Memoirs of the American Mathematical Society

Number 281

Andrew Majda

The existence of multi-dimensional shock fronts

Published by the

AMERICAN MATHEMATICAL SOCIETY

Providence, Rhode Island, USA

May 1983 · Volume 43 · Number 281 (fourth of 5 numbers)

MEMOIRS of the American Mathematical Society

This journal is designed particularly for long research papers (and groups of cognate papers) in pure and applied mathematics. It includes, in general, longer papers than those in the TRANSACTIONS.

Mathematical papers intended for publication in the Memoirs should be addressed to one of the editors. Subjects, and the editors associated with them, follow:

MEMOIRS are printed by photo-offset from camera-ready copy fully prepared by the authors. Prospective authors are encouraged to request booklet giving detailed instructions regarding reproduction copy. Write to Editorial Office, American Mathematical Society, P.O. Box 6248, Providence, Rhode Island 02940. For general instructions, see last page of Memoir.

SUBSCRIPTION INFORMATION. The 1983 subscription begins with Number 272 and consists of six mailings, each containing one or more numbers. Subscription prices for 1983 are $104.00 list; $52.00 member. Each number may be ordered separately; *please specify number* when ordering an individual paper. For prices and titles of recently released numbers, refer to the New Publications sections of the **NOTICES** of the American Mathematical Society.

BACK NUMBER INFORMATION. For back issues see the AMS Catalogue of Publications.

TRANSACTIONS of the American Mathematical Society

This journal consists of shorter tracts which are of the same general character as the papers published in the MEMOIRS. The editorial committee is identical with that for the MEMOIRS so that papers intended for publication in this series should be addressed to one of the editors listed above.

Subscriptions and orders for publications of the American Mathematical Society should be addressed to American Mathematical Society, P. O. Box 1571, Annex Station, Providence, R. I. 02901. *All orders must be accompanied by payment.* Other correspondence should be addressed to P. O. Box 6248, Providence, R. I. 02940.

MEMOIRS of the American Mathematical Society (ISSN 0065-9266) is published bimonthly (each volume consisting usually of more than one number) by the American Mathematical Society at 201 Charles Street, Providence, Rhode Island 02904. Second Class postage paid at Providence, Rhode Island 02940. Postmaster: Send address changes to Memoirs of the American Mathematical Society, American Mathematical Society, P. O. Box 6248, Providence, RI 02940.

TABLE OF CONTENTS

Library of Congress Cataloging in Publication Data

Majda, Andrew, 1949-
The existence of multi-dimensional shock fronts.

(Memoirs of the American Mathematical Society, ISSN 0065-9266 ; no. 281)
Bibliography: p.
1. Shock waves. 2. Differential equations, Hyperbolic--Numerical solutions. 3. Conservation laws (Physics) I. Title. II. Series.
QA3.A57 no. 281 [QA927] 510s [531'.1133] 83-3725
ISBN 0-8218-2281-0

Abstract

The short-time existence of discontinuous shock front solutions of a system of conservation laws in several space variables is proved below under suitable hypotheses. These shock front solutions are nonlinear progressing wave solutions associated with the nonlinear wave fields. The results developed here apply to the equations of compressible fluid flow in two or three space variables with standard equations of state where the initial data can have shock discontinuities of arbitrary strength which lie on a given smooth initial surface with arbitrary geometry. These shock front solutions are constructed via a classical iteration scheme so that the shock-fronts for the physical equations are extremely stable and do not exhibit a "loss of derivatives." Two of the main tools used in the proof of convergence are estimates from Friedrichs' theory of positive symmetric systems and the linearized stability theory for shock fronts developed recently by the author. The convergence proof for the iteration scheme utilizes exponentially weighted square-integrable norms in space-time together with a judicious choice of these weights as the time interval tends to zero.

AMOS(MOS) subject classification. Primary 76L05; 35L65
Secondary 35B40; 35A40

Key words and phrases. Hyperbolic conservation laws, multi-dimensional shock fronts, stability, mixed problems.

§0. INTRODUCTION

Most of the interesting phenomena in classical inviscid continuum mechanics are described by solutions of the hyperbolic system of conservation laws,

$$(0.1) \qquad \frac{\partial u}{\partial t} + \sum_{j=1}^{N} \frac{\partial}{\partial x_j} F_j(u) = 0 , \quad t > 0$$

where $x = (x_1, \ldots , x_N) \in R^N$, $u = {}^t(u_1, \ldots , u_M)$, and $F(u) \equiv (F_j(u))$, $1 \leqslant j \leqslant N$, are smooth nonlinear mappings defined on an open subset of R^M and mapping back into R^M with $A(u) \equiv (A_j(u))$, $1 \leqslant j \leqslant N$, and $A_j(u) \equiv \partial F_j/\partial u$, the corresponding $M \times M$ Jacobian matrices. Following Friedrichs, we express the hyperbolicity of the system in (0.1) by the assumption that there is a smoothly varying, positive definite, symmetric matrix, $A_0(u)$, with

$$(0.2) \qquad A_0(u)A_j(u) \text{ symmetric} , \quad 1 \leqslant j \leqslant N .$$

Important physical applications where the systems in (0.1) arise and (0.2) is satisfied include gas dynamics, shallow water thoery, and magnetofluid dynamics where, of course, $N = 1, 2,$ or 3 . In fact, all the physically inter-

Received by the editor: June 10, 1981

Partially supported by N.S.F. Grant #8102360.

esting hyperbolic conservation laws known to this author satisfy the additional condition in (0.2)

Despite the abundance of practical applications of the above systems of conservation laws, remarkably little is known rigorously regarding the solutions of these physical systems in two or three space dimensions. In [8], Kato proved the local existence in time of general smooth solutions of (0.1) provided that (0.2) is satisfied and $u_0(x)$ is a sufficiently smooth function -- Kato required $u_0(x) \in H^s(R^N)$ for $s > \frac{N}{2} + 1$ where, here and below, $H^s(G)$ denotes the standard Sobolev space of order s defined on the domain, G. When $N = 1$, the theory of discontinuous solutions (even globally in time) of (0.1) is well-understood as a consequence of the work of Glimm in [7]. In that work, one of the main ideas involved using the explicit structure of the nonlinear progressing wave solutions of (0.1) in a single space variable given by the solution of the Riemann problem and constructed by Lax in [9]. Lax also discusses, for weak waves, the fashion in which the nonlinear shock waves and rarefaction waves generalize the progressing wave solutions of the linearized version of (0.1) in a single space variable.

In this paper, we prove the short-time existence of multi-dimensional "shock front" solutions of (0.1) under structural hypotheses always satisfied by the physical equations of compressible fluid flow with standard equations of state and initial shock strengths of arbitrary size. These shock front solutions are discontinuous, multi-dimensional, nonlinear progressing wave solutions of (0.1) associated with the nonlinear wave fields of (0.1). For gas dynamics, the only previous results regarding the existence of discontinuous solutions of (0.1) for $N = 2, 3$, are the explicit spherically symmetric shock fronts with spherically symmetric initial data described, for example, in [4] and constructed by using scaling laws.

To describe the general shock front solutions which we construct below, for the moment, we consider the behavior of discontinuous solutions in multidimensions of the linear hyperbolic equations,

$$\frac{\partial u}{\partial t} + \sum_{j=1}^{N} A_j \frac{\partial u}{\partial x_j} = 0 \ . \tag{0.3}$$

We regard the equation in (0.3) as the linearization of (0.1) about a given constant state. The simplest intuitive discontinuous solutions of (0.3) are piecewise smooth weak solutions which jump across wave fronts which are characteristic space-time hypersurfaces of (0.3) (see [3]). Analogously, shock front solutions of (0.1) are piecewise smooth weak solutions of (0.1) with the following structure:

(0.4) There are a C^2 space-time hypersurface $S(t)$, defined in (x, t) space for $T \geqslant t \geqslant 0$ with space-time normal, $(n_t, n_1, \dots, n_N) = (n_t, n_x)$, and two C^1 vector-valued functions, $u^+(x, t)$ and $u^-(x, t)$, defined on respective domains, G^+ and G^-, on either side of this hypersurface so that

$$\frac{\partial u^+}{\partial t} + \sum_{j=1}^{N} A_j(u^+) \frac{\partial u^+}{\partial x_j} = 0 \quad \text{in} \quad G^+ \ ,$$

$$\frac{\partial u^-}{\partial t} + \sum_{j=1}^{N} A_j(u^-) \frac{\partial u^-}{\partial x_j} = 0 \quad \text{in} \quad G^- \ .$$

Furthermore, in order the define a weak solution of (0.1), the boundary values of u^+, u^- restricted to the hypersurface, $S(t)$, cannot jump in an arbitrary fashion but are constrained to satisfy the Rankine Hugoniot conditions,

$$n_t(u^+ - u^-)\big|_S + \sum_{j=1}^{N} n_j(F_j(u^+) - F_j(u^-))\big|_S = 0 \ . \tag{0.5}$$

Also, unlike the situation described above for linear progressing waves, for shock waves,

(0.6) the surface, S, is noncharacteristic for (0.1).

We call discontinuous piecewise smooth solutions of (0.1) satisfying (0.4) - (0.6), shock-front solutions. Obviously, unlike the linear case, the surface S is not known in advance and must be determined as part of the solution of the problem; thus, the equations in (0.4) and (0.5) describe a multi-dimensional, highly nonlinear, free boundary value problem for a system of quasi linear hyperbolic equations.

Before stating our main theorems, we describe the classes of discontinuous initial data, u_0, for (0.1) for which we construct the above shock-front solutions. In a precise sense described in §§1 and 2 below, we assume that the initial data (called shock-front data below) is piecewise smooth with a jump discontinuity across a smooth compact hypersurface, M_0 which defines the intial position for the shock front, $S(t)$. We denote the inside of M_0 by Ω_- and the outside of M_0 by Ω_+ with corresponding pieces of the initial data given by u_0^-, u_0^+ defined on Ω_-, Ω_+ respectively. We parametrize M_0 by α and let $n(\alpha) = (n_1(\alpha), \ldots, n_N(\alpha))$ denote the outward normal to M_0. Obviously the main restriction satisfied by the shock front initial data is that necessarily (0.5) must be satisfied at $t = 0$. Therefore,

(0.7) There is a smooth function $\sigma(\alpha)$ defined on M_0 so that

$$-\sigma(\alpha)(u_0^+(\alpha) - u_0^-(\alpha)) + \sum_{j=1}^{N} n_j(\alpha)(F_j(u_0^+(\alpha)) - F_j(u_0^-(\alpha)) = 0 ,$$

$$\text{for all} \quad \alpha \in M_0 .$$

We also impose other restrictions on the initial data, u_0^+, u_0^-, which also occur in the standard theory of hyperbolic mixed problems (see [13]),

(0.8) For a given integer s, the compatibility conditions up to order $s - 1$ are satisfied on M_0 by (u_0^+, u_0^-).

To describe these conditions, we impose the requirement from (0.6) initially at $t = 0$; thus,

(0.9) $\det \tilde{A}(\alpha) \neq 0$ for $\alpha \in M_0$ where

$$\tilde{A}(\alpha) = \begin{pmatrix} A(u_0^+)\cdot n(\alpha) - \sigma I & 0 \\ 0 & -(A(u_0^-)\cdot n(\alpha) - \sigma I) \end{pmatrix},$$

and it follows that (see §1)

(0.10) $\tilde{A}(\alpha)$ has exactly $m - 1$ positive real eigenvalues with $P(\alpha)$ the corresponding smooth projection.

Let $\partial j/\partial n^j$ denote j-th order normal differentiation to M_0. In §2, we prove that to satisfy the compatibility conditions in (0.8),

(0.11) $$(I - P) \frac{\partial^j}{\partial n^j} \begin{pmatrix} u_0^+ \\ u_0^- \end{pmatrix}\Bigg|_{M_0} \text{ can be prescribed arbitrarily and}$$

$$P \frac{\partial^j}{\partial n^j} \begin{pmatrix} u_0^+ \\ u_0^- \end{pmatrix}\Bigg|_{M_0} \text{ is then uniquely determined, } 1 \leq j \leq s - 1 .$$

Thus, $m + 1$ of the $2m$ traces, $\left.\frac{\partial^j}{\partial n^j}\begin{pmatrix} u_0^+ \\ u_0^- \end{pmatrix}\right|_{M_0}$, can be prescribed arbitrarily and (0.8) is then guaranteed for $1 \leqslant j \leqslant s - 1$. (See §2 below.) In particular, once the main condition in (0.7) is satisfied, the conditions in (0.8) are automatically guaranteed for a wide class of initial data. Without further comment, for technical convenience, we always assume below that u_0^+ satisfies u_0^+ is constant for $|x| \geqslant R_0$. However, by finite propagation speed for the Cauchy problem, this restriction is easily removed.

Next, we state our main theorem for the simplest multi-dimensional physical system, the 3×3 system describing isentriopic compressible fluid flow in two space dimensions given by

$$(0.12) \qquad \frac{\partial}{\partial t}\begin{pmatrix} \rho w_1 \\ \rho w_2 \\ \rho \end{pmatrix} + \frac{\partial}{\partial x_1}\begin{pmatrix} \rho w_1^2 + p \\ \rho w_1 w_2 \\ \rho w_1 \end{pmatrix} + \frac{\partial}{\partial x_2}\begin{pmatrix} \rho w_1 w_2 \\ \rho w_2^2 + p \\ \rho w_2 \end{pmatrix} = \begin{pmatrix} 0 \\ 0 \\ 0 \end{pmatrix}$$

where $w = {}^t(w_1, w_2)$ is the fluid velocity, ρ is the density, and $p(\rho)$ is a well-defined function of ρ with $p'(\rho) > 0$ and determined by an equation of state.

<u>THEOREM 1</u>. (Shock Fronts for Isentropic Flow). Assume that M_0 is an arbitrary smooth curve and that the initial data (w_0^-, ρ_0^-), (w_0^+, ρ_0^+) belong to H^s of the respective domains, Ω_- and $\Omega_+ \cap \{|x| < R_0\}$ for some fixed $s \geqslant 10$. Also assume that there is a function $\sigma(\alpha) \in H^s(M_0)$ so that (0.7) and (0.8) are satisfied for $1 \leqslant s \leqslant 8$ together with the entropy condition,

$$(0.13) \qquad w_0^+ \cdot n(\alpha) + (p_\rho(\rho^+))^{1/2} < \sigma(\alpha) < w_0^- \cdot n(\alpha) + (p_\rho(\rho^-))^{1/2} ,$$

and the stability condition,

$$(0.14) \qquad \frac{[p(\rho)]}{[\rho]} < p_\rho(\rho^-) + (w_0^-\cdot n - \sigma)^2 .$$

Then, there is a C^2 hypersurface, $S(t)$, defined for $0 \leqslant t \leqslant T$ with T sufficiently small so that a piecewise C^1 shock front solution of the compressible fluid equations in (0.12) exists with the above initial data and satisfies (0.4), (0.5).

REMARK. We have already proved in §3 of [12] that the stability condition in (0.14) is always satisfied for compressive shocks of arbitrary strength provided $p(\rho)$ is a convex function of ρ. In particular, for the ideal gas equations of state, $p(\rho) = A\rho^r$, $A > 0$, $r > 1$. In (0.13), we have described a shock front advancing into Ω_+; by changing the signs in (0.13), (0.14), appropriately, we can treat, just as simply a shock advancing into Ω_-.

The reader might suspect that the condition in (0.7) is a severe restriction on the geometry of M_0 and the boundary values of (u_0^+, u_0^-) restricted to M_0. Next, we remark that this is not the situation for the physical equations of compressible fluid flow. For simplicity we assume that $p(\rho)$ is a convex function of ρ so that (0.14) is always satisfied. The conditions in (0.7) and (0.13), (0.14) are satisfied if and only if

(0.15) (1) $w_0^+(\alpha) - (w_0^+\cdot n(\alpha))n(\alpha) \equiv w_0^-(\alpha) - (w_0^-\cdot n(\alpha))n(\alpha)$;

(2) $(w_0^+(\alpha)\cdot n(\alpha), \rho^+)$ lies along the two (one) shock curve with speed $\sigma(\alpha)$ emanating from $(w_0^-(\alpha)\cdot n(\alpha), \rho^-)$ for the one-dimensional fluid equations

$$\frac{\partial\rho}{\partial t} + \frac{\partial}{\partial x}\rho w = 0 ,$$

$$\frac{\partial\rho w}{\rho t} + \frac{\partial}{\partial x}(\rho w^2 + p(\rho)) = 0 .$$

By classical one-dimensional facts (see [4]), (1) and (2) can be used to generate large classes of initial data satisfying the hypotheses in (0.7), (0.13), (0.14) independent of the geometry of M_0 and with arbitrary shock strength.

Finally, we state the general theorem which we prove in this paper. Theorem 1 and also Theorem 2 below are immediate Corollaries of this theorem by the results for the physical examples proved in §3 of our earlier paper, [12].

THEOREM (The Existence of Shock Fronts). Assume

(1) The general system from (0.1) satisfies the structural conditions in §1.

(2) The initial data, $u_0^-(x)$, $u_0^+(x)$ belong to $H^{s+1}(\Omega_-)$, $H^{s+1}(\Omega_+ \cap \{|x| < R_0\})$ for a fixed integer s with $s \geqslant 2[\frac{n}{2}] + 7$.

(3) There is a $\sigma(\alpha) \in H^{s+1}(M_0)$ so that the initial data also satisfies (0.7), (0.8) for $1 \leqslant j \leqslant s - 1$.

(4) The planar constant coefficient shock front problem defined by $(u_0^+(\alpha), u_0^-(\alpha), \sigma(\alpha))$ is uniformly stable for any $\alpha \in M_0$ (see §1 and [12]).

Then, for a sufficiently short-time, T , with $0 \leqslant t \leqslant T$, there is a hypersurface $S(t)$ in H^{s+1} and smooth functions $u^+(x, t)$, $u^-(x, t)$ belonging to $H^s(G^-)$, $H^s(G^+ \cap \{x \,|\, |x| < C_0R_0\})$ respectively, with u^+ constant for $|x| > C_0R_0$ so that (0.4), (0.5) are satisfied and (u^+, u^-) have the initial data, $(u_0^+(x), u_0^-(x))$. Thus, (u^+, u^-) define a classical shock front solution of (0.1) with the given initial data.

We also state a second corollary of the above theorem which applies to construct shock front folutions for the full Euler equations describing compressible flow in three space dimensions. We consider solutions of the Euler equations,

(0.16) $$\frac{\partial\rho}{\partial t} + \operatorname{div}(\rho w) = 0 \;;$$

$$\frac{\partial}{\partial t}(\rho w_i) + \sum_{j=1}^{3} \frac{\partial}{\partial x_j}(\rho w_i w_j + \delta_{ij} p) = 0 \;, \quad i = 1, 2, 3 \;;$$

$$\frac{\partial}{\partial t}(\rho E) + \sum_{j=1}^{3} \frac{\partial}{\partial x_j}(\rho w_j E + p w_j) = 0 \;.$$

Where $E = \frac{1}{2}|w|^2 + e(p, \rho)$, $e(p, \rho)$ is the specific internal energy, and $w = {}^t(w_1, w_2, w_3)$ is the fluid velocity. We assume a general equation of state, $p(\tau, S)$, where $\tau = \frac{1}{\rho}$ is the specific volume, and the entropy S , together with the temperature, T , is defined through the second law of thermodynamics,

$$TdS - pd\tau = de \;.$$

We require $-\frac{dp}{d\tau}(\tau, S) > 0$ and introduce the quantity, ℓ , defined by

$$\ell = 2 - \frac{M^2(\tau^+ - \tau^-)}{T^-} p_S \Big|_{(\tau^-, S^-)} \;,$$

where M^2 is the square of the Mach number behind the shock, i.e.,

$$M^2 = \frac{[p]}{[\tau]} \frac{dp}{d\tau}\Big|_{\tau^-} \;.$$

THEOREM 2 (Shock Fronts for the Compressibel Euler Equations). Assume that M_0 is an arbitrary smooth compact surface in R^3 and that the initial data (ρ^+, w^+, E^+), (ρ^-, w^-, E^-) satisfy the smoothness conditions and compatibility conditions of the above theorem for $n = 3$, i.e. $s = 10$, etc.

for some appropriate shock speed function $\sigma(\alpha)$. Also assume that these data satisfy the entropy condition,

$$(0.17) \qquad w^{+}\cdot n + \left(\frac{dp}{d\rho}\bigg|_{(\rho^{+},S^{+})}\right)^{1/2} < \sigma(\alpha) < w^{-} n + \left(\frac{dp}{d\rho}\bigg|_{(\rho^{-},S^{-})}\right)^{1/2} ,$$

and also the stability condition,

$$(0.18) \qquad (\ell - 1) + M^{2}(1 - \frac{\tau^{+}}{\tau^{-}}) > 0 , \quad \alpha \in M_{0} .$$

Then, a shock front solution with the smoothness properties of the above Theorem exists for the full Euler equations of three-dimensional compressible flow in (0.16) with the above initial data. In particular, the condition in (0.18) is always satisfied for compressive shocks of any strength in an ideal gas with equation of state, $p(\tau, S) = \exp(\frac{S}{c_v})\tau^{-\gamma}$, $\gamma > 1$ and for sufficiently weak shocks for general equations of state.

Similar remarks as in (0.15) can be used for the Euler equations to generate large classes of shock-front initial data with arbitrary strength satisfying (0.7), (0.17), and (0.18) independent of the geometry of M_0. Theorem 2 is an immediate Corollary of our main Theorem and Proposition 3 of §3 in our earlier paper, [12].

Next, we discuss the proof of our main theorem and the contents of the remainder of this paper. In outline, our proof is a straightforward adaptation of the standard proof of short-time existence for smooth solutions of the Cauchy problem; however, the technical details are quite different due to the unusual features of the above problem. In §§1 and 2, we develop the structural conditions on the system needed for the proof of the Theorem, discuss the compatibility conditions from (0.8), and reformulate the problem so that all functions have a common domain of definition. We also build an approximate solution in §2 which satisfies (0.4) and (0.5) to order t^{s-1} at $t = 0$. The main arguments in the paper are contained in §§3-5. In §3, we define the iteration scheme. In §4, we prove the convergence of this iteration scheme. The iteration scheme is a classical one and our method of proof verifies that multi-dimensional shock fronts for the physical equations do not exhibit the "loss of derivatives" phenomena. The iteration scheme is based on a linearization which crucially utilizes the theory of linearized stability for shock fronts developed by the author in [12]. As opposed to the use of uniform norms in time for the Cauchy problem ([8]), here, we need L^2 weighted norms for functions and their derivatives in space-time such as

$$\int_0^T \int_G e^{-2\eta t} |F|^2 \, ds \, dt ,$$

where G is M_0, $\Omega_\pm$, etc. This is necessary precisely because of the new phenomena studied here -- we need to treat interactions between corrections to the shock surface and the interior equations. We use standard fixed point iteration for the interior equations but it is crucial that we apply Newton's method pointwise on $M_0 \times [0, T]$ for the nonlinear boundary conditions to deduce "high norm boundedness." Also, it is important that we prove a priori estimates in the above hyperbolic weighted norms independent of η, for η large, for the appropriate linearized problems in a high Sobolev norm; then, we make a judicious choice of $\eta(T) = O(T^{-\beta})$ as $T \longrightarrow 0$, β fixed with $0 < \beta < 1$ which implies high norm boundedness (see the detailed remarks in §4). In §5, we develop the main a priori estimates in high Sobolev norms for the linearized shock front problem mentioned above. The existence of solutions and the main a priori estimates in the zero norm follow from the theory developed in [12] and suitable approximation of the coefficients. The estimates in the high norms are deduced from this zero norm estimate by differentiating the equations tangentially, estimating normal derivatives directly on $M_0 \times [0, T]$ via tangential derivatives, and then unsing Friedrichs' main energy estimates ([5]) for positive symmetric systems to estimate the normal derivatives in the interior -- in particular, essential use is made of the hypothesis in (0.2). The straightforward and more elegant argument of Rauch (see [14]) for estimating interior derivatives in space-time for general hyperbolic systems implies an illusory loss of derivatives which we avoid by using the stronger boundary estimates valid for positive symmetric systems in §5. In the appendix, we discuss modifications of well-known nonlinear calculus inequalities which we need in the η-weighted norms throughout §§3-5.

In a series of interesting papers, ([10], [11]) Li Da-qian and Yu Wen-ci have discussed general free boundary value problems for hyperbolic equations in a single space variable including the full perturbed Riemann problem and steady supersonic flow in two dimensions past a curved wedge (also see Schaeffer [15]). In these cases, the interior hyperbolic equations can be treated by integration along characteristics and the linearized theory developed in [12] degenerates to coupled systems of scalar hyperbolic equations and ordinary differential equations. This is the only other work known to the author on hyperbolic free boundary problems. The work in [11] also applies for a treatment of spherically symmetric shock fronts with spherically symmetric initial data. Of course, if one is interested in even the slightest non-spherically symmetric perturbation of these shock fronts, the only theory known to the author which treats this case rigorously is developed here.

Finally, we point out to the reader that the extremely technical arguments developed in [12] are not needed specifically in this paper. Only a knowledge of the introduction and the results quoted in §§2 and 3 of that paper are necessary to read this paper. Also, for a first reading, it is helpful to make the stronger assumptions in §§1 and 2 that the initial data has $s + 2$ classical derivatives.

§1. STRUCTURAL CONDITIONS AND SHOCK FRONT INITIAL DATA: SOME PRELIMINARY FACTS

First, we remark that all of the structural assumptions which we make below have been verified for the physical equations of compressible fluid flow (defined in (0.12) and (0.16)) in §3 of [12]. We have already made our first assumption in (0.2) above. We also require that the initial data, $u_0^+(x)$, $u_0^-(x)$ lies in the domain of hyperbolicity for (0.1), i.e., for the physical examples, the initial density is strictly positive. The above condition implies that there are positive constants, δ , $c > 0$ so that

$$(1.1) \qquad c^{-1}I \leq A_0(u) \leq cI$$

for all vector values, $u \in R^m$, satisfying

$$(1.2) \qquad |u - u_0^{\pm}(x)| < \delta \quad \text{for some} \quad x \in R^N .$$

Such a condition is also necessary for the existence results in [8] regarding the Cauchy problem. We also assume here and in the remainder of the paper that the boundary values $(u_0^+(\alpha),\ u_0^-(\alpha),\ \sigma(\alpha))$ for the shock front initial data belong to $H^{s+1}(M_0)$ and satisfy

$$(1.3) \qquad -\sigma(\alpha)(u_0^+(\alpha) - u_0^-(\alpha)) + n(\alpha)\cdot(F(u_0^+(\alpha)) - F(u_0^-(\alpha) = 0 .$$

In fact, our point of view in this section and §2 always assumes that we are given functions $(u_0^+(\alpha),\ u_0^-(\alpha),\ \sigma(\alpha))$ satisfying (1.3) and lying in a domain of hyperbolicity for (0.1) so that (1.1) and (1.2) are satisfied for $\alpha \in M_0$.

Then, in §2, we show how to construct all initial data $u_0^+(x)$, $u_0^-(x)$ belonging to $H^{s+1}(\Omega_\pm)$, satisfying $u_0^+|_{M_0} = u_0^+(\alpha)$, $u_0^-|_{M_0} = u_0^-(\alpha)$ and also the compatibility conditions in (0.8). For $s > \frac{n}{2}$, by continuity, we can then guarantee (1.1) and (1.2) for all x in a neighborhood of M_0. Without further comment, we extend these values of u_0^+, u_0^- away from this neighborhood so that (1.1) and (1.2) are satisfied since the same considerations are needed for the Cauchy problem for smooth data which we regard below as completely understood. Furthermore, the strategy described above is consistent with our construction of shock front data in (0.15) for the physical equations.

We assume that the jump conditions in (1.3) are associated with a given p-shock independent of $\alpha \in M_0$. Thus, if $\lambda_j(u_0^\pm(\alpha)) = \lambda_j$, $1 \leqslant j \leqslant M$ are the eigenvalues of

$$A(u_0^\pm)\cdot n(\alpha) = \sum_{j=1}^{N} A_j(u_0^\pm)n_j(\alpha) \tag{1.4}$$

ordered with $\lambda_1^\pm \leqslant \lambda_2^\pm \leqslant \dots \leqslant \lambda_M^\pm$, we assume that there is a p with $1 \leqslant p \leqslant M$ so that

$$\lambda_{p-1}^\pm < \lambda_p^\pm < \lambda_{p+1}^\pm \quad , \tag{1.5}$$

and that Lax's shock inequalities are satisfied; thus,

$$\lambda_p(u_0^-(\alpha)) > \sigma(\alpha) > \lambda_p(u_0^+(\alpha)) \quad , \tag{1.6}$$

$$\lambda_{p+1}(u_0^+(\alpha)) > \sigma(\alpha) > \lambda_{p-1}(u_0^-(\alpha))$$

for all $\alpha \in M_0$. Under the condition in (1.5), we let $P^+(\alpha)$, $P^-(\alpha)$ denote, respectively, the smoothly varying projections onto the subspaces spanned by the eigenvectors associated with the eigenvalues, $\lambda^+_{p+1} \leq \dots \leq \lambda^+_M$ of $A(u^+_0)\cdot n(\alpha)$ and $\lambda^-_1 \leq \dots \leq \lambda^-_{p-1}$ of $A(u^-_0)\cdot n(\alpha)$. By using the matrix resolvent formula and Lemma A-2 of the appendix, (also see appendix B of [12] for this type of argument) we observe that $P^+(\alpha)$, $P^-(\alpha)$ belong to $H^{s+1}(M_0)$ provided that $(\sigma(\alpha), u^+_0(\alpha), u^-_0(\alpha))$ belongs to $H^{s+1}(M_0)$ and $s > \frac{n-1}{2}$.

Our next structural assumption is that the technical block structure condition stated in Appendix B of [12] is valid at any fixed point, $u^{\pm}_0(\alpha)$, $\alpha \in M_0$, for the perturbed operators with the form,

$$\frac{\partial}{\partial t} + \sum_{j=1}^{N} A_j(u^{\pm}_0(\alpha) + v) \frac{\partial}{\partial x_j} ,$$

provided v satisfies

(1.7) $$|v| < \delta .$$

Such an assumption is automatic for any strictly hyperbolic system like the isentropic equations in (0.12) and in §3 of [12] we have verified the above block structure condition explicitly for the compressible Euler equations in (0.16) where strict hyperbolicity fails.

Our final structural condtion on the initial data, $(u^+_0(\alpha), u^-_0(\alpha), \sigma(\alpha))$ is assumption (4) of our main theorem. Thus,

(1.8) The planar shock front problems associated with $(u^+_0(\alpha), u^-_0(\alpha), \sigma(\alpha))$ are uniformly for every $\alpha \in M_0$.

The significance of the condition in (1.8) and the block structure assumption above (1.7) is that Theorems 1 and 2 of §2 in [12] apply to the linearized shock front problems developed in [12]. These unusual boundary value problems are the basic linear problem in the nonlinear iteration scheme described in §3. More precisely, the assumption in (1.8) guarantees that there is a $\delta > 0$ and smaller than δ in (1.2), (1.7) so that, provided that the smooth coefficients, $(u^+, u^-, \beta, \nabla_{tan}\beta, \beta_t)$ satisfy,

(1.9)
$$|u^+(x, t) - u_0^+(x)| + |u^-(x, t) - u_0^-(x)| + |\beta(\alpha, t)| + |\nabla_{tan}\beta(\alpha, t) + |\beta_t(\alpha, t) - \sigma(\alpha)| < \delta$$

for $-\infty < t < +\infty$ and also the trivial condition in (2.13) of [12], the theory in [12] applies to the following linear problem: We define coefficients by

(1.10)
$$b_0 = u^+ - u^- ;$$

$$b_1 = (F(u_0^+) - F(u_0^-))\,(I + \beta W(\alpha))^{-1} ;$$

$$b_2 = -(I + \beta W(\alpha))^{-2}\nabla_{tan}\beta\cdot(F(u_0^+) - F(u_0^-)) ;$$

$$M(v^+, v^-) = [(n(\alpha) - (1 + \beta W(\alpha))^{-1}\nabla_{tan}\beta)\cdot A(u^+) - \frac{\partial\beta}{\partial t}]v^+ - [(n(\alpha) - (1 + \beta W(\alpha))^{-1}\nabla_{tan}\beta)\cdot A(u^-) - \frac{\partial\beta}{\partial t}]v^- ;$$

where $W(\alpha)$ is the Weingarten map at $\alpha \in M_0$ and ∇_{tan} is the gradient defined by the Riemannian metric induced on M_0 by the Euclidean inner product. We form the boundary operator,

(1.11)
$$B(v^+, v^-, \phi) \equiv b_0\phi_t + b_1\nabla_{tan}\phi + b_2\phi + M(v^+, v^-)$$

$$\text{on } M_0 \times (-\infty, \infty) .$$

Looking ahead, we also consider the interior operators $L^+(u^+, \beta)$, $L^-(u^-, \beta)$ defined in (2.5) below.

(1.12) When (1.9) is satisfied, Theorems 1 and 2 of [12] are valid for the linearized shock front boundary problem

$$L^+(u^+, \beta)v^+ = F_+ \quad \text{in} \quad \Omega_+ \times (-\infty, \infty) \; ;$$

$$L^-(u^-, \beta)v^- = F_- \quad \text{in} \quad \Omega_- \times (-\infty, \infty) \; ;$$

$$\mathcal{B}(v^+, v^-, \phi) = g \quad \text{on} \quad M_0 \times (-\infty, \infty) \; .$$

Finally, we conclude this section with some remarks regarding conditions guaranteeing that there are functions $(u_0^+(\alpha), u_0^-(\alpha), \sigma(\alpha)) \in H^{s+1}(M_0)$ satisfying (1.3). For the physical equations in (A.12) or (A.16) with standard equations of state, we use the prescription in (0.15) and also the remarks in [4]. It is well-known that the shock speed, σ , and the state ahead of the shock, u^+ , define a unique state behind the shock, $\mathcal{U}^-(u^+, \sigma)$, so that (1.3) is satisfied and also (0.13) or (0.17) for the given system in either (0.12) or (0.16). Here $\mathcal{U}^-(u^+, \sigma)$ is a smoothly varying function of u^+, σ . Given $u^+(\alpha), \sigma(\alpha)$ belonging to $H^{s+1}(M_0)$, we define $u^-(\alpha)$ by the recipe

$$u^-(\alpha) = \mathcal{U}^-(u^+(\alpha), \sigma(\alpha)) \; ;$$

then by A-2 of the appendix, $u^-(\alpha) \in H^{s+1}(M_0)$ provided $s > \frac{n-1}{2}$. For the general system in (0.1), in addition to (1.5) and (1.16), assume that

(1.13) $\lambda_p(u, n)$ is a genuinely nonlinear eigenvalue of

$$A(n, u) \equiv \sum_{j=1}^{N} A_j(u)n_j \quad \text{for} \quad u \in \mathcal{V} \quad \text{and} \quad |n| = 1 .$$

Assume that $u_0^+(\alpha) \in \mathcal{V}$, $u_0^+(\alpha)$, $\sigma(\alpha) \in H^{s+1}(M_0)$, $s > \frac{n-1}{2}$, and $|\sigma(\alpha) - \lambda_p(u_0^+(\alpha), n(\alpha))|$ is small. Then it follows from Lax's construction in [9] using the implicit function theorem that there is a unique $u^-(\alpha)$ satisfying (1.3) with $|u^-(\alpha) - u^+(\alpha)|$ sufficiently small. By the same remarks as in the above, $u^-(\alpha) \in H^{s+1}(M_0)$ too.

§2. THE MAP TO A FIXED DOMAIN, COMPATIBILITY CONDITIONS, AND AN APPROXIMATE SOLUTION

In this section and in the remainder of this paper, we assume the basic hypothesis from (1.12) is satisfied for the system of conservation laws with the given shock-front initial data.

2.A Reformulation by Mapping to a Fixed Domain

We define a change of variables which depends on the unknown shock front and has the obvious technical advantage of reformulating the problem so that all functions in the iteration scheme have a common fixed domain of definition. Recall that the initial position of the shock front at $t = 0$ is the compact smooth hypersurface, M_0, of R^N with the variable vector α parametrizing M_0 and $n(\alpha)$, the outward normal to M_0. The mapping

$$x(\alpha, \tau) = \alpha + \tau n(\alpha) \tag{2.1}$$

parametrizes a tubular neighborhood of M_0 provided that $|\tau| < \delta_1$ so that given any x with $d(x, M_0) < \delta_2$ ($d(\cdot, S)$ is the distance of a point to the set S) there is a unique $\alpha(x) \in M_0$ well-defined through the equation in (2.1). Consider any C^2 function, $\beta(\alpha, t)$, defined on $M_0 \times (-\infty, \infty)$ with $\beta(\alpha, 0) = 0$. We consider the change of variables,

$$(2.2) \qquad \tilde{x} = \chi_t(x, \beta) \equiv x - \rho(x)\beta(\alpha(x), t)n(\alpha(x)) \; ;$$

$$\tilde{t} = t \; ,$$

where ρ is a fixed smooth cut-off function with $\rho \equiv 1$ for $d(x, M_0) < \frac{\delta_2}{2}$ and vanishing for $d(x, M_0) > \delta_2$. There is a $\delta_3 > 0$ sufficiently small so that $\chi_t(x, \beta)$ is a well-defined diffeomorphism of R^N to R^N provided that $\beta(\alpha, t)$ is an arbitrary C^2 function satisfying

(2.3) $$\sup_{\substack{\alpha \in M_0 \\ -\infty < t < -\infty}} |\beta(\alpha, t)| + |\nabla_{\tan}\beta(\alpha, t)| < \delta_3 .$$

Here and below, $\nabla_{\tan}$ denotes the gradient of a function defined on M_0 with respect to the natural Riemannian metric induced on M_0 as a submanifold of R^N. In the process of defining the iteration scheme in the next sections, we will restrict furhter the constant δ_3 in (2.3) a fixed finite number of times. We denote the inverse mapping to χ_t by $\Phi_t(\tilde{x}, \beta) \equiv \chi_t^{-1}(\tilde{x}, \beta)$.

Now, parametrize the position of the C^2 shock surface $S(t)$ (to be constructed below) for sufficiently short times, $0 \leq t \leq T_0$, by a function, $\tilde{\beta}(\alpha, t)$ so that

(2.4) $$\tilde{\beta}(\alpha, 0) = 0 ;$$

$$S(t) = \{\alpha + \tilde{\beta}(\alpha, t)n(\alpha) | \alpha \in M_0\} ,$$

where the conditions in (2.3) are satisfied for $0 \leq t \leq T_0$. Corresponding to (2.2), we introduce the change of variables,

$$x = \Phi_{\tilde{t}}(\tilde{x}, \tilde{\beta}) ;$$

$$t = \tilde{t} ,$$

and define

$$u^{\pm}(\tilde{x}, \tilde{t}) \equiv u^{\pm}(\Phi_{\tilde{t}}(\tilde{x}, \tilde{\beta}), \tilde{t})$$

where $u^{\pm}$ are the C^1 solutions defined on each side of the shock surface. One easily computes that $u^+(\tilde{x}, \tilde{t})$, $u^-(\tilde{x}, \tilde{t})$ satisfy the equations,

$$\text{(2.5)} \qquad L^+(u^+, \tilde{\beta})u^+ = 0 \quad \text{in} \quad \Omega_+ \times [0, T_0] \quad ;$$

$$L^-(u^-, \tilde{\beta})u^- = 0 \quad \text{in} \quad \Omega_- \times [0, T_0] \quad ,$$

where

$$L^{\pm} = \frac{\partial}{\partial \tilde{t}} + \sum_{j=1}^{N} A_j^{\pm} \frac{\partial}{\partial \tilde{x}_j}$$

and

$$\text{(2.6)} \qquad A_j^{\pm}(u^{\pm}, \tilde{\beta})$$

$$\equiv [A_j(u^{\pm}) - \sum_{k=1}^{N} A_k(u^{\pm})c_{jk}(\tilde{x}, \tilde{t}, \tilde{\beta})] - c_{jk}(\tilde{x}, \tilde{t}, \tilde{\beta})I \quad .$$

The coefficients, $c_{j,k}(\tilde{x}, \tilde{t}, \tilde{\beta})$ are given by the formulae,

$$\text{(2.7)} \qquad c_{j,k}(\tilde{x}, \tilde{t}, \tilde{\beta}) \equiv \frac{\partial}{\partial x_k} (\rho\tilde{\beta}n_j)\Big|_{(\phi_{\tilde{t}}(\tilde{x},\tilde{\beta}),\tilde{t})} \quad ,$$

$$1 \leqslant j, k \leqslant N \ , \quad 0 \leqslant \tilde{t} \leqslant T_0 \quad ;$$

$$c_{k,0}(\tilde{x}, \tilde{t}, \tilde{\beta}, \frac{\partial\tilde{\beta}}{\partial t}) \equiv \frac{\partial\tilde{\beta}}{\partial t} \rho n_k\Big|_{(\Phi_{\tilde{t}}(\tilde{x},\tilde{\beta}),\tilde{t})} \quad ,$$

$$1 \leqslant k \leqslant N \ , \quad 0 \leqslant \tilde{t} \leqslant T_0 \quad ;$$

where $\tilde{\beta}$ is the composite function $\tilde{\beta}(\alpha(x), t)$. The Rankine-Hugoniot jump conditions from (0.7) are transformed into the boundary conditions on the fixed surface $M_0 \times [0, T_0]$ given by

(2.8)
$$G(u^+, u^-, \tilde{\beta}, \frac{\partial\tilde{\beta}}{\partial t}, \nabla_{\tan})\Big|_{M_0\times[0,T]}$$
$$\equiv \frac{d\tilde{\beta}}{dt}(u^+ - u^-) - n(\alpha)\cdot(F(u^+) - F(u^-))$$
$$+ (I + \tilde{\beta}W(\alpha))^{-1}\nabla_{\tan}\tilde{\beta}\cdot(F(u^+) - F(u^-))\Big|_{M_0\times[0,T]}$$
$$= 0 ,$$

where $W(\alpha)$ is the Weingarten map at $\alpha \in M_0$, mapping the tangent space into itself. The formula in (2.8) follows from the elementary fact that under the transformation below (2.4), the space-time normal to the shock front (n_t, n_x) transforms to a vector parallel to

$$\left(-\frac{d\tilde{\beta}}{dt}, n(\alpha) - (1 + \tilde{\beta}W(\alpha))^{-1}\nabla_{\tan}\tilde{\beta}\right) .$$

The following proposition is immediate by reversing the above steps:

PROPOSITION 2.1 (Reformulation on a Fixed Domain). Assume that there is a function $\tilde{\beta} \in C^2(M_0 \times (-\infty, \infty))$ and functions $(u^+(\tilde{x}, \tilde{t}), u^-(\tilde{x}, \tilde{t})) \in C^1(\Omega_+ \times (-\infty, \infty)) \oplus C^1(\Omega_- \times (-\infty, \infty))$ so that $\tilde{\beta}$ satisfies (2.3) and up to some sufficiently short time T_0, the equations

$$L^+(u^+, \tilde{\beta})u^+ = 0 \quad \text{in} \quad \Omega_+ \times [0, T_0] ,$$

$$L^-(u^-, \tilde{\beta})u^- = 0 \quad \text{in} \quad \Omega_- \times [0, T_0] ,$$

are satisfied together with the boundary conditions from (2.8) on $M_0 \times [0, T_0]$, then $S(t) = \{\alpha + \tilde{\beta}(\alpha, t)n(\alpha) | 0 \leq t \leq T_0\}$ defines a C^2 shock front and $u^+(\chi_t(x, \beta), t)$, $u^-(\chi_t(x, \beta), t)$ define functions on each side of this hypersurface which solve the shock front problem for $0 \leq t \leq T_0$.

In the remainder of this paper, we only study the equivalent problem in (2.7), (2.8) and drop the tildas for notational convenience.

2.B Derivation of the Higher Order Compatibility Conditions

The derivation of the higher order compatibility conditions follows the analogous derivation for standard hyperbolic mixed problems (see [13]) utilizing formal Cauchy-Kowaleski computations. Here we assume that there is a smooth solution (u^+, u^-, β) of the equations in (2.5) and (2.8) and derive the formal compatibility conditions up to order $s - 1$, s a given positive integer, which must be satisfied by the initial data. In the next subsection, we translate these formal conditions into a quantitative statement about classes of initial data which satisfy these compatibility conditions.

The zero-order compatibility conditions arise from evaluating the expression in (2.8) at $t = 0$; thus, necessarily,

$$\beta(\alpha, 0) = 0, \quad \frac{\partial \beta}{\partial t}\bigg|_{t=0} = \sigma(\alpha) , \tag{2.9}$$

where $\sigma(\alpha)$ is the initial shock speed function defined in (0.7). We compute relations among $(\partial^{k+1}/\partial t^{k+1})\beta(\alpha, 0)$ and $(\partial^k u^{\pm}/\partial t^k)(\alpha, 0)$ recursively by differentiating (2.8) and evaluating at $t = 0$. By utilizing (2.9) we see that these formulae are complicated but have the inductive pattern

$$\frac{\partial^{k+1}\beta}{\partial t^{k+1}}(u_0^+ - u_0^-) + (n(\alpha)\cdot A(u_0^+) - \sigma(\alpha))\frac{\partial^k u^+}{\partial t^k} - (n(\alpha)\cdot A(u_0^-) - \sigma(\alpha))\left.\frac{\partial^k u^-}{\partial t^k}\right|_{t=0} = H_k \tag{2.10}$$

for $1 \leqslant k \leqslant s - 1$ and $\alpha \in M_0$.

Here H_k is a complicated highly nonlinear function which involves only $\left.\frac{\partial^\ell \beta}{\partial t^\ell}\right|_{t=0}$, $0 \leqslant \ell \leqslant k$ and $\left.\frac{\partial^\ell u^\pm}{\partial t^\ell}\right|_{t=0}$, $0 \leqslant \ell \leqslant k - 1$, for $\alpha \in M_0$. By a standard computation using (2.5) - (2.7), we compute that for $1 \leqslant k \leqslant s - 1$,

$$\left.\frac{\partial^k u^\pm}{\partial t^k}\right|_{M_0\times\{0\}} = (-1)^k(n(\alpha)\cdot A(u_0^\pm) - \sigma I)^k \left.\frac{\partial^k u_0^\pm}{\partial n^k}\right|_{M_0} + J_k^\pm \, , \tag{2.11}$$

where $J_k^\pm$ is another complicated nonlinear function which involves only tangential derivatives to M_0 of order at most $k - \ell$ of the quantities $\left.\frac{\partial^\ell u_0^\pm}{\partial n^\ell}\right|_{M_0}$ for $0 \leqslant \ell \leqslant k - 1$ together with tangential derivatives up to order $k - \ell$ of $\left.\frac{\partial^\ell \beta}{\partial t^\ell}\right|_{t=0}$ for $0 \leqslant \ell \leqslant k$. Here, $\partial^\ell/\partial n^\ell$ denotes ℓ-th order differentiation normal to M_0. By substituting (2.11) into (2.10), we derive formally that initial data for a smooth solution of (2.5) and (2.8) should satisfy the <u>higher order compatibility conditions</u> up to order $s - 1$ at $t = 0$ given by

$$\left.\frac{\partial^{k+1}\beta}{\partial t^{k+1}}\right|_{t=0}(u_0^+ - u_0^-) + (-1)^k(n(\alpha)\cdot A(u_0^+) - \sigma(\alpha)I)^{k+1}\left.\frac{\partial^k u_0^\pm}{\partial n^k}\right|_{M_0} - (-1)^k(n(\alpha)\cdot A(u_0^-) - \sigma(\alpha)I)^{k+1}\left.\frac{\partial^k u^-}{\partial n^k}\right|_{M_0} = I_k \tag{2.12}$$

for $1 \leqslant k \leqslant s - 1$ where I_k is a well-determined function involving only the tangential derivatives up to order $k - \ell$ of $\left.\frac{\partial^\ell \beta}{\partial t^\ell}\right|_{t=0}$, $0 \leqslant \ell \leqslant k$, together with the tangential derivatives up to order $k - \ell$ of the quantities, $\left.\frac{\partial^\ell u_0^\pm}{\partial n^\ell}\right|_M$. For a given $\alpha \in M_0$ and a given k , the compatibility conditions in (2.12) involve M equations for the $2M + 1$ unknowns, $\{\partial^{k+1}\beta/\partial t^{k+1}, \partial^k u_0^+/\partial n^k, \partial^k u_0^-/\partial n^k\}$. The following algebraic lemma indicates that $(I - P^+(\alpha)) \left.\frac{\partial^k u_0^+}{\partial n^k}\right|_{M_0}$ and $(I - P^-(\alpha)) \left.\frac{\partial^k u_0^-}{\partial n^k}\right|_{M_0}$ can be prescribed arbitrarily and then $P^+(\alpha) \left.\frac{\partial^k u_0^+}{\partial n^k}\right|_{M_0}$, $P^-(\alpha) \left.\frac{\partial^k u_0^+}{\partial n^k}\right|_{M_0}$, and $\left.\frac{\partial^{k+1}}{\partial t^{k+1}} \beta\right|_{t=0}$ are uniquely determined by the equation in (2.12).

LEMMA 2.1. Under the main assumption in (1.8), if (v^+, v^-) are vectors in R^{2M} with $P^+v_+ = v_+$, $P^-v_- = v_-$ and β is a given constant. Suppose

$$M((\beta, v^+, v^-)) \equiv \beta(u_0^+ - u_0^-) + (-1)^k(n(\alpha)\cdot(u_0^+) - \sigma I)^{k+1} v_+$$

$$- (-1)^k(n(\alpha)\cdot(u_0^-) - \sigma I)^{k+1} v_- = 0 ;$$

then $(\beta, v^+, v^-) = 0$.

PROOF OF LEMMA 2.1. Consider the basis for the M-dimensional vector space of (β, v^+, v^-) that satisfy $P^+v_+ = v_+$, $P^-v_- = v_-$ given by the vectors

$$\{(1, 0, 0)\} \cup \{(0, r_j^+, 0)\}_{j=p+1}^{M} \cup \{(0, 0, r_j^-)\}_{j=1}^{p-1}$$

corresponding to the given p-shock where

$$n(\alpha)\cdot A(u_0^+)r_j^+ = \lambda_j(u_0^+)r_j^+ , \qquad p+1 \leqslant j \leqslant M ;$$

$$n(\alpha)\cdot A(u_0^-)r_j^- = \lambda_j(u_0^-)r_j^- , \qquad 1 \leqslant j \leqslant p-1 ;$$

then the corresponding $M \times M$ determinant is given explicitly by

$$\text{(2.13)} \qquad \det(M(1, 0, 0)), M((0, r_j^+, 0), M((0, 0, r_j^-)))$$

$$= (-1)^{k(M+1)} \prod_{p+1 \leqslant j \leqslant M} (\lambda_j^+ - \sigma)^{k+1} \prod_{1 \leqslant j \leqslant p-1} (\lambda_j^- - \sigma)^{k+1}$$

$$\times \det(r_M^+, \ldots , r_{M-p+1}^+, u_0^+ - u_0^-, r_{M-p-1}^-, \ldots , r_1^-) .$$

Since the shock front is uniformly stable to perturbations in multi-dimensions by our main assumption in (1.1), one expects that it is also stable to one-dimensional perturbations; in particular, by Proposition 3.1 in [12], the determinant on the right hand side of (2.13) is non-zero. From this fact, and (1.6), since the shock front is noncharacteristic so the other factors are also nonzero, the conclusion in Lemma 2.1 follows immediately.

2.C Large Classes of Initial Data Satisfying the Compatibility Conditions

For s an integer with $s > \frac{n}{2} + 1$, we assume that we are given functions $u_0^+(\alpha)$, $\sigma(\alpha)$, $u_0^-(\alpha)$ belonging to $H^{s+1}(M_0)$ and satisfying, for $\alpha \in M_0$,

$$\text{(2.14)} \qquad -\sigma(\alpha)(u_0^+(\alpha) - u_0^-(\alpha)) + n(\alpha)\cdot(F(u_0^+(\alpha) - F(u_0^-(\alpha))) = 0 .$$

For an arbitrary curved geometry for M_0 , we have shown how to construct such initial data for the physical examples including shock wave fronts of arbitrary strength in the introduction and also for sufficiently weak shocks for genuinely nonlinear wave fields in a general system in §1. The next proposition guarantees that large classes of initial data can be generated so that (2.14) and the compatibility equations in (2.12) are satisfied.

PROPOSITION 2.2. Assume that s is an integer with $s > \frac{n}{2} + 1$ and $(u_0^+(\alpha), u_0^-(\alpha), \sigma(\alpha)) \in H^{s+1}(M_0)$ satisfy (2.14). Assume $g_\ell^\pm(\alpha) \in H^{s+1-\ell}(M_0)$, $\ell = 1, \dots , s - 1$ are arbitrary functions satisfying $P^\pm g_\ell^\pm(\alpha) = 0$. Then there are $(u_0^+(x), u_0^-(x), \beta(\alpha, t))$ in $H^{s+1}(\Omega_+) \oplus H^{s+1}(\Omega_-) \oplus H^{s+2}_{comp}(M_0 \times (-\infty, \infty))$ so that

(1) $$u_0^\pm\Big|_{M_0} = u_0^\pm(\alpha) , \quad \beta(\alpha, 0) = 0 , \quad \frac{\partial\beta}{\partial t}(\alpha, 0) = \sigma(\alpha) ,$$

$$(I - P^\pm) \left.\frac{\partial^\ell u_0}{\partial n^\ell}\right|_{M_0} = g_\ell^\pm , \quad 1 \leq \ell \leq s - 1 .$$

(2) $\left.\dfrac{\partial^{k+1}\beta}{\partial t^{k+1}}\right|_{t=0}$ $\left.\dfrac{\partial^k u_0^\pm}{\partial n^k}\right|_{M_0}$ satisfy the compatibility conditions in (2.12) for $1 \leq k \leq s - 1$. Furthermore:

(3) Choose a fixed pair of functions $u_0^+(x), u_0^-(x)$ satisfying (1) and (2) above; then the initial data $u_0^+(x) + v_0^+(x), u_0^-(x) + v_0^-(x)$ also satisfies (1) and (2) with the same β provided that $v_0^+(x), v_0^-(x)$ are arbitrary functions in $H^{s+1}(\Omega_\pm) \cap H_0^s(\Omega_\pm)$.

REMARK. Tacitly, we have assumed above that $u_0^\pm + v_0^\pm$ defines coefficients satisfying (1.2) and remaining in the domain of hyperbolicity for $L^\pm$.

This requires trivial modifications of $u_0^{\pm} + v_0^{\pm}$ far away from the shock front, M_0.

Given (1) and (2) in the above proposition, (3) is obvious. Parts (1) and (2) of Proposition 2.2 follow from Lemma 2.1 and the following: consider the functions, I_k, defined in (2.12) where the arguments are evaluated on more general functions of the form $\nabla_{\tan}^j v_\ell^{\pm}$, $\nabla_{\tan}^j a_\ell$, $|j| + \ell \leq k$, $\ell < k$ instead of $\nabla_{\tan}^j \frac{\partial^\ell}{\partial n^\ell} u_0^{\pm}$, $\nabla_{\tan}^j \frac{\partial^\ell}{\partial t^\ell}\Big|_{t=0}$, etc.

LEMMA 2.2. Under the assumptions in Proposition 2, if $v_\ell^{\pm}$ and a_ℓ satisfy $v_\ell \in H^{s+1-\ell}$, $a_\ell \in H^{s+1-\ell}$, then there are finite constants C_k so that

$$\|I_k\|_{H^{s+1-K}(M_0)} \leq C_k .$$

We postpone a discussion of the tedious proof of Lemma 2.2 until the Appendix since we need more complicated and essentially similar facts using the same ideas in later sections.

REMARKS. If a variant of Lemma 2.2 could be established for non-integer s, we could use the weaker assumption $u_0^{\pm}(\alpha)$, $\sigma(\alpha) \in H^{s+1/2}(M_0)$ in (2.14). This would give precise results compatible with the trace theorem for functions in $H^{s+1}(\Omega_{\pm})$. By a simple induction argument on k, Lemma 2.1, Lemma 2.2, and the Banach algebra properties of $H^s(M_0)$, $s > \frac{n-1}{2}$, there are uniquely determined functions, $(h_\alpha^+(\alpha), h_\ell^-(\alpha), a_\ell(\alpha)) \in H^{s+1-\ell}(M_0)$, $1 \leq \ell \leq s-1$ with

$$P^{\pm}h^{\pm}(\alpha) = h_\ell^{\pm}(\alpha) ,$$

$$v_\ell^{\pm}(\alpha) \equiv h_\ell^{\pm}(\alpha) + g_\ell^{\pm}(\alpha) ,$$

and so that the $s - 1$ compatibility equations in (2.12) have the solutions

$$a_K(\alpha)(u^+ - u^-) + (-1)^k(n(\alpha)\cdot A(u_0^+) - \sigma(\alpha)I)^{k+1}v_k^+$$

$$- (-1)^k(n(\alpha)\cdot A(u_0^-) - \sigma(\alpha)I)^{k+1}v_k^- = I_k(\nabla^j_{\tan}v^{\pm}_\ell, \nabla^j_{\tan}a_\ell) ;$$

$$1 \leqslant k \leqslant s - 1 , \quad |j| + \ell \leqslant k , \quad |\ell| < k .$$

By the trace theorem, there exists $\beta(\alpha, t) \in H^{s+2}_{comp}(M_0 \times (-\infty, \infty))$ and $u^+(x)$, $u^-(x)$ belonging to $H^{s+1}(\Omega_+)$, $H^{s+1}(\Omega_-)$ respectively so that

$$\beta(\alpha, 0) = 0 ,$$

$$\left.\frac{\partial\beta}{\partial t}\right|_{t=0} = \sigma(\alpha) ,$$

$$\left.\frac{\partial^{\ell+1}}{\partial t^{\ell+1}}\right|_{t=0} = a_\ell(\alpha) , \quad \ell = 1, \dots , s - 1 ,$$

$$\left.\frac{\partial^\ell u_0^{\pm}}{\partial n^\ell}\right|_{M_0} = v^{\pm}_\ell(\alpha) .$$

This completes the construction for (1) and (2) in the above Proposition.

2.D Construction of an Approximate Solution

Our objective here is to construct an approximate solution of (2.4) and (2.8) which vanishes to order $s - 1$ at $t = 0$ and satisfies the linearized well-posedness condition from (1.9) for all points on $M_0 \times (-\infty, \infty)$ with the smaller constant $\frac{\delta}{2}$ replacing δ on the right hand side of (1.9).

For the moment, we ignore the additional requirement from (1.9) and solve (2.5) and (2.8) to order $s-1$ at $t = 0$. For any fixed integer s with $s > \frac{n}{2} + 1$, we consider any of the shock front initial data $(u_0^+, u_0^-) \in H^{s+1}(\Omega_+ \cap \{x \mid |x| < R\}) + H^{s+1}(\Omega_-)$ and denote by $\tilde{\beta}_0 \in H^{s+2}_{comp}(M_0 \times (-\infty, \infty))$, the corresponding approximate shock front function constructed to satisfy the compatibility conditions in (2.12) for $1 \leq k \leq s-1$. First, we construct

$$\tilde{u}_0^+(x, t) \in \bigcap_{j=0}^{s+1} C^j((-\infty, \infty), H^{s+1-j}(\Omega_+ \cap |x| < CR)) ,$$

$$\tilde{u}_0^-(x, t) \in \bigcap_{j=0}^{s+1} C^j((-\infty, \infty)), H^{s+1-j}(\Omega_-))$$

so that

$$L^+(\tilde{u}_0^+, \tilde{\beta}_0)\tilde{u}_0^+ = \tilde{F}_0^+ ; \tag{2.15}$$

$$L^-(\tilde{u}_0^-, \tilde{\beta}_0)\tilde{u}_0^- = \tilde{F}_0^- ;$$

$$G(\tilde{u}_0^+, \tilde{u}_0^-, \tilde{\beta}_0, (\tilde{\beta}_0)_t, \nabla_{tan}\tilde{\beta}_0) = \tilde{g}_0 ;$$

where $\tilde{F}_0^+, \tilde{F}_0^- \in H^s_{comp}(\overline{\Omega}_\pm \times [-T, T])$ for any $T > 0$, $\tilde{g}_0 \in H^s_{loc}(M_0 \times (-\infty, \infty))$, and

$$\partial_t^j \tilde{F}_0^\pm \Big|_{t=0} = 0 , \quad 0 \leq j \leq s-1 ; \tag{2.16}$$

$$\partial_t^j g_0 \Big|_{t=0} = 0 , \quad 0 \leq j \leq s-1 .$$

The function $\tilde{\beta}_0 \in H^{s+2}_{comp}(M_0 \times (-\infty, \infty))$ has already been determined from our discussion above. We write the operators, $L^{\pm}(\tilde{u}_0^{\pm}, \tilde{\beta}_0)$ in the form, $\frac{\partial}{\partial t} - G^{\pm}$, and set

$$m_K^{\pm} = \left.\frac{\partial^k \tilde{u}_0^{\pm}}{\partial t^k}\right|_{t=0}, \quad 0 \leqslant k \leqslant s \ .$$

The formal Cauchy-Kowaleski computation yields the recursion relations,

$$(2.17) \qquad m_k^{\pm} = \sum_{i=0}^{k-1} \binom{k-1}{i} G_i^{\pm}(0) m_{k-1-i}$$

$$\equiv L_k^{\pm}(m_0^{\pm}, \ldots, m_{k-1}^{\pm}, \tilde{\beta}_0), \quad 1 \leqslant k \leqslant 5 \ ,$$

where $G_0^{\pm} = G^{\pm}$ and $G_i = \sum_{j=1}^{N} (-1) \frac{\partial^i}{\partial t^i} (A_j^{\pm}) \frac{\partial}{\partial x_j}$. Analogous to Lemma 2.2, we have the following fact with a proof which we similarly postpone discussion until the Appendix.

LEMMA 2.3. If $\tilde{\beta}_0 \in H^{s+2}_{comp}(M_0 \times (-\infty, \infty))$ and $m_j^{\pm} \in H^{s+1-j}$, $1 \leqslant j \leqslant k - 1$, then $L_k^{\pm} \in H^{s+1-k}$ for $1 \leqslant k \leqslant s$, thus $m_k^{\pm} \in H^{s+1-k}$.

As a consequence of Lemma 2.3 and a simple induction with k, we only need to find $\tilde{u}_0^{\pm}$ with

$$\left.\frac{\partial^k u_0}{\partial t^k}\right|_{t=0} = m_k^{\pm}, \quad 0 \leqslant k \leqslant s \ ,$$

and with the regularity stated above (2.15). First, extend $m_k^{\pm}$ across M_0 by the Lion's extension procedure so that these functions are defined in all of space. For simplicity, define $m_{s+1}^{\pm} = 0$ and let P denote a scalar, constant coefficient, strictly hyperbolic operator of order $s+2$ with t, a time-like direction. Choose $\tilde{u}_0^{\pm}$ to satisfy the scalar Cauchy problem,

$$P\tilde{u}_0^{\pm} = 0 \; ;$$

$$\left.\frac{\partial^k u_0}{\partial t^k}\right|_{t=0} = m_k^{\pm} , \quad 0 \leqslant k \leqslant s+1 ,$$

component-wise.

Standard energy estimates guarantee that $\tilde{u}_0^{\pm} \in \bigcap_{j=0}^{s+1} C^j((-\infty, \infty),$ $H^{s+1-j}(\Omega_{\pm}))$, etc. Furthermore, since $s > \frac{n}{2} + 1$, it follows from Sobolev's lemma that there is a sufficiently small T_0 so that

$$\text{(2.18)} \qquad \max_{\substack{|t| \leqslant T_0 \\ x \in \Omega^+}} |\tilde{u}_0^+(x, t) - u_0^+(x)| + \max_{\substack{|t| \leqslant T_0 \\ x \in \Omega^-}} |\tilde{u}_0^-(x, t) - u_0^-(x)|$$

$$+ \max_{\substack{|t| \leqslant T_0 \\ \alpha \in M_0}} (|\tilde{\beta}| + |\nabla_{\tan}\tilde{\beta}|) + \max_{\substack{|t| \leqslant T_0 \\ \alpha \in M_0}} \left|\frac{\partial\tilde{\beta}}{\partial t} - \sigma(\alpha)\right| < \frac{\delta}{2} ,$$

where with a slight abuse of notation $\delta = \min(\delta_3, \delta)$ where δ_3 is defined in (2.3) above and δ has been defined in (1.9). We claim that the approximation $(\tilde{u}_0^+, \tilde{u}_0^-, \tilde{\beta}_0)$ also satisfies the condition in (2.15) and (2.16); this follows from the fact that (u_0^+, u_0^-) are shock front initial data so that $(u_0^+, u_0^-, \tilde{\beta}_0)$ satisfy the compatibility conditions in (2.12) together with a repetition of the calculations in (2.9) - (2.12) utilizing the Banach algebra properties of $H^s(M_0)$ for $s > \frac{n}{2} + 1$ -- we omit the details here since they can be justified by repeating the argument of Lemma 2.2.

Our next step is to take this approximate solution and modify the coefficients in a trivial fashion for $|t| > T_0/2$ so that the hypotheses of Theorem 2 in [12] are satisfied globally in time together with (2.18). We consider $\beta_0 = \tilde{\rho}(t)\tilde{\beta}_0$ and the associated transformation, $\chi_t(x, \tilde{\rho}(t)\tilde{\beta}_0)$, defined in (2.2) where $\tilde{\rho}(t)$ is a positive smooth cut-off function with $\tilde{\rho}(t) \equiv 1$ for $|t| < \frac{1}{2} T_0$ and $\tilde{\rho}(t) \equiv 0$ for $|t| > T_0$. We define extended coefficients from (2.7) via the formulae,

$$c_{j,k} = \frac{\partial}{\partial x_k}(\rho\tilde{\rho}\tilde{\beta}_0 n_j)\Big|_{(\Phi_t(\tilde{x},\tilde{\rho}(t)\tilde{\beta}_0),\tilde{t})}, \qquad 1 \leqslant j,\ k \leqslant N \ ;$$

$$c_{k,0} = \left(\frac{\partial\tilde{\beta}_0}{\partial t}\tilde{\rho} + (I - \tilde{\rho})\sigma_\gamma\right)\rho n_k\Big|_{(\Phi_t(\tilde{x},\tilde{\rho}(t)\tilde{\beta}_0),\tilde{t})}, \qquad 1 \leqslant k \leqslant N \ .$$

We also consider the boundary conditions in (2.8) and (1.11) with $\partial\tilde{\beta}_0/\partial t$ replaced by $\tilde{\rho}(\partial\beta_0/\partial t) + (1 - \tilde{\rho})\sigma_\gamma(\alpha)$ where σ_γ is a mollification of $\sigma(\alpha)$. After extending the initial data, u_0^+, u_0^- across M_0 by Lion's reflection, we let $u_{0,\gamma}^+$, $u_{0,\gamma}^-$ denote the standard mollification of $u_0^\pm$ in R^N where γ is the mollification parameter. We define extended values for u_0^+, u_0^- via the formulae

$$u_0^\pm = \tilde{\rho}\tilde{u}_0^\pm + (I - \tilde{\rho})u_{0,\gamma}^\pm \ . \tag{2.20}$$

From (2.19) and (2.20), it is clear that the requirement in (9) is satisfied for all time for the operators $L^\pm(u_0^\pm, \beta_0)$ and the associated linearized boundary problems in (2.20) by merely choosing γ sufficiently small. Furthermore, from (2.19) and (2.20), all the requirements of Theorem 2 of [12] regarding these coefficients for $|t| > T_0$ are satisfied. By summarizing all of the above steps we have the following.

PROPOSITION 2.3 (The Approximate Solution). For $s > \frac{n}{2} + 1$, s an integer. Under the hypotheses of Proposition 2.2 on the initial data, there are functions $\beta_0 \equiv \tilde{\rho}(t)\tilde{\beta}_0 \in H^{s+2}_{\text{comp}}(M_0 \times (-\infty, \infty))$ and (u_0^+, u_0^-) defined in (2.20) so that

$$\text{(1)}\quad u_0^+ \in \bigcap_{j=0}^{s+1} C^j((-\infty,\infty),\ H^{s+1-j}(\Omega_+ \cap |x| < 2R)\ ;$$

$$\cap$$

$$u_0^- \in \bigcap_{j=0}^{s+1} C^j((-\infty,\infty),\ H^{s+1-j}(\Omega_-)\ .$$

(2) (u_0^+, u_0^-, β_0) is an approximate solution of (2.5) and (2.8) in the sense that

$$L^+(u_0^+, \beta_0)u_0^+ = F_+ \quad \text{in} \quad \Omega^+ \times (-\infty, \infty)\ ;$$

$$L^-(u_0^-, \beta_0)u_0^- = F_- \quad \text{in} \quad \Omega^- \times (-\infty, \infty)\ ;$$

$$G(u_0^+, u_0^-, \beta_0, \nabla_{\tan}, \beta_0, (\beta_0)_t) = g_0 \quad \text{in} \quad M_0 \times (-\infty, \infty)\ ,$$

where $F_+, F_- \in H^s_{\text{comp}}(\Omega_\pm \times [-T_0, T_0])$, $g_0 \in H^s_{\text{loc}}(M_0 \times (-\infty, \infty))$, and

$$\partial_t^j F\Big|_{t=0} = 0\ , \quad 0 \leqslant j \leqslant s-1\ ;$$

$$\partial_t^j g\Big|_{t=0} = 0\ , \quad 0 \leqslant j \leqslant s-1\ .$$

(3) With the extended definition of the coefficients in and below (2.19), (u_0^+, u_0^-, β_0) satisfy all the conditions in (1.9) so that the linearized boundary value problem is well-posed. Furthermore,

all the hypotheses of Theorem 2 of [12] are valid for the extended coefficients for $|t| > T_0$.

REMARK. We shall continue to denote by $L^{\pm}(u^{\pm}, \beta)$, and $\mathcal{B}(u^{+}, u^{-}, \beta, \beta_t, \nabla_{tan}\beta)$ the operators with extended coefficients defined in (2.19) and below wherever (u^{+}, u^{-}, β) has the form $(u_0^{+}, u_0^{-}, \beta_0) + (v^{+}, v^{-}, \phi)$ where $\phi|_{t=0} = 0$ and

$$\left.\frac{\partial^k v^{\pm}}{\partial t^k}\right|_{t=0} = \left.\frac{\partial^{k+1}\phi}{\partial t^{k+1}}\right|_{t=0} = 0 \ , \qquad 0 \leqslant k \leqslant s \ ,$$

by utilizing the same fixed function $\tilde{\rho}(t)$ as a cut-off. Of course, by Proposition 2.1, we are only constructing a solution of the shock front problem for $|t| < \frac{1}{2} T_0$ so our construction already has been limited to sufficiently short times.

Without further comments, we always assume tacitly for the remainder of the paper that we are dealing with shock front initial data satisfying the uniform stability condition in (1.9) so that Proposition 2.2 and Proposition 2.3 have been applied to construct the corresponding initial data and approximate solution.

§3. THE ITERATION SCHEME

Given the shock front initial data, we let (u_0^+, u_0^-, β_0) be the corresponding approximate solution of (2.5) and (2.8) constructed in Proposition 2.3 where according to the hypotheses of our main theorem, we fix s as an integer with $s \geqslant 2[\frac{n}{2}] + 7$. We define functions inductively with the form,

$$\text{(3.1)} \qquad \left.\begin{aligned} u_n^+ &= u_0^+ + v_n^+ \\ u_n^- &= u_0^- + v_n^- \\ \beta_n &= \beta_0 + \phi_n \end{aligned}\right\} \quad n = 0, 1, 2, \ldots$$

and introduce the notiation,

$$\text{(3.2)} \qquad V_n = (v_n^+, v_n^-, \phi_n) \ .$$

We set $V_0 = (0, 0, 0)$ and below we indicate how to construct V_{n+1} inductively given V_n.

First, we introduce the standard hyperbolic L^2 weighted Sobolev norms,

$$|F^+|_{0,\eta,T}^2 = \int_0^T \int_{\Omega_+} e^{-2\eta t} |F^+|^2 \, dx \, dt \quad ;$$

$$|F^-|_{0,\eta,T}^2 = \int_0^T \int_{\Omega_+} e^{-2\eta t} |F^-|^2 \, dx \, dt \quad ;$$

$$\langle g\rangle^2_{0,\eta,T} = \int_0^T \int_{M_0} e^{-2\eta t}|g|^2 \, ds\, dt \quad ,$$

for $\eta \geqslant 1$ and also the corresponding higher order weighted Sobolev norms. These higher order wieghted Sobolev norms by introducing a fixed partition of unity subordinate to co-ordinate transformations mapping M_0 locally to the half-space $x_N = 0$ and introducing the weighted higher derivative norms corresponding to (3.3) and defined on a half-space by

$$(3.4) \quad \langle g\rangle^2_{s,\eta,T} = \sum_{|\alpha_1|+|\alpha_2|+|\alpha_3|=s} \int_0^T \int_{x_N=0} |\eta|^{2|\alpha_1|} e^{-2\eta t}|D_x^{\alpha_2} D_t^{\alpha_3} g|^2 \quad ;$$

$$|F^+|^2_{s,\eta,T} = \sum_{K=0}^{s} \int_0^\infty \langle D^K_{x_N} F^+\rangle^2_{s-K,\eta,T} \, dx_N \quad ;$$

$$|F^-|^2_{s,\eta,T} = \sum_{K=0}^{s} \int_{-\infty}^0 \langle D^K_{x_N} F^-\rangle^2_{s-K,\eta,T} \, dx_N \quad .$$

Here $x' = (x_1, \ldots, x_{N-1})$ corresponds to the tangential directions to $x_N = 0$. We use the notation, $\langle g\rangle^2_{s,\eta}$, etc. when $T = +\infty$ and also we denote the corresponding conventional unweighted Sobolev norms of order s by $|v|^2_{s,T}$, $|v^{\pm}|^2_{s,T}$ defined on the respective regions $M_0 \times [0, T]$ and $\Omega_{\pm} \times [0, T]$. For the regions $M_0 \times (-\infty, \infty)$ and $\Omega_{\pm} \times (-\infty, \infty)$, we use the notations, $\langle v\rangle^2_s$, $|v^{\pm}|^2_s$ for the standard Sobolev norms of order s. We introduce the following norm for functions, V, where $V = (v^+, v^-, \phi)$,

$$(3.5) \quad |||V|||^2_{s,\eta,T} \equiv \langle\phi\rangle^2_{s+1,\eta,T} + \sum_{j=0}^{s} \langle\frac{\partial^j v^+}{\partial n^j}\rangle^2_{s-j,\eta,T} + \sum_{j=0}^{s} \langle\frac{\partial^j v^-}{\partial n^j}\rangle^2_{s-j,\eta,T}$$

$$+ \eta\left[|v^+|^2_{s,\eta,T} + |v^-|^2_{s,\eta,T}\right] \quad .$$

We also introduce the analogous unweighted Sobolev norms, $|||V|||^2_{s,T}$; here an equivalent norm is defined by setting $\eta \equiv I$ in (3.5).

For the iterative construction, we assume by induction that V_n satisfies the following conditions:

(1) V_n vanishes for $t < 0$ and $t > T_0$ where T_0 is defined above (2.18) and also for $|x| > cR_0$.

(2) $C_s|||V_n|||^2_{s,T_0} \leq \frac{\delta}{4}$ where δ is the fixed positive number in (2.18), s is a fixed integer with $s \geq 2[\frac{n}{2}] + 7$ and $C_s^{1/2}$ is the Sobolev embedding constant estimating V_n in

$$C((\Omega_+ \times [0, T_0] \oplus C(\Omega_- \times [0, T_0] \oplus C^1(M_0 \times [0, T_0]) \tag{3.6}$$

by $|||V_n|||_{s,T_0}$.

(3) $\partial_t^j v_n^{\pm}\big|_{t=0} = 0 , \quad 0 \leq j \leq s - 1$;

$\partial_t^{\ell} \phi_n\big|_{t=0} = 0 , \quad 0 \leq \ell \leq s$.

V_0 trivially satisfies the conditions in (3.6) and below, we show how to define V_{n+1} satisfying (3.6) provided that V_n also satisfies these conditions. We need the following fact:

<u>LEMMA 3.1</u>. Assume that T satisfies, $0 < T \leq T_0/2$. Assume that $V = (v^+, v^-, \phi)$ satisfies, $|||V|||^2_{s,\eta,T} < \infty$ and also

$$(3.7)\qquad \partial_t^j v^{\pm}\Big|_{t=0} = 0 \ , \quad 0 \leqslant j \leqslant s-1 \ ;$$

$$\partial_t^{\ell}\phi\Big|_{t=0} = 0 \ , \quad 0 \leqslant \ell \leqslant s \ ;$$

then there exists an extended function, $E_T V$, so that

(1) $E_T V = v$ for $0 \leqslant t \leqslant T$;

(2) $E_T V = 0 \begin{cases} \text{for } t < 0 \\ \text{for } t > T_0 \end{cases}$;

(3) $|||E_T V|||^2_{\tilde{s},\eta,T_0} \leqslant C_s |||V|||^2_{\tilde{s},\eta,T}$, $0 \leqslant \tilde{s} \leqslant s$;

where C_s is independent of T and $\eta \geqslant 1$ and depends only on the integer s.

<u>REMARK</u>. Without a condition like (3.7), the estimate in (3) is false and the constant $C(T)$ will blow up as $T \longrightarrow 0$. Lemma 3.1 is an exercise proved by a repetition of the standard proof of the Lion's extension lemma ([6]) so we omit the details. Proceeding formally for the moment (the subsequent lemmas justify these formal considerations), we define the iteration scheme. In the interior, we use conventional fixed point iteration. Thus, we set

$$(3.8)\qquad F^{\pm}_{n+1} \equiv \begin{cases} 0 & t < 0 \\ -L^{\pm}(u^{\pm}_n, \beta_n)u^{\pm}_0 \ , & t > 0 \end{cases} ,$$

and choose $v^{\pm}_{n+1}$, $n = 0, 1, 2, \dots$, to satisfy the

(3.9) Interior Iteration Scheme

$$L^+(u_n^+, \beta_n)v_{n+1}^+ = F_{n+1}^+ \quad \text{in} \quad \Omega_+ \times (-\infty, \infty) \ ,$$

$$v_{n+1}^+ \text{ vanishes for } t < 0 \ ;$$

$$L^-(u_n^-, \beta_n)v_{n+1}^- = F_{n+1}^- \quad \text{in} \quad \Omega_- \times (-\infty, \infty) \ ,$$

$$v_{n+1}^- \text{ vanishes for } t < 0 \ .$$

For defining the iteration scheme on the boundary, recall that we would like to find u^+, u^-, β so that the nonlinear boundary condtions from (2.8) are satisfied, i.e.,

(3.10)

$$G(u^+, u^-, \beta, \frac{\partial\beta}{\partial t}, \nabla_{tan}\beta) \equiv \frac{d\beta}{dt}(u^+ - u^-) - n(\alpha)\cdot(F(u^+) - F(u^-))$$

$$+ (1 + \beta W(\alpha))^{-1}\nabla_{tan}\beta\cdot(F(u^+) - F(u^-))$$

$$= 0 \quad \text{on} \quad M_0 \times [0, T] \ .$$

We introduce the large vector, $w = (u^+, u^-, \beta, \frac{\partial\beta}{\partial t}, \nabla_{tan}\beta)$; at each point, $(\alpha, t) \in M_0 \times [0, T]$ for sufficiently small T we would like to satisfy the nonlinear equation, $G(w) = 0$. We do this by applying Newton's method pointwise on the boundary. Abstractly, this leads to the iteration scheme,

$$d_w G(w_n)(w_{n+1} - w_n) = -G(w_n) \ , \tag{3.11}$$

where $d_w G$ is just the ordinary vector calculus differential of G with respect to all arguments. If we set $w_n \equiv w_0 + \tilde{w}_n$, $\tilde{w}_0 = 0$, then (3.11) becomes

(3.12) $$d_w G(w_n)\tilde{w}_{n+1} = -G(w_n) + d_w G(w_n)\tilde{w}_n \ .$$

The well-known quadratic character of Newton's method in finite dimensions implies the following two useful estimates:

(3.13)
$$\text{a)} \quad -G(w_n) + d_w G(w_n)\tilde{w}_n = -G(w_0) + O(|\tilde{w}_n|^2) \ ;$$
$$\text{b)} \quad d_w G(w_n)(\tilde{w}_{n+1} - \tilde{w}_n) = O(|\tilde{w}_n - \tilde{w}_{n-1}|^2) \ ,$$

for any smooth function $G(w)$. We apply the abstract discussion in (3.11) - (3.13) to the specific nonlinear function in (3.10) to arrive at the

(3.14) <u>Iteration Scheme on the Boundary</u>

$$\left(\frac{d\phi_{n+1}}{dt} - \frac{d\phi_n}{dt}\right)(u_n^+ - u_n^-) + \frac{d\beta_n}{dt}((v_{n+1}^+ - v_n^+) - (v_{n+1}^- - v_n^-))$$
$$- [n(\alpha)\cdot A(u_n^+)(v_{n+1}^+ - v_n^+) - n(\alpha)\cdot A(u_n^-)(v_{n+1}^- - v_n^-)]$$
$$+ (1 + \beta_n W(\alpha))^{-1}\nabla_{\tan}(\phi_{n+1} - \phi_n)\cdot(F(u_n^+) - F(u_n^-))$$
$$+ (1 + \beta_n W(\alpha))^{-1}\nabla_{\tan}(\beta_n)\cdot(A(u_n^+)(v_{n+1}^+ - v_n^+) - A(u_n^-)(v_{n+1}^- - v_n^-)$$
$$- (\phi_{n+1} - \phi_n)[1 + \beta_n W(\alpha)]^{-2} W(\alpha)\nabla_{\tan}\beta_n\cdot(F(u_n^+) - F(u_n^-))$$
$$= -G_+(u_n^+, u_n^-, \beta_n, \frac{\partial\beta_n}{\partial t}, \nabla_{\tan}\beta_b) \qquad \text{on } M_0 \times (-\infty, \infty) \ ;$$

$$\phi_{n+1} \text{ vanishes for } t < 0 \ ,$$

where

$$G_+(u^+, u^-, \beta, \frac{\partial\beta}{\partial t}, \nabla_{\tan}\beta) = \begin{cases} G, & t \geqslant 0 \\ 0, & t < 0 \end{cases},$$

with G already defined in (3.10).

By introducing the notation corresponding to (1.10) above given by

(3.15) (1) $$M_n(v^+, v^-) \equiv ((1 + \beta_n W(\alpha))^{-1}\nabla_{\tan}\beta_n - n(\alpha)\cdot A(u_n^+)v^+$$

$$+ \frac{d\beta_n}{dt} I)v^+ - ((1 + \beta_n W(\alpha)^{-1}\nabla_{\tan}\beta_n - n(\alpha))\cdot A(u_n^-) + \frac{d\beta_n}{dt} I)v^- ;$$

(2) $b_{0,n} = u_n^+ - u_n^-$;

(3) for any tangent vector, T, to M_0 :

$$b_{1,n}\cdot T \equiv (I + \beta_n W(\alpha))^{-1}(T)\cdot(F(u_n^+) - F(u_n^-)) \quad ;$$

(4) $b_{2,n} \equiv -(I + \beta_n W(\alpha))^{-2} W(\alpha)\nabla_{\tan}\beta_n\cdot(F(u_n^+) - F(u_n^-))$;

we define B_n via

(3.16) $$B_n(v^+, v^-, \phi) \equiv b_{0,n}\phi_t + b_{1,n}\cdot\nabla_{\tan}\phi + M_n(v^+, v^-) + b_{2,n}\phi \ .$$

Then, the iteration scheme on the boundary from (3.14) can be written concisely as

(3.17) <u>Boundary Conditions for Iteration Scheme</u>

$$B_n(v_{n+1}^+, v_{n+1}^-, {}_{n+1})$$

$$= -G_+(u_n^+, u_n^-, \beta_n, (\beta_n)_t, \nabla_{\tan}\beta_n) + B_n(v_n^+, v_n^-, \phi_n) \equiv g_{n+1}$$

on $M_0 \times (-\infty, \infty)$; ϕ_{n+1} vanishes for $t < 0$.

To show that the equations in (3.9) and (3.17) have a unique solution, we need the following main fact regarding this linear problem. §5 is devoted to the detailed proof of this result.

PROPOSITION 3.1 (The Main Linear Estimate). Assume that (v_n^+, v_n^-, ϕ_n) satisfy all the conditions in (3.6) for $s \geq 2[\frac{n}{2}] + 7$, s a fixed integer. Assume that F_+, F_- belong to $H^s(\Omega_\pm \times (-\infty, \infty))$, $g \in H^s(M_0 \times (-\infty, \infty))$, that these functions vanish for $t < 0$, and also that they satisfy

$$\partial_t^j F\Big|_{t=0} = \partial_t^j g\Big|_{g=0} = 0 , \qquad 0 \leq j \leq s - 1 .$$

Consider the linear boundary value problem,

$$(3.18) \qquad \begin{aligned} L^+(u_n^+, \beta_n)v^+ &= F^+ \quad \text{in} \quad \Omega_+ \times (-\infty, \infty) \; ; \\ L^-(u_n^-, \beta_n)v^- &= F^- \quad \text{in} \quad \Omega_- \times (-\infty, \infty) \; ; \\ B_n(v^+, v^-, \phi) &= g \quad \text{in} \quad M_0 \times (-\infty, \infty) \; ; \\ (v^+, v^-, \phi) &= V \quad \text{vanishes for} \quad t < 0 \; . \end{aligned}$$

Then the boundary value problem in (3.18) has a unique solution, V, with $|||V|||^2_{s,\eta,T}$ finite for any $T > 0$. There are fixed constants C_1 and C_2 independent of (v_n^+, v_n^-, ϕ_n) satisfying (3.6) so that this unique solution satisfies the estimates,

$$(3.19) \qquad |||V|||^2_{s,\eta,T} \leq C_1 \left(\frac{|F^+|^2_{s,\eta,T} + |F^-|^2_{s,\eta,T}}{\eta} + \langle g \rangle^2_{s,\eta,T} \right)$$

for $\eta > C_2$ provided that T satisfies $0 < T < T_0$. The triple V also satisfies

$$\partial_t^k v^{\pm}\Big|_{t=0} = 0 \ , \quad 0 \leq K \leq s-1 \ ; \tag{3.20}$$

$$\partial_t^{\ell}\phi\Big|_{t=0} = 0 \ , \quad 0 \leq \ell \leq s \ .$$

Furthermore, the same estimate as in (3.19) is valid for solutions of (3.18) with $s = 0$.

The following lemma is proved in a similar fashion to Lemmas 2.2 and 2.3 discussed in the Appendix. However, a more precise version of this lemma will already be used in §5 so we won't discuss the proof of Lemma 3.2 until the Appendix.

LEMMA 3.2. Assume that V_n satisfies the conditions in (3.6), then the functions $F^{\pm}_{n+1}$, g_{n+1} defined in (2.8_ and (3.17) satisfy

$$F^{\pm}_{n+1} \in H^s(\Omega_{\pm} \times [0, T_0]) \ ;$$

$$g_{n+1} \in H^s(M_0 \times [0, T_0]) \ ,$$

with the compatibility conditions,

$$\partial_t^j F^{\pm}_{n+1}\Big|_{t=0} = \partial_t^j g_{n+1}\Big|_{t=0} = 0$$

for $0 \leq j \leq s-1$.

Finally, assuming Proposition 3.1 and Lemma 3.2, we show how to define V_{n+1} in the iteration scheme so that all the conditions in (3.6) are satisfied. We let $(v^+_{n+1}, v^-_{n+1}, \phi_{n+1})$ denote the unique solution of the boundary value problem in (3.9) and (3.17). It follows from Proposition 3.1 and Lemma 3.2 that $|||(v^+_{n+1}, v^-_{n+1}, \phi_{n+1})|||^2_{s,\eta,T_0/2}$ is finite and

$$\partial^j_t v^{\pm}_{n+1}\Big|_{t=0} = 0 , \quad 0 \leqslant j \leqslant s-1 ; \tag{3.21}$$

$$\partial^{\ell}_t \phi_{n+1}\Big|_{t=0} = 0 , \quad 0 \leqslant \ell \leqslant s .$$

Thus, $(v^+_{n+1}, v^-_{n+1}, \phi_{n+1})$ vanishes for $t < 0$ and satisfies condition (3) of (3.6). To define V_{n+1} we choose a fixed number ε_0 lying inside the interval of ε_0 satisfying the fixed condition,

$$C_s \varepsilon_0 < \frac{\delta}{8} , \tag{3.22}$$

where $\frac{\delta}{4}$ is the constant in (2) of (3.6) and $C_s^{1/2}$ is the product of the constant in (3) of Lemma 3.1 (actually the constant in (3) of Lemma 3.1 when $\eta \equiv 1$ is the exact value needed in (3.22)) and the Sobolev embedding constant estimating the $C(\Omega_{\pm} \times [0, T_0])$, $C^1(M_0 \times [0, T_0])$ norms in terms of $|||V|||_{s,T_0}$. A judicious choice of ε_0 satisfying (3.22) will be made later in §4 in the convergence proof. Consider the time, $\mathcal{T}_{n+1}$, defined by

$$\mathcal{T}_{n+1} = \min\{T \,|\, |||(v^+_{n+1}, v^-_{n+1}, \phi_{n+1})|||^2_{s,T} \geqslant \varepsilon_0\} , \tag{3.23}$$

and let T_{n+1} be given by

$$T_{n+1} = \min\{\frac{T_0}{2}, \mathcal{T}_{n+1}\} . \tag{3.24}$$

We define V_{n+1} by the formula,

$$V_{n+1} \equiv E_{T_{n+1}}(v^+_{n+1}, v^-_{n+1}, \phi_{n+1}) \quad , \tag{3.25}$$

where E_T is the extension operator from Lemma 3.1. Then, it follows from Lemma 3.1 and (3.21), (3.22), above V_{n+1} satisfies all three of the conditions in (3.6) provided that V_n did. Thus, by induction, since $V_0 = (0, 0, 0)$, the successive iterates $\{V_n\}_{n=0}^{\infty}$ have been well-defined by (3.25) and satisfy all of the conditions in (3.6). In the next section, often we will abuse notation slightly by omitting the extension operator in (3.25) and calling V_{n+1}, the triple $(v^+_{n+1}, v^-_{n+1}, \phi_{n+1})$.

§4. CONVERGENCE OF THE ITERATION SCHEME

In order to prove the convergence of the iteration shceme, we utilize the general strategy for proving the short-time existence of solutions for quasi-linear hyperbolic systems ([8]). However, our choice of norms is not standard and is dictated by the nature of the iteration scheme defined in (3.9) and (3.17) -- in particular, by the interaction between the free surface corrections, ϕ_{n+1}, and the interior solutions, $v^{\pm}_{n+1}$. In proving the boundedness in a high norm, we utilize the norms, $|||\cdot|||_{s,\eta(T),T}$ where $\eta(T)$ has the form, $\eta(T) = c_0 T^{-\beta}$, β fixed with $0 < \beta < 1$. To prove the contraction stimate in a low norm, we use the norm, $|||\cdot|||_{0,\eta_0,T}$ with η_0 fixed and sufficiently large.

Recall that until the time T_{n+1}, the term, V_{n+1} in the iteration scheme is defined by $(v^{+}_{n+1}, v^{-}_{n+1}, \phi_{n+1})$ where this triple satisfies the linear boundary value problem,

$$(4.1) \qquad \begin{aligned} L^{+}(u^{+}_{n}, \beta_{n})v^{+}_{n+1} &= F^{+}_{n+1} \quad \text{in} \quad \Omega_{+} \times (-\infty, \infty) \ ; \\ L^{-}(u^{-}_{n}, \beta_{n})v^{-}_{n+1} &= F^{-}_{n+1} \quad \text{in} \quad \Omega_{-} \times (-\infty, \infty) \ ; \\ \mathcal{B}_{n}(v^{+}_{n+1}, v^{-}_{n+1}, \phi_{n+1}) &= g_{n+1} \quad \text{on} \quad M_{0} \times (-\infty, \infty) \ ; \\ (v^{+}_{n+1}, v^{-}_{n+1}, \phi_{n+1}) & \ \text{vanishes for} \ t < 0 \ ; \end{aligned}$$

where F^{+}_{n+1}, F^{-}_{n+1}, and g_{n+1} are defined in (3.8) and (3.17) respectively. The boundedness in the high norm must necessarily incorporate the condition that there is a fixed time $T^{*} < T_0/2$ and a judicious choice of ε_0

satisfying (3.22) so that the solution $(v^+_{n+1}, v^-_{n+1}, \phi_{n+1})$ to (4.1) satisfies

$$(4.2) \qquad |||(v^+_{n+1}, v^-_{n+1}, \phi_{n+1})|||^2_{s,T*} \leq \varepsilon_0 \ , \qquad n = 0, ., 2, 3, \ldots$$

When (4.2) is satisfied, we conclude from (3.24) that necessarily T_{n+1} satisfies

$$T_{n+1} \geq T_* \ , \qquad n = 0, 1, 2, 3 \ldots$$

Otherwise, the fixed point of (3.9) and (3.17) to be established for some time $T_{**} > 0$ would not define a solution of (2.5) and (2.8) for $0 < t < T_*$ as we require in Proposition 2.1. The following two propositions are the two basic facts needed for convergence.

<u>PROPOSITION 4.1</u> (Boundedness in the High Norm). Fix the integer s with $s \geq 2[\frac{n}{2}] + 7$. There are a fixed number ε_0 satisfying (3.22), a fixed number β, $0 < \beta < 1$, and a $T_* > 0$ so that with $\eta(T) \equiv c_0 T^{-\beta}$ and $c_0 \equiv 2^{\beta} T_0^{-\beta} C_2$ where C_2 is the constant in Proposition 3.1, the function $(v^+_{n+1}, v^-_{n+1}, \phi_{n+1})$ solving the boundary value problem in (4.1) satisfies the stronger estimate,

$$(4.3) \quad |||(v^+_{n+1}, v^-_{n+1}, \phi_{n+1})|||^2_{s,\eta(T_*),T_*} \leq \varepsilon_0 \ ; \qquad n = 0, 1, 2, 3, \ldots$$

More precisely, ε_0, β, and T_* are chosen in the fashion dictated in (4.15) - (4.20) below. In particular, ε_0 can be restricted initially to satisfy the condition in (3.22).

<u>PROPOSITION 4.2</u> (The Contraction in the Low Norm). There are fixed constants C_1 and C_2 depending essentially only on the constant δ from (3.6) and ε_0 from Proposition 4.1 so that for $\eta > C_2$ and $T < T_*$, the itera-

tion scheme defined in (3.25) satisfies the estimates,

$$(4.4)\qquad |||v_{n+1} - v_n|||^2_{0,\eta,T} \leqslant C_1\left(\frac{1}{\eta} + T^2\right)|||v_n - v_{n-1}|||^2_{0,\eta,T} .$$

Next, for completeness, we finish the proof of the main theorem given Proposition 4.1 and Proposition 4.2. The proof follows standard arguments. From (4.3), by looking back at the definition of the norms in (3.4) and (3.5), we conclude that when $\eta(T) = c_0T^{-\beta}$, there is a fixed constant c_0 so that

$$(4.5)\qquad |||(v^+_{n+1}, v^-_{n+1}, \phi_{n+1})|||^2_{s,T_*}$$

$$\leqslant C_0|||(v^+_{n+1}, v^0_{n+1}, \phi_{n+1})|||^2_{s,\eta(T_*),T_*}$$

$$\leqslant C_0\varepsilon_0 ,$$

where trivially ε_0 also has been constructed below in the proof of Proposition 4.1 to satisfy the fixed constraint $C_sC_0\varepsilon_0 < \frac{\delta}{4}$. Pick a number $\alpha < 1$ and choose η_0 so that $C_1\eta_0^{-1} < \alpha^2/2$ and then choose $T_{**} \leqslant T_*$ so that $C_1T^2_{**} < \alpha^2/2$; then Proposition 4.2 implies the estimate

$$(4.6)\qquad |||v_{n+1} - v_n|||_{0,\eta_0,T_{**}} \leqslant \alpha|||v_n - v_{n-1}|||_{0,\eta_0,T_{**}} ;$$

$$n = 1, 2, 3, \ldots , \quad \text{with } \alpha \text{ fixed} , \quad 0 < \alpha < 1 .$$

The estimates in (4.5) and (4.6) imply by routine arguments that there is a $V \equiv (v^+, v^-, \phi)$ with

$$(4.7)\qquad |v^+|_{s,T_{**}} + |v^-|_{s,T_{**}} + \langle v^+\rangle_{s,T_{**}} + \langle v^-\rangle_{s,T_{**}} + \langle\phi\rangle_{s+1,T_{**}}$$

finite so that

$$(4.8) \qquad |||V_{n+1} - V_n|||_{0,T_{**}} \longrightarrow 0 \quad \text{as} \quad n \longrightarrow \infty \ .$$

By utilizing the Sobolev space interpolation inequalities, (4.5), and (4.8) in a standard fashion (see [8]), we improve the conclusion in (4.8) to

$$(4.9) \qquad V_{n+1} \longrightarrow V \quad \text{as} \quad n \longrightarrow \infty \quad \text{in}$$

$$H^{s-\varepsilon}(\Omega_+ \times [0, T_{**}]) \oplus H^{s-\varepsilon}(\Omega_- \times [0, T_{**}]) \oplus H^{s+1-\varepsilon}(M_0 \times [0, T_{**}])$$

for any fixed $\varepsilon > 0$. By Sobolev's lemma, since $s > \frac{(N+1)}{2} + 1$, we conclude that

$$(4.10) \qquad V_{n+1} \longrightarrow V \quad \text{as} \quad n \longrightarrow \infty \quad \text{in}$$

$$C^1(\overline{\Omega}_+ \times [0, T_{**}]) \oplus C^1(\overline{\Omega}_- \times [0, T_{**}]) \oplus C^2(M_0 \times [0, T_{**}]) \ .$$

We use (3.1), (3.9), (3.17), the remark contained in (4.2), and straightforwardly pass to the limit with (4.10) to deduce that

$$u^+ = u_0^+ + v^+ \ ;$$

$$u^- = u_0^- + V^- \ ;$$

$$\beta = \beta_0 + \phi \ ,$$

is a classical solution of (2.5) and (2.8) for $0 \leq t \leq T_{**}$. By applying Proposition 2.1 and using the transformation $\chi_t(x, \beta)$, we see that we have constructed a classical solution of the shock front problem for $0 \leq t \leq T_{**}$.

Furthermore, by applying Lemma A-4 of [12], we see that $u^+(\chi_t(x, \beta), t)$, $u^-(\chi_t(x, \beta), t)$ and $\beta(\alpha, t)$ have the required additional regularity as stated in the main theorem. Below, it is always understood that C is a fixed a priori constant depending only on the quantities in (3.6) and can vary from relation to relation.

4.A High Norm Boundedness -- The Proof of Proposition 4.1

Since $(v_0^+, v_0^-, \phi_0) = (0, 0, 0)$, by induction, we only need to prove that there are fixed constants ε_0, β, and T_* satisfying the requirements of Proposition 4.1 so that with $\eta(T) = c_0 T^{-\beta}$, if

$$|||(v_n^+, v_n^-, \phi_n)|||^2_{s,\eta(T_*),T_*} \leq \varepsilon_0 ,$$

then

$$|||(v_{n+1}^+, v_{n+1}^-, \phi_{n+1})|||^2_{s,\eta(T_*),T_*} \leq \varepsilon_0 , \tag{4.11}$$

where $(v_{n+1}^+, v_{n+1}^-, \phi_{n+1})$ is the solution of (4.1). For simplicity in exposition and without loss of generality, we now assume that the constant T_0 defined in (2.18) satisfies $T_0 < 1$.

First, we remark that for $0 \leq \beta \leq 1$ and $0 \leq t \leq T \leq T_0 \leq 1$, we have

$$e^{-2c_0} \leq e^{-2c_0 t T^{-\beta}} \leq 1 .$$

Therefore, with $\eta(T) = c_0 T^{-\beta}$, we have the inequalities,

$$(4.12)\qquad (1)\ \ |||(v^+, v^-, \phi)|||^2_{s,T} \leqslant \tilde{C}|||(v^+, v^-, \phi)|||^2_{s,\eta(T),T}\ ;$$

$$(2)\ \ |F^{\pm}|^2_{s,\eta(T),T} \leqslant C \sum_{j=0}^{s} T^{-2(s-j)\beta}|F^{\pm}|^2_{j,T}\ ;$$

$$(3)\ \ \langle g\rangle^2_{s,\eta(T),T} \leqslant C \sum_{j=0}^{s} T^{-2(s-j)\beta}\langle g\rangle^2_{j,T}\ ;$$

where $\tilde{C}$, C are fixed a priori constants provided T varies over $0 \leqslant T \leqslant T_0 \leqslant 1$. The facts in (4.12) follow easily from applying the above remark to the definitions of the norms in (3.3), (3.4). The main estimate in the proof of Proposition 4.2 is the set of a priori inequalities contained in the following:

LEMMA 4.1. Consider the terms g_{n+1}, $F^{\pm}_{n+1}$ defined in (3.8) and (3.17) respectively. Then there exists a fixed constant C, depending only on δ in (3.6) and $T_0 \leqslant 1$, so that for $0 \leqslant T \leqslant T_0$ and β with $0 \leqslant \beta \leqslant \frac{3}{2(s+1)}$, we have the a priori estimates,

$$(4.13)\qquad \left.\begin{array}{l} (1)\ \ |F^{\pm}_{n+1}|^2_{s,\eta(T),T} \leqslant C + CT^{\frac{3}{s}-2\beta} \\[2ex] (2)\ \ \langle g_{n+1}\rangle^2_{s,\eta(T),T} \leqslant C|||V_n|||^4_{s,T} \\[2ex] \qquad + CT^{\frac{3}{s}-2\beta} + C\langle g_1\rangle^2_{s,T} \end{array}\right\} \text{ for } T \leqslant T_n \text{ where}$$

$$(3)\ \ g_1 \equiv G_+(u_0^+, u_0^-, \beta_0, (\beta_0)_t, \nabla_{\tan}\beta_0)\ .$$

Here G is the nonlinear function defined in (3.10) and G_+ has been defined in (3.14).

REMARK. We must use the norms $|||\cdot|||^2_{s,\eta(T),T}$ precisely because the estimat in (1) of (4.13) involves a fixed constant without a small factor of time; the introduction of the norms with $\eta(T)$ allows small factors involving T to multiply these terms on the right hand side of the main linear estimate of Proposition 3.1 which we use below. Typical terms involving the constant C arise from the H^s norm of the interaction of the shock front, β_n, and the interior equations; for example, consider the term contributing to F^+_{n+1} given by, $\rho n_j(\partial\beta_n/\partial t)(\partial u^+_0/\partial x_j)$, then we have

$$\sum_{|\alpha|=s} \left| D^\alpha \left(\rho n_j \frac{\partial\beta_n}{\partial t} \frac{\partial u^+_0}{\partial x_j} \right) \right|^2_{0,T} \leq C\langle\beta_n\rangle^2_{s+1,T} \quad ,$$

and this term does not seem to have a small factor of T preceding it even by utilizing stronger "semigroup" estimates for u^+, u^-.

Assuming Lemma 4.1 for the moment, we complete the proof of Proposition 4.1. From (4.1), Lemma 4.1, and Proposition 3.1, we deduce that, for $0 \leq T \leq T_n \leq T_0/2$

$$(4.14) \qquad |||(v^+_{n+1}, v^-_{n+1}, \phi_{n+1})|||^2_{s,\eta(T),T}$$

$$\leq C(\eta(T))^{-1}(|F^+_{n+1}|^2_{s,\eta(T),T} + |F^-_{n+1}|^2_{s,\eta(T),T})$$

$$+ C\langle g_{n+1}\rangle^2_{s,\eta(T),T}$$

$$\leq \{C(T^\beta + T^{\frac{3}{s}-\beta} + T^{\frac{3}{s}-2\beta})\} + \{C|||v_n|||^4_{s,T}\} + \{C\langle g_1\rangle^2_{s,T}\}$$

$$\equiv \{1\} + \{2\} + \{3\} \quad .$$

Recall that in order to satisfy (3.22), we always chose ε_0 to satisfy the fixed restriction

$$C_s\tilde{C}\varepsilon_0 < \frac{\delta}{8} , \tag{4.15}$$

where C_s, δ are the constants in (3.22) and $\tilde{C}$ is the constant in (1) of (4.12). To treat the term {2}, we also choose ε_0 so that

$$\tilde{C}^2 C\varepsilon_0 < \frac{1}{3} , \tag{4.16}$$

where C is the constant appearing in {2} above. Set $\beta \equiv \frac{1}{s}$. Then trivially we have

$$\{1\} \leqslant CT^{\frac{1}{s}} \quad \text{for} \quad s \geqslant 2 . \tag{4.17}$$

With this fixed choice of ε_0 made in (4.15), (4.16), choose T_1 so that with C the constant in (4.17),

$$CT_1^{\frac{1}{s}} = \frac{1}{3}\varepsilon_0 . \tag{4.18}$$

Recall that g_1 is a fixed function with $\langle g_1 \rangle^2_{s,T_0}$ finite; thus, given ε_0, there is a T_3 so that

$$\langle g_1 \rangle^2_{s,T_3} \leqslant \frac{\varepsilon_0}{3} . \tag{4.19}$$

Finally, determine T_* by

$$T_* = \min\{T_1, T_3\} . \tag{4.20}$$

Now, let's assume with the above choices of ε_0, β, and T_*, that the induction hypotheses in (4.11) is satisfied for (v_n^+, v_n^-, ϕ_n) so that in particular, $T_n \geqslant T_*$, then (2) of (4.13) and (4.15) - (4.20) guarantee that for the time, T_*, $\{1\} + \{2\} + \{3\} \leqslant \varepsilon_0$. Thus, the right hand side of (4.14) is less than ε_0 and by (1(of (4.12), we deduce the conclusion in (4.11); also, (4.15) guarantees that (3.22) is satisfied so that $T_{n+1} \leqslant T_*$.

REMARK. The proof crucially depends upon the use of a superlinear boundary iteration scheme such as Newton's method (see (3.11)) to yield the estimate in (2) of (4.13). If only $|||V_n|||^2_{s,T}$ appeared on the right hand side of (4.13), the above argument would fail.

PROOF OF LEMMA 4.1. First, let's remark that for the fixed regions, $\Omega_\pm \times [0, T_0]$ and $M_0 \times [0, T_0]$, we have the well-known Sobolev space interpolation inequalities,

$$(4.12) \qquad |F_\pm|^2_{j,T_0} \leqslant C_s |F_\pm|_{s,T_0}^{2(\frac{j}{s})} |F_\pm|_{0,T_0}^{2(1-\frac{j}{s})}, \qquad 0 \leqslant j \leqslant s \ ;$$

$$\langle g\rangle^2_{j,T_0} \leqslant C_s \langle g\rangle_{s,T_0}^{2(\frac{j}{s})} \langle g\rangle_{0,T_0}^{2(1-\frac{j}{s})}, \qquad 0 \leqslant j \leqslant s \ .$$

Assume that $F_\pm$, g belong to $H^s(\Omega_\pm \times [0, T_0])$ and $H^s(M_0 \times [0, T_0])$ and satisfy,

$$(4.22) \qquad \partial_t^j F_\pm\Big|_{t=0} = \partial_t^j g\Big|_{t=0} = 0 \ , \qquad 0 \leqslant j \leqslant s - 1 \ .$$

Then, by applying (4.21) to $E_T F_\pm$, $E_T g$ where E_T is the extension operator in Lemma 2.1, we obtain the useful lemma,

LEMMA 4.2. Assume that (4.22) is satisfied then there is a constant, C_s, independent of T with $0 \leqslant T \leqslant T_0/2$ so that

$$(4.23) \qquad (1) \quad |F_\pm|^2_{j,T} \leqslant C_s |F_\pm|_{s,T}^{2(\frac{j}{s})} \, |F_\pm|_{0,T}^{2(1-\frac{j}{s})} \, , \qquad 0 \leqslant j \leqslant s \; ;$$

$$(2) \quad \langle g \rangle^2_{j,T} \leqslant C_s \langle g \rangle_{s,T}^{2(\frac{j}{s})} \langle g \rangle_{0,T}^{2(1-\frac{j}{s})} \, , \qquad 0 \leqslant j \leqslant s \; .$$

For the proof of Lemma 4.1 and Lemma 4.3 below, by the definition of the iteration scheme, we can assume that V_n satisfies all the conditions in (3.6) independent of ε_0 satisfying the restriction in (3.22). In particular, from (2) of (3.6), we have

$$(4.24) \qquad C_s |||V_n|||^2_{s,T_0} \leqslant \frac{\delta}{4} \, .$$

Also, since $s > \frac{n+1}{2} + 1$, we have from Sobolev's lemma and (4.24), the estimate,

$$(4.25) \qquad |u_n^\pm|^2_{C^1(\bar{\Omega}_\pm \times [0,T_0])} + |\beta_n|^2_{C^2(M_0 \times [0,T_0])} \leqslant C \, .$$

To establish the estimate in (1) of Lemma 4.1, first, we derive the estimate

$$(4.26) \qquad |F^\pm_{n+1}|^2_{s,T_0} \leqslant C \, .$$

Looking back at the definition of $F^\pm_{n+1}$ in (3.8) and also (2.5) - (2.7), we only need to verify that

(4.27) $$\left|A_k(u_n^{\pm})(c_{j,k}(x, t, \beta_n) \frac{\partial u_0^{\pm}}{\partial x_j}\right|_{s,T_0} \leqslant C ,$$

since a similar argument is valid for the term,

$$c_{j,0}(x, t, \beta_n) \frac{\partial u_0^{\pm}}{\partial x_j} .$$

However, from (A-1) - (A-5) of the Appendix and the definition of $c_{j,k}$ in (2.7), we have

(4.28) (1) $$|c_{j,k}(x, t, \beta_n)|^2_{s,T_0} \leqslant C\langle\beta_n\rangle^2_{s+1,T_0} \leqslant C ;$$

(2) $$\left|\frac{\partial u_0}{\partial x_j}\right|^2_{s,T_0} \leqslant C|\psi u_0^{\pm}|^2_{s+1,T_0} \leqslant C ;$$

(3) $$|\psi A_k(u_n^{\pm})|^2_{s,T_0} \leqslant C ;$$

where ψ is a fixed C_0^{∞} function with $\psi \equiv 1$ for $|x| < CR_0$ and supp $\psi \subseteq (C + 1)R_0$. However, (4.27), (4.28), and the Banach algebra properties of H^s, $s > \frac{(n + 1)}{2}$ (see (A-1) of the Appendix) clearly imply (4.26). We also claim that

(4.29) $$|F^{\pm}_{n+1}|^2_{0,T} \leqslant CT^3 .$$

To derive (4.2), we observe that it follow from (4.25) that $F^{\pm}_{n+1}$ has bounded support and belongs to $C^1(\overline{\Omega}_{\pm} \times [0, T_0])$. Furthermore, from Lemma 3.2 and (4.25),

$$F^{\pm}_{n+1}(x, 0) = 0 , \quad x \in \Omega_{\pm} \tag{4.30}$$

$$|F^{\pm}_{n+1}|_{C^1(\overline{\Omega}_{\pm}\times[0,T_0])} \leqslant C$$

so that by Taylor's theorem with remainder,

$$|F^{\pm}_{n}(x, t)| \leqslant Ct , \quad 0 \leqslant t \leqslant T \leqslant \frac{T_0}{2} . \tag{4.31}$$

By integrating (4.31) over $\Omega_{\pm} \times [0, T]$ for $T \leqslant T_0/2$ and utilizing the fact that $F^{\pm}_{n+1}$ has support in a fixed bounded set, we obtain the inequality in (4.29). To derive the a priori estimate in (1) of Lemma 4.1, we use (2) of (4.12), (4.26), (4.29), Lemma 2.3, and interpolate by (1) of Lemma 4.3 to obtain, for $0 \leqslant T \leqslant T_0/2 < 1/2$,

$$\begin{aligned} &|F^{\pm}_{n+1}|^2_{s,\eta(T),T} \\ &\leqslant C \sum_{j=0}^{s} T^{-2(s-j)\beta} |F_{n+1}|^2_{j,T} \\ &\leqslant C \sum_{j=0}^{s} T^{-2(s-j)\beta+3(1-\frac{j}{s})} \\ &\leqslant C + CT^{\frac{3}{s}-2\beta} , \end{aligned} \tag{4.32}$$

where we have used $0 \leqslant \beta \leqslant \frac{3}{2(s+1)}$ in the final step.

To verify (2) of Lemma 4.1, we introduce the notation,

$$w_0 = (u_0^+, u_0^-, \beta_0, (\beta_0)_t, \nabla_{tan}\beta_0) \quad ;$$

$$\tilde{w}_0 = (v_n^+, v_n^-, \phi_n, (\phi_n)_t, \nabla_{tan}\phi_n) \quad .$$

Then the definition of g_{n+1} in (3.17), the construction of the boundary iteration scheme by Newton's method so that (a) of (3.13) is valid, and Taylor's theorem with remainder imply that

$$g_{n+1} = -G_+(w_0) + C(w_0, \tilde{w}_n)\tilde{w}_n^2 \quad ,$$

where $C(w_0, \tilde{w}_n)$ is a smooth matrix function of its arguments provided (3.22) is satisfied; the expression $\tilde{w}_n^2$, is symbolic notation for a quadratic function of all variables, $\tilde{w}_n$, and $G_+(w_0)$ is the fixed function in (3.14). From (A-9) of the Appendix, we derive that for $0 \leqslant T \leqslant T_n$,

$$(4.33) \qquad \langle C(w_0, \tilde{w}_n)\tilde{w}_n^2\rangle_{s,T}^2 \leqslant C\langle \tilde{w}_n\rangle_{s,T}^4 \leqslant C|||v_n|||_{s,T}^4 \quad ,$$

thus,

$$(4.34) \qquad \langle g_{n+1}\rangle_{s,T}^2 \leqslant C\langle G_+(w_0)\rangle_{s,T}^2 + C|||v_n|||_{s,T}^4 \quad .$$

Also, from Lemma 3.2 and (4.25), $g_{n+1} \in C^1(M_0 \times [0, T_0])$ with

$$g_{n+1}(\alpha, 0) = 0 \quad ;$$

$$\langle g_{n+1}\rangle_{C^1(M_0\times[0,T_0])} \leqslant C \quad ,$$

so that analogous to (4.30) - (4.31), we obtain

(4.35) $$\langle g_{n+1} \rangle^2_{0,T} \leqslant CT^3 \quad ,$$

so that by using (2) of Lemma 4.3, (3) in (4.12), (4.34), (4.35) and repeating a similar argument as in (4.32) we deduce (2) of Lemma 4.1.

4.B The Proof of Proposition 4.2

With the choice of ε_0 from Proposition 4.1, we compute that for $T \leqslant T_*$, $V_{n+1} - V_n$ satisfies the boundary value problem,

(4.36) (1) $$L^+(u_n^+, \beta_n)(v_{n+1}^+ - v_n^+)$$

$$= F_{n+1}^+ - F_n^+ + (L^+(u_{n-1}^+, \beta_{n-1}) - L^+(u_n^+, \beta_n))v_n^+ \equiv \tilde{F}_{n+1}^+$$

$$\text{in } \Omega^+ \times [0, T_*] \quad ;$$

(2) $$L^-(u_n^-, \beta_n)(v_{n+1}^- - v_n^-)$$

$$= F_{n+1}^- - F_n^- + (L^-(u_{n-1}^-, \beta_{n-1}) - L^-(u_n^-, \beta_n))v_n^- \equiv \tilde{F}_{n+1}^-$$

$$\text{in } \Omega^- \times [0, T_*] \quad ;$$

(3) $$\mathcal{B}_n(v_{n+1}^+ - v_n^+, v_{n+1}^- - v_n^-, \phi_{n+1} - \phi_n)$$

$$= g_{n+1} - g_n + (\mathcal{B}_{n-1} - \mathcal{B}_n)(v_n^+, v_n^-, \phi_n) \equiv \tilde{g}_{n+1}$$

$$\text{on } M_0 \times [0, T_*] \quad .$$

$$V_{n+1} - V_n \text{ vanishes for } t < 0 \quad .$$

For the moment, we assume the following:

LEMMA 4.3. Under the hypotheses of Proposition 4.2, for $0 \leqslant T \leqslant T_*$, the terms $(\tilde{F}^+_{n+1}, \tilde{F}^-_{n+1}, \tilde{g}_{n+1})$ satisfy the estimates

$$(4.37)\qquad (1)\quad |\tilde{F}_{n+1}|^2_{0,\eta,T_*} \leqslant C_1 |||V_n - V_{n-1}|||^2_{0,\eta,T} \;;$$

$$(2)\quad \langle \tilde{g}_{n+1}\rangle^2_{0,\eta,T} \leqslant C_1 T^2 |||V_n - V_{n-1}|||^2_{0,\eta,T} \;;$$

for $\eta > C_2$.

The conclusion of Proposition 4.2 follows immedaitely from Lemma 4.3 and a direct application of the main linear estimate from Proposition 3.1 when $s = 0$ to the boundary value problem in (4.36). It remains to establish the

PROOF OF LEMMA 4.3. Under the hypotheses of Lemma 4.2, from (4.25) we have the estimates,

$$(4.38)\qquad |u^{\pm}_j|^2_{C^1(\bar{\Omega}\times[0,T_*)} + |\beta_j|^2_{C^2(M_0\times[0,T_*])} \leqslant C$$

$$\text{for } j = 0, 1, 2, \ldots$$

which imply, by a repetition of the argument in (4.30), (4.31) using (3.6)(3) that for $0 \leqslant t \leqslant T_*$,

$$(4.39)\qquad |v^{\pm}_j(x, t)| + |\phi_j(\alpha, t)| + |\nabla_{\tan}\phi_j(\alpha, t)| + |(\phi_j)_t(\alpha, t)| \leqslant Ct \;, \qquad j = 1, 2, 3, \ldots$$

To estimate (2) in Lemma 4.3, we use the quadratic estimate for Newton's method on the boundary as expressed in (3.13) together with (4.39) to estimate that pointwise on $M_0 \times [0, T_*]$,

$$(4.40) \qquad |d_w G(w_n)\tilde{w}_{n+1} - \tilde{w}_n)|^2(\alpha, t)$$

$$= |\mathcal{B}_n((v^+_{n+1} - v^+_n, v^-_{n+1} - v^-_n, \phi_{n+1} - \phi_n))|^2(\alpha, t)$$

$$= |\tilde{g}_{n+1}|^2(\alpha, t)$$

$$\leqslant C|\tilde{w}_n - \tilde{w}_{n-1}|^4(\alpha, t)$$

$$\leqslant C(|v^\pm_n - v^\pm_{n-1}|^2 + |\phi_n - \phi_{n-1}|^2 + |\nabla_{\tan}\phi_n - \nabla_{\tan}\phi_{n-1}|^2$$

$$+ |(\phi_n)_t - (\phi_{n-1})_t|^2)^2(\alpha, t)$$

$$\leqslant Ct^2(|v^\pm_n - v^\pm_{n-1}|^2 + |\phi_n - \phi_{n-1}|^2 + |\nabla_{\tan}\phi_n - \nabla_{\tan}\phi_{n-1}|^2$$

$$+ |(\phi_n)_t - (\phi_{n-1})_t|^2)(\alpha, t) \quad .$$

By multiplying (4.40) by $e^{-\eta t}$ and integrating over $M_0 \times [0, T)$ with $T < T_*$ we obtain the estimate in (2) of Lemma 4.3.

The estimate for (2) in Lemma 4.3 is also completely straightforward so we will be terse. The typical term contributing to $F_{n+1} - F_n$ has the form, (reversing the roles of x and $\tilde{x}$ in (2.6) and (2.7))

$$(4.41) \qquad \left[A_k(u^\pm_n)\, \frac{\partial}{\partial \tilde{x}_k}\, (\rho\beta_n n_j)\Big|_{(\Phi^t(x,\beta_n),t)} \frac{\partial u^\pm_0}{\partial x_j}\right]$$

$$- \left[A_k(u^\pm_{n-1})\, \frac{\partial}{\partial \tilde{x}_k}\, (\rho\beta_{n-1} n_j)\Big|_{(\Phi^t(x,\beta_{n-1}),t)} \frac{\partial u^\pm_0}{\partial x_j}\right] \quad .$$

From the definition in (2.2) and (4.38), $\Phi^t(x, \beta_j)$, $j = 0, 1, \ldots$ is a C^2 transformation of R^N to R^N satisfying

$$|\Phi^t(x, \beta_{n-1}) - \Phi^t(x, \beta_n)| \leq C|\phi_n - \phi_{n-1}|(\alpha(x), t) \quad . \tag{4.42}$$

If we use (4.42) in (4.41) together with the mean-value theorem in (4.38), by straightforward estimates adding and subtracting terms, we dominate the square of the absolute value of the left hand side of (4.41) by

$$C|v_n^{\pm} - v_{n-1}^{\pm}|^2 + C\rho(\tilde{x})(\nabla_{tan}(\phi_n - \phi_{n-1})|^2(\alpha(x), t)|_{x=\Phi^t(x,\beta_{n-1})} \tag{4.43}$$

$$+ C\rho(\tilde{x})|_{\tilde{x}=\Phi^t(x,\beta_{n-1})} |\phi_n - \phi_{n-1}|^2(\alpha(x), t) \quad .$$

Multiplying (4.41) and (4.43) by $e^{-\eta t}$ and integrating from $[0, T]$, we obtain easily that

$$|F_{n+1}^{\pm} - F_n^{\pm}|^2_{0,\eta,T} \leq C|||v_n - v_{n-1}|||^2_{0,\eta,T} \quad . \tag{4.44}$$

The contribution to $\tilde{F}_{n+1}$ from

$$(L^{\pm}(u_{n-1}^{\pm}, \beta_{n-1}) - L^{\pm}(u_n^{\pm}, \beta_n^{\pm}))v_n^{\pm}$$

has a similar estimate as in (4.41) - (4.43) once we remark that it follows from (4.38) that for $0 \leq t \leq T$

$$\sup\left|\frac{\partial v_n^{\pm}}{\partial x_j}\right|(x, t) \leq C \quad .$$

This completes the proof of Lemma 4.3.

§5. THE MAIN LINEAR ESTIMATE

Our objective here is to prove Proposition 3.1. Estimates derived by integration by parts following Friedrichs' theory of positive symmetric systems ([5]) and the linear theory developed in [12] play a crucial role in the proof -- in particular, essential use is made of the terms $\langle \partial^\ell v^\pm/\partial n^\ell \rangle^2_{s-\ell,\eta,T}$ contributing to $|||V|||^2_{s,\eta,T}$. For convenience we restate a more general version of Proposition 3.1 which we prove below.

PROPOSITION 5.1. Consider the operators, $L^+(u^+, \beta)$, $L^-(u^-, \beta)$, $\mathcal{B}(u^+, u^-, \beta)$ defined in (2.5) - (2.7) and (3.15) with $u^\pm = u_0^\pm + \tilde{v}^\pm$, $\beta = \beta_0 + \tilde{\phi}$ where $\tilde{V} = (\tilde{v}^+, \tilde{v}^-, \tilde{\phi})$ satisfies (1) and (2) of (3.6) where s is an integer with $s \geqslant [\frac{n}{2}] + 7$. Consider the boundary value problem for $V = (v^+, v^-, \phi)$,

(5.1)
$$L^+(u^+, \beta)v^+ = F^+ \quad \text{in} \quad \Omega_+ \times (-\infty, \infty) \ ;$$

$$L^-(u^-, \beta)v^- = F^- \quad \text{in} \quad \Omega_- \times (-\infty, \infty) \ ;$$

$$\mathcal{B}(u^+, u^-, \beta)(v^+, v^-, \phi) = g \quad \text{on} \quad M_0 \times (-\infty, \infty) \ ;$$

$$V \text{ vanishes for } t \leqslant 0 \ ,$$

where F^+, F^- are functions in $H^s(\Omega_\pm \times (-\infty, \infty))$ and $g \in H^s(M_0 \times (-\infty, \infty))$, all vanishing for $t < 0$ and $t > T_0$ and satisfying

(5.2)
$$\partial_t^j F^\pm\Big|_{t=0} = \partial_t^j g\Big|_{t=0} = 0 \ , \quad 0 \leqslant j \leqslant s - 1 \ .$$

Then there exists constants C_1 and C_2 depending only upon (1) and (2) of (3.6) so that the unique strong solution to (5.1) has $|||v|||^2_{s,\eta,T}$ finite and satisfies the estimates,

$$(5.3)\qquad |||v|||^2_{s,\eta,T} \leqslant C_1\left(\frac{|F^+|^2_{s,\eta,T} + |F^-|^2_{s,\eta,T}}{\eta} + \langle g\rangle^2_{s,\eta,T}\right)$$

for $\eta \geqslant C_2$ provided that $0 \leqslant T \leqslant T_0$ and also the conditions

$$(5.4)\qquad 0 = \partial_t^j v^{\pm}\Big|_{t=0} \quad , \quad 0 \leqslant j \leqslant s - 1 \quad ;$$

$$0 = \partial_t^{\ell}\phi\Big|_{t=0} \quad , \quad 0 \leqslant \ell \leqslant s \quad .$$

A similar estimate is valid when $s = 0$ for $F^{\pm} \in L^2(\Omega_{\pm} \times [0, T_0])$, $g \in L^2(M_0 \times [0, T_0])$.

At the end of this section, we sketch a proof of the following lemma:

LEMMA 5.1. Under the hypotheses of Proposition 5.1 for the coefficients (u^+, u^-, β) , there is a sequence of smooth functions, $(u^+_{\varepsilon}, u^-_{\varepsilon}, \beta_{\varepsilon})$ so that for $0 \leqslant \varepsilon \leqslant \varepsilon_0$

(5.5) (1) $(u^+_{\varepsilon}, u^-_{\varepsilon}, \beta_{\varepsilon}) \equiv (u^+_0, u^-_0, \beta_0)$ for $|x| > (C + 1)R$ or $|t| > T_0 + 1$.

(2) $|||(u^+_{\varepsilon}, u^-_{\varepsilon}, \beta_{\varepsilon}) - (u^+, u^-, \beta)|||^2_{s,2T_0} \longrightarrow 0$ as $\varepsilon \longrightarrow 0$; in particular, ε_0 can be chosen small enough so that $|||(u^+_{\varepsilon}, u^-_{\varepsilon}, \beta_{\varepsilon}) - (u^+_0, u^-_0, \beta_0)|||^2_{s,2T_0} \leqslant C + 1$ provided that $|||v|||^2_{s,T_0} \leqslant C$.

(3) The inequality in (2.18) is satisfied with $\frac{\delta}{2}$ replaced by $\frac{3\delta}{4}$ for $0 < \varepsilon \leqslant \varepsilon_0$.

The significance of Lemma 5.1 is that we only need to prove Proposition 5.1 under the additional assumption that

$$\text{(5.6)} \qquad \text{the coefficients } (u^+, u^-, \beta) \text{ are smooth.}$$

We consider the boundary value problem in (5.1) using the operator $L^+(u_\varepsilon^+, \beta_\varepsilon)$, $L^-(u_\varepsilon^-, \beta_\varepsilon)$ and $\mathcal{B}(u_\varepsilon^+, u_\varepsilon^-, \beta_\varepsilon)$. Then, under the additional assumption of (5.6) and Lemma 5.1, we prove that the conclusion of Proposition 5.1 is satisfied for these operators; furthermore, by conditions (1) - (3) of Lemma 5.1, we shall see that the constants C_1 and C_2 in (5.3) are independent of ε for $0 < \varepsilon \leqslant \varepsilon_0$. Therefore, the estimate in (5.3) is uniform over the family $(L_\varepsilon^+, L_\varepsilon^-, \beta_\varepsilon)$ for (F^+, F^-, g) fixed. Also, the subspace defined by (5.4) is preserved under weak limits in the norm, $|||\cdot|||_{s,\eta,T}$; thus, we can pass to the limit and verify the conclusions of Proposition 5.1 for the general operator satisfying the hypotheses of that proposition. Without further comment, we use the additional assumption in (5.6) for the remainder of the section.

Under the assumptions on the coefficients, $(u_\varepsilon^+, u_\varepsilon^-, \beta_\varepsilon)$ as discussed in Lemma 5.1 and (5.6), in Theorem 1 and Theorem 2 of [12], we have already proved that provided $F^\pm$ and g satisfy the hypotheses of Proposition 5.1, there is a unique strong solution, V , to the boundary value problem in (5.1) with V belonging to $H^s(\overline{\Omega}_+ \times [0, T_0]) \oplus H^s(\overline{\Omega}_- \times [0, T_0]) \oplus H^{s+1}(M_0 \times [0, T_0])$ and satisfying (5.4) (the condition in (5.4) is not stated explicitly in Theorem 1 of [12] but follows implicitly from the argument in §5). In fact,

(5.7) when $F^{\pm}$ and g are smooth and vanish in a neighborhood of $t = 0$, this unique strong solution is smooth and vanishes in a neighborhood of $t = 0$.

It remains to prove the estimates in (5.3) with constants C_1, C_2 having the dependence asserted in Proposition 5.1 provided that the coefficients $(u_\varepsilon^+, u_\varepsilon^-, \beta_\varepsilon)$ are smooth and have the properties of (1) - (3) in Lemma 5.1. (Below we shall omit the subscript epsilon.) In deriving the estimates in (5.3), we can assume, that $F^{\pm}$ and g satisfy the condtions in (5.7) and then pass to the limit.

Following standard arguments ([12]), we utilize a fixed partition of unity flattening the boundary, M_0, so that we only need to derive the estimates in the half-spaces $x_N \begin{smallmatrix} < 0 \\ > 0 \\ = 0 \end{smallmatrix}$ with $x' = (x_1, \dots, x_{N-1})$ the tangential variables, for a smooth solution (v^+, v^-, ϕ) and $F^{\pm}$, g satisfying (5.7) for the corresponding boundary value problem,

$$\tilde{L}^+ v^+ = F^+ \quad \text{in} \quad x_N > 0 \; ; \tag{5.8}$$

$$\tilde{L}^- v^- = F^- \quad \text{in} \quad x_N < 0 \; ;$$

$$\tilde{B}(v^+, v^-, \phi) = g \quad \text{on} \quad x_N = 0 \; ;$$

$$(v^+, v^-, \phi) \text{ vanishes for } t < 0 \; ;$$

where

$$\tilde{L}^{\pm} = \frac{\partial}{\partial t} + \sum_{j=1}^{N} \tilde{A}_j \frac{\partial}{\partial x_j} \; ; \quad \tilde{A}_j^{\pm} = \sum_k a_{j,k}(x) A_k^{\pm}(u^{\pm}, \beta) \; , \tag{5.9}$$

with $A_k^{\pm}(u^{\pm}, \beta)$ already defined in (2.6) and the $a_{j,k}(x)$ smooth coefficients depending only on the fixed coordinate transformations. Similarly, the coefficients of the boundary operator, $\tilde{\mathcal{B}}$, are unchanged except for the expressions involving $\nabla_{\tan}$ which are replaced by $(\nabla_{\tan})_i = \sum_j g_{ij}(x') \frac{\partial}{\partial x_j}$ where g_{ij} is a uniformly positive, symmetric matrix depending on the co-ordinate transformation -- we will denote the corresponding transformed coefficients for the boundary conditions, $\tilde{\mathcal{B}}$, in (5.8) by $\tilde{b}_0$, $\tilde{b}_{1,i}(x, u^+, u^-)$, $1 \leqslant i \leqslant n - 1$, $\tilde{M}(u_+, u_-, \beta)$, etc., without confusion. Furthermore, it follows from (3.6) and (1.9) that we can assume that,

$$|\det(\tilde{A}_N^{\pm})| \geqslant C > 0 . \tag{5.10}$$

Here and below in this section, C will denote generic <u>a priori</u> constant with the dependence described for C_1 and C_2 in (5.3) and permitted for $(u_\varepsilon^+, u_\varepsilon^-, \beta_\varepsilon)$ in Lemma 5.1.

The following lemma is an immediate consequence of Theorem 2 in [12]. This is the only lemma of this paper which requires crucial use of the assumption, $s \geqslant 2[\frac{n}{2}] + 7$ -- all other lemmas can be proved under the weaker requirements, $s > \frac{n + 1}{2} + 1$.

<u>LEMMA 5.2</u>. If $\hat{V} = (\hat{v}_+, \hat{v}_-, \hat{\phi})$ is a smooth solution of

$$\tilde{L}^+\hat{v}_+ = \hat{F}_+ \quad \text{in} \quad x_N > 0 ;$$

$$\tilde{L}^-\hat{v}_- = \hat{F}_- \quad \text{in} \quad x_N < 0 ;$$

$$\tilde{\mathcal{B}}(\hat{v}_+, \hat{v}_-, \hat{\phi}) = \hat{g} \quad \text{on} \quad x_N = 0 ;$$

$$(\hat{v}_+, \hat{v}_-, \hat{\phi}) \ \text{vanishes for} \ t < 0 ;$$

then there are constants, C_1, C_2 so that for $0 \leq T \leq T_0$,

(5.12) $$|||\hat{v}|||^2_{0,\eta,T} \leq C_1 \left(\frac{|\hat{F}_+|^2_{0,\eta,T} + |\hat{F}_-|^2_{0,\eta,T}}{\eta} + \langle \hat{g} \rangle^2_{0,\eta,T} \right),$$

for $\eta \geq C_2$.

We differentiate (5.8) in directions purely tangential to $x_N = 0$ and introduce

(5.13) $$v^{\pm}_{\alpha_0,\alpha'} = D_t^{\alpha_0} D_x^{\alpha'} v_{\pm} \; ; \qquad \phi_{\alpha_0,\alpha'} = D_t^{\alpha_0} D_x^{\alpha'} \phi$$

for $|\alpha_0| + |\alpha'| \leq s$. From (5.8), it follows that the equations

(5.14) $$\tilde{L}^{\pm} v^{\pm}_{\alpha_0,\alpha'} = F^{\pm}_{\alpha_0,\alpha'} \quad \text{in} \quad \begin{matrix} x_N > 0 \\ x_N < 0 \end{matrix},$$

$$\tilde{B}(v^{+}_{\alpha_0,\alpha'}, v^{-}_{\alpha_0,\alpha'}, \phi_{\alpha_0,\alpha'}) = g_{\alpha_0,\alpha'} \quad \text{on} \quad x_N = 0$$

are satisfied for these derivatives where

(5.15) $$F^{\pm}_{\alpha_0,\alpha'}$$

$$\equiv D_t^{\alpha_0} D_x^{\alpha'} F^{\pm} - \left[D_t^{\alpha_0} D_x^{\alpha'} \left(\sum_{j=1}^{N} \tilde{A}_j \frac{\partial v^{\pm}}{\partial x_j} \right) - \sum_{j=1}^{N} \tilde{A}^{\pm}_j \frac{\partial}{\partial x_j} (D_t^{\alpha_0} D_x^{\alpha'} v^{\pm}) \right]$$

and similarly

$$g_{\alpha_0,\alpha'} \equiv D_t^{\alpha_0} D_x^{\alpha'} g - [D_t^{\alpha_0} D_x^{\alpha'}(\tilde{b}_0\phi_t) - \tilde{b}_0(D_t^{\alpha_0} D_x^{\alpha'}\phi_t)]$$

$$- \left| D_t^{\alpha_0} D_x^{\alpha'} \left(\sum_{i=1}^{N-1} \tilde{b}_{1,i} \frac{\partial\phi}{\partial x_i} \right) - \sum_{i=1}^{N-1} \tilde{b}_{1,i} \frac{\partial}{\partial x_i} (D_t^{\alpha_0} D_x^{\alpha'})\phi) \right|$$

$$- [D_t^{\alpha_0} D_x^{\alpha'}(\tilde{M}(v^+, v^-)) - \tilde{M}(D_t^{\alpha_0} D_x^{\alpha'} v^+, D_t^{\alpha_0} D_x^{\alpha''} v^-)]$$

$$- [D_t^{\alpha_0} D_x^{\alpha''}(\tilde{b}_2\phi) - \tilde{b}_2 D_t^{\alpha_0} D_x^{\alpha'}\phi] \quad .$$

Setting $V_{\alpha_0,\alpha'} = (v^+_{\alpha_0,\alpha'}, v^-_{\alpha_0,\alpha'}, \phi_{\alpha_0,\alpha'})$, we apply Lemma 5.2 to obtain the estimates,

$$(5.17) \qquad |||V_{\alpha_0,\alpha'}|||^2_{0,\eta,T} \leqslant C_1 \sum_{\pm} \frac{|F^{\pm}_{\alpha_0,\alpha'}|^2_{0,\eta,T}}{\eta} + \langle g_{\alpha_0,\alpha'} \rangle^2_{0,\eta,T} \quad ,$$

for $|\alpha_0| + |\alpha'| \leqslant s$ and $\eta > C_2$, $0 \leqslant T \leqslant T_0$.

The following normal derivative estimate is proved in the third section of the Appendix. Here we make crucial use of the finiteness of the L^2-norms, $\langle \partial j \tilde{v}^{\pm}/\partial x_N^j \rangle_{s-j,T}$, for the coefficients and the constants depend essentially on these norms. If we did not use the norms and merely applied the trace theorem or used the elegant technique in [14], we would "lose" $\frac{1}{2}$ derivative as regards dependence on the coefficients.

<u>LEMMA 5.3</u> (The Normal Derivative Lemma). Assume that $\tilde{L}^{\pm}$ has coefficients satisfying Lemma 5.1.

If $v^{\pm}$ is a smooth solution of

$$\tilde{L}^{\pm} v^{\pm} = F_{\pm} \quad \text{in} \quad \begin{matrix} x_n > 0 \\ x_N < 0 \end{matrix} ,$$

$$v^{\pm} \text{ vanishes for } t < 0 ,$$

and s is an integer with $s \geqslant \frac{(n+1)}{2} + 7$, we have the estimate,

$$\sum_{\pm, j=0}^{s} \langle \partial j v^{\pm}/\partial x_N^j \rangle^2_{s-j,\eta,T} \leqslant C \left(\sum_{\pm} \langle v^{\pm} \rangle^2_{s,\eta,T} + \sum_{\pm} \frac{|F_{\pm}|^2_{s,\eta,T}}{\eta} \right) \tag{5.18}$$

for $0 \leqslant t \leqslant T_0$. Here C depends upon $\sum_{j=0}^{s} \langle \partial j u^{\pm}/\partial x_N^j \rangle^2_{s-j,T_0}$, $\langle \beta \rangle^2_{s+1,T_0}$ and (3) of (5.5)

Multiplying (5.17) by the appropriate powers of η yields

$$\sum_{|\alpha_0|+|\alpha_0'| \leqslant s} |\eta|^{2(s-(|\alpha_0|+|\alpha'|))} |||v_{\alpha_0,\alpha'}|||^2_{0,\eta,T} \tag{5.19}$$

$$\leqslant C_I \left[\sum_{\pm, |\alpha_0|+|\alpha'| \leqslant s} \frac{|\eta|^{2(s-(|\alpha_0|+|\alpha'|))} |F^{\pm}_{\alpha_0,\alpha'}|^2_{0,\eta,T}}{\eta} \right.$$

$$\left. + \sum_{|\alpha_0|+|\alpha'| \leqslant s} |\eta|^{2(s-(|\alpha_0|+|\alpha'|))} \langle g_{\alpha_0,\alpha'} \rangle^2_{0,\eta,T} \right] .$$

On the other hand, (5.18) implies that

$$(5.20)\qquad \sum_{\pm, j=0}^{s} \langle \partial^j v^{\pm}/\partial x_N^j \rangle_{s-j,\eta,T}^2$$

$$\leqslant C\left\{ \sum_{|\alpha_0|+|\alpha'|\leqslant s} |\eta|^{2(s-(|\alpha_0|+|\alpha'|))} |||V_{\alpha_0,\alpha'}|||_{0,\eta,T}^2 + \sum_{\pm} \frac{|F_{\pm}|_{s,\eta,T}^2}{\eta} \right\} ,$$

so that from (5.19) and (5.20), we deduce that

(5.21)

$$\sum_{\pm} \langle \partial^j v^{\pm}/\partial x_N^j \rangle_{s-j,\eta,T}$$

$$+ \sum_{|\alpha_0|+|\alpha'|\leqslant s} |\eta|^{2(s-(|\alpha_0|+|\alpha'|))} |||V_{\alpha_0,\alpha'}|||_{0,\eta,T}^2$$

$$\leqslant C\left\{ \sum_{\pm,|\alpha_0|+|\alpha'|\leqslant s} \frac{|\eta|^{2(s-|\alpha_0|+|\alpha'|)} |F^{\pm}_{\alpha_0,\alpha'}|_{0,\eta,T}^2}{\eta} + \sum_{\pm} \frac{|F_{\pm}|_{s,\eta,T}^2}{\eta} + \sum_{|\alpha_0|+|\alpha'|\leqslant s} |\eta|^{2(s-|\alpha_0|+|\alpha'|)} \langle g_{\alpha_0,\alpha'} \rangle_{0,\eta,T}^2 \right\} .$$

We still need to estimate, $(\partial^j/\partial x_N^j) D_t^{\alpha_0} D_x^{\alpha'} v^{\pm}$ in $\Omega_{\pm} \times [0, T]$ with $|j| + |\alpha_0| + |\alpha'| \leqslant s$ to include all the terms in the definition of $|||\cdot|||_{s,\eta,T}^2$. Here, we use the symmetrizability assumption from (0.2) and well-known special estimates for Friedrich's positive symmetric systems derived by integration by parts. We have the following:

LEMMA 5.4 (The Energy Estimate for Symmetrized Systems). Assume that $\hat{v}^{\pm}$ is a smooth solution of

$$L^{\pm}\hat{v} = \hat{F}^{\pm} \quad \text{in} \quad x_N > 0 \quad (x_N < 0) \quad ;$$

$$\hat{v}^{\pm} \quad \text{vanishes for} \quad t < 0 \quad .$$

Then there are constants, C_1 and C_2 so that for $\eta > C_2$, we have the estimate

$$(5.22) \qquad |\hat{v}^{\pm}|^2_{0,\eta,T} \leq C_1\left(\frac{|\hat{F}^{\pm}|^2_{0,\eta,T}}{\eta} + \langle \hat{v}^{\pm}\rangle^2_{0,\eta}\right)$$

$$\text{for} \quad 0 \leq T \leq T_0 \quad .$$

PROOF OF LEMMA 5.4 (for completeness). Under the assumption in (1.1), (1.2), there is a positive symmetric matrix, $A_0^{\pm}(u^{\pm})$ so that

$$A_0^{\pm}\tilde{A}_j^{\pm} \quad \text{is symmetric,} \quad j = 1, \ldots, N \quad ,$$

and $cI \leq A_0^{\pm} \leq CI$. Set $\hat{w}^{\pm} = e^{-\eta t}\hat{v}^{\pm}$, then

$$A_0^{\pm}\hat{w}_t + \eta A_0^{\pm}\hat{w} + \sum A_0^{\pm}A_j^{\pm}\frac{\partial \hat{w}^{\pm}}{\partial x_j} = A_0^{\pm}\hat{F}^{\pm} \quad .$$

If we multiply by $\hat{w}^{\pm}$ and apply Green's formula over the space-time domain, $\{x_N \gtrless 0\} \times [0, T]$ we obtain (5.22) in standard fashion.

REMARK. We have included the above sketch merely to indicate where we have used the symmetrizability assumption in our argument explicitly.

Next, we apply the above estimate in (5.22) to the term, $v^{\pm}_{\alpha_0,\alpha',j} \equiv D_t^{\alpha_0} D_x^{\alpha'} D_{x_N}^{j} v^{\pm}$ where $|\alpha_0| + |\alpha'| + |j| \leq s$. We observe from (5.21) that we have already estimated $\langle v^{\pm}_{\alpha_0,\alpha',j} \rangle^2_{0,\eta,T}$ in (5.19). Analogous to (5.14),

$$\tilde{L}^{\pm} v^{\pm}_{\alpha_0,\alpha',j} = F^{\pm}_{\alpha_0,\alpha',j}$$

where

$$F^{\pm}_{\alpha_0,\alpha',j}$$

$$= D_t^{\alpha_0} D_x^{\alpha'} D_{x_N}^{j} F^{\pm} - \left[D_t^{\alpha_0} D_x^{\alpha'} D_{x_N}^{j} \left(\sum_{K=1}^{N} \tilde{A}_k^{\pm} \frac{\partial v^{\pm}}{\partial x_k} \right) - \sum_{k=1}^{N} \tilde{A}_k^{\pm} \frac{\partial}{\partial x_k} \left(D_t^{\alpha_0} D_x^{\alpha'} D_{x_N}^{j} v^{\pm} \right) \right] .$$

By applying the estimate in (5.22) to $v^{\pm}_{\alpha_0,\alpha',j}$, multiplying by $\eta^{2(s-|\alpha_0|-|\alpha'|+|j|)}$, and then by using (5.21) and summing, we obtain that

$$\text{(5.24)} \quad |||v|||^2_{s,\eta,T} \leq C_1 \left(\frac{|F_+|^2_{s,\eta,T} + |F_-|^2_{s,\eta,T}}{\eta} + \langle g \rangle^2_{s,\eta,T} \right) + \{1\} + \{2\}$$

$$\text{for } \eta > C_2 ,$$

where the term $\{1\}$ is defined by

(5.25)

$$\{1\} = \frac{C}{\eta} \sum_{\substack{\pm, 1 \leq k \leq N \\ |\alpha_0|+|\alpha'|+|j| \leq s}} \eta^{2(s-|\alpha_0|-|\alpha'|-|j|)}$$

$$\times \left| e^{-\eta t} \left[D_t^{\alpha_0} D_x^{\alpha'} D_{x_N}^{j} \left(\tilde{A}_k^{\pm} \frac{\partial v_{\pm}}{\partial x_k} \right) - \tilde{A}_k^{\pm} \left(D_t^{\alpha_0} D_x^{\alpha'} D_{x_N}^{j} \frac{\partial v_{\pm}}{\partial x_k} \right) \right] \right|^2_{0,T} ,$$

and

$$\{2\} = C \sum_{|\alpha_0|+|\alpha'|\leqslant s} \eta^{2(s-|\alpha_0|-|\alpha'|)} \left\langle e^{-\eta t}(g_{\alpha_0,\alpha'} - D_t^{\alpha_0} D_x^{\alpha'} g)\right\rangle^2_{0,T} ,$$

where $g_{\alpha_0,\alpha'}$ has already been defined in (5.16). To estimate these terms, we use the following lemma which is a variant of well-known nonlinear commutator estimates adapted to the norms with weights in η. The proof of this lemma is postponed until the second section of the Appendix.

<u>LEMMA 5.5</u> (The Nonlinear Commutator Estimates).

(1) Assume $u \in H^s(\Omega_\pm \times [0, T_0]$, $v \in H^s(\Omega_\pm \times [0, T_0])$ and $\partial_t^j v|_{t=0} = 0$, $0 \leqslant j \leqslant s - 1$, with $s > \frac{n+1}{2} + 1$, s integer; then for $\tilde{\alpha} = |\alpha_0| + |\alpha'| + |j| \leqslant s$,

$$\left| e^{-\eta t}[D_t^{\alpha_0} D_x^{\alpha'} D_{x_N}^j (uv) - uD_t^{\alpha_0} D_{x_N}^{\alpha'} D_{x_N}^j v]\right|^2_{0,T} \leqslant C|u|^2_{s,T_0} |v|^2_{\tilde{\alpha}-1,\eta,T} .$$

(2) Assume $u \in H^s(M_0 \times [0, T_0])$, $v \in H^s(M_0 \times [0, T_0])$ with $\partial_t^j v|_{t=0} = 0$, $0 \leqslant j \leqslant s - 1$; then, for $|\alpha_0| + |\alpha'| = \hat{\alpha} \leqslant s$,

$$\left\langle e^{-\eta t}[D_t^{\alpha_0} D_x^{\alpha'}(uv) - uD_t^{\alpha_0} D_x^{\alpha} v]\right\rangle^2_{0,T} \leqslant C\langle u\rangle^2_{s,T_0} \langle v\rangle^2_{\hat{\alpha}-1,\eta,T}$$

for $0 \leqslant T \leqslant T_0/2$ and $\eta \geqslant 1$.

Below, we will also use the elementary inequalities

(5.27) (1) $|e^{-\eta t}v|_{\tilde{s},T} \leq C|v|_{\tilde{s},\eta,T}$,

$$\langle e^{-\eta t}v\rangle_{\tilde{s},T} \leq C\langle v\rangle_{\tilde{s},\eta,T} .$$

(2) For any integer θ with $0 \leq \theta \leq s$,

$$\eta^{s-\theta}\langle v\rangle_{\theta,\eta,T} \leq \langle v\rangle_{s,\eta,T} ,$$

$$\eta^{s-\theta}|v|_{\theta,\eta,T} \leq |v|_{s,\eta,T} .$$

(3) $\left|\frac{\partial v}{\partial x_j}\right|_{\tilde{s}-1,\eta,T} \leq |v|_{\tilde{s},\eta,T}$,

$$\sum_{j=1}^{N-1}\left\langle\frac{\partial\phi}{\partial x_j}\right\rangle_{\tilde{s}-1,\eta,T} + \left\langle\frac{\partial\phi}{\partial t}\right\rangle_{\tilde{s}-1,\eta,T} \leq \langle\phi\rangle_{\tilde{s},\eta,T} .$$

By (1) of Lemma 5.5, (A-2) of the Appendix, and (3) in (5.27), we estimate $\{1\}$ in (5.25) by

$$\{1\} \leq \sum_{\substack{\pm,1\leq l\leq N\\ 0\leq\tilde{\alpha}\leq s}} \frac{C}{\eta}\eta^{2(s-\tilde{\alpha})}\left|\frac{\partial v^{\pm}}{\partial x_k}\right|^2_{\tilde{\alpha}-1,\eta,T} \tag{5.28}$$

$$\leq \frac{C}{\eta}(|v^+|^2_{s,\eta,T} + |v^-|^2_{s,\eta,T}) .$$

By looking back at (5.16), we observe that we can use (2) of Lemma 5.5, (A-2) of the Appendix, and (2) in (5.27) to estimate the term, $\{2\}$, defined in (5.26) by

(5.29)

$$\{2\} \leq C\left(\sum_{0\leq j\leq N-1} \eta^{2(s-\hat{\alpha})} \langle \phi_{x_j} \rangle^2_{\hat{\alpha}-1,\eta,T} \right.$$

$$\left. + \sum_{0\leq\hat{\alpha}\leq s} \eta^{2(s-\hat{\alpha})} (\langle \phi_t \rangle^2_{\hat{\alpha}-1,\eta,T} + \langle v^+ \rangle^2_{\hat{\alpha}-1,\eta,T} + \langle v^- \rangle^2_{\hat{\alpha}-1,\eta,T}) \right)$$

$$\leq \frac{C}{\eta^2} (\langle\langle \phi \rangle^2_{s+1,\eta,T} + \langle v^+ \rangle^2_{s,\eta,T} + \langle v^- \rangle^2_{s,\eta,T}) \quad .$$

From (5.28) and (5.29), the error terms in (5.24) satisfy the estimates,

(5.30)
$$\{1\} + \{2\} \leq \frac{C}{\eta^2} |||V|||^2_{s,\eta,T} \quad .$$

Thus, if we pick $\eta \geq \eta_0$ a priori depending on C in (5.30) so that $C/\eta_0^2 \leq 1/2$, then these terms can be absorbed on the right hand side in (5.42) and we have derived the main a priori estimate in (5.3) of Proposition 5.1 where C_1 is twice the constant C appearing in (5.24).

We finish this section by sketching the

PROOF OF LEMMA 5.1. We only develop the construction for $(\tilde{\rho}(t)u^+$, $\tilde{\rho}(t)u^-$, $\tilde{\rho}(t)\beta(t))$ where $\tilde{\rho}(t)$ is the time cut-off used in (2.20) above since identical considerations as developed there guarantee (1) of Lemma 5.1. (We omit $\tilde{\rho}$ below for notational convenience.) The construction of β_ε is trivial since by standard mollification, we build a smooth β_ε with

$$\langle \beta - \beta_\varepsilon \rangle^2_{s+1} \longrightarrow 0 \quad \text{as} \quad \varepsilon \longrightarrow 0 \quad .$$

The construction of u^+_ε, u^-_ε is slightly more complex. For u^-, we set

$f_j = (\partial^j u^- / \partial n^j)\big|_{M\times(-\infty,\infty)}$, $0 \leq j \leq s$. For the moment we assume the existence of an extension operator, T , with the following properties:

(1) T maps $\bigoplus_{j=0}^{s} H^s(M_0 \times (-\infty, \infty))$ to $H^s(\Omega_- \times (-\infty, \infty))$.

(2) $|T(f_j)|_s^2 \leq C \sum_{j=0}^{s} \langle f_j \rangle_{s-j}^2$.

(3) If (f_j) is smooth, $T(f_j)$ is smooth.

(4) $\dfrac{\partial^j}{\partial n^j} T(f_j)\Big|_{M_0\times(-\infty,\infty)} = f_j$, $0 \leq j \leq s$.

Using the extension operator T with the properties in (5.31), we can decompose any u^- with bounded support and

$$\sum_{j=0}^{s} \left\langle \frac{\partial^j u^-}{\partial x_N^j} \right\rangle_{s-j}^2 + |u^-|_s^2$$

finite into

(5.32) $$u^- = T(f_j) + v ,$$

where v belongs to $H_0^s(\Omega_- \times (-\infty, \infty))$ and so that with $f_j = (\partial^j u^- / \partial n^j)\big|_{M_0\times(-\infty,\infty)}$, (4) is satisfied. Define u_ε^- by

(5.33) $$u_\varepsilon^- = T(f_{j,\varepsilon}) + v_\varepsilon ,$$

where $f_{j,\varepsilon}$ is a smooth mollification of f_j so that

$$(5.34)\qquad \sum_{j=0}^{s} \langle f_{j,\varepsilon} - f_j \rangle_{s-j}^2 \leq \varepsilon ,$$

and also v_ε belongs to $C_0^\infty(\Omega_- \times (-\infty, \infty))$ with

$$(5.35)\qquad |v_\varepsilon - v|_s^2 \leq \varepsilon .$$

Then, u_ε^- is a smooth function by (3) of (5.31) and it follows immediately from (5.31)(2), (5.32), and (5.33) that

$$(5.36)\qquad \sum_{j=0}^{s} \left\langle \frac{\partial^j}{\partial n^j} (u_\varepsilon^- - u^-) \right\rangle_{s-j}^2 + |u_\varepsilon^- - u^-|_s^2 \leq C\varepsilon .$$

Since a similar construction applies for u^+, the conditions in (2) of Lemma 5.1 are satisfied for ε sufficiently small as a consequence of (5.36), and an obvious application of Sobolev's lemma guarantees that (3) of Lemma 5.1 is satisfied for sufficiently small ε provided that $s > \frac{(n+1)}{2}$.

To construct the extension operator, by a standard partition of unity arguemnt, we only need to build T for the half-space $x_N \geq 0$ with the additional property that given R_0 sufficiently small, there is an $R_1 \leq R_0$ so that

$$(5.37)\qquad \operatorname{supp}(f_j) \subseteq \{|x'| < R_1\} \text{ guarantees}$$

$$\operatorname{supp} T(f_j) \subseteq \{x | x_N \geq 0 ,\ |x| \leq R_0\} .$$

Choose L to be a given constant coefficient strictly hyperbolic operator of order $s + 1$ with x_N a given time-like direction and let $\tilde{T}(f_j) = \tilde{u}$ be the solution of Cauchy problem,

$$L(\tilde{u}) = 0 \ ;$$

$$\left.\frac{\partial^{\ell}\tilde{u}}{\partial x_N^{\ell}}\right|_{x_N=0} = f_{\ell} \ , \qquad 0 \leqslant \ell \leqslant s \ ;$$

then $\tilde{u}$ belongs to $C^j([0, \infty), H^{s-j}(R_{N-1}))$ for $0 \leqslant j \leqslant s$ and satisfies (3) and (4) of (5.31). We define

$$T(f_j) \equiv \psi\tilde{u} \ ,$$

where ψ is a fixed smooth cut-off function. The basic energy estimates for $s + 1$ order constant coefficient strictly hyperbolic equations guarantee the estimate in (2) and because L has finite propagation speed, it is clear that the cut-off, ψ, can be chosen guaranteeing (5.37) and (4) of (5.31).

Appendix

NONLINEAR CALCULUS INEQUALITIES ON SOBOLEV SPACES WEIGHTED WITH TIME

A.A A Summary of Standard Calculus Inequalities

In proving all the inequalities of the Appendix which are neededfor the preceding developments, we shall use the basic calculus inequalities stated in Lemma A.1. These facts are proved by results in [2] and [8] once we make the following comments. The proofs in the Appendix of [2] apply only for a bounded domain; however, they trivially extend to the case treated below since the diffeomorphisms of R^N to R^N, $\Phi^t(x, \beta)$, used in this paper satisfy the additional conditons,

(A-0) $$\Phi_t(x, \beta) \equiv x \quad \text{for} \quad |x| > 2R_0 \quad \text{and} \quad |t| > 2T_0 \quad .$$

Also, in treating the unbounded region, $\Omega_+ \times [0, T_0]$, all functions, $F_\pm$, $c_{j,K}$, etc. used in the preceding sections have support in the fixed bounded region, $|x| \leqslant CR_0$ by the uniform finite propagation speed of all Cauchy problems for operators with coefficients satisfying (1.9). Thus, our tacit assumption below is that all functions, f_i^+, etc. defined on $\Omega^+ \times [0, T_0]$ satisfy $\text{supp}\, f_i^+ \subseteq \{(x, t) \,|\, |x| \leqslant CR_0\}$ and all constants depend on the size of this fixed region. Also, the proofs in [8] are for the case, $\Omega = R^{N+1}$. However, the results below in (A-1) follow from [8] by utilizing a fixed Lion's extension operator and quoting the results in [8] for R^{N+1}.

LEMMA A-1 (Standard Calculus Inequalities).

(1) Suppose s_j are integers with $s_j \geq r$, $j = 1, \ldots, k$ and $\min_{1\leq h\leq k} \min_{j_1<\ldots<j_h} \{s_{j_1} + \ldots + s_{j_h} - (h-1)([\frac{n+1}{2}]+1)\} = r \geq 0$, then

$$\text{(A-1)} \qquad \Big|\prod_{i=1}^{k} f_i^{\pm}\Big|^2_{r,T_0} \leq C \prod_{i=1}^{k} |f_i^{\pm}|^2_{s_i,T_0} ,$$

$$\langle \prod_{i=1}^{k} g_i \rangle^2_{r,T_0} \leq C \prod_{i=1}^{k} \langle g_i \rangle^2_{s_i,T_0} .$$

(2) Assume $g(\alpha, t, w)$ is smooth for $|w - w_0| < \delta$, $F_{\pm}(x, t, w)$ is smooth for $|w - w_0| < \delta$, and $F_{+}(x, t, w)$ vanishes for $|x| > (C_0 + 1)R_0$. If $w_0 - w \in H^s$ of the respective domains, $\Omega_- \times [0, T_0]$, $\Omega_+ \times [0, T_0]$, $M_0 \times [0, T_0]$, $s > \frac{N+1}{2}$, and $|w_0 - w| < \frac{3}{4}\delta$, then

$$\text{(A-2)} \qquad |F_{\pm}(x, t, w)|^2_{s,T_0} \leq C|F_{\pm}|^2_{C^s(|w-w_0|\leq\frac{3}{4})} (1 + |w|^s_{s,T_0})^2 ,$$

$$\langle g \rangle^2_{s,T_0} \leq C|g|^2_{C^s(|w-w_0|\leq\frac{3}{4}\delta)} (1 + \langle w \rangle^s_{s,T_0})^2 .$$

(3) Assume $\tilde{\Phi}$ is a diffeomorphism of R^{N+1} to R^{N+1} with the form $(x, t) \longmapsto (\Phi_t(x), t)$ and that (A-0) is satisfied. If F belongs to $H^s(R^N \times [0, T_0])$ and $s > \frac{N+1}{2} + 1$, then

$$\text{(A-3)} \qquad |F \circ \tilde{\Phi}|_{s,T_0} \leq C|F|_{s,T_0} \left(\frac{1}{\inf|d\Phi_t|^{1/2}} (|\tilde{\Phi}_t - I|^s_s + 1\right) .$$

REMARK. The estimate in (1) above includes the well-known fact that $H^s(\tilde{\Omega})$, $\tilde{\Omega} \subseteq R^{N+1}$, is a Banach algebra for $s > \frac{N+1}{2}$.

A DISCUSSION OF (4.28) AND (4.33) IN LEMMA 4.1. First, we discuss the estimates in (4.28). Recall that the condition from (4.24), $\|V_n\|^2_{s,T_0} \leqslant C_s^{-1} \frac{\delta}{4}$, is satisfied. From (4.24) and the definition of $\chi_t(x, \beta)$ in (2.2), $\tilde{\chi} = (\chi_t(x, \beta_n), t)$ is a C^1 diffeomorphism in H^{s+1} satisfying (A-0) and also

$$\text{(A-4)} \qquad |\chi_t - 1|^2_{x+1,2T_0} \leqslant C \ ,$$

$$|d\chi_t| \geqslant C > 0 \ .$$

(Recall that satisfying (4.24) guarantees (2.18) which in turn guarantees the condition in (2.3)). The formulae from the inverse function theorem and Lemma A-1 (see [2]) guarantees the same properties for $\tilde{\chi}^{-1}$ which is the map $\tilde{\Phi} = (\Phi_t(x, \beta_n), t)$. From (1) of (4.28)

$$c_{j,k}(x, t, \beta_n) = \frac{\partial}{\partial \tilde{x}_K} (\rho\beta_n n_j)|_{\tilde{\Phi}} \ .$$

Thus, by (3) of Lemma A-1 and the above discussion,

$$\text{(A-5)} \qquad |c_{j,k}(x, t, \beta_n)|^2_{s,T_0} \leqslant C\left|\frac{\partial}{\partial \tilde{x}_K}(\rho\beta_n n_j)\right|^2_{s,T_0}$$

$$\leqslant C\langle\beta_n\rangle^2_{s+1,T_0}$$

as required in (1) of (4.28). The estimate in (3) of (4.28) is an immediate application of (3) of Lemma A-1, while (2) in (4.28) is obvious since u_0^+

is constant for $|x| > 2R_0$.

To verify the estimate in (4.33), we again recall that (4.24) is satisfied and this guarantees that (3.22) is satisfied. In fact, (4.24) and (3.22) guarantee that

$$\text{(A-6)} \qquad \sup|E_T\tilde{w}_n| \leq \frac{\delta}{8} , \quad \text{for any} \quad 0 \leq T \leq T_n ,$$

where E_T is the Lion's extension operator defined in Lemma 3.1 at $t = T$. Also recall that $\tilde{w}_n$ satisfies for $0 \leq j \leq s - 1$,

$$\text{(A-7)} \qquad \partial_t^j \tilde{w}_n |_{t=0} = 0 .$$

The term, $C(w_0, \tilde{w}_n)\tilde{w}_n^2$, is the restriction to $0 \leq t \leq T$ of

$$\text{(A-8)} \qquad C(w_0, E_T\tilde{w}_n)(E_T\tilde{w}_n)^2 .$$

Since $s \geq 2[\frac{n}{2}] + 7$ and (A-6) is satisfied, by (1) and (2) of Lemma A-1, for $0 \leq T \leq T_n$, we have

$$\begin{aligned} \text{(A-9)} \qquad \langle C(w_0, \tilde{w}_n)\tilde{w}_n^2\rangle_{s,T}^2 &\leq \langle C(w_0, E_{T_n}\tilde{w}_n)(E_{T_n}\tilde{w}_n)^2\rangle_{s,T_0}^2 \\ &\leq c\langle C(w_0, E_{T_n}\tilde{w}_n)\rangle_{s,T_0}^2 \langle E_{T_n}\tilde{w}_n\rangle_{s,T_0}^4 \\ &\leq \alpha\langle \tilde{w}_n\rangle_{s,T}^4 . \end{aligned}$$

In the last line above, we have used (A-7) and Lemma 3.1. The inequality in (A-9) implies the estimate in (4.33).

A.B Nonlinear Commutator Estimates -- The Proof of Lemma 5.5

We only need to discuss the proof of (1) of Lemma 5.5 since (2) follows in the same fashion. With $s \geq |\tilde{\alpha}| = |\alpha_0| + |\alpha'| + |j|$ and the obvious symbolic notation for the inequality in (1) of Lemma 5.5, we have

(A-10)
$$|e^{-\eta t}[D^{\tilde{\alpha}}(uv) - uD^{\tilde{\alpha}}v]|^2_{0,T} \leq C \sum_{0 \leq |\tilde{\beta}| \leq |\tilde{\alpha}|-1} |D^{\tilde{\alpha}-\tilde{\beta}}uD^{\tilde{\beta}}v|^2_{0,\eta,T}$$
$$\leq C \sum_{\substack{0 \leq |\tilde{\beta}| \leq |\tilde{\alpha}|-I \\ 0 \leq |\tilde{\gamma}| \leq |\tilde{\beta}|}} \eta^{2|\tilde{\gamma}|} |D^{\tilde{\alpha}-\tilde{\beta}}uD^{\tilde{\beta}-\tilde{\gamma}}w|^2_{0,T} \quad ,$$

with $w \equiv e^{-\eta t}v$.

We estimate each of the individual terms on the right hand side of (A-10) by using E_T and (1) of Lemma A-1. Thus,

(A-11)
$$\eta^{2|\tilde{\gamma}|}|D^{\tilde{\alpha}-\tilde{\beta}}uD^{\tilde{\beta}-\tilde{\gamma}}w|^2_{0,T} \leq \eta^{2|\tilde{\gamma}|}|D^{\tilde{\alpha}-\tilde{\beta}}uD^{\tilde{\beta}-\tilde{\gamma}}E_Tw|^2_{0,T_0} \quad ,$$

and choosing $s_1 = s - |\tilde{\alpha}| + |\tilde{\beta}|$, $s_2 = |\tilde{\alpha}| - |\tilde{\beta}| - 1$, we have $s_1 \geq 0$, $s_2 \geq 0$ since $0 \leq |\tilde{\beta}| \leq |\tilde{\alpha}| - 1$ and $s - 1 \geq [\frac{n+1}{2}] + 1$, so by (1) of Lemma A-1 and Lemma 3.1,

(A-12)
$$\eta^{2|\tilde{\gamma}|}|D^{\tilde{\alpha}-\tilde{\beta}}uD^{\tilde{\beta}-\tilde{\gamma}}E_Tw|^2_{0,T_0} \leq \eta^{2|\tilde{\gamma}|}C|D^{\tilde{\alpha}-\tilde{\beta}}u|^2_{s-|\tilde{\alpha}|+|\tilde{\beta}|,T_0}|D^{\tilde{\beta}-\tilde{\gamma}}E_Tw|^2_{|\tilde{\alpha}|-|\tilde{\beta}|-1,T_0}$$
$$\leq \eta^{2|\tilde{\gamma}|}C|u|^2_{s,T_0}|E_Tw|^2_{|\tilde{\alpha}|-|\tilde{\gamma}|-1,T_0}$$
$$\leq C|u|^2_{s,T_0}\eta^{2|\tilde{\gamma}|}|v|^2_{|\tilde{\alpha}|-|\tilde{\gamma}|-1,\eta,T_0}$$
$$\leq C|u|^2_{s,T_0}|v|^2_{|\tilde{\alpha}|-1,\eta,T} \quad .$$

In the last step of (A-12), we have used (2) of (5.27). The estimates in (A-10) through (A-12) yield the proof of (1) in Lemma 5.5 as required there.

For the proof of Lemma 5.3 in the next subsection, we will use the following lemma with a proof similar to the above.

LEMMA A-2.

(1) Assume $u \in H^{s+k-p}(\Omega_\pm \times [0, T_0])$, $v \in H^{\ell-1-k}(\Omega_\pm \times [0, T_0])$, $\partial_t^j v|_{t=0} = 0$, $0 \leqslant j \leqslant \ell - 1 - k - 1$ where the integers k, p, ℓ, s satisfy

(A-13) $$0 \leqslant k \leqslant p < p + 1 \leqslant \ell \leqslant s, \quad s > [(\frac{n+1}{2})] + 1 ;$$

then

(A-14) $$|uv|_{\ell-(p+1),\eta,T} \leqslant C|u|_{s+k-p,T_0}|v|_{\ell-1-k,\eta,T}$$

$$\text{for } 0 \leqslant T \leqslant T_0/2 .$$

(2) Assume $u, v \in H^s(M_0 \times [0, T_0])$ and that they satisfy the conditions above (A-13) with the same integers satisfying (A-13), then

(A-15) $$\langle uv\rangle_{\ell-(p+1),\eta,T} \leqslant C\langle u\rangle_{s+k-p,T_0}\langle v\rangle_{\ell-1-k,\eta,T} .$$

PROOF OF LEMMA A-2. Analogous to (A-10), we have

(A-16) $$|uv|_{\ell-(p+1),\eta,T}$$

$$\leqslant C \sum_{\substack{0\leqslant|\gamma|\leqslant|\beta| \\ 0\leqslant|\beta|\leqslant|R| \\ 0\leqslant|R|\leqslant\ell-(p+1)}} \eta^{\ell-(p+1+|R|)+|\gamma|}|D^{R-\beta}uD^{\beta-\gamma}E_T w|_{0,T_0} .$$

where $w = e^{-\eta t}v$. We apply (1) of Lemma A-1 again to each term individually on the right hand side of (A-16). We choose $s_1 = s + k - p - |R| + |\beta|$, $s_2 = p + |R| - k - |\beta|$, so that by (A-13), $s_1 \geq 0$, $s_2 \geq 0$ and $s_1 + s_2 \geq [\frac{n+1}{2}] + 1$, therefore,

$$(A\text{-}17) \qquad |D^{R-\beta}uD^{\beta-\gamma}E_T w|_{0,T_0} \leq C|u|_{s+k-p,T_0}|D^{\beta-\gamma}E_T w|_{p+|R|-k-|\beta|,T_0}$$

$$\leq C|u|_{s+k-p,T_0}|v|_{p+|R|-k-|\gamma|,\eta,T} .$$

By using (2) of (5.27), we have

$$(A\text{-}18) \qquad \eta^{\ell-(p+1+|R|)+|\gamma|}|v|_{p+|R|-k-|\gamma|,\eta,T} \leq |v|_{\ell-k-1,\eta,T} ,$$

and (A-16) - (A-18) yield the inequality in (A-14).

A.C <u>Estimates for Normal Derivatives -- Lemma 5.3 and Lemmas 2.2, 2.3, and 3.2</u>

Our first objective here is to prove the most important of the above lemmas, Lemma 5.3. From (5.9) and (5.10), we can solve for the normal derivatives on $x_N = 0$ by (with $\tilde{A}_0^{\pm} = I$ and $\partial/\partial t$ corresponding to $\partial/\partial x_0$)

$$(A\text{-}19) \qquad \left.\frac{\partial v^{\pm}}{\partial x_N}\right|_{x_N=0} = -(\tilde{A}_N^{\pm})^{-1}\left.\left(\sum_{j=0}^{N-1} A_j \frac{\partial v^{\pm}}{\partial x_j}\right)\right|_{x_N=0} + (\tilde{A}_N^{\pm})^{-1}F_{\pm}\Big|_{x_N=0} .$$

If we define $m_k^{\pm}$ by $m_K^{\pm} = \left.\frac{\partial^k v^{\pm}}{\partial x_N^k}\right|_{x_N=0}$, $1 \leq k \leq s$, by induction we see that

$$m_{p+1}^{\pm} = \sum_{i=0}^{p} \binom{p}{i} \tilde{G}_i^{\pm} m_{p-i} + \sum_{i=0}^{p} \binom{p}{i} \tilde{H}_i^{\pm} \left. \frac{\partial^{p-i} F_{\pm}}{\partial x_N^{p-i}} \right|_{x_N=0} ,$$

where $\tilde{G}_i^{\pm}$ are the first order tangential operators with the coefficients

(A-20)
$$\tilde{G}_0^{\pm} = \sum_{j=0}^{N-1} \tilde{G}_{j,0}^{\pm} \Big|_{x_N=0} \frac{\partial}{\partial x_j} ;$$

$$\tilde{G}_{j,0}^{\pm} = -(A_N^{\pm})^{-1}(\tilde{A}_j^{\pm}) ;$$

$$\tilde{G}_\ell^{\pm} = \sum_{j=0}^{N-1} \tilde{G}_{j,\ell} \frac{\partial}{\partial x_j} ;$$

$$\tilde{G}_{j,\ell}^{\pm} = \left. \frac{\partial^\ell G_{j,0}}{\partial x_N^\ell} \right|_{x_N=0} ;$$

and $\tilde{H}_i^{\pm} \equiv \tilde{G}_{0,i}^{\pm}$.

Analogous to Lemma 2.2 and 2.3, we have the following lemma with a proof which we postpone until later in this Appendix.

LEMMA A-3. Under the assumptions in Lemma 5.1 on the coefficients, (u^+, u^-, β) , $\tilde{G}_{j,\ell}^{\pm}$, $\tilde{H}_\ell^{\pm}$ belong to $H^{s-\ell}(M_0 \times [0, T_0])$ for $0 \leqslant \ell \leqslant s$ and

(A-21)
$$\sum_{j=0}^{N-1} \langle \tilde{G}_{j,\ell}^{\pm} \rangle_{s-\ell,T_0} \leqslant C \left\{ 1 + \left(\langle \beta \rangle_{s+1} + \sum_{k=0}^{s-\ell-1} \left\langle \frac{\partial^k u^{\pm}}{\partial x_N^k} \right\rangle_{s-k,T_0} \right)^s \right\} \leqslant C .$$

Next, using Lemma A-3, we complete the proof of Lemma 5-3. We need to estimate

$$\langle m^{\pm}_{p+1} \rangle_{s-(p+1),\eta,T}$$

for $0 \leqslant p \leqslant s-1$. By induction, we assume the conclusion of Lemma 5.3 is valid for $1 \leqslant i \leqslant p$ and establish the conclusion for $p+1$. Let's apply Lemma A-2 with $\ell = s$, $K = p-i$, $0 \leqslant i \leqslant p$, then

(A-22)

$$\begin{aligned}\langle \tilde{G}^{\pm}_{i} m_{p-i} \rangle^2_{s-(p+1),\eta,T} &\leqslant C \sum_{j=0}^{N-1} \langle \tilde{G}^{\pm}_{j,i} \rangle^2_{s-i} \left\langle \frac{\partial m_{p-i}}{\partial x_j} \right\rangle^2_{s-1-(p-i),\eta,T} \\ &\leqslant C \langle m_{p-i} \rangle^2_{s-(p-i),\eta,T} \\ &\text{for } s-(p+1) \geqslant 0 ,\end{aligned}$$

where we have used (A-21) in the last inequality in (A-22). Similarly, we have

(A-23)

$$\begin{aligned}&\sum_{i=0}^{p} \left\langle \tilde{H}^{\pm}_{i} \left. \frac{\partial^{p-i} F_{\pm}}{\partial x_N^{p-i}} \right|_{x_N=0} \right\rangle^2_{s-(p+1),\eta,T} \\ &\leqslant C \sum_{i=0}^{p} \left\langle \frac{\partial^{p-i} F_{\pm}}{\partial x_N^{p-i}} \right\rangle^2_{s-1-(p-i),\eta,T} \\ &\leqslant \frac{C}{\eta} \sum_{i=0}^{p} \left\langle \frac{\partial^{p-i} F_{\pm}}{\partial x_N^{p-i}} \right\rangle^2_{s-\frac{1}{2}-(p-i),\eta,T} \\ &\leqslant \frac{C}{\eta} |F_{\pm}|^2_{s,\eta,T} .\end{aligned}$$

In the last two inequalities, we have applied (5.27)(2) (also valid for non-integer spaces) and the trace theorem in η-weighted norms to the function $E_T F_\pm$ in a familiar fashion as in several arguments above. A proof of the trace theorem and (5.27)(2) in the non-integer case follows easily from the discussion by Beals in [1] where he studies similar η-weighted spaces. From (A-22), (A-23), and the induction hypotheses,

$$\left\langle \frac{\partial^{p+1} v_\pm}{\partial x_N^{p+1}} \right\rangle^2_{s-(p+1),\eta,T} \leq C\left\{\langle v_\pm \rangle^2_{s,\eta,T} + \frac{1}{\eta} |F_\pm|^2_{s,\eta,T}\right\} .$$

This completes the proof of Lemma 5.3.

The key fact for the proof of Lemma A-3 and also Lemma 2.2 and 2.3 (with $M_0 \times [0, T_0]$ replaced by M_0, $\Omega_\pm$ respectively) is the following:

(A-24) Suppose $f_k \in H^{s-k}(M_0 \times [0, T_0])$, $1 \leq k \leq j$ with $s - j \geq 0$, $s \geq [\frac{n}{2}] + 1$ and α_K satisfies, $\prod_{i=1}^{j} k\alpha_k \leq j$, then

$$\langle \prod_{k=1}^{j} (f_k)^{\alpha_k} \rangle_{s-j,T_0} \leq C \prod_{k=1}^{j} \langle f_k \rangle^{\alpha_k}_{s-k} ..$$

To prove the fact in (A-24) we apply (1) of Lemma A-1. Thus, we must find a lower bound for

$$\text{(A-25)} \qquad \sum_{k=1}^{j} \theta_k(s - k) - \left(\sum_{k=1}^{j} \theta_k - 1\right)\left([\tfrac{n}{2}] + 1\right) ,$$

where $\sum_{k=1}^{j} \theta_k \geq 1$ and $\theta_k \leq \alpha_k$. Since

$$\sum_{j=1}^{j} \theta_k k \leqslant \sum_{k=1}^{j} \alpha_k k \leqslant j ,$$

$$\left(\sum_{k=1}^{j} \theta_k\right) s - \left(\sum_{k=1}^{j} \theta_k - 1\right)\left([\tfrac{n}{2}] + 1\right) \geqslant s$$

under the assumption in (A-24), a lower bound for (A-25) is always $s - j$ -- this implies the conclusion in (A-24). It is obvious that a vector analogue of (A-24) is proved in the same way with only more complicated notation. To prove Lemma A-3, we set $f_k^{\pm} = \left(\left.\frac{\partial^k u^{\pm}}{\partial x_N^k}\right|_{x_N=0} , \nabla_{\tan}^k \beta , \nabla_{\tan}^k \beta_t , \nabla_{\tan}^{k+1} \beta \right)$. It follows from (A-20) that the coefficients, $\tilde{G}_{j,\ell}^{\pm}$, $0 \leqslant j \leqslant N - 1$ have the form

(A-26) $$\tilde{G}_{j,\ell}^{\pm} = \sum_{\vec{\alpha} \in J} \tilde{g}_{\vec{\alpha},j}^{\pm} \prod_{k=1}^{\ell} (f_k^{\pm})^{\alpha_k} ,$$

where $g_{\vec{\alpha},j}$ are smooth functions of $(u^{\pm}, \beta, \nabla_{\tan}\beta, \beta_t)$ and $\vec{\alpha} \in R^{\ell}$, $J = \{\vec{\alpha} \mid \sum_{k=1}^{\ell} k\alpha_k \leqslant \ell\}$. Then,

(A-27) $$\langle \tilde{G}_{j,\ell}^{\pm} \rangle_{s-\ell,T_0} \leqslant C \sum_{\vec{\alpha} \in J} \langle \tilde{g}_{\vec{\alpha},j}^{\pm} \rangle_{s,T_0} \langle \prod_{k=1}^{\ell} (f_k^{\pm})^{\alpha_k} \rangle_{s-\ell,T_0} .$$

By applying (2) of Lemma A-1 to $\tilde{g}_{\vec{\alpha},j}^{\pm}$ and (A-24), we get the conclusion in (A-21). If we look back at the statements in Lemma 2.2 and Lemma 2.3, then it is clear that these lemmas are proved in the same fashion as in the proof of Lemma 5.3 above. In fact, the proofs are simpler since no η-weighted norms are needed so we omit any further discussion here. Also, the conclusions in Lemma 3.2 are obvious provided that (β_n, u_n^+, u_n^-) are smooth functions with the form in (3.1) so all that remains is to justify the conclusion of these formal calculations. However, since $F_{n+1}^{\pm} \in H^s(\Omega_{\pm} \times [0, T_0])$, by the trace theorem, $\partial_t^j F_{n+1}^{\pm}\big|_{t=0} \in H^{s-j-\frac{1}{2}}(\Omega_{\pm})$ for $0 \leqslant j \leqslant s - 1$. By using a similar argument as in (A-24) - (A-27) above, the recursion calculations as in (2.17) for $\partial_t^j F_{n+1}\big|_{t=0}$ can be justified in $H^{s-j-1}(\Omega_{\pm})$ for $0 \leqslant j \leqslant s - 1$. In this fashion, we can deduce that $\partial_t^j F_{n+1}\big|_{t=0} = 0$, $0 \leqslant j \leqslant s - 1$. Once again, for brevity we omit the straightforward details.

BIBLIOGRAPHY

[1] R. Beals, "Hyperbolic Equations and Systems with Multiple Characteristics," Arch. Rat. Mech. Anal., 48 (1972), pp. 123-152.

[2] J.P. Bourguignon and H. Brezis, "Remarks on the Euler Equation," J. Fcnl. Anal., 15 (1974), pp. 341-363.

[3] R. Courant and D. Hilbert, Methods of Mathematical Physics, Vol. II, Wiley-Interscience, New York, 1961.

[4] R. Courant and K.O. Friedrichs, Supersonic Flow and Shock Waves, Springer-Verlag, New York, 1948.

[5] K.O. Friedrichs, "Symmetric Positive Linear Differential Equations," Comm. Pure Appl. Math., 11 (1958), pp. 333-418.

[6] A. Friedman, Partial Differential Equations, Academic Press, New York, 1969.

[7] J. Glimm, "Solutions in the Large for Nonlinear Hyperbolic Systems of Equations," Comm. Pure Appl. Math., 18 (1965), pp. 697-715.

[8] T. Kato, "The Cauchy Problem for Quasi-Linear Symmetric Hyperbolic Systems," Arch. Rat. Mech. Anal., 58 (1975), 181-205.

[9] P.D. Lax, "Hyperbolic Systems of Conservation Laws, II," Comm. Pure Appl. Math., 10 (1957), pp. 537-556.

[10] Li Da-qian and Yu Wen-Ci, "Some Existence Theorems for Quasi-Linear Hyperbolic Systems of Partial Differential Equations in Two Independent Variables, II," Scientia Sinica, 13 (1964), pp. 551-564.

[11] Li Da-qian and Yu Wen-ci, "The Local Solvability of Boundary Value Problems for Quasilinear Hyperbolic Systems," (to appear).

[12] A. Majda, "The Stability of Multi-Dimensional Shock Fronts: A New Problem for Linear Hyperbolic Equations," (to appear in Memoirs of A.M.S.).

[13] J.B. Rauch and F.J. Massey, "Differentiability of Solutions to Hyperbolic Initial-Boundary Value Problems," Trans. Amer. Math. Soc., 189 (1974), pp. 303-318.

[14] J.B. Rauch, " L_2 is a Continuable Condition for Kreiss' Mixed Problems," Comm. Pure Appl. Math., 25 (1972), pp. 265-285.

[15] D.G. Schaeffer, "Supersonic Flow Past a Nearly Straight Wedge," Duke Math. J., 43 (1976), pp. 637-670.

Department of Mathematics
University of California
Berkeley, CA 94720

General instructions to authors for
PREPARING REPRODUCTION COPY FOR MEMOIRS

For more detailed instructions send for AMS booklet, "A Guide for Authors of Memoirs." Write to Editorial Offices, American Mathematical Society, P. O. Box 6248, Providence, R. I. 02940.

MEMOIRS are printed by photo-offset from camera copy fully prepared by the author. This means that, except for a reduction in size of 20 to 30%, the finished book will look exactly like the copy submitted. Thus the author will want to use a good quality typewriter with a new, medium-inked black ribbon, and submit clean copy on the appropriate model paper.

Model Paper, provided at no cost by the AMS, is paper marked with blue lines that confine the copy to the appropriate size. Author should specify, when ordering, whether typewriter to be used has PICA-size (10 characters to the inch) or ELITE-size type (12 characters to the inch).

Line Spacing – For best appearance, and economy, a typewriter equipped with a half-space ratchet – 12 notches to the inch – should be used. (This may be purchased and attached at small cost.) Three notches make the desired spacing, which is equivalent to 1-1/2 ordinary single spaces. Where copy has a great many subscripts and superscripts, however, double spacing should be used.

Special Characters may be filled in carefully freehand, using dense black ink, or INSTANT ("rub-on") LETTERING may be used. AMS has a sheet of several hundred most-used symbols and letters which may be purchased for $5.

Diagrams may be drawn in black ink either directly on the model sheet, or on a separate sheet and pasted with rubber cement into spaces left for them in the text. Ballpoint pen is *not* acceptable.

Page Headings (Running Heads) should be centered, in CAPITAL LETTERS (preferably), at the top of the page – just above the blue line and touching it.

LEFT-hand, EVEN-numbered pages should be headed with the AUTHOR'S NAME;

RIGHT-hand, ODD-numbered pages should be headed with the TITLE of the paper (in shortened form if necessary).

Exceptions: PAGE 1 and any other page that carries a display title require NO RUNNING HEADS.

Page Numbers should be at the top of the page, on the same line with the running heads.

LEFT-hand, EVEN numbers – flush with left margin;

RIGHT-hand, ODD numbers – flush with right margin.

Exceptions: PAGE 1 and any other page that carries a display title should have page number, centered below the text, on blue line provided.

FRONT MATTER PAGES should be numbered with Roman numerals (lower case), positioned below text in same manner as described above.

MEMOIRS FORMAT

It is suggested that the material be arranged in pages as indicated below.
Note: Starred items (*) are requirements of publication.

Front Matter (first pages in book, preceding main body of text).

Page i – *Title, *Author's name.

Page iii – Table of contents.

Page iv – *Abstract (at least 1 sentence and at most 300 words).

*1980 Mathematics Subject Classifications represent the primary and secondary subjects of the paper. For the classification scheme, see Annual Subject Indexes of MATHEMATICAL REVIEWS beginning in December 1978.

Key words and phrases, if desired. (A list which covers the content of the paper adequately enough to be useful for an information retrieval system.)

Page v, etc. – Preface, introduction, or any other matter not belonging in body of text.

Page 1 – Chapter Title (dropped 1 inch from top line, and centered).

Beginning of Text.

Footnotes: *Received by the editor date.

Support information – grants, credits, etc.

Last Page (at bottom) – Author's affiliation.

ABCDEFGHIJ–AMS–89876543

Memoirs of the American Mathematical Society

Number 282

John Hempel

Intersection calculus on surfaces with applications to 3-manifolds

Published by the

AMERICAN MATHEMATICAL SOCIETY

Providence, Rhode Island, USA

May 1983 · Volume 43 · Number 282 (end of volume)

MEMOIRS of the American Mathematical Society

This journal is designed particularly for long research papers (and groups of cognate papers) in pure and applied mathematics. It includes, in general, longer papers than those in the TRANSACTIONS.

Mathematical papers intended for publication in the Memoirs should be addressed to one of the editors. Subjects, and the editors associated with them, follow:

MEMOIRS are printed by photo-offset from camera-ready copy fully prepared by the authors. Prospective authors are encouraged to request booklet giving detailed instructions regarding reproduction copy. Write to Editorial Office, American Mathematical Society, P.O. Box 6248, Providence, Rhode Island 02940. For general instructions, see last page of Memoir.

SUBSCRIPTION INFORMATION. The 1983 subscription begins with Number 272 and consists of six mailings, each containing one or more numbers. Subscription prices for 1983 are $104.00 list; $52.00 member. Each number may be ordered separately; ***please specify number*** when ordering an individual paper. For prices and titles of recently released numbers, refer to the New Publications sections of the **NOTICES** of the American Mathematical Society.

BACK NUMBER INFORMATION. For back issues see the AMS Catalogue of Publications.

TRANSACTIONS of the American Mathematical Society

This journal consists of shorter tracts which are of the same general character as the papers published in the MEMOIRS. The editorial committee is identical with that for the MEMOIRS so that papers intended for publication in this series should be addressed to one of the editors listed above.

Subscriptions and orders for publications of the American Mathematical Society should be addressed to American Mathematical Society, P. O. Box 1571, Annex Station, Providence, R. I. 02901. *All orders must be accompanied by payment.* Other correspondence should be addressed to P. O. Box 6248, Providence, R. I. 02940.

MEMOIRS of the American Mathematical Society (ISSN 0065-9266) is published bimonthly (each volume consisting usually of more than one number) by the American Mathematical Society at 201 Charles Street, Providence, Rhode Island 02904. Second Class postage paid at Providence, Rhode Island 02940. Postmaster: Send address changes to Memoirs of the American Mathematical Society, American Mathematical Society, P. O. Box 6248, Providence, RI 02940.

TABLE OF CONTENTS

ABSTRACT

If a group, T, acts on a surface, S, there is a pairing

$$H_1(S) \times H_1(S) \to \mathbb{Z}T$$

called the Reidemeister pairing. We use properties of this pairing, in the case that S/T is a Heegaard splitting surface for a 3-manifold, M, to prove a duality theorem for the fundamental group of M and, in a relative form, for the fundamental group system. We also give criteria for the reducibility of the Heegaard splitting. In the process we review material about group and module presentations as related by the free differential calculus with emphasis on its geometric interpretation and give explicit formulae for computing the Reidemeister pairing.

AMS(MOS) SUBJECT CLASSIFICATIONS (1980): 55M05, 57M05, 57M10, 57N05, 57N10.

Library of Congress Cataloging in Publication Data

Hempel, John, 1935-
Intersection calculus on surfaces with applications to 3-manifolds.

(Memoirs of the American Mathematical Society, ISSN 0065-9266 ; no. 282)
Bibliography: p.
1. Three-manifolds (Topology) 2. Duality theory (Mathematics) 3. Calculus. 4. Surfaces. I. Title. II. Series.
QA3.A57 no. 282 [QA614.5] 510s [514'.3] 83-3724
ISBN 0-8218-2282-9

INTRODUCTION

Given a surface, S, and a normal subgroup N of $\pi_1(S)$ there is a pairing $\phi: N \times N \to \mathbb{Z}(\pi_1(S)/N)$, the *Reidemeister pairing* (section 3) which, when S is a Heegaard splitting surface of a 3-manifold (possibly with boundary), reflects interesting properties of the manifold.

We study this pairing and give two applications to 3-manifold theory. The first gives properties of the homology of the fundamental group system of a 3-manifold (or equivalently of the relative homology modules of its covering spaces). The second deals with the question of whether a given Heegaard splitting of a 3-manifold can be "reduced" which has bearing on the Poincaré conjecture and other aspects of the homeomorphism problem for 3-manifolds.

Given a regular covering space $p:\widetilde{M} \to M$ the homology groups $H_q(\widetilde{M})$ are modules over the group ring, $\mathbb{Z}T$, of the group, T, of covering transformations. The module structure provides useful information about the spaces M, $\widetilde{M}$, and the action of T on $\widetilde{M}$. For $q = 1$ the module $H_1(\widetilde{M})$ is completely determined by the fundamental group, $\pi_1(M)$, and the representation $\pi_1(M) \to T$. More generally, for $B \subset M$ the module $H_1(\widetilde{M},\widetilde{B})$ is determined by the fundamental group system $\{i_*:\pi_1(B) \to \pi_1(M)\}$ consisting of the inclusion induced maps of the fundamental groups of the components of B into $\pi_1(M)$. The relationship can be expressed in terms of the free differential calculus developed by Fox [F_1-F_5] in several different ways [Tr], [Cr_1], [Cr_2]. We prefer one which stresses the topological significance of the free derivatives and which shows that the Jacobian matrix of free derivatives for a suitable presentation of $\pi_1(M)$ projects in $\mathbb{Z}T$ to a matrix for the boundary map of cellular chain modules $\partial:C_2(\widetilde{M}) \to C_1(\widetilde{M})$. This fact appears not to be universally recognized and we have included an exposition of it in Chapter 2. Our treatment includes the relative case necessary to consider group systems.

We are interested in the case in which M is a compact, 3-manifold. For a large class of these, the P^2-irreducible, sufficiently large 3-manifolds, the total peripheral group system $\{i_*:\pi_1(\partial M) \to \pi_1(M)\}$ is a complete invariant of the topological type of M [W_1], [He_1], [Ev], [Tu], [Sw]. Thus for the classification problem for 3-manifolds it becomes important to find effective ways of distinguishing between peripheral group systems of 3-manifolds as well as for distinguishing these from arbitrary group systems. For this last problem it is natural to expect that, because of the relation between fundamental group systems and homology modules, that in dimension three duality would impose restrictions on the allowable group systems. This is indeed the case -- the symmetry of the Alexander polynomial of a knot [S] and generalizations there of [Bℓ], [To], [Mi] provide examples of this. We extend these results by proving that for M a compact oriented 3-manifold and $\partial M = B_1 \cup B_2$ a decomposition of M as a union of surfaces that the group systems $\{\pi_1(B_1) \to \pi_1(M)\}$ and $\{\pi_1(B_2) \to \pi_1(M)\}$ (more precisely the associated joined systems as defined in Chapter 1) have presentations which are dual. From this we have a corresponding duality between the homology modules $H_1(\widetilde{M},\widetilde{B}_1)$ and $H_1(\widetilde{M},\widetilde{B}_2)$. We stress that this applies to arbitrary (e.g. non abelian) covering groups, T, and that these results apply regarding the module structure over the integral group ring $\mathbb{Z}T$ not just over the more tractable rational group ring $\mathbb{Q}T$. However in the case of the universal abelian covering they imply and extend the known symmetry properties of the Alexander invariants.

We note that a rather different approach to duality in $H_*(\widetilde{M},\widetilde{B})$ is taken in [K]. Also I have been informed that V. G. Tureav has also obtained similar results on dual presentations of $\pi_1(M)$ in the case $\partial M = \phi$ [Tur].

To obtain the dual presentations we take a Heegaard splitting surface $S \subset M$ (which must be chosen with some care in the case $\partial M \neq \phi$). We have in the natural way two presentations of the corresponding group systems which are the desired ones. Proof of duality involves consideration of the regular covering $p:\widetilde{S} \to S$ corresponding to $N = \ker(i_*:\pi_1(S) \to \pi_1(M))$ and with $i_*\pi_1(S) < \pi_1(M)$ as covering group. We have an operator valued intersection pairing (the Reidemeister pairing)

$$N \times N \to H_1(\widetilde{S}) \times H_1(\widetilde{S}) \to \mathbb{Z}\pi_1(M)$$

which in skew Hermetian relative to the conjugation $\overline{\sum m_i g_i} = \sum m_i g_i^{-1}$. This pairing can be computed in terms of free derivatives. A formula for this is developed in Chapter 3. The formula provides a comparison of the Jacobian matrices of the two presentations via the Reidemeister pairing. The desired duality is a reflection of the Hermetian character of this pairing.

From a different point of view, our formula for the Reidemeister pairing proves a means of expressing the integral intersection pairing $H_1(\widetilde{S}) \times H_1(\widetilde{S}) \to \mathbb{Z}$ (for any regular covering $\widetilde{S} \to S$) by computations in the group ring of covering transformations. This has applications to the problem of determining when a normal subgroup of $\pi_1(S)$ contains a simple element and in particular when a given Heegaard splitting of a 3-manifold can be reduced (and thus to the Poincaré conjecture).

This material is developed in Chapter 5 where we prove generalization of theorems of Maskit [Ma] and Papakyriakopoulos [P].

In Chapter 6 we compute some examples illustrating some applications of our theorems.

CHAPTER 1: GROUP SYSTEMS AND PAIRS

Throughout we will work in the category of CW complexes except where further restrictions (e.g. to simplicial complexes) are made explicit.

We are interested in studying pairs (M,B) in which B need not be connected, (e.g., where M is a manifold and $B = \partial M$). This causes some problems with fundamental groups because of base point considerations. In order to resolve these problems one is led to the definition of a fundamental group system.

A *group system* is an indexed collection $\{\alpha_i : G_i \to G\}$ of group homomorphisms with common target group G. A *morphism of group systems* is a collection

$$\{\phi : G \to H,\ \phi_i : G_i \to H_{j(i)}\}$$

of homomorphisms such that for each i,

$$\beta_{j(i)} \circ \phi_i = \gamma_i \circ \phi \circ \alpha_i$$

for some inner automorphism γ_i of H. For a CW pair (M,B) with M connected and B having components $\{B_i\}$, the *fundamental group system* is the system $\{\alpha_i : \pi_1(B_i) \to \pi_1(M)\}$ where α_i are induced by inclusion and reference to a common base point. The motivation for these somewhat awkward definitions is first that the fundamental group system is well defined up to isomorphism of group systems and second that the following result is valid (c.f. [Hm, lemma 13.8]).

1.1 PROPOSITION. *Let* (M,B) *and* (N,C) *be* CW *pairs with* $\pi_q(N) = 0$ *for* $q \geq 2$. *Then a homomorphism* $\phi : \pi_1(M) \to \pi_1(N)$ *is induced by a map* $f : (M,B) \to (N,C)$ *of pairs if and only if* ϕ *extends to a homomorphism of the fundamental group systems.*

If each $\beta_j : \pi_1(C_j) \to \pi_1(N)$ is monic the situation is simpler: ϕ is induced by a map of pairs if and only if $\phi(\alpha_i(\pi_1(B_i))$ is conjugate to a subgroup of some $\beta_j(\pi_1(C_j))$ for each i. Thus we need, in this case, only consider a group together with a collection of conjugacy classes of subgroups.

Received by the editors October 3, 1980. Research supported in part by NSF grant MCS 78-06116. This paper contains material presented to the Society in an invited address at the Houston, Texas, meeting April 8, 1978.

It is natural to try to simplify even further by replacing a group system by a system with a single homomorphism or, perhaps, even by a group pair. One approach is taken in [Tr] where the *fused system* corresponding to a group system $\{\alpha_i : G_i \to G\}$ is defined to be the system $\{\beta : H \to G\}$ where H is the free product of the G_i and β is induced by the α_i. This is not entirely satisfactory for two reasons.

First there is the fact that isomorphic group systems need not have isomorphic fusings. Even the conjugacy class of the image group $\beta(H)$ need not be independent of the isomorphism class of the system. Let, for example, G be a knot group. Then G can be generated by finitely many "meridians" all of which are conjugate. Thus if G_i are the cyclic groups generated by these meridians, then within the isomorphism class of the system $\{G_i \hookrightarrow G\}$ the corresponding fusings can have images ranging from a cyclic subgroup to all of G.

Second it is shown in [Tr] that the homology of a group system (as defined therein) maps isomorphically to the homology of the corresponding fused system in dimensions two or more but only maps onto in dimensions zero and one. These are the dimensions of our particular interest; so we seek a sharper definition.

We restrict ourselves to fundamental group systems. Let (M,B) be a CW pair with M connected. If $B \neq \emptyset$ we add a new 0-cell, m_0, and for each component B_i of B we add a 1-cell a_i joining m_0 to a 0-cell of B_i to obtain a pair

$$(M^*,B^*) = (M \cup \cup a_i \cup m_0,\ B \cup \cup a_i \cup m_0).$$

If $B = \emptyset$ we let $(M^*,B^*) = (M,m_0)$ for some 0-cell m_0 of M; so, by convention, we never have $B^* = \emptyset$. We call (M^*,B^*) the *joined pair* associated to the pair (M,B), $\{i_* : \pi_1(B^*) \to \pi_1(M^*)\}$ the *joined fundamental group system* of (M,B), and $(\pi_1(M^*), i_*\pi_1(B^*))$ the *joined fundamental group pair* of (M,B).

The justification for these definitions is the validity of theorem 2.5. In example 6.3 we give some calculations illustrating the distinction between the fused and joined fundamental group systems.

We note the following

1.2 PROPOSITION.

(i) $\pi_1(M^*) \cong \pi_1(M) * F$ *where* F *is a free group of rank* $\overline{\beta}_0(B)$ *(= rank* $\overline{H}_0(B)$*),*

(ii) $\pi_1(B^*) \cong \mathop{\ast}_i \pi_1(B_i)$,

(iii) if each map $\pi_1(B_i) \to \pi_1(M)$ *is monic then so is* $i_* : \pi_1(B^*) \to \pi_1(M^*)$,

(iv) If $\rho:M^* \to M$ *is a retraction, then* $\{\rho_* \circ i_* : \pi_1(B^*) \to \pi_1(M)\}$ *is the fusing of some system in the isotopy class of the fundamental group system of* (M,B), *and all such fusings have this form, and*

(v) $H_q(M,B) \cong H_q(M^*,B^*)$ *unless* $q = 0$ *and* $B = \emptyset$.

CHAPTER 2. FREE CALCULUS; GROUP AND MODULE PRESENTATIONS

In this chapter we discuss presentations for groups, group systems, and modules. We briefly review the free differential calculus developed by Fox [F_1-F_5] and show how it is used to relate a presentation for the fundamental group system of a pair (M,B) to presentations of the homology modules, $H_1(\widetilde{M},\widetilde{B})$, of its regular coverings. This material is not new. It is developed in a more algebraic form in [Tr], [Cr_2]. We give a different development depending more on topology -- in particular clarifying the topological significance of free derivatives -- and being perhaps more suitable for topological applications. We feel that this approach may also clarify certain matters, e.g., the use of the joined fundamental group system (as defined in chapter 1) in the relative case and the distinction between the modules $H_1(\widetilde{M},p^{-1}(m_0))$ and $H_1(\widetilde{M})$ (lemma 2.8, 2.10). We include a brief account of the elementary ideals associated with presentations.

Let $F = F(X)$, be a free group with a specified free basis X, and $\mathbb{Z}F$ be the integral group ring of F.

For each $x_j \in X$ there is a mapping

$$\frac{\partial}{\partial x_j} : F \to \mathbb{Z}F$$

called the j-th *free derivative* and determined by the conditions

$$\frac{\partial x_i}{\partial x_j} = \delta_{i,j}$$

$$\frac{\partial (uv)}{\partial x_j} = \frac{\partial u}{\partial x_j} + u \frac{\partial v}{\partial x_j}$$

From these it follows

$$\frac{\partial u^{-1}}{\partial x_j} = -u^{-1} \frac{\partial u}{\partial x_j}$$

and in general

2.1 If $u = x_{j_1}^{\varepsilon_1} x_{j_2}^{\varepsilon_2} \ldots x_{j_n}^{\varepsilon_n}$ $(\varepsilon_k = \pm 1)$,

then

$$\frac{\partial u}{\partial x_j} = \sum_{k=1}^{n} \varepsilon_k \, \delta_{j_k,j} \, x_{j_1}^{\varepsilon_1} \cdots x_{j_{k-1}}^{\varepsilon_{k-1}} x_{j_k}^{(\varepsilon_k-1)/2}$$

Any group is (in many ways) isomorphic to a quotient of a free group. A *presentation* for a group, G, is a triple $(X:R:\eta)$ where X is a set, $R \subset F(X)$ and $\eta:F(X) \to G$ is an epimorphism whose kernel is the smallest normal subgroup of $F(X)$ containing R. When specific mention of η is unnecessary, we will suppress it from the notation.

The concept of presentation extends naturally to group systems [F5]. We will only need this for a system $\{\alpha:G_0 \to G\}$ consisting of a single homomorphism. In this case a *presentation of the group system* is

$$(X;\ X_0:\ R;\ R_0:\ \eta;\ \eta_0)$$

where $(X;\ R;\ \eta)$ and $(X_0:\ R_0:\ \eta_0)$ are presentations for G and G_0 respectively, $X_0 \subset X$, $R_0 \subset R$, and $\alpha \circ \eta_0 = \eta | F(X_0)$.

If we have a finite presentation

$$P = (x_1,\ldots,x_{n+t};\ x_{n+1},\ldots,x_{n+t}:\ r_1,\ldots,r_{m+s};\ r_{m+1},\ldots,r_{m+s}:\ \eta;\ \eta_0)$$

of $\alpha:G_0 \to G$ then $\operatorname{def} P = n-m$ is the *deficiency* of P and the matrix

$$J(P) = \eta \begin{bmatrix} \frac{\partial r_1}{\partial x_1} & \cdots & \frac{\partial r_1}{\partial x_n} \\ \vdots & & \vdots \\ \frac{\partial r_m}{\partial x_1} & & \frac{\partial r_m}{\partial x_n} \end{bmatrix}$$

with entries in $\mathbb{Z}G$ is called the *Jacobian* of the presentation.

Two finite presentations for a given group, or more generally for a given homomorphism, are related by a sequence of Teitze transformations [F2,F5]. It is straightforward to check the corresponding effect on their Jacobians. We summarize this in

2.2 THEOREM (4.4 of [F5]) *Any two finite presentations of a homomorphism* $\alpha:G_0 \to G$ *have Jacobian matrices which are equivalent under the equivalence relation generated by*

(1) permuting rows or columns,

(2) adding a new row which is a left linear combination,

over $\mathbb{Z}G$, *of the existing rows, and*

(3) *replacing the matrix* J *by*

$$\begin{bmatrix} & & 0 \\ & J & \vdots \\ & & 0 \\ x & \ldots\ x & 1 \end{bmatrix}$$

where all the elements of the new row, except for the last, may be arbitrary elements of $\mathbb{Z}G$.

The following operations are derivable from (1), (2), and (3) (see [F_2])

(4) *add to any row a left linear combination, over* $\mathbb{Z}G$, *of other rows or add to any column a right linear combination of other columns,*

(5) *left multiply any row or right multiply any column by a unit of* $\mathbb{Z}G$.

We note that equivalence is not completely symmetric in rows and columns. In order to symmetrize it would be necessary (and sufficient) to allow the adjunction of a new column of zeros. This would correspond to adding to the presentation a new generator for G, but no new relations. Thus the symmetrized equivalence classes of Jacobian matrices would be invariants only of the "stable isomorphism" type of the corresponding group or group homomorphism -- where stabilizing allows taking the free product with a free group. Some of our results, e.g., theorem 4.1, could be stated in terms of stable equivalence, but we prefer the somewhat sharper forms given.

2.3 PROPOSITION

(i) *The Jacobian of a presentation of a homomorphism* $\alpha: G_0 \to G$ *depends only on the group pair* $(G, \alpha(G_0))$.

(ii) *If* $P = (x_1,\ldots,x_n : r_1,\ldots,r_m : \eta)$ *is a presentation of* G, $w_1,\ldots,w_t \in F(x_1,\ldots,x_n)$ *and* H *is the subgroup of* G *generated by* $\{\eta(w_1),\ldots,\eta(w_t)\}$ *then there is a presentation* P' *for* $\{i : H \to G\}$ *with*

$$J(P') = \begin{bmatrix} J(P) \\ \eta\left(\frac{\partial w_k}{\partial x_j}\right) \end{bmatrix} \qquad \begin{matrix} 1 \leqslant k \leqslant t \\ 1 \leqslant j \leqslant n. \end{matrix}$$

(iii) *If* N *is the normal closure in* G *of* H *and* $\nu: G \to G/N$ *is projection then* $\nu(J(P'))$ *is the Jacobian of a presentation of* G/N.

PROOF: If $(X; X_0: R; R_0: \eta, \eta_0)$ presents $\alpha: G_0 \to G$, then for some $S \subset \operatorname{Ker} \eta_0$ $(X_0: R_0 \cup S: \alpha \circ \eta_0)$ presents $\alpha(G\ )$ and $(X; X_0: R \cup S; R_0 \cup S: \eta; \alpha \circ \eta_0)$ presents $i: \alpha(G_0) \to G$. The first and last presentations have identical Jacobians; since they only involve $\frac{\partial r}{\partial x}$ for $r \in R - R_0 = (R \cup S) - (R_0 \cup S)$. This proves (i).

For (ii) we see that

$$P' = (x_1, \ldots, x_{n+t}; x_{n+1}, \ldots, x_{n+t}: r_1, \ldots, r_m, w_1 x_{n+1}^{-1}, \ldots, w_t x_{n+t}^{-1},$$

$$s_1, \ldots, s_k; s_1, \ldots, s_k: \bar{\eta}; \bar{\eta} | F(x_{n+1}, \ldots, x_{n+t}))$$

presents $i: H \to G$ for some $s_i \in F(x_{n+1}, \ldots, x_{n+t})$ where $\bar{\eta}$ is the extension of η defined by $\bar{\eta}(x_{n+i}) = \eta(w_i)$. Since

$$\frac{\partial(w_k x_{n+k}^{-1})}{\partial x_j} = \frac{\partial w_k}{\partial x_j}$$

if $j \leqslant n$, it follows that $J(P')$ is as described.

Part (iii) follows from the observation that

$$(x_1, \ldots, x_n: r_1, \ldots, r_m, w_1, \ldots, w_t: \nu \circ \eta)$$

presents G/N.

A presentation for a (left) module, A, over a ring Λ is a triple

$$[X : R : \eta]$$

where $R \subset \Lambda^X$ (the free Λ-module on the set X) and $\eta: \Lambda^X \to A$ is an epimorphism whose kernel is the Λ-submodule generated by R.

If $X = \{x_1, \ldots, x_n\}$ and $R = \{r_1, \ldots, r_m\}$ then for each i

$r_i = \sum_{j=1}^{n} \lambda_{ij} x_j$ for some $\lambda_{ij} \in \Lambda$. The $m \times n$ matrix (λ_{ij}) is called the *presentation matrix* for A corresponding to the given presentation.

2.4 THEOREM. *If Λ is any ring with 1 then any two finite presentation matrices for isomorphic Λ modules are equivalent in precisely the same sense as in theorem 2.2. Conversely equivalent matrices over Λ present isomorphic Λ modules.*

PROOF. Operation (1), (2), or (3) applied to the matrix L, of a presentation $[X:R:\eta]$ of A correspond to reordering bases, adding to R an element of $\ker \eta$, or adding to X a new generator and extending η to map this generator into A. Clearly equivalent matrices present isomorphic modules.

Conversely if $[X:R:\eta]$, $[X':R':\eta']$ present modules A and A' respectively then an isomorphism $\phi:A \to A'$ would be covered by a homomorphism $p: \Lambda^X \to \Lambda^{X'}$ and ϕ^{-1} would be covered by $q: \Lambda^{X'} \to \Lambda^X$.

By type (3) operations applied to $[X:R:\eta]$ we expand X to $X \cup X'$ and expand R to $R_1 = R \cup \{-q(x') + x'; x' \in X'\}$. Next we enlarge R_1 to $R_2 = R_1 \cup R'$. This is a type (2) operation; since for $r' \in R'$, $r' = q(r') + r' - q(r')$ is a linear combination of elements of R_1. Then we enlarge R_2 to $R_3 = R_2 \cup \{-p(x) + x: x \in X\}$. This is again of type (2); since for $x \in X$ $x = x-q(p(x)) + q(p(x)) - p(x)$ is a linear combination of elements of R_3. By symmetry the resulting presentation is also equivalent to $[X':R':\eta']$. In matrix form

$$L \equiv \begin{bmatrix} L & 0 \\ -Q & I \\ 0 & L' \\ I & -P \end{bmatrix} \equiv L'$$

Thus we see that the same equivalence relation applies to Jacobian matrices of group presentations and to presentation matrices for modules.

The common ground for these concepts is provided by the following

2.5 THEOREM. *Let* (M,B) *be a finite* CW *pair with* M *connected and let* $p: \widetilde{M} \to M$ *be a regular covering space with covering group* T *and quotient map* $\theta: \pi_1(M) \to T$. *Put* $\widetilde{B} = p^{-1}(B)$. *Then if* K *is the Jacobian matrix for any finite presentation of the associated joined group system* $\{i_*: \pi_1(B^*) \to \pi_1(M^*)\}$ *then for* $\theta^*:\pi_1(M^*) \to T$ *an extension of* θ, θ^*K *is a presentation matrix of*

$$H_1(\widetilde{M},p^{-1}(m_0)) \quad \textit{if} \quad B = \emptyset$$

$$H_1(\widetilde{M},\widetilde{B}) \quad \textit{if} \quad B \neq \emptyset$$

PROOF. We note that the relation between $H_1(\widetilde{M},p^{-1}(m_0))$ and $H_1(\widetilde{M})$ will be given by 2.8 and 2.10.

Choose a retraction $\rho:M^* \to M$ then $p:\widetilde{M} \to M$ extends to a covering $p^*: \widetilde{M}^* \to M^*$ uniquely determined by the condition that ρ is covered by a retraction $\tilde{\rho}:\widetilde{M}^* \to M$. Put $\theta^* = \theta \circ \rho_*$. By excision

$$H_1(\widetilde{M}^*,p^{-1}(B^*)) = \begin{cases} H_1(\widetilde{M},p^{-1}(m_0)); & \text{if } B = \emptyset \\ H_1(\widetilde{M},\widetilde{B}); & \text{if } B \neq \emptyset \end{cases}$$

Thus it suffices to assume that B is connected and nonempty. Without changing the homotopy type of the pair (M,B) we may first expand B to contain the zero skeleton of M and then collapse a maximal tree in (the new) B. Thus we may assume that M has a single 0-cell m_0.

By 2.2 and 2.4 the result is independent of the presentations considered. To determine particular presentations we will make choices of

(i) Characteristic maps $f_i^n:(I^n,0) \to (M,m_0)$ for the n-cells, e_i^n of M, and

(ii) a base point $\widetilde{m}_0 \in p^{-1}(m_0)$.

We will denote a choice of this data by D. It determines uniquely a presentation P_D for $\{i_*:\pi_1(B) \to \pi_1(M))\}$, a quotient map $\theta_D:\pi_1(M) \to T$ and free $\mathbb{Z}T$ bases $\{\widetilde{e}_i^n\}$ for the cellular chain modules $C_n(\widetilde{M})$ according to the following rules.

$$P_D = (x_1,\ldots,x_{n+t};\, x_{n+1},\ldots,x_{n+t}:\, r_1,\ldots,r_{m+s};\, r_{m+1},\ldots,r_{m+s})$$

where $x_i = [f_i^1]$, $r_j = [f_j^2\circ\omega]$ where ω represents a fixed generator of $\pi_1(\partial I^2,0)$, and the indexing is chosen so that $e_{n+1}^1,\ldots,e_{n+t}^1$ and $e_{m+1}^2,\ldots,e_{m+s}^2$ are respectively the 1 and 2-cells of B.

Let $\alpha:(I,\partial I) \to (M,m_0)$ and $\widetilde{\alpha}$ be the lifting of α with $\widetilde{\alpha}(1) = \widetilde{m}_0$, then θ_D is determined by the condition that $\theta_D([\alpha])(\widetilde{m}_0) = \widetilde{\alpha}(1)$.

Finally $\widetilde{e}_i^n = \widetilde{f}_i^n(\mathrm{Int}\, I^n)$ where $\widetilde{f}_i^n$ is the lifting of f_i^n satisfying $\widetilde{f}_i^n(0) = \widetilde{m}_0$.

In order to complete the proof we must understand the geometric significance of the free derivatives. This is given as

2.6 LEMMA. *Given* $u \in \pi_1(M^{(1)}) = F(x_1,\ldots,x_{n+t})$ *let* $du \in C_1(\widetilde{M})$ *be the chain determined by the lifting, which begins at* $\widetilde{m}_0$, *of the edge path representing* u. *Then*

$$du = \sum_{j=1}^{n+t} \theta\eta\left(\frac{\partial u}{\partial x_j}\right) \widetilde{e}_j^1.$$

PROOF. If $u = x_{j_1}^{\varepsilon_j} \ldots x_{j_r}^{\varepsilon_r}$ $(\varepsilon_k = \pm 1)$ then clearly

$$du = \sum_k \varepsilon_k t_k \tilde{e}^1_{j_k}$$

for some $t_k \in T$. If α_k is any path in $\tilde{M}$ from $\tilde{m}_0$ to $t_k(\tilde{m}_0)$ then $t_k = \theta([p\circ\alpha_k])$ according to our description of $\theta = \theta_D$. We can take α_k to be the lifting of the edge path corresponding to the initial segment of u preceeding the occurrence of $x_{j_k}^{\varepsilon_k}$ if $\varepsilon_k = +1$ or including $x_{j_k}^{\varepsilon_k}$ if $\varepsilon_k = -1$. That is to say

$$[p\circ\alpha_k] = \eta\left(x_{j_1}^{\varepsilon_1} \cdots x_{j_{k-1}}^{\varepsilon_{k-1}} x_{j_k}^{(\varepsilon_k-1)/2}\right).$$

The proof is completed by comparing with formula 2.1.

The map $C_n(\tilde{M}) \to C_n(\tilde{M},\tilde{B})$ sends the cells of $\tilde{B}$ to 0. Since $\partial e_i^2 = dr_i$, 2.6 immediately gives

2.7 LEMMA. *If* J *is the Jacobian matrix of the presentation* P_D *of* $\{i_*:\pi_1(B) \to \pi_1(M)\}$, *then* $\theta_D(J)$ *is the matrix of* $\partial:C_2(\tilde{M},\tilde{B}) \to C_1(\tilde{M},\tilde{B})$ *in terms of the bases* $\{e_i^2\}$ *and* $\{e_j^1\}$ *determined by* D.

Now since $\tilde{B}$ contains the 0-skeleton of $\tilde{M}$, $Z_1(\tilde{M},\tilde{B}) = C_1(\tilde{M},\tilde{B})$. Thus $H_1(M,B) = \operatorname{coker} \partial$ and the proof of the theorem 2.5 is complete.

One point of this argument deserves attention. In reducing to the case where B is connected we arbitrarily chose a retraction $\rho:M^* \to M$ in order to induce a covering $p^*:\tilde{M}^* \to M^*$ and a corresponding representation $\theta^* = \theta\circ\rho_*:\pi_1(M^*) \to T$. It is not the case that ρ_* maps the Jacobian of a presentation for $\{i_*:\pi_1(B^*) \to \pi_1(M^*)\}$ to the Jacobian of a presentation for $\{\rho_*\circ i_*:\pi_1(B^*) \to \pi_1(M)\}$. The latter depends on the choice of ρ (c.f. example 6.3). This illustrates the distinction between fusing and joinings of group systems.

Let $\varepsilon:\mathbb{Z}T \to \mathbb{Z}$ be the *augmentation map* defined by

$$\varepsilon(\sum n_i t_i) = \sum n_i \quad (n_i \in \mathbb{Z},\ t_i \in T)$$

and let $\mathfrak{a} = \ker \varepsilon$ be the *augmentation ideal* of $\mathbb{Z}T$. We have:

2.8 LEMMA. *There is an exact sequence over* $\mathbb{Z}T$:

$$0 \to H_1(\widetilde{M}) \to H_1(\widetilde{M},p^{-1}(m_0)) \xrightarrow{\partial_*} \mathcal{G} \to 0$$

Moreover

(i) this sequence always splits over $\mathbb{Z}$, *but*

(ii) it splits over $\mathbb{Z}T$ *if and only if there is a commutative diagram*

$$\begin{array}{ccc} \pi_1(M) & \xrightarrow{\ \theta\ } & T \\ & \searrow \quad \swarrow \phi & \\ & \dfrac{\pi_1(M)}{p_*[\pi_1(\widetilde{M}),\ \pi_1(\widetilde{M}))]} & \end{array}$$

PROOF. As before we may assume that M has a single 0-cell m_0. Now $H_0(p^{-1}(m_0)) \cong \mathbb{Z}T$ where $1 \in \mathbb{Z}T$ corresponds to the homology class of $\widetilde{m}_0$. The given sequence is just the homology sequence of the pair $(\widetilde{M},p^{-1}(m_0))$.

As an abelian group $\mathcal{G}$ is freely generated by $\{(t-1):t \in T-\{1\}\}$. Thus we can define an abelian group homomorphism

$$\sigma': \mathcal{G} \to C_1(\widetilde{M},\ p^{-1}(m_0))$$

by $\sigma'(t-1) = du_t$ where $u_t \in F(x_1,\ldots,x_n)$ is such that $\theta(\eta(u_t)) = t$, and $d:F(x_1,\ldots,x_n) \to C_1(\widetilde{M})$ is the map of lemma 2.6. By 2.6 du_t is a 1-chain whose boundary in $C_0(p^{-1}(m_0))$ is $tm_0 - m_0$ which corresponds to $t-1$ in $\mathbb{Z}T$. That is to say σ' induces a splitting $\sigma:\mathcal{G} \to H_1(\widetilde{M},p^{-1}(m_0))$ of the sequence.

For $s,t \in T$, $s(t-1) = (st-1)-(s-1)$. Since σ is $\mathbb{Z}$-linear, $\sigma(s(t-1)) = \sigma(st-1)-\sigma(s-1)$. Thus σ will be $\mathbb{Z}T$-linear if for all s,t, $du_{st} - du_s - sdu_t$ is null homologous. But, by computing with free derivatives, $du_{st} - du_s - sdu_t = du_{st} - d(u_su_t) = d(u_{st}u_t^{-1}u_s^{-1})$.

Now suppose we are given a commutative diagram as in (ii). Then $\phi(\theta(\eta(u_{st}u_t^{-1}u_s^{-1}))) = \phi(1) = 1$; so $\eta(u_{st}u_t^{-1}u_s^{-1}) \in p_*[\pi_1(\widetilde{M}),\ \pi_1(\widetilde{M})]$. So the edge loop representing $u_{st}u_t^{-1}u_s^{-1}$ lifts to a loop, ω, representing an element of the commutator subgroup of $\pi_1(\widetilde{M})$. By 2.6 $d(u_{st}u_t^{-1}u_s^{-1})$ is a cycle representing the homology class of ω -- which is 0. Thus σ is $\mathbb{Z}T$-linear.

Conversely if $\sigma:\mathcal{G} \to H_1(M,\ p^{-1}(m_0))$ is any $\mathbb{Z}T$-linear splitting of the sequence we can define $\phi:T \to \pi_1(M)/p_*[\pi_1(M),\ \pi_1(M)]$ as follows. For $t \in T$ let $u_t \in F(x_1,\ldots,x_n)$ be chosen so that the cycle du_t represent $\sigma(t-1)$.

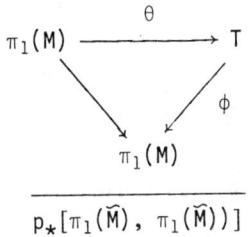

This is possible since any $c \in C_1(\widetilde{M})$, with $\partial c = t-1$ is in the image of d. We let $\phi(t)$ be the coset of $\eta(u_t)$. If du_t is homologous to du'_t then $\eta(u_t^{-1}u'_t) \in p_*[\pi_1(\widetilde{M}), \pi_1(\widetilde{M})]$; so ϕ is well defined. That ϕ is a homomorphism follows from the $\mathbb{Z}T$-linearity of σ by reversing the previous argument.

If $H_1(\widetilde{M}) = 0$ (e.g. if $\pi_1(M) = T$) then $H_1(\widetilde{M},p^{-1}(m_0)) \cong \mathfrak{a}$. Thus we have

2.9 COROLLARY. *The Jacobian matrix for any finite presentation of a group* T *is a presentation matrix for the augmentation ideal of* $\mathbb{Z}T$.

In the classical cases of this subject, arising from knot theory, T is infinite cyclic. In this situation $\mathfrak{a} \cong \mathbb{Z}T$ and the sequence of 2.8 splits over $\mathbb{Z}T$; so

$$H_1(\widetilde{M}) \cong H_1(\widetilde{M},p^{-1}(x_0)) \oplus \mathbb{Z}T$$

and $\theta(J)$ with a column of zeros added is a presentation matrix for $H_1(\widetilde{M})$.

However in the more general context (even when T is abelian) the distinction between $H_1(\widetilde{M})$ and $H_1(\widetilde{M},p^{-1}(m_0))$ is greater, but a description of $H_1(\widetilde{M})$ is still obtainable from $\theta(J)$.

As abelian groups

$$H_1(\widetilde{M},p^{-1}(m_0)) \cong H_1(\widetilde{M}) \oplus \mathbb{Z}^{o(T)-1}$$

and a presentation (infinite in case $o(T) = \infty$)) for the former can be obtained from θJ be extending everything in terms of bases over $\mathbb{Z}$ (c.f. [F_3]).

In order to describe the $\mathbb{Z}T$-module structure of $H_1(\widetilde{M})$ it clearly suffices to:

(i) Find generators for $Z_1(\widetilde{M}) = \ker(\partial : C_1(\widetilde{M}) \to \mathfrak{a})$.

(ii) Rewrite the relations $\sum_j \frac{\partial r_i}{\partial x_j} \tilde{e}_j = 0$ in terms of these new generators, and

(iii) add relations in the new generators necessary to present $Z_1(\widetilde{M})$ (which will not, in general, be a free $\mathbb{Z}T$ module).

We illustrate this in the case of a finitely generated abelian group

$$T = \langle t_1 : t_1^{n_1} \rangle \times \dots \times \langle t_r : t_r^{n_r} \rangle \times \langle t_{r+1} : \underline{\quad} \rangle \times \dots \times t_s : \underline{\quad} \rangle$$

We may suppose that

$$\theta(x_i) = t_i \quad \text{for} \quad 1 \leqslant i \leqslant s$$

for this can always be arranged by enlarging M to a complex which collapses to M. This clearly implies $\partial\tilde{e}_i = t_i - 1$ $(1 \leqslant i \leqslant s)$. Thus if E is the submodule of $C_1(\tilde{M})$ generated by $\{\tilde{e}_1,\ldots,\tilde{e}_s\}$, then ∂ maps E onto $\mathfrak{a}$.

For $i = s+1,\ldots,n$ let $c_i = \tilde{e}_i - a_i$ where $a_i \in E$ is chosen so that $\partial\tilde{e}_i = \partial a_i$.

2.10 THEOREM. *As a* $\mathbb{Z}T$ *module* $H_1(\tilde{M})$ *is presented with generators:*

$$c_i \qquad ; \qquad s+1 \leqslant i \leqslant n$$

$$f_{i,j} = (t_i-1)\tilde{e}_j - (t_j-1)\tilde{e}_i \qquad ; \qquad 1 \leqslant i < j \leqslant s$$

$$g_i = (1 + t_i + \ldots + t_i^{n_i-1})\tilde{e}_i ; \qquad 1 \leqslant i \leqslant r$$

and relations:

$$\sum_j \theta\left(\frac{\partial r_i}{\partial x_j}\right)\tilde{e}_j = 0 \qquad 1 \leqslant i \leqslant m$$

rewritten in terms of the new generators as well as

$$(t_i-1)g_i = 0; \qquad i \leqslant r$$

$$(t_k-1)f_{i,j} + (t_i-1)f_{j,k} + (t_j-1)f_{k,i} = 0; \quad i,j,k \text{ distinct}$$

$$(t_j-1)g_i + (1 + t_i + \ldots + t_i^{n_i-1})f_{i,j} = 0; \quad i \leqslant r$$

where we have used $f_{i,j}$ *to denote* $-f_{j,i}$ *if* $i > j$.

PROOF. It is clear that the c_i's represent free generators for $Z_1(\tilde{M})/E \cap Z_1(\tilde{M})$. Thus it suffices to show that the remaining generators and the relations involving only these generators present $E \cap Z_1(\tilde{M})$. For this let

$$X = L_{n_1} \times \ldots \times L_{n_r} \times S^1 \times \ldots \times S^1$$

where L_{n_i} is a 3-dimensional lens space with $\pi_1(L_{n_i}) = Z_{n_i}$ and let $\tilde{X}$ be the universal cover of X. Thus $\pi_1(X) \cong T$ and $H_1(\tilde{X}) = H_2(\tilde{X}) = 0$. We give each factor the "natural" cell structure and give X the resulting product cell structure. This is done in such a way that in the cellular chain complex

$$C_3(\widetilde{X}) \xrightarrow{\partial} C_2(\widetilde{X}) \xrightarrow{\partial} C_1(\widetilde{X}) \xrightarrow{\partial} \mathbb{G}$$

the elements $\widetilde{e}_1,\ldots,\widetilde{e}_s$ correspond precisely to the appropriate lifts of the 1-cells of X, the proposed generators for $E \cap Z_1(\widehat{M})$ correspond to the boundaries of the lifts of the 2-cells of X and the proposed relations correspond to the lifts of the 3-cells of $\widetilde{X}$. The conclusion follows from the exactness of the sequence.

While presentation matrices (Jacobian matrices) provide the most complete information about the corresponding module (group pair), it is customary to consider successively weaker, but more tractable, invariants -- the "elementary ideals" and "Alexander polynomials". We conclude with a discussion of these invariants.

Let Λ be a commutative ring with unit and let P be a matrix over Λ with n columns and m rows. For any $k \in \mathbb{Z}$ the ideal in Λ generated by all $(n-k) \times (n-k)$ minor determinants of P is called the k-th *elementary ideal* of P and is denoted $\mathcal{E}_k(P)$. By convention $\mathcal{E}_k(P) = (0)$ if $m < n-k \leqslant n$ or $k < 0$ and $\mathcal{E}_k(P) = (1)$ if $k \geqslant n$. Thus we have a chain

$$(0) \subseteq \mathcal{E}_1(P) \subseteq \ldots \subseteq \mathcal{E}_n(P) \subseteq (1)$$

of ideals. It is straightforward to show that $\mathcal{E}_k(P)$ depends only on the equivalence class of P. Thus if H is any finitely presented Λ module we define $\mathcal{E}_k(H)$ to be $\mathcal{E}_k(P)$ for P a presentation matrix for H. Finally if (M,B) is a CW-pair and $\theta:\pi_1(M) \to T$ is any representation to an abelian group T, then, with $\Lambda = \mathbb{Z}T$, we put $\mathcal{E}_k(M,B;\theta) = \mathcal{E}_k(H_1(\widehat{M},\widehat{B}))$, where $\widehat{M}$ is the regular covering of M corresponding to $\ker\theta$. By 2.5 we have

2.11 THEOREM. *If* $B \neq \emptyset$, *then for* J *the Jacobian matrix of any finite presentation of* $i_*:\pi_1(B^*) \to \pi_1(M^*)$ *we have* $\mathcal{E}_k(M,B;\theta) = \mathcal{E}_k(J)$.

Note there is a distinction between $\mathcal{E}_k(M,m_0;\theta)$ and $\mathcal{E}_k(M;\theta)$; the latter can be computed with the help of 2.10.

If Λ is a factorization domain, we let $\Delta_k(\ldots)$ be any generator of the smallest principal ideal of Λ containing $\mathcal{E}_k(\ldots)$ -- with the understanding that it is only defined up to multiplication by a unit of Λ we will call $\Delta_k(\ldots)$ the k-th *Alexander polynomial* of $(\ldots)$.

If Λ is a principal ideal domain, then any finitely generated Λ module is completely determined by its chain of elementary ideals. This is the content of the structure theorem for such modules. This is not true in greater generality. The following lemmas summarize some of the properties which we use in later sections. For more details and related results see [Bℓ], [F_2], [F_5], [Cr_3], [Cr,S], [Hi_1], [Hi_2].

2.12 LEMMA. *If* $K \subset H$ *are finitely presented* Λ *modules, then for all* i,j,k

(i) $\mathcal{E}_k(H) \subset \mathcal{E}_k(H/K)$,

(ii) $\mathcal{E}_i(K)\mathcal{E}_j(H/K) \subset \mathcal{E}_{i+j}(H)$, *and*

(iii) *if* $H = K \oplus H/K$ *then* $\mathcal{E}_k(H) = \sum_{i+j=k} \mathcal{E}_i(K)\mathcal{E}_j(H/K)$.

PROOF. A presentation matrix for H extends, by adding rows, to one for H/K.

If P and Q are presentation matrices for K and H/K respectively, then H has a presentation matrix of the form

$$\begin{bmatrix} P & 0 \\ R & Q \end{bmatrix}$$

where $R = 0$ if $H = K \oplus H/K$.

The above clearly implies, in case Λ is a factorization domain, that $\Delta_k(H/K)$ divides $\Delta_k(H)$. We also have

2.13 LEMMA. (4.2 of [Bℓ]) *If* Λ *is a unique factorization domain and* $K \subset H$ *are finitely presented* Λ *modules then for every* k $\Delta_k(K)$ *divides* $\Delta_k(H)$.

PROOF. By induction on rank (H/K) it suffices to consider the case in which H has a presentation matrix $P = (p_{ij})$ $1 \leqslant i \leqslant m$, $1 \leqslant j \leqslant n$ relative to generators $h_1,\dots,h_n$ and relations $r_i = \sum_j p_{ij}h_j$, and that K is generated by $h_1,\dots,h_{n-1}$.

Let $s = n-k$ and X be an $s \times s$ minor of P so that $\det X$ is a generator of $\mathcal{E}_k(H)$. Let $p_i = (p_{i1},\dots,p_{in})$ be the ith row vector of P and suppose that X is a minor of

$$\cdot \begin{bmatrix} p_{i_1} \\ \vdots \\ p_{i_s} \end{bmatrix}$$

Let $d = \gcd\{p_{i_1,n},\ldots,p_{i_s,n}\}$ and put $p_{i_j,n} = q_j d$, $j = 1,\ldots,s$. Now $q_1^{s-1}\det X = \det X'$ where X' is an $s \times s$ minor of

$$P' = \begin{bmatrix} p_{i_1} \\ q_1 p_{i_2} \\ q_1 p_{i_s} \end{bmatrix}$$

which is equivalent to

$$P'' = \begin{bmatrix} p_{i_1} \\ q_1 p_{i_2} - q_2 p_{i_1} \\ \cdot \\ \cdot \\ \cdot \\ q_1 p_{i_s} - q_s p_{i_1} \end{bmatrix} = \begin{bmatrix} \cdots & p_{i_1,n} \\ Q & 0 \\ & \cdot \\ & \cdot \\ & \cdot \\ & 0 \end{bmatrix}$$

where Q is a matrix of relations in $h_1,\ldots,h_{n-1}$ and thus extends, by adding rows, to a presentation matrix for K.

Thus we have shown that $q_1^{s-1}\det X \in \mathcal{E}_k(P'') = \mathcal{E}_k(Q) \subset \mathcal{E}_k(K)$. So $\Delta_k(K)$ divides $q_1^{s-1}\det X$. The same argument shows that $\Delta_k(K)$ divides $q_j^{s-1}\det X$, $j = 1,\ldots,s$. Since $\gcd\{q_1,\ldots,q_s\} = 1$; $\Delta_k(K)$ divides $\det X$ and the proof is complete.

Note that the stronger assertion that $\mathcal{E}_k(H) \subset \mathcal{E}_k(K)$ is false. Examples are supplied in case H is free and K is a non-free submodule.

For example let $T = (t_1,t_2:[t_1,t_2])$ be free abelian of rank two and $\Lambda = \mathbb{Z}T$. Then $H = \Lambda$ is free so $\mathcal{E}_0(H) = (o)$, $\mathcal{E}_1(H) = 1$. But if K is the augmentation ideal of Λ, then according to 2.9 $\mathcal{E}_1(K) = (t_1-1,\ t_2-1)$.

2.14 LEMMA. *If* $\mathfrak{u} = \{\lambda \in \Lambda : \lambda H = 0\}$ *is the annihilating ideal of* H *and* $\rho = $ *rank* H *then*

$$\mathfrak{u}^{\rho} \subset \mathcal{E}_0(H) \subset \mathfrak{u}$$

Thus if Λ *is an integral domain,* H *is a torsion module if and only if* $\mathcal{E}_0(H) \neq (0)$.

PROOF. Let $P = (p_{ij})$ be a ρ-columned presentation matrix for H. If $\lambda_1, \ldots, \lambda_\rho \in \mathfrak{U}$ then

$$\begin{bmatrix} \lambda_1 & & 0 \\ & \ddots & \\ 0 & & \lambda_\rho \\ & P & \end{bmatrix}$$

also presents H. Thus $\mathfrak{U}^\rho \subset \mathcal{E}_0(H)$.

Now if $h_1, \ldots, h_\rho$ are generators of H corresponding to the columns of P and $X = (x_{ij})$ is any $\rho \times \rho$ minor of P, then for each i $\sum_j x_{ij} h_j = 0$. So for any k, putting X_{ik} the cofactor of x_{ik} we have

$$0 = \sum_i \sum_j x_{ij} X_{ik} h_j = \sum_j (\sum_i x_{ij} X_{ik}) h_j =$$

$$\sum_j (\delta_{jk} \det X) h_j = (\det X) h_k .$$

So $\mathcal{E}_0(H) \subset \mathfrak{U}$.

By definition H (being finitely generated) is a torsion module if and only if $\mathfrak{U} \neq (0)$. But if Λ is an integral domain then $\mathfrak{U} \neq (0)$ if and only if $\mathfrak{U}^\rho \neq (o)$ and the second part follows.

2.15 LEMMA (4.10 of [Bℓ]). *Let* H *be a finitely presented module over a Notherian unique factorization domain* Λ, *let* $T = \{h \in H: \lambda h = 0$ for some $\lambda \in \Lambda - \{0\}\}$ *be the torsion submodule of* H *and let* $\mu = min\{k : \Delta_k(H) \neq 0\}$. *Then*

(i) $\Delta_k(H) = \Delta_{k-\mu}(T)$ *for all* k, *and*

(ii) μ *is the rank of any maximal free submodule of* H.

PROOF. Let F be a maximal free submodule of H and put $\nu = \text{rank } F < \infty$. Let H′ be the submodule generated by $F \cup T$, and note that $H' = F \oplus T$. By 2.12 $\Delta_{k-\nu}(T) = \Delta_k(H')$ which, by 2.13, divides $\Delta_k(H)$. By 2.12 (ii) $\Delta_k(H)$ divides $\Delta_{k-\nu}(T) \cdot \Delta_\nu(H/T)$. Since F is maximal and H is finitely generated, there is some $\lambda \in \Lambda - \{0\}$ such that $\lambda H \subset F$. Since H/T is torsion free, the map $h \mapsto \lambda h$ induces an embedding of H/T into F. By 2.13 $\Delta_\nu(H/T)$ divides $\Delta_\nu(F) = 1$.

Thus $\Delta_\nu(H/T) = 1$ and $\Delta_k(H) = \Delta_{k-\nu}(T)$. In particular $\Delta_k(H) = 0$ if $k < \nu$ and $\Delta_\nu(H) = \Delta_0(T) \neq 0$ by 2.14. So $\nu = \mu$ and the proof is complete.

2.16 LEMMA. *If* M *is a finite* CW *complex and* θ *is a representation of* $\pi_1(M)$ *onto a free abelian group,* T, *then for all* k

$$\Delta_{k-1}(M;\theta) = \Delta_k(M,m_0:\theta).$$

If T *is cyclic then*

$$\mathcal{E}_{k-1}(M;\theta) = \mathcal{E}_k(M,m_0;\theta).$$

PROOF. We have the exact sequence

$$0 \to H_1(\tilde{M}) \to H_1(\tilde{M},p^{-1}(m_0)) \to \mathfrak{a} \to 0$$

from 2.8. If T is cyclic then $\mathfrak{a} \cong \mathbb{Z}T$, the sequence splits, and the conclusion follows from 2.12.

In general, since $\mathfrak{a}$ is torsion free $H_1(\tilde{M})$ and $H_1(\tilde{M},p^{-1}(m_0))$ have the same torsion submodule. By 2.15 $\Delta_{k-i}(M;\theta) = \Delta_k(M,m_0;\theta)$ for some $i \geq 1$. But $\Delta_k(M,m_0;\theta)$ divides $\Delta_{k-1}(M;\theta)\cdot\Delta_1(\mathfrak{a})$. Again by 2.13, $\Delta_1(\mathfrak{a})$ divides $\Delta_1(\mathbb{Z}T) = 1$; so $\Delta_1(\mathfrak{a}) = 1$, $i = 1$, and the proof is complete.

We note that the above lemma explains the discrepancy in indexing the Alexander polynomials between the group theoretic and module theoretic approaches. We also note that the corresponding statement about the elementary ideals is not true when rank $T > 1$. There is still a close relationship between these ideals. See [Cr_3], [Cr,S], [Hi_1], [Hi_2] for further information along these lines.

CHAPTER 3. REIDEMEISTER PAIRING

If a group, T, acts on oriented n-manifold, M, there are pairings $H_q(M) \times H_{n-q}(M) \to \mathbb{Z}T$ which have been studied by Reidemeister [R_1, R_2]. We are most interested in the case $n = 2$, $q = 1$ where this pairing can also be viewed as a pairing on the fundamental group.

Throughout this section S will denote a compact, connected, oriented surface, N will be a normal subgroup of $\pi_1(S,s_0)$, $p:\tilde{S} \to S$ will be associated regular covering, and T will be the group of covering transformations. As in Chapter 2, an exact sequence

$$1 \to N \to \pi_1(S) \overset{\theta}{\to} T \to 1$$

is determined once a choice of base point $\tilde{s}_0 \in p^{-1}(s_0)$ is made.

We let $\langle\ ,\ \rangle: H_1(\tilde{S}) \times H_1(\tilde{S}) \to \mathbb{Z}$ be the integral intersection pairing Recall that $H_*(\tilde{S})$ is a $\mathbb{Z}T$ module with respect to the action defined by $t\alpha = t_*(\alpha)$ $t \in T$, $\alpha \in H_*(\tilde{S})$.

We define pairings:

$$\Phi: H_1(\tilde{S}) \times H_1(\tilde{S}) \to \mathbb{Z}T$$

$$\phi: N \times N \to \mathbb{Z}T$$

by

$$\Phi(\alpha,\beta) = \sum_{t\in T} \langle\alpha,t\beta\rangle t$$

$$\phi(u,v) = \Phi(h(u), h(v))$$

where $h: N \to H_1(\tilde{S})$ is the composition of "lifting at $\tilde{s}_0$" and the Hurewicz map. We call these maps the *Reidemeister pairings* on homology and homotopy respectively.

Recall that we have a conjugation on $\mathbb{Z}T$ defined by

$$\overline{\sum_i n_i t_i} = \sum_i n_i t_i^{-1} \quad n_i \in \mathbb{Z}, t_i \in T.$$

3.1 PROPOSITION. Φ *and* ϕ *are skew Hermetian bilinear. That is*

$$\text{(i)}\quad \Phi(\alpha,\beta) = -\overline{\Phi(\beta,\alpha)} \qquad \alpha,\beta \in H_1(\tilde{S}),$$

$$\phi(u,v) = -\overline{\phi(v,u)} \qquad u,v \in N,$$

$$\text{(ii)}\quad \Phi(\sum_i \lambda_i\alpha_i , \sum_j \mu_j\beta_j) = \sum_{i,j} \lambda_i\, \Phi(\alpha_i,\beta_j)\bar{\mu}_j$$

$$\alpha_i,\beta_j \in H_1(\tilde{S}) \qquad \lambda_i,\mu_j \in \mathbb{Z}T, \text{ and}$$

$$\phi(\prod_i a_iu_ia_i^{-1}, \prod_j b_jv_jb_j^{-1}) = \sum_{i,j} \theta(a_i)\phi(u_i,v_j)\overline{\theta(b_j)}$$

$$u_i,v_j \in N \quad a_i,b_j \in \pi_1(S).$$

PROOF. Note that $h(uv) = h(u) + h(v)$ and that $h(aua^{-1}) = \theta(a)h(u)$. That is, h is a module homomorphism from the multiplicative $\pi_1(S)$ module N to the additive $\mathbb{Z}T$ module $H_1(\tilde{S})$. Thus the properties for ϕ follow from those for Φ. The conjugate linearity in the second variable follows from the linearity in the first (which is obvious for Φ) and from (i).

Finally

$$\Phi(\alpha,\beta) = \sum_{t\in T} \langle\alpha,t\beta\rangle t = \sum_{t\in T} \langle t^{-1}\alpha,\beta\rangle t = -\sum_{t\in T} \langle\beta,t^{-1}\alpha\rangle t = -\sum_{t\in T} \langle\beta,t\alpha\rangle t^{-1}$$

$$= -\overline{\Phi(\beta,\alpha)}.$$

We are working towards a formula for expressing ϕ in terms of free derivatives. It is natural to give this in two forms. The first, which facilitates the proof, is expressed in terms of a pair of "dual" bases for $\pi_1(S)$. The second expresses the formula in terms of a single basis. We first develope the relationship between these bases.

Let D be the disk shown in figure 1. We give D the right hand orientation. There is an orientation preserving identification map $D \to S$ which makes the indicated edge identifications. The curves $a_1,\ldots,a_{2g+k}$ represent a set of "standard" generators (which we denote by the same symbols) for $\pi_1(S,s_0)$. Note that g is the genus of S and k is the number of components of ∂S. In terms of these generators $\pi_1(S)$ is presented with the single relation

$$[a_1,a_2] \ldots [a_{2g-1},a_{2g}]\, a_{2g+1} \ldots a_{2g+k} = 1.$$

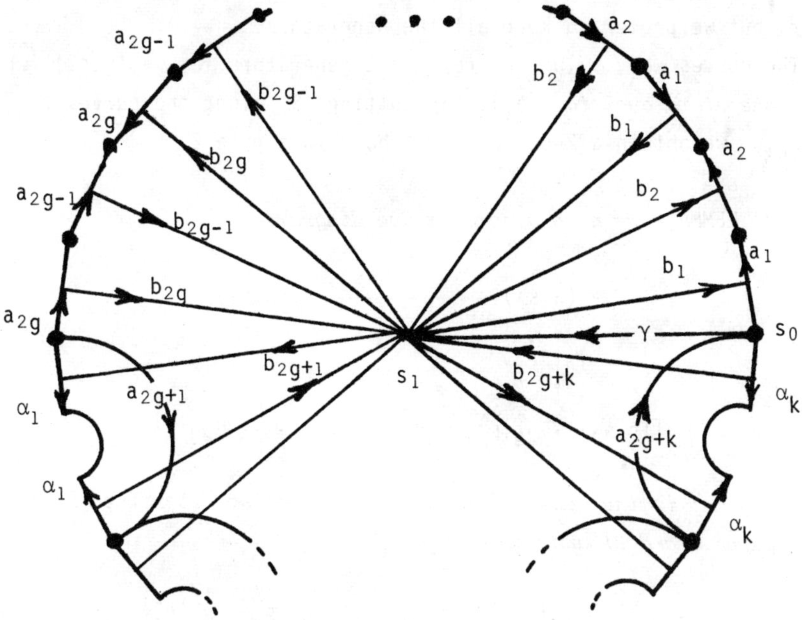

Figure 1

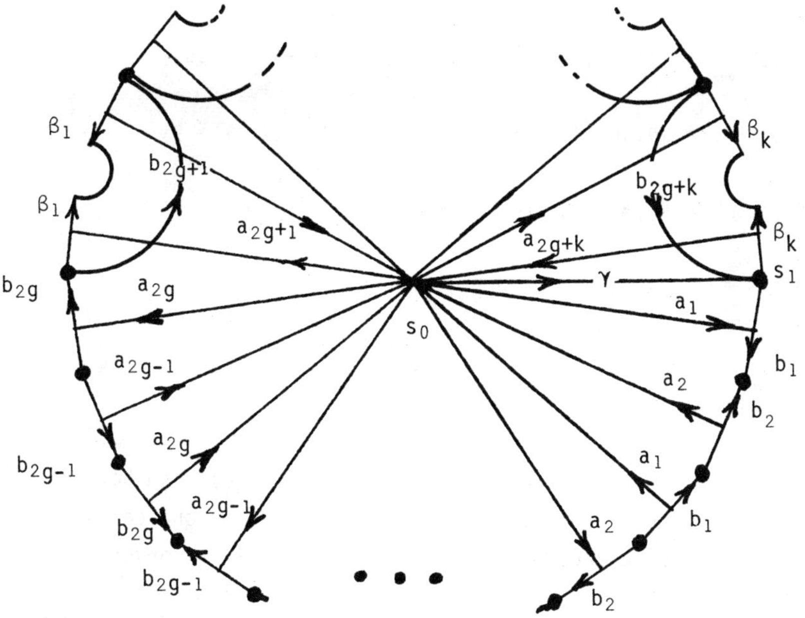

Figure 2

If $k > 0$ then $\pi_1(S)$ is free and this relation can be used to eliminate one generator, but we prefer to keep all the generators.

The curves $b_1,\dots,b_{2g+k}$ represent generators for $\pi_1(S,s_1)$ which we call the *generators dual to* $\{a_i\}$. By cutting S along the curves $b_1,\dots,b_{2g+k}$ we obtain a 2-cell E as shown in figure 2.

3.2 LEMMA. *We have a commutative diagram*

$$\begin{array}{ccc} \pi_1(S,s_0) & \longrightarrow & \pi_1(S,s_1) \\ \uparrow{\scriptstyle i_0} & & \uparrow{\scriptstyle i_1} \\ F(a_1,\dots,a_{2g+k}) & \xrightarrow{\ f\ } & F(b_1,\dots,b_{2g+k}) \end{array}$$

where the top map is the base point change isomorphism determined by the path γ *(see figures 1 and 2) and where:*

$$f(a_{2i-1}) = Q_{2i-1}b_{2i}Q_{2i-1}^{-1}, \quad f^{-1}(b_{2i-1}) = P_{2i-1}a_{2i}^{-1}P_{2i-1}^{-1}; \quad i \leqslant g$$

$$f(a_{2i}) = Q_{2i}b_{2i-1}^{-1}Q_{2i}^{-1}, \quad f^{-1}(b_{2i}) = P_{2i}a_{2i-1}P_{2i}^{-1}; \quad i \leqslant g$$

$$f(a_j) = Q_jb_j^{-1}Q_j^{-1}, \quad f^{-1}(b_j) = P_ja_j^{-1}P_j^{-1} \quad j > 2g$$

$$P_{2i-1} = [a_1,a_2]\ \dots\ [a_{2i-3},a_{2i-2}]\ a_{2i-1}; \quad i \leqslant g$$

$$Q_{2i-1} = [b_1,b_2^{-1}]\ \dots\ [b_{2i-3},b_{2i-2}^{-1}]\ b_{2i-1}; \quad i \leqslant g$$

$$P_{2i} = P_{2i-1}a_{2i}, \quad Q_{2i} = Q_{2i-1}\,b_{2i}^{-1} \quad i \leqslant g$$

$$P_j = [a_1,a_2]\ \dots\ [a_{2g-1},a_{2g}]\ a_{2g+1}\ \dots\ a_j$$

$$Q_j = [b_1,b_2^{-1}]\ \dots\ [b_{2g-1},b_{2g}^{-1}]\ b_{2g+1}\ \dots\ b_j; \quad j > 2g.$$

PROOF. Homotopies between $\gamma b_i\gamma^{-1}$ and $f^{-1}(b_i)$ and between $\gamma^{-1}a_i\gamma$ and $f(a_i)$ can be seen by examining figures 1 and 2; or one can show that $f\circ f^{-1} = f^{-1}\circ f = 1$ by direct computation. For this it is convenient to verify inductively that

$$f(P_i) = \begin{cases} b_iQ_{i+1}^{-1} \\ Q_i^{-1} \end{cases} \qquad f^{-1}(Q_i) = \begin{cases} a_iP_{i+1}^{-1}; & i \text{ odd and } i < 2g \\ P_i^{-1} & ; \text{ otherwise.} \end{cases}$$

Let

$$N_0 = i_0^{-1}(N) < F_0 = F(a_1,\dots,a_{2g+k}),$$

$$N_1 = i_1^{-1}(N) < F_1 = F(b_1,\dots,b_{2g+k}),$$

and put $\theta_j = \theta\circ i_j : F_j \to T;\quad j = 0,1.$

3.3 THEOREM.

(i) For $(u,v) \in N_1 \times N_0$

$$\phi(i_1u,i_0v) = \sum_{j=1}^{2g+k} \theta_1\left(\frac{\partial u}{\partial b_j}\right)\ \theta_0(\mu_j)\ \overline{\theta_0\left(\frac{\partial v}{\partial a_j}\right)}$$

where

$$\mu_j = \begin{cases} P_j a_j^{-1}; & j \leqslant 2g \\ P_j(a_j^{-1} - a_j^{-2}); & j > 2g. \end{cases}$$

(ii) For $(u,v) \in N_0 \times N_0$

$$\phi(i_0u,i_0v) = \theta_0\left(\sum_{i,j=1}^{2g+k} \frac{\partial u}{\partial a_i}\ \lambda_{ij}\ \overline{\frac{\partial v}{\partial a_j}}\right)$$

where

$$\begin{bmatrix} \lambda_{2i-1,2i-1} & \lambda_{2i-1,2i} \\ \lambda_{2i,2i-1} & \lambda_{2i,2i} \end{bmatrix} = \begin{bmatrix} 1 - a_{2i-1} & a_{2i-1}a_{2i}^{-1} \\ 1-a_{2i-1}^{-1}-a_{2i} & 1 - a_{2i}^{-1} \end{bmatrix};\quad i \leqslant g$$

$\lambda_{i,i} = 1-a_i;\quad i > 2g$ *and in all other cases*

$$\lambda_{i,j} = \begin{cases} (1-a_i)(1-a_j^{-1}); & j < i \\ 0; & j > i \end{cases}$$

(iii) If we use (ii) to extend ϕ *to* $\hat{\phi}: F_0 \times F_0 \to \mathbb{Z}T$ *by*

$\hat{\phi}(u,v) = \theta_0\left(\sum_{ij} \frac{\partial u}{\partial a_i}\ \lambda_{ij}\ \overline{\frac{\partial v}{\partial a_j}}\right)$, *then*

$$\hat{\phi}(u,v) + \overline{\hat{\phi}(v,u)} = \theta_0((1-u)(1-v^{-1})).$$

In particular $u \in N_0$ *if and only if* $\hat{\phi}(u,v) = -\overline{\hat{\phi}(v,u)}$ *for all* $v \in F_0$.

PROOF. The loops $a_1,\dots,a_{2g+k}$ respectively $b_1,\dots,b_{2g+k}$, determine cell structures on their underlying 1-complexes which we denote by K_0 and K_1. The liftings $\tilde{a}_i$, $\tilde{b}_i$ beginning at $\tilde{s}_0$ and $\tilde{s}_1$ respectively represent free $\mathbb{Z}T$ bases for $C_1(\tilde{K}_0)$, and $C_1(\tilde{K}_1)$. We choose $\tilde{s}_0$ and $\tilde{s}_1$ to cobound a lifting of γ.

We construct a commutative diagram

$$\begin{array}{ccccc}
N \times N & & \xrightarrow{\quad\phi\quad} & & \mathbb{Z}T \\
\uparrow{\scriptstyle i_1 \times i_0} & & C_1(\tilde{K}_1) \times C_1(\tilde{K}_0) & \xrightarrow{\hat{\Phi}} & \uparrow{\scriptstyle \Phi} \\
 & & \cup & & \\
N_1 \times N_0 & \xrightarrow{d_1 \times d_0} & Z_1(\tilde{K}_1) \times Z_1(\tilde{K}_0) & \longrightarrow & H_1(\tilde{S}) \times H_1(\tilde{S})
\end{array}$$

where, according to lemma 2.6

$$d_1 u = \sum_i \theta_1\left(\frac{\partial u}{\partial b_i}\right) \tilde{b}_i$$

$$d_0 v = \sum_i \theta_0\left(\frac{\partial v}{\partial a_j}\right) \tilde{a}_j.$$

Since elements of $C_1(\tilde{K}_0)$ and $C_1(\tilde{K}_1)$ intersect transversely, there is an integral intersection pairing $C_1(\tilde{K}_1) \times C_1(\tilde{K}_0) \to \mathbb{Z}$ which we use to define $\hat{\Phi}$ in direct analogy to Φ.

Thus

$$\phi(i_1 u, i_0 v) = \hat{\Phi}\left(\sum \theta_1\left(\frac{\partial u}{\partial b_i}\right)\tilde{b}_i,\ \sum \theta_0\left(\frac{\partial v}{\partial a_j}\right)\tilde{a}_j\right) =$$

$$\sum_{i,j} \theta_1\left(\frac{\partial u}{\partial b_i}\right) \hat{\Phi}(\tilde{b}_i,\tilde{a}_j)\ \overline{\theta_1\left(\frac{\partial v}{\partial a_j}\right)}$$

If $i \neq j$ then $\hat{\Phi}(\tilde{b}_i,\tilde{a}_j) = 0$. If $i = j \leqslant 2g$ then $\langle \tilde{b}_i, t\tilde{a}_j \rangle \neq 0$ for exactly one $t \in T$ (in which case $\langle \tilde{b}_i, t\tilde{a}_j \rangle = +1$ by our choice of orientation). To determine this value of t, let ω be the path in ∂D which begins at the initial point of γ, transverses in the counterclockwise direction, and ends at the first occurrence of the initial point of a_i. If $g: D \to \tilde{S}$ is the lifting of the identification map $D \to S$ such that $g(s_1) = \tilde{s}_1$ then $g(\omega(0)) = \tilde{s}_0$

and $g(\omega(1)) = t\cdot\tilde{s}_0$. But then $t = \theta[p\circ g\circ\omega] = \theta_0(P_j a_j^{-1})$.

If $i > 2g$ then $\tilde{b}_i$ meets two translates of $\tilde{a}_i$ -- one with intersection number +1, the other with intersection number -1. The corresponding translates are determined as above to finish the proof of (i).

For (ii) take $(u,v) \in N_0 \times N_0$. Now $\phi(i_0 u, i_0 v) = \phi(i_1 f(u), i_0 v)$ where $f:F_0 \to F_1$ is the isomorphism of lemma 3.2. From part (i) we have

$$\phi(i_1 f(u), i_0 v) = \sum_j \theta_1\left(\frac{\partial f(u)}{\partial b_j}\right) \theta_0(\mu_j)\, \overline{\theta_0\left(\frac{\partial v}{\partial a_j}\right)}.$$

There is a chain rule for free derivatives (which can be derived from formula 2.1, c.f. 2.6 of $[F_1]$) which gives:

$$\frac{\partial f(u)}{\partial b_j} = \sum_i f\left(\frac{\partial u}{\partial a_i}\right) \frac{\partial f(a_i)}{\partial b_j}.$$

Thus

$$\theta_1\left(\frac{\partial f(u)}{\partial b_j}\right) = \sum_i \theta_0\left(\frac{\partial u}{\partial a_i}\, f^{-1}\left(\frac{\partial f(a_i)}{\partial b_j}\right)\right),$$

and

$$\phi(i_0 u, i_0 v) = \theta_0\left(\sum_{i,j} \frac{\partial u}{\partial a_i}\, \lambda_{ij}\, \overline{\frac{\partial v}{\partial a_j}}\right),$$

where

$$\lambda_{i,j} = f^{-1}\left(\frac{\partial f(a_i)}{\partial b_j}\right)\mu_j.$$

One can verify that the values of λ_{ij} are as stated by using the formulas of lemma 3.2.

For (iii) we compute, using (ii), that $\lambda_{ij} + \overline{\lambda_{ji}} = (1-a_i)(1-a_j^{-1})$. Thus

$$\hat{\phi}(u,v) + \overline{\hat{\phi}(v,u)} = \theta\left(\sum_i \frac{\partial u}{\partial a_i}(1-a_i)\cdot \overline{\sum_j \frac{\partial v}{\partial a_j}(1-a_j)}\right).$$

But

$$1-u = \sum_i \frac{\partial u}{\partial a_i}(1-a_i)$$

is an identity in F_0. A proof of this is easily established by inducting on the length of u as a word in the a_i's (c.f. 2.3 of $[F_1]$). Thus $\hat{\phi}(u,v) + \overline{\hat{\phi}(v,u)} = \theta_0((1-u)(1-v^{-1}))$.

Finally, if $u \in N_0$ then $\theta_0(u) = 1$; so $\hat{\phi}(u,v) + \overline{\hat{\phi}(v,u)} = 0$.

Conversely if $\hat{\phi}(u,v) = -\overline{\hat{\phi}(v,u)}$ for all v, then putting $v = u$ we have

$$0 = \theta_0((1-u)(1-u^{-1})) = 2-\theta_0(u) - \theta_0(u)^{-1};$$

which can happen only if $\theta_0(u) = \theta_0(u)^{-1} = 1$.

CHAPTER 4. DUALITY

Through this chapter M will be a compact, connected, oriented 3-manifold with (possibly empty) boundary, and $\partial M = B_1 \cup B_2$ will be a decomposition into compact (possibly empty, possibly not connected) 2-manifolds with $B_1 \cap B_2 = \partial B_1 = \partial B_2$. Let $(G_k, H_k) = (\pi_1(M_k^*), i_*\pi_1(B_k^*))$ denote the joined fundamental group pair of the pair (M, B_k); $k = 1,2$.

4.1 THEOREM. *Let* F_k *be the free group of rank* $\overline{\beta}_0(B_k) = $ *rank* $\overline{H}_0(B_k)$ *and let* $g_k = $ *genus* B_k. *Then there are presentations* P_1 *for* $(G_1,H_1)*(F_2,1)$ *and* P_2 *for* $(G_2,H_2)*(F_1,1)$ *whose Jacobian matrices satisfy*

$$J(P_1) = -\overline{J(P_2)}^{tr}$$

and such that $\operatorname{def} P_1 = -\operatorname{def} P_2 = g_2 - g_1$.

PROOF. We first show that we may assume that B_k is connected; $k = 1,2$. Let $\hat{M}$ be a 3-manifold obtained from M by attaching 3-cells C_1, C_2 by identifying disjoint 2-cells in ∂C_k with 2-cells in B_k -- one for each component of B_k. This is done in such a way that $\hat{B}_k = B_k \cup \partial C_k - \operatorname{Int}(B_k \cap C_k)$ is a connected surface with the same genus, g_k, as B_k, and such that there are isomorphisms

$$(\pi_1(\hat{M}), i_*\pi_1(\hat{B}_1)) \cong (G_1,H_1)*(F_2,1)$$

$$(\pi_1(\hat{M}), i_*\pi_1(\hat{B}_2)) \cong (G_2,H_2)*(F_1,1)$$

Note that $\pi_1(\hat{B}_k)$ is not isomorphic to H_k, but its image in $\pi_1(\hat{M})$ is isomorphic to H_k.

Thus the validity of the theorem for $(\hat{M}; \hat{B}_1, \hat{B}_2)$ yields its proof in general. For the remainder of the proof we revert to our original notation and assume that B_k is connected and, hence, that $F_k = 1$; $k = 1,2$. We proceed to find "dual" presentations for (G_1,H_1) and (G_2,H_2). These will be the two presentations corresponding to a Heegaard splitting of M -- properly chosen to treat the relative case. In case one or both of B_1, B_2 is empty the proof is somewhat simpler. However, to be consistent we assume that both B_1 and B_2 are non-empty. This can always be achieved by removing the interior of a 3-cell from $\operatorname{Int} M$ and making its boundary the appropriate B_k. This changes none of the data since $\overline{\beta}_0(\emptyset) = \overline{\beta}_0(S^2) = 0 = g(\emptyset) = g(S^2)$.

We want to express $M = V_1 \cup V_2$ where V_k is a cube with handles and $S = V_1 \cap V_2 = \partial V_1 \cap \partial V_2$ is a connected 2-manifold and such that

(i) $E_1 = V_2 \cap B_1$ and $E_2 = V_1 \cap B_2$ are 2-cells in $\text{Int } B_1$ and $\text{Int } B_2$ respectively,

(ii) there are disjoint, properly embedded 2-cells $D_{k,1},\ldots,D_{k,n_k+t_k}$ in V_k which cut V_k to a 3-cell and such that $D_{k,j} \subset \text{Int } S$ for $1 \le j \le n_k$ and

$$D_{k,n_k+1} \cap \overline{B_k - E_k},\ldots,D_{k,n_k+t_k} \cap \overline{B_k - E_k}$$

are 1-cells which cut $\overline{B_k - E_k}$ to a 2-cell.

To make this construction we choose a triangulation $\mathfrak{T}$ of M containing each of B_1, B_2, and $\partial B_1 = \partial B_2$ as full subcomplexes. We let $\mathfrak{T}^*$ be the dual cell complex.

The union of all 1-simplexes of $\mathfrak{T}$ in ∂M which are disjoint from B_1 is connected. Let A_2 be a maximal tree in this 1-complex. Let A_1 be a maximal tree in the union of all 1-cells of $\mathfrak{T}^*$ in ∂M which are disjoint from B_2. Let X_1 be the union of A_2, ∂B_1, and all 1-simplexes of $\mathfrak{T}$ which are not contained in B_2 or which do not intersect A_1. Let X_2 be the maximal subcomplex of $\mathfrak{T}^*$ which is disjoint from X_1. Then each of X_k and $X_k \cap B_k$ is connected $(k = 1,2)$, and for $j \ne k$ $X_k \cap \text{Int } B_j = A_j$. We let V_k be the star neighborhood of X_k in $\mathfrak{T}''$. Then V_k is a cube with handles, $M = V_1 \cup V_2$, and $S = V_1 \cap V_2 = \partial V_1 \cap \partial V_2$ is a 2-manifold.

Now $V_2 \cap B_1$ is a regular neighborhood in B_1 of A_1 and is therefore a 2-cell. Also $V_1 \cap B_2$ is a regular neighborhood in B_2 of $\partial B_2 \cap A_2$. We can adjust, by an isotopy of M, so that $V_1 \cap \text{Int } B_2$ is also a 2-cell, and we assume, without changing notation, that this has been done. Thus (i) follows.

To establish (ii) note that we can choose cutting disks $D_{1,1}, D_{1,2},\ldots$ for V_1 to be the intersection of V_1 with the 2-cells of $\mathfrak{T}^*$ which are dual to the 1-simplexes of X_1 which are not in some maximal tree, T_1, of X_1. We can choose T_1 so that $A_2 \subset T_1$ and so that $T_1 \cap B_1$ is a maximal tree in $X_1 \cap B_1$ (which is connected). This choice of the $D_{1,j}$'s, properly ordered, satisfies condition (ii) for V_1.

For V_2, we let T_2 be a maximal tree in X_2 such that $T_2 \cap B_2$ is a maximal tree in $X_2 \cap B_2$ and $A_1 \subset T_2$. For each 1-cell, σ, of $X_2 - T_2$ let τ be a 2-simplex of $\mathfrak{T}$ dual to σ. If σ is not in ∂M, τ is uniquely determined. If $\sigma \subset \partial M$ then σ is dual to a 1-simplex τ' of $\mathfrak{T}$ in ∂M and we can choose τ to be any 2-simplex of $\mathfrak{T}$ containing τ' and not contained in ∂M. The component of $\tau \cap V_2$ which intersects σ will serve as one of the cutting disk $D_{2,j}$.

Note that when we cut V_k along the $D_{k,j}$'s S is cut into a punctured 2-sphere. This shows that S is connected.

For $k = 1,2$ we can choose a standard set of generators $\{a_{k,j}\}$, $j \leqslant 2(n_k+t_k)$, for $\pi_1(\partial V_k)$ where the loops $a_{k,2j}$ are freely homotopic in ∂V_k to the curves $\partial D_{k,j}$. Moreover we may assume that for $j \leqslant 2n_k$ $a_{k,j}$ is a loop in S and that $\{a_{k,1},\ldots,a_{k,2n_k}\}$ extends to a standard set of generators for $\pi_1(S)$. Finally we may assume that the loops $a_{k,2n_k+1}, a_{k,2n_k+3},\ldots$ represent generators for $i_*(\pi_1(B_k-E_k)) < \pi_1(V_k)$.

Consider the following diagram in which all maps are induced by inclusion

$$\begin{array}{ccccc}
\pi_1(\partial V_1) & \longleftarrow & \pi_1(S) & \longrightarrow & \pi_1(\partial V_2) \\
\Big\downarrow \mu_1 & \overset{\nu_1}{\swarrow} & \Big\downarrow \eta & \overset{\nu_2}{\searrow} & \Big\downarrow \mu_2 \\
\pi_1(V_1) & \underset{\eta_1}{\longrightarrow} & \pi_1(M) & \underset{\eta_2}{\longleftarrow} & \pi_1(V_2)
\end{array}$$

Put $x_{k,j} = \mu_k(a_{k,2j-1})$. Then $\pi_1(V_k)$ is freely generated by $\{x_{k,1},\ldots,x_{k,n_k+t_k}\}$ and $\{\eta_k(x_{n_k+1}),\ldots,\eta_k(x_{n_k+t_k})\}$ generates image $(\pi_1(B_k) \to \pi_1(M)) = H_k$.

By first pushing across the open 2-cell

$$\mathrm{Int}(B_2-E_2) - (D_{2,n_2+1} \cup \ldots \cup D_{2,n_2+t_2})$$

we get a sequence of collapses

$$M \searrow V_1 \cup D_{2,1} \cup \ldots \cup D_{2,n_2+t_2} \cup E_2 \searrow$$

$$V_1 \cup D_{2,1} \cup \ldots \cup D_{2,n_2} \cup E_2.$$

Thus $\pi_1(M)$ is presented with generators $\{x_{1,1},\ldots,x_{1,n_1+t_1}\}$ and with relators any elements in the conjugacy classes determined by $\partial D_{2,1},\ldots,\partial D_{2,n_2},\partial E_2$. For technical reasons we choose

$$r_{1,i} = \nu_1(a_{2,2i-1}a_{2,2i}a^{-1}_{2,2i-1}); \quad i \leqslant n_2$$

and let r_{1,n_2+1} be any word in $x_{1,n_1+1},\ldots,x_{1,n_1+t_1}$ representing ∂E_2.

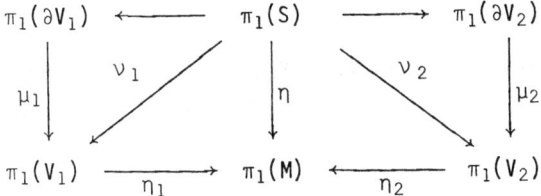

Then $P_1 =$

$$(x_{1,1},\ldots,x_{1,n_1+t_1};x_{1,n_1+1},\ldots,x_{1,n_1+t_1}:r_{1,1},\ldots,r_{1,n_2+1};r_{1,n_2+1})$$

is a presentation for $i_*:\pi_1(B_1) \to \pi_1(M)$. The Jacobian matrix

$$J(P_1) = \eta_1\left(\frac{\partial r_{1,i}}{\partial x_{1,j}}\right) \begin{array}{l} 1 \leqslant i \leqslant n_2 \\ 1 \leqslant j \leqslant n_1 \end{array}$$

depends only on the pair (G_1,H_1).

In the same way, by putting

$$r_{2,i} = \nu_2(a_{1,2i-1}a_{1,2i}a^{-1}_{1,2i-1}); \quad i \leqslant n_1$$

we get a presentation P_2 for $i_*:\pi_1(B_2) \to \pi_1(M)$ whose Jacobian is

$$J(P_2) = \eta_2\left(\frac{\partial r_{2,i}}{\partial x_{2,j}}\right) \begin{array}{l} 1 \leqslant i \leqslant n_1 \\ 1 \leqslant j \leqslant n_2 \end{array} .$$

We now apply theorem 3.3 where $N = \ker(\eta:\pi_1(S) \to \pi_1(M)$ and $\phi:N \times N \to \mathbb{Z}(\pi_1(S)/N) \subset \mathbb{Z}\pi_1(M)$ is the associated Reidemeister pairing.

If $i \leqslant 2n_2$ and $j \leqslant 2n_1$ then

$$\phi(a_{2,2i-1}a_{2,2i}a^{-1}_{2,2i-1},a_{1,2j-1}a_{1,2j}a^{-1}_{1,2j-1})$$

$$= \phi(a_{2,2i-1}a_{2,2i}a^{-1}_{2,2i-1},a_{1,2j})\eta(a^{-1}_{1,2j-1})$$

$$= \eta\left(\sum_{p,q} \frac{\partial(a_{2,2i-1}a_{2,2i}a^{-1}_{2,2i-1})}{\partial a_{1,p}} \lambda_{p,q} \overline{\frac{\partial a_{1,2j}}{\partial a_{1,q}}}\right)\eta(a^{-1}_{1,2j-1})$$

$$= \eta\left(\sum_{p} \frac{\partial(a_{2,2i-1}a_{2,2i}a^{-1}_{2,2i-1})}{\partial a_{1,p}} \lambda_{p,2j}\right) \eta(a^{-1}_{1,2j-1})$$

$$= \eta\left(\frac{\partial r_{1,i}}{\partial x_{1,j}}\right) .$$

The last equality uses the fact that $\eta(a_{1,2j}) = 1$ combined with the computation of $\lambda_{p,q}$ from 3.3 which shows that

$$\eta(\lambda_{p,2j}) = \begin{cases} \eta(a_{1,2j-1}); & p = 2j \\ 0 & ; \text{ otherwise} \end{cases}$$

By the skew Hermetian property of ϕ,

$$\phi(a_{2,2i-1}a_{2,2i}a_{2,2i-1}^{-1}, a_{1,2j-1}a_{1,2j}a_{1,2j-1}^{-1})$$

$$= \overline{-\phi(a_{1,2j-1}a_{1,2j}a_{1,2j-1}^{-1}, a_{2,2i-1}a_{2,2i}a_{2,2i-1}^{-1})}$$

which by the same sort of computation is equal to $-\overline{\eta_2(\frac{\partial r_{2,j}}{\partial x_{2,i}})}$.

Thus we have shown that $J(P_1) = -\overline{J(P_2)}^{tr}$.

For the final part of the theorem we note that since

$$M \searrow V_1 \cup D_{2,1} \cup \ldots \cup D_{2,n_2} \cup E_2,\ \chi(M) = 2 - n_1 - t_1 + n_2.$$

Similarly $\chi(M) = 2 - n_2 - t_2 + n_1$; so $2(n_1 - n_2) = t_2 - t_1$. Since $\overline{B_k - E_k}$ is cut to a 2-cell by t_k arcs, $\chi(B_k) = 2 - t_k$. Since $\partial B_1 = \partial B_2$,

$$g_2 - g_1 = \tfrac{1}{2}(\chi(B_1) - \chi(B_2)) = \tfrac{1}{2}(t_2 - t_1) = n_1 - n_2.$$

This is, by definition, def P_1 (= -def P_2).

The above theorem yields, immediately, relations between the elementary ideals as defined in Section 2.

4.2 COROLLARY. *Let* $\theta : \pi_1(M) \to T$ *be a representation of* $\pi_1(M)$ *onto an abelian group* T.

(i) If $B_1 \neq \emptyset \neq B_2$, *then for all* i $\mathcal{E}_i(M, B_1; \theta) = \overline{\mathcal{E}}_{i+n}(M, B_2; \theta)$ where $n = \tfrac{1}{2}(\chi(B_2) - \chi(B_1))$.

(ii) If $\partial M \neq \emptyset$, *then for all* i $\mathcal{E}_i(M, m_0; \theta) = \overline{\mathcal{E}}_{i+n}(M, \partial M; \theta)$ *where* $n = \tfrac{1}{2}\chi(\partial M) - 1 = \chi(M) - 1$.

(iii) If $\partial M = \emptyset$, *then for all* i $\mathcal{E}_i(M, m_0; \theta) = \overline{\mathcal{E}}_i(M, m_0; \theta)$.

PROOF. We use the notation of theorem 4.1. A presentation P_k' for (G_k,H_k) can be extended to a presentation P_k'' for $(G_k,H_k) * (F_j,1)$ by adding r_j generators and no relations (r_j = rank $F_j = \overline{\beta}_0(B_j)$). Thus $\mathcal{E}_i(M,B_k;\theta) = \mathcal{E}_i(\theta J(P_k')) = \mathcal{E}_{i+r_j}(\theta J(P_k'')) = \mathcal{E}_{i+r_j}(\theta J(P_k))$. By 4.1 $\mathcal{E}_n(\theta J(P_1)) = \overline{\mathcal{E}}_{n-(g_2-g_1)}(\theta J(P_2))$. Combining we have

$$\mathcal{E}_i(M,B_1;\theta) = \mathcal{E}_{i+(r_2-r_1)-(g_2-g_1)}(M,B_2;\theta).$$

Since $\partial B_1 = \partial B_2$ and since neither B_k is empty,

$$(r_2-r_1) - (g_2-g_1) = \tfrac{1}{2}(\chi(B_2) - \chi(B_1)).$$

For part (ii) let C be a 3-cell in Int M. Put $M' = M\text{-Int } C$, $B_1' = \partial C$, $B_2' = \partial M$. Part (i) applied to $(M';B_1',B_2')$ gives part (ii).

Part (iii) follows directly from the conclusion that, for M closed, $\pi_1(M)$ has a pair of dual presentations.

This is the strongest form of symmetry. Fox and Torres [FT] proved that the group of any link in S^3 has a pair of presentations which are dual in the somewhat weaker sense; whereas Blanchfield [Bℓ] has shown that the group of a link in any 3-manifold has symmetric Alexander polynomials. This suggests that the amount of symmetry in the absolute invariants decreases with the increase in complexity of the boundary (see example 6.4). In the case of tori boundary it is possible to compare the absolute invariants with certain symmetric relative invariants. The next two corollaries illustrate this.

4.3 COROLLARY. (5.6 of [Bℓ]). *If the components of* ∂M *are all tori and if* $\theta:\pi_1(M) \to T$ *is any representation onto a free abelian group,* T, *then for all* $i(\Delta_i(M;\theta)) = (\overline{\Delta_i(M;\theta)})$.

PROOF. We can express $\partial M = B_1 \cup B_2$ where the components of B_k are annuli, $\partial B_1 = \partial B_2$, and $i_*:H_1(B_k) \to H_1(M)$ is monic. By 4.2

$$\mathcal{E}_i(M,B_1;\theta) = \overline{\mathcal{E}}_i(M,B_2;\theta)$$

for all i. But there is isotopy of M taking B_1 to B_2. Thus $\mathcal{E}_i(M,B_1;\theta) = \overline{\mathcal{E}}_i(M,B_1;\theta)$.

It suffices to prove the case in which $T = H_1(M)/\text{torsion}$ and the corresponding cover $p:\widetilde{M} \to M$ is the maximal free abelian covering. Thus each component of $\widetilde{B}_1 = p^{-1}(B_1)$ is simply connected. Thus we have an exact sequence

$$0 \to H_1(\widetilde{M}) \to H_1(\widetilde{M},\widetilde{B}_1) \xrightarrow{\partial*} H_0(\widetilde{B}_1) \to \mathbb{Z} \to 0$$

By 2.12 $\Delta_i(M,B_1;\theta)$ divides $\Delta_i(M;\theta)\cdot\Delta_0(\text{image } \partial_*)$ and by 2.13 $\Delta_i(M;\theta)$ divides $\Delta_i(M,B_1;\theta)$. Thus $\Delta_i(M,B_1;\theta) = \Delta_i(M;\theta)\cdot X$ for some divisor, X, of $\Delta_0(\text{image } \partial_*)$.

Now $H_0(\widetilde{B}_1)$ is a direct sum of cyclic modules:

$$H_0(\widetilde{B}_1) = \bigoplus_i \mathbb{Z}T/(1-z_i).$$

By 2.13 X divides $\Delta_0(H_0(B_1)) = (1-z_1) \ldots (1-z_n)$.

If $z_i = w_i^{q_i}$ where w_i is primitive then the prime divisors of $1-z_i$ are the cyclotomic polynomials in w_i. These are symmetric. But $\Delta_i(M,B_1;\theta)$ is also symmetric; so $\Delta_i(M;\theta)$ is also and the proof is complete.

We say that a compact, orientable 3-manifold M is a *boundary link complement* if the components $C_1,\ldots,C_k$ of ∂M are tori and if there are pairwise disjoint surfaces $S_1,\ldots,S_k$ in M with $S_i \cap \partial M = S_i \cap C_i$ a simple closed curve which is not contractible in C_i. Any knot space (in a homology 3-sphere) is such a manifold. For M as above, $T = \text{image } (H_1(\partial M) \to H_1(M))$ is free abelian of rank k and is a direct summand of $H_1(M)$. The natural projection $\theta:\pi_1(M) \to T$ is called the *boundary representation.*

4.4 COROLLARY. *Let* M *be a boundary link complement,* $\theta:\pi_1(M) \to T$ *be the boundary representation, and* $p:\widetilde{M} \to M$ *the regular covering corresponding to* ker θ. *Then there is a square matrix* P *over* $\mathbb{Z}T$ *such that both* P *and* $\overline{P}^{tr}$ *present* $H_1(M,p^{-1}(m_0)) \oplus (\mathbb{Z}T)^{k-1}$ $(k = \beta_0(\partial M))$. *In particular* $\mathcal{E}_i(M,m_0;\theta) = \overline{\mathcal{E}}_i(M,m_0;\theta)$ *for all* i.

PROOF. As in 4.3 we write $\partial M = B_1 \cup B_2$; where the components of B_i are annuli. Only in this case we take B_1 to be a regular neighborhood of $\cup\partial S_i$. Now $M\text{-}\cup S_i$ lifts homeomorphically to $\widetilde{M}$. Thus image $(H_1(\widetilde{B}_1)\to H_1(\widetilde{M}))=0$, and we have an exact sequence

$$0 \to H_1(\widetilde{M},p^{-1}(m_0)) \to H_1(\widetilde{M},\widetilde{B}_1) \to H_0(\widetilde{B}_1,p^{-1}(m_0)) \to 0$$

But each component of B_1 projects homeomorphically; so

$$H_0(B_1, p^{-1}(m_0)) \cong (\mathbb{Z}T)^{k-1}$$

and the sequence splits. The conclusion follows by applying 4.1 to $(M;B_1,B_2)$.

Let $\mu(M,B;\theta) = \min\{k : \Delta_k(M,B;\theta) \neq 0\}$ (c.f. lemma 2.15). We have immediately from 4.2

4.5 COROLLARY. *Let* $\partial M = B_1 \cup B_2$ *as in theorem 4.1. Then*

(i) If $B_1 \neq \emptyset \neq B_2$ *then* $\mu(M,B_1;\theta) \geq \frac{1}{2}(\chi(B_1)-\chi(B_2))$, *and*

(ii) If $\partial M \neq \emptyset$ *then* $\mu(M,m_0;\theta) \geq 1-\frac{1}{2}\chi(\partial M)$.

CHAPTER 5. REDUCIBILITY OF HEEGAARD SPLITTINGS

In [P] Papkyriakopoulos developed a method, using free differential calculus combined with a "planarity theorem" of Maskit [Ma], for determining whether a Heegaard splitting of a 3-manifold is reducible.

In this section we present Papkyriakopoulos' ideas from a somewhat different point of view which depends on properties of the Reidemeister pairing already developed and combine this with a generalization of Maskit's theorem using the theory of ends to give further results in this direction.

We restrict our attention to closed, orientable, 3-manifolds. A *Heegaard splitting* of such a manifold, M, is a pair (M,S) where S is a surface in M separating M into two cubes with handles. The *genus of the splitting* is the genus of S.

The *connected sum,* $(M_1,S_1) \# (M_2,S_2)$, of two Heegaard splittings is constructed by choosing a 3-cell $B_i \subset M_i$ such that $B_i \cap S_i$ is a 2-cell, $C_i (i=1,2)$ and putting

$$(M_1,S_1) \# (M_2,S_2) = (\overline{M_1-B_1},\overline{S_1-C_1}) \cup (\overline{M_2-B_2},\overline{S_2-C_2})$$
$$(\partial B_1,\partial C_1) = (\partial B_2,\partial C_2)$$

A Heegaard splitting (M,S) is *reducible* if $(M,S) = (M_1,S_1) \# (M_2,S_2)$ where $(M_i,S_i) \neq (S^3,S^2)$; $i = 1,2$. The following theorems are pertinent.

5.1 THEOREM. (Haken [Ha]) *If* $M = M_1 \# M_2$, *where* $M_i \neq S^3$ $(i = 1,2)$, *and* (M,S) *is any Heegaard splitting of* M, *then* $(M,S) = (M_1,S_1) \# (M_2,S_2)$ *for some splittings* (M_i,S_i) *of* M_i.

5.2 THEOREM. (Waldhausen [W_2]) *Every Heegaard splitting of* S^3 *of genus* >1 *is reducible.*

Now let (M,S) be a Heegaard splitting of some 3-manifold M. We will denote by V_1 and V_2 the cubes with handles such that $M = V_1 \cup V_2$ and $V_1 \cap V_2 = \partial V_1 = \partial V_2 = S$ and we will put $N_i = \ker\{\mu_i : \pi_1(S) \to \pi_1(V_i)\}$.

5.3 LEMMA. *If* $g(S) > 1$ *then* (M,S) *is reducible if and only if some nontrivial element of* $N_1 \cap N_2$ *is represented by a simple loop in* S.

PROOF. If α is a simple loop representing $1 \neq [\alpha] \in N_1 \cap N_2$, then we may assume α separates S; otherwise the commutator $[\alpha,\beta]$, where β is a simple loop meeting α transversely is a single point, is a nontrivial (since $g(S) > 1$) element of $N_1 \cap N_2$ represented by a separating simple loop.

By Dehn's lemma α bounds 2-cells $D_i \subset V_i$. The 2-sphere $D_1 \cup D_2$ realizes a reduction of (M,S) and every reduction is of this form.

Note that

$$N_1 \cap N_2 \;=\; \ker\{\mu_1 \times \mu_2 : \pi_1(S) \to \pi_1(V\ ) \times \pi_1(V\ )\}$$

The maps, $\mu_1 \times \mu_2$, called *splitting homomorphisms*, have been studied by Stallings [St_1] and Jaco [J_1].

Note also that by Van Kampen's theorem

$$\pi_1(M) = \pi_1(S)/N_1N_2.$$

It is never the case that $N_1 \cap N_2 = 1$ (if $g(S) > 1$) for N_i is not cyclic, and if $n_1 \in N_1$ and $n_2 \in N_2$ are not powers of a third element of $\pi_1(S)$, then $1 \neq [n_1,n_2] \in N_1 \cap N_2$.

Thus we are interested in criteria for determining when a normal subgroup of $\pi_1(S)$ "contains" a simple loop. Our result will be stated in terms of end theory. The needed results are given below.) See [E] for an exposition of this theory.

For X a locally compact CW complex, the *number of ends* of X, $e(X)$, is the supremum, taken over all finite subcomplexes K of X, of the number of components of $X-K$ which are not relatively compact. For $p:\widetilde{X} \to X$ a regular covering space with X a compact CW complex, the number $e(\widetilde{X})$ depends only on the group, G, of covering transformations and the *number of ends of* G, $e(G)$, is unambiguously defined to be $e(\widetilde{X})$. The only possibilities for $e(G)$ are $0,1,2,\infty$.

Maskit [Ma] has shown that a normal subgroup, N, of $\pi_1(S)$ contains a power of a simple loop if and only if N contains some normal subgroup, K, of $\pi_1(S)$ whose corresponding regular covering space is planar (embeds in $\mathbb{R}^2$). Since a planar, nonsimply connected surface without boundary has more than one end, the following generalizes Maskit's theorem.

5.4 THEOREM. *Let* S *be a closed surface and* N *be a normal subgroup of* $\pi_1(S)$. *Then there is a simple loop* α *in* S *such that* $1 \neq [\alpha]^n \in N$ *for some* n *if and only if* N *contains a normal subgroup* K *of* $\pi_1(S)$ *such that* $e(\pi_1(S)/K) > 1$.

PROOF. Necessity follows from [Ma] which shows that the covering of S corresponding to the normal closure of $[\alpha]^n$ (α a simple loop) is planar.

So suppose that N contains a normal subgroup, K, of $\pi_1(S)$ with $e(\pi_1(S)/K) > 1$. If $e(\pi_1(S)/K) = \infty$, then according to Stallings (theorem 5.A.9 of [St]) either

(i) $\pi_1(S)/K \cong G_1 \underset{A}{*} G_2$ - an amalgamated free product with amalgamating subgroup, A, a finite, proper subgroup of both G_1 and G_2, or

(ii) $\pi_1(S)/K \cong G_A \leftrightarrows \emptyset$ - an HNN group with bonding subgroup, A, a finite, proper subgroup of G.

If $e(\pi_1(S)/K) = 2$, then by lemma 4.1 of [Wa] $\pi_1(S)/K$ is an extension of a finite group by either $\mathbb{Z}$ or by $\mathbb{Z}_2 * \mathbb{Z}_2$.

In any case we can construct a complex X (with $\pi_1(X) = \pi_1(S)/K$, or $\mathbb{Z}$, or $\mathbb{Z}_2 * \mathbb{Z}_2$) and a map $f:S \to X$ so that

(i) X contains a subcomplex Y with a neighborhood homeomorphic to $Y \times [-1,1]$,

(ii) $\pi_1(Y)$ is finite,

(iii) $j_*:\pi_1(X-Y) \to \pi_1(X)$ is not epic,

(iv) $f_*:\pi_1(S) \to \pi_1(X)$ is epic, and

(v) $\ker f_*$ contains K as a subgroup of finite index.

We may assume that f is transverse with respect to Y; so that each component of $f^{-1}(Y)$ is a simple closed curve. If each of these components contracts in S, there is a surface $S_0 \subset S$ such that $S-S_0$ is a finite disjoint union of 2-cells and $f(S_0) \subset X-Y$. Thus we have a commutative diagram

$$\begin{array}{ccc} \pi_1(S_0) & \xrightarrow{\ i_*\ } & \pi_1(S) \\ \Big\downarrow (f|S_0)_* & & \Big\downarrow f_* \\ \pi_1(X-Y) & \xrightarrow{\ j_*\ } & \pi_1(X) \end{array}$$

This gives a contradiction since i_* and f_* are epic whereas j_* is not.

Thus some component of $f^{-1}(Y)$ determines a simple loop α in S such that $1 \neq [\alpha] \in \pi_1(S)$. Since $\pi_1(Y)$ is finite, $[\alpha]^m \in \ker f_*$ for some $m > 0$. Since K has finite index in $\ker f_*$, $[\alpha]^n \in K \subset N$ for some $n \geqslant m$ and the proof is complete.

5.5 COROLLARY. *The Heegaard splitting* (M,S) $(g(S) > 1)$ *is reducible if and only if* $N_1 \cap N_2$ *contains a normal subgroup,* K, *of* $\pi_1(S)$ *with* $e(\pi_1(S)/K) > 1$.

PROOF. This follows immediately from 5.3 and 5.4 with the observation that, by the loop theorem, $[\alpha]^n \in N_i$ for $n \neq 0$ implies $[\alpha] \in N_i$.

Note that a surface with empty boundary has more than one end if and only if it contains a simple closed curve which separates the surface but does not bound a compact subsurface. The intersection number of this curve with any other curve is zero. Thus we have proved

5.6 LEMMA. *If* S *is a closed surface,* K *is a normal subgroup of* $\pi_1(S)$, *and* $p:\tilde{S} \to S$ *is the associated regular covering, then* $e(\pi_1(S)/K) > 1$ *if and only if the intersection pairing* $\langle\, ,\, \rangle : H_1(\tilde{S}) \times H_1(\tilde{S}) \to \mathbb{Z}$ *is singular.*

Note, by comparison, that $\tilde{S}$ is planar if and only if $\langle\, ,\, \rangle$ is trivial.

By use of the Reidemeister pairing we can express the intersection pairing on $\tilde{S}$ in terms of the group structure of $(\pi_1(S),K)$.

We consider the case in which K is the normal closure of a finite set $\{u_1,\ldots,u_n\}$ of elements of $\pi_1(S)$ written in terms of a standard set, $\{a_1,\ldots,a_{2g}\}$, of generators for $\pi_1(S)$. As in section 3 we have the Reidemeister pairings:

$$\begin{array}{ccc} H_1(\tilde{S}) \times H_1(\tilde{S}) & \xrightarrow{\quad\Phi\quad} & \\ \uparrow h\times h & & \mathbb{Z}(\pi_1(S)/K) \\ K \times K & \xrightarrow{\quad\phi\quad} & \end{array}$$

and the projection $\theta:\pi_1(S) \to \pi_1(S)/K$. $\tilde{S}$ is planar if and only if $\phi(u_i,u_j) = 0$ for all i,j. By contrast we have

5.7 THEOREM. $e(\pi_1(S)/K) > 1$ *if and only if there is an element*

$$(\alpha_1,\ldots,\alpha_n) \in (\mathbb{Z}(\pi_1(S)/K))^n$$

such that

(*i*) *for each* j $\sum_i \alpha_i\phi(u_i,u_j) = 0$, *and*

(*ii*) *for some* j, $\sum_i \alpha_i\theta(\frac{\partial u_i}{\partial a_j})$ *is not divisible by*

$$\theta(\frac{\partial r}{\partial a_j}); \text{ where } r = [a_1,a_2] \ldots [a_{2g-1},a_{2g}].$$

PROOF. By 5.6, $e(\pi_1(S)/K) > 1$ if and only if there exists $w \in K$ such that $\langle h(w),z\rangle = 0$ for all $z \in H_1(\tilde{S})$ and such that $h(w) \neq 0$. By 3.1 $\langle h(w),z\rangle = 0$ for all z if and only if $\Phi(h(w),h(u_j)) = \phi(w,u_j) = 0$ for all j. If $w = \prod_k c_k u_{m_k}^{p_k} c_k^{-1}$, then

$$\phi(w,u_j) = \sum_k p_k \theta(c_k)\phi(u_{m_k},u_j) = \sum_i \alpha_i \phi(u_i,u_j),$$

$$\text{where } \alpha_i = \sum_{m_k=i} p_k \theta(c_k) \in \mathbb{Z}(\pi_1(S)/K).$$

Moreover, given $\alpha_i \in \mathbb{Z}(\pi(S)/K)$ there exists $w \in K$ such that

$$\phi(w,u_j) = \sum_i \alpha_i \phi(u_i,u_j).$$

Thus condition (i) of the theorem is equivalent to finding $w \in K$ such that $\langle h(w),z\rangle = 0$ for all z.

By 2.6 $h(w)$ is the homology class of the cycle

$$dw = \sum_j \theta\left(\frac{\partial w}{\partial a_j}\right)e_j = \sum_j \left(\sum_i \alpha_i \theta\left(\frac{\partial u_i}{\partial a_j}\right)\right)e_j.$$

But $\pi_1(S)$ is presented with the single relator, r, corresponding to the 2-cell of S. Thus by 2.7, the boundaries, $B_1(\tilde{S})$, are generated by $\sum_j \theta\left(\frac{\partial r}{\partial a_j}\right)e_j$. Thus condition (ii) is equivalent to $h(w) \neq 0$.

We conclude by noting that, because of the formula of theorem 3.3, the computations required to apply the above theorem can be made explicit modulo a solution to the word problem in $\pi_1(S)/K$. The difficulty in trying to apply this theorem, via 5.5, to decide on the reducibility of a Heegaard splitting is that one must deal with arbitrary $K < N_1 \cap N_2$. In particular, regarding the Poincaré conjecture, we have

5.8 THEOREM. *The 3-dimensional Poincaré conjecture is equivalent to the following statement.*

If S *is a closed orientable surface of genus* $g > 1$ *and* N_i *is a normal subgroup of* $\pi_1(S)$ *with* $\pi_1(S)/N_i$ *free of rank* g; $i = 1,2$, *and* $\pi_1(S) = N_1N_2$, *then* $N_1 \cap N_2$ *contains a normal subgroup,* K, *of* $\pi_1(S)$ *with* $e(\pi_1(S)/K) > 1$.

PROOF. By Waldhausen's theorem (5.2) the Poincaré conjecture is equivalent to the statement that every Heegaard splitting of a closed, simply connected 3-manifold is reducible. By Jaco, [J_1], all such pairs N_1, N_2 are realized by Heegaard splittings of 3-manifolds, which, because $N_1N_2 = \pi_1(S)$ are simply connected. Thus the theorem follows from corollary 5.5.

CHAPTER 6. EXAMPLES

The groups of 6.1 below are probably the simplest examples of finitely presented, nonhopfian groups [B,S] and naturally attracted attention relative to the conjecture that fundamental groups of 3-manifolds are hopfian (c.f. [Hm, pg. 175]). Several proofs exist to show that they are not fundamental groups of 3-manifolds [He_2], [J_2], [K]. We illustrate that this follows easily from the duality developed here.

6.1 PROPOSITION. *The group*

$$G_{p,q} = \langle x,y : xy^p x^{-1} = y^q \rangle, \quad pq \neq 0,$$

is the fundamental group of a 3-manifold if and only if $|p| = |q|$.

PROOF. Sufficiency follows easily by construction.

For necessity suppose $\pi_1(M) = G_{p,q}$ for some 3-manifold M and suppose $|p| \neq |q|$. By [Sc] we may assume M is compact. Since $G_{p,q}$ is torsion free and is neither cyclic nor a nontrivial free product it follows from the sphere and projective plane theorems that $\pi_2(M) = 0$. Thus M is aspherical and so $H_i(M) = H_i(G_{p,q})$ for all i. For $p \neq q$ $H_2(G_{p,q}) = 0$. Thus $H_2(M,\partial M)$ is torsion free and it follows that M is orientable.

Let $T = \langle t: \ \rangle$ and $\theta:\pi_1(M) \to T$ be given by $\theta(x) = t$, $\theta(y) = 1$. One easily computes that $\Delta_1(M,m_0;\theta) = pt-q$.

By 5.4 ∂M consists of tori; so by 4.3 we must have

$$(\Delta_1(M,m_0:\theta)) = \overline{(\Delta_1(M,m_0:\theta))}.$$

This is a contradiction if $|p| \neq |q|$.

To illustrate the effect of duality in distinguishing peripheral group systems of 3-manifolds we consider the "torus knot" groups:

$$H_{p,q} = \langle x,y : x^p = y^q \rangle. \quad (p,q) = 1$$

It is known (c.f. 13.11 of [Hm]) that $H_{p,q}$ is the fundamental group of several, nonhomeomorphic, 3-manifolds. It is easy to show that each such

3-manifold must have boundary a single, incompressible torus. Thus the peripheral subgroup is a maximal, rank two free abelian subgroup of $H_{p,q}$. Thus we investigate those subgroups generated by a pair z,w of elements. We suppose that $z = x^p$ (which generates the center of $H_{q,p}$). We let $\theta: H_{p,q} \to \langle t: \rangle$ be abelianization -- given by $\theta(x) = t^q$ and $\theta(y) = t^p$.

6.2 PROPOSITION. *Let* K *be the subgroup of* $H_{p,q}$ *generated by* $z = x^p$ *and* $w \in H_{p,q}$. *If* $(H_{p,q},K)$ *is the peripheral group pair of some 3-manifold,* M, *then*

$$\left(\frac{t^{pq}-1}{t-1}, \frac{t^p-1}{t-1}\,\theta\,\frac{\partial w}{\partial x}, \frac{t^q-1}{t-1}\,\theta\,\frac{\partial w}{\partial x}\right) = (1).$$

For $w = x^a y^b$ *this condition becomes* $(a,p) = 1$ *and* $(b,q) = 1$ *which is also sufficient.*

PROOF. Let $f_n(t) = \dfrac{t^n - 1}{t - 1}$. For $r = x^p y^{-q}$ we compute from the given presentation for $H_{p,q}$ that

$$\left(\theta\,\frac{\partial r}{\partial x}, \theta\,\frac{\partial r}{\partial y}\right) = (f_p(t^q), -t^{p-q} f_q(t^p))$$

that

$$\Delta_1(M,m_0;\theta) = f_{pq}(t)/f_p(t)f_q(t)$$

and that

$$\mathcal{E}_1(M,m_0;\theta) = (\Delta_1(M,m_0;\theta)).$$

By 2.3 there is a presentation for the pair $(H_{p,q},K)$ whose Jacobian maps under θ to

$$\theta\begin{bmatrix} \frac{\partial r}{\partial x} & \frac{\partial r}{\partial y} \\ \frac{\partial z}{\partial x} & \frac{\partial z}{\partial y} \\ \frac{\partial w}{\partial x} & \frac{\partial w}{\partial y} \end{bmatrix} = \begin{bmatrix} f_p(t^q) & -t^{p-q}f_q(t^p) \\ f_p(t^q) & 0 \\ \theta\,\frac{\partial w}{\partial x} & \theta\,\frac{\partial w}{\partial y} \end{bmatrix}$$

Thus $\mathcal{E}_0(M,\partial M;\theta) =$

$$\left(f_p(t^q)f_q(t^p), f_p(t^q)\,\theta\,\frac{\partial w}{\partial y}, f_q(t^p)\,\theta\,\frac{\partial w}{\partial x}\right) =$$

$$(\Delta_1(M,m_0;\theta))\left(f_{pq}(t), f_p(t)\,\theta\,\frac{\partial w}{\partial y}, f_q(t)\,\theta\,\frac{\partial w}{\partial x}\right).$$

By 4.2, 4.3 and 2.16

$$\mathcal{E}_1(M,m_0;\theta) = \mathcal{E}_0(M,\partial M;\theta)$$

and the conclusion follows.

If $w = x^a y^b$ then $\theta \frac{\partial w}{\partial x} = f_a(t^q)$, $\theta \frac{\partial w}{\partial x} t^{aq} f_b(t^q)$ and the above conclusion forces $(a,p) = 1 = (b,q)$. Conversely if this holds one can construct M as in 13.11 of [Hm].

6.3 EXAMPLE. *Fused vs joined group systems.*

It should be clear from the remarks in section 1 that there is a distinction between *fused* and *joined* fundamental groups systems. We wish to illustrate this difference at the level of Jacobian matrices and elementary ideals.

So let (M,B) be a compact CW pair with M connected and B having components $B_1,\dots,B_p$. Let (M^*,B^*) be the corresponding joined pair. Let $P = (x_1,\dots,x_n: r_1,\dots,r_m)$ be a presentation of $\pi_1(M)$ and let $\theta:\pi_1(M) \to T$ be a given representation.

As in theorem 2.5 we choose a retraction $\rho:M^* \to M$ (arbitrarily) inducing a representation $\theta^* = \theta\circ\rho:\pi_1(M^*) \to T$. We choose generators $z_1,\dots,z_{p-1}$ for $\pi_1(M^*) = \pi_1(M) * F(z_1,\dots,z_{p-1})$. It is convenient, but not necessary, to make this choice so that $\rho_*(z_i) = 1$ for each i.

We now choose elements $u_{i,j} \in F(x_1,\dots,x_n)$, $1 \leqslant i \leqslant p-1$, $1 \leqslant j \leqslant n_i$, so that the elements $\{z_i u_{i,j} z_i^{-1}\}$ represent generators for $\pi_1(B^*)$.

Note that

$$\frac{\partial}{\partial x_k}(z_i u_{i,j} z_i^{-1}) = z_i \frac{\partial u_{i,j}}{\partial x_k}$$

$$\frac{\partial}{\partial z_k}(z_i u_{i,j} z_i^{-1}) = \delta_{ik}(1 - z_i u_{i,j} z_i^{-1}).$$

Thus, according to 2.3, there is a presentation, P', for $\{i_*:\pi_1(B^*) \to \pi_1(M^*)\}$ whose Jacobian matrix is

$$J(P') = \left[\begin{array}{c|c} J(P) & 0 \\ \hline z_i \frac{\partial u_{i,j}}{\partial x_k} & 1 - z_i u_{i,j} z_j^{-1} \end{array}\right]$$

By 2.5 $\theta^{*}J(P')$ is a presentation matrix for $H_1(\widetilde{M},\widetilde{B})$ (where $\widetilde{M}$ corresponds to $\ker\theta$). In particular $\theta^{*}J(P')$ is an invariant of $(M,B;\theta)$.

Now the elements $\{u_{i,j}\}$ represent generators for one particular fusing of the fundamental group system (determined by ρ) and all fusing arise in this way. Again by 2.3 there is a presentation, P'', of this fused system whose Jacobian is

$$J(P'') = \begin{bmatrix} J(P) \\ \\ \frac{\partial u_{i,j}}{\partial x_k} \end{bmatrix}$$

and, see below, may depend on the choice of fusing.

To see the distinction in a computational way observe that if we replace each $u_{i,j}$ by a conjugate, $\alpha_i u_{i,j}\alpha_i^{-1}$, $\alpha_i \in F(x_1,\ldots,x_n)$ the effect on $\theta^{*}J(P')$ is, essentially, to add to each of the first n columns linear combinations of the remaining columns. This does not change the equivalence class of $\theta^{*}J(P')$ but may change the class of $\theta J(P'')$.

Examples in which a change actually occurs are easy to come by. For one suppose $\pi_1(M) = F(x,y)$ (M a cube with handles) and that B consists of two curves corresponding to the generators. Then an arbitrary fusing of the system produces, up to conjugacy, a subgroup of $F(x,y)$ generated by x and $\alpha y\alpha^{-1}$. Taking $T = H_1(M)$ one computes that the 1st elementary ideal in $\mathbb{Z}T$ of the fused system is

$$(\theta(\alpha + (1-y)\frac{\partial\alpha}{\partial y}))$$

which clearly depends on α. For example it is (1) if $\alpha = 1$ and is $(1-y^2 + xy^2)$ if $\alpha = (xy)^2$.

6.4 EXAMPLE. *A non symmetric ideal*

To illustrate that the symmetry of the elementary ideals established in Section 4 need not hold in case ∂M contains a surface of genus greater than one let M be the 3-manifold obtained by attaching a 2-handle to a genus three cube with handles along the curve shown in figure 3. Then

$$\pi_1(M) = \langle x,y,z: x^2yz^2xy^{-2}z = 1\rangle$$

Let $\theta:\pi(M) \to \langle s,t: [s,t] = 1\rangle$ be given by $\theta(x) = s$, $\theta(y) = s^3t^3$, $\theta(z) = t$.

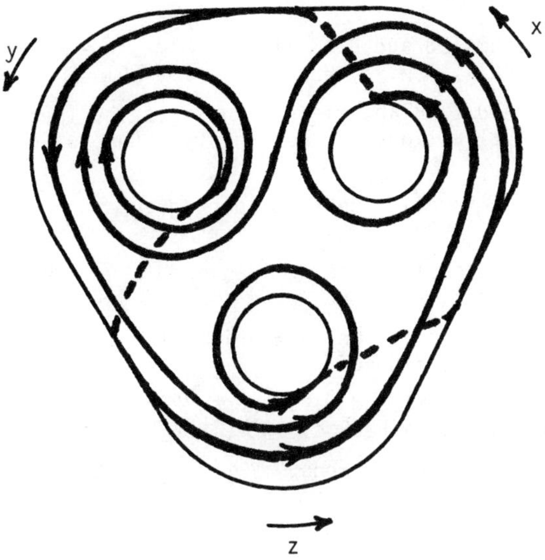

Figure 3

One has

$$\mathcal{E}_2(M,m_0;\theta) = (1+s+s^5t^5,\ 1+s^3t^3-s^2t,\ 1+s^5t^4+s^5t^5).$$

Define $\psi:\mathbb{Z}T \to Z_5$ by $\psi(S) = 2$, $\psi(s^{-1}) = 3$, $\psi(t) = \psi(t^{-1}) = 1$. Then $\mathcal{E}_2(M,m_0;\theta) \subset \ker\psi$ but $\overline{\mathcal{E}_2(M,m_0;\theta)} \not\subset \ker\psi$; so $\mathcal{E}_2(M,m_0;\theta) \neq \overline{\mathcal{E}_2(M,m_0;\theta)}$.

REFERENCES

[B,S] G. Baumslag and D. Solitar, "Some two-generator, one-relator non hopfian groups," Bull. A.M.S., 68 (1962), 199-201.

[Bℓ] Richard C. Blanchfield, "Intersection theory of manifolds with operators with applications to knot theory," Ann. of Math. 65 (1957), 340-356.

[Cr_1] R. H. Crowell, "Corresponding group and module sequences," Nagoya Math. J. 19 (1961), 27-40.

[Cr_2] R. H. Crowell, "The derived module of a homomorphism," Advances in Math. 6 (1971), 210-238.

[Cr_3] R. H. Crowell, "On the annihilator of a knot module," Proc. Amer. Math. Soc. 15 (1964), 696-700.

[Cr,S] R. H. Crowell and D. Strangs, "On the elementary ideals of link modules," Trans. Amer. Math. Soc. 142 (1969), 93-109.

[E] D.B.A. Epstein, "Ends," *Topology of 3-manifolds*, Prentice Hall (1962), 110-118.

[Ev] B. Evans, "Boundary respecting maps of 3-manifolds," Pacific J. Math., 42 (1972), 639-655.

[F_1] R. H. Fox, "Free differential calculus I, Derivation in the free group ring," Ann. of Math. 57 (1953), 547-560.

[F_2] R. H. Fox, "Free differential calculus II, The isomorphism problem for groups," Ann. of Math. 59 (1954), 196-210.

[F_3] R. H. Fox, "Free differential calculus III, Subgroups," Ann. of Math. 64 (1956), 407-419.

[F_4] K. T. Chen, R. H. Fox, and R. C. Lyndon, "Free differential calculus IV, The quotient groups of the lower central series," Ann. of Math. 68 (1958), 81-95.

[F_5] R. H. Fox, "Free differential calculus V, The Alexander matrices re-examined," Ann. of Math. 71 (1960), 408-422.

[F_6] R. H. Fox, "The homology characters of the cyclic coverings of knots of genus one," Ann. of Math. 71 (1960), 187-196.

[F,T] R. H. Fox, and G. Torres, "Dual presentations of the group of a knot," Ann. of Math. 59 (1954), 211-218.

[Go] L. Goeritz, "Die Bettishchen Zahlen der Zyklische Überlagerungsraüme der Knotenaussenraüme," Amer. J. Math. 56 (1934), 194-198.

[G] C. M. Gordon, "Knots whose branched cyclic covers have periodic homology," Trans. Amer. Math. Soc. 168 (1972), 357-370.

[Ha] Wolfgang Haken, "Some results on surfaces in 3-manifolds," *Studies in Modern Topology,* Math. Assoc. of America, distributed by Prentice Hall (1968), 39-98.

[He_1] Wolfgang Heil, "On P^2-irreducible 3-manifolds," Bull. Amer. Math. Soc. 75 (1963), 772-775.

[He_2] Wolfgang Heil, "Some finitely presented non 3-manifold groups," Proc. Amer. Math. Soc. 53 (1975), 497-500.

[Hm] John Hempel, *3-manifolds, Annals of Math Studies* No. 86, Princeton Univ. Press (1976).

[Hi_1] Jonathan Hillman, "Longitudes of a link and principality of an Alexander ideal," Proc. Amer. Math. Soc. 72 (1978), 370-374.

[Hi_2] Jonathan Hillman, "Alexander polynomials, annihilator ideals and the Steinitz-Fox invariant," Proc. Lond. Math. Soc (3) 45 (1982), 31-48.

[J_1] William Jaco, "Heegaard splittings and splitting homomorphisms," Trans. Amer. Math. Soc. 144 (1969), 365-379.

[J_2] William Jaco, "Roots, relations, and centralizers in 3-manifold groups," *Lecture Notes in Math,* No. 483, Springer-Verlag (1975), 283-309.

[K] Akio Kawauchi, "on quadratic forms of 3-manifolds," Inventiones Math 43 (1977), 177-198.

[Ma] Bernard Maskit, "A theorem on planar covering surfaces with applications to 3-manifolds," Ann. of Math. 81 (1965), 341-355.

[Mi] John Milnor, "A duality theorem for Reidemeister torsion," Ann. of Math. 76 (1962), 137-147.

[P] C. D. Papakyriakopoulos, "Planar regular coverings of orientable, closed surfaces," *Knots groups and 3-manifolds, Ann. of Math. Studies* No. 84, Princeton, Univ. Press (1975), 261-292.

[$P\ell$] Antonio Plans, "Contribution to the study of the cyclic ramified coverings corresponding to a knot," Revista Acad. Ci. Madrid 47 (1953), 161-193.

[R_1] K. Reidemeister, "Homotopiegruppen von Komplexen," Abh. Math. Sem. Hamburgischen Univ. 10 (1935), 211-215.

[R_2] K. Reidemeister, "Complexes and homotopy chains," Bull. Amer. Math. Soc. 56 (1950), 297-307.

[Sc] G. P. Scott, "Finitely generated 3-manifold groups are finitely presented," J. Lond. Math. Soc. (2) 6 (1973), 437-440.

[S] H. Seifert, "Uber das Geschlect von knoten," Math. Ann. 110 (1934), 571-592.

[St_1] John Stallings, "How not to prove the Poincaré conjecture," *Ann. of Math Study* No. 60 (1966), 83-88.

[St_2] John Stallings, *Group Theory and three-dimensional manifolds,* Yale University Press (1971).

[Sw] G. A. Swarup, "Boundary preserving maps of 3-manifolds," Proc. Amer. Math. Soc. 78 (1980), 291-294.

[To] G. Torres, "On the Alexander polynomial," Ann. of Math. 57 (1953), 57-89.

[Tr] H. F. Trotter, "Homology of group systems with applications to knot theory," Ann. of Math. 76 (1962), 464-498.

[Tu] Thomas Tucker, "Boundary reducible 3-manifolds and Waldhausen's theorem," Mich. J. Math 20 (1973), 321-327.

[Tur] V. B. Turaev, "The fundamental groups of manifolds and Poincaré complexes," (Russian), Mat. Sb., 38 (1981), 255-270.

[W_1] F. Waldhausen, "On irreducible 3-manifolds which are sufficiently large," Ann. of Math. 87 (1968), 56-88.

[W_2] F. Waldhausen, "Heegaard-Zerlengungen der 3-sphäre," Topology 7 (1968), 195-203.

[Wa] C. T. C. Wall, "Poincaré complexes I," Ann. of Math. 86 (1967), 213-245.

RICE UNIVERSITY, HOUSTON, TEXAS

General instructions to authors for
PREPARING REPRODUCTION COPY FOR MEMOIRS

For more detailed instructions send for AMS booklet, "A Guide for Authors of Memoirs." Write to Editorial Offices, American Mathematical Society, P. O. Box 6248, Providence, R. I. 02940.

MEMOIRS are printed by photo-offset from camera copy fully prepared by the author. This means that, except for a reduction in size of 20 to 30%, the finished book will look exactly like the copy submitted. Thus the author will want to use a good quality typewriter with a new, medium-inked black ribbon, and submit clean copy on the appropriate model paper.

Model Paper, provided at no cost by the AMS, is paper marked with blue lines that confine the copy to the appropriate size. Author should specify, when ordering, whether typewriter to be used has PICA-size (10 characters to the inch) or ELITE-size type (12 characters to the inch).

Line Spacing – For best appearance, and economy, a typewriter equipped with a half-space ratchet – 12 notches to the inch – should be used. (This may be purchased and attached at small cost.) Three notches make the desired spacing, which is equivalent to 1-1/2 ordinary single spaces. Where copy has a great many subscripts and superscripts, however, double spacing should be used.

Special Characters may be filled in carefully freehand, using dense black ink, or INSTANT ("rub-on") LETTERING may be used. AMS has a sheet of several hundred most-used symbols and letters which may be purchased for $5.

Diagrams may be drawn in black ink either directly on the model sheet, or on a separate sheet and pasted with rubber cement into spaces left for them in the text. Ballpoint pen is *not* acceptable.

Page Headings (Running Heads) should be centered, in CAPITAL LETTERS (preferably), at the top of the page – just above the blue line and touching it.

LEFT-hand, EVEN-numbered pages should be headed with the AUTHOR'S NAME;

RIGHT-hand, ODD-numbered pages should be headed with the TITLE of the paper (in shortened form if necessary).

Exceptions: PAGE 1 and any other page that carries a display title require NO RUNNING HEADS.

Page Numbers should be at the top of the page, on the same line with the running heads.

LEFT-hand, EVEN numbers – flush with left margin;

RIGHT-hand, ODD numbers – flush with right margin.

Exceptions: PAGE 1 and any other page that carries a display title should have page number, centered below the text, on blue line provided.

FRONT MATTER PAGES should be numbered with Roman numerals (lower case), positioned below text in same manner as described above.

MEMOIRS FORMAT

It is suggested that the material be arranged in pages as indicated below.
Note: Starred items (*) are requirements of publication.

Front Matter (first pages in book, preceding main body of text).

Page i – *Title, *Author's name.

Page iii – Table of contents.

Page iv – *Abstract (at least 1 sentence and at most 300 words).

*1980 Mathematics Subject Classifications represent the primary and secondary subjects of the paper. For the classification scheme, see Annual Subject Indexes of MATHEMATICAL REVIEWS beginning in December 1978.

Key words and phrases, if desired. (A list which covers the content of the paper adequately enough to be useful for an information retrieval system.)

Page v, etc. – Preface, introduction, or any other matter not belonging in body of text.

Page 1 – Chapter Title (dropped 1 inch from top line, and centered).

Beginning of Text.

Footnotes: *Received by the editor date.
Support information – grants, credits, etc.

Last Page (at bottom) – Author's affiliation.

ABCDEFGHIJ–AMS–89876543

RAYMOND

Memoirs of the American Mathematical Society

Number 283

M. Scott Osborne
and Garth Warner

The Selberg trace formula III: Inner product formulae (Initial considerations)

Published by the
AMERICAN MATHEMATICAL SOCIETY
Providence, Rhode Island, USA

July 1983 · Volume 44 · Number 283 (first of three numbers) · ISSN 006

MEMOIRS of the American Mathematical Society

This journal is designed particularly for long research papers (and groups of cognate papers) in pure and applied mathematics. It includes, in general, longer papers than those in the TRANSACTIONS.

Mathematical papers intended for publication in the Memoirs should be addressed to one of the editors. Subjects, and the editors associated with them, follow:

MEMOIRS are printed by photo-offset from camera-ready copy fully prepared by the authors. Prospective authors are encouraged to request booklet giving detailed instructions regarding reproduction copy. Write to Editorial Office, American Mathematical Society, P.O. Box 6248, Providence, Rhode Island 02940. For general instructions, see last page of Memoir.

SUBSCRIPTION INFORMATION. The 1983 subscription begins with Number 272 and consists of six mailings, each containing one or more numbers. Subscription prices for 1983 are $104.00 list; $52.00 member. Each number may be ordered separately; *please specify number* when ordering an individual paper. For prices and titles of recently released numbers, refer to the New Publications sections of the **NOTICES** of the American Mathematical Society.

BACK NUMBER INFORMATION. For back issues see the AMS Catalogue of Publications.

TRANSACTIONS of the American Mathematical Society

This journal consists of shorter tracts which are of the same general character as the papers published in the MEMOIRS. The editorial committee is identical with that for the MEMOIRS so that papers intended for publication in this series should be addressed to one of the editors listed above.

Subscriptions and orders for publications of the American Mathematical Society should be addressed to American Mathematical Society, P. O. Box 1571, Annex Station, Providence, R. I. 02901. *All orders must be accompanied by payment.* Other correspondence should be addressed to P. O. Box 6248, Providence, R. I. 02940.

MEMOIRS of the American Mathematical Society (ISSN 0065-9266) is published bimonthly (each volume consisting usually of more than one number) by the American Mathematical Society at 201 Charles Street, Providence, Rhode Island 02904. Second Class postage paid at Providence, Rhode Island 02940. Postmaster: Send address changes to Memoirs of the American Mathematical Society, American Mathematical Society, P. O. Box 6248, Providence, RI 02940.

CONTENTS

Abstract: In this memoir, we lay the foundations for the study of inner product formulae, one of the key technical preliminaries in the derivation of the Selberg trace formula.

1980 Mathematics Subject Classification

Primary: 10D40, 32N10

Secondary: 22E40, 22E46

Key Words and Phrases: Selberg trace formula, Eisenstein series, Eisenstein systems, inner product formulae, automorphic forms, truncation.

Library of Congress Cataloging in Publication Data

Osborne, M. Scott, 1946-
The Selberg trace formula III.

(Memoirs of the American Mathematical Society, ISSN 0065-9266 ; no. 283)
Bibliography: p.
1. Selberg trace formula. 2. Eisenstein series.
3. Automorphic forms. I. Warner, Garth, 1940- .
II. Title. III. Series.
QA3.A57 no. 283 [QA241] 510s [515.9] 83-3918
ISBN 0-8218-2283-7

First printing, July 1983
Corrected printing, August 1983

§1. INTRODUCTION

This is the third in a projected series of papers in which we plan to come to grips with the Selberg trace formula, the ultimate objective being a reasonably explicit expression. In the present work, we take up the problem of obtaining a formula for the (L^2) inner product of two truncated Eisenstein series which, as one knows, constitutes a key step in the derivation of the Selberg trace formula itself. Of course, the main issue in this circle of ideas is the case of Eisenstein series associated with residual forms, the formula in the case of cuspidal Eisenstein series being an acquired fact (cf. [2]). It turns out that, for various reasons, the general problem is fairly difficult, requiring, as it does, a considerable amount of preparation, a portion of which will be found here. On the one hand, therefore, this article is, to a certain extent, anticipatory in character; on the other hand, when applied to certain special cases, it becomes definitive.

Received by the editors October 2, 1981. The research of both authors was supported in part by the National Science Foundation.

As always, the theory centers on a reductive Lie group G and a non-uniform lattice Γ in G, both satisfying the usual conditions. Associated with the pair (G,Γ) is a formal apparatus of considerable complexity, namely, the theory of Eisenstein systems. In §§2-3 we summarize those facts from this theory pertinent for our purposes, adding here and there certain related results. §4 is technical in nature, being devoted to a study of the singularities of cuspidal Eisenstein series or at least the ones which play a role in the later going. The inner product formula for truncated residual Eisenstein series at the first level is the subject of §5, the upshot being a result which is not only of intrinsic interest but also is quite suggestive as to what to expect in general. For the applications, it is eventually necessary to integrate over the unitary spectrum meaning, then, that the parameter in the inner product formula of §5 must be moved to the imaginary axis; this is done in §6, it actually being necessary to start with the cuspidal case which, until now, had never been discussed in complete generality. The first step toward a general formula is taken in §7, the principle embodied therein being, in fact, the point of departure for our next paper on this topic. We terminate in §8 with some miscellaneous observations.

It would not be possible to make this paper totally self-contained. Accordingly, as a general reference and suggested overall introduction to the subject, we shall use our monograph:

The Theory of Eisenstein Systems, Academic Press, N.Y., 1981.

Throughout the sequel, the title of this work has been abbreviated to TES.

§2. CALCULUS OF Hom - ⊗

The purpose of this § is to establish notation and recall certain definitions and facts from the formal apparatus prefacing the theory of Eisenstein systems. It is definitely expected that the reader is not unfamiliar with this theory, as spelled out in TES, to which we refer for omitted proofs and a complete discussion. However, those points which are new in the sense that they were not explicitly taken up by us in TES, primarily because they were not needed there, will, of course, be dealt with in all necessary detail here. Generally speaking, in the sequel our policy will be to use without specific mention the various and sundry rules of calculation to be found below.

Let G be a reductive Lie group, Γ a lattice in G, both subject to the usual conditions. Let (P,S) be a Γ-cuspidal split parabolic subgroup of G with split component A. Let $\mathfrak{X}$ be an affine subspace of the complexification of $\check{\mathfrak{a}}$ -- then we shall say that $\mathfrak{X}$ is admissible if $\mathfrak{X}$ can be represented as an intersection of hyperplanes of the form

$\check{\lambda} = c(\check{\lambda}\epsilon\Sigma_P(\mathfrak{g},\check{\mathfrak{a}}))$.

Fix an admissible affine subspace $\mathfrak{X}$. If

$$\mathfrak{X} = \cap (\check{\lambda} = c),$$

then $\mathfrak{X}$ admits a unique decomposition

$$\mathfrak{X} = \mathfrak{X}^{\sim} \oplus X$$

where

$$\mathfrak{X}^{\sim} = \cap (\check{\lambda} = 0)$$

and X is a vector which is orthogonal to $\mathfrak{X}^{\sim}$, the normal translation in $\mathfrak{X}$. We shall write $\mathfrak{X}^{\perp}$ for the orthogonal complement of $\mathfrak{X}^{\sim}$. By $S_{\mathfrak{X}}$, we then understand the symmetric algebra over $\mathfrak{X}^{\perp}$. There is a unique conjugate linear isomorphism

$$\begin{cases} *: S_{\mathfrak{X}} \to S_{\mathfrak{X}} \\ u \mapsto u^* \end{cases}$$

of $S_{\mathfrak{X}}$ with itself such that $\Lambda^* = -\bar{\Lambda}$ if Λ belongs to $\mathfrak{X}^{\perp}$.

Let V be a finite dimensional complex Hilbert space. Form

$$\begin{cases} S_{\mathfrak{X}} \otimes V \\ \mathrm{Hom}(S_{\mathfrak{X}}, V). \end{cases}$$

There is a natural pairing

$$S_{\mathfrak{X}} \otimes V \times \mathrm{Hom}(S_{\mathfrak{X}}, V) \to \mathbf{C}$$

which is linear in the first variable and conjugate linear in the second variable, characterized by the condition

$$(u \otimes v, T) = (v, T(u^*)).$$

Form

$$\begin{cases} S_{\mathfrak{X}} \otimes \check{V} \\ \mathrm{Hom}(S_{\mathfrak{X}}, V). \end{cases}$$

There is a natural pairing

$$S_{\mathfrak{X}} \otimes \check{V} \times \mathrm{Hom}(S_{\mathfrak{X}}, V) \to \mathbf{C}$$

which is linear in both variables, characterized by the condition

$$\langle u \otimes v, T \rangle = \langle \check{v}, T(u) \rangle.$$

LEMMA *Suppose that* Λ *is a linear function on* $S_{\mathfrak{X}} \otimes V$ -- *then there is a* T_Λ *in* $\mathrm{Hom}(S_{\mathfrak{X}}, V)$ *such that*

$$\Lambda(Q) = (Q, T_\Lambda) \qquad (Q \in S_{\mathfrak{X}} \otimes V).$$

LEMMA (bis) *Suppose that* Λ *is a linear function on* $S_{\mathfrak{X}} \otimes \check{V}$ -- *then there is a* T_Λ *in* $\mathrm{Hom}(S_{\mathfrak{X}}, V)$ *such that*

$$\Lambda(Q) = \langle Q, T_\Lambda \rangle \qquad (Q \in S_{\mathfrak{X}} \otimes V).$$

One can identify $\mathrm{Hom}(S_{\mathfrak{X}},V)$ with the space of formal power series over $\mathfrak{X}^{\perp}$ with coefficients in V. In this interpretation, we can speak of the order of an element T: $\mathrm{ord}(T)$ is the order of the term of lowest degree which actually occurs in the power series expansion of T. This being so, a linear transformation from $\mathrm{Hom}(S_{\mathfrak{X}},V)$ to another vector space is said to be of degree n iff it annihilates all the terms of order $>n$ but does not annihilate every term of order $=n$.

LEMMA *Suppose that* Λ *is a linear function on* $\mathrm{Hom}(S_{\mathfrak{X}},V)$ -- *then* Λ *is of finite degree iff there is a* Q_Λ *in* $S_{\mathfrak{X}} \otimes V$ *such that*

$$\Lambda(T) = \overline{(Q_\Lambda, T)} \qquad (T \in \mathrm{Hom}(S_{\mathfrak{X}},V)).$$

LEMMA (bis) *Suppose that* Λ *is a linear function on* $\mathrm{Hom}(S_{\mathfrak{X}},V)$ -- *then* Λ *is of finite degree iff there is a* Q_Λ *in* $S_{\mathfrak{X}} \otimes \check{V}$ *such that*

$$\Lambda(T) = \langle Q_\Lambda, T \rangle \qquad (T \in \mathrm{Hom}(S_{\mathfrak{X}},V)).$$

REMARK Suppose given

$$\begin{cases} (P',S';A') \\ (P'',S'';A''), \end{cases}$$

with associated data

$$\begin{cases} (\check{\mathfrak{a}}', \mathfrak{X}', V') \\ (\check{\mathfrak{a}}'', \mathfrak{X}'', V'') . \end{cases}$$

Let

$$\nabla : \mathrm{Hom}(S_{\mathfrak{X}'}, V') \to S_{\mathfrak{X}''} \otimes V''$$

be a linear transformation of finite degree -- then there exists a unique linear transformation

$$\nabla^* : \mathrm{Hom}(S_{\mathfrak{X}''}, V'') \to S_{\mathfrak{X}'} \otimes V'$$

of finite degree such that

$$(\nabla T', T'') = \overline{(\nabla^* T'', T')}$$

for all

$$\begin{cases} T' \in \mathrm{Hom}(S_{\mathfrak{X}'}, V') \\ T'' \in \mathrm{Hom}(S_{\mathfrak{X}''}, V''). \end{cases}$$

We shall, accordingly, refer to ∇^* as the adjoint of ∇.

There is a certain profit to be made in regarding

$$\begin{cases} S_{\mathfrak{X}} \otimes V \quad \text{or} \quad S_{\mathfrak{X}} \otimes \check{V} \\ \mathrm{Hom}(S_{\mathfrak{X}}, V) \end{cases}$$

as locally convex topological vector spaces. Let us agree to write $S_{\mathfrak{X}}^n$ for the terms of degree $\leq n$ -- then

$$S_{\mathfrak{X}} \otimes V = \bigcup_n S_{\mathfrak{X}}^n \otimes V.$$

Since a given $S_{\mathfrak{X}}^n \otimes V$ is finite dimensional, we can treat it as a Banach space. This done, equip $S_{\mathfrak{X}} \otimes V$ with the corresponding inductive limit topology -- then $S_{\mathfrak{X}} \otimes V$ is an LB-space. The topological dual of

$$S_{\mathfrak{X}} \otimes V \quad \text{or} \quad S_{\mathfrak{X}} \otimes \check{V}$$

is the same as its algebraic dual, each being identified with $\mathrm{Hom}(S_{\mathfrak{X}},V)$ (cf. supra). Under the strong topology, $\mathrm{Hom}(S_{\mathfrak{X}},V)$ is a Fréchet space; in fact, $\mathrm{Hom}(S_{\mathfrak{X}},V)$ is also a Montel space, this being the case of

$$S_{\mathfrak{X}} \otimes V \quad \text{or} \quad S_{\mathfrak{X}} \otimes \check{V}.$$

Write

$$\mathrm{Hom}^n(S_{\mathfrak{X}},V)$$

for the set of elements in $\mathrm{Hom}(S_{\mathfrak{X}},V)$ of order $\geq n+1$ -- then $\mathrm{Hom}^n(S_{\mathfrak{X}},V)$ is a closed subspace of finite codimension in $\mathrm{Hom}(S_{\mathfrak{X}},V)$. The linear functions on $\mathrm{Hom}(S_{\mathfrak{X}},V)$ which are of finite degree can be viewed, alternatively, as the continuous linear functions on $\mathrm{Hom}(S_{\mathfrak{X}},V)$, thus the topological dual of $\mathrm{Hom}(S_{\mathfrak{X}},V)$ can be identified with

$$S_{\mathfrak{X}} \otimes V \quad \text{or} \quad S_{\mathfrak{X}} \otimes \check{V}.$$

That this is indeed so is a consequence of the following result which also serves to put the Remark above into proper perspective.

LEMMA *Suppose given*

$$\begin{cases} (P',S';A') \\ (P'',S'';A''), \end{cases}$$

with associated data

$$\begin{cases} (\check{a}',\mathfrak{X}',V') \\ (\check{a}'',\mathfrak{X}'',V''). \end{cases}$$

Let

$$\nabla: \mathrm{Hom}(S_{\mathfrak{X}'},V') \to S_{\mathfrak{X}''} \otimes V''$$

be a linear transformation. Consider the conditions:

(i) $\exists n$ st $\nabla | \mathrm{Hom}^n(S_{\mathfrak{X}'},V') = 0$;

(ii) $\exists n$ st $\mathrm{Im}(\nabla) \subset S^n_{\mathfrak{X}''} \otimes V''$;

(iii) ∇ *is continuous*;

(iv) ∇ *has finite rank*.

Then

$$\text{(iii)} \Leftrightarrow \text{(i)} \underset{\not\Leftarrow}{\Rightarrow} \text{(iv)} \Leftrightarrow \text{(ii)}.$$

PROOF (ii) $\Longrightarrow$ (iv): This is clear.

(iv) $\Longrightarrow$ (ii): We have

$$\mathrm{Im}(\nabla) = \bigcup_{n=1}^{\infty} (\mathrm{Im}(\nabla) \cap S^n_{\mathfrak{X}''} \otimes V''),$$

therefore, by dimension,

$$\mathrm{Im}(\nabla) = \mathrm{Im}(\nabla) \cap S^n_{\mathfrak{X}''} \otimes V''$$

for some n.

(i) $\Longrightarrow$ (iii): There is a factorization

$$\begin{array}{ccc} \mathrm{Hom}(S_{\mathfrak{X}'}, V') & \rightarrow & \mathrm{Hom}(S_{\mathfrak{X}'}, V')/\mathrm{Hom}^n(S_{\mathfrak{X}'}, V') \\ & \searrow^{\nabla} & \downarrow \nabla^n \\ & & S_{\mathfrak{X}''} \otimes V'' \end{array}$$

of ∇. Since the quotient space is finite dimensional and Hausdorff, ∇^n is continuous.

(iii) $\Longrightarrow$ (ii): We have

$$\mathrm{Hom}(S_{\mathfrak{X}'}, V') = \bigcup_{n=1}^{\infty} \nabla^{-1}(S^n_{\mathfrak{X}''} \otimes V''),$$

therefore, by the Baire category theorem,

$$\nabla^{-1}(S^n_{\mathfrak{X}''} \otimes V'')$$

has an interior point for some n. But then, being a subspace,

$$\mathrm{Hom}(S_{\mathfrak{X}'}, V') = \nabla^{-1}(S^n_{\mathfrak{X}''} \otimes V''),$$

which implies, of course, that

$$\mathrm{Im}(\nabla) \subset S^n_{\mathfrak{X}''} \otimes V''.$$

(iii) $\Longrightarrow$ (i): Thanks to '(iii) $\Longrightarrow$ (ii)', $\mathrm{Im}(\nabla)$ is a Banach space. On the other hand, the topology of $\mathrm{Hom}(S_{\mathfrak{X}'},V')$ can be described by a countable family of semi-norms, the nth such having kernel $\mathrm{Hom}^n(S_{\mathfrak{X}'},V')$. Accordingly, $\|\nabla(?)\|$ is bounded by a positive constant times the nth semi-norm $(\exists n)$, thus, for this n,

$$\nabla|\mathrm{Hom}^n(S_{\mathfrak{X}'},V') = 0,$$

as desired.

(iv) $\not\Longrightarrow$ (i): Fix a non-zero vector $v'' \in V''$. For every n, choose a

$$T'_n \in \mathrm{Hom}^n(S_{\mathfrak{X}'},V') - \mathrm{Hom}^{n+1}(S_{\mathfrak{X}'},V').$$

Define a complex valued functional Λ' on the algebraic span of the T'_n by the rule

$$\Lambda'(\Sigma c_n T'_n) = \Sigma c_n.$$

Extend Λ' linearly to all of $\mathrm{Hom}(S_{\mathfrak{X}'},V')$. Put

$$\nabla(T') = \Lambda'(T')v'.$$

Then ∇ has finite rank but

$$\nabla|\mathrm{Hom}^n(S_{\mathfrak{X}'},V') \neq 0 \quad (\forall n).$$

Hence the lemma. //

Let $\mathfrak{Y}$ be an admissible affine subspace which is contained in $\mathfrak{X}$, so that

$$\mathfrak{Y} = \mathfrak{Y}^{\sim} \oplus Y$$

with $\mathfrak{Y}^{\sim}$ contained in $\mathfrak{X}^{\sim}$. Call $S_{\mathfrak{Y}-\mathfrak{X}}$ the symmetric algebra over the orthogonal complement of $\mathfrak{X}^{\perp}$ in $\mathfrak{Y}^{\perp}$ -- then

$$S_{\mathfrak{Y}} = S_{\mathfrak{Y}-\mathfrak{X}} \otimes S_{\mathfrak{X}}.$$

If

$$\begin{cases} u_o \in S_{\mathfrak{Y}-\mathfrak{X}} \\ T \in \mathrm{Hom}(S_{\mathfrak{Y}}, V), \end{cases}$$

then there is determined an element

$$u_o \vee T \in \mathrm{Hom}(S_{\mathfrak{X}}, V)$$

characterized by the condition that

$$(u_o \vee T)(u) = T(u_o \otimes u) \qquad (u \in S_{\mathfrak{X}}).$$

One has

$$\begin{cases} (Q, u_o \vee T) = (u_o^* \otimes Q, T) & (Q \in S_{\mathfrak{X}} \otimes V) \\ \langle Q, u_o \vee T \rangle = \langle u_o \otimes Q, T \rangle & (Q \in S_{\mathfrak{X}} \otimes \check{V}). \end{cases}$$

The discussion thus far has centered on $(\check{\mathfrak{a}}, \mathfrak{X}, V)$. We shall now inject another ingredient into our considerations. Let X be a linear subspace of

$\check{\mathfrak{a}}$ -- then X is said to be $\mathfrak{X}$-admissible if X is contained in $\mathfrak{X}^{\sim}$.

Fix an $\mathfrak{X}$-admissible subspace X. Obviously: $\mathfrak{X}^{\perp} \subset X^{\perp}$.

LEMMA <u>There is a canonical injection</u> $u \mapsto \partial_u$ <u>of</u> $S_{\mathfrak{X}}$ <u>into the algebra of holomorphic differential operators on</u> $X^{\perp}$ <u>such that</u>

$$\partial_{\Lambda} f(\Lambda_o) = \frac{df}{dt}(\Lambda_o + t\Lambda)\Big|_{t=0}$$

<u>for all</u> Λ <u>in</u> $\mathfrak{X}^{\perp}$.

Fix a point Λ_o in $X^{\perp}$ and an open neighborhood N_o of Λ_o.

Let

$$\Phi: N_o \to V$$

be a holomorphic function. By $d\Phi(\Lambda_o)$, we mean that element of $\mathrm{Hom}(S_{\mathfrak{X}}, V)$ specified by the rule

$$d\Phi(\Lambda_o)(u) = \partial_u \Phi(\Lambda_o) \qquad (u \in S_{\mathfrak{X}}).$$

In the formal power series picture, $d\Phi(\Lambda_o)$ is obtained by expanding

$$\Phi_{\Lambda_o} \qquad (\Phi_{\Lambda_o}(\Lambda) = \Phi(\Lambda_o + \Lambda))$$

about the origin. Observe that $d\Phi$ may be thought of as a map

$$d\Phi: N_o \to \mathrm{Hom}(S_{\mathfrak{X}}, V)$$

with the property that for every $u \in S_{\mathfrak{X}}$,

$$d\Phi(?:u):N_0 \to V$$

is holomorphic on N_0. This, however, is entirely equivalent to saying that

$$d\Phi:N_0 \to \mathrm{Hom}(S_{\mathfrak{X}},V)$$

is weakly holomorphic.

Let

$$F:N_0 \to \mathrm{Hom}(S_{\mathfrak{X}},V)$$

be a weakly holomorphic map. Let, as before, $\mathfrak{Y}$ be an admissible affine subspace which is contained in $\mathfrak{X}$. Assume: X is $\mathfrak{Y}$-admissible. We shall then denote by $d_0F(\Lambda_0)$ that element of $\mathrm{Hom}(S_{\mathfrak{Y}},V)$ given by the prescription

$$d_0F(u_0 \otimes u) = \partial_{u_0} F(\Lambda_0:u).$$

REMARK It is easy to check that

$$d_0(d\Phi)(\Lambda_0) = d\Phi(\Lambda_0).$$

The notation, however, is deceptive. On the left, $d\Phi$ is viewed as an element of $\mathrm{Hom}(S_{\mathfrak{X}},V)$ while, on the right, $d\Phi$ is viewed as an element of $\mathrm{Hom}(S_{\mathfrak{Y}},V)$. No confusion will arise, though, the context always dictating the appropriate interpretation.

Let again

$$\Phi:N_o \to V$$

be a holomorphic function -- then, as we have seen above,

$$d\Phi:N_o \to \mathrm{Hom}(S_{\mathfrak{X}},V)$$

is weakly holomorphic. For certain applications, it is useful to know that $d\Phi$ is actually strongly continuous. To convince ourselves of this point, we need only establish the following generality.

PROPOSITION M *A weakly holomorphic map from an open subset of* C^n *to a Montel space is strongly continuous.*

One can, in reality, get away with far less, viz.:

LEMMA *Let* X *be a Hausdorff space whose topology is compactly generated; let* Y *be a Montel space. Suppose that* $f:X \to Y$ *is weakly continuous -- then* f *is strongly continuous.*

PROOF Let C be a strongly closed subset of Y -- then it need only be shown that $f^{-1}(C)$ is a closed subset of X. For this purpose, let K be an arbitrary compact subset of X -- then $f(K)$ is

$$\begin{array}{ccccccc} & \nearrow & \text{weakly closed} & \Longrightarrow & \text{strongly closed} & \searrow & \\ \text{weakly compact} & & & & & & \text{strongly compact} . \\ & \searrow & \text{weakly bounded} & \Longrightarrow & \text{strongly bounded} & \nearrow & \end{array}$$

Because

$$(f(K), \text{strong topology}) \rightarrow (f(K), \text{weak topology})$$

is a continuous map of compact Hausdorff spaces, it is a homeomorphism. Thus $C \cap f(K)$ is weakly closed, and so

$$f^{-1}(C) \cap K = f^{-1}(C \cap f(K)) \cap K$$

is the intersection of two closed sets in X, hence is closed. As the topology on X is, by hypothesis, compactly generated, it follows that $f^{-1}(C)$ is a closed subset of X. //

Let V_1 and V_2 be finite dimensional complex Hilbert spaces -- then $\text{Hom}(V_1,V_2)$ is again a finite dimensional complex Hilbert space. Let

$$f:\mathcal{N}_o \rightarrow \text{Hom}(V_1,V_2)$$

be a holomorphic function. One then has that

$$df(\Lambda_o) \in \text{Hom}(S_{\mathfrak{X}},\text{Hom}(V_1,V_2)),$$

i.e., that

$$df(\Lambda_o) \in \text{Hom}(S_{\mathfrak{X}} \otimes V_1,V_2),$$

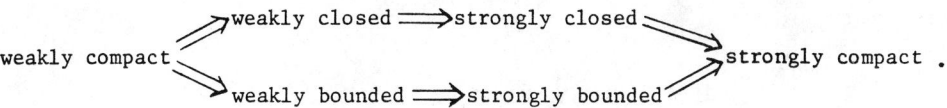

i.e., that

$$df(\Lambda_o) \in \mathrm{Hom}(V_1, \mathrm{Hom}(S_{\mathfrak{X}}, V_2)).$$

The composition of any element in $\mathrm{Hom}(S_{\mathfrak{X}}, V_1)$ with $df(\Lambda_o)$ provides us with an element of

$$\mathrm{Hom}(S_{\mathfrak{X}}, \mathrm{Hom}(S_{\mathfrak{X}}, V_2))$$

or still, with an element of

$$\mathrm{Hom}(S_{\mathfrak{X}} \otimes S_{\mathfrak{X}}, V_2).$$

The Hopf algebra map

$$S_{\mathfrak{X}} \to S_{\mathfrak{X}} \otimes S_{\mathfrak{X}}$$

then induces a morphism of restriction

$$\mathrm{Hom}(S_{\mathfrak{X}} \otimes S_{\mathfrak{X}}, V_2) \to \mathrm{Hom}(S_{\mathfrak{X}}, V_2).$$

There is, therefore, a natural map

$$\mathrm{Hom}(S_{\mathfrak{X}}, V_1) \to \mathrm{Hom}(S_{\mathfrak{X}}, V_2),$$

thereby determining an element

$$d_{\mathrm{Hom}} f(\Lambda_o) \in \mathrm{Hom}(\mathrm{Hom}(S_{\mathfrak{X}}, V_1), \mathrm{Hom}(S_{\mathfrak{X}}, V_2)).$$

Let us now go back to the beginning of this discussion. Starting from the fact that

$$df(\Lambda_o) \in \mathrm{Hom}(S_{\mathfrak{X}}, \mathrm{Hom}(V_1, V_2)),$$

i.e., that

$$df(\Lambda_o) \in \mathrm{Hom}(S_{\mathfrak{X}} \otimes V_1, V_2),$$

we obtain an element in

$$\mathrm{Hom}(S_{\mathfrak{X}} \otimes S_{\mathfrak{X}} \otimes V_1, S_{\mathfrak{X}} \otimes V_2)$$

by tensoring $df(\Lambda_o)$ on the left with the identity map on $S_{\mathfrak{X}}$. Tensoring the Hopf algebra map

$$S_{\mathfrak{X}} \to S_{\mathfrak{X}} \otimes S_{\mathfrak{X}}$$

on the right with the identity map on V_1 yields an element in

$$\mathrm{Hom}(S_{\mathfrak{X}} \otimes V_1, S_{\mathfrak{X}} \otimes S_{\mathfrak{X}} \otimes V_1).$$

Combine these two maps under the composition on

$$\mathrm{Hom}(S_{\mathfrak{X}} \otimes V_1, S_{\mathfrak{X}} \otimes S_{\mathfrak{X}} \otimes V_1) \times \mathrm{Hom}(S_{\mathfrak{X}} \otimes S_{\mathfrak{X}} \otimes V_1, S_{\mathfrak{X}} \otimes V_2),$$

thereby determining an element

$$d_{\otimes} f(\Lambda_o) \in \mathrm{Hom}(S_{\mathfrak{X}} \otimes V_1, S_{\mathfrak{X}} \otimes V_2).$$

There is a calculus for

$$\begin{cases} d_{\mathrm{Hom}} \\ d_{\otimes} \end{cases},$$

the elements of which are developed in the following lemmas.

Agreeing to employ the usual multiindex notation, we may view $S_{\mathfrak{X}}$ as comprised of all formal polynomials

$$\Sigma a_I x^I,$$

$S_{\mathfrak{X}} \otimes V$ then consisting of all formal polynomials

$$\Sigma v_I x^I \qquad (v_I \in V).$$

It is then convenient to regard $\mathrm{Hom}(S_{\mathfrak{X}}, V)$ as made up of all formal power series

$$\Sigma v^I x_I \qquad (v^I \in V)$$

where

$$x_I(x^J) = \begin{cases} I! & \text{if } I = J \\ 0 & \text{if } I \neq J. \end{cases}$$

Let

$$\Phi : N_0 \to V$$

be a holomorphic function -- then

$$\begin{cases} d\Phi : N_0 \to \mathrm{Hom}(S_{\mathfrak{X}}, V) \\ d\Phi(\Lambda_0)(x^I) = \partial^{|I|}\Phi/\partial x^I|_{\Lambda_0} \end{cases} .$$

Let

$$f : N_0 \to \mathrm{Hom}(V_1, V_2)$$

be a holomorphic function -- then

$$\begin{cases} d_{\mathrm{Hom}} f(\Lambda_o):\mathrm{Hom}(S_{\mathfrak{X}},V_1) \to \mathrm{Hom}(S_{\mathfrak{X}},V_2) \\ \\ d_{\otimes} f(\Lambda_o):S_{\mathfrak{X}} \otimes V_1 \to S_{\mathfrak{X}} \otimes V_2 \ , \end{cases}$$

or still, upon explication,

$$\begin{cases} d_{\mathrm{Hom}} f(\Lambda_o)(v^I x_I) = \sum\limits_{J \geq I} \binom{J}{I} \partial^{|J-I|} f/\partial x^{J-I}|_{\Lambda_o} (v^I) x_J \\ d_{\otimes} f(\Lambda_o)(v_I x^I) = \sum\limits_{J \leq I} \binom{I}{J} \partial^{|I-J|} f/\partial x^{I-J}|_{\Lambda_o} (v_I) x^J . \end{cases}$$

These formulas are at the basis of the verifications infra.

Associated with the holomorphic function

$$f:N_o \to \mathrm{Hom}(V_1,V_2)$$

is the transpose map

$$f^t:N_o \to \mathrm{Hom}(\check{V}_2,\check{V}_1),$$

itself also holomorphic. Accordingly,

$$d_{\mathrm{Hom}} f^t(\Lambda_o):\mathrm{Hom}(S_{\mathfrak{X}},\check{V}_2) \to \mathrm{Hom}(S_{\mathfrak{X}},\check{V}_1).$$

On the other hand,

$$d_{\otimes} f(\Lambda_o):S_{\mathfrak{X}} \otimes V_1 \to S_{\mathfrak{X}} \otimes V_2 \ .$$

For the tensor spaces, there is no distinction between algebraic dual and topological dual. Therefore $d_{\otimes} f(\Lambda_o)$ admits a transpose

$$d_{\otimes} f(\Lambda_o)^t:\mathrm{Hom}(S_{\mathfrak{X}},\check{V}_2) \to \mathrm{Hom}(S_{\mathfrak{X}},\check{V}_1) \ ,$$

of necessity continuous. This said, it can then be shown without difficulty that

$$d_{\mathrm{Hom}} f^t(\Lambda_o) = d_{\otimes} f(\Lambda_o)^t \quad .$$

Here is a corollary: $d_{\mathrm{Hom}} f(\Lambda_o)$ is continuous. [Proof: $f = (f^t)^t$.] By comparison, the continuity of $d_{\otimes} f(\Lambda_o)$ is, of course, automatic.

In general, any continuous

$$D \in \mathrm{Hom}(\mathrm{Hom}(S_{\mathfrak{X}}, V_1), \mathrm{Hom}(S_{\mathfrak{X}}, V_2))$$

has an adjoint

$$D^* \in \mathrm{Hom}(S_{\mathfrak{X}} \otimes V_2, S_{\mathfrak{X}} \otimes V_1)$$

characterized by the condition

$$(D^*(u \otimes v_2), T_1) = (u \otimes v_2, DT_1)$$

whereas, in general, any

$$D \in \mathrm{Hom}(S_{\mathfrak{X}} \otimes V_1, S_{\mathfrak{X}} \otimes V_2)$$

has an adjoint

$$D^* \in \mathrm{Hom}(\mathrm{Hom}(S_{\mathfrak{X}}, V_2), \mathrm{Hom}(S_{\mathfrak{X}}, V_1))$$

characterized by the condition

$$(D(u \otimes v_1), T_2) = (u \otimes v_1, D^*T_2).$$

LEMMA *Let*

$$\begin{cases} f:N_o \to \mathrm{Hom}(V_1,V_2) \\ g:N_o \to \mathrm{Hom}(V_2,V_1) \end{cases}$$

be holomorphic functions. Suppose that

$$f(\Lambda_o)^* = g(-\overline{\Lambda}_o).$$

Then

$$\begin{cases} d_{\mathrm{Hom}} f(\Lambda_o)^* = d_{\otimes} g(-\overline{\Lambda}_o) \\ d_{\mathrm{Hom}} g(-\overline{\Lambda}_o) = d_{\otimes} f(\Lambda_o)^*. \end{cases}$$

[Needless to say, it is necessary to impose an obvious condition on N_o.]

There are also rules relating the ordinary differential with d_{Hom} or $d_{\otimes}$. The d_{Hom} statement is transparent, the $d_{\otimes}$ statement less so involving, as it does, the trace map

$$\mathrm{tr}_V:\check{V}\otimes V \to \mathbf{C} \qquad (V = V_1 \text{ or } V_2).$$

LEMMA *Let*

$$\begin{cases} \Phi:N_o \to V_1 \\ f:N_o \to \mathrm{Hom}(V_1,V_2) \end{cases}$$

be holomorphic. Form

$$f\Phi:N_o \to V_2 .$$

Then

$$d_{Hom} f(\Lambda_o)(d\Phi(\Lambda_o)) = d(f\Phi)(\Lambda_o).$$

LEMMA *Let*

$$\begin{cases} f: N_o \to \mathrm{Hom}(V_1, V_2) \\ \Phi: N_o \to \check{V}_2 \end{cases}$$

be holomorphic. Form

$$\Phi f: N_o \to \check{V}_1.$$

Then

$$\mathrm{tr}_{V_1} \circ (d(\Phi f)(\Lambda_o) \otimes 1_{V_1}) = \mathrm{tr}_{V_2} \circ (d\Phi(\Lambda_o) \otimes 1_{V_2}) \circ d_{\otimes} f(\Lambda_o).$$

There is a one dimensional variant of this last result, useful in practice, which does not involve the trace map, viz.: Given holomorphic functions

$$\begin{cases} \phi: N_o \to C \\ \Phi: N_o \to V, \end{cases}$$

so

$$\begin{cases} d_{\otimes}\phi(\Lambda_o): S_{\mathfrak{X}} \to S_{\mathfrak{X}} \\ \quad d\Phi(\Lambda_o): S_{\mathfrak{X}} \to V, \end{cases}$$

we then have

$$d(\phi\Phi)(\Lambda_o) = d\Phi(\Lambda_o)\circ d_{\otimes}\phi(\Lambda_o).$$

Composites also behave correctly.

LEMMA *Let*

$$\begin{cases} f:N_o \to \mathrm{Hom}(V_1,V_2) \\ g:N_o \to \mathrm{Hom}(V_2,V_3) \end{cases}$$

be holomorphic functions -- then

$$d_{\mathrm{Hom}}(g\circ f)(\Lambda_o) = d_{\mathrm{Hom}}g(\Lambda_o)\circ d_{\mathrm{Hom}}f(\Lambda_o)$$

and

$$d_{\otimes}(g\circ f)(\Lambda_o) = d_{\otimes}g(\Lambda_o)\circ d_{\otimes}f(\Lambda_o).$$

There is one final point in this circle of ideas which should be mentioned.

LEMMA *There is a canonical injection* $u \mapsto p_u$ *of* $S_{\mathfrak{X}}$ *into the algebra of polynomial functions on* $\check{\mathfrak{X}}^{\perp}$ *such that*

$$p_{\Lambda} = \Lambda$$

for all Λ *in* $\mathfrak{X}^{\perp}$.

§3. EISENSTEIN SYSTEMS

The purpose of this § is to review the geometric, as opposed to analytic, aspects of the theory of Eisenstein systems, concluding with an important deduction, alluded to but not explicitly formulated in TES, to which we shall need to appeal in the sequel. In this connection, it seems best first, in order to prevent any misunderstanding, to set up the relevant notation completely. This procedure will, we hope, serve a two-fold purpose, providing, on the one hand, a convenient review for the initiated reader and, on the other hand, a capsule summary for the uninitiated reader. It is to be observed that while the definitions are given in maximum generality, i.e., per the daggered picture, we shall, for the most part (but not always), actually only need them per the initial picture.

Let G be a reductive Lie group, Γ a lattice in G, both subject to the usual conditions. We shall then agree to employ without comment the customary conventions as regards this situation.

Let $\mathcal{C}$ be an association class of Γ-cuspidal split parabolic subgroups of G, $\mathcal{C}_i$ and $\mathcal{C}_j$ G-conjugacy classes in $\mathcal{C}$. By the symbol

$$\underset{\sim}{W}(\mathcal{C}_j,\mathcal{C}_i),$$

we understand the set of equivalence classes in

$$\coprod_{\mathcal{C}_j,\mathcal{C}_i} W(A_j,A_i) \quad \begin{cases} (P_j,S_j;A_j) \in \mathcal{C}_j \\ (P_i,S_i;A_i) \in \mathcal{C}_i \end{cases}$$

where

$$\begin{cases} w'_{ji} \in W(A'_j,A'_i) \\ w''_{ji} \in W(A''_j,A''_j) \end{cases}$$

are declared equivalent iff

$$w''_{ji} = I(P''_j|A''_j:P'_j|A'_j)\circ w'_{ji}\circ I(P'_i|A'_i:P''_i|A''_i).$$

Observe that if

$$\underset{\sim}{w}_{ji} \in \underset{\sim}{W}(\mathcal{C}_j,\mathcal{C}_i) \quad\text{and}\quad \begin{cases} (P_j,S_j;A_j) \in \mathcal{C}_j \\ (P_i,S_i;A_i) \in \mathcal{C}_i, \end{cases}$$

then $\underset{\sim}{w}_{ji} \cap W(A_j,A_i)$ is a singleton, call it $\underset{\sim}{w}_{ji}(P_j|A_j;P_i|A_i)$ (or w_{ji}).

Let $\mathcal{C}$, $\mathcal{C}_0$ be association classes of Γ-cuspidal split parabolic subgroups of G, $\mathcal{C}_k$, $\mathcal{C}_{i_0}$ and $\mathcal{C}_{j_0}$ G-conjugacy classes in $\mathcal{C}$, $\mathcal{C}_0$. Suppose that

$$\mathcal{C}_k \succeq \begin{cases} \mathcal{C}_{i_o} \\ \mathcal{C}_{j_o} \end{cases} .$$

Let

$$\begin{cases} (P_{i_o}, S_{i_o}; A_{i_o}) \in \mathcal{C}_{i_o} \\ (P_{j_o}, S_{j_o}; A_{j_o}) \in \mathcal{C}_{j_o} \end{cases} .$$

Then there exist unique elements

$$\begin{cases} (P_i, S_i; A_i) \\ (P_j, S_j; A_j) \end{cases} \in \mathcal{C}_k$$

such that

$$\begin{cases} (P_i, S_i; A_i) \succeq (P_{i_o}, S_{i_o}; A_{i_o}) \\ (P_j, S_j; A_j) \succeq (P_{j_o}, S_{j_o}; A_{j_o}) \end{cases} .$$

This being so, let

$$W^{\dagger}_{\mathcal{C}_k}(A_{j_o}, A_{i_o})$$

be the subset of $W(A_{j_o}, A_{i_o})$ comprised of those $w^{\dagger}_{j_o i_o}$ such that

$$w^{\dagger}_{j_o i_o} | A_i = I(P_j | A_j : P_i | A_i).$$

Let then

$$\underset{\sim}{W}^{\dagger}_{\mathcal{C}_k}(\mathcal{C}_{j_o}, \mathcal{C}_{i_o})$$

be the subset of $\underset{\sim}{W}(\mathcal{C}_{j_o}, \mathcal{C}_{i_o})$ formed of those $\underset{\sim}{w}^{\dagger}_{j_o i_o}$ which factor through the triangle

$$\begin{array}{ccc} \underset{\sim}{W}(\mathcal{C}_{j_0}, \mathcal{C}_{i_0}) & \longrightarrow & W(A_{j_0}, A_{i_0}) \\ & \searrow & \uparrow \\ & & W^{\dagger}_{\mathcal{C}_k}(A_{j_0}, A_{i_0}) \end{array}$$

for all choices on the right.

REMARK The data

$$\begin{cases} (P_i, S_i; A_i) \succeq (P_{i_0}, S_{i_0}; A_{i_0}) \\ (P_j, S_j; A_j) \succeq (P_{j_0}, S_{j_0}; A_{j_0}) \end{cases}$$

determines, in the usual way, triples

$$\begin{cases} (P^{\dagger}_{i_0}, S^{\dagger}_{i_0}; A^{\dagger}_{i_0}) \\ (P^{\dagger}_{j_0}, S^{\dagger}_{j_0}; A^{\dagger}_{j_0}) \end{cases}$$

per

$$\begin{cases} M_i \\ M_j. \end{cases}$$

Because M_i and M_j may be distinct groups, one cannot, in general, ask whether $(P^{\dagger}_{i_0}, S^{\dagger}_{i_0})$ and $(P^{\dagger}_{j_0}, S^{\dagger}_{j_0})$ are associate. It is for this reason that one is forced to introduce the set $W^{\dagger}_{\mathcal{C}_k}(A_{j_0}, A_{i_0})$. If, however,

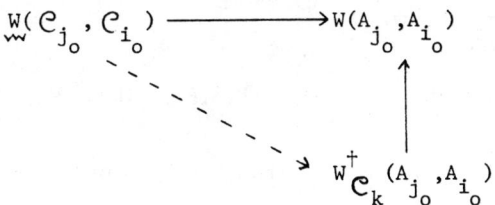

$$\begin{cases} (P_{i_o}, S_{i_o}; A_{i_o}) \\ (P_{j_o}, S_{j_o}; A_{j_o}) \end{cases}$$

admit a common dominant in $\mathcal{C}_k$, say $(P,S;A)$, then $W^{\dagger}_{\mathcal{C}_k}(A_{j_o}, A_{i_o})$ can be identified with $W(A^{\dagger}_{j_o}, A^{\dagger}_{i_o})$, a set, though, which may be empty (since the relation of association is not necessarily preserved under the daggering procedure).

Let $\mathcal{C}$ be an association class of Γ-cuspidal split parabolic subgroups of G, $\mathcal{C}_i$ a G-conjugacy class in $\mathcal{C}$. By an equivariant system $\mathfrak{X}$ of admissible affine subspaces attached to $\mathcal{C}_i$ we mean a map $\mathfrak{X}$ which assigns to each

$$(P,S;A) \in \mathcal{C}_i$$

a non-empty admissible affine subspace $\mathfrak{X}(P,A)$ of the complexification of $\mathfrak{a}$ subject to the following compatibility condition:

For all

$$\begin{cases} (P_1, S_1; A_1) \\ (P_2, S_2; A_2) \end{cases} \in \mathcal{C}_i,$$

$$\mathfrak{X}(P_2, A_2) = I(P_2|A_2 : P_1|A_1)\mathfrak{X}(P_1, A_1).$$

An $\mathfrak{X}$-admissible subspace X of $\mathfrak{X}$ is an equivariant system of linear subspaces attached to $\mathcal{C}_i$ such that for each

$$(P,S;A) \in \mathcal{C}_i,$$

$X(P,A)$ is $\mathfrak{X}(P,A)$-admissible. The terms

$$\begin{cases} \text{'dimension of } \mathfrak{X}\text{'} \\ \text{'dimension of } X\text{'} \end{cases}$$

are to be used in the obvious way. If $\mathcal{C}_j$ is another G-conjugacy class in $\mathcal{C}$, then one can associate with each element $\underset{\sim}{w}_{ji} \in \underset{\sim}{W}(\mathcal{C}_j, \mathcal{C}_i)$ an equivariant system $\mathfrak{X}_{\underset{\sim}{w}_{ji}}$ of admissible affine subspaces attached to $\mathcal{C}_j$ by requiring that

$$\mathfrak{X}_{\underset{\sim}{w}_{ji}}(P_j, A_j) = \overline{-\underset{\sim}{w}_{ji}(P_j|A_j : P_i|A_i)\mathfrak{X}(P_i, A_i)} \, .$$

Fix an equivariant system $\mathfrak{X}$ of admissible affine subspaces together with an $\mathfrak{X}$-admissible subspace X. Let

$$(P,S;A) \in \mathcal{C}_i.$$

The conventions and notations set down in §2 are then applicable here. In particular, we can write

$$\mathfrak{X}(P,A) = \mathfrak{X}(P,A)^{\sim} \oplus X(P,A)$$

and consider

$$S_{\mathfrak{X}(P,A)}.$$

Let δ be a K-type and $\mathcal{O}$ an M-type -- then $E_{cus}(\delta,\mathcal{O})$ is a finite dimensional Hilbert space. Specializing the earlier discussion, we can say that there is a natural pairing

$$S_{\mathfrak{X}(P,A)} \otimes E_{cus}(\delta,\mathcal{O}) \times \mathrm{Hom}(S_{\mathfrak{X}(P,A)}, E_{cus}(\delta,\mathcal{O})) \to \mathbf{C}$$

which is linear in the first variable and conjugate linear in the second variable, characterized by the condition

$$(u \otimes \Phi, T) = (\Phi, T(u^*)),$$

and a natural pairing

$$S_{\mathfrak{X}(P,A)} \otimes E_{cus}(\delta,\mathcal{O})^{\vee} \times \mathrm{Hom}(S_{\mathfrak{X}(P,A)}, E_{cus}(\delta,\mathcal{O})) \to \mathbf{C}$$

which is linear in both variables, characterized by the condition

$$\langle u \otimes \check{\Phi}, T\rangle = \langle \check{\Phi}, T(u)\rangle .$$

Since a polynomial function on a subspace of $\mathfrak{a}$ may always be regarded as a function on G, the presence of the $\mathfrak{X}(P,A)$-admissible subspace $X(P,A)$ implies that there is a map

$$S_{\mathfrak{X}(P,A)} \otimes E_{cus}(\delta,\mathcal{O}) \to \mathrm{Fnc}(G)$$

sending $u \otimes \Phi$ to $p_u \cdot \Phi$. We hardly need point out that this map depends on $X(P,A)$.

Fix an association class $\mathcal{C}_o$ of Γ-cuspidal split parabolic subgroups of G. Let $\mathcal{C}_{i_o}$ be a G-conjugacy class in $\mathcal{C}_o$, $\mathfrak{X}$ an equivariant system of admissible affine subspaces attached to $\mathcal{C}_{i_o}$. Suppose that $\mathcal{C} \succeq \mathcal{C}_{i_o}$ -- then the set of all $(P,S;A_i)$ in $\mathcal{C}$ for which there exists a $(P_{i_o},S_{i_o};A_{i_o})$ in $\mathcal{C}_{i_o}$ such that

$$(P,S;A_i) \succeq (P_{i_o},S_{i_o};A_{i_o})$$

is itself a union of G-conjugacy classes in $\mathcal{C}$, call it $\mathcal{C}(i_o)$. Let $\mathcal{C}_k$ be a G-conjugacy class in $\mathcal{C}$ which is contained in $\mathcal{C}(i_o)$. One may then define an equivariant system $X_{\mathcal{C}_k}$ of linear subspaces attached to $\mathcal{C}_{i_o}$ as follows. Let $(P_{i_o},S_{i_o};A_{i_o})$ be a member of $\mathcal{C}_{i_o}$ -- then there exists a unique element $(P,S;A_i)$ in $\mathcal{C}_k$ such that

$$(P,S;A_i) \succeq (P_{i_o},S_{i_o};A_{i_o}).$$

We write $\mathfrak{X}(P_{i_o},A_{i_o})^\dagger$ for the orthogonal projection of $\mathfrak{X}(P_{i_o},A_{i_o})$ onto the complexification of $\check{\mathfrak{a}}_{i_o}^\dagger$. This said, let

$$X_{\mathcal{C}_k}(P_{i_o},A_{i_o}) = \check{\mathfrak{a}}_i.$$

It is easy to check that $X_{\mathcal{C}_k}$ does in fact have the required properties. Write

$$\mathcal{C}(i_o;\mathfrak{X})$$

for those G-conjugacy classes $\mathcal{C}_k$ in $\mathcal{C}(i_o)$ for which $X_{\mathcal{C}_k}$ is $\mathfrak{X}$-admissible

Of course, $\mathcal{C}(i_o;\mathfrak{X})$ could be empty. Let $\mathcal{C}_{j_o}$ be another G-conjugacy class in $\mathcal{C}_o$. Suppose that there exists $\mathcal{C}_k$ in $\mathcal{C}(i_o;\mathfrak{X})$ such that $\mathcal{C}_k \succeq \mathcal{C}_{j_o}$ -- then we can form $\underset{\sim}{W}^{\dagger}_{\mathcal{C}_k}(\mathcal{C}_{j_o},\mathcal{C}_{i_o})$ which, as we recall, is a subset of $\underset{\sim}{W}(\mathcal{C}_{j_o},\mathcal{C}_{i_o})$. Each $\underset{\sim}{w}^{\dagger}_{j_o i_o}$ in $\underset{\sim}{W}^{\dagger}_{\mathcal{C}_k}(\mathcal{C}_{j_o},\mathcal{C}_{i_o})$ thus takes $\mathfrak{X}$ to another equivariant system $\mathfrak{X}_{\underset{\sim}{w}^{\dagger}_{j_o i_o}}$. If we agree that two elements of $\underset{\sim}{W}^{\dagger}_{\mathcal{C}_k}(\mathcal{C}_{j_o},\mathcal{C}_{i_o})$ are to be regarded as equivalent when their action on $\mathfrak{X}$ coincides, then we obtain a set of equivalence classes

$$\underset{\sim}{W}^{\dagger}_{\mathcal{C}_k}(\mathfrak{X};\mathcal{C}_{j_o},\mathcal{C}_{i_o}).$$

Let

$$\begin{cases}(P_{i_o},S_{i_o};A_{i_o}) \in \mathcal{C}_{i_o}\\(P_{j_o},S_{j_o};A_{j_o}) \in \mathcal{C}_{j_o}.\end{cases}$$

Denote by

$$W^{\dagger}_{\mathcal{C}_k}(\mathfrak{X};A_{j_o},A_{i_o})$$

the set of distinct linear transformations from $\mathfrak{X}(P_{i_o},A_{i_o})$ into the complexification of $\mathfrak{a}_{j_o}$ obtained by restricting the elements of $W^{\dagger}_{\mathcal{C}_k}(A_{j_o},A_{i_o})$ to $\mathfrak{X}(P_{i_o},A_{i_o})$ -- then there is a canonical bijection

$$\underset{\sim}{W}^{\dagger}_{\mathcal{C}_k}(\mathfrak{X};\mathcal{C}_{j_o},\mathcal{C}_{i_o}) \to W^{\dagger}_{\mathcal{C}_k}(\mathfrak{X};A_{j_o},A_{i_o}).$$

We write

$$\mathfrak{X}(P_{j_o},A_{j_o})_{w^\dagger_{j_o i_o}} = \overline{-w^\dagger_{j_o i_o}\mathfrak{X}(P_{i_o},A_{i_o})} \quad (w^\dagger_{j_o i_o} \in W^\dagger_{\mathcal{C}_k}(\mathfrak{X};A_{j_o},A_{i_o})).$$

REMARK Suppose that

$$(P,S:A_i) \succeq (P_{i_o},S_{i_o};A_{i_o}).$$

The daggering procedure then determines a split parabolic subgroup $(P^\dagger_{i_o},S^\dagger_{i_o})$ of M_i with split component $A^\dagger_{i_o}$ which, of course, is Γ_{M_i}-cuspidal. There are orthogonal decompositions

$$\begin{cases} \mathfrak{a}_{i_o} = \mathfrak{a}^\dagger_{i_o} \oplus \mathfrak{a}_i \\ \check{\mathfrak{a}}_{i_o} = \check{\mathfrak{a}}^\dagger_{i_o} \oplus \check{\mathfrak{a}}_i \end{cases}.$$

$\mathfrak{X}(P_{i_o},A_{i_o})^\dagger$ is an admissible affine subspace of the complexification of $\check{\mathfrak{a}}^\dagger_{i_o}$ with normal translation $X(P_{i_o},A_{i_o})$. Conventionally, an admissible hyperplane in $\mathfrak{X}(P_{i_o},A_{i_o})^\dagger$ is an intersection of the form

$$\mathfrak{X}(P_{i_o},A_{i_o})^\dagger \cap (\check{\lambda}^\dagger = c) \quad (\check{\lambda}^\dagger \in \Sigma_{P^\dagger_{i_o}}(\mathfrak{m}_i,\check{\mathfrak{a}}^\dagger_{i_o})).$$

Consider, then, those association classes $\mathcal{C}$ such that

(i) $\mathcal{C} \succeq \mathcal{C}_{i_o}$;

(ii) $\mathcal{C}(i_o;\mathfrak{X}) \neq \emptyset$.

Fix a K-type δ and an orbit type $\underset{\sim}{\mathcal{O}}_o$. The assignment to all

$$\begin{cases} (P,S;A_i) \in \mathcal{C}(i_o;\mathfrak{X}) \\ (P_{i_o},S_{i_o};A_{i_o}) \in \mathcal{C}_{i_o} \end{cases}$$

standing in the relation

$$(P,S;A_i) \succeq (P_{i_o},S_{i_o};A_{i_o})$$

of a complex valued function

$$E(\mathfrak{X}:P|A_i:P_{i_o}|A_{i_o}:T_{i_o}:\Lambda_{i_o}^{\dagger}:x)$$

on

$$\mathrm{Hom}(S_{\mathfrak{X}(P_{i_o},A_{i_o})},\mathcal{E}_{cus}(\delta,\mathcal{O}_{i_o})) \times \mathfrak{X}(P_{i_o},A_{i_o})^{\dagger} \times G$$

such that

$$\begin{cases} E(\mathfrak{X}:P|A_i:P_{i_o}|A_{i_o}:c_1T_{i_o}^1 + c_2T_{i_o}^2:\Lambda_{i_o}^{\dagger}:x) \\ = c_1E(\mathfrak{X}:P|A_i:P_{i_o}|A_{i_o}:T_{i_o}^1:\Lambda_{i_o}^{\dagger}:x) + c_2E(\mathfrak{X}:P|A_i:P_{i_o}|A_{i_o}:T_{i_o}^2:\Lambda_{i_o}^{\dagger}:x) \\ E(\mathfrak{X}:P|A_i:P_{i_o}|A_{i_o}:T_{i_o}:\Lambda_{i_o}^{\dagger}:x) = 0 \quad \text{if ord } (T_{i_o})>>0(\text{uniformly in } (\Lambda_{i_o}^{\dagger},x)) \\ E(\mathfrak{X}:P|A_i:P_{i_o}|A_{i_o}:T_{i_o}:\Lambda_{i_o}^{\dagger}:x\gamma a_i n) \\ = E(\mathfrak{X}:P|A_i:P_{i_o}|A_{i_o}:T_{i_o}:\Lambda_{i_o}^{\dagger}:x) \qquad (\gamma a_i n \in (\Gamma\cap P)\cdot A_i\cdot N) \end{cases}$$

PLUS the assignment to all

$$\begin{cases} (P,S;A_i) \in \mathcal{C}(i_o;\mathfrak{X}) \\ (P_{i_o},S_{i_o};A_{i_o}) \in \mathcal{C}_{i_o} \end{cases} \qquad \begin{cases} (P,S;A_j) \in \mathcal{C}(i_o;\mathfrak{X}) \\ (P_{j_o},S_{j_o};A_{j_o}) \in \mathcal{C}_{j_o} \end{cases}$$

standing in the relation

$$\begin{cases} (P,S;A_i) \succeq (P_{i_o},S_{i_o};A_{i_o}) \\ (P,S;A_j) \succeq (P_{j_o},S_{j_o};A_{j_o}) \end{cases}$$

of a function

$$\nabla(\mathfrak{X}:P|(A_j,A_i):P_{j_o}|A_{j_o}:P_{i_o}|A_{i_o}:w^{\dagger}_{j_o i_o}:\Lambda^{\dagger}_{i_o})$$

on

$$W^{\dagger}_{\mathcal{C}_k}(\mathfrak{X};A_{j_o},A_{i_o}) \times \mathfrak{X}(P_{i_o},A_{i_o})^{\dagger}$$

which is a linear transformation from

$$\mathrm{Hom}(S_{\mathfrak{X}(P_{i_o},A_{i_o})},E_{cus}(\delta,\mathcal{O}_{i_o}))$$

to

$$S_{\mathfrak{X}(P_{j_o},A_{j_o})_{w^{\dagger}_{j_o i_o}}} \otimes E_{cus}(\delta,\mathcal{O}_{j_o})$$

such that

$$(\nabla(\mathfrak{X}:P|(A_j,A_i):P_{j_o}|A_{j_o}:P_{i_o}|A_{i_o}:w^{\dagger}_{j_o i_o}:\Lambda^{\dagger}_{i_o})T_{i_o},T_{j_o}) = 0$$

if

$$\begin{cases} \mathrm{ord}(T_{i_o}) >> 0 \\ \mathrm{ord}(T_{j_o}) >> 0 \end{cases}$$

is said to be an Eisenstein system belonging to $\mathfrak{X}$ provided that certain assumptions are met in a non-trivial way.

These assumptions are listed in TES. They involve conditions of meromorphicity, equivariance, compatibility, transitivity, and negligibility, as well as a representation theoretic condition. There is little to be gained by reproducing all of them in extenso here. We shall, accordingly, recall only the two which will be needed in what follows, namely the hypotheses of meromorphicity and negligibility.

The supposition of meromorphicity is simply this. Working first with the E-functions, suppose that

$$(P,S;A_i) \succeq (P_{i_0},S_{i_0};A_{i_0}).$$

We then assume that

$$E(\mathfrak{X}:P|A_i:P_{i_0}|A_{i_0}:T_{i_0}:\Lambda_{i_0}^{\dagger}:x)$$

is a meromorphic function of $\Lambda_{i_0}^{\dagger}$ in $\mathfrak{X}(P_{i_0},A_{i_0})^{\dagger}$, whose singularities lie along admissible hyperplanes. Write

$$D(\mathfrak{X}:E:P|A_i:P_{i_0}|A_{i_0})$$

for the set of all $\Lambda_{i_0}^{\dagger}$ in $\mathfrak{X}(P_{i_0},A_{i_0})^{\dagger}$ which admit a neighborhood $\mathcal{N}_{i_0}^{\dagger}$ with the property that

$$E(\mathfrak{X}:P|A_i:P_{i_0}|A_{i_0}:T_{i_0}:?:x)$$

is holomorphic on $\mathcal{N}_{i_0}^{\dagger}$ for all T_{i_0} and all x. We then assume that

$$E(\mathfrak{X}:P|A_i:P_{i_o}|A_{i_o}:T_{i_o}:\Lambda_{i_o}^{\dagger}:x)$$

is a differentiable function of $(\Lambda_{i_o}^{\dagger},x)$ on

$$D(\mathfrak{X}:E:P|A_i:P_{i_o}|A_{i_o}) \times G.$$

Finally, if $\mathfrak{S}_{M_i}$ be a Siegel domain associated with a Γ_{M_i}-percuspidal parabolic subgroup P_{M_i} of M_i, it will then be required that

$$|E(\mathfrak{X}:P|A_i:P_{i_o}|A_{i_o}:T_{i_o}:\Lambda_{i_o}^{\dagger}:km_i)| \le C\cdot\|T_{i_o}\|\cdot\Xi_{P_{M_i}}(m_i)^r \quad ((k,m_i)\in K\times\mathfrak{S}_{M_i}),$$

C and r being uniform on compacta in $D(\mathfrak{X}:E:P|A_i:P_{i_o}|A_{i_o})$. As for the ∇-functions, suppose that

$$\begin{cases}(P,S;A_i)\succeq(P_{i_o},S_{i_o};A_{i_o})\\(P,S;A_j)\succeq(P_{j_o},S_{j_o};A_{j_o}).\end{cases}$$

We then assume that

$$(\nabla(\mathfrak{X}:P|(A_j,A_i):P_{j_o}|A_{j_o}:P_{i_o}|A_{i_o}:w_{j_o i_o}^{\dagger}:\Lambda_{i_o}^{\dagger})T_{i_o},T_{j_o})$$

is a meromorphic function of $\Lambda_{i_o}^{\dagger}$ in $\mathfrak{X}(P_{i_o},A_{i_o})^{\dagger}$, whose singularities lie along admissible hyperplanes. Write

$$D(\mathfrak{X}:\nabla:P|A_i:P_{i_o}|A_{i_o})$$

for the set of all $\Lambda_{i_o}^{\dagger}$ in $\mathfrak{X}(P_{i_o},A_{i_o})^{\dagger}$ which admit a neighborhood N_{i_o} with the property that

$$(\nabla(\mathfrak{X}:P|(A_j,A_i):P_{j_o}|A_{j_o}:P_{i_o}|A_{i_o}:w^{\dagger}_{j_o i_o}:?)T_{i_o},T_{j_o})$$

is holomorphic on $N^{\dagger}_{i_o}$ for all $w^{\dagger}_{j_o i_o}$ and all T_{i_o},T_{j_o}. While no a priori assumptions are made vis-a-vis this set, certain a posteriori conclusions can be drawn; cf. infra.

Turning to the supposition of negligibility, suppose that

$$\begin{cases} (P,S;A_i) \succeq (P_{i_o},S_{i_o};A_{i_o}) \\ (P,S;A_j) \succeq (P_{j_o},S_{j_o};A_{j_o}). \end{cases}$$

We then assume that

$$E_{K\times P^{\dagger}_{j_o}}(\mathfrak{X}:P|A_i:P_{i_o}|A_{i_o}:T_{i_o}:\Lambda^{\dagger}_{i_o}:km_j)$$

$$= \sum_{w^{\dagger}_{j_o i_o}\in W^{\dagger}_{\mathcal{C}_k}(\mathfrak{X};A_{j_o},A_{i_o})}$$

$$\times\ a^{\dagger}_{j_o}(m_j)^{w^{\dagger}_{j_o i_o}\Lambda^{\dagger}_{i_o}}\cdot(\nabla(\mathfrak{X}:P|(A_j,A_i):P_{j_o}|A_{j_o}:P_{i_o}|A_{i_o}:w^{\dagger}_{j_o i_o}:\Lambda^{\dagger}_{i_o})T_{i_o})(km_j).$$

On the other hand, suppose that $(P,S;A)$ is a dominant successor of some other Γ-cuspidal split parabolic subgroup of G which is not in $\mathcal{C}_o$, say

$$(P,S;A) \succeq (P',S';A').$$

We then assume that

$$E_{K\times'P}(\mathfrak{X}:P|A:P_{i_o}|A_{i_o}:T_{i_o}:\Lambda^{\dagger}_{i_o}:km) \sim 0.$$

[An empty sum will, by convention, be set equal to zero. In this connection, observe that $(P_{i_0}^\dagger, S_{i_0}^\dagger)$ and $(P_{j_0}^\dagger, S_{j_0}^\dagger)$ could very well lie in distinct association classes when projected into $P/A_iN = P/A_jN$; but this happens iff $W_{\mathcal{C}_k}^\dagger(\mathfrak{X}; A_{j_0}, A_{i_0})$ is empty.]

REMARK To say that an Eisenstein system belonging to $\mathfrak{X}$ is non-trivial means exactly this: There are triples

$$\begin{cases} (P, S; A_i) \\ (P_{i_0}, S_{i_0}; A_{i_0}) \end{cases}$$

standing in the relation

$$(P, S; A_i) \succeq (P_{i_0}, S_{i_0}; A_{i_0})$$

for which

$$E(\mathfrak{X} : P|A_i : P_{i_0}|A_{i_0} : ? : ? : ?) \neq 0 .$$

It is necessary to make such an assumption in order to draw certain geometric and group theoretic conclusions (see TES).

The assumptions which define an Eisenstein system are, of course, abstracted from those encountered in the study of Eisenstein series associated with cusp forms. The ambient group is $K \times M_i$ (or M_i) rather than just G itself. This degree of generality is needed for induction arguments. It should be noted that the E-functions, viewed on $K \times M_i$ (or M_i), are automorphic forms per $\{1\} \times \Gamma_{M_i}$ (or Γ_{M_i}).

EXAMPLE (THE CANONICAL EISENSTEIN SYSTEM) Let $\mathfrak{X}$ be the equivariant system of admissible affine subspaces attached to $\mathcal{C}_{i_0}$ via the prescription

$$\mathfrak{X}(P_{i_0}, A_{i_0}) = \text{complexification of } \check{\mathfrak{a}}_{i_0}.$$

In this case, then, $\mathcal{C}(i_0;\mathfrak{X}) = \mathcal{C}(i_0)$. Moreover

$$S_{\mathfrak{X}(P_{i_0}, A_{i_0})} = \text{scalars},$$

so the map $T_{i_0} \mapsto T_{i_0}(1)$ sets up an isomorphism

$$\mathrm{Hom}(S_{\mathfrak{X}(P_{i_0}, A_{i_0})}, \mathcal{E}_{cus}(\delta, \mathcal{O}_{i_0})) \rightarrow \mathcal{E}_{cus}(\delta, \mathcal{O}_{i_0}).$$

This said, suppose that

$$(P, S; A_i) \succeq (P_{i_0}, S_{i_0}; A_{i_0}).$$

Put

$$E(\mathfrak{X}:P|A_i:P_{i_0}|A_{i_0}:T_{i_0}:\Lambda_{i_0}^{\dagger}:x)$$

$$= \sum_{\gamma \in \Gamma \cap P/\Gamma \cap P_{i_0}} a_{i_0}(x\gamma)^{(\Lambda_{i_0}^{\dagger} - \rho_{i_0}^{\dagger})} \cdot T_{i_0}(1)(x\gamma).$$

Then

$$E(\mathfrak{X}:P|A_i:P_{i_0}|A_{i_0}:T_{i_0}:\Lambda_{i_0}^{\dagger}:km_i)$$

$$= \sum_{\gamma \in \Gamma_{M_i}/\Gamma_{M_i} \cap P_{i_0}} a_{i_0}^{\dagger}(m_i)^{(\Lambda_{i_0}^{\dagger} - \rho_{i_0}^{\dagger})} \cdot T_{i_0}(1)(km_i\gamma),$$

the Eisenstein series

$$E(K \times P_{i_0}^{\dagger} \,|\, \{1\} \times A_{i_0}^{\dagger} : T_{i_0}(1) : \Lambda_{i_0}^{\dagger} : (k, m_i))$$

on $K \times M_i$ attached to $T_{i_0}(1)$. Suppose now that

$$\begin{cases} (P,S;A_i) \succeq (P_{i_0}, S_{i_0}; A_{i_0}) \\ (P,S;A_j) \succeq (P_{j_0}, S_{j_0}; A_{j_0}). \end{cases}$$

Put

$$\nabla(\mathfrak{X}:P\,|\,(A_j,A_i):P_{j_0}|A_{j_0}:P_{i_0}|A_{i_0}:w_{j_0 i_0}^{\dagger}:\Lambda_{i_0}^{\dagger})T_{i_0}$$

$$= 1 \otimes \Phi,$$

where

$$\Phi(x) = a_{j_0}(x)^{(\rho_{j_0}^{\dagger} - w_{j_0 i_0}^{\dagger}\Lambda_{i_0}^{\dagger})} \cdot \sum_{\gamma \in \Gamma \cap N_{j_0} \backslash \Gamma(w_{j_0 i_0}^{\dagger})/\Gamma \cap P_{i_0}}$$

$$\times \int_{N_{j_0}/N_{j_0} \cap \gamma N_{i_0} \gamma^{-1}} a_{i_0}(xn_{j_0}\gamma)^{(\Lambda_{i_0}^{\dagger} - \rho_{i_0}^{\dagger})} \cdot T_{i_0}(1)(xn_{j_0}\gamma)\, d_{N_{j_0}}(n_{j_0}).$$

For simplicity, let us suppose that $A_i = A_j$ (it is not difficult to reduce the general case to this one) -- then we have

$$\Phi(km_i) = a_{j_0}^{\dagger}(m_i)^{(\rho_{j_0}^{\dagger} - w_{j_0 i_0}^{\dagger}\Lambda_{i_0}^{\dagger})} \cdot \sum_{\gamma \in \Gamma_{M_i} \cap N_{j_0}^{\dagger} \backslash \Gamma_{M_i}(w_{j_0 i_0}^{\dagger})/\Gamma_{M_i} \cap P_{i_0}^{\dagger}}$$

$$\times \int_{N_{j_0}^{\dagger}/N_{j_0}^{\dagger} \cap \gamma N_{i_0}^{\dagger} \gamma^{-1}} a_{i_0}^{\dagger}(m_i n_{j_0}^{\dagger}\gamma)^{(\Lambda_{i_0}^{\dagger} - \rho_{i_0}^{\dagger})} \cdot T_{i_0}(1)(km_i n_{j_0}^{\dagger}\gamma)\, d_{N_{j_0}^{\dagger}}(n_{j_0}^{\dagger}),$$

or still

$$\Phi(km_i) = (c_{cus}(P^{\dagger}_{j_o}|A^{\dagger}_{j_o}:P^{\dagger}_{i_o}|A^{\dagger}_{i_o}:w^{\dagger}_{j_o i_o}:\Lambda^{\dagger}_{i_o})T_{i_o}(1))(km_i).$$

That this data does in fact constitute an Eisenstein system is a consequence of the theory of Eisenstein series associated with cusp forms.

On the basis of the assumptions alone, it is possible to relate the singularities of the E-functions with those of the ∇-functions. In this connection, recall that we have introduced sets

$$\begin{cases} D(\mathfrak{X}:E:P|A_i:P_{i_o}|A_{i_o}) \\ D(\mathfrak{X}:\nabla:P|A_i:P_{i_o}|A_{i_o}). \end{cases}$$

Our assumptions imply that both of these sets are open and dense in $\mathfrak{X}(P_{i_o},A_{i_o})^{\dagger}$ and that the complement of either one is a locally finite set of hyperplanes, which will be referred to as the singular hyperplanes of E or ∇, as the case may be.

PROPOSITION E <u>Let</u> $\{E,\nabla\}$ <u>be an Eisenstein system belonging to</u> $\mathfrak{X}$ -- <u>then the singularities of</u> E <u>are contained in those of</u> ∇. <u>In precise terms, suppose that</u>

$$(P,S;A_i) \succeq (P_{i_o},S_{i_o};A_{i_o}) .$$

Let H_s be a singular hyperplane of

$$E(\mathfrak{X}:P|A_i:P_{i_o}|A_{i_o}:T_{i_o}:\Lambda_{i_o}^{\dagger}:x).$$

Then there exists a triple $(P_{j_o},S_{j_o};A_{j_o})$, with

$$(P,S;A_j) \succeq (P_{j_o},S_{j_o};A_{j_o}),$$

and an element $w_{j_o i_o}^{\dagger} \in W_{\mathcal{C}_k}^{\dagger}(\mathfrak{X};A_{j_o},A_{i_o})$ such that H_s is a singular hyperplane of

$$\nabla(\mathfrak{X}:P|(A_j,A_i):P_{j_o}|A_{j_o}:P_{i_o}|A_{i_o}:w_{j_o i_o}^{\dagger}:\Lambda_{i_o}^{\dagger}).$$

COROLLARY Let the notation and hypotheses be as above. Then

$$D(\mathfrak{X}:\nabla:P|A_i:P_{i_o}|A_{i_o}) \subset D(\mathfrak{X}:E:P|A_i:P_{i_o}|A_{i_o}).$$

By a singular hyperplane of $W_{\mathcal{C}_k}^{\dagger}(\mathfrak{X};A_{i_o},A_{i_o})$ in $\mathfrak{X}(P_{i_o},A_{i_o})^{\dagger}$, we mean a set of the form

$$H_w = \{\Lambda_{i_o}^{\dagger}:w_{i_o i_o}^{\dagger}\Lambda_{i_o}^{\dagger} = \Lambda_{i_o}^{\dagger}\ (w_{i_o i_o}^{\dagger} \in W_{\mathcal{C}_k}^{\dagger}(\mathfrak{X};A_{i_o},A_{i_o}),w_{i_o i_o}^{\dagger} \neq 1)\}.$$

Let

$$D(\mathfrak{X}:W:P|A_i:P_{i_o}|A_{i_o})$$

be the complement of the union of the singular hyperplanes of

$W_{\mathcal{C}_k}^{\dagger}(\mathfrak{X};A_{i_o},A_{i_o})$ in $\mathfrak{X}(P_{i_o},A_{i_o})^{\dagger}$.

PROPOSITION ∇ *Let* $\{E,\nabla\}$ *be an Eisenstein system belonging to* $\mathfrak{X}$ *-- then the singularities of* ∇ *are contained in those of* E *and* W. *In precise terms, suppose that*

$$\begin{cases} (P,S;A_i) \succeq (P_{i_o},S_{i_o};A_{i_o}) \\ (P,S;A_j) \succeq (P_{j_o},S_{j_o};A_{j_o}). \end{cases}$$

Let H_s *be a singular hyperplane of*

$$\nabla(\mathfrak{X}:P|(A_j,A_i):P_{j_o}|A_{j_o}:P_{i_o}|A_{i_o}:w^{\dagger}_{j_o i_o}:\Lambda^{\dagger}_{i_o})T_{i_o}.$$

Then H_s *is either a singular hyperplane of*

$$E(\mathfrak{X}:P|A_i:P_{i_o}|A_{i_o}:T_{i_o}:\Lambda^{\dagger}_{i_o}:x)$$

or else is a singular hyperplane of $W^{\dagger}_{\mathcal{C}_k}(\mathfrak{X};A_{i_o},A_{i_o})$.

COROLLARY *Let the notation and hypotheses be as above. Then*

$$D(\mathfrak{X}:E:P|A_i:P_{i_o}|A_{i_o}) \cap D(\mathfrak{X}:W:P|A_i:P_{i_o}|A_{i_o})$$

is contained in

$$D(\mathfrak{X}:\nabla:P|A_i:P_{i_o}|A_{i_o}).$$

The equivariant systems of admissible affine subspaces and their associated Eisenstein systems, which actually figure in the theory, satisfy certain ancillary conditions. As they are of definite importance, we had best recall them.

Keeping to the earlier assumptions and notations, write

$$\mathcal{C}_o = \coprod_{i_o} \mathcal{C}_{i_o} .$$

Suppose there is attached to each $\mathcal{C}_{i_o}$ a collection

$$\underset{\sim}{\mathfrak{X}}_{i_o} = \{\mathfrak{X}\}$$

of equivariant systems of admissible affine subspaces of dimension $\mathfrak{d}$. One sets

$$\underset{\sim}{\mathfrak{X}}(\mathcal{C}_o) = \coprod_{i_o} \underset{\sim}{\mathfrak{X}}_{i_o} .$$

Let now $\mathcal{C}$ be another association class. If, as usual, $\mathcal{C}_k$ is a G-conjugacy class in $\mathcal{C}$, then we put

$$\underset{\sim}{\mathfrak{X}}_{i_o}(\mathcal{C}_k) = \{\mathfrak{X} : \mathcal{C}_k \subset \mathcal{C}(i_o;\mathfrak{X})\}$$

and form

$$\underset{\sim}{\mathfrak{X}}(\mathcal{C}_k;\mathcal{C}_o) = \coprod_{i_o} \underset{\sim}{\mathfrak{X}}_{i_o}(\mathcal{C}_k),$$

a subset of $\underset{\sim}{\mathfrak{X}}(\mathcal{C}_o)$ which could, of course, be empty. It will be convenient to introduce an equivalence relation in $\underset{\sim}{\mathfrak{X}}(\mathcal{C}_k;\mathcal{C}_o)$. For this purpose, suppose that

$$\mathcal{C}_k \subset \mathcal{C}(i_o;\mathfrak{X}) \cap \mathcal{C}(j_o;\mathfrak{Y}) \quad \text{with} \begin{cases} \mathfrak{X} & \text{per } \mathcal{C}_{i_o} \\ \mathfrak{Y} & \text{per } \mathcal{C}_{j_o} \end{cases}.$$

Let

$$\underset{\sim}{W}^{\dagger}_{\mathcal{C}_k}(\mathfrak{Y},\mathfrak{X};\mathcal{C}_{j_o},\mathcal{C}_{i_o})$$

be the subset of $\underset{\sim}{W}^{\dagger}_{\mathcal{C}_k}(\mathfrak{X};\mathcal{C}_{j_o},\mathcal{C}_{i_o})$ consisting of those transformations $\underset{\sim}{w}^{\dagger}_{j_o i_o}$ such that $\mathfrak{X}_{\underset{\sim}{w}^{\dagger}_{j_o i_o}} = \mathfrak{Y}$. In terms of specific data, if

$$\begin{cases} (P_{i_o},S_{i_o};A_{i_o}) \in \mathcal{C}_{i_o} \\ (P_{j_o},S_{j_o};A_{j_o}) \in \mathcal{C}_{j_o} \end{cases}$$

and if

$$W^{\dagger}_{\mathcal{C}_k}(\mathfrak{Y},\mathfrak{X};A_{j_o},A_{i_o})$$

is the subset of $W^{\dagger}_{\mathcal{C}_k}(\mathfrak{X};A_{j_o},A_{i_o})$ consisting of those transformations $w^{\dagger}_{j_o i_o}$ such that

$$\mathfrak{X}(P_{j_o},A_{j_o})_{w^{\dagger}_{j_o i_o}} = \mathfrak{Y}(P_{j_o},A_{j_o}),$$

then there is a canonical bijection

$$\underset{\sim}{W}^{\dagger}_{\mathcal{C}_k}(\mathfrak{Y},\mathfrak{X};\mathcal{C}_{j_o},\mathcal{C}_{i_o}) \to W^{\dagger}_{\mathcal{C}_k}(\mathfrak{Y},\mathfrak{X};A_{j_o},A_{i_o}).$$

This said, elements

$$\begin{cases} \mathfrak{X} \ \text{per} \ \mathcal{C}_{i_0} \\ \mathfrak{Y} \ \text{per} \ \mathcal{C}_{j_0} \end{cases}$$

in $\underset{\sim}{\mathfrak{X}}(\mathcal{C}_k;\mathcal{C}_0)$ are declared equivalent iff

$$\underset{\sim}{W}^{\dagger}_{\mathcal{C}_k}(\mathfrak{Y},\mathfrak{X};\mathcal{C}_{j_0},\mathcal{C}_{i_0})$$

is non-empty.

Let $\mathfrak{X}$ be an equivariant system of admissible affine subspaces attached to $\mathcal{C}_{i_0}$ -- then by $\mathrm{Dis}(\mathfrak{X})$ we understand the $\mathfrak{X}$-admissible subspace of $\mathfrak{X}$ defined as follows. Let $(P_{i_0},S_{i_0};A_{i_0})$ be a member of $\mathcal{C}_{i_0}$ -- then

$$\mathrm{Dis}(\mathfrak{X})(P_{i_0},A_{i_0})$$

is the distinguished subspace of $\check{\mathfrak{a}}_{i_0}$ of maximal dimension which is contained in $\mathfrak{X}(P_{i_0},A_{i_0})^{\sim}$. In the event that the complexification of

$$\mathrm{Dis}(\mathfrak{X})(P_{i_0},A_{i_0})$$

is actually equal to $\mathfrak{X}(P_{i_0},A_{i_0})^{\sim}$, then $\mathfrak{X}$ is termed principal.

There corresponds to $\mathrm{Dis}(\mathfrak{X})$, in a natural way, an association class $\mathcal{C}(\mathrm{Dis}(\mathfrak{X}))$, containing a G-conjugacy class $\mathcal{C}_k(\mathrm{Dis}(\mathfrak{X}))$ such that

$$\mathcal{C}_k(\mathrm{Dis}(\mathfrak{X})) \succeq \mathcal{C}_{i_0}.$$

One can write

$$\check{\mathfrak{a}}_{i_0} = \mathrm{Dis}(\mathfrak{X})(P_{i_0},A_{i_0})^{\dagger} \oplus \mathrm{Dis}(\mathfrak{X})(P_{i_0},A_{i_0}).$$

Let

$$(P_{Dis(\mathfrak{X})}, S_{Dis(\mathfrak{X})}; A^{i}_{Dis(\mathfrak{X})})$$

denote the parabolic data determined by $Dis(\mathfrak{X})(P_{i_o}, A_{i_o})$. We then have

$$(P_{Dis(\mathfrak{X})}, S_{Dis(\mathfrak{X})}; A^{i}_{Dis(\mathfrak{X})}) \succeq (P_{i_o}, S_{i_o}; A_{i_o}),$$

so the daggering procedure determines parabolic data

$$(P^{i_o}_{Dis(\mathfrak{X})}, S^{i_o}_{Dis(\mathfrak{X})}; A^{i_o}_{Dis(\mathfrak{X})})$$

in $M^{i}_{Dis(\mathfrak{X})}$.

What follows is a list of possible geometric conditions which could be imposed on $\underset{\sim}{\mathfrak{X}}_{i_o}$.

GEOM: I (LOCAL FINITENESS) $\underset{\sim}{\mathfrak{X}}_{i_o}$ is locally finite in the sense that for all $(P_{i_o}, S_{i_o}; A_{i_o})$ in $\mathcal{C}_{i_o}$ and every compact subset ω_{i_o} of the complexification of $\check{\mathfrak{a}}_{i_o}$, the set

$$\{\mathfrak{X} \in \underset{\sim}{\mathfrak{X}}_{i_o} : \mathfrak{X}(P_{i_o}, A_{i_o}) \cap \omega_{i_o} \neq \emptyset\}$$

is finite.

GEOM: II (REAL COMPACTNESS) $\underset{\sim}{\mathfrak{X}}_{i_o}$ is real compact in the sense that for all $(P_{i_o}, S_{i_o}; A_{i_o})$ in $\mathcal{C}_{i_o}$, the set

$$\{\mathrm{Re}(X(P_{i_o},A_{i_o})):\mathfrak{X}\in\underset{\sim}{\mathfrak{X}}_{i_o}\}$$

has compact closure.

GEOM: III (CONICAL CONTAINMENT) $\underset{\sim}{\mathfrak{X}}_{i_o}$ is conically contained in the sense that for all $(P_{i_o},S_{i_o};A_{i_o})$ in $\mathcal{C}_{i_o}$,

$$-\mathrm{Re}(X(P_{i_o},A_{i_o}))\in\mathcal{D}_{i_o}{}_{P_{\mathrm{Dis}(\mathfrak{X})}}(\mathfrak{a}^{\vee i_o}_{\mathrm{Dis}(\mathfrak{X})})$$

for every $\mathfrak{X}$ belonging to $\underset{\sim}{\mathfrak{X}}_{i_o}$.

There is also a group theoretic requirement which might be invoked per $\underset{\sim}{\mathfrak{X}}_{i_o}$.

GR For every $\mathfrak{X}$ belonging to $\underset{\sim}{\mathfrak{X}}_{i_o}$, there exists an element

$$\underset{\sim}{w}^{o}_{i_o}\in\underset{\sim}{W}(\mathfrak{X},\mathfrak{X};\mathcal{C}_{i_o},\mathcal{C}_{i_o})$$

such that for all $(P_{i_o},S_{i_o};A_{i_o})$ in $\mathcal{C}_{i_o}$,

$$\underset{\sim}{w}^{o}_{i_o}(P_{i_o}|A_{i_o}:P_{i_o}|A_{i_o})|\mathfrak{X}(P_{i_o},A_{i_o})^{\sim}$$

is the identity map.

[Set, for brevity,

$$w^{o}_{i_o}=\underset{\sim}{w}^{o}_{i_o}(P_{i_o}|A_{i_o}:P_{i_o}|A_{i_o}).$$

Needless to say, $w^{o}_{i_o}\in W(A_{i_o},A_{i_o})$. In passing, note that

$$w_{i_o}^o(X(P_{i_o},A_{i_o})) = \overline{-X(P_{i_o},A_{i_o})},$$

so $w_{i_o}^o$, if it exists at all, is unique.]

So far we have said nothing about Eisenstein systems. Let us now suppose that for each i_o there is an Eisenstein system $\{E,\nabla\}$ belonging to each $\mathfrak{X}$ in $\underset{\sim}{\mathfrak{X}}_{i_o}$. We consider three conditions.

E-S:I Let $\mathfrak{X}\in\underset{\sim}{\mathfrak{X}}_{i_o}(\mathcal{C}_k)$. Suppose that

$$(P,S;A_i) \succeq \begin{cases} (P_{i_o},S_{i_o};A_{i_o}) \\ (P_{j_o},S_{j_o};A_{j_o}) \end{cases}.$$

Let $w_{j_o i_o}^{\dagger}\in W_{\mathcal{C}_k}^{\dagger}(\mathfrak{X};A_{j_o},A_{i_o})$; let $\underset{\sim}{w}_{j_o i_o}^{\dagger}\in \underset{\sim}{W}_{\mathcal{C}_k}^{\dagger}(\mathfrak{X};\mathcal{C}_{j_o},\mathcal{C}_{i_o})$ correspond to $w_{j_o i_o}^{\dagger}$ under the canonical bijection

$$\underset{\sim}{W}_{\mathcal{C}_k}^{\dagger}(\mathfrak{X};\mathcal{C}_{j_o},\mathcal{C}_{i_o}) \rightarrow W_{\mathcal{C}_k}^{\dagger}(\mathfrak{X};A_{j_o},A_{i_o}).$$

Then:

$$\mathfrak{X}_{\underset{\sim}{w}_{j_o i_o}^{\dagger}} \notin \mathfrak{X}_{j_o}(\mathcal{C}_k)$$

$$\Longrightarrow$$

$$\nabla(\mathfrak{X}:P|(A_i,A_i):P_{j_o}|A_{j_o}:P_{i_o}|A_{i_o}:w_{j_o i_o}^{\dagger}:\Lambda_{i_o}^{\dagger}) = 0.$$

E-S:II Let $\mathfrak{X}\in\underset{\sim}{\mathfrak{X}}_{i_o}(\mathcal{C}_k)$. Suppose that

$$(P,S;A_i) \succeq \begin{cases} (P_{i_0},S_{i_0};A_{i_0}) \\ (P_{j_0},S_{j_0};A_{j_0}) . \end{cases}$$

Let $w^{\dagger}_{j_0 i_0} \in W^{\dagger}_{\mathcal{C}_k}(\mathfrak{X};A_{j_0},A_{i_0})$; let $\underset{\sim}{w}^{\dagger}_{j_0 i_0} \in \underset{\sim}{W}^{\dagger}_{\mathcal{C}_k}(\mathfrak{X};\mathcal{C}_{j_0},\mathcal{C}_{i_0})$ correspond to $w^{\dagger}_{j_0 i_0}$ under the canonical bijection

$$\underset{\sim}{W}^{\dagger}_{\mathcal{C}_k}(\mathfrak{X};\mathcal{C}_{j_0},\mathcal{C}_{i_0}) \to W^{\dagger}_{\mathcal{C}_k}(\mathfrak{X};A_{j_0},A_{i_0}).$$

Then: The adjoint

$$\nabla(\mathfrak{X}:P|(A_i,A_i):P_{j_0}|A_{j_0}:P_{i_0}|A_{i_0}:w^{\dagger}_{j_0 i_0}:\Lambda^{\dagger}_{i_0})^*$$

of

$$\nabla(\mathfrak{X}:P|(A_i,A_i):P_{j_0}|A_{j_0}:P_{i_0}|A_{i_0}:w^{\dagger}_{j_0 i_0}:\Lambda^{\dagger}_{i_0})$$

is given by

$$\nabla(\mathfrak{X}_{\underset{\sim}{w}^{\dagger}_{j_0 i_0}}:P|(A_i,A_i):P_{i_0}|A_{i_0}:P_{j_0}|A_{j_0}:(w^{\dagger}_{j_0 i_0})^{-1}:-w^{\dagger}_{j_0 i_0}\bar{\Lambda}^{\dagger}_{i_0}),$$

provided

$$\mathfrak{X}_{\underset{\sim}{w}^{\dagger}_{j_0 i_0}} \in \mathfrak{X}_{\underset{\sim}{j}_0}(\mathcal{C}_k).$$

E-S:III Let $\mathfrak{X} \in \mathfrak{X}_{\underset{\sim}{i}_0}(\mathcal{C}_k)$. Suppose that $\underset{\sim}{p}^{\dagger} = \{p^{\dagger}_{j_0}\}$ is a completely equivariant system of polynomials, i.e., to each j_0 there is assigned a polynomial $p^{\dagger}_{j_0} \in C[\check{\mathfrak{a}}^{\dagger}_{j_0}]$ such that for all triples

$$\begin{cases} (P_{j_o^1}, S_{j_o^1}; A_{j_o^1}) \\ (P_{j_o^2}, S_{j_o^2}; A_{j_o^2}) \end{cases}$$

standing in the relation

$$(P,S;A_i) \succeq \begin{cases} (P_{j_o^1}, S_{j_o^1}; A_{j_o^1}) \\ (P_{j_o^2}, S_{j_o^2}; A_{j_o^2}) \end{cases}$$

one has

$$p^{\dagger}_{j_o^2}(w^{\dagger}_{j_o^2 j_o^1}\Lambda^{\dagger}_{j_o^1}) = p^{\dagger}_{j_o^1}(\Lambda^{\dagger}_{j_o^1})$$

for all

$$w^{\dagger}_{j_o^2 j_o^1} \in W^{\dagger}_{\mathcal{C}_k}(A_{j_o^2}, A_{j_o^1}).$$

View $p^{\dagger}_{j_o}$ as a holomorphic function on the complexification of $\mathfrak{a}^{\dagger}_{j_o}$ with values in

$$\mathrm{Hom}(\mathcal{E}_{cus}(\delta,\mathcal{O}_{j_o}), \mathcal{E}_{cus}(\delta,\mathcal{O}_{j_o}))$$

by means of scalar multiplication. Suppose that

$$(P,S;A_i) \succeq \begin{cases} (P_{i_o}, S_{i_o}; A_{i_o}) \\ (P_{j_o}, S_{j_o}; A_{j_o}). \end{cases}$$

Then:

$$\nabla(\mathfrak{X}:P|(A_i,A_i):P_{j_o}|A_{j_o}:P_{i_o}|A_{i_o}:w^{\dagger}_{j_o i_o}:\Lambda^{\dagger}_{i_o})(d_{Hom}p^{\dagger}_{i_o}(\Lambda^{\dagger}_{i_o})(T_{i_o}))$$

$$= d_{\otimes}p^{\dagger}_{j_o}(w^{\dagger}_{j_o i_o}\Lambda^{\dagger}_{i_o})(\nabla(\mathfrak{X}:P|(A_i,A_i):P_{j_o}|A_{j_o}:P_{i_o}|A_{i_o}:w^{\dagger}_{j_o i_o}:\Lambda^{\dagger}_{i_o})T_{i_o}).$$

Ostensibly, E-S:I, II, III do not explicitly involve the E-functions. This, however, is illusory (cf. TES).

Some idea as to the significance of the conditions supra is contained in the following result.

PROPOSITION $\mathfrak{X}$ Suppose that for each i_o, $\underset{\sim}{\mathfrak{X}}_{i_o}$ is subject to

$$\begin{cases} \text{Geom: I} \\ \text{Geom: II,} \quad \text{Gr} \\ \text{Geom: III} \end{cases}$$

and

E-S: I, II, III.

Then $\underset{\sim}{\mathfrak{X}}(\mathcal{C}_o)$ is a finite set. Moreover, the normal translation in any member of $\underset{\sim}{\mathfrak{X}}(\mathcal{C}_o)$ is real. Finally, given $\mathcal{C}_k$, every equivalence class

$$\mathcal{E} \subset \underset{\sim}{\mathfrak{X}}(\mathcal{C}_k;\mathcal{C}_o)$$

contains at least one principal element.

Implicit in the proof of this proposition is a fact fundamental to the entire theory.

SCHOLIUM <u>Let</u> $\mathcal{E}$ <u>be an equivalence class in</u> $\underset{\sim}{\mathfrak{X}}(\mathcal{C}_k;\mathcal{C}_o)$; <u>let</u> $\mathfrak{X}$ <u>be a principal element of</u> $\mathcal{E}$, <u>say</u> $\mathfrak{X}\in\underset{\sim}{\mathfrak{X}}_{i_o}(\mathcal{C}_k)$ -- <u>then</u>

$$E(\mathfrak{X}:P_{\mathrm{Dis}(\mathfrak{X})}|A^i_{\mathrm{Dis}(\mathfrak{X})}:P_{i_o}|A_{i_o}:T_{i_o}:X(P_{i_o},A_{i_o}):?)$$

<u>is square integrable on</u> $K\times M^i_{\mathrm{Dis}(\mathfrak{X})}|\{1\}\times\Gamma_{M^i_{\mathrm{Dis}(\mathfrak{X})}}$ <u>for any</u>

$$T_{i_o}\in\mathrm{Hom}(S_{\mathfrak{X}(P_{i_o},A_{i_o})},\mathcal{E}_{cus}(\delta,\mathcal{O}_{i_o})).$$

[This is the point of departure for the spectral decomposition of $L^2(G/\Gamma)$; cf. TES.]

All the Eisenstein systems of interest satisfy the various assumptions set forth above.

It is suggestive to view the next statement as a summary of the 'functional calculus' for the E-functions and the ∇-functions.

FNC CAL <u>Suppose that for each</u> i_o, $\underset{\sim}{\mathfrak{X}}_{i_o}$ <u>is subject to</u>

$$\begin{cases} \text{Geom: I} \\ \text{Geom: II,} \quad \text{Gr} \\ \text{Geom: III} \end{cases}$$

and

E - S: I, II, III.

Let $\underset{\sim}{p}^{\dagger} = \{p_{j_o}^{\dagger}\}$ be a completely equivariant system of polynomials.

Let $\mathfrak{X} \in \underset{\sim}{\mathfrak{X}}(\mathcal{C}_k; \mathcal{C}_o)$, say $\mathfrak{X} \in \underset{\sim}{\mathfrak{X}}_{i_o}(\mathcal{C}_k)$ -- then

$$E(\mathfrak{X}:P|A_i:P_{i_o}|A_{i_o}:d_{\mathrm{Hom}} p_{i_o}^{\dagger}(\Lambda_{i_o}^{\dagger})(T_{i_o}):\Lambda_{i_o}^{\dagger}:x)$$

$$= p_{i_o}^{\dagger}(\Lambda_{i_o}^{\dagger})E(\mathfrak{X}:P|A_i:P_{i_o}|A_{i_o}:T_{i_o}:\Lambda_{i_o}^{\dagger}:x)$$

and

$$\nabla(\mathfrak{X}:P|(A_i,A_i):P_{j_o}|A_{j_o}:P_{i_o}|A_{i_o}:w_{j_o i_o}^{\dagger}:\Lambda_{i_o}^{\dagger})(d_{\mathrm{Hom}} p_{i_o}^{\dagger}(\Lambda_{i_o}^{\dagger})(T_{i_o}))$$

$$= p_{i_o}^{\dagger}(\Lambda_{i_o}^{\dagger})\nabla(\mathfrak{X}:P|(A_i,A_i):P_{j_o}|A_{j_o}:P_{i_o}|A_{i_o}:w_{j_o i_o}^{\dagger}:\Lambda_{i_o}^{\dagger})T_{i_o}.$$

We had remarked in TES that the second of these rules could be used to place a priori restrictions on the range of ∇ but did not enter into detail there. It will now be necessary to make matters explicit.

PROPOSITION 3.1 Suppose that for each i_o, $\underset{\sim}{\mathfrak{X}}_{i_o}$ is subject to

$$\begin{cases} \text{Geom: I} & \\ \text{Geom: II,} & \text{Gr} \\ \text{Geom: III} & \end{cases}$$

and

E - S: I, II, III.

Let $\mathfrak{X}$ *be an element of* $\mathfrak{X}_{i_0}$ *with the following property:*

$$\forall w_{i_0} \in W(A_{i_0})$$

$$w_{i_0} \neq 1 \Rightarrow w_{i_0} | \mathfrak{X}(P_{i_0}, A_{i_0}) \neq \mathrm{ID}.$$

Then

$$\mathrm{Im}(\nabla(\mathfrak{X}:G|(\{1\},\{1\}):P_{j_0}|A_{j_0}:P_{i_0}|A_{i_0}:w_{j_0 i_0}:\Lambda_{i_0}))$$

is contained in

$$C \otimes \mathcal{E}_{cus}(\delta, \mathcal{O}_{j_0}) \hookrightarrow S_{\mathfrak{X}(P_{j_0}, A_{j_0})_{w_{j_0 i_0}}} \otimes \mathcal{E}_{cus}(\delta, \mathcal{O}_{j_0}).$$

We shall preface the proof with a lemma, it being understood that the assumptions are as in the statement of the proposition itself.

LEMMA 3.2 *Let*

$$T_{i_0} \in \mathrm{Hom}(S_{\mathfrak{X}(P_{i_0}, A_{i_0})}, \mathcal{E}_{cus}(\delta, \mathcal{O}_{i_0})).$$

Suppose that $T_{i_0}(1) = 0$ -- *then*

$$\nabla(\mathfrak{X}:G|(\{1\},\{1\}):P_{j_0}|A_{j_0}:P_{i_0}|A_{i_0}:w_{j_0 i_0}:\Lambda_{i_0})T_{i_0} = 0$$

Setting aside the proof of the lemma for the moment, let us pass to the proof of our proposition. It is clear that $\mathfrak{X}_{w_{j_0 i_0}}$ satisfies the same conditions per $W(A_{j_0})$ as does $\mathfrak{X}$ per $W(A_{i_0})$. If

$$\mathfrak{X}_{\underset{\sim}{w}_{j_o i_o}} \notin \mathfrak{X}_{\underset{\sim}{j}_o},$$

then, in view of E - S:I,

$$\nabla(\mathfrak{X}:G|(\{1\},\{1\}):P_{j_o}|A_{j_o}:P_{i_o}|A_{i_o}:w_{j_o i_o}:\Lambda_{i_o}) = 0,$$

and so the assertion is trivially true in this case. Consider, therefore, the other possibility:

$$\mathfrak{X}_{\underset{\sim}{w}_{j_o i_o}} \in \mathfrak{X}_{\underset{\sim}{j}_o}.$$

Bearing in mind E - S: II, to complete the proof it need only be shown that if

$$T_{j_o} \in \mathrm{Hom}(S_{\mathfrak{X}(P_{j_o},A_{j_o})_{w_{j_o i_o}}}, \mathcal{E}_{cus}(\delta,\mathcal{O}_{j_o}))$$

has the property that $T_{j_o}(1) = 0$, then

$$\nabla(\mathfrak{X}_{\underset{\sim}{w}_{j_o i_o}}:G|(\{1\},\{1\}):P_{i_o}|A_{i_o}:P_{j_o}|A_{j_o}:w_{j_o i_o}^{-1}:-w_{j_o i_o}\bar{\Lambda}_{i_o})T_{j_o} = 0.$$

But this is precisely the gist of Lemma 3.2 (in the j_o-picture).

The proof of Lemma 3.2 depends on a well-known elementary fact which may be formulated as follows.

LEMMA <u>Let</u> W <u>be a finite group of automorphisms of</u> $\mathbf{C}^n$. <u>Fix an integer</u> $N \geq 0$ <u>and complex numbers</u> $c_I(0 \leq |I| \leq N)$. <u>Suppose that</u> z_o <u>is a point</u>

in C^n *with the property that* $w \cdot z_o \neq z_o \ \forall w \in W (w \neq 1)$ -- *then there exists a* W-*invariant polynomial* p_o *such that*

$$\partial^{|I|} p_o / \partial z^I \big|_{z_o} = c_I \qquad (0 \leq |I| \leq N).$$

[We shall omit the verification, it being but a simple variant on standard themes of Harish-Chandra and Steinberg.]

PROOF OF LEMMA 3.2 To begin with, observe that we need only establish our assertion on a non-empty open subset of $\mathfrak{X}(P_{i_o}, A_{i_o})$. This said, for any $w_{i_o} \neq 1$ in $W(A_{i_o})$, put

$$\mathfrak{X}(P_{i_o}, A_{i_o} : w_{i_o}) = \{\Lambda_{i_o} \in \mathfrak{X}(P_{i_o}, A_{i_o}) : w_{i_o} \Lambda_{i_o} = \Lambda_{i_o}\}.$$

Then $\mathfrak{X}(P_{i_o}, A_{i_o} : w_{i_o})$ is a closed, nowhere dense subset of $\mathfrak{X}(P_{i_o}, A_{i_o})$, hence, thanks to the Baire category theorem,

$$\mathfrak{X}(P_{i_o}, A_{i_o}) - \bigcup_{w_{i_o} \neq 1} \mathfrak{X}(P_{i_o}, A_{i_o} : w_{i_o})$$

is a non-empty open subset of $\mathfrak{X}(P_{i_o}, A_{i_o})$, call it

$$\mathcal{O}_{\mathfrak{X}}(P_{i_o}, A_{i_o}).$$

Fix a point $\Lambda_{i_o}^o \in \mathcal{O}_{\mathfrak{X}}(P_{i_o}, A_{i_o})$ at which $\nabla(\ldots : \Lambda_{i_o}^o)$ is defined. Let T_{i_o} be an element of

$$\mathrm{Hom}(S_{\mathfrak{X}(P_{i_o}, A_{i_o})}, \mathcal{E}_{cus}(\delta, \mathcal{O}_{i_o}))$$

such that $T_{i_o}(1) = 0$. Since

$$\mathrm{Hom}(S_{\mathfrak{X}(P_{i_o}, A_{i_o})}, \mathcal{E}_{cus}(\delta, \mathcal{O}_{i_o}))$$

can be identified with

$$\mathrm{Hom}(S_{\mathfrak{X}(P_{i_o}, A_{i_o})}, C) \otimes \mathcal{E}_{cus}(\delta, \mathcal{O}_{i_o}),$$

upon taking linear combinations, it can be supposed that

$$T_{i_o} \in \mathrm{Hom}(S_{\mathfrak{X}(P_{i_o}, A_{i_o})}, C) \otimes \Phi_{i_o}.$$

Choose a positive integer $N >> 0$ so as to guarantee the vanishing of the ∇'s whenever $\mathrm{ord}(?) > N$. Determine complex numbers $c_I (0 \leq |I| \leq N)$ by the requirement

$$T_{i_o}(x^I) = c_I \otimes \Phi_{i_o}.$$

In view of the above lemma, there exists a polynomial p_{i_o}, invariant under $W(A_{i_o})$, such that

$$\partial^{|I|} p_{i_o} / \partial \Lambda_{i_o}^I \Big|_{\Lambda_{i_o}^o} = c_I \qquad (0 \leq |I| \leq N).$$

Because $c_o = 0$ by assumption, we have, in particular,

$$p_{i_o}(\Lambda^o_{i_o}) = 0.$$

Define $T^o_{i_o}$ by the prescription

$$\begin{cases} T^o_{i_o}(1) = \Phi_{i_o} \\ T^o_{i_o}(x^I) = 0 \qquad (|I| \geq 1). \end{cases}$$

Then

$$\mathrm{ord}[(d_{\mathrm{Hom}} p_{i_o}(\Lambda^o_{i_o})T^o_{i_o}) - T_{i_o}] > N.$$

Call $\underset{\sim}{p} = \{p_{j_o}\}$ the completely equivariant system of polynomials associated in the canonical way with p_{i_o}. Taking into account the functional calculus supra, we have now

$$\nabla(\mathfrak{X}:G|(\{1\},\{1\}):P_{j_o}|A_{j_o}:P_{i_o}|A_{i_o}:w_{j_o i_o}:\Lambda^o_{i_o})T_{i_o}$$

$$= \nabla(\mathfrak{X}:G|(\{1\},\{1\}):P_{j_o}|A_{j_o}:P_{i_o}|A_{i_o}:w_{j_o i_o}:\Lambda^o_{i_o})(d_{\mathrm{Hom}} p_{i_o}(\Lambda^o_{i_o})T^o_{i_o})$$

$$= p_{i_o}(\Lambda^o_{i_o})\nabla(\mathfrak{X}:G|(\{1\},\{1\}):P_{j_o}|A_{j_o}:P_{i_o}|A_{i_o}:w_{j_o i_o}:\Lambda^o_{i_o})T^o_{i_o}$$

$$= 0,$$

thereby completing the proof. //

Keeping still to the hypotheses on $\mathfrak{X}$ set forth in Proposition 3.1, consider the diagram

$$\begin{array}{ccc}
\operatorname{Hom}(S_{\mathfrak{X}(P_{i_0},A_{i_0})}, \mathcal{E}_{cus}(\delta,\mathcal{O}_{i_0})) & \xrightarrow{\nabla(\mathfrak{X}:\ldots)} & S_{\mathfrak{X}(P_{j_0},A_{j_0})^{w_{j_0 i_0}}} \otimes \mathcal{E}_{cus}(\delta,\mathcal{O}_{j_0}) \\
T_{i_0} \mapsto T_{i_0}(1) \Big\downarrow & & \Big\uparrow \Phi_{j_0} \mapsto 1 \otimes \Phi_{j_0} \\
\mathcal{E}_{cus}(\delta,\mathcal{O}_{i_0}) & \cdots\cdots\cdots\cdots > & \mathcal{E}_{cus}(\delta,\mathcal{O}_{j_0})
\end{array}$$

Since $\nabla(\mathfrak{X}:\ldots)$ vanishes on $\operatorname{Ker}(\downarrow)$ and takes values in $\operatorname{Im}(\uparrow)$, the fact that

$$\begin{cases} \downarrow \text{ is surjective} \\ \uparrow \text{ is injective} \end{cases}$$

implies that there is one and only one arrow $\cdots>$ filling in the diagram, call it

$$\nabla_{cus}(\mathfrak{X}:G|(\{1\},\{1\}):P_{j_0}|A_{j_0}:P_{i_0}|A_{i_0}:w_{j_0 i_0}:\Lambda_{i_0}) \quad .$$

There is also an analogue for E-functions. Thus put

$$\begin{aligned} &E_{cus}(\mathfrak{X}:G|\{1\}:P_{i_0}|A_{i_0}:T_{i_0}(1):\Lambda_{i_0}:?) \\ &= E(\mathfrak{X}:G|\{1\}:P_{i_0}|A_{i_0}:T_{i_0}:\Lambda_{i_0}:?). \end{aligned}$$

Of course, to justify the definition it is necessary to bear in mind the negligibility assumption as well as the familiar principle that a slowly increasing differentiable function on G/Γ, which is negligible along every

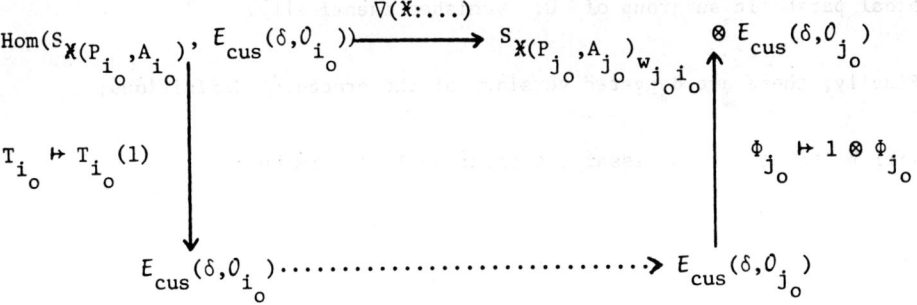

Γ-cuspidal parabolic subgroup of G, vanishes identically.

Finally, there are daggered versions of the preceding definitions; they will be used, when necessary, without explicit comment.

§4. ON THE SINGULARITIES OF CUSPIDAL EISENSTEIN SERIES

The purpose of this § is to analyze the geometry of the singularities of cuspidal Eisenstein series which meet the tube over the negative chamber. In the terminology of amalgamations (cf. TES, Chap. 5), these are the hyperplanes which arise in the construction of Amal (2). The results obtained here will be used in the next § when we take up the inner product formula for truncated residual Eisenstein series of level one.

Agreeing to employ the usual notations, consider

$$\begin{cases} E(P_{i_0}|A_{i_0}:\Phi_{i_0}:\Lambda_{i_0}:x) & (\Phi_{i_0} \in \mathcal{E}_{cus}(\delta,\mathcal{O}_{i_0})) \\ c_{cus}(P_{j_0}|A_{j_0}:P_{i_0}|A_{i_0}:w_{j_0i_0}:\Lambda_{i_0}), & \end{cases}$$

the data which are the object of study in the theory of Eisenstein series associated with cusp forms or still, the ingredients of the canonical Eisenstein system (with $\mathcal{C}_k = \{G\}$, of course). Suppose that H_s is a singular hyperplane of

$$E(P_{i_0}|A_{i_0}:\Phi_{i_0}:\Lambda_{i_0}:x).$$

Then H_s is admissible, say

$$H_s:\check{\lambda} = c \quad (\exists\check{\lambda} \in \Sigma_{P_{i_0}}(\mathfrak{g},\check{\mathfrak{a}}_{i_0})).$$

Because

$$E(P_{i_o} | A_{i_o} : \Phi_{i_o} : \Lambda_{i_o} : x)$$

is non-singular along the imaginary axis, there are two possibilities, viz:

$$\begin{cases} \mathrm{Re}(c) < 0 \\ \mathrm{Re}(c) > 0. \end{cases}$$

Observe that in either case the normal translation in H_s is $(c/2)\cdot\lambda$.

LEMMA 4.1 $H_s \cap (-\mathcal{C}_{P_{i_o}}(\check{a}_{i_o}) + \sqrt{-1}\check{a}_{i_o}) \neq \emptyset$ <u>iff</u> $\mathrm{Re}(c) < 0$.

PROOF Let $\Lambda_{i_o} \in \check{a}_{i_o} + \sqrt{-1}\check{a}_{i_o}$ -- then

$$\Lambda_{i_o} \in H_s \cap (-\mathcal{C}_{P_{i_o}}(\check{a}_{i_o}) + \sqrt{-1}\check{a}_{i_o})$$

iff

$$\mathrm{Re}(\Lambda_{i_o}) \in -\mathcal{C}_{P_{i_o}}(\check{a}_{i_o}) \text{ and } \check{\lambda}(\Lambda_{i_o}) = c.$$

But

$$\check{\lambda}(\mathrm{Re}(\Lambda_{i_o})) = \mathrm{Re}(c).$$

Consequently,

$$\mathrm{Re}(\Lambda_{i_o}) \in -\mathcal{C}_{P_{i_o}}(\check{a}_{i_o}) \text{ and } \check{\lambda}(\Lambda_{i_o}) = c$$

$$\Longrightarrow \check{\lambda}(\mathrm{Re}(\Lambda_{i_o})) < 0 \Longrightarrow \mathrm{Re}(c) < 0.$$

Turning to the converse, it must now be shown that if $\operatorname{Re}(c) < 0$, then

$$H_s \cap (-\mathcal{C}_{P_{i_0}}(\check{a}_{i_0}) + \sqrt{-1}\check{a}_{i_0}) \neq \emptyset.$$

If this were untrue, then we would have

$$(\operatorname{Re}(c/2)\cdot\lambda + \operatorname{Ker}(\check{\lambda})) \cap -\mathcal{C}_{P_{i_0}}(\check{a}_{i_0}) = \emptyset$$

or still

$$(\lambda + \operatorname{Ker}(\check{\lambda})) \cap \mathcal{C}_{P_{i_0}}(\check{a}_{i_0}) = \emptyset,$$

implying, therefore, that

$$(t\lambda + \operatorname{Ker}(\check{\lambda})) \cap \mathcal{C}_{P_{i_0}}(\check{a}_{i_0}) = \emptyset$$

for all $t > 0$. On the other hand, since $-\lambda$ is a negative root, it must necessarily be the case that

$$(-\lambda + \operatorname{Ker}(\check{\lambda})) \cap \mathcal{C}_{P_{i_0}}(\check{a}_{i_0}) = \emptyset,$$

hence again

$$(-t\lambda + \operatorname{Ker}(\check{\lambda})) \cap \mathcal{C}_{P_{i_0}}(\check{a}_{i_0}) = \emptyset$$

for all $t > 0$. As it is obvious that

$$\operatorname{Ker}(\check{\lambda}) \cap \mathcal{C}_{P_{i_0}}(\check{a}_{i_0}) = \emptyset,$$

we can thus say that

$$t \in R \Longrightarrow (t\lambda + \mathrm{Ker}(\check{\lambda})) \cap \mathcal{C}_{P_{i_0}}(\check{a}_{i_0}) = \emptyset.$$

However,

$$\bigcup_{t \in R} (t\lambda + \mathrm{Ker}(\check{\lambda})) = \check{a}_{i_0} \supset \mathcal{C}_{P_{i_0}}(\check{a}_{i_0}),$$

so, upon intersecting both sides with $\mathcal{C}_{P_{i_0}}(\check{a}_{i_0})$, it follows that $\mathcal{C}_{P_{i_0}}(\check{a}_{i_0}) = \emptyset$, a contradiction. //

Let us agree now to place ourselves in the setting of the main result of TES, namely Theorem 5.12. This done, consider $\mathfrak{X}_{i_0}^{(2)}$ -- then each element H of $\mathfrak{X}_{i_0}^{(2)}$ is a completely equivariant system of admissible hyperplanes, there being a characterization

$$H \in \mathfrak{X}_{i_0}^{(2)}$$
$$\Longleftrightarrow H(P_{i_0}, A_{i_0}) \cap (-\mathcal{C}_{P_{i_0}}(\check{a}_{i_0}) + \sqrt{-1}\check{a}_{i_0}) \neq \emptyset .$$

It is to be noted that the normal translation $H(P_{i_0}, A_{i_0})$ of any such H is necessarily real (cf. Proposition $\mathfrak{X}$ (§3)).

In terms of specific data, these facts admit the following interpretation. Suppose that H_s is a singular hyperplane of

$$E(P_{i_0} | A_{i_0} : \Phi_{i_0} : \Lambda_{i_0} : x)$$

with the property that

$$H_s \cap (-\mathcal{C}_{P_{i_0}}(\check{a}_{i_0}) + \sqrt{-1}\check{a}_{i_0}) \neq \emptyset.$$

Then

$$H_s = H(P_{i_0}, A_{i_0}) \quad (\exists H \in \underset{\sim}{\mathfrak{X}}_{i_0}^{(2)})$$

and

$$\begin{cases} \mathrm{Re}(c) < 0 \\ \\ \mathrm{Im}(c) = 0. \end{cases}$$

In addition, and, as it turns out, this is a point of considerable importance, the singularity of

$$E(P_{i_0}|A_{i_0}:\Phi_{i_0}:\Lambda_{i_0}:x)$$

along H_s must be simple. Indeed, thanks to Proposition E (§3), H_s is also a singular hyperplane of

$$c_{cus}(P_{j_0}|A_{j_0}:P_{i_0}|A_{i_0}:w_{j_0 i_0}:\Lambda_{i_0})$$

for some j_0. Accordingly, it need only be shown that the corresponding singularity for this particular c-function is simple. To facilitate the verification, let us pass to the Eisenstein system picture. Take $\mathfrak{X}$ per the canonical Eisenstein system -- then we have (cf. §3)

$$\nabla(\mathfrak{X}:G|(\{1\},\{1\}):P_{j_o}|A_{j_o}:P_{i_o}|A_{i_o}:w_{j_o i_o}:\Lambda_{i_o})T_{i_o}$$

$$= 1 \otimes (c_{cus}(P_{j_o}|A_{j_o}:P_{i_o}|A_{i_o}:w_{j_o i_o}:\Lambda_{i_o})T_{i_o}(1)).$$

Put $\Lambda_\perp = -\lambda/||\lambda||$ -- then $\Lambda_\perp$ is a real unit normal to $\mathcal{H}(P_{i_o},A_{i_o})$ in $\mathfrak{X}(P_{i_o},A_{i_o})$ belonging, moreover, to $-\mathcal{D}_{P_{i_o}}(\check{a}_{i_o})^-$. On $\mathcal{H}(P_{i_o},A_{i_o})$, the Laurent expansion gives

$$\nabla(\mathfrak{X}:G|(\{1\},\{1\}):P_{j_o}|A_{j_o}:P_{i_o}|A_{i_o}:w_{j_o i_o}:\Lambda_{i_o} + \zeta\Lambda_\perp)$$

$$= \sum_{n=-N}^{\infty} \nabla_n(w_{j_o i_o}:\Lambda_{i_o})\zeta^n,$$

the ∇_n being meromorphic functions. Our contention amounts to the claim that $N = 1$. We shall argue by contradiction. Assume, therefore, that $N > 1$ -- then ∇_{-N} is not identically zero, say

$$\nabla_{-N}(w_{j_o i_o}:\Lambda_{i_o})T_{i_o}^o \neq 0.$$

Noting that

$$S_{\mathcal{H}(P_{i_o},A_{i_o})} \simeq C[\Lambda_\perp],$$

define an element

$$T_{i_o} \in \mathrm{Hom}(S_{\mathcal{H}(P_{i_o},A_{i_o})}, \mathcal{E}_{cus}(\delta,\mathcal{O}_{i_o}))$$

by the prescription

$$\begin{cases} T_{i_o}(\Lambda_\perp^n) = 0 & (n \neq N-1) \\ T_{i_o}(\Lambda_\perp^{N-1}) = T_{i_o}^o(1). \end{cases}$$

On the basis of the residual definitions (cf. Chap. 5 in TES), we then have

$$\nabla(H:G|(\{1\},\{1\}):P_{j_o}|A_{j_o}:P_{i_o}|A_{i_o}:w_{j_o i_o}:\Lambda_{i_o})T_{i_o}$$

$$= \sum_{n=0}^{N-1}\sum_{k=0}^{n} \frac{1}{k!(n-k)!}\cdot (w_{j_o i_o}\Lambda_\perp)^{n-k} \otimes \nabla_{-n-1}(w_{j_o i_o}:\Lambda_{i_o})((\Lambda_\perp)^k \vee T_{i_o})$$

$$= \frac{1}{(N-1)!}\cdot 1 \otimes \nabla_{-N}(w_{j_o i_o}:\Lambda_{i_o})T_{i_o}^o \neq 0.$$

However, by construction, $\underset{\sim}{\mathfrak{X}}_{i_o}^{(2)}$ is subject to

$$\begin{cases} \text{Geom: I} \\ \text{Geom: II} \quad , \quad \text{Gr} \\ \text{Geom: III} \end{cases}$$

and

$$\text{E - S : I, II, III.}$$

Furthermore, it is clear that

$$\forall w_{i_o} \in W(A_{i_o})$$

$$w_{i_o} \neq 1 \Longrightarrow w_{i_o}|H(P_{i_o},A_{i_o}) \neq \text{ID},$$

since $H(P_{i_o},A_{i_o})$ is a hyperplane not passing through the origin, hence has for its linear span the full complexification of $\check{a}_{i_o}$. But then, as

$N > 1$, $T_{i_o}(1) = 0$, so, in view of Lemma 3.2,

$$\nabla(H:G|(\{1\},\{1\}):P_{j_o}|A_{j_o}:P_{i_o}|A_{i_o}:w_{j_o i_o}:\Lambda_{i_o})T_{i_o} = 0,$$

a contradiction.

There is one final remark which should be made in connection with the foregoing. The given singular hyperplane H_s for

$$E(P_{i_o}|A_{i_o}:\Phi_{i_o}:\Lambda_{i_o}:x)$$

may or may not be a singular hyperplane for

$$c_{cus}(P_{j_o}|A_{j_o}:P_{i_o}|A_{i_o}:w_{j_o i_o}:\Lambda_{i_o}),$$

it being understood, of course, that j_o is now arbitrary. If, as before, we write

$$H_s = H(P_{i_o}, A_{i_o}),$$

then, in terms of the ambient normal translation, there are two possibilities, viz:

(1) $w_{j_o i_o} H(P_{i_o}, A_{i_o}) \in \mathcal{D}_{P_{j_o}}(\check{a}_{j_o})^-$;

(2) $w_{j_o i_o} H(P_{i_o}, A_{i_o}) \in -\mathcal{D}_{P_{j_o}}(\check{a}_{j_o})^-$.

[That these are in fact the <u>only</u> possibilities can be seen by noting that $H(P_{i_o}, A_{i_o})$ is proportional to an element of $\Sigma_{P_{i_o}}(\mathfrak{g}, \mathfrak{a}_{i_o})$.]

RE-(1): In this case, H_s is either a singular hyperplane for

$$c_{cus}(P_{j_o}|A_{j_o}:P_{i_o}|A_{i_o}:w_{j_o i_o}:\Lambda_{i_o})$$

of order one or else

$$c_{cus}(P_{j_o}|A_{j_o}:P_{i_o}|A_{i_o}:w_{j_o i_o}:\Lambda_{i_o})$$

is meromorphic along H_s. [Observe that H_s cannot be a singular hyperplane for

$$W_{\{G\}}(\mathfrak{X};A_{i_o},A_{i_o}) = W(A_{i_o}).$$

Taking into account Proposition ∇(§3), our assertion thus follows from what has been said above.]

RE-(2): In this case, H_s is not a singular hyperplane of

$$c_{cus}(P_{j_o}|A_{j_o}:P_{i_o}|A_{i_o}:w_{j_o i_o}:\Lambda_{i_o}),$$

that is,

$$c_{cus}(P_{j_o}|A_{j_o}:P_{i_o}|A_{i_o}:w_{j_o i_o}:\Lambda_{i_o})$$

is meromorphic along H_s. [Here is the point. The vector

$$-w_{j_o i_o} H(P_{i_o}, A_{i_o})$$

is the normal translation per

$$H(P_{j_o}, A_{j_o}) w_{j_o i_o} .$$

Accordingly, in view of our hypothesis,

$$H_{w_{j_o i_o}} \not\subset \mathfrak{X}_{j_o}^{(2)} .$$

It therefore follows from ES:I that

$$\nabla(H:G|(\{1\},\{1\}):P_{j_o}|A_{j_o}:P_{i_o}|A_{i_o}:w_{j_o i_o}:\Lambda_{i_o}) = 0,$$

implying, then, that

$$c_{cus}(P_{j_o}|A_{j_o}:P_{i_o}|A_{i_o}:w_{j_o i_o}:\Lambda_{i_o})$$

is meromorphic along H_s.]

§5. INNER PRODUCTS AND TRUNCATION: LEVEL ONE EISENSTEIN SERIES

The purpose of this § is to obtain a formula for the (L^2) inner product of two truncated Eisenstein series associated with residual forms of level one. The analogous problem for truncated Eisenstein series associated with cusp forms, i.e., the level zero situation, has already been treated by us in [2]. Of course, one will ultimately want a formula at any level but it seems worthwhile to consider the level one case initially since it can be handled by direct methods; this will not be true of the general case. Philosophically, the situation here is not unlike that encountered in TES (cf. Chap. 4): The relevant results can be established via a procedure which immediately suggests itself but, unfortunately, while the results themselves can be generalized, the methods cannot. In this connection, it is interesting to note that the difficulty is eventually the same, namely simple singularities vs. arbitrary singularities.

To begin with, it will first be necessary to review some facts from the general theory of Eisenstein series; see TES, Chap. 5. Thus let (P,S) be a Γ-cuspidal split parabolic subgroup of G with special split component A, $P = M\cdot A\cdot N$ the corresponding Langlands decomposition of P. Introduce, as usual, the space $\mathcal{E}(\delta,\mathcal{O})$ -- then

$$E(\delta,\mathcal{O}) = E_{cus}(\delta,\mathcal{O}) \oplus E_{res}(\delta,\mathcal{O})$$

and one may form the Eisenstein series

$$E(P|A:\Phi:\Lambda:x) = \sum_{\gamma\in\Gamma/\Gamma\cap P} a_{x\gamma}^{(\Lambda-\rho)}\cdot\Phi(x\gamma)$$

for any $\Phi\in E(\delta,\mathcal{O})$. Write, in an obvious notation,

$$\Phi = \Phi_{cus} + \Phi_{res} .$$

Then

$$E(P|A:\Phi:\Lambda:x)$$

is the sum of

$$\begin{cases} E(P|A:\Phi_{cus}:\Lambda:x) \\ E(P|A:\Phi_{res}:\Lambda:x) . \end{cases}$$

Needless to say, it is the 'residual constituent'

$$E(P|A:\Phi_{res}:\Lambda:x)$$

with which we shall primarily be concerned.

Let $\mathcal{C}_k$ be the G-conjugacy class in the association class $\mathcal{C}$ containing (P,S;A) -- then the space $E(\delta,\mathcal{O})$ admits a decomposition

$$\sum_{\mathcal{C}_k \succeq \mathcal{C}_o} \oplus\, E(\delta,\mathcal{O}:\mathcal{C}_o) .$$

In this connection, it should be kept in mind that $\mathcal{C}_o$ is a summation variable which can take, in particular, the value $\mathcal{C}$. We have

$$E_{cus}(\delta,\mathcal{O}) = E(\delta,\mathcal{O}:\mathcal{C})$$

while $E_{res}(\delta,\mathcal{O})$ is the sum of what remains. Let $\Phi \in E(\delta,\mathcal{O})$ -- then

$$\Phi = \sum_{\mathcal{C}_0} \Phi_{\mathcal{C}_0} \quad (\Phi_{\mathcal{C}_0} \in E(\delta,\mathcal{O}:\mathcal{C}_0)),$$

so

$$E(P|A:\Phi:\Lambda:x) = \sum_{\mathcal{C}_0} E(P|A:\Phi_{\mathcal{C}_0}:\Lambda:x).$$

Again,

$$E(P|A:\Phi_{cus}:\Lambda:x)$$

is the same as

$$E(P|A:\Phi_{\mathcal{C}}:\Lambda:x)$$

while

$$E(P|A:\Phi_{res}:\Lambda:x)$$

is the sum of what remains. These remarks make it clear that an arbitrary Eisenstein series is composed of other Eisenstein series associated with the elements the various $E(\delta,\mathcal{O}:\mathcal{C}_0)$. The latter Eisenstein series are said to be of level

$$\text{rank}(\mathcal{C}_0) - \text{rank}(\mathcal{C}).$$

Of course, it may very well happen that for distinct $\mathcal{C}_0'$ and $\mathcal{C}_0''$,

$$\text{rank}(\mathcal{C}_0') = \text{rank}(\mathcal{C}_0''),$$

i.e., the level does not uniquely determine the association class. On the other hand, the Eisenstein series of level zero are precisely those which are associated with cusp forms.

Keeping always to the notations of TES, the space $E(\delta,\mathcal{O}:\mathcal{C}_0)$, upon explication, appears as the orthogonal projection of $E(\delta,\mathcal{O})$ onto

$$\sum_{\underset{\sim}{\mathcal{O}}_0} \oplus\, E(\delta,\underset{\sim}{\mathcal{O}}_0;\underset{\sim}{\mathfrak{X}}:P|A),$$

where, in fact, $E(\delta,\mathcal{O})$ is preserved by the orthogonal projection onto each

$$E(\delta,\underset{\sim}{\mathcal{O}}_0;\underset{\sim}{\mathfrak{X}}:P|A).$$

Let $\Phi_{\mathcal{C}_0} \in E(\delta,\mathcal{O}:\mathcal{C}_0)$ -- then, in terms of the orbit types $\underset{\sim}{\mathcal{O}}_0$,

$$\Phi_{\mathcal{C}_0} = \sum_{\underset{\sim}{\mathcal{O}}_0} \Phi_{\mathcal{C}_0,\underset{\sim}{\mathcal{O}}_0} ,$$

from which it follows that

$$E(P|A:\Phi_{\mathcal{C}_0}:\Lambda:x) = \sum_{\underset{\sim}{\mathcal{O}}_0} E(P|A:\Phi_{\mathcal{C}_0,\underset{\sim}{\mathcal{O}}_0}:\Lambda:x).$$

Later on, we shall refine the description on the right even further but it will suffice for now.

With this preparation, let us pass to the problem with which we shall be concerned, setting it up in a precise manner and then making some more or less immediate reductions. It will be supposed throughout that the reader is acquainted with the truncation operator $Q^{\underset{\sim}{H}}$ and its basic properties; cf [2].

Let (P_1,S_1), (P_2,S_2) be Γ-cuspidal split parabolic subgroups of G with special split components A_1, A_2. Let

$$\begin{cases} \delta_1,\delta_2 & \text{be } K\text{-types} \\ \mathcal{O}_1,\mathcal{O}_2 & \text{be } M_1,M_2\text{-types.} \end{cases}$$

Let

$$\begin{cases} \Phi_1 \in \mathcal{E}(\delta_1,\mathcal{O}_1) \\ \Phi_2 \in \mathcal{E}(\delta_2,\mathcal{O}_2) \end{cases} .$$

Simply put, then, our problem is to compute in explicit terms the inner product

$$(Q^{\underset{\sim}{H}}E(P_1|A_1:\Phi_1:\Lambda_1:?),\ Q^{\underset{\sim}{H}}E(P_2|A_2:\Phi_2:\Lambda_2:?))$$

for all $\underset{\sim}{H}\in\underset{\sim}{\mathfrak{a}}_Q$.

There is no harm in supposing outright that $\delta_1 = \delta_2$, say δ, since otherwise

$$(Q^{\underset{\sim}{H}}E(P_1|A_1:\Phi_1:\Lambda_1:?),\ Q^{\underset{\sim}{H}}E(P_2|A_2:\Phi_2:\Lambda_2:?)) = 0 \qquad (\underset{\sim}{H}\in\underset{\sim}{\mathfrak{a}}_Q) .$$

Additional reductions can be made with the help of the discussion supra. For this purpose, let

$$\begin{cases} \mathcal{C}_1 \\ \mathcal{C}_2 \end{cases}$$

be the association classes containing

$$\begin{cases} (P_1, S_1; A_1) \\ (P_2, S_2; A_2) \end{cases}$$

and let

$$\begin{cases} \mathcal{C}_{k_1} \\ \mathcal{C}_{k_2} \end{cases}$$

be the corresponding G-conjugacy classes thus determined. Write

$$\begin{cases} E(P_1|A_1:\Phi_1:\Lambda_1:x) = \sum\limits_{\mathcal{C}_{k_1} \succeq \mathcal{C}_o} E(P_1|A_1:\Phi^1_{\mathcal{C}_o}:\Lambda_1:x) \\ E(P_2|A_2:\Phi_2:\Lambda_2:x) = \sum\limits_{\mathcal{C}_{k_2} \succeq \mathcal{C}_o} E(P_2|A_2:\Phi^2_{\mathcal{C}_o}:\Lambda_2:x) \end{cases}$$

where, of course,

$$\begin{cases} \Phi^1_{\mathcal{C}_o} \in \mathcal{E}(\delta, \mathcal{O}_1 : \mathcal{C}_o) \\ \Phi^2_{\mathcal{C}_o} \in \mathcal{E}(\delta, \mathcal{O}_2 : \mathcal{C}_o), \end{cases}$$

being the components in the canonical decomposition of Φ_1, Φ_2 per the $\mathcal{C}_o$. In general, a $\mathcal{C}_o$ which is dominated by $\mathcal{C}_{k_1}$ need not be dominated by $\mathcal{C}_{k_2}$ and vice-versa. [Note: This is true if $\mathcal{C}_1 = \mathcal{C}_2$, as can be seen from Lemma 8 in the Appendix to Chap. 2 of TES.] Let

$$\mathrm{Dom}_o(\mathcal{C}_{k_1},\mathcal{C}_{k_2}) = \{\mathcal{C}_o : \mathcal{C}_{k_1} \succeq \mathcal{C}_o\} \cap \{\mathcal{C}_o : \mathcal{C}_{k_2} \succeq \mathcal{C}_o\} .$$

Owing to the theory of wave-packets,

$$(Q^{\underline{H}}E(P_1|A_1:\Phi^1_{\mathcal{C}_o'}:\Lambda_1:?),\ Q^{\underline{H}}E(P_2|A_2:\Phi^2_{\mathcal{C}_o''}:\Lambda_2:?)) = 0 \qquad (\underline{H}\in\underline{a}_Q)$$

if $\mathcal{C}_o' \neq \mathcal{C}_o''$. Consequently,

$$(Q^{\underline{H}}E(P_1|A_1:\Phi_1:\Lambda_1:?),\ Q^{\underline{H}}E(P_2|A_2:\Phi_2:\Lambda_2:?))$$

is the sum over the

$$\mathcal{C}_o \in \mathrm{Dom}_o(\mathcal{C}_{k_1},\mathcal{C}_{k_2})$$

of the

$$(Q^{\underline{H}}E(P_1|A_1:\Phi^1_{\mathcal{C}_o}:\Lambda_1:?),\ Q^{\underline{H}}E(P_2|A_2:\Phi^2_{\mathcal{C}_o}:\Lambda_2:?))$$

which makes it clear that we need only deal with the latter.

Assume, henceforth, that

$$\begin{cases} \Phi_1 \in \mathcal{E}(\delta,\mathcal{O}_1:\mathcal{C}_o) \\ \Phi_2 \in \mathcal{E}(\delta,\mathcal{O}_2:\mathcal{C}_o) \end{cases}$$

where

$$\begin{cases} \mathcal{C}_{k_1} \succeq \mathcal{C}_o \\ \mathcal{C}_{k_2} \succeq \mathcal{C}_o \ . \end{cases}$$

Decompose Φ_1, Φ_2 in terms of the orbit types $\underline{\mathcal{O}}_o$:

$$\begin{cases} \Phi_1 = \sum\limits_{\underset{\sim}{\mathcal{O}}_o} \Phi^1_{\underset{\sim}{\mathcal{O}}_o} \\ \\ \Phi_2 = \sum\limits_{\underset{\sim}{\mathcal{O}}_o} \Phi^2_{\underset{\sim}{\mathcal{O}}_o} \end{cases} .$$

Again, it is the case that

$$(Q^H_{\sim}E(P_1|A_1:\Phi^1_{\underset{\sim}{\mathcal{O}}'_o}:\Lambda_1:?),\ Q^H_{\sim}E(P_2|A_2:\Phi^2_{\underset{\sim}{\mathcal{O}}''_o}:\Lambda_2:?)) = 0 \qquad (\underset{\sim}{H}\in\underset{\sim}{a}_Q)$$

if $\underset{\sim}{\mathcal{O}}'_o \neq \underset{\sim}{\mathcal{O}}''_o$. This means that there are no mixed terms and

$$(Q^H_{\sim}E(P_1|A_1:\Phi_1:\Lambda_1:?),\ Q^H_{\sim}E(P_2|A_2:\Phi_2:\Lambda_2:?))$$

is the sum over the $\underset{\sim}{\mathcal{O}}_o$ of the

$$(Q^H_{\sim}E(P_1|A_1:\Phi^1_{\underset{\sim}{\mathcal{O}}_o}:\Lambda_1:?),\ Q^H_{\sim}E(P_2|A_2:\Phi^2_{\underset{\sim}{\mathcal{O}}_o}:\Lambda_2:?)).$$

Taking into account the preceding remarks, we see that there is no loss of generality in actually assuming that

$$\begin{cases} \Phi_1 \in E(\delta,\mathcal{O}_1:\mathcal{C}_o) \cap E(\delta,\underset{\sim}{\mathcal{O}}_o;\underset{\sim}{\mathfrak{X}}:P_1|A_1) \\ \\ \Phi_2 \in E(\delta,\mathcal{O}_2:\mathcal{C}_o) \cap E(\delta,\underset{\sim}{\mathcal{O}}_o;\underset{\sim}{\mathfrak{X}}:P_2|A_2) . \end{cases}$$

As will be recalled below, thanks to the theory of Eisenstein systems, Φ_1 and Φ_2 then admit direct characterizations.

At this juncture, we had best remind ourselves that there exists an explicit inner product formula in the level zero situation, i.e., when $\mathcal{C}_1 = \mathcal{C}_2$, say $\mathcal{C}$, thence $\mathcal{C}_o = \mathcal{C}$, so that

$$\begin{cases} \Phi_1 \in \mathcal{E}_{cus}(\delta, \mathcal{O}_1) \\ \Phi_2 \in \mathcal{E}_{cus}(\delta, \mathcal{O}_2) \end{cases} . \qquad (\mathcal{O}_1 \text{ and } \mathcal{O}_2 \text{ associate})$$

This is in fact the thrust of Theorem 9.6 in [2]. Before stating the result, we shall set up the relevant notation. Thus let $P_i (1 \leq i \leq r)$ be a set of representatives for $G\backslash\mathcal{C}$ -- then

$$\mathcal{C} = \coprod_i \mathcal{C}_i$$

where $\mathcal{C}_i = G\cdot\{P_i\} \cap \mathcal{C}$. Let $P_{i\mu}(1 \leq \mu \leq r_i)$ be a set of representatives for $\Gamma\backslash\mathcal{C}_i$ -- then

$$\{P_{i\mu} : 1 \leq i \leq r, \ 1 \leq \mu \leq r_i\}$$

is a set of representatives for $\Gamma\backslash\mathcal{C}$. The symbols

$$\begin{cases} W(A_{i\mu}, A_1), \ W(A_{i\mu}, A_2) \\ \Pi_{\lambda_{i\mu}}, \ I_{P_{i\mu}}(\underset{\sim}{H}) \end{cases}$$

are then to be assigned the usual meanings.

THEOREM 5.1 *Let*

$$\begin{cases} \Phi_1 \in \mathcal{E}_{cus}(\delta, \mathcal{O}_1) \\ \Phi_2 \in \mathcal{E}_{cus}(\delta, \mathcal{O}_2) \end{cases} . \qquad (\mathcal{O}_1 \text{ and } \mathcal{O}_2 \text{ associate})$$

Then, for all $\underset{\sim}{H} \in \underset{\sim}{\mathfrak{a}}_Q$, *the inner product*

$$(Q_{\underset{\sim}{H}}E(P_1|A_1:\Phi_1:\Lambda_1:?),\ Q_{\underset{\sim}{H}}E(P_2|A_2:\Phi_2:\Lambda_2:?))$$

is equal to the sum

$$(-1)^{\ell}\cdot \mathrm{vol}(\mathcal{C})\cdot \sum_{i=1}^{r}\sum_{\mu=1}^{r_i}\sum_{w_{i\mu:2}\in W(A_{i\mu},A_2)}\sum_{w_{i\mu:1}\in W(A_{i\mu},A_1)}$$

of the product of

$$\exp(\langle I_{P_{i\mu}}(\underset{\sim}{H}),\ w_{i\mu:2}\bar{\Lambda}_2 + w_{i\mu:1}\Lambda_1\rangle)$$

$$\times\ (1/\prod_{\lambda_{i\mu}}(w_{i\mu:2}\bar{\Lambda}_2 + w_{i\mu:1}\Lambda_1,\lambda_{i\mu}))$$

with

$$(c_{cus}(P_{i\mu}|A_{i\mu}:P_1|A_1:w_{i\mu:1}:\Lambda_1)\Phi_1,\ c_{cus}(P_{i\mu}|A_{i\mu}:P_2|A_2:w_{i\mu:2}:\Lambda_2)\Phi_2).$$

[Note: Here ℓ and $\mathrm{vol}(\mathcal{C})$ are as always. Furthermore, all split components are special.]

There is a suggestive way to interpret the conclusion of the preceding theorem. To this end, it will be convenient to introduce some notation which will also be helpful later on. Let $(P,S;A)$ be an element of $\mathcal{C}$ -- then by $\mathrm{Exp}(\mathfrak{a})$ we understand the exponentials on $\mathfrak{a}$. Given

$$\begin{cases} \Lambda',\Lambda'' \ \text{in}\ \check{\mathfrak{a}} + \sqrt{-1}\check{\mathfrak{a}} \\ \Phi',\Phi'' \ \text{in}\ \mathcal{E}(\delta,\mathcal{O}), \end{cases}$$

write

$$\mathfrak{K}(\mathcal{C}:P|A:e^{\langle ?,\Lambda'\rangle}\cdot\Phi',\ e^{\langle ?,\Lambda''\rangle}\cdot\Phi'':\underset{\sim}{H})$$

for

$$\frac{e^{\langle I_P(\underset{\sim}{H}),\Lambda' + \bar{\Lambda}''\rangle}}{\prod_\lambda(\Lambda' + \bar{\Lambda}'',\lambda)} \cdot (\Phi',\Phi'')$$

and then extend

$$\mathcal{K}(\mathfrak{C}:P|A:\ldots,\ldots:\underset{\sim}{H})$$

by linearity to

$$\mathrm{Exp}(\mathfrak{a}) \otimes \mathcal{E}(\delta,\mathcal{O}).$$

Returning to the theorem, let us now note that the constant terms

$$\begin{cases} E_{P_{i\mu}}(P_1|A_1:\Phi_1:\Lambda_1:?) \in \mathrm{Exp}(\mathfrak{a}_{i\mu}) \otimes \mathcal{E}_{cus}(\delta,\mathcal{O}_{i\mu}) \\ E_{P_{i\mu}}(P_2|A_2:\Phi_2:\Lambda_2:?) \in \mathrm{Exp}(\mathfrak{a}_{i\mu}) \otimes \mathcal{E}_{cus}(\delta,\mathcal{O}_{i\mu}) \end{cases}.$$

Accordingly, for all $\underset{\sim}{H} \in \underset{\sim}{\mathfrak{a}}_Q$, the inner product

$$(Q^{\underset{\sim}{H}}E(P_1|A_1:\Phi_1:\Lambda_1:?),\ Q^{\underset{\sim}{H}}E(P_2|A_2:\Phi_2:\Lambda_2:?))$$

is equal to

$$(-1)^{\ell} \cdot \mathrm{vol}(\mathfrak{C})$$

times

$$\mathcal{K}(\mathfrak{C}:P_{i\mu}|A_{i\mu}:E_{P_{i\mu}}(P_1|A_1:\Phi_1:\Lambda_1:?),\ E_{P_{i\mu}}(P_2|A_2:\Phi_2:\Lambda_2:?):\underset{\sim}{H}),$$

summed over i and μ.

Dropping all assumptions of cuspidality, we have already remarked that the essential problem in the development of the theory centers on the explicit computation of the inner product

$$(\underset{\sim}{Q}^{H}E(P_1|A_1:\Phi_1:\Lambda_1:?),\ \underset{\sim}{Q}^{H}E(P_2|A_2:\Phi_2:\Lambda_2:?)),$$

where

$$\begin{cases} \Phi_1 \in \mathcal{E}(\delta,\mathcal{O}_1:\mathcal{C}_o) \cap \mathcal{E}(\delta,\underset{\sim}{\mathcal{O}}_o;\underset{\sim}{\mathfrak{X}}:P_1|A_1) \\ \Phi_2 \in \mathcal{E}(\delta,\mathcal{O}_2:\mathcal{C}_o) \cap \mathcal{E}(\delta,\underset{\sim}{\mathcal{O}}_o;\underset{\sim}{\mathfrak{X}}:P_2|A_2) \end{cases},$$

the association class $\mathcal{C}_o$ being subject to the condition

$$\begin{cases} \mathcal{C}_{k_1} \succeq \mathcal{C}_o \\ \mathcal{C}_{k_2} \succeq \mathcal{C}_o \end{cases}.$$

It will not be possible in the present paper to make the determination in all generality; this will be considered elsewhere. We shall, however, take an important first step by establishing the sought for result under the supposition that

$$\begin{cases} \operatorname{rank}(\mathcal{C}_o) - \operatorname{rank}(\mathcal{C}_1) = 1 \\ \operatorname{rank}(\mathcal{C}_o) - \operatorname{rank}(\mathcal{C}_2) = 1 \end{cases}.$$

Let us fix, for the moment, an association class $\mathcal{C}$, an element

$$(P,S;A) \in \mathcal{C}_k \subset \mathcal{C},$$

and an association class $\mathcal{C}_0$ dominated by $\mathcal{C}$. The elements

$$\Phi \in \mathcal{E}(\delta,\mathcal{O}:\mathcal{C}_0) \cap \mathcal{E}(\delta,\underset{\sim}{\mathcal{O}}_0;\underset{\sim}{\mathfrak{X}}:P|A)$$

can then be explicated as follows (cf. Chap. 6 of TES). One knows that

$$\mathcal{E}(\delta,\underset{\sim}{\mathcal{O}}_0;\underset{\sim}{\mathfrak{X}}:P|A) = \sum_{\mathfrak{X}} \mathcal{E}(\delta,\underset{\sim}{\mathcal{O}}_0;\mathfrak{X}:P|A).$$

Here, $\sum_{\mathfrak{X}}$ stands for a sum over those $\mathfrak{X}$ in $\underset{\sim}{\mathfrak{X}}(\mathcal{C}_0;\ \mathrm{rank}(\mathcal{C}))$ for which $x_{\mathcal{C}_k}$ is $\mathfrak{X}$-admissible. Fix such an $\mathfrak{X}$, say $\mathfrak{X} \in \underset{\sim}{\mathfrak{X}}_{i_0}(\mathrm{rank}(\mathcal{C}))$ -- then $\mathfrak{X}$ is principal with

$$\begin{cases} P_{\mathrm{Dis}(\mathfrak{X})} = P \\ A_{\mathrm{Dis}(\mathfrak{X})} = A. \end{cases}$$

Consider

$$E(\mathfrak{X}:P|A:P_{i_0}|A_{i_0}:T_{i_0}:X(P_{i_0},A_{i_0}):x).$$

These functions appear in the Scholium formulated in §3. By definition, the space of such is

$$\mathcal{E}(\delta,\underset{\sim}{\mathcal{O}}_0;\mathfrak{X}:P|A).$$

This said, take a

$$\Phi \in \mathcal{E}(\delta,\mathcal{O}:\mathcal{C}_0) \cap \mathcal{E}(\delta,\underset{\sim}{\mathcal{O}}_0;\underset{\sim}{\mathfrak{X}}:P|A).$$

Write

$$\Phi = \Sigma\ \Phi_{\mathfrak{X}},$$

where

$$\Phi_{\mathfrak{X}} \in \mathcal{E}(\delta, \underset{\sim}{\mathcal{O}}_0; \mathfrak{X}:P|A).$$

Let

$$T_{i_0}(\Phi:\mathfrak{X}) \in \mathrm{Hom}(S_{\mathfrak{X}(P_{i_0},A_{i_0})}, \mathcal{E}_{cus}(\delta, \mathcal{O}_{i_0}))$$

be defined by the requirement that

$$\Phi_{\mathfrak{X}} = E(\mathfrak{X}:P|A:P_{i_0}|A_{i_0}:T_{i_0}(\Phi:\mathfrak{X}):X(P_{i_0},A_{i_0}):?).$$

Then it can be shown that

$$E(P|A:\Phi_{\mathfrak{X}}:\Lambda:x)$$

$$= E(\mathfrak{X}:G|\{1\}:P_{i_0}|A_{i_0}:T_{i_0}(\Phi:\mathfrak{X}):\Lambda+X(P_{i_0},A_{i_0}):x).$$

In other words, the Eisenstein series

$$E(P|A:\Phi:\Lambda:x)$$

attached to Φ is a finite linear combination of certain Eisenstein system functions canonically associated with the data at hand. So far, the difference

$$\mathrm{rank}(\mathcal{C}_0) - \mathrm{rank}(\mathcal{C})$$

has been arbitrary. Let us now assume that it is unity -- then, as it is a question of co-dimension one, to reflect this point, we change the notation, replacing the $\mathfrak{X}$'s by $\mathfrak{H}$'s, the focus being on

$$E(\mathfrak{H}:G|\{1\}:P_{i_o}|A_{i_o}:T_{i_o}:\Lambda_{i_o}:x)$$

$$(T_{i_o} \in \mathrm{Hom}(S_{\mathfrak{H}(P_{i_o},A_{i_o})}, \mathcal{E}_{cus}(\delta,\mathcal{O}_{i_o})))$$

and the associated

$$\nabla(\mathfrak{H}:G|(\{1\},\{1\}):P_{j_o}|A_{j_o}:P_{i_o}|A_{i_o}:w_{j_o i_o}:\Lambda_{i_o})$$

or still, as is permissible (see the definition at the end of §3),

$$\nabla_{cus}(\mathfrak{H}:G|(\{1\},\{1\}):P_{j_o}|A_{j_o}:P_{i_o}|A_{i_o}:w_{j_o i_o}:\Lambda_{i_o}).$$

Lest there be any confusion, we explicitly point out that

$$\Lambda_{i_o} \in \mathfrak{H}(P_{i_o},A_{i_o}),$$

hence, in particular, Λ_{i_o} could be

$$\Lambda + \mathfrak{H}(P_{i_o},A_{i_o}).$$

Let $\Lambda_\perp$ be the ubiquitous real unit normal to $\mathfrak{H}(P_{i_o},A_{i_o})$ -- then, in view of what has been said in §4, we have

$$E(\mathfrak{H}:G|\{1\}:P_{i_o}|A_{i_o}:T_{i_o}:\Lambda_{i_o}:x)$$

$$= \lim_{s\to 0} s\cdot E(P_{i_o}|A_{i_o}:T_{i_o}(1):\Lambda_{i_o} + s\Lambda_\perp:x)$$

and

$$\nabla_{cus}(\mathfrak{H}:G|(\{1\},\{1\}):P_{j_o}|A_{j_o}:P_{i_o}|A_{i_o}:w_{j_o i_o}:\Lambda_{i_o})$$

$$= \lim_{s\to 0} s\cdot c_{cus}(P_{j_o}|A_{j_o}:P_{i_o}|A_{i_o}:w_{j_o i_o}:\Lambda_{i_o} + s\Lambda_{\perp}),$$

it being a question here of simple singularities.

Coming back to our problem, namely that of calculating the inner product

$$(Q_{\sim}^{H}E(P_1|A_1:\Phi_1:\Lambda_1:?),\ Q_{\sim}^{H}E(P_2|A_2:\Phi_2:\Lambda_2:?)),$$

where

$$\begin{cases} \Phi_1 \in E(\delta,\mathcal{O}_1:\mathcal{C}_o) \cap E(\delta,\underset{\sim}{\mathcal{O}}_o;\underset{\sim}{\mathfrak{X}}:P_1|A_1) \\ \Phi_2 \in E(\delta,\mathcal{O}_2:\mathcal{C}_o) \cap E(\delta,\underset{\sim}{\mathcal{O}}_o;\underset{\sim}{\mathfrak{X}}:P_2|A_2), \end{cases}$$

under the supposition that the common dominant $\mathcal{C}_o$ is subject to the condition

$$\begin{cases} \mathrm{rank}(\mathcal{C}_o) - \mathrm{rank}(\mathcal{C}_1) = 1 \\ \mathrm{rank}(\mathcal{C}_o) - \mathrm{rank}(\mathcal{C}_2) = 1, \end{cases}$$

decompose

$$\begin{cases} \Phi_1 \ \text{ per } \ \sum_{\mathfrak{X}} E(\delta,\underset{\sim}{\mathcal{O}}_o;\mathfrak{X}:P_1|A_1) \\ \Phi_2 \ \text{ per } \ \sum_{\mathfrak{X}} E(\delta,\underset{\sim}{\mathcal{O}}_o;\mathfrak{X}:P_2|A_2). \end{cases}$$

Now, as above, change the notation, writing $\mathfrak{H}$ in place of $\mathfrak{X}$ -- then we see that it will be enough to explicate the inner product

$$(Q^{H}_{\sim}E(H_1:G|\{1\}:P_{i_o^1}|A_{i_o^1}:T_{i_o^1}:\Lambda_{i_o^1}:?),\quad Q^{H}_{\sim}E(H_2:G|\{1\}:P_{i_o^2}|A_{i_o^2}:T_{i_o^2}:\Lambda_{i_o^2}:?)).$$

Here

$$\begin{cases} H_1 \in \underset{\sim}{H}(\mathcal{C}_o;\mathrm{rank}(\mathcal{C}_o)-1) : X_{\mathcal{C}_{k_1}} \quad H_1\text{-admissible} \\ H_2 \in \underset{\sim}{H}(\mathcal{C}_o;\mathrm{rank}(\mathcal{C}_o)-1) : X_{\mathcal{C}_{k_2}} \quad H_2\text{-admissible,} \end{cases}$$

the specific placement being

$$\begin{cases} H_1 \in \underset{\sim}{H}_{i_o^1}(\mathrm{rank}(\mathcal{C}_o)-1) \\ H_2 \in \underset{\sim}{H}_{i_o^2}(\mathrm{rank}(\mathcal{C}_o)-1). \end{cases}$$

Then, of course,

$$\begin{cases} T_{i_o^1} \in \mathrm{Hom}(S_{H_1}(P_{i_o^1},A_{i_o^1}),\ E_{cus}(\delta,\mathcal{O}_{i_o^1})) \\ T_{i_o^2} \in \mathrm{Hom}(S_{H_2}(P_{i_o^2},A_{i_o^2}),\ E_{cus}(\delta,\mathcal{O}_{i_o^2})) . \end{cases}$$

Write

$$\begin{aligned} &E(H_1:G|\{1\}:P_{i_o^1}|A_{i_o^1}:T_{i_o^1}:\Lambda_{i_o^1}:x) \\ &= \lim_{s\to 0} s\cdot E(P_{i_o^1}|A_{i_o^1}:T_{i_o^1}(1):\Lambda_{i_o^1} + s\Lambda_{\perp}^{1}:x) \end{aligned}$$

and

$$E(H_2:G|\{1\}:P_{i_o^2}|A_{i_o^2}:T_{i_o^2}:\Lambda_{i_o^2}:x)$$

$$= \lim_{t\to 0} t\cdot E(P_{i_o^2}|A_{i_o^2}:T_{i_o^2}(1):\Lambda_{i_o^2} + t\Lambda_\perp^2:x).$$

Assuming that s and t are real, as we may with no loss of generality, our inner product thus admits a representation as a limit, viz.:

$$\lim_{\substack{s\to 0\\ t\to 0}} st\cdot(\ldots,\ldots)$$

where $(\ldots,\ldots)$ is itself an inner product, viz:

$$(\underset{\sim}{Q}^H E(P_{i_o^1}|A_{i_o^1}:T_{i_o^1}(1):\Lambda_{i_o^1} + s\Lambda_\perp^1:?),\ \underset{\sim}{Q}^H E(P_{i_o^2}|A_{i_o^2}:T_{i_o^2}(1):\Lambda_{i_o^2} + t\Lambda_\perp^2:?)),$$

this entity being computed in turn by Theorem 5.1 as the sum

$$(-1)^{\ell_o}\cdot\mathrm{vol}(\mathcal{C}_o)\cdot\sum_{i_o=1}^{r_o}\ \sum_{\mu_o=1}^{r_{i_o}}\ \sum_{w_{i_o\mu_o}:i_o^2\in W(A_{i_o\mu_o},A_{i_o^2})}\ \sum_{w_{i_o\mu_o}:i_o^1\in W(A_{i_o\mu_o},A_{i_o^1})}$$

of the product of

$$\exp(\langle I_{P_{i_o\mu_o}}(\underset{\sim}{H}), w_{i_o\mu_o:i_o^2}(\bar\Lambda_{i_o^2} + t\Lambda_\perp^2) + w_{i_o\mu_o:i_o^1}(\Lambda_{i_o^1} + s\Lambda_\perp^1)\rangle)$$

$$\times\ (1/\prod_{\lambda_{i_o\mu_o}}(w_{i_o\mu_o:i_o^2}(\bar\Lambda_{i_o^2} + t\Lambda_\perp^2) + w_{i_o\mu_o:i_o^1}(\Lambda_{i_o^1} + s\Lambda_\perp^1),\ \lambda_{i_o\mu_o}))$$

with

$$(c_{cus}(P_{i_o\mu_o}|A_{i_o\mu_o}:P_{i_o^1}|A_{i_o^1}:w_{i_o\mu_o:i_o^1}:\Lambda_{i_o^1} + s\Lambda_\perp^1)T_{i_o^1}(1),$$

$$c_{cus}(P_{i_o\mu_o}|A_{i_o\mu_o}:P_{i_o^2}|A_{i_o^2}:w_{i_o\mu_o:i_o^2}:\Lambda_{i_o^2} + t\Lambda_\perp^2)T_{i_o^2}(1)).$$

Observe that singularities in this formula can arise in two ways:

(1) Singularities in the root denominator;

(2) Singularities in the c-functions.

We shall start the analysis with a simple lemma.

LEMMA <u>The product</u>

$$\prod_{\lambda_{i_o\mu_o}} (w_{i_o\mu_o:i_o^2}(\bar{\Lambda}_{i_o^2} + t\Lambda_\perp^2) + w_{i_o\mu_o:i_o^1}(\Lambda_{i_o^1} + s\Lambda_\perp^1), \lambda_{i_o\mu_o})$$

<u>has a limit as</u> $s,t \to 0$ <u>which is identically zero for all</u>

$$\begin{cases} \Lambda_{i_o^1} \in \mathcal{H}_1(P_{i_o^1}, A_{i_o^1}) \\ \Lambda_{i_o^2} \in \mathcal{H}_2(P_{i_o^2}, A_{i_o^2}) \end{cases}$$

<u>iff there exists</u>

$$\lambda_{i_o\mu_o} \in \Sigma^o_{P_{i_o\mu_o}}(\mathfrak{g}, \mathfrak{a}_{i_o\mu_o})$$

<u>such that</u>

$$\begin{cases} \mathcal{H}_1(P_{i_o\mu_o}, A_{i_o\mu_o})_{w_{i_o\mu_o:i_o^1}} = (\check{\lambda}_{i_o\mu_o} = c_{i_o\mu_o}) \\ \qquad\qquad\qquad\qquad\qquad\qquad\qquad\qquad (\exists c_{i_o\mu_o}) \\ \mathcal{H}_2(P_{i_o\mu_o}, A_{i_o\mu_o})_{w_{i_o\mu_o:i_o^2}} = (\check{\lambda}_{i_o\mu_o} = -c_{i_o\mu_o}). \end{cases}$$

PROOF If, in the limit, the product vanishes identically, then there is at least one $\lambda_{i_o\mu_o}$ with the property that

$$(w_{i_o\mu_o:i_o^2}\bar{\Lambda}_{i_o^2} + w_{i_o\mu_o:i_o^1}\Lambda_{i_o^1}, \lambda_{i_o\mu_o}) = 0$$

for all

$$\begin{cases} \Lambda_{i_o^1} \in \mathcal{H}_1(P_{i_o^1}, A_{i_o^1}) \\ \Lambda_{i_o^2} \in \mathcal{H}_2(P_{i_o^2}, A_{i_o^2}) \end{cases} .$$

Fix such a $\lambda_{i_o\mu_o}$ -- then we claim that

$$\begin{cases} \mathcal{H}_1(P_{i_o\mu_o}, A_{i_o\mu_o})^{\sim}_{w_{i_o\mu_o:i_o^1}} = \operatorname{Ker}(\check{\lambda}_{i_o\mu_o}) \\ \mathcal{H}_2(P_{i_o\mu_o}, A_{i_o\mu_o})^{\sim}_{w_{i_o\mu_o:i_o^2}} = \operatorname{Ker}(\check{\lambda}_{i_o\mu_o}). \end{cases}$$

Consider, e.g., the $\mathcal{H}_1$-assertion. Recalling that the relevant normals are real, we have

$$(w_{i_o\mu_o:i_o^1}H_1(P_{i_o^1},A_{i_o^1}) + w_{i_o\mu_o:i_o^1}\tilde{\Lambda}_{i_o^1},\lambda_{i_o\mu_o})$$

$$= -(w_{i_o\mu_o:i_o^2}H_2(P_{i_o^2},A_{i_o^2}),\lambda_{i_o\mu_o})$$

for any

$$\tilde{\Lambda}_{i_o^1} \in \mathcal{H}_1(P_{i_o^1},A_{i_o^1})^{\sim} .$$

But since $\tilde{\Lambda}_{i_o^1}$ can take, in particular, the value zero, it follows that

$$\mathcal{H}_1(P_{i_o\mu_o},A_{i_o\mu_o})^{\sim}_{w_{i_o\mu_o:i_o^1}} = \mathrm{Ker}(\check{\lambda}_{i_o\mu_o}),$$

as desired. The verification of the $\mathcal{H}_2$-assertion is similar. On the other hand,

$$w_{i_o\mu_o:i_o^1}H_1(P_{i_o^1},A_{i_o^1}) + w_{i_o\mu_o:i_o^2}H_2(P_{i_o^2},A_{i_o^2})$$

is orthogonal to both $\lambda_{i_o\mu_o}$ and the intersection of

$$\mathcal{H}_1(P_{i_o\mu_o},A_{i_o\mu_o})^{\sim}_{w_{i_o\mu_o:i_o^1}}$$

with

$$\mathcal{H}_2(P_{i_o\mu_o},A_{i_o\mu_o})^{\sim}_{w_{i_o\mu_o:i_o^2}} ,$$

i.e., is orthogonal to both $\lambda_{i_o\mu_o}$ and $\mathrm{Ker}(\check{\lambda}_{i_o\mu_o})$, hence is null. Consequently,

$$\mathcal{H}_1(P_{i_o\mu_o}, A_{i_o\mu_o})w_{i_o\mu_o:i_o^1} = -\mathcal{H}_2(P_{i_o\mu_o}, A_{i_o\mu_o})w_{i_o\mu_o:i_o^2} .$$

Because the various steps in this argument are evidently reversible, we are done. //

The preceding lemma can also be formulated in a qualitative manner. For this purpose, introduce (cf. §3)

$$\underset{\sim}{W}_{\{G\}}(\mathcal{H}_2,\mathcal{H}_1; \mathcal{C}_{i_o^2}, \mathcal{C}_{i_o^1}),$$

$\mathcal{H}_1$ and $\mathcal{H}_2$ being termed equivalent iff

$$\underset{\sim}{W}_{\{G\}}(\mathcal{H}_2,\mathcal{H}_1; \mathcal{C}_{i_o^2}, \mathcal{C}_{i_o^1}) \neq \emptyset$$

Our lemma can then be interpreted to read: The product

$$\prod_{\lambda_{i_o\mu_o}} \cdots (\exists \begin{cases} w_{i_o\mu_o:i_o^2} \\ w_{i_o\mu_o:i_o^1} \end{cases})$$

has a limit as $s,t \to 0$ which is identically zero for all

$$\begin{cases} \Lambda_{i_o^1} \in \mathcal{H}_1(P_{i_o^1}, A_{i_o^1}) \\ \Lambda_{i_o^2} \in \mathcal{H}_2(P_{i_o^2}, A_{i_o^2}) \end{cases}$$

iff $\mathcal{H}_1$ and $\mathcal{H}_2$ are equivalent.

To reflect these circumstances, put

$$\begin{cases} W^{\pm}_{H_1}(A_{i_o\mu_o},A_{i_o^1}) = \{w_{i_o\mu_o:i_o^1} \in W(A_{i_o\mu_o},A_{i_o^1}) : w_{i_o\mu_o:i_o^1}H_1(P_{i_o^1},A_{i_o^1}) \in \pm\mathfrak{D}_{P_{i_o\mu_o}}(\check{\mathfrak{a}}_{i_o\mu_o})^-\} \\ W^{\pm}_{H_2}(A_{i_o\mu_o},A_{i_o^2}) = \{w_{i_o\mu_o:i_o^2} \in W(A_{i_o\mu_o},A_{i_o^2}) : w_{i_o\mu_o:i_o^2}H_2(P_{i_o^2},A_{i_o^2}) \in \pm\mathfrak{D}_{P_{i_o\mu_o}}(\check{\mathfrak{a}}_{i_o\mu_o})^-\}. \end{cases}$$

Then we have

$$\begin{cases} W(A_{i_o\mu_o},A_{i_o^1}) = W^{+}_{H_1}(A_{i_o\mu_o},A_{i_o^1}) \coprod W^{-}_{H_1}(A_{i_o\mu_o},A_{i_o^1}) \\ W(A_{i_o\mu_o},A_{i_o^2}) = W^{+}_{H_2}(A_{i_o\mu_o},A_{i_o^2}) \coprod W^{-}_{H_2}(A_{i_o\mu_o},A_{i_o^2}). \end{cases}$$

Accordingly, split the sum

$$\sum_{W(A_{i_o\mu_o},A_{i_o^2})} \sum_{W(A_{i_o\mu_o},A_{i_o^1})}$$

into three pieces

$$\begin{cases} \sum_{W^{+}_{H_2}(A_{i_o\mu_o},A_{i_o^2})} \sum_{W^{+}_{H_1}(A_{i_o\mu_o},A_{i_o^1})} \\ \sum_{W^{+}_{H_2}(A_{i_o\mu_o},A_{i_o^2})} \sum_{W^{-}_{H_1}(A_{i_o\mu_o},A_{i_o^1})} + \sum_{W^{-}_{H_2}(A_{i_o\mu_o},A_{i_o^2})} \sum_{W^{+}_{H_1}(A_{i_o\mu_o},A_{i_o^1})} \\ \sum_{W^{-}_{H_2}(A_{i_o\mu_o},A_{i_o^2})} \sum_{W^{-}_{H_1}(A_{i_o\mu_o},A_{i_o^1})}. \end{cases}$$

Bearing in mind points (1) and (2) detailed near the end of §4, we can now say the following.

(a) The first piece will contain no singularities from the root denominator but there will be potential (simple) singularities from both of the c-functions.

(b) The second piece will contain singularities from the root denominator iff H_1 and H_2 are equivalent and there will be potential (simple) singularities from one of the c-functions.

(c) The third piece will contain no singularities from the root denominator and no singularities from the c-functions.

It is clear, therefore, that the contribution from the first piece is null. As for the contribution from the first piece, its determination is immediate in that the limit as s and t approach zero of

$$st \cdot \sum_{i_o=1}^{r_o} \sum_{\mu_o=1}^{r_{i_o}} \sum_{w_{i_o\mu_o:i_o^2} \in W_{H_2}^+(A_{i_o\mu_o}, A_{i_o^2})} \sum_{w_{i_o\mu_o:i_o^1} \in W_{H_1}^+(A_{i_o\mu_o}, A_{i_o^1})} \cdots$$

is the product of

$$\exp(\langle I_{P_{i_o\mu_o}}(H), w_{i_o\mu_o:i_o^2} \bar{\Lambda}_{i_o^2} + w_{i_o\mu_o:i_o^1} \Lambda_{i_o^1} \rangle)$$

$$\times (1/\prod_{\lambda_{i_o\mu_o}} (w_{i_o\mu_o:i_o^2} \bar{\Lambda}_{i_o^2} + w_{i_o\mu_o:i_o^1} \Lambda_{i_o^1}, \lambda_{i_o\mu_o}))$$

with

$$(\nabla_{cus}(H_1:G|(\{1\},\{1\}):P_{i_o\mu_o}|A_{i_o\mu_o}:P_{i_o^1}|A_{i_o^1}:w_{i_o\mu_o:i_o^1}:\Lambda_{i_o^1})T_{i_o^1}(1),$$

$$\nabla_{cus}(H_2:G|(\{1\},\{1\}):P_{i_o\mu_o}|A_{i_o\mu_o}:P_{i_o^2}|A_{i_o^2}:w_{i_o\mu_o:i_o^2}:\Lambda_{i_o^2})T_{i_o^2}(1)),$$

summed, of course, over i_o, μ_o and

$$\begin{cases} w_{i_o\mu_o:i_o^2} \in W_{H_2}^{+}(A_{i_o\mu_o}, A_{i_o^2}) \\ \\ w_{i_o\mu_o:i_o^1} \in W_{H_1}^{+}(A_{i_o\mu_o}, A_{i_o^1}) \; . \end{cases}$$

The negligibility supposition on the Eisenstein system functions implies, in the case at hand, that

$$\begin{cases} E_{P_{i_o\mu_o}}(H_1:G|\{1\}:P_{i_o^1}|A_{i_o^1}:T_{i_o^1}:\Lambda_{i_o^1}:?) \\ \\ E_{P_{i_o\mu_o}}(H_2:G|\{1\}:P_{i_o^2}|A_{i_o^2}:T_{i_o^2}:\Lambda_{i_o^2}:?) \end{cases}$$

can be calculated in terms of

$$\begin{cases} \nabla_{cus}(H_1:G|(\{1\},\{1\}):P_{i_o\mu_o}|A_{i_o\mu_o}:P_{i_o^1}|A_{i_o^1}:w_{i_o\mu_o:i_o^1}:\Lambda_{i_o^1})T_{i_o^1}(1) \\ \nabla_{cus}(H_2:G|(\{1\},\{1\}):P_{i_o\mu_o}|A_{i_o\mu_o}:P_{i_o^2}|A_{i_o^2}:w_{i_o\mu_o:i_o^2}:\Lambda_{i_o^2})T_{i_o^2}(1) \end{cases}$$

via a sum over all

$$\begin{cases} w_{i_o\mu_o:i_o^1} \in W_{\{G\}}(H_1;A_{i_o\mu_o}, A_{i_o^1}) \\ \\ w_{i_o\mu_o:i_o^2} \in W_{\{G\}}(H_2;A_{i_o\mu_o}, A_{i_o^2}) \; . \end{cases}$$

But here

$$\begin{cases} W_{\{G\}}(H_1;A_{i_o\mu_o},A_{i_o^1}) = W(A_{i_o\mu_o},A_{i_o^1}) \\ W_{\{G\}}(H_2;A_{i_o\mu_o},A_{i_o^2}) = W(A_{i_o\mu_o},A_{i_o^2}) \end{cases} .$$

Furthermore (cf. §4),

$$\begin{cases} w_{i_o\mu_o:i_o^1} \in W_{H_1}^-(A_{i_o\mu_o},A_{i_o^1}) \\ \Longrightarrow \nabla_{cus}(H_1:\ldots:w_{i_o\mu_o:i_o^1}:\Lambda_{i_o^1}) = 0 \end{cases}$$

and

$$\begin{cases} w_{i_o\mu_o:i_o^2} \in W_{H_2}^-(A_{i_o\mu_o},A_{i_o^2}) \\ \Longrightarrow \nabla_{cus}(H_2:\ldots:w_{i_o\mu_o:i_o^2}:\Lambda_{i_o^2}) = 0 \end{cases} .$$

Bringing in $\mathcal{K}(\mathcal{C}_o:P_{i_o\mu_o}|A_{i_o\mu_o}:\ldots)$, the contribution from the first piece is thus seen to take the form

$$\mathcal{K}(\mathcal{C}_o:P_{i_o\mu_o}|A_{i_o\mu_o}:E_{P_{i_o\mu_o}}(H_1:\ldots:T_{i_o^1}:\Lambda_{i_o^1}:?),E_{P_{i_o\mu_o}}(H_2:\ldots:T_{i_o^2}:\Lambda_{i_o^2}:?):\underset{\sim}{H}),$$

summed over i_o and μ_o, multiplied by $(-1)^{\ell_o}\cdot\mathrm{vol}(\mathcal{C}_o)$.

To complete the calculation of the inner product

$$(\underset{\sim}{Q}^{H}E(H_1:G|\{1\}:P_{i_o^1}|A_{i_o^1}:T_{i_o^1}:\Lambda_{i_o^1}:?),\underset{\sim}{Q}^{H}E(H_2:G|\{1\}:P_{i_o^2}|A_{i_o^2}:T_{i_o^2}:\Lambda_{i_o^2}:?)),$$

we must determine the contribution from the second piece. This, however, turns out to be a good deal more troublesome.

Choose $k_{i_o\mu_o} \in K$ with the property that $P_{i_o\mu_o} = k_{i_o\mu_o} P_{i_o} k_{i_o\mu_o}^{-1}$ -- then $A_{i_o\mu_o} = k_{i_o\mu_o} A_{i_o} k_{i_o\mu_o}^{-1}$. [Note: In this connection, let us keep in mind that it is a question throughout of special split components.] One has, in obvious notation,

$$\begin{cases} W(A_{i_o\mu_o}, A_{i_o^1}) = k_{i_o\mu_o} \cdot W(A_{i_o}, A_{i_o^1}) \\ W(A_{i_o\mu_o}, A_{i_o^2}) = k_{i_o\mu_o} \cdot W(A_{i_o}, A_{i_o^2}). \end{cases}$$

Therefore the products

$$\begin{cases} \prod_{\lambda_{i_o\mu_o}} (k_{i_o\mu_o} w_{i_o:i_o^2}(\bar{\Lambda}_{i_o^2} + t\Lambda_\perp^2) + k_{i_o\mu_o} w_{i_o:i_o^1}(\Lambda_{i_o^1} + s\Lambda_\perp^1), \lambda_{i_o\mu_o}) \\ \prod_{\lambda_{i_o}} (w_{i_o:i_o^2}(\bar{\Lambda}_{i_o^2} + t\Lambda_\perp^2) + w_{i_o:i_o^1}(\Lambda_{i_o^1} + s\Lambda_\perp^1), \lambda_{i_o}), \end{cases}$$

qua functions of $\Lambda_{i_o^1}$ and $\Lambda_{i_o^2}$, are equal for all

$$\begin{cases} w_{i_o:i_o^1} \in W(A_{i_o}, A_{i_o^1}) \\ w_{i_o:i_o^2} \in W(A_{i_o}, A_{i_o^2}). \end{cases}$$

Consequently, if i_o is fixed, then

$$1/\prod_{\lambda_{i_o\mu_o}}$$

can be pulled through the sum over μ_o to

$$1/\prod_{\lambda_{i_o}} .$$

The Weyl sum

$$\sum_{W^+_{H_2}(A_{i_o\mu_o}, A_{i_o^2})} \sum_{W^-_{H_1}(A_{i_o\mu_o}, A_{i_o^1})} + \sum_{W^-_{H_2}(A_{i_o\mu_o}, A_{i_o^2})} \sum_{W^+_{H_1}(A_{i_o\mu_o}, A_{i_o^1})}$$

can in turn be written as

$$\sum_{W^+_{H_2}(A_{i_o}, A_{i_o^2})} \sum_{W^-_{H_1}(A_{i_o}, A_{i_o^1})} + \sum_{W^-_{H_2}(A_{i_o}, A_{i_o^2})} \sum_{W^+_{H_1}(A_{i_o}, A_{i_o^1})}$$

provided, of course, that $k_{i_o\mu_o}$ is inserted at the appropriate places. Not all terms here will contribute. To isolate those that do, given $\lambda_{i_o} \in \Sigma^o_{P_{i_o}}(\mathfrak{g}, \mathfrak{a}_{i_o})$, write

$$W(\lambda_{i_o} : H_2 \times H_1)$$

for the set of all pairs

$$(w_{i_o:i_o^2}, w_{i_o:i_o^1})$$

in

$$W^{\pm}_{H_2}(A_{i_o}, A_{i_o^2}) \times W^{\mp}_{H_1}(A_{i_o}, A_{i_o^1})$$

such that

$$\left\{\begin{array}{l} H_1(P_{i_o},A_{i_o})_{w_{i_o:i_o^1}} = (\check{\lambda}_{i_o} = c_{i_o}) \\ \\ H_2(P_{i_o},A_{i_o})_{w_{i_o:i_o^2}} = (\check{\lambda}_{i_o} = -c_{i_o}) \end{array}\right. \quad (\exists c_{i_o}) .$$

Observe that if H_1 and H_2 are not equivalent, then, of necessity,

$$W(\lambda_{i_o}:H_2 \times H_1) = \emptyset$$

for all λ_{i_o}; on the other hand, if H_1 and H_2 are equivalent, then, of necessity,

$$W(\lambda_{i_o}:H_2 \times H_1) \neq \emptyset$$

for some λ_{i_o}. The

$$W(\lambda_{i_o}:H_2 \times H_1)$$

are pairwise disjoint; in general,

$$\coprod_{\lambda_{i_o}} W(\lambda_{i_o}:H_2 \times H_1)$$

will be a proper subset of

$$W_{H_2}^{\pm}(A_{i_o},A_{i_o^2}) \times W_{H_1}^{\mp}(A_{i_o},A_{i_o^1}) .$$

Nevertheless, the sub-sum

$$\sum_{\lambda_{i_o}} \sum_{W(\lambda_{i_o}:H_2 \times H_1)}$$

contains all the information relevant to our problem. Here is the point. Only one of the two c-functions has a potential singularity. Such a singularity, being simple, is wiped out by taking $s = t$ unless there is also a singularity in the root denominator, thus explaining the introduction of the

$$W(\lambda_{i_o} : \mathfrak{H}_2 \times \mathfrak{H}_1).$$

Thanks to the preceding remarks, our position can now be reinforced, in that we must calculate the limit as s and t approach zero of

$$st \cdot \sum_{i_o=1}^{r_o} \sum_{\lambda_{i_o} \in \Sigma^o_{P_{i_o}}(\mathfrak{g}, \mathfrak{a}_{i_o})} \sum_{(w_{i_o:i_o^2}, w_{i_o:i_o^1}) \in W(\lambda_{i_o} : \mathfrak{H}_2 \times \mathfrak{H}_1)}$$

$$\Big[\prod_{\lambda \in \Sigma^o_{P_{i_o}}(\mathfrak{g}, \mathfrak{a}_{i_o})} (w_{i_o:i_o^2}(\bar{\Lambda}_{i_o^2} + t\Lambda_\perp^2) + w_{i_o:i_o^1}(\Lambda_{i_o^1} + s\Lambda_\perp^1), \lambda)\Big]^{-1}$$

times the sum over μ_o of the product of

$$\exp(\langle I_{P_{i_o\mu_o}}(\underset{\sim}{H}), k_{i_o\mu_o} w_{i_o:i_o^2}(\bar{\Lambda}_{i_o^2} + t\Lambda_\perp^2) + k_{i_o\mu_o} w_{i_o:i_o^1}(\Lambda_{i_o^1} + s\Lambda_\perp^1)\rangle)$$

with

$$(c_{cus}(P_{i_o\mu_o}|A_{i_o\mu_o} : P_{i_o^1}|A_{i_o^1} : k_{i_o\mu_o} w_{i_o:i_o^1} : \Lambda_{i_o^1} + s\Lambda_\perp^1)T_{i_o^1}(1),$$

$$c_{cus}(P_{i_o\mu_o}|A_{i_o\mu_o} : P_{i_o^2}|A_{i_o^2} : k_{i_o\mu_o} w_{i_o:i_o^2} : \Lambda_{i_o^2} + t\Lambda_\perp^2)T_{i_o^2}(1)),$$

all multiplied by $(-1)^{\ell_o} \cdot \mathrm{vol}(\mathcal{C}_o)$.

To this end, fix i_o and λ_{i_o}. Put

$$\begin{cases} W^+(\lambda_{i_o}:H_2\times H_1) = \{(w_{i_o:i_o^2}, w_{i_o:i_o^1}) : w_{i_o:i_o^1}H_1(P_{i_o^1}, A_{i_o^1}) \in \mathcal{D}_{P_{i_o}}(\check{a}_{i_o})^-\} \\ W^-(\lambda_{i_o}:H_2\times H_1) = \{(w_{i_o:i_o^2}, w_{i_o:i_o^1}) : w_{i_o:i_o^1}H_1(P_{i_o^1}, A_{i_o^1}) \in -\mathcal{D}_{P_{i_o}}(\check{a}_{i_o})^-\} . \end{cases}$$

Then

$$W(\lambda_{i_o}:H_2\times H_1) = W^+(\lambda_{i_o}:H_2\times H_1) \coprod W^-(\lambda_{i_o}:H_2\times H_1).$$

Call $w_{\lambda_{i_o}}$ the root reflection associated with λ_{i_o} (secured by GR) -- then still

$$W^-(\lambda_{i_o}:H_2\times H_1)$$

$$= \{(w_{\lambda_{i_o}} w_{i_o:i_o^2}, w_{\lambda_{i_o}} w_{i_o:i_o^1}) : (w_{i_o:i_o^2}, w_{i_o:i_o^1}) \in W^+(\lambda_{i_o}:H_2\times H_1)\} .$$

Utilizing this decomposition of $W(\lambda_{i_o}:H_2\times H_1)$, we must therefore explicate

$$\lim_{\substack{s\to 0\\ t\to 0}} st\cdot \sum_{(w_{i_o:i_o^2}, w_{i_o:i_o^1}) \in W^+(\lambda_{i_o}:H_2\times H_1)} (\mathrm{I} + \mathrm{II})$$

where I is

$$\Big[\prod_{\lambda\in\Sigma^o_{P_{i_o}}(\mathfrak{g},\mathfrak{a}_{i_o})} (w_{i_o:i_o^2}(\bar{\Lambda}_{i_o^2} + t\Lambda_\perp^2) + w_{i_o:i_o^1}(\Lambda_{i_o^1} + s\Lambda_\perp^1), \lambda)\Big]^{-1}$$

times the sum over μ_o of the product of

$$\exp(\langle I_{P_{i_o\mu_o}}(\underset{\sim}{H}), k_{i_o\mu_o} w_{i_o:i_o^2}(\bar{\Lambda}_{i_o^2} + t\Lambda_\perp^2) + k_{i_o\mu_o} w_{i_o:i_o^1}(\Lambda_{i_o^1} + s\Lambda_\perp^1)\rangle)$$

with

$$(c_{cus}(P_{i_o\mu_o}|A_{i_o\mu_o}:P_{i_o^1}|A_{i_o^1}:k_{i_o\mu_o}w_{i_o:i_o^1}:\Lambda_{i_o^1} + s\Lambda^1)T_{i_o^1}(1),$$

$$c_{cus}(P_{i_o\mu_o}|A_{i_o\mu_o}:P_{i_o^2}\,A_{i_o^2}:k_{i_o\mu_o}w_{i_o:i_o^2}:\Lambda_{i_o^2} + t\Lambda^2)T_{i_o^2}(1)),$$

all multiplied by $(-1)^{\ell_o}\cdot\mathrm{vol}(\mathcal{C}_o)$, while II is

$$\Big[\prod_{\lambda\in\Sigma^o_{P_{i_o}}(\mathfrak{g},\mathfrak{a}_{i_o})}(w_{\lambda_{i_o}}w_{i_o:i_o^2}(\bar{\Lambda}_{i_o^2} + t\Lambda_\perp^2) + w_{\lambda_{i_o}}w_{i_o:i_o^1}(\Lambda_{i_o^1} + s\Lambda_\perp^1),\ \lambda)\Big]^{-1}$$

times the sum over μ_o of the product of

$$\exp(\langle I_{P_{i_o\mu_o}}(\underset{\sim}{H}),k_{i_o\mu_o}w_{\lambda_{i_o}}w_{i_o:i_o^2}(\bar{\Lambda}_{i_o^2} + t\Lambda_\perp^2) + k_{i_o\mu_o}w_{\lambda_{i_o}}w_{i_o:i_o^1}(\Lambda_{i_o^1} + s\Lambda_\perp^1)\rangle)$$

with

$$(c_{cus}(P_{i_o\mu_o}|A_{i_o\mu_o}:P_{i_o^1}|A_{i_o^1}:k_{i_o\mu_o}w_{\lambda_{i_o}}w_{i_o:i_o^1}:\Lambda_{i_o^1} + s\Lambda_\perp^1)T_{i_o^1}(1),$$

$$c_{cus}(P_{i_o\mu_o}|A_{i_o\mu_o}:P_{i_o^2}|A_{i_o^2}:k_{i_o\mu_o}w_{\lambda_{i_o}}w_{i_o:i_o^2}:\Lambda_{i_o^2} + t\Lambda_\perp^2)T_{i_o^2}(1)),$$

all multiplied by $(-1)^{\ell_o}$ $\mathrm{vol}(\mathcal{C}_o)$. In reality, it will be more convenient to evaluate

$$\lim_{t\to 0}\lim_{s\to t} st\cdot \sum_{(w_{i_o:i_o^2},w_{i_o:i_o^1})\in W^+(\lambda_{i_o}:\mathfrak{H}_2\times\mathfrak{H}_1)} (I + II).$$

Naturally, no loss of generality is entailed in so doing.

The final result turns out to be fairly simple. Before we state it, let us recall that one may attach to λ_{i_o}, in the usual way, a Γ-cuspidal parabolic

$P_{\lambda_{i_o}} \succeq P_{i_o}$, the special split component of $P_{\lambda_{i_o}}$ being, of course,

$$A_{\lambda_{i_o}} = \exp(\mathrm{Ker}(\lambda_{i_o})).$$

Denote by

$$\mathcal{C}(\lambda_{i_o})$$

the association class to which $P_{\lambda_{i_o}}$ belongs.

LEMMA 5.2 <u>The limit</u>

$$\lim_{t\to 0} \lim_{s\to t} st \cdot \sum_{(w_{i_o:i_o^2}, w_{i_o:i_o^1}) \in W^+(\lambda_{i_o}:\mathfrak{H}_2 \times \mathfrak{H}_1)} (I + II)$$

<u>exists and is equal to</u>

$$(-1)^{\ell_o - 1} \cdot \mathrm{vol}(\mathcal{C}(\lambda_{i_o})) \cdot \sum_{(w_{i_o:i_o^2}, w_{i_o:i_o^1}) \in W^+(\lambda_{i_o}:\mathfrak{H}_2 \times \mathfrak{H}_1)}$$

$$\Big[\prod_{\lambda \neq \lambda_{i_o}} (w_{i_o:i_o^2}\bar{\Lambda}_{i_o^2} + w_{i_o:i_o^1}\Lambda_{i_o^1}, \lambda)\Big]^{-1}$$

<u>times the sum over</u> μ_o <u>of the product of</u>

$$\exp(\langle I_{P_{i_o\mu_o}}(\underset{\sim}{H}), k_{i_o\mu_o} w_{i_o:i_o^2}\bar{\Lambda}_{i_o^2} + k_{i_o\mu_o} w_{i_o:i_o^1}\Lambda_{i_o^1}\rangle)$$

<u>with</u>

$$(\nabla_{cus}(\mathfrak{H}_1:G|(\{1\},\{1\}):P_{i_o\mu_o}|A_{i_o\mu_o}:P_{i_o^1}|A_{i_o^1}:k_{i_o\mu_o}w_{i_o:i_o^1}:\Lambda_{i_o^1})T_{i_o^1}(1),$$

$$c_{cus}(P_{i_o\mu_o}|A_{i_o\mu_o}:P_{i_o^2}|A_{i_o^2}:k_{i_o\mu_o}w_{i_o:i_o^2}:\Lambda_{i_o^2})T_{i_o^2}(1)).$$

[Note: The symbol $\prod_{\lambda \neq \lambda_{i_o}}$ stands for a product over the $\lambda \neq \lambda_{i_o}$ in $\Sigma^o_{P_{i_o}}(\mathfrak{g}, \mathfrak{a}_{i_o})$.]

In making the evaluation, we might just as well fix an element

$$(w_{i_o:i_o^2}, w_{i_o:i_o^1}) \in W^+(\lambda_{i_o}:H_2 \times H_1).$$

This done, let us first explain how the term

$$\mathrm{vol}(\mathcal{C}(\lambda_{i_o}))$$

arises. Consider, per I, the product

$$\prod_{\lambda \in \Sigma^o_{P_{i_o}}(\mathfrak{g}, \mathfrak{a}_{i_o})} \cdots \quad .$$

Pull out the factor corresponding to $\lambda = \lambda_{i_o}$ -- then (cf. supra)

$$\begin{aligned}
&(w_{i_o:i_o^2}(\bar{\Lambda}_{i_o^2} + t\Lambda_\perp^2) + w_{i_o:i_o^1}(\Lambda_{i_o^1} + s\Lambda_\perp^1), \lambda_{i_o}) \\
&= (tw_{i_o:i_o^2}\Lambda_\perp^2 + sw_{i_o:i_o^1}\Lambda_\perp^1, \lambda_{i_o}).
\end{aligned}$$

But, in view of the definitions,

$$\begin{cases}
w_{i_o:i_o^1}\Lambda_\perp^1 = \dfrac{\lambda_{i_o}}{\|\lambda_{i_o}\|} \\[2ex]
w_{i_o:i_o^2}\Lambda_\perp^2 = \dfrac{-\lambda_{i_o}}{\|\lambda_{i_o}\|}
\end{cases} .$$

Thus our inner product is equal to

$$(s-t)\cdot \|\lambda_{i_o}\| .$$

On the other hand, per II, the associated contribution is evidently just the opposite, viz.

$$(t-s)\cdot \|\lambda_{i_o}\| .$$

Since the relevant factors actually occur in the denominator, we come up with

$$\mathrm{vol}(\mathcal{C}_o)/\|\lambda_{i_o}\|$$

which, after a moments reflection, is seen to be precisely

$$\mathrm{vol}(\mathcal{C}(\lambda_{i_o})).$$

To compute

$$\lim_{s\to t} st\cdot \ldots,$$

we shall need to invoke L'Hôspital's rule. Accordingly, it must be shown that

$$\Big[\prod_{\lambda\neq\lambda_{i_o}} (w_{i_o:i_o^2}(\bar{\Lambda}_{i_o^2} + t\Lambda_{\perp}^2) + w_{i_o:i_o^1}(\Lambda_{i_o^1} + t\Lambda_{\perp}^1),\lambda)\Big]^{-1}$$

times the sum over μ_o of the product of

$$\exp(\langle I_{P_{i_o\mu_o}}(\underset{\sim}{H}), k_{i_o\mu_o} w_{i_o:i_o^2}(\bar{\Lambda}_{i_o^2} + t\Lambda_{\perp}^2) + k_{i_o\mu_o} w_{i_o:i_o^1}(\Lambda_{i_o^1} + s\Lambda_{\perp}^1)\rangle)$$

with

$$(c_{cus}(P_{i_o\mu_o}|A_{i_o\mu_o}:P_{i_o^1}|A_{i_o^1}:k_{i_o\mu_o}w_{i_o:i_o^1}:\Lambda_{i_o^1}+t\Lambda_\perp^1)T_{i_o^1}(1),$$

$$c_{cus}(P_{i_o\mu_o}|A_{i_o\mu_o}:P_{i_o^2}|A_{i_o^2}:k_{i_o\mu_o}w_{i_o:i_o^2}:\Lambda_{i_o^2}+t\Lambda_\perp^2)T_{i_o^2}(1))$$

LESS

$$\Big[\prod_{\lambda\neq\lambda_{i_o}}(w_{\lambda_{i_o}}w_{i_o:i_o^2}(\bar\Lambda_{i_o^2}+t\Lambda_\perp^2)+w_{\lambda_{i_o}}w_{i_o:i_o^1}(\Lambda_{i_o^1}+t\Lambda_\perp^1),\lambda)\Big]^{-1}$$

times the sum over μ_o of the product of

$$\exp(\langle I_{P_{i_o\mu_o}}(\underset{\sim}{H}),k_{i_o\mu_o}w_{\lambda_{i_o}}w_{i_o:i_o^2}(\bar\Lambda_{i_o^2}+t\Lambda_\perp^2)+k_{i_o\mu_o}w_{\lambda_{i_o}}w_{i_o:i_o^1}(\Lambda_{i_o^1}+t\Lambda_\perp^1)\rangle)$$

with

$$(c_{cus}(P_{i_o\mu_o}|A_{i_o\mu_o}:P_{i_o^1}|A_{i_o^1}:k_{i_o\mu_o}w_{\lambda_{i_o}}w_{i_o:i_o^1}:\Lambda_{i_o^1}+t\Lambda_\perp^1)T_{i_o^1}(1),$$

$$c_{cus}(P_{i_o\mu_o}|A_{i_o\mu_o}:P_{i_o^2}|A_{i_o^2}:k_{i_o\mu_o}w_{\lambda_{i_o}}w_{i_o:i_o^2}:\Lambda_{i_o^2}+t\Lambda_\perp^2)T_{i_o^2}(1)).$$

is

$$\equiv 0.$$

Since

$$w_{i_o:i_o^1}\Lambda_\perp^1+w_{i_o:i_o^2}\Lambda_\perp^2=0,$$

the outside quotients are the same. The seemingly more serious part

$$\sum_{\mu_o}\exp\cdot(c_{cus},c_{cus})-\sum_{\mu_o}\exp\cdot(c_{cus},c_{cus})$$

has actually figured earlier in a different context (see §9 of [2]), the desired nullity being a consequence of the functional equations for the c-functions.

Legitimately, therefore,

$$\lim_{s\to t} \frac{1}{s-t}\cdot\Big\{\Big[\prod_{\lambda\neq\lambda_{i_o}} \cdots\Big]^{-1}\cdot\sum_{\mu_o}\cdots - \Big[\prod_{\lambda\neq\lambda_{i_o}}\cdots\Big]^{-1}\cdot\sum_{\mu_o}\cdots\Big\}$$

is

$$\frac{\partial}{\partial s}\cdot\Big\{\Big[\prod_{\lambda\neq\lambda_{i_o}} \cdots\Big]^{-1}\cdot\sum_{\mu_o}\cdots - \Big[\prod_{\lambda\neq\lambda_{i_o}}\cdots\Big]^{-1}\cdot\sum_{\mu_o}\cdots\Big\}\Big|_{s=t}\,.$$

One term which arises from the product rule is

$$\Big[\prod_{\lambda\neq\lambda_{i_o}} (w_{i_o:i_o^2}(\bar\Lambda_{i_o^2}+t\Lambda_\perp^2)+w_{i_o:i_o^1}(\Lambda_{i_o^1}+t\Lambda_\perp^1),\lambda)\Big]^{-1}$$

times the sum over μ_o of the product of

$$\exp(\langle I_{P_{i_o\mu_o}}(\underset{\sim}{H}),k_{i_o\mu_o}w_{i_o:i_o^2}(\bar\Lambda_{i_o^2}+t\Lambda_\perp^2)+k_{i_o\mu_o}w_{i_o:i_o^1}(\Lambda_{i_o^1}+t\Lambda_\perp^1)\rangle)$$

with

$$(\frac{\partial}{\partial s}c_{cus}(P_{i_o\mu_o}|A_{i_o\mu_o}:P_{i_o^1}|A_{i_o^1}:k_{i_o\mu_o}w_{i_o:i_o^1}:\Lambda_{i_o^1}+s\Lambda_\perp^1)\Big|_{s=t}T_{i_o^1}(1),$$
$$c_{cus}(P_{i_o\mu_o}|A_{i_o\mu_o}:P_{i_o^2}|A_{i_o^2}:k_{i_o\mu_o}w_{i_o:i_o^2}:\Lambda_{i_o^2}+t\Lambda_\perp^2)T_{i_o^2}(1)).$$

When this expression is multiplied by t^2 and the limit

$$\lim_{t\to 0}$$

is taken, an appeal to the generality

$$\lim_{t\to 0} t^2\cdot(\frac{d}{ds}\sum_{n=-1}^{\infty} a_n s^n)\Big|_{s=t} = -a_{-1}$$

implies that what remains is precisely

$$\Big[\prod_{\lambda\neq\lambda_{i_o}} (w_{i_o:i_o^2}\bar{\Lambda}_{i_o^2} + w_{i_o:i_o^1}\Lambda_{i_o^1},\lambda)\Big]^{-1}$$

times the sum over μ_o of the product of

$$\exp(\Big\langle I_{P_{i_o\mu_o}}(\underset{\sim}{H}), k_{i_o\mu_o} w_{i_o:i_o^2}\bar{\Lambda}_{i_o^2} + k_{i_o\mu_o} w_{i_o:i_o^1}\Lambda_{i_o^1}\Big\rangle)$$

with

$$-(\nabla_{cus}(H_1:G|(\{1\},\{1\}):P_{i_o\mu_o}|A_{i_o\mu_o}:P_{i_o^1}|A_{i_o^1}:k_{i_o\mu_o}w_{i_o:i_o^1}:\Lambda_{i_o^1})T_{i_o^1}(1),$$

$$c_{cus}(P_{i_o\mu_o}|A_{i_o\mu_o}:P_{i_o^2}|A_{i_o^2}:k_{i_o\mu_o}w_{i_o:i_o^2}:\Lambda_{i_o^2})T_{i_o^2}(1)),$$

which, up to a minus sign, is the ambient term in our lemma. Of course, the minus sign itself is taken care of by

$$(-1)^{\ell_o-1}.$$

What about the other terms? The claim is that they must vanish when we pass to the limit. That this is in fact the case can be seen by remarking that it is either a question of a simple pole (and we are multiplying by t^2) or else the relevant ∇_{cus} is null (as follows from the definitions and the way matters have been arranged). Lemma 5.2 is thereby demonstrated.

This is a good point at which to formulate a statement of recapitulation.

PROPOSITION 5.3 *The inner product*

$$(Q_{\sim}^{H}E(H_1:G|\{1\}:P_{i_o^1}|A_{i_o^1}:T_{i_o^1}:\Lambda_{i_o^1}:?),\ Q_{\sim}^{H}E(H_2:G|\{1\}:P_{i_o^2}|A_{i_o^2}:T_{i_o^2}:\Lambda_{i_o^2}:?))$$

is equal to

$$(-1)^{\ell_o}\cdot\mathrm{vol}(\mathcal{C}_o)$$

times the sum over

$$\begin{cases} i_o, 1 \le i_o \le r_o \\ \mu_o, 1 \le \mu_o \le r_{i_o} \end{cases}$$

of

$$\mathcal{K}(\mathcal{C}_o:P_{i_o\mu_o}|A_{i_o\mu_o}:E_{P_{i_o\mu_o}}(H_1:\ldots:T_{i_o^1}:\Lambda_{i_o^1}:?),E_{P_{i_o\mu_o}}(H_2:\ldots:T_{i_o^2}:\Lambda_{i_o^2}:?):\underset{\sim}{H})$$

plus the sum over

$$\begin{cases} i_o, 1 \le i_o \le r_o \\ \lambda_{i_o} \in \Sigma_{P_{i_o}}^{o}(\mathfrak{g},\mathfrak{a}_{i_o}) \\ (w_{i_o:i_o^2}, w_{i_o:i_o^1}) \in W^{+}(\lambda_{i_o}:H_2\times H_1) \end{cases}$$

of

$$(-1)^{\ell_o-1}\cdot\mathrm{vol}(\mathcal{C}(\lambda_{i_o}))\cdot\Big[\prod_{\lambda\neq\lambda_{i_o}}(w_{i_o:i_o^2}\bar{\Lambda}_{i_o^2}+w_{i_o:i_o^1}\Lambda_{i_o^1},\lambda)\Big]^{-1}$$

times the sum over μ_o *of the product of*

$$\exp(\langle I_{P_{i_o\mu_o}}(\underset{\sim}{H}),k_{i_o\mu_o}w_{i_o:i_o^2}\bar{\Lambda}_{i_o^2}+k_{i_o\mu_o}w_{i_o:i_o^1}\Lambda_{i_o^1}\rangle)$$

with

$$(\nabla_{\text{cus}}(H_1:G|(\{1\},\{1\}):P_{i_o\mu_o}|A_{i_o\mu_o}:P_{i_o^1}|A_{i_o^1}:k_{i_o\mu_o}w_{i_o:i_o^1}:\Lambda_{i_o^1})T_{i_o^1}(1),$$

$$c_{\text{cus}}(P_{i_o\mu_o}|A_{i_o\mu_o}:P_{i_o^2}|A_{i_o^2}:k_{i_o\mu_o}w_{i_o:i_o^2}:\Lambda_{i_o^2})T_{i_o^2}(1)).$$

[Note: Here, it is understood that the truncation parameter $\underset{\sim}{H}$ belongs to $\underset{\sim}{\mathfrak{a}}_Q$.]

REMARK It should be kept in mind that the primary case of interest is when

$$\begin{cases} T_{i_o^1} = T_{i_o^1}(\Phi_1:H_1) \\ T_{i_o^2} = T_{i_o^2}(\Phi_2:H_2) \end{cases}$$

for

$$\begin{cases} \Phi_1 \in E(\delta,\mathcal{O}_1:\mathcal{C}_o) \cap E(\delta,\underset{\sim}{\mathcal{O}}_o;\underset{\sim}{H}:P_1|A_1) \\ \Phi_2 \in E(\delta,\mathcal{O}_2:\mathcal{C}_o) \cap E(\delta,\underset{\sim}{\mathcal{O}}_o;\underset{\sim}{H}:P_2|A_2). \end{cases}$$

The parameters may then be specialized to

$$\begin{cases} \Lambda_{i_o^1} = \Lambda_1 + H_1(P_{i_o^1},A_{i_o^1}) \\ \Lambda_{i_o^2} = \Lambda_2 + H_2(P_{i_o^2},A_{i_o^2}). \end{cases}$$

Let us recall that the initial data consists of two association classes $\mathcal{C}_1$ and $\mathcal{C}_2$ admitting $\mathcal{C}_c$ as the common dominant subject to the condition

$$\begin{cases} \text{rank } (\mathcal{C}_o) - \text{rank } (\mathcal{C}_1) = 1 \\ \text{rank } (\mathcal{C}_o) - \text{rank } (\mathcal{C}_2) = 1. \end{cases}$$

It has been noted earlier that if H_1 and H_2 are not equivalent, then, of necessity,

$$W(\lambda_{i_o} : H_2 \times H_1) = \emptyset$$

for all λ_{i_o}. Since an empty sum is conventionally null, there can be no substantive contribution from the second part in this case. On the other hand, if H_1 and H_2 are equivalent, then as

$$\begin{cases} H_1 \\ H_2 \end{cases}$$

are principal with

$$\begin{cases} P_{\text{Dis}(H_1)} = P_1 \\ A_{\text{Dis}(H_1)} = A_1 \end{cases} \qquad \begin{cases} P_{\text{Dis}(H_2)} = P_2 \\ A_{\text{Dis}(H_2)} = A_2 \end{cases},$$

it follows that

$$\begin{cases} (P_1, S_1; A_1) \\ (P_2, S_2; A_2) \end{cases}$$

are associate, hence that

$$\mathcal{C}_1 = \mathcal{C}_2 .$$

There is more to be said about the contribution from the second part. On the basis of the preceding remarks, when treating it, we might just as well suppose that $\mathcal{C}_1 = \mathcal{C}_2$, say $\mathcal{C}$. However, before we embark on this journey, which requires a considerably deeper and more subtle analysis, let us formally state as a theorem what has been learned in the case when $\mathcal{C}_1 \neq \mathcal{C}_2$.

THEOREM 5.4 *Let $\mathcal{C}_1$ and $\mathcal{C}_2$ be distinct association classes of Γ-cuspidal split parabolic subgroups of G. Let $\mathcal{C}_0$ be an association class of Γ-cuspidal split parabolic subgroups of G, dominated by both $\mathcal{C}_1$ and $\mathcal{C}_2$ and subject to the condition*

$$\begin{cases} \text{rank}\,(\mathcal{C}_0) - \text{rank}\,(\mathcal{C}_1) = 1 \\ \text{rank}\,(\mathcal{C}_0) - \text{rank}\,(\mathcal{C}_2) = 1. \end{cases}$$

Let

$$\begin{cases} \Phi_1 \in E(\delta, \mathcal{O}_1 : \mathcal{C}_0) \\ \Phi_2 \in E(\delta, \mathcal{O}_2 : \mathcal{C}_0). \end{cases}$$

Then, for all $\underset{\sim}{H} \in \underset{\sim}{\mathfrak{a}}_Q$, the inner product

$$(Q^{\underset{\sim}{H}} E(P_1 | A_1 : \Phi_1 : \Lambda_1 : ?),\ Q^{\underset{\sim}{H}} E(P_2 | A_2 : \Phi_2 : \Lambda_2 : ?))$$

is equal to

$$(-1)^{\ell_0} \cdot \text{vol}(\mathcal{C}_0)$$

<u>times</u>

$$\mathcal{K}(\mathcal{C}_o : P_{i_o\mu_o} | A_{i_o\mu_o} : E_{P_{i_o\mu_o}}(P_1|A_1:\Phi_1:\Lambda_1:?),\ E_{P_{i_o\mu_o}}(P_2|A_2:\Phi_2:\Lambda_2:?):\underset{\sim}{H}),$$

<u>summed over</u> i_o <u>and</u> μ_o.

We shall now proceed to the study of the second expression figuring in Proposition 5.3. In so doing, we can and will assume that $\mathcal{H}_1$ and $\mathcal{H}_2$ are equivalent with

$$\begin{cases} (P_1, S_1; A_1) \\ \\ (P_2, S_2; A_2) \end{cases} \in \mathcal{C} \ .$$

At the outset, it should be noted that there is present an apparent asymmetry vis-a-vis ∇_{cus} and c_{cus}. In the end, this turns out to be a mere illusion. But the justification is not immediate in that there are some preliminary questions that must be dealt with, the answers to which depending ultimately on the most profound aspects of the theory of Eisenstein systems.

The association class $\mathcal{C}$ being given, fix an element

$$(P_i, S_i; A_i) \in \mathcal{C}_i \subset \mathcal{C} \ .$$

Keeping to the assumption that $\mathcal{C}_o$ is an association class dominated by $\mathcal{C}$

with

$$\text{rank}\ (\mathcal{C}_o) - \text{rank}\ (\mathcal{C}) = 1,$$

let

$$H \in \underset{\sim}{H}(\mathcal{C}_o;\ \text{rank}\ (\mathcal{C}_o) - 1) : \chi_{\mathcal{C}_i}\ H\text{-admissible}$$

where, specifically,

$$H \in \underset{\sim}{H}_{i_o}\ (\text{rank}\ (\mathcal{C}_o) - 1).$$

Then H is principal with

$$\begin{cases} P_{\text{Dis}(H)} = P_i \\ A_{\text{Dis}(H)} = A_i. \end{cases}$$

Suppose that j_o is another index lying between 1 and r_o. Let

$$w_{j_o i_o} \in W_{\{G\}}(H; A_{j_o}, A_{i_o})$$

or still, let

$$w_{j_o i_o} \in W(A_{j_o}, A_{i_o}).$$

Then we may form

$$H_{\underset{\sim}{w}_{j_o i_o}},$$

which will be assumed to belong to

$$\underset{\sim}{H}_{j_o}\ (\text{rank}\ (\mathcal{C}_o) - 1)$$

and which we shall take, in addition, to be principal. The latter assumption implies that

$$\mathfrak{H}(P_{j_o}, A_{j_o})^{\sim}_{w_{j_o i_o}} = \mathrm{Ker}(\check{\lambda}_{j_o})$$

for some λ_{j_o} in $\Sigma^o_{P_{j_o}}(\mathfrak{g}, \mathfrak{a}_{j_o})$, the associated element $w^o_{j_o}$ per Gr (cf. §3) being the root reflection $w_{\lambda_{j_o}}$. Thus

$$P_{\mathrm{Dis}(\mathfrak{H}_{w_{j_o i_o}})} = P_{\lambda_{j_o}},$$

say P_j, so that

$$(P_j, S_j; A_j) \succeq (P_{j_o}, S_{j_o}; A_{j_o}),$$

determining, therefore, a G-conjugacy class $\mathcal{C}_j \subset \mathcal{C}$. Let, as usual (cf. TES), $\mathcal{C}_{j_o}(P_j|A_j)$ be the subset of $\mathcal{C}_{j_o}$ consisting of those elements which are dominated predecessors of $(P_j, S_j; A_j)$. Fix a set

$$\{(P_{j_o\nu_o}, S_{j_o\nu_o}; A_{j_o\nu_o})\}$$

of representatives for

$$\Gamma \cap P_j \backslash \mathcal{C}_{j_o}(P_j|A_j).$$

Needless to say, the triple $(P_{j_o}, S_{j_o}; A_{j_o})$ is to be regarded as the base point of the data. Choose, accordingly, $k_{j_o\nu_o} \in K$ with the property that $k_{j_o\nu_o}$

conjugates

$$(P_{j_o}, S_{j_o}; A_{j_o})$$

to

$$(P_{j_o\nu_o}, S_{j_o\nu_o}; A_{j_o\nu_o}).$$

We remark, in passing, that parabolics $P_{j_o\nu_o}$ corresponding to distinct indices ν_o cannot be Γ-conjugate; on the other hand, the $P_{j_o\nu_o}$ need not fill out $\Gamma\backslash\mathcal{C}_{j_o}$. This will not be any cause for concern, though (cf. infra).

Our immediate objective is to study

$$\nabla_{cus}(H:G|(\{1\},\{1\}):P_{j_o}|A_{j_o}:P_{i_o}|A_{i_o}:w_{j_o i_o}:\Lambda_{i_o})$$

in the light of the present particular circumstances. To this end, consider

$$\nabla_{cus}(H_{w_{j_o i_o}}:G|(\{1\},\{1\}):P_{j_o}|A_{j_o}:P_{j_o\nu_o}|A_{j_o\nu_o}:w_{\lambda_{j_o}}k_{j_o\nu_o}^{-1}:k_{j_o\nu_o}\Lambda_{j_o}),$$

which, by definition, is a linear transformation

$$E_{cus}(\delta,\mathcal{O}_{j_o\nu_o}) \to E_{cus}(\delta,\mathcal{O}_{j_o})$$

of a certain prescribed type, depending, a priori, on the parameter Λ_{j_o}. In fact, however,

$$\nabla_{cus}(H_{w_{j_o i_o}}:G|(\{1\},\{1\}):P_{j_o}|A_{j_o}:P_{j_o\nu_o}|A_{j_o\nu_o}:w_{\lambda_{j_o}}k_{j_o\nu_o}^{-1}:k_{j_o\nu_o}\Lambda_{j_o})$$

is a constant, being precisely (cf. TES, Chap. 5)

$$\nabla_{cus}(H_{\underset{\sim}{w}_{j_o i_o}}:P_j|(A_j,A_j):P_{j_o}|A_{j_o}:P_{j_o\nu_o}|A_{j_o\nu_o}:w_{\lambda_{j_o}}k^{-1}_{j_o\nu_o}:?_{j_o\nu_o}) ,$$

where

$$?_{j_o\nu_o} = -k_{j_o\nu_o}w_{j_o i_o}H(P_{i_o},A_{i_o})$$

is the normal of $H_{\underset{\sim}{w}_{j_o i_o}}$ per $(P_{j_o\nu_o},A_{j_o\nu_o})$. Call it

$$\nabla_{cus}(H:(w_{\lambda_{j_o}},w_{j_o i_o}):\nu_o)$$

for short.

PROPOSITION 5.5 <u>Suppose that</u> H <u>and</u> $H_{\underset{\sim}{w}_{j_o i_o}}$ <u>satisfy the hypotheses set down above -- then</u>

$$\nabla_{cus}(H:G|(\{1\},\{1\}):P_{j_o}|A_{j_o}:P_{i_o}|A_{i_o}:w_{j_o i_o}:\Lambda_{i_o})$$

<u>is equal to</u>

$$\sum_{\nu_o}^{(j)}\nabla_{cus}(H:(w_{\lambda_{j_o}},w_{j_o i_o}):\nu_o)\circ c_{cus}(P_{j_o\nu_o}|A_{j_o\nu_o}:P_{i_o}|A_{i_o}:k_{j_o\nu_o}w_{\lambda_{j_o}}w_{j_o i_o}:\Lambda_{i_o}).$$

Here, the symbol

$$\sum_{\nu_o}^{(j)}$$

stands for a sum over the ν_o per

$$\Gamma\cap P_j\backslash \mathcal{C}_{j_o}(P_j|A_j).$$

Before we give the proof, there are two things which should be noticed.

(1) $H(P_{i_o}, A_{i_o})$ is not a singular hyperplane for

$$c_{cus}(P_{j_o\nu_o}|A_{j_o\nu_o}:P_{i_o}|A_{i_o}:k_{j_o\nu_o}w_{\lambda_{j_o}}w_{j_oi_o}:\Lambda_{i_o})$$

Indeed,

$$k_{j_o\nu_o}w_{\lambda_{j_o}}w_{j_oi_o}H(P_{i_o}\ A_{i_o}) \in -\mathfrak{D}_{P_{j_o\nu_o}}(\check{\underline{a}}_{j_o\nu_o})^-,$$

thus our assertion follows from what was said near the end of §4.

(2) The composition of linear transformations supra is meaningful.

Indeed, it is a question of putting together arrows

$$\begin{cases} E_{cus}(\delta, \mathcal{O}_{i_o}) \to E_{cus}(\delta, \mathcal{O}_{j_o\nu_o}) \\ E_{cus}(\delta, \mathcal{O}_{j_o\nu_o}) \to E_{cus}(\delta, \mathcal{O}_{j_o}) \end{cases}$$

to get a map

$$E_{cus}(\delta, \mathcal{O}_{i_o}) \to E_{cus}(\delta, \mathcal{O}_{j_o}).$$

PROOF OF PROPOSITION 5.5 We have

$$\nabla_{cus}(H:G|(\{1\},\{1\}):P_{j_o}|A_{j_o}:P_{i_o}|A_{i_o}:w_{j_oi_o}:\Lambda_{i_o})$$

$$= \lim_{s\to 0} s\cdot c_{cus}(P_{j_o}|A_{j_o}:P_{i_o}|A_{i_o}:w_{j_oi_o}:\Lambda_{i_o} + s\Lambda_{\perp}).$$

Owing to the functional equations for the c-functions,

$$c_{cus}(P_{j_o}|A_{j_o}:P_{i_o}|A_{i_o}:w_{j_oi_o}:\Lambda_{i_o} + s\Lambda_{\perp})$$

is the sum

$$\sum_{\nu_o}^{(j)}$$

of

$$c_{cus}(P_{j_o}|A_{j_o}:P_{j_o\nu_o}|A_{j_o\nu_o}:w_{\lambda_{j_o}}k_{j_o\nu_o}^{-1}:k_{j_o\nu_o}w_{\lambda_{j_o}}w_{j_oi_o}(\Lambda_{i_o}+s\Lambda_{\perp}))$$

composed with

$$c_{cus}(P_{j_o\nu_o}|A_{j_o\nu_o}:P_{i_o}|A_{i_o}:k_{j_o\nu_o}w_{\lambda_{j_o}}w_{j_oi_o}:\Lambda_{i_o}+s\Lambda_{\perp}).$$

We hasten to add that, strictly speaking, the functional equations involve a sum per $\Gamma\backslash\mathcal{C}_{j_o}$; but, for any other such $P_{j_o\nu_o}$, of necessity

$$c_{cus}(P_{j_o}|A_{j_o}:P_{j_o\nu_o}|A_{j_o\nu_o}:w_{\lambda_{j_o}}k_{j_o\nu_o}^{-1}:k_{j_o\nu_o}w_{\lambda_{j_o}}w_{j_oi_o}(\Lambda_{i_o}+s\Lambda_{\perp}))$$

would be null, hence need not be explicitly taken into account. In view of the first point noted at the beginning,

$$\lim_{s\to 0} c_{cus}(P_{j_o\nu_o}|A_{j_o\nu_o}:P_{i_o}|A_{i_o}:k_{j_o\nu_o}w_{\lambda_{j_o}}w_{j_oi_o}:\Lambda_{i_o}+s\Lambda_{\perp})$$
$$= c_{cus}(P_{j_o\nu_o}|A_{j_o\nu_o}:P_{i_o}|A_{i_o}:k_{j_o\nu_o}w_{\lambda_{j_o}}w_{j_oi_o}:\Lambda_{i_o}).$$

As for the other term,

$$w_{\lambda_{j_o}}w_{j_oi_o}\Lambda_{\perp}$$

is a real unit normal to $\mathfrak{H}(P_{j_o},A_{j_o})_{w_{j_oi_o}}$ in $\mathfrak{X}(P_{j_o},A_{j_o})$, so

$$\lim_{s\to 0} s\cdot c_{cus}(P_{j_o}|A_{j_o}:P_{j_o\nu_o}|A_{j_o\nu_o}:w_{\lambda_{j_o}}k^{-1}_{j_o\nu_o}:k_{j_o\nu_o}w_{\lambda_{j_o}}w_{j_oi_o}(\Lambda_{i_o}+s\Lambda_\perp))$$

$$= \nabla_{cus}(H_{\underset{\sim}{w}_{j_oi_o}}:G|(\{1\},\{1\}):P_{j_o}|A_{j_o}:P_{j_o\nu_o}|A_{j_o\nu_o}:w_{\lambda_{j_o}}k^{-1}_{j_o\nu_o}:k_{j_o\nu_o}w_{\lambda_{j_o}}w_{j_oi_o}\Lambda_{i_o})$$

$$= \nabla_{cus}(H:(w_{\lambda_{j_o}},w_{j_oi_o}):\nu_o).$$

Summing over ν_o then finishes the proof. //

Keeping for the time being to the i - j notation, suppose that $\mathcal{O}_i$ and $\mathcal{O}_j$ are associate orbits -- then, for each $w_{ji} \in W(A_j,A_i)$, there is defined a linear transformation

$$c_{res}(P_j|A_j:P_i|A_i:w_{ji}:\Lambda_i)$$

taking

$$E(\delta,\mathcal{O}_i:\mathcal{C}_o) \cap E(\delta,\underset{\sim}{\mathcal{O}}_o;\underset{\sim}{H}:P_i|A_i)$$

to

$$E(\delta,\mathcal{O}_j:\mathcal{C}_o) \cap E(\delta,\underset{\sim}{\mathcal{O}}_o;\underset{\sim}{H}:P_j|A_j).$$

For the work to follow, we shall need the exact definition which, of course, may be found in TES (cf. Chap. 6) but only in disguise, being masked, as it were, by the boldface picture.

Since a given

$$\Phi_i \in E(\delta,\mathcal{O}_i:\mathcal{C}_o) \cap E(\delta,\underset{\sim}{\mathcal{O}}_o;\underset{\sim}{H}:P_i|A_i)$$

can be written as a sum

$$\Phi_i = \Sigma\, \Phi_H ,$$

where

$$\Phi_H \in \mathcal{E}(\delta,\underset{\sim}{\mathcal{O}}_o;H:P_i|A_i),$$

there is no loss of generality in supposing that actually

$$\Phi_H = E(H:P_i|A_i:P_{i_o}|A_{i_o}:T_{i_o}(\Phi_i:H):H(P_{i_o},A_{i_o}):?),$$

the notation being as before. In particular, therefore,

$$T_{i_o}(\Phi_i:H) \in \mathrm{Hom}(S_{H(P_{i_o},A_{i_o})},\mathcal{E}_{cus}(\delta,\mathcal{O}_{i_o})).$$

Take now an element $w_{ji} \in W(A_j,A_i)$ -- then we may form, as in Chap. 6 of TES, $[\underset{\sim}{w}_{ji}]$, determining thereby, per specific data, $w_{j_o i_o}$ and $w_{\lambda_{j_o}} w_{j_o i_o}$. This said, let us agree to write

$$\nabla(H:(w_{\lambda_{j_o}},w_{j_o i_o}):\nu_o',\nu_o)$$

in place of

$$\nabla(H_{\underset{\sim}{w}_{j_o i_o}}:G|(\{1\},\{1\}):P_{j_o\nu_o'}|A_{j_o\nu_o'}:P_{j_o\nu_o}|A_{j_o\nu_o}:k_{j_o\nu_o'}w_{\lambda_{j_o}}k_{j_o\nu_o}^{-1}:?_{j_o\nu_o}),$$

where (cf. supra)

$$?_{j_o\nu_o} = - k_{j_o\nu_o} w_{j_o i_o} H(P_{i_o},A_{i_o}).$$

Owing to Proposition 5.4 in TES, there exist

$$T_{j_o\nu_o}(\Phi_i:H:\Lambda_i) \in \mathrm{Hom}(S_{H(P_{j_o\nu_o},A_{j_o\nu_o})w_{j_oi_o}}, E_{cus}(\delta,\mathcal{O}_{j_o\nu_o}))$$

such that for all ν_o',

$$\nabla(H:G|(\{1\},\{1\}):P_{j_o\nu_o'}|A_{j_o\nu_o'}:P_{i_o}|A_{i_o}:w_{j_oi_o}:\Lambda_i + H(P_{i_o},A_{i_o}))T_{i_o}(\Phi_i:H)$$

$$= \sum_{\nu_o}^{(j)} \nabla(H:(w_{\lambda_{j_i}},w_{j_oi_o}):\nu_o',\nu_o)T_{j_o\nu_o}(\Phi_i:H:\Lambda_i).$$

It is then the case that

$$c_{res}(P_j|A_j:P_i|A_i:w_{ji}:\Phi_i)\Phi_H$$

is equal to

$$\sum_{\nu_o}^{(j)} E(H_{w_{j_oi_o}}:P_j|A_j:P_{j_o\nu_o}|A_{j_o\nu_o}:T_{j_o\nu_o}(\Phi_i:H:\Lambda_i):H_{w_{j_oi_o}}(P_{j_o\nu_o},A_{j_o\nu_o}):?).$$

Thanks to Proposition 5.5, this expression can also be written in terms of E_{cus} (see the definition at the end of §3), namely as

$$\sum_{\nu_o}^{(j)} E_{cus}(H_{w_{j_oi_o}}:P_j|A_j:P_{j_o\nu_o}|A_{j_o\nu_o}:\Phi_{j_o\nu_o}(\Phi_i:H:\Lambda_i):H_{w_{j_oi_o}}(P_{j_o\nu_o},A_{j_o\nu_o}):?),$$

where

$$\Phi_{j_o\nu_o}(\Phi_i:H:\Lambda_i)$$

$$= c_{cus}(P_{j_o\nu_o}|A_{j_o\nu_o}:P_{i_o}|A_{i_o}:k_{j_o\nu_o}w_{\lambda_{j_o}}w_{j_oi_o}:\Lambda_i + H(P_{i_o},A_{i_o}))T_{i_o}(\Phi_i:H)(1).$$

REMARK The reader will observe that ∇_{cus} has been utilized systematically throughout the entire discussion; its companion E_{cus}, on the other hand, has not entered in at all, primarily because no particular simplification would have resulted. However, there is ample justification for its introduction here, the reason being the virtual trivialization of the proof of Lemma 5.6 below.

The necessary preparation is now complete. Anticipating the applications, make the following changes in the notation:

$$(P_i,S_i;A_i) \begin{array}{l} \nearrow (P_1,S_1;A_1) \\ \\ \searrow (P_2,S_2;A_2) \end{array} \qquad \left(i_o \begin{array}{l} \nearrow i_o^1 \\ \\ \searrow i_o^2 \end{array} \right)$$

$$(P_j,S_j;A_j) \to (P_i,S_i;A_i) \qquad (j_o \to i_o, \nu_o \to \mu_o).$$

$\mathcal{H}$ is replaced, accordingly, by $\mathcal{H}_1$ or $\mathcal{H}_2$, $\mathcal{H}_1$ and $\mathcal{H}_2$ being taken equivalent. The element w_{ji} then becomes

$$\begin{cases} w_{i1} \in W(A_i,A_1) \\ \\ w_{i2} \in W(A_i,A_2), \end{cases}$$

thence

$$\begin{cases} [\underset{\sim}{w}_{i1}] \to w_{i_o:i_o^1} \\ \\ [\underset{\sim}{w}_{i2}] \to w_{i_o:i_o^2} \end{cases} \quad \text{and} \quad \lambda_{i_o},$$

where

$$(w_{i_o:i_o^2}, w_{i_o:i_o^1}) \in W^+(\lambda_{i_o}:H_2 \times H_1),$$

as is permissible.

LEMMA 5.6 Let

$$\begin{cases} \Phi_1 \in E(\delta, \mathcal{O}_1 : \mathcal{C}_o) \cap E(\delta, \mathcal{O}_o; H:P_1|A_1) \\ \Phi_2 \in E(\delta, \mathcal{O}_2 : \mathcal{C}_o) \cap E(\delta, \mathcal{O}_o; H:P_2|A_2). \end{cases}$$

Then the inner product

$$(c_{res}(P_i|A_i:P_1|A_1:w_{i1}:\Lambda_1)\Phi_{H_1}, c_{res}(P_i|A_i:P_2|A_2:w_{i2}:\Lambda_2)\Phi_{H_2})$$

is equal to the sum

$$\sum_{\mu_o}^{(i)}$$

of the inner products

$$(\nabla_{cus}(H_1:G|(\{1\},\{1\}):P_{i_o\mu_o}|A_{i_o\mu_o}:P_{i_o^1}|A_{i_o^1}:k_{i_o\mu_o}w_{i_o:i_o^1}:\Lambda_1+H_1(P_{i_o^1},A_{i_o^1}))T_{i_o^1}(\Phi_1:H_1)(1)$$

$$c_{cus}(P_{i_o\mu_o}|A_{i_o\mu_o}:P_{i_o^2}|A_{i_o^2}:k_{i_o\mu_o}w_{i_o:i_o^2}:\Lambda_2+H_2(P_{i_o^2},A_{i_o^2}))T_{i_o^2}(\Phi_2:H_2)(1)).$$

[This is immediate. Simply unravel the first term of the second inner product via Proposition 5.5 and then use the definitions to get the first inner product via Proposition 5.9 (E and ∇) from TES.]

In essence, we now have our hands on a generic term of the second expression figuring in Proposition 5.3. To finish up, we need only establish the proper notation and make a few simple remarks.

Let us recall that it is a question here of two association classes $\mathcal{C}$ and $\mathcal{C}_o$ with $\mathcal{C} \succeq \mathcal{C}_o$, where

$$\text{rank}\ (\mathcal{C}_o) - \text{rank}\ (\mathcal{C}) = 1.$$

Repeating our notational principles, let $P_i (1 \leq i \leq r)$ be a set of representatives for $G \backslash \mathcal{C}$ -- then

$$\mathcal{C} = \coprod_i \mathcal{C}_i$$

where $\mathcal{C}_i = G \cdot \{P_i\} \cap \mathcal{C}$. Let $P_{i\mu} (1 \leq \mu \leq r_i)$ be a set of representatives for $\Gamma \backslash \mathcal{C}_i$ -- then

$$\{P_{i\mu} : 1 \leq i \leq r,\ 1 \leq \mu \leq r_i\}$$

is a set of representatives for $\Gamma \backslash \mathcal{C}$. The same notation is employed at the $\mathcal{C}_o$-level by appending a sub-zero whenever appropriate, as has been done systematically from the beginning. Owing to Lemma 8 in the Appendix to Chap. 2 of TES,

$$\forall\, \mathcal{C}_i \quad \exists\, \mathcal{C}_{i_o} \quad \text{st} \quad \mathcal{C}_i \succeq \mathcal{C}_{i_o}.$$

Therefore the set

$$\mathrm{Dom}_o(\mathcal{C}_i) = \{\mathcal{C}_{i_o} : \mathcal{C}_i \succeq \mathcal{C}_{i_o}\}$$

is never empty. Of course, it can happen that

$$\mathrm{Dom}_o(\mathcal{C}_{i'}) \cap \mathrm{Dom}_o(\mathcal{C}_{i''}) \neq \emptyset$$

for distinct $\mathcal{C}_{i'}, \mathcal{C}_{i''}$, a potential source of confusion. Note, too, that there may very well be a $\mathcal{C}_{i_o}$ which belongs to no $\mathrm{Dom}_o(\mathcal{C}_i)$. These points recorded, given the index i, choose, for each index i_o that $\mathcal{C}_i \succeq \mathcal{C}_{i_o}$, a set

$$\{(P_{i;i_o\mu_o}, S_{i;i_o\mu_o}; A_{i;i_o\mu_o})\}$$

of representatives for $\Gamma\backslash\mathcal{C}_{i_o}$ where we assume, as we may, that each

$$(P_{i;i_o\mu_o}, S_{i;i_o\mu_o}; A_{i;i_o\mu_o})$$

is a dominated predecessor of some $(P_{i\mu}, S_{i\mu}; A_{i\mu})$, thence (cf supra)

$$\Gamma\backslash\mathcal{C}_{i_o} = \coprod_{\mu} \Gamma\cap P_{i\mu}\backslash\mathcal{C}_{i_o}(P_{i\mu}|A_{i\mu}),$$

a decomposition which depends on i. However, as is in Chap. 6 of TES, we can and will assume that matters have been so arranged that for distinct indices i' and i'', $P_{i';i_o\mu_o}$ is Γ-conjugate to $P_{i'';i_o\mu_o}$.

Again, let

$$\begin{cases} \Phi_1 \in E(\delta, \mathcal{O}_1 : \mathcal{C}_o) \cap E(\delta, \underset{\sim}{\mathcal{O}}_o ; \underset{\sim}{H} : P_1 | A_1) \\ \Phi_2 \in E(\delta, \mathcal{O}_2 : \mathcal{C}_o) \cap E(\delta, \underset{\sim}{\mathcal{O}}_o ; \underset{\sim}{H} : P_2 | A_2). \end{cases} \quad (\mathcal{O}_1 \text{ and } \mathcal{O}_2 \text{ associate})$$

Then Lemma 5.6 provides us with a formula for the inner product

$$(c_{res}(P_{i\mu}|A_{i\mu}:P_1|A_1:w_{i\mu:1}:\Lambda_1)\Phi_{H_1},\ c_{res}(P_{i\mu}|A_{i\mu}:P_2|A_2:w_{i\mu:2}:\Lambda_2)\Phi_{H_2}),$$

namely as a sum

$$\sum_{\mu_o}^{(i\mu)}$$

of other inner products

$$(\nabla_{cus}(\ldots)T_{i_o^1}(\Phi_1:H_1)(1), c_{cus}(\ldots)T_{i_o^2}(\Phi_2:H_2)(1)).$$

Needless to say, the $(P_i, S_i; A_i)$ appearing there has been replaced by the $(P_{i\mu}, S_{i\mu}; A_{i\mu})$ appearing here, which is certainly permissible. Consider now the sum

$$(-1)^{\ell} \cdot \mathrm{vol}(\mathcal{C}) \cdot \sum_{i=1}^{r} \sum_{\mu=1}^{r_i} \sum_{w_{i\mu:2} \in W(A_{i\mu}, A_2)} \sum_{w_{i\mu:1} \in W(A_{i\mu}, A_1)}$$

of the product of

$$\exp(\langle I_{P_{i\mu}}(\underset{\sim}{H}),\ w_{i\mu:2}\bar{\Lambda}_2 + w_{i\mu:1}\Lambda_1\rangle)$$

$$\times\ (1/\prod_{\lambda_{i\mu}} (w_{i\mu:2}\bar{\Lambda}_2 + w_{i\mu:1}\Lambda_1, \lambda_{i\mu}))$$

with

$$(c_{res}(P_{i\mu}|A_{i\mu}:P_1|A_1:w_{i\mu:1}:\Lambda_1)\Phi_{H_1}, c_{res}(P_{i\mu}|A_{i\mu}:P_2|A_2:w_{i\mu:2}:\Lambda_2)\Phi_{H_2}).$$

Our problem will be to compare this sum with the sum associated with the second part of the inner product formula in Proposition 5.3. To begin with, if for some index i_o, the G-conjugacy class $\mathcal{C}_{i_o}$ is not dominated by any $\mathcal{C}_i$, then there is no contribution. Consequently, that part of the sum can be ignored. This said, fix the index i. Suppose that

$$\mathcal{C}_{i_o} \in \mathrm{Dom}_o(\mathcal{C}_i).$$

Then, per the 'base point'

$$(P_{i;i_o}, S_{i;i_o}; A_{i;i_o})$$

of the data, there is determined a simple root $\lambda_{i;i_o}$ with the property that

$$P_{\lambda_{i;i_o}} = P_i,$$

implying, therefore, that

$$\mathcal{C}(\lambda_{i;i_o}) = \mathcal{C}.$$

On the basis of our definitions, it then follows that the sum

$$(-1)^{\ell}\cdot\mathrm{vol}(\mathcal{C})\cdot\sum_{\mu=1}^{r_i}\sum_{w_{i\mu:2}\in W(A_{i\mu},A_2)}\sum_{w_{i\mu:1}\in W(A_{i\mu},A_1)}$$

of the product of

$$\exp(\langle I_{P_{i\mu}}(\underset{\sim}{H}),\ w_{i\mu:2}\bar{\Lambda}_2 + w_{i\mu:1}\Lambda_1\rangle)$$

$$\times\quad (1/\prod_{\lambda_{i\mu}}(w_{i\mu:2}\bar{\Lambda}_2 + w_{i\mu:1}\Lambda_1,\lambda_{i\mu}))$$

with

$$(c_{res}(P_{i\mu}|A_{i\mu}:P_1|A_1:w_{i\mu:1}:\Lambda_1)\Phi_{H_1},\ c_{res}(P_{i\mu}|A_{i\mu}:P_2|A_2:w_{i\mu:2}:\Lambda_2)\Phi_{H_2})$$

is the same as the sum

$$\begin{cases} i_o : \mathcal{C}_{i_o} \in \mathrm{Dom}_o(\mathcal{C}_i) \\ \mu_o : P_{i;i_o\mu_o} \preceq P_{i\mu} \\ (w_{i;i_o:i_o^2},\ w_{i;i_o:i_o^1}) \in W^+(\lambda_{i;i_o}:H_2 \times H_1) \end{cases}$$

of

$$(-1)^{\ell_o - 1}\cdot\mathrm{vol}(\mathcal{C}(\lambda_{i;i_o}))\cdot\Big[\prod_{\lambda\neq\lambda_{i;i_o}}(w_{i;i_o:i_o^2}\bar{\Lambda}_{i_o^2} + w_{i;i_o:i_o^1}\Lambda_{i_o^1},\lambda)\Big]^{-1}$$

times the sum over μ_o of the product of

$$\exp(\langle I_{P_{i;i_o\mu_o}}(\underset{\sim}{H}),k_{i;i_o\mu_o}w_{i;i_o:i_o^2}\bar{\Lambda}_{i_o^2} + k_{i;i_o\mu_o}w_{i;i_o:i_o^1}\Lambda_{i_o^1}\rangle)$$

with

$$(\nabla_{cus}(H_1:G|(\{1\},\{1\}):P_{i;i_o\mu_o}|A_{i;i_o\mu_o}:P_{i_o^1}|A_{i_o^1}:k_{i;i_o\mu_o}w_{i;i_o:i_o^1}:\Lambda_{i_o^1})T_{i_o^1}(1),$$

$$c_{cus}(P_{i;i_o\mu_o}|A_{i;i_o\mu_o}:P_{i_o^2}|A_{i_o^2}:k_{i;i_o\mu_o}w_{i;i_o:i_o^2}:\Lambda_{i_o^2})T_{i_o^2}(1)),$$

where

$$\begin{cases} T_{i_o^1} = T_{i_o^1}(\Phi_1:H_1) \\ T_{i_o^2} = T_{i_o^2}(\Phi_2:H_2) \end{cases}$$

the parameters $\Lambda_{i_o 1}$, $\Lambda_{i_o 2}$ being connected with the parameters Λ_1, Λ_2 via the relations

$$\begin{cases} \Lambda_{i_o 1} = \Lambda_1 + H_1(P_{i_o 1}, A_{i_o 1}) \\ \Lambda_{i_o 2} = \Lambda_2 + H_2(P_{i_o 2}, A_{i_o 2}). \end{cases}$$

Needless to say, the minor details as regards $\prod$ and exp have been deliberately ignored since an analogous elaboration may be found in §9 of [2].

If the index is changed, from i to j, say, then at first it seems that a problem might be present in that the sums supra at the $\mathcal{C}_o$-level conceivably could overlap. But this can't happen: The simple root $\lambda_{j:j_o}$, when conjugated, is not the same as $\lambda_{i:i_o}$ ($j_o = i_o$, of course). The pertinent sums are therefore disjoint. Moreover, in an obvious sense, they exhaust all that is relevant.

Before we summarize our conclusions we had best recall that the sum

$$\sum_{w_{ji} \in W(A_j, A_i)} a_j(x)^{w_{ji}\Lambda_i} \cdot (c_{res}(P_j|A_j : P_i|A_i : w_{ji} : \Lambda_i)\Phi_i)(x)$$

does not give, in general, the full constant term

$$E_{P_j}(P_i|A_i : \Phi_i : \Lambda_i : x)$$

along P_j for the Eisenstein series $E(P_i|A_i : \Phi_i : \Lambda_i : x)$ but rather only part of it, say

$$E^{w}_{P_j}(P_i|A_i:\Phi_i:\Lambda_i:x),$$

the so-called weak constant term of the Eisenstein series $E(P_i|A_i:\Phi_i:\Lambda_i:x)$ along P_j.

In this connection, let us now note that the weak constant terms

$$\begin{cases} E^{w}_{P_{i\mu}}(P_1|A_1:\Phi_1:\Lambda_1:?) \in \text{Exp}(\mathfrak{a}_{i\mu}) \otimes \mathcal{E}_{res}(\delta,\mathcal{O}_{i\mu}) \\ E^{w}_{P_{i\mu}}(P_2|A_2:\Phi_2:\Lambda_2:?) \in \text{Exp}(\mathfrak{a}_{i\mu}) \otimes \mathcal{E}_{res}(\delta,\mathcal{O}_{i\mu}). \end{cases}$$

We have proved:

THEOREM 5.7 *Let $\mathcal{C}$ be an association class of Γ-cuspidal split parabolic subgroups of G. Let $\mathcal{C}_o$ be an association class of Γ-cuspidal split parabolic subgroups of G, dominated by $\mathcal{C}$ and subject to the condition*

$$\text{rank}(\mathcal{C}_o) - \text{rank}(\mathcal{C}) = 1.$$

Let

$$\begin{cases} \Phi_1 \in \mathcal{E}(\delta,\mathcal{O}_1:\mathcal{C}_o) \\ \Phi_2 \in \mathcal{E}(\delta,\mathcal{O}_2:\mathcal{C}_o) \end{cases} \quad (\mathcal{O}_1 \text{ and } \mathcal{O}_2 \text{ associate}).$$

Then, for all $H \in \mathfrak{a}_Q$, the inner product

$$(Q^{H}E(P_1|A_1:\Phi_1:\Lambda_1:?),\ Q^{H}E(P_2|A_2:\Phi_2:\Lambda_2:?))$$

is equal to

$$(-1)^{\ell_o}\cdot \mathrm{vol}(\mathcal{C}_o)$$

times

$$Ж(\mathcal{C}_o : P_{i_o\mu_o} | A_{i_o\mu_o} : E_{P_{i_o\mu_o}}(P_1|A_1:\Phi_1:\Lambda_1:?),\ E_{P_{i_o\mu_o}}(P_2|A_2:\Phi_2:\Lambda_2:?):\underset{\sim}{H}),$$

summed over i_o and μ_o, PLUS

$$(-1)^{\ell}\cdot \mathrm{vol}(\mathcal{C})$$

times

$$Ж(\mathcal{C} : P_{i\mu} | A_{i\mu} : E^w_{P_{i\mu}}(P_1|A_1:\Phi_1:\Lambda_1:?),\ E^w_{P_{i\mu}}(P_2|A_2:\Phi_2:\Lambda_2:?):\underset{\sim}{H}),$$

summed over i and μ.

§6. PASSAGE TO THE IMAGINARY AXIS

The purpose of this § is to take the inner product formulas of the preceding §, in both the level zero and level one situations, and determine the form which they assume when the parameters are equal and pure imaginary. Such a determination is absolutely essential for the eventual application of this machinery to the Selberg trace formula; cf. [1] for the Γ-rank one case.

We shall begin by studying the level zero situation; as will become apparent in due course, certain features of the analysis can then be carried over to the level one situation without substantial change. Keeping to the notation of Theorem 5.1, the set-up is this. Fix a

$$(P,S;A) \in \mathcal{C},$$

the split component A being, as always special. Let

$$\Phi',\Phi'' \in \mathcal{E}_{cus}(\delta,\mathcal{O}).$$

Then, as we know, for all $\underset{\sim}{H} \in \underset{\sim}{a}_Q$, the inner product

$$(Q^{\underset{\sim}{H}}E(P|A:\Phi':\Lambda':?),\ Q^{\underset{\sim}{H}}E(P|A:\Phi'':\Lambda'':?))$$

is equal to

$$(-1)^{\ell}\cdot \mathrm{vol}(\mathcal{C})$$

times

$$\mathbb{K}(\mathcal{C}:P_{i\mu}|A_{i\mu}:E_{P_{i\mu}}(P|A:\Phi':\Lambda':?),\ E_{P_{i\mu}}(P|A:\Phi'':\Lambda'':?):\underset{\sim}{H}),$$

summed over i and μ. Briefly put, our problem now is to see what happens when $\Lambda' = \Lambda''$, say Λ, $\Lambda \in \sqrt{-1}\,\check{\mathfrak{a}}$. Naturally, some care must be exercised, the equality supra being only in the sense of meromorphic functions. Tacitly, therefore, we suppose throughout that the points Λ', Λ'' are in general position (cf. Chap. 5 of TES).

There is a slightly different way to write the inner product formula which turns out to be more convenient for our immediate purposes. Let i be any index between 1 and r. Given

$$w', w'' \in W(A_i, A),$$

write

$$\text{ю}_i(\Lambda', \Lambda''; w', w''; \Phi', \Phi'' : \underset{\sim}{H})$$

for

$$\sum_{\mu=1}^{r_i} \exp(\langle I_{P_{i\mu}}(\underset{\sim}{H}), k_{i\mu} w''\bar{\Lambda}'' + k_{i\mu} w'\Lambda'\rangle)$$

$$\times\ (c_{cus}(P_{i\mu}|A_{i\mu}:P|A:k_{i\mu}w':\Lambda')\Phi',\ c_{cus}(P_{i\mu}|A_{i\mu}:P|A:k_{i\mu}w'':\Lambda'')\Phi'').$$

Then, for all $\underset{\sim}{H} \in \underset{\sim}{\mathfrak{a}}_Q$, the inner product

$$(Q^{\underset{\sim}{H}}E(P|A:\Phi':\Lambda':?),\ Q^{\underset{\sim}{H}}E(P|A:\Phi'':\Lambda'':?))$$

is equal to

$$(-1)^{\ell} \cdot \mathrm{vol}(\mathfrak{C})$$

times

$$\sum_{i=1}^{r} \sum_{w' \in W(A_i, A)} \sum_{w'' \in W(A_i, A)} \frac{\mathcal{H}_i(\Lambda', \Lambda''; w', w''; \Phi', \Phi'': \underset{\sim}{H})}{\prod_{\lambda_i} (w''\bar{\Lambda}'' + w'\Lambda', \lambda_i)} .$$

For the sake of simplicity, we shall henceforth suppress Φ', Φ'', and $\underset{\sim}{H}$ from the notation, adding them in again only in the end.

We explicitly remark that $\mathcal{H}_i$ are meromorphic functions of $(\Lambda', \bar{\Lambda}'')$, all singularities being of the form

$$H_s \times (\check{a} + \sqrt{-1}\,\check{a}) \quad \text{or} \quad (\check{a} + \sqrt{-1}\,\check{a}) \times H_s .$$

In addition, the functional equations satisfied by the c-functions imply that the $\mathcal{H}_i$ admit an important property of invariance per the ambient simple roots. Indeed, let $\lambda_i \in \Sigma_{P_i}^{o}(\mathfrak{g}, \mathfrak{a}_i)$ -- then one may attach to λ_i, in the usual way, a Γ-cuspidal parabolic $P_{\lambda_i} \succeq P_i$. On the other hand, there is associated with the corresponding simple reflection w_{λ_i} another Γ-cuspidal parabolic P_j in $\mathcal{C}$, itself a dominated precedessor of P_{λ_i}. [Note: Strictly speaking P_j may not be one of our fixed representatives for $G\backslash\mathcal{C}$ but there is no real harm in pretending that it is.] We recall that

$$w_{\lambda_i} \in W(A_j^{\dagger}, A_i^{\dagger}),\ w_{\lambda_i}(\lambda_i) = -\lambda_j, \quad \text{say},$$

so

$$-w_{\lambda_i}(\lambda_i) = \lambda_j \in \Sigma_{P_j}^{o}(\mathfrak{g},\mathfrak{a}_j).$$

In terms of this data, then,

$$\text{IO}_i(\Lambda',\Lambda";w',w") = \text{IO}_j(\Lambda',\Lambda";w_{\lambda_i}w',w_{\lambda_i}w")$$

provided

$$(w"\bar{\Lambda}" + w'\Lambda',\lambda_i) = 0.$$

Such invariance carries with it certain implications of non-singularity which are best exploited by re-parameterizing our problem. Thus given $w \in W(A)$, put

$$T(\Lambda',\Lambda";w) = \sum_{i=1}^{r} \sum_{w_i \in W(A_i,A)} \frac{\text{IO}_i(\Lambda',\Lambda";w_i,w_i w)}{\prod_{\lambda_i}(w_i w\bar{\Lambda}" + w_i\Lambda',\lambda_i)} .$$

Then, for all $\underset{\sim}{H} \in \underset{\sim}{\mathfrak{a}}_{\underset{\sim}{Q}}$, the inner product

$$(Q^{\underset{\sim}{H}}E(P|A:\Phi':\Lambda':?),\ Q^{\underset{\sim}{H}}E(P|A:\Phi":\Lambda":?))$$

is equal to

$$(-1)^{\ell}\cdot\text{vol}(\mathcal{C})\cdot\sum_{w\in W(A)} T(\Lambda',\Lambda";w) .$$

Furthermore, and this is the point, the potential singularity of $T(\Lambda',\Lambda";w)$ along any set of the form

$$\{(\Lambda',\Lambda"):(w_i w\bar{\Lambda}" + w_i\Lambda',\lambda_i) = 0\}$$

is evidently removable.

Suppose that $\Lambda'' = \Lambda \in \sqrt{-1}\,\check{\mathfrak{a}}$ is in general position. Let $\Lambda' = \Lambda + \Lambda_o$, Λ_o being a real variable which, in practice, will be sent to zero. Now fix a $w \in W(A)$ -- then we ask: What is

$$\lim_{\Lambda_o \to 0} T(\Lambda + \Lambda_o, \Lambda; w)?$$

The answer, of course, depends on w itself. To get some insight into the problem, we shall consider the two simplest cases first. This analysis will provide the requisite background for tackling the general case, the investigation of which is somewhat more complicated.

Since Λ is pure imaginary, a generic term of the root denominator in

$$T(\Lambda + \Lambda_o, \Lambda; w)$$

has the form

$$(w_i \Lambda_o + w_i(1-w)\Lambda, \lambda_i).$$

The limit in question then depends upon the rank of $1-w$. Accordingly, for all

$$\begin{cases} w_i \in W(A_i, A) \\ w \in W(A) \end{cases},$$

let

$$\Sigma^o_{P_i}(w_i, w) = \{\lambda_i \in \Sigma^o_{P_i}(\mathfrak{g}, \mathfrak{a}_i) : \check{\lambda}_i \,|\, \mathrm{Im}(w_i(1-w)) \neq 0\}.$$

We shall then agree to denote by

$$\iota(w_i, w)$$

the cardinality of the set

$$\Sigma_{P_i}^{o}(\mathfrak{g}, \mathfrak{a}_i) - \Sigma_{P_i}^{o}(w_i, w).$$

The simplest case of all is when $1-w$ is non-singular. Because $(1-w)\Lambda$ is still in general position, we have

$$\begin{aligned}&\lim_{\Lambda_o \to 0} T(\Lambda + \Lambda_o, \Lambda; w)\\ &\qquad = T(\Lambda, \Lambda; w)\end{aligned}$$

so there is no difficulty here. In this connection, note that $\iota(w_i, w) = 0$ for all choices of w_i.

The next simplest case is when the rank of $1-w$ is equal to $\ell-1$. We shall treat it in detail since what is learned will eventually be needed for the discussion of the general case. To begin with, note that for each w_i there is at most one simple root λ_i with the property that

$$(w_i(1-w)\Lambda, \lambda_i)$$

vanishes identically. In other words,

$$\text{rank}(1-w) = \ell-1 \Longrightarrow \iota(w_i, w) = \begin{cases} 0 \\ 1 \end{cases}.$$

Split the sum defining

$$T(\Lambda + \Lambda_o, \Lambda; w)$$

into two parts, viz.:

$$\begin{cases} \sum_i \sum_{w_i : \iota(w_i, w) = 0} \\ \sum_i \sum_{w_i : \iota(w_i, w) = 1.} \end{cases}$$

It is then immediate that

$$\lim_{\Lambda_o \to 0} \sum_i \sum_{w_i : \iota(w_i, w)=0} \frac{\mathfrak{P}_i(\Lambda + \Lambda_o, \Lambda; w_i, w_i w)}{\prod_{\lambda_i} (w_i \Lambda_o + w_i(1-w)\Lambda, \lambda_i)}$$

$$= \sum_i \sum_{w_i : \iota(w_i, w)=0} \frac{\mathfrak{P}_i(\Lambda, \Lambda; w_i, w_i w)}{\prod_{\lambda_i} (w_i(1-w)\Lambda, \lambda_i)}.$$

As for the other part, corresponding to each w_i with $\iota(w_i, w) = 1$ there is exactly one λ_i such that

$$(w_i(1-w)\Lambda, \lambda_i)$$

is identically zero. Determining the index j per the simple reflection w_{λ_i}, put $w_j = w_{\lambda_i} w_i$ -- then, in an obvious sense, our problem becomes symmetrical, the investigation of the limit hinging on the behavior of the pairs

$$\frac{\mathfrak{P}_i(\Lambda + \Lambda_o, \Lambda; w_i, w_i w)}{\prod_{\lambda_i} (w_i \Lambda_o + w_i(1-w)\Lambda, \lambda_i)} + \frac{\mathfrak{P}_j(\Lambda + \Lambda_o, \Lambda; w_j, w_j w)}{\prod_{\lambda_j} (w_j \Lambda_o + w_j(1-w)\Lambda, \lambda_j)}$$

as $\Lambda_o \to 0$. [Note: The sum is in fact non-singular qua a function of Λ_o as Λ_o passes through the origin.] Replacing Λ_o by $t\Lambda_o$, L'Hôspital's rule may be applied to infer that

$$\lim_{t\to 0}\left[\frac{\mathrm{IO}_i(\Lambda + t\Lambda_o, \ldots)}{\prod_{\lambda_i}(tw_i\Lambda_o + \ldots)} + \frac{\mathrm{IO}_j(\Lambda + t\Lambda_o, \ldots)}{\prod_{\lambda_j}(tw_j\Lambda_o + \ldots)}\right]$$

is equal to

$$\frac{1}{(w_i\Lambda_o, \lambda_i)}$$

times

$$\frac{d}{dt}\left[\frac{\mathrm{IO}_i(\Lambda + t\Lambda_o, \Lambda; w_i, w_i w)}{\prod_{\lambda\neq\lambda_i}(tw_i\Lambda_o + w_i(1-w)\Lambda, \lambda)}\right]_{t=0}$$

plus

$$\frac{1}{(w_j\Lambda_o, \lambda_j)}$$

times

$$\frac{d}{dt}\left[\frac{\mathrm{IO}_j(\Lambda + t\Lambda_o, \Lambda; w_j, w_j w)}{\prod_{\lambda\neq\lambda_j}(tw_j\Lambda_o + w_j(1-w)\Lambda, \lambda)}\right]_{t=0}.$$

Both of these derivatives can be computed by explicit calculation; this offers no difficulty, hence need not be reproduced. We remark that, in particular, one term which arises is the familiar 'logarithmic derivative' of the c-function, as is to be expected. Putting everything together, then, leads to the conclusion that

$$\lim_{\Lambda_o\to 0} \sum_i \sum_{w_i:\iota(w_i,w)=1} \frac{Ю_i(\Lambda+\Lambda_o,\Lambda;w_i,w_iw)}{\prod_{\lambda_i}(w_i\Lambda_o+w_i(1-w)\Lambda,\lambda_i)}$$

exists and equals

$$\sum_i \sum_{w_i:\iota(w_i,w)=1} Я_i^{(1)}(\Lambda;w_i,w_iw) ,$$

where by definition, anticipating the notation below,

$$Я_i^{(1)}(\Lambda;w_i,w_iw)$$

is

$$\frac{1}{(w_i\Lambda_o,\lambda_i)}$$

times

$$\frac{d}{dt}\left[\frac{Ю_i(\Lambda+t\Lambda_o,\Lambda;w_i,w_iw)}{\prod_{\lambda\neq\lambda_i}(tw_i\Lambda_o+w_i(1-w)\Lambda,\lambda)}\right]_{t=0} .$$

We pass now to the general case. Suppose that $\mathrm{rank}(1-w) = \ell-\kappa$, $0\le\kappa\le\ell$. Split the sum defining

$$T(\Lambda+\Lambda_o,\Lambda;w)$$

into $\kappa+1$ parts, viz.:

$$\begin{cases} \displaystyle\sum_i \sum_{w_i : \iota(w_i, w) = 0} \\ \displaystyle\sum_i \sum_{w_i : \iota(w_i, w) = 1} \\ \vdots \\ \displaystyle\sum_i \sum_{w_i : \iota(w_i w) = \kappa.} \end{cases}$$

The study of

$$\lim_{\Lambda_o \to 0} T(\Lambda + \Lambda_o, \Lambda; w)$$

then depends upon consideration of each limit

$$\lim_{\Lambda_o \to 0} \sum_i \sum_{w_i : \iota(w_i, w) = k} \frac{\text{Ю}_i(\Lambda + \Lambda_o, \Lambda; w_i, w_i w)}{\prod_{\lambda_i} (w_i \Lambda_o + w_i(1-w)\Lambda, \lambda_i)} \qquad (0 \le k \le \kappa)$$

separately. To this end, let

$$\text{Я}_i^{(k)}(\Lambda; w_i, w_i w)$$

stand for

$$\frac{1}{k!} \cdot \frac{1}{\prod_{\lambda_i \in \Sigma_{P_i}^o(\mathfrak{g}, \mathfrak{a}_i) - \Sigma_{P_i}^o(w_i, w)} (w_i \Lambda_o, \lambda_i)}$$

times

$$\frac{d^k}{dt^k}\left[\frac{\text{Ю}_i(\Lambda + t\Lambda_o, \Lambda; w_i, w_i w)}{\prod_{\lambda_i \in \Sigma^o_{P_i}(w_i,w)} (tw_i\Lambda_o + w_i(1-w)\Lambda, \lambda_i)}\right]_{t=0} .$$

Fix k, $0 \le k \le \kappa$ -- then we claim that

$$\lim_{\Lambda_o \to 0} \sum_i \sum_{w_i : \iota(w_i,w)=k} \frac{\text{Ю}_i(\Lambda + \Lambda_o, \Lambda; w_i, w_i w)}{\prod_{\lambda_i} (w_i\Lambda_o + w_i(1-w)\Lambda, \lambda_i)}$$

exists and equals

$$\sum_i \sum_{w_i : \iota(w_i,w)=k} \text{Я}_i^{(k)}(\Lambda; w_i, w_i w) .$$

The cases $k=0$, $k=1$ have been dealt with above so we can assume that $k > 1$. A remark on methodology is then in order. When $k=1$, the limit was determined by an application of L'Hôspital's rule. In general, an approach along these lines will work provided we replace 'distinguished pairs' by a more comprehensive notion.

Consider the set $\mathcal{P}_k(w)$ of pairs

$$(i, w_i) \quad (\iota(w_i, w) = k).$$

Suppose that

$$\lambda_i \in \Sigma^o_{P_i}(\mathfrak{g}, \mathfrak{a}_i) - \Sigma^o_{P_i}(w_i, w) \qquad (\iota(w_i, w) = k).$$

Then

$$(i,w_i) \in \mathcal{P}_k(w) \Longrightarrow (j,w_{\lambda_i} w_i) \in \mathcal{P}_k(w).$$

That being, let us agree to term the pairs

$$\begin{cases} (i,w_i) \\ (j,w_{\lambda_i} w_i) \end{cases}$$

w-preequivalent. The equivalence relation generated by w-preequivalence splits $\mathcal{P}_k(w)$ into equivalence classes $\mathcal{E}$, meaning that

$$\sum_{(i,w_i) \in \mathcal{P}_k(w)}$$

can be written as

$$\sum_{\mathcal{E}} \sum_{(i,w_i) \in \mathcal{E}} .$$

Fix $\mathcal{E}$ -- then $\mathcal{E}$ contains enough reflections to ensure that

$$\sum_{(i,w_i) \in \mathcal{E}}$$

is holomorphic in t near the origin. But

$$\sum_{(i,w_i) \in \mathcal{E}}$$

admits an alternative expression as t^{-k} times another sum, each term of which is holomorphic in t near the origin. Therefore the sum, with t^{-k} pulled out, must vanish to order k at $t=0$, thus the limit claim is a consequence of a k-fold application of L'Hôspital's rule.

Our conclusions can be summarized as follows.

THEOREM 6.1 *Suppose that* Λ *is pure imaginary and in general position. Let*

$$\Phi', \Phi'' \in \mathcal{E}_{cus}(\delta, \mathcal{O}).$$

Then, for all $\underset{\sim}{H} \in \underset{\sim}{a}_Q$, *the inner product*

$$(Q^{\underset{\sim}{H}}E(P|A:\Phi':\Lambda:?),\ Q^{\underset{\sim}{H}}E(P|A:\Phi'':\Lambda:?))$$

is equal to the sum

$$(-1)^{\ell} \cdot \text{vol}(\mathcal{C}) \cdot \sum_{w \in W(A)} \sum_{k=0}^{\ell_w} \sum_{i} \sum_{w_i : \iota(w_i, w) = k}$$

of the

$$\mathcal{A}_i^{(k)}(\Lambda; w_i, w_i w; \Phi', \Phi'' : \underset{\sim}{H}).$$

[Note: We have written ℓ_w for

$$\ell\text{-rank}(1-w).$$

In addition, Φ', Φ'', and $\underset{\sim}{H}$ have been put back into the notation per $\mathcal{A}_i^{(k)}$ (cf. the comment made at the very beginning).]

In passing, we remark that a result slightly more general than Theorem 6.1 is actually valid. Indeed, let

$$\begin{cases} (P_1,S_1;A_1) \\ (P_2,S_2;A_2) \end{cases}$$

be two members of the same G-conjugacy class $\mathcal{C}_i$, say, in $\mathcal{C}$, $(P_i,S_i;A_i)$ then being, as usual, a fixed member of $\mathcal{C}_i$. Fix elements k_1,k_2 of K such that

$$\begin{cases} k_1(P_i,S_i;A_i)k_1^{-1} = (P_1,S_1;A_1) \\ k_2(P_i,S_i;A_i)k_2^{-1} = (P_2,S_2;A_2). \end{cases}$$

Let

$$\begin{cases} \Phi_1 \in E_{cus}(\delta,\mathcal{O}_1) \\ \Phi_2 \in E_{cus}(\delta,\mathcal{O}_2). \end{cases} \quad (\mathcal{O}_1 \text{ and } \mathcal{O}_2 \text{ associate})$$

Supposing now that $\Lambda_i \in \sqrt{-1}\check{a}_i$ is in general position, a simple variant on the preceding considerations leads at once to an inner product formula for

$$(Q^{\underset{\sim}{H}}E(P_1|A_1:\Phi_1:k_1\Lambda_i:?),\ Q^{\underset{\sim}{H}}E(P_2|A_2:\Phi_2:k_2\Lambda_i:?)) \quad (\underset{\sim}{H}\in\underset{\sim}{a}_Q).$$

We omit the (obvious) details.

The hypothesis of 'general position' in the foregoing can be weakened considerably, a point of some importance. Retaining the earlier notation, fix a $w\in W(A)$ -- then any singularity of

$$T(\Lambda',\Lambda'';w) = \sum_{i=1}^{r} \sum_{w_i \in W(A_i,A)} \frac{\wp_i(\Lambda',\Lambda'';w_i,w_i w)}{\prod_{\lambda_i} (w_i w\bar{\Lambda}'' + w_i \Lambda', \lambda_i)}$$

which is not removable is necessarily a singularity of some $\wp_i$. As for the removable singularities, given such and supposing that $\Lambda' = \Lambda''$, say Λ, is pure imaginary and in general position, the value of the implied continuation is the sum

$$\sum_{k=0}^{\ell_w} \sum_{i} \sum_{w_i : \iota(w_i,w)=k}$$

of the

$$\mathcal{R}_i^{(k)}(\Lambda;w_i,w_i w) \quad .$$

Needless to say, this conclusion is the main step in the proof of Theorem 6.1. To formulate a generalization, per an arbitrary pair

$$\begin{cases} \Lambda' \in \check{\mathfrak{a}} + \sqrt{-1}\,\check{\mathfrak{a}} \\ \Lambda'' \in \check{\mathfrak{a}} + \sqrt{-1}\,\check{\mathfrak{a}} \end{cases},$$

let

$$\Sigma_{P_i}^{o}(\Lambda',\Lambda'';w_i,w)$$

stand for the λ_i in $\Sigma_{P_i}^{o}(\mathfrak{g},\mathfrak{a}_i)$ such that

$$(w_i w\bar{\Lambda}'' + w_i \Lambda', \lambda_i) \neq 0.$$

We shall then agree to denote by

$$\iota(\Lambda',\Lambda'';w_i,w)$$

the cardinality of the set

$$\Sigma_{P_i}^{o}(\mathfrak{g},\mathfrak{a}_i) - \Sigma_{P_i}^{o}(\Lambda',\Lambda'';w_i,w).$$

Assuming that $k = \iota(\Lambda',\Lambda'';w_i,w)$, write

$$\mathfrak{R}_i^{(k)}(\Lambda',\Lambda'';w_i,w_iw)$$

for

$$\frac{1}{k!}\cdot\frac{1}{\prod_{\lambda_i\in\Sigma_{P_i}^{o}(\mathfrak{g},\mathfrak{a}_i)-\Sigma_{P_i}^{o}(\Lambda',\Lambda'';w_i,w)}(w_i\Lambda_o,\lambda_i)}$$

times

$$\frac{d^k}{dt^k}\left[\frac{\mathfrak{P}_i(\Lambda'+t\Lambda_o,\Lambda'';w_i,w_iw)}{\prod_{\lambda_i\in\Sigma_{P_i}^{o}(\Lambda',\Lambda'';w_i,w)}(w_iw\bar{\Lambda}''+w_i\Lambda'+tw_i\Lambda_o,\lambda_i)}\right]_{t=0}.$$

Finally, put

$$\ell_w(\Lambda',\Lambda'') = \max \iota(\Lambda',\Lambda'';w_i,w).$$

The requisite notation established, suppose still that the singularity at (Λ',Λ'') is removable -- then this time the value of the implied continuation is the sum

$$\sum_{0\le k\le \ell_w(\Lambda',\Lambda'')} \sum_i \sum_{w_i:\iota(\Lambda',\Lambda'';w_i,w)=k}$$

of the

$$\mathcal{R}_i^{(k)}(\Lambda',\Lambda'';w_i,w_iw) .$$

That this is in fact the case can be seen by employing a simple variant on the preceding theme. Accordingly, we need not insist upon a formal proof in extenso. We remark only that the point is again to study

$$\lim_{\Lambda_o\to 0} T(\Lambda'+\Lambda_o,\Lambda'';w)$$

which, as before, is done by splitting the sum defining

$$T(\Lambda' + \Lambda_o, \Lambda''; w)$$

into its constituent parts

$$\sum_i \sum_{w_i : \iota(\Lambda', \Lambda''; w_i, w) = k},$$

where $0 \le k \le \ell_w(\Lambda', \Lambda'')$. We then fix the index k and prove that

$$\lim_{\Lambda_o \to 0} \sum_i \sum_{w_i : \iota(\Lambda', \Lambda''; w_i, w) = k} \frac{Ю_i(\Lambda' + \Lambda_o, \Lambda''; w_i, w_i w)}{\prod_{\lambda_i} (w_i w \bar{\Lambda}'' + w_i(\Lambda' + \Lambda_o), \lambda_i)}$$

exists and equals the sum of the

$$Я_i^{(k)}(\Lambda', \Lambda''; w_i, w_i w).$$

In so doing, we use a suitable equivalence relation (cf. infra) and L'Hôspitals rule, except, of course, that (Λ', Λ'') appears rather than (Λ, Λ).

In other words:

THEOREM 6.1 (bis) <u>Suppose that</u>

$$\begin{cases} \forall i, \quad 1 \le i \le r \\ \forall w_i, \quad w_i \in W(A_i, A) \\ \forall w, \quad w \in W(A), \end{cases}$$

$\text{Ю}_i(?,?;\ldots)$ is holomorphic at (Λ',Λ''). Let

$$\Phi',\Phi'' \in \mathcal{E}_{cus}(\delta,\mathcal{O}).$$

Then, for all $H \in \mathfrak{a}_Q$, the inner product

$$(Q^H E(P|A:\Phi':\Lambda':?),\ Q^H E(P|A:\Phi'':\Lambda'':?))$$

is equal to the sum

$$(-1)^{\ell} \cdot \mathrm{vol}(\mathcal{C}) \cdot \sum_{w \in W(A)} \ \sum_{0 \le k \le \ell_w(\Lambda',\Lambda'')} \ \sum_i \ \sum_{w_i : \iota(\Lambda',\Lambda'';w_i,w)=k}$$

of the

$$\mathcal{R}_i^{(k)}(\Lambda',\Lambda'';w_i,w_i w;\Phi',\Phi'':H) .$$

It is to be noted that this theorem applies in the important special case when $\Lambda' = \Lambda''$, say Λ, $\Lambda \in \sqrt{-1}\check{\mathfrak{a}}$, providing, therefore, an inner product formula at any point on the imaginary axis. Consequently, the requirement of 'general position' is not necessary; on the other hand, there is a trade off in that the parameter sets then depend on Λ too.

Consider now arbitrary pairs

$$\begin{cases} (P_1,S_1;A_1) \in \mathcal{C} \\ (P_2,S_2;A_2) \in \mathcal{C} . \end{cases}$$

Let

$$\begin{cases} \Phi_1 \in E_{cus}(\delta,\mathcal{O}_1) \\ \Phi_2 \in E_{cus}(\delta,\mathcal{O}_2). \end{cases} \qquad (\mathcal{O}_1 \text{ and } \mathcal{O}_2 \text{ associate})$$

Then Theorem 5.1 provides us with a formula for the inner product

$$(Q^{H}_{\sim}E(P_1|A_1:\Phi_1:\Lambda_1:?),\ Q^{H}_{\sim}E(P_2|A_2:\Phi_2:\Lambda_2:?)).$$

Since

$$\begin{cases} (P_1,S_1;A_1) \\ (P_2,S_2;A_2) \end{cases}$$

need not be in the same G-conjugacy class in $\mathcal{C}$, it is generally meaningless even to attempt to move the parameters Λ_1 and Λ_2 to some common imaginary axis. Nevertheless, it is possible to rework the formula so as to bring it into line with Theorems 6.1 and 6.1(bis). We should stress outright that there is a pressing reason for doing this. Indeed, at a crucial juncture during the level one discussion infra we shall have to appeal to the present, modified version.

Actually, thanks to what has been said so far, the necessary generalization is quite formal, it being basically a question of altering the notation so as to reflect the current circumstances. Let i be any index between 1 and r.

Given

$$\begin{cases} w_{i:1} \in W(A_i, A_1) \\ w_{i:2} \in W(A_i, A_2), \end{cases}$$

write

$$\text{Ю}_i(\Lambda_1, \Lambda_2; w_{i:1}, w_{i:2}; \Phi_1, \Phi_2 : \underset{\sim}{H})$$

for

$$\sum_{\mu=1}^{r_i} \exp(\langle I_{P_{i\mu}}(\underset{\sim}{H}), w_{i\mu:2}\bar{\Lambda}_2 + w_{i\mu:1}\Lambda_1\rangle)$$

$$\times (c_{cus}(P_{i\mu}|A_{i\mu}:P_1|A_1:w_{i\mu:1}:\Lambda_1)\Phi_1, c_{cus}(P_{i\mu}|A_{i\mu}:P_2|A_2:w_{i\mu:2}:\Lambda_2)\Phi_2).$$

Then, for all $\underset{\sim}{H} \in \underset{\sim}{\mathfrak{a}}_Q$, the inner product

$$({}_Q^{\underset{\sim}{H}}E(P_1|A_1:\Phi_1:\Lambda_1:?), {}_Q^{\underset{\sim}{H}}E(P_2|A_2:\Phi_2:\Lambda_2:?))$$

is equal to

$$(-1)^{\ell} \cdot \text{vol}(\mathcal{C})$$

times

$$\sum_{i=1}^{r} \sum_{w_{i:1} \in W(A_i, A_1)} \sum_{w_{i:2} \in W(A_i, A_2)} \frac{\text{Ю}_i(\Lambda_1, \Lambda_2; w_{i:1}, w_{i:2}; \Phi_1, \Phi_2 : \underset{\sim}{H})}{\prod_{\lambda_i}(w_{i:2}\bar{\Lambda}_2 + w_{i:1}\Lambda_1, \lambda_i)}.$$

Omitting henceforth Φ_1, Φ_2, and $\underset{\sim}{H}$ from the notation, we shall also set $A = A_1$ and thus agree to write w_i in place of $w_{i:1}$. Given $w_{12} \in W(A_1, A_2)$, put

$$T(\Lambda_1,\Lambda_2;w_{12}) = \sum_{i=1}^{r} \sum_{w_i \in W(A_i,A)} \frac{Ю_i(\Lambda_1,\Lambda_2;w_i,w_iw_{12})}{\prod_{\lambda_i}(w_iw_{12}\bar{\Lambda}_2 + w_i\Lambda_1,\lambda_i)} .$$

Then, for all $\underset{\sim}{H} \in \underset{\sim}{\mathfrak{a}}_Q$, the inner product

$$(\underset{\sim}{Q}^{H}E(P_1|A_1:\Phi_1:\Lambda_1:?),\ \underset{\sim}{Q}^{H}E(P_2|A_2:\Phi_2:\Lambda_2:?))$$

is equal to

$$(-1)^{\ell} \cdot \mathrm{vol}(\mathcal{C}) \cdot \sum_{w_{12} \in W(A_1,A_2)} T(\Lambda_1,\Lambda_2;w_{12}) .$$

Each constituent here can be analyzed exactly as above. Thus fix a w_{12}. Suppose that

$$Ю_i(?,?;w_i,w_iw_{12})$$

is holomorphic at (Λ_1,Λ_2) for all i and w_i -- then, in obvious notation, the value of the implied continuation is the sum

$$\sum_{0 \le k \le \ell_{w_{12}}(\Lambda_1,\Lambda_2)} \sum_{i} \sum_{w_i : \iota(\Lambda_1,\Lambda_2;w_i,w_{12})=k}$$

of the

$$Я_i^{(k)}(\Lambda_1,\Lambda_2;w_i,w_iw_{12}) .$$

Granted this remark, the modified inner product formula alluded to supra follows upon performing the evident summation.

As for the proof, given Λ_1, Λ_2, our objective will be to introduce an equivalence relation

$$R(\Lambda_1,\Lambda_2;w_{12})$$

into the disjoint union

$$\coprod_{i=1}^{r} (\{i\} \times W(A_i,A)).$$

The full sum

$$\sum_{i=1}^{r} \sum_{w_i \in W(A_i,A)}$$

can then be written as a sum over the equivalence classes E of the partial sums

$$\sum_{(i,w_i)\in E} \quad .$$

Hence, in a self-explanatory notation, we have

$$T(?,?;w_{12}) = \sum_{E} T(?,?;w_{12}:E).$$

Assuming now that

$$\text{Ю}_i(?,?;w_i,w_iw_{12})$$

is holomorphic at (Λ_1,Λ_2) for all i and w_i, the value of the implied continuation per each E is the sum

$$\sum_{0\leq k\leq \ell_{w_{12}}(\Lambda_1,\Lambda_2)} \quad \sum_{(i,w_i)\in E} \quad \sum_{w_i:\iota(\Lambda_1,\Lambda_2;w_i,w_{12})=k}$$

of the

$$\mathcal{A}_i^{(k)}(\Lambda_1,\Lambda_2;w_i,w_iw_{12}) \quad .$$

Turning to the definition of

$$R(\Lambda_1,\Lambda_2;w_{12}),$$

suppose that

$$\lambda_i \in \Sigma_{P_i}^{o}(\mathfrak{g},\mathfrak{a}_i) - \Sigma_{P_i}^{o}(\Lambda_1,\Lambda_2;w_i,w_{12}).$$

Then, as at the beginning, we determine the corresponding simple reflection w_{λ_i} and associated data. Typical pairs

$$\begin{cases} (i,w_i) \\ (j,w_{\lambda_i}w_i) \end{cases}$$

are said to be $(\Lambda_1,\Lambda_2;w_{12})$-- preequivalent. Since

$$\lambda_j \in \Sigma^o_{P_j}(\mathfrak{g},\mathfrak{a}_j) - \Sigma^o_{P_j}(\Lambda_1,\Lambda_2;w_{\lambda_i}w_i,w_{12})$$

and $w_{\lambda_j} = w_{\lambda_i}^{-1}$, it is clear that $(\Lambda_1,\Lambda_2;w_{12})$ -- preequivalence is a symmetric relation. In general, arbitrary pairs

$$\begin{cases} (i,w_i) \\ (j,w_j) \end{cases}$$

are said to be $(\Lambda_1,\Lambda_2;w_{12})$ -- equivalent if there exist

$$\begin{cases} k_1,\ldots,k_n \\ w_{k_1},\ldots,w_{k_n} \end{cases}$$

such that

$$\begin{cases} i = k_1,\ w_i = w_{k_1} \\ j = k_n,\ w_j = w_{k_n} \end{cases}$$

and $\forall\nu = 1,\ldots,n-1$,

$$(k_\nu,w_{k_\nu})$$

is $(\Lambda_1,\Lambda_2;w_{12})$ -- preequivalent to

$$(k_{\nu+1},w_{k_{\nu+1}}).$$

By

$$R(\Lambda_1,\Lambda_2;w_{12})$$

we then understand $(\Lambda_1,\Lambda_2;w_{12})$ -- equivalence, the latter being, evidently, the equivalence relation generated by $(\Lambda_1,\Lambda_2;w_{12})$ -- preequivalence.

The putative continuability assertion is then manifest, each equivalence class E containing the simple reflections necessary to make the argument go.

The level one situation has still to be addressed. Maintaining the notation of Theorem 5.7, fix a

$$(P,S;A) \in \mathcal{C},$$

the split component A being, as always, special. Let

$$\Phi',\Phi'' \in E(\delta,\mathcal{O}:\mathcal{C}_0).$$

Then, as we know, for all $\underset{\sim}{H} \in \underset{\sim}{\mathfrak{a}}_Q$, the inner product

$$(Q^{\underset{\sim}{H}}E(P|A:\Phi':\Lambda':?),\ Q^{\underset{\sim}{H}}E(P|A:\Phi'':\Lambda'':?))$$

is equal to

$$(-1)^{\ell_0}\cdot\text{vol}(\mathcal{C}_0)$$

times

$$\mathcal{K}(\mathcal{C}_0:P_{i_0\mu_0}|A_{i_0\mu_0}:E_{P_{i_0\mu_0}}(P|A:\Phi':\Lambda':?),\ E_{P_{i_0\mu_0}}(P|A:\Phi'':\Lambda'':?):\underset{\sim}{H}),$$

summed over i_0 and μ_0, plus

$$(-1)^{\ell}\cdot\text{vol}(\mathcal{C})$$

times

$$Ж(\mathcal{C}:P_{i\mu}|A_{i\mu}:E^{w}_{P_{i\mu}}(P|A:\Phi':\Lambda':?),\ E^{w}_{P_{i\mu}}(P|A:\Phi'':\Lambda'':?):\underset{\sim}{H}),$$

summed over i and μ. Once again, our problem is to see what happens when $\Lambda' = \Lambda''$, say Λ, $\Lambda \in \sqrt{-1}\,\check{\mathfrak{a}}$ (subject, of course, to the usual provisos). Interestingly enough, this time the difficulties lie with the first term, the analysis of the second term being exactly the same as that per the level zero situation.

Indeed, the weak constant terms

$$\begin{cases} E^{w}_{P_{i\mu}}(P|A:\Phi':\Lambda':?) \\ E^{w}_{P_{i\mu}}(P|A:\Phi'':\Lambda'':?) \end{cases}$$

of the Eisenstein series

$$\begin{cases} E(P|A:\Phi':\Lambda':?) \\ E(P|A:\Phi'':\Lambda'':?) \end{cases}$$

are computed in terms of c_{res}. Furthermore, c_{res} satisfies the same type of functional equations as does c_{cus}. Accordingly, we may introduce exactly as before

$$\begin{cases} Ю_i \\ Я_i^{(k)} \end{cases},$$

where, naturally, c_{cus} is now replaced by c_{res}. This done, we then find that the sum over i and μ of

$$Ж(\mathcal{C}:P_{i\mu}|A_{i\mu}:E^{w}_{P_{i\mu}}(P|A:\Phi':\Lambda:?),\ E^{w}_{P_{i\mu}}(P|A:\Phi'':\Lambda:?):\underset{\sim}{H})$$

is equal to the sum

$$\sum_{w\in W(A)} \sum_{k=0}^{\ell_w} \sum_{i} \sum_{w_i:\iota(w_i,w)=k}$$

of the

$$Я_i^{(k)}(\Lambda;w_i,w_iw;\Phi',\Phi'':\underset{\sim}{H}) \quad .$$

Here, for simplicity of statement, we take Λ in general position.

There remains the task of analyzing the contribution from $Ж(\mathcal{C}_o:\ldots)$. It will be best to make first an assumption about Φ' and Φ''. For the application this will entail no loss of generality. So, start off with

$$\Phi',\Phi'' \in \mathcal{E}(\delta,\mathcal{O}:\mathcal{C}_o) \cap \mathcal{E}(\delta,\underset{\sim}{\mathcal{O}}_o;\underset{\sim}{H}:P|A).$$

Then, as in TES (cf. Chap. 6), write

$$\mathcal{E}(\delta,\underset{\sim}{\mathcal{O}}_o;\underset{\sim}{H}:P|A) = \sum_{\mathfrak{E}} \oplus\ \mathcal{E}(\delta,\underset{\sim}{\mathcal{O}}_o;\mathfrak{E}:P|A),$$

the $\mathfrak{E}$ being equivalence classes in

$$\underset{\sim}{H}(\mathcal{C}_o;\ \text{rank}(\mathcal{C}_o) - 1).$$

Fix one such, say $\mathfrak{E}_o$. By definition,

$$E(\delta,\underset{\sim}{\mathcal{O}}_0;\mathbf{E}_0:P|A) = \sum_{H\in \mathbf{E}_0} E(\delta,\underset{\sim}{\mathcal{O}}_0;H:P|A),$$

the H in the sum on the right subject to the requirement that $\chi_{\mathcal{C}_k}$ be H-admissible. In this connection, bear in mind that

$$(P,S;A)\in \mathcal{C}_k \subset \mathcal{C}.$$

Our assumption then will be that

$$\Phi',\Phi'' \in E(\delta,\mathcal{O}:\mathcal{C}_0) \cap E(\delta,\underset{\sim}{\mathcal{O}}_0;\mathbf{E}_0:P|A).$$

This being the case, decompose

$$\begin{cases} \Phi' & \text{per} \quad E(\delta,\underset{\sim}{\mathcal{O}}_0;\mathbf{E}_0:P|A) \\ \Phi'' & \text{per} \quad E(\delta,\underset{\sim}{\mathcal{O}}_0;\mathbf{E}_0:P|A). \end{cases}$$

Then, in the notation of §5, we see that our problem is to study

$$\lim_{\Lambda_0\to 0}\ \lim_{s,t\to 0}\ st\cdot \sum_{i_0=1}^{r_0} \sum_{w_{i_0:i_0^1}\in W(A_{i_0},A_{i_0^1})} \sum_{w_{i_0:i_0^2}\in W(A_{i_0},A_{i_0^2})}$$

of

$$\mathfrak{D}_{i_0}(\Lambda + H_1(P_{i_0^1},A_{i_0^1}) + \Lambda_0 + s\Lambda_\perp^1, \Lambda + H_2(P_{i_0^2},A_{i_0^2}) + t\Lambda_\perp^2, w_{i_0:i_0^1}, w_{i_0:i_0^2};\Phi_{i_0^1},\Phi_{i_0^2}:\underset{\sim}{H})$$

divided by

$$\prod_{\lambda_{i_0}} (w_{i_0:i_0^2}(\bar{\Lambda} + H_2(P_{i_0^2},A_{i_0^2}) + t\Lambda_\perp^2) + w_{i_0:i_0^1}(\Lambda + H_1(P_{i_0^1},A_{i_0^1}) + \Lambda_0 + s\Lambda_\perp^1),\lambda_{i_0}).$$

Here

$$\begin{cases} \Phi_{i_o^1} = T_{i_o^1}(\Phi':H_1)(1) \in E_{cus}(\delta, \mathcal{O}_{i_o^1}) \\ \\ \Phi_{i_o^2} = T_{i_o^2}(\Phi'':H_2)(1) \in E_{cus}(\delta, \mathcal{O}_{i_o^2}) . \end{cases}$$

Of course, we are not really interested in the full sum over

$$\begin{cases} W(A_{i_o}, A_{i_o^1}) \\ \\ W(A_{i_o}, A_{i_o^2}) \end{cases}$$

but rather only in the partial sum over

$$\begin{cases} W_{H_1}^{+}(A_{i_o}, A_{i_o^1}) \\ \\ W_{H_2}^{+}(A_{i_o}, A_{i_o^2}). \end{cases}$$

In order to exploit the inherent symmetries, it will be necessary to expand the partial sum. Needless to say, we must make sure that any new terms give 0 in the limit. This is easy enough. In fact, to get a contribution to $\mathfrak{K}(\mathbb{C}:\ldots)$, of necessity

$$w_{i_o:i_o^1}H_1(P_{i_o^1}, A_{i_o^1}) = -\, w_{i_o:i_o^2}H_2(P_{i_o^2}, A_{i_o^2}) .$$

On the other hand,

$$\begin{cases} w_{i_o:i_o^1} \in W_{H_1}^+(A_{i_o}, A_{i_o^1}) \\ w_{i_o:i_o^2} \in W_{H_2}^+(A_{i_o}, A_{i_o^2}) \end{cases}$$

$$\Longrightarrow$$

$$\begin{cases} w_{i_o:i_o^1} H_1(P_{i_o^1}, A_{i_o^1}) \in \mathfrak{D}_{P_{i_o}}(\check{\mathfrak{a}}_{i_o})^- \\ w_{i_o:i_o^2} H_2(P_{i_o^2}, A_{i_o^2}) \in \mathfrak{D}_{P_{i_o}}(\check{\mathfrak{a}}_{i_o})^- , \end{cases}$$

and so

$$w_{i_o:i_o^1} H_1(P_{i_o^1}, A_{i_o^1}) \neq - w_{i_o:i_o^2} H_2(P_{i_o^2}, A_{i_o^2}).$$

Consequently, to determine the contribution to $\mathfrak{K}(\mathcal{C}_o:\ldots)$, it is permissible to replace the sum over

$$\begin{cases} W_{H_1}^+(A_{i_o}, A_{i_o^1}) \\ W_{H_2}^+(A_{i_o}, A_{i_o^2}) \end{cases}$$

by a sum over all the pairs

$$(w_{i_o:i_o^1}, w_{i_o:i_o^2}) \in W(A_{i_o}, A_{i_o^1}) \times W(A_{i_o}, A_{i_o^2})$$

for which

$$w_{i_o:i_o^1} H_1(P_{i_o^1}, A_{i_o^1}) \neq - w_{i_o:i_o^2} H_2(P_{i_o^2}, A_{i_o^2}).$$

We then make a by now familiar re-parametrization. Thus call

$$W_o(A_{i_o^1}, A_{i_o^2})$$

the subset of $W(A_{i_o^1}, A_{i_o^2})$ consisting of the

$$w_{12} \in W(A_{i_o^1}, A_{i_o^2})$$

such that

$$H_1(P_{i_o^1}, A_{i_o^1}) \neq - w_{12}H_2(P_{i_o^2}, A_{i_o^2}).$$

In this notation, our problem, in revised form, is to study

$$\lim_{\Lambda_o \to 0} \ \lim_{s,t \to 0} \ st \cdot \sum_{w_{12} \in W_o(A_{i_o^1}, A_{i_o^2})}$$

of

$$T(\Lambda + H_1(P_{i_o^1}, A_{i_o^1}) + \Lambda_o + s\Lambda_\perp^1, \ \Lambda + H_2(P_{i_o^2}, A_{i_o^2}) + t\Lambda_\perp^2; w_{12}).$$

In passing, let us observe that we are working at the $\mathcal{C}_o$-level, hence are free to make use of the corresponding cuspidal theory (as developed above).

Some restrictions on Λ are, of course, going to be required. To begin wi let us remind ourselves that

$$\Lambda \in \sqrt{-1}\,\check{\mathfrak{a}} \subset \begin{cases} \check{\mathfrak{a}}_{i_o^1} + \sqrt{-1}\,\check{\mathfrak{a}}_{i_o^1} \\ \check{\mathfrak{a}}_{i_o^2} + \sqrt{-1}\,\check{\mathfrak{a}}_{i_o^2} \end{cases} .$$

Since any H figuring in the definition of

$$E(\delta, \mathcal{O}_o; \mathfrak{E}_o : P|A)$$

is principal, it follows from Proposition 5.11 in TES that $E(H:\ldots)$ is holomorphic at $\Lambda + H(P_{i_o}, A_{i_o})$. Unfortunately, this does not guarantee that $\Lambda + H(P_{i_o}, A_{i_o})$ lies on no other singular hyperplane of the associated cuspidal Eisenstein series $E(P_{i_o}|A_{i_o}:\ldots)$ besides $H(P_{i_o}, A_{i_o})$ itself. Accordingly, we shall simply make the blanket assumption that only those Λ will be considered which have this property whatever be the H (per $\mathfrak{E}_o$). Denote by

$$\sqrt{-1}\,\check{\mathfrak{a}}(\mathfrak{E}_o)$$

the set of all such Λ. Because the number of H involved is finite, it is clear that $\sqrt{-1}\,\check{\mathfrak{a}}(\mathfrak{E}_o)$ is a set of full measure in $\sqrt{-1}\,\check{\mathfrak{a}}$.

Now fix a

$$w_{12} \in W_o(A_{i_o^1}, A_{i_o^2}).$$

Given $\Lambda \in \sqrt{-1}\,\check{\mathfrak{a}}(E_o)$, put

$$\begin{cases} \Lambda_1 = \Lambda + H_1(P_{i_o^1}, A_{i_o^1}) \\ \Lambda_2 = \Lambda + H_2(P_{i_o^2}, A_{i_o^2}). \end{cases}$$

Replace $A_{i_o^1}$ by A_o (say) -- then

$$T(\Lambda_1 + \Lambda_o + s\Lambda_\perp^1, \Lambda_2 + t\Lambda_\perp^2; w_{12})$$

is defined by

$$\sum_{i_o=1}^{r_o} \sum_{w_{i_o} \in W(A_{i_o}, A_o)} \frac{\mathbf{O}_{i_o}(\Lambda_1 + \Lambda_o + s\Lambda_\perp^1, \Lambda_2 + t\Lambda_\perp^2; w_{i_o}, w_{i_o} w_{12})}{\prod_{\lambda_{i_o}} (w_{i_o} w_{12}(\bar{\Lambda}_2 + t\Lambda_\perp^2) + w_{i_o}(\Lambda_1 + s\Lambda_\perp^1), \lambda_{i_o})}.$$

To compute

$$\lim_{\Lambda_o \to 0} \lim_{s,t \to 0} st,$$

it is not necessary to sum over all pairs in

$$\coprod_{i_o=1}^{r_o} (\{i_o\} \times W(A_{i_o}, A_o))$$

but rather only over the subset $W_o(w_{12})$ consisting of the (i_o, w_{i_o}) such that

$$\begin{cases} w_{i_o} \in W_{H_1}^+(A_{i_o}, A_{i_o^1}) \\ w_{i_o} w_{12} \in W_{H_2}^+(A_{i_o}, A_{i_o^2}). \end{cases}$$

To take advantage of this remark, let $\mathcal{W}_o(w_{12})$ be the set of $(\Lambda_1,\Lambda_2;w_{12})$ -- equivalence classes determined by $W_o(w_{12})$ -- then

$$\sum_{(i_o,w_{i_o})\in W_o(w_{12})}$$

can be exchanged for

$$\sum_{(i_o,w_{i_o})\in \mathcal{W}_o(w_{12})}$$

without altering the situation in any essential way. This being so, let E_o be an equivalence class in $\mathcal{W}_o(w_{12})$ -- then we have

$$\lim_{\Lambda_o\to 0}\left[\lim_{s,t\to 0} st\cdot\sum_{(i_o,w_{i_o})\in E_o}\frac{\mathfrak{D}_{i_o}(\Lambda_1+\Lambda_o+s\Lambda_\perp^1,\Lambda_2+t\Lambda_\perp^2;w_{i_o},w_{i_o}w_{12})}{\prod_{\lambda_{i_o}}(w_{i_o}w_{12}(\bar{\Lambda}_2+t\Lambda_\perp^2)+w_{i_o}(\Lambda_1+s\Lambda_\perp^1),\lambda_{i_o})}\right]$$

$$= \text{value of extended holomorphic function at } (0,0,0),$$

or still

$$\lim_{s\to 0} s^2\cdot\left[\lim_{\Lambda_o\to 0}\sum_{(i_o,w_{i_o})\in E_o}\frac{\mathfrak{D}_{i_o}(\Lambda_1+\Lambda_o+s\Lambda_\perp^1,\Lambda_2+t\Lambda_\perp^2;w_{i_o},w_{i_o}w_{12})}{\prod_{\lambda_{i_o}}(w_{i_o}w_{12}(\bar{\Lambda}_2+s\Lambda_\perp^2)+w_{i_o}(\Lambda_1+s\Lambda_\perp^1),\lambda_{i_o})}\right].$$

In turn,

$$\lim_{\Lambda_o\to 0}\sum_{(i_o,w_{i_o})\in E_o}\cdots$$

is the sum

$$\sum_{0\le k\le \ell_{w_{12}}(\Lambda_1+s\Lambda_\perp^1,\Lambda_2+s\Lambda_\perp^2)} \sum_{(i_o,w_{i_o})\in E_o} \sum_{w_{i_o}:\iota(\Lambda_1+s\Lambda_\perp^1,\Lambda_2+s\Lambda_\perp^2;w_{i_o},w_{12})=k}$$

of the

$$\mathcal{A}_{i_o}^{(k)}(\Lambda_1+s\Lambda_\perp^1,\Lambda_2+s\Lambda_\perp^2;w_{i_o},w_{i_o}w_{12})$$

The dependence on s in the last expression can be partially eliminated. In fact, since the real part of

$$\begin{cases} \Lambda_1+s\Lambda_\perp^1 \text{ is proportional to } H_1(P_{i_o^1},A_{i_o^1}) \\ \Lambda_2+s\Lambda_\perp^2 \text{ is proportional to } H_2(P_{i_o^2},A_{i_o^2}) , \end{cases}$$

and since H_1 and H_2 are equivalent, it is clear that

$$(w_{i_o}w_{12}(\bar{\Lambda}_2+s\Lambda_\perp^2)+w_{i_o}(\Lambda_1+s\Lambda_\perp^1),\lambda_{i_o})\neq 0$$

iff

$$(w_{i_o}w_{12}\bar{\Lambda}_2+w_{i_o}\Lambda_1,\lambda_{i_o})\neq 0.$$

Therefore

$$\Sigma_{P_{i_o}}^{o}(\Lambda_1+s\Lambda_\perp^1,\Lambda_2+s\Lambda_\perp^2;w_{i_o},w_{12})$$

$$=\Sigma_{P_{i_o}}^{o}(\Lambda_1,\Lambda_2;w_{i_o},w_{12}) .$$

Consequently, we are reduced to explicating the limit as $s \to 0$ of s^2 times the sum

$$\sum_{0 \leq k \leq \ell_{w_{12}}(\Lambda_1, \Lambda_2)} \sum_{(i_o, w_{i_o}) \in E_o} \sum_{w_{i_o} : \iota(\Lambda_1, \Lambda_2; w_{i_o}, w_{12}) = k}$$

of the

$$\mathcal{A}_{i_o}^{(k)}(\Lambda_1 + s\Lambda_\perp^1, \Lambda_2 + s\Lambda_\perp^2; w_{i_o}, w_{i_o} w_{12}) .$$

For this purpose, take s complex -- then the limit is simply the residue at $s = 0$ of

$$s \cdot \mathcal{A}_{i_o}^{(k)}(\Lambda_1 + s\Lambda_\perp^1, \Lambda_2 + \bar{s}\Lambda_\perp^2; w_{i_o}, w_{i_o} w_{12}) ,$$

call it

$$Y_{i_o}^{(k)}(\Lambda_1, \Lambda_2; w_{i_o}, w_{i_o} w_{12})$$

for short.

Noting that

$$W_o(w_{12}) = \coprod E_o,$$

we can thus say, by way of recapitulation, that

$$\lim_{\Lambda_o \to 0} \lim_{s,t \to 0} st \cdot T(\Lambda_1 + \Lambda_o + s\Lambda_\perp^1, \Lambda_2 + t\Lambda_\perp^2; w_{12})$$

exists and is equal to the sum

$$\sum_{0\le k\le \ell_{w_{12}}(\Lambda_1,\Lambda_2)} \sum_{(i_o,w_{i_o})\in \mathcal{W}_o(w_{12})} \sum_{w_{i_o}:\iota(\Lambda_1,\Lambda_2;w_{i_o},w_{12}) = k}$$

of the

$$Y_{i_o}^{(k)}(\Lambda_1,\Lambda_2;w_{i_o},w_{i_o}w_{12}) \quad .$$

The computation, in toto, can then be completed by making one last summation, namely, over the

$$w_{12} \in W_o(A_{i_o 1},A_{i_o 2}) \ .$$

If now we proceed to add up each of the contributions from all possible pairs (H_1,H_2), then the result is an explicit formula for the sum over i_o and μ_o of

$$\mathbb{K}(\mathcal{C}_o:P_{i_o\mu_o}|A_{i_o\mu_o}:E_{P_{i_o\mu_o}}(P|A:\Phi':\Lambda:?),\ E_{P_{i_o\mu_o}}(P|A:\Phi'':\Lambda:?):\underset{\sim}{H}).$$

Here,

$$\Phi',\Phi'' \in \mathcal{E}(\delta,\mathcal{O}:\mathcal{C}_o) \cap \mathcal{E}(\delta,\underset{\sim}{\mathcal{O}}_o;E_o:P|A).$$

Moreover,

$$\Lambda \in \sqrt{-1}\ \check{\mathfrak{a}}(E_o).$$

Admittedly, the final formula is complicated; on the other hand, it will not survive into the trace formula itself, a point for which we should certainly be grateful. Since a systematic discussion would take us too far afield, we shall settle, at the moment, for a partial justification of this assertion. Each

$$Y_{i_o}^{(k)}(\Lambda_1,\Lambda_2;w_{i_o},w_{i_o}w_{12})$$

depends not only on the indicated parameters but also on Φ_{i_o1}, Φ_{i_o2}, and $\underset{\sim}{H}$. With Λ (and hence Λ_1,Λ_2) fixed, we claim that

$$|Y_{i_o}^{(k)}(\Lambda_1,\Lambda_2;w_{i_o},w_{i_o}w_{12})| \to 0$$

as $\underset{\sim}{H} \to -\infty$. [Note: We omit for now any discussion of uniformities in the decay.] The argument runs as follows. Consider the set $S(w_{12})$ of all pairs (i_o,w_{i_o}) in

$$\coprod_{i_o=1}^{r_o} (\{i_o\} \times W(A_{i_o},A_o))$$

such that

$$w_{i_o}H_1(P_{i_o1},A_{i_o1}) + w_{i_o}w_{12}H_2(P_{i_o2},A_{i_o2}) \in \mathcal{D}_{P_{i_o}}(\check{a}_{i_o})^- .$$

Then it is clear that

$$S(w_{12}) \supset W_o(w_{12}) \ .$$

In addition, it is not difficult to check that $S(w_{12})$ is $R(\Lambda_1,\Lambda_2;w_{12})$-stable, i.e., is a union of $(\Lambda_1,\Lambda_2;w_{12})$--equivalence classes. Thus, it is actually true that

$$S(w_{12}) \supset \mathcal{W}_o(w_{12}),$$

and so

$$\forall w_{12} \in W_o(A_{i_o 1},A_{i_o 2}),\ \forall (i_o,w_{i_o}) \in \mathcal{W}_o(w_{12}):$$

$$\operatorname{Re}(w_{i_o}\Lambda_1 + w_{i_o}w_{12}\bar{\Lambda}_2)$$

$$= w_{i_o}H_1(P_{i_o 1},A_{i_o 1}) + w_{i_o}w_{12}H_2(P_{i_o 2},A_{i_o 2}) \in \mathcal{D}_{P_{i_o}}(\check{a}_{i_o})^- \ .$$

Consequently,

$$|\exp(\langle I_{P_{i_o\mu_o}}(\underset{\sim}{H}),k_{i_o\mu_o}(w_{i_o}w_{12}\bar{\Lambda}_2 + w_{i_o}\Lambda_1)\rangle)| \to 0$$

as $\underset{\sim}{H} \to -\infty$. Taking into account the derivative formula for a residue (and the definitions) then leads at once to our contention.

The preceding state of affairs will be denoted simply by writing

$$o(\Lambda:\underset{\sim}{H}) \ .$$

All told, therefore:

THEOREM 6.2 *Suppose that* Λ *is pure imaginary and in general position. Let*

$$\Phi', \Phi'' \in E(\delta, \mathcal{O}: \mathfrak{C}_o) \cap E(\delta, \underset{\sim}{\mathcal{O}}_o; E_o : P|A).$$

Suppose further that

$$\Lambda \in \sqrt{-1}\, \check{\mathfrak{a}}(E_o)\ .$$

Then, for all $\underset{\sim}{H} \in \underset{\sim}{\mathfrak{a}}_Q$, *the inner product*

$$(Q^{\underset{\sim}{H}}E(P|A:\Phi':\Lambda:?),\ Q^{\underset{\sim}{H}}E(P|A:\Phi'':\Lambda:?))$$

is equal to the sum

$$(-1)^{\ell} \cdot \mathrm{vol}(\mathfrak{C}) \cdot \sum_{w \in W(A)} \sum_{k=0}^{\ell_w} \sum_{i} \sum_{w_i : \iota(w_i, w)=k}$$

of the

$$\mathcal{R}_i^{(k)}(\Lambda; w_i, w_i w; \Phi', \Phi'' : \underset{\sim}{H})$$

PLUS

$$o(\Lambda:\underset{\sim}{H}).$$

§7. EXTENSION OF AN EARLIER RESULT

The purpose of this § is to obtain a formula for the inner product of an Eisenstein system function with a truncated cuspidal Eisenstein series. This formula will be the point of departure for later work. It is a generalization of the result established by us in [2] which dealt with cuspidal Eisenstein series only. Furthermore, the proof given here is very different since it makes no use of functional equations, depending, instead, on still another inner product formula (from the theory of wave-packets), to be recalled below.

We shall have need of some preliminaries of an analytic nature whose significance will become apparent in short order.

Introduce, as usual, the space $S(G/\Gamma)$ of slowly increasing functions on G/Γ -- then

$$S(G/\Gamma) = \bigcup_r S_r(G/\Gamma),$$

$S_r(G/\Gamma)$ the set of slowly increasing functions on G/Γ with exponent of growth r. Note that the union can be taken over all r less than some fixed r, e.g. -1. Keep in mind, too, that $S_r(G/\Gamma)$ is, in a natural way, a Banach space. Let, still, $S_r^\infty(G/\Gamma)$ be the space of slowly increasing differentiable functions on G/Γ with exponent of growth r such that for every right invariant differential operator D on G, Df is also slowly increasing with exponent of growth r -- then $S_r^\infty(G/\Gamma)$ is a Fréchet space.

Let $\mathcal{D}$ be an open subset of C^n. Let

$$\Phi : \mathcal{D} \to S_r^\infty(G/\Gamma)$$

be a map. We shall assume that

$$\Phi \in C^\infty(\mathcal{D} \times G/\Gamma)$$

and, in addition, that

$$\forall x \in G, \quad \Lambda \mapsto \Phi(x:\Lambda) \qquad (\Lambda \in \mathcal{D})$$

is holomorphic. It then follows from the Schwartz kernel theorem that, for every D, the assignment

$$\begin{cases} \mathcal{D} \to C^\infty(G/\Gamma) \\ \Lambda \mapsto D\Phi(?:\Lambda) \end{cases}$$

is weakly holomorphic, hence strongly continuous, $C^\infty(G/\Gamma)$ being a Montel space (cf. Proposition M (§2)). We would then like to conclude that Φ itself is weakly holomorphic (qua a map from $\mathcal{D}$ to $S_r^\infty(G/\Gamma)$, of course). To quote the generality infra, it is necessary to know that Φ is, at least, weakly continuous. To ensure this, it will be supposed that

$$\forall \text{ compact subset } K \subset \mathcal{D},$$

$$D\Phi(K) \text{ is bounded in } S_r(G/\Gamma)$$

whatever D be. [That this condition does in fact force the weak continuity of Φ requires a small approximation argument which, however, can be omitted, similar principles having been employed several times by us in [2].]

LEMMA *Let* X *and* Y *be locally convex quasi-complete topological vector spaces. Assume that*

$$X \subset Y \quad \textit{with} \quad \iota : X \hookrightarrow Y \quad \textit{continuous}.$$

Let $\mathcal{D}$ *be an open subset of* C^n; *let* $\Phi:\mathcal{D} \to X$ *be a weakly continuous map such that* $\Phi:\mathcal{D} \to Y$ *is weakly holomorphic* -- *then* $\Phi:\mathcal{D} \to X$ *is weakly holomorphic.*

[Note: Taking

$$\begin{cases} X = S_r^\infty(G/\Gamma) \\ Y = C^\infty(G/\Gamma) \end{cases}$$

completes the discussion supra.]

PROOF Let X^*_Φ be the subset of X^* comprised of those T with the property that

$$\Lambda \mapsto \langle \Phi(\Lambda), T \rangle \qquad (\Lambda \in \mathcal{D})$$

is holomorphic. We must prove that $X^*_\Phi = X^*$. Since $\Phi:\mathcal{D} \to X$ is weakly continuous, X^*_Φ is closed in the Mackey topology $\tau(X^*,X)$. Indeed, if $T_o \notin X^*_\Phi$, then there exists a Λ_o and a polydisk $\Delta_o \ni \Lambda_o$ such that

$$\langle \Phi(\Lambda_0), T_0 \rangle \neq \frac{1}{(2\pi)^n} \int_{\partial\Delta_0} \cdots .$$

Let K be the weakly closed convex hull of $\Phi(\{\Lambda_0\} \cup \partial\Delta_0)$ -- then, by Krein's theorem, K is a (convex) weakly compact subset of X. Use K to separate T_0 and X^*_Φ. This point settled, it remains only to show that X^*_Φ is $\tau(X^*,X)$-dense. By hypothesis, X^*_Φ contains $\iota^*(Y^*)$, the latter being weak*-dense in X^*, i.e., $\sigma(X^*,X)$-dense (Hahn-Banach). Since $\sigma(X^*,X)$-dense subspaces are dense in any topology on X^* yielding X as the dual and since, by Mackey's theorem, this includes $\tau(X^*,X)$, it must therefore be the case that X^*_Φ is dense. //

Now bring in the truncation operator $\underset{\sim}{Q}^{\underset{\sim}{H}}(\underset{\sim}{H} \in \underset{\sim}{\mathfrak{a}})$. Referring to [2] for the particulars, let us recall only that

$$\forall \underset{\sim}{H} \in \underset{\sim}{\mathfrak{a}}_Q,$$

$$\begin{cases} \underset{\sim}{Q}^H \circ \underset{\sim}{Q}^H = \underset{\sim}{Q}^H, \quad (\underset{\sim}{Q}^H)^* = \underset{\sim}{Q}^H \\ \underset{\sim}{Q}^H(S_r^\infty(G/\Gamma)) \subset R(G/\Gamma), \end{cases}$$

$R(G/\Gamma)$ the space of rapidly decreasing functions on G/Γ.

Let again $\mathcal{D}$ be an open subset of $\mathbf{C}^n$. Let

$$\Phi : \mathcal{D} \to \bigcup_r S_r^\infty(G/\Gamma)$$

be a map. Write $\mathcal{D}_r$ for the interior of

$$\{\Lambda \in \mathcal{D} : \Phi(\Lambda) \in S_r^\infty(G/\Gamma)\}.$$

We then assume that

$$\mathcal{D} = \bigcup_r \mathcal{D}_r .$$

Let

$$\Phi_r = \Phi | \mathcal{D}_r .$$

Supposing that Φ_r satisfies the same general conditions as the 'Φ' at the beginning, it follows that

$$\Phi_r : \mathcal{D}_r \to S_r^\infty(G/\Gamma)$$

is weakly holomorphic. Owing to the closed graph theorem,

$$\underset{\sim}{Q}^H : S_r^\infty(G/\Gamma) \to R(G/\Gamma)$$

is continuous. Form the triangle

$$\begin{array}{ccc} \mathcal{D}_r & \xrightarrow{\ \Phi_r\ } & S_r^\infty(G/\Gamma) \\ & \ddots & \Big\downarrow \underset{\sim}{Q}^H \\ & & R(G/\Gamma) \end{array}$$

The dotted arrow, being the composition, is therefore itself weakly holomorphic. But $R(G/\Gamma)$ is independent of r. Consequently,

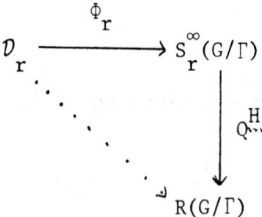

$$\begin{cases} \mathcal{D} \to R(G/\Gamma) \\ \Lambda \mapsto Q^{H}_{\sim}\Phi(?:\Lambda) \end{cases}$$

is weakly holomorphic.

Agreeing to use subscripts to denote two situations per supra, we could then say that

$$\begin{cases} \mathcal{D}_1 \times \mathcal{D}_2 \to C \\ (\Lambda_1,\Lambda_2) \mapsto (Q^{H}_{\sim}\Phi_1(?:\Lambda_1),\ Q^{H}_{\sim}\Phi_2(?:\bar{\Lambda}_2)) \end{cases}$$

is separately holomorphic, hence, by Hartog's theorem, holomorphic.

The analytic preparation established, let us pass to the inner product formula. Fix an association class $\mathcal{C}_o$ of Γ-cuspidal split parabolic subgroups of G. Let $\mathcal{C}_{i_o}$ be a G-conjugacy class in $\mathcal{C}_o$, $\mathfrak{X}$ an equivariant system of admissible affine subspaces attached to $\mathcal{C}_{i_o}$. Suppose given an Eisenstein system $\{E,\nabla\}$ belonging to $\mathfrak{X}$. There is then attached to each

$$(P_{i_o}, S_{i_o}; A_{i_o}) \in \mathcal{C}_{i_o}$$

an E-function

$$E(\mathfrak{X}:G|\{1\}:P_{i_o}|A_{i_o}:T_{i_o}:\Lambda_{i_o}:x).$$

Here, of course, we are in the special case when

$$\mathcal{C}_k = \{G\},\quad P = G,\quad A_i = \{1\}.$$

Let $\mathcal{C}_{j_o}$ be another G-conjugacy class in $\mathcal{C}_o$ -- then with any

$$(P_{j_o}, S_{j_o}; A_{j_o}) \in \mathcal{C}_{j_o}$$

there is associated an Eisenstein series

$$E(P_{j_o}|A_{j_o}:\Phi_{j_o}:\Lambda_{j_o}:x).$$

In this connection, we had best remind ourselves that

$$\begin{cases} T_{i_o} \in \operatorname{Hom}(S_{\mathfrak{X}(P_{i_o},A_{i_o})}, \mathcal{E}_{cus}(\delta,\mathcal{O}_{i_o})) \\ \Phi_{j_o} \in \mathcal{E}_{cus}(\delta,\mathcal{O}_{j_o}). \end{cases}$$

It will be assumed that $\mathcal{O}_{i_o}$ and $\mathcal{O}_{j_o}$ belong to the same orbit type, $\underset{\sim}{\mathcal{O}}_o$, say. Now observe that the general properties of E-functions and Eisenstein series serve to ensure that the preceding considerations are applicable, the concomitant consequence being that the inner product

$$(Q_{\sim}^{H}E(\mathfrak{X}:G|\{1\}:P_{i_o}|A_{i_o}:T_{i_o}:\Lambda_{i_o}:?),\ Q_{\sim}^{H}E(P_{j_o}|A_{j_o}:\Phi_{j_o}:\Lambda_{j_o}:?))$$

is a holomorphic function of $(\Lambda_{i_o}, \overline{\Lambda}_{j_o})$ so long as Λ_{i_o} (or Λ_{j_o}) is confined to a suitable open domain. Our intention is to obtain an explicit formula for this inner product. Since in the end it is a question of an equality of meromorphic functions, there is no loss of generality in the beginning by restricting Λ_{i_o} (or Λ_{j_o}) to a relatively compact open domain and taking Λ_{j_o} very negative.

For simplicity, we shall suppose that the ambient split components infra are special.

Due to the idempotence and self-adjointness of $Q_{\sim}^{H}$, we need only explicitly evaluate

$$(E(\mathfrak{X}:G|\{1\}:P_{i_o}|A_{i_o}:T_{i_o}:\Lambda_{i_o}:?),\ Q_{\sim}^{H}E(P_{j_o}|A_{j_o}:\Phi_{j_o}:\Lambda_{j_o}:?)).$$

In turn, this leads us to the point of our procedure. For, as will be recalled, we found in [2] a way of writing

$$Q_{\sim}^{H}E(P_{j_o}|A_{j_o}:\Phi_{j_o}:\Lambda_{j_o}:?)$$

in terms of (generalized) wave-packets, actually a finite sum thereof. What has to be done, then, is to take each of the individual summands and compute its inner product with

$$E(\mathfrak{X}:G|\{1\}:P_{i_o}|A_{i_o}:T_{i_o}:\Lambda_{i_o}:?).$$

The desired result would then be a formal consequence of Lemma 5.7 from TES provided the E-function was L^2 (which it need not be) and the (generalized) wave-packets were really wave-packets (which they need not be). We must therefore proceed with care, making sure as we go along that the relevant estimates are valid.

The expression for

$$Q_{\sim}^{H}E(P_{j_o}|A_{j_o}:\Phi_{j_o}:\Lambda_{j_o}:?)$$

alluded to above (cf. Proposition 9.5 in [2]) can only be written down after some additional notation has been introduced. Let P_{k_o} $(1 \le k_o \le r_o)$ be a set of representatives for $G\backslash\mathcal{C}_o$ -- then

$$\mathcal{C}_o = \coprod_{k_o} \mathcal{C}_{k_o}$$

where $\mathcal{C}_{k_o} = G\cdot\{P_{k_o}\} \cap \mathcal{C}_o$. Let $P_{k_o\xi_o}$ $(1 \le \xi_o \le r_{k_o})$ be a set of representatives for $\Gamma\backslash\mathcal{C}_{k_o}$ -- then

$$\{P_{k_o\xi_o} : 1 \le k_o \le r_o,\ 1 \le \xi_o \le r_{k_o}\}$$

is a set of representatives for $\Gamma\backslash\mathcal{C}_o$. This said, under the supposition that Λ_{j_o} is sufficiently negative,

$$Q\overset{H}{\sim}E(P_{j_o}|A_{j_o}:\Phi_{j_o}:\Lambda_{j_o}:?)(x)$$

is equal to

$$\sum_{k_o=1}^{r_o}\ \sum_{\xi_o=1}^{r_{k_o}}\ \sum_{w_{k_o\xi_o:j_o} \in W(A_{k_o\xi_o},A_{j_o})} \Theta_{\Phi_{k_o\xi_o}}(w_{k_o\xi_o:j_o}:\Lambda_{j_o}:?)(x).$$

Here,

$$\Theta_{\Phi_{k_o\xi_o}}(w_{k_o\xi_o:j_o}:\Lambda_{j_o}:?)(x)$$

is the series

$$\sum_{\gamma_{k_o\xi_o} \in \Gamma/\Gamma\cap P_{k_o\xi_o}} \varphi_{k_o\xi_o}(w_{k_o\xi_o:j_o}:\Lambda_{j_o}:x\gamma_{k_o\xi_o}),$$

where, by definition,

$$\varphi_{k_o\xi_o}(w_{k_o\xi_o:j_o}:\Lambda_{j_o}:x)$$

is equal to

$$\exp(\langle H_{P_{k_o\xi_o}|A_{k_o\xi_o}}(x),\ w_{k_o\xi_o:j_o}\Lambda_{j_o}-\rho_{k_o\xi_o}\rangle)$$

$$\times\ (-1)^{\#(F(w_{k_o\xi_o:j_o}))}\tau_{*,*}(F(w_{k_o\xi_o:j_o}):I_{P_{k_o\xi_o}}(\underset{\sim}{H}) - H_{P_{k_o\xi_o}|A_{k_o\xi_o}}(x))$$

$$\times\ (c_{cus}(P_{k_o\xi_o}|A_{k_o\xi_o}:P_{j_o}|A_{j_o}:w_{k_o\xi_o:j_o}:\Lambda_{j_o})\Phi_{j_o})(x).$$

Formally, therefore, the inner product

$$(E(\mathfrak{X}:G|\{1\}:P_{i_o}|A_{i_o}:T_{i_o}:\Lambda_{i_o}:?),\ Q^{\underset{\sim}{H}}E(P_{j_o}|A_{j_o}:\Phi_{j_o}:\Lambda_{j_o}:?))$$

is equal to the sum over

$$\begin{cases} k_o, & 1\le k_o\le r_o \\ \xi_o, & 1\le\xi_o\le r_{k_o} \\ w_{k_o\xi_o:j_o}, w_{k_o\xi_o:j_o} & \in W(A_{k_o\xi_o}, A_{j_o}) \end{cases}$$

of the inner products

$$(E(\mathfrak{X}:G|\{1\}:P_{i_o}|A_{i_o}:T_{i_o}:\Lambda_{i_o}:?),\Theta_{\Phi_{k_o\xi_o}}(w_{k_o\xi_o:j_o}:\Lambda_{j_o}:?)).$$

Since Λ_{i_o} is confined to a relatively compact open domain, the exponent of growth for the E-function is uniform, call it $r(<-1)$. The idea now will be to show that if Λ_{j_o} is very negative (but compactly confined), then the

$$\Theta_{\Phi_{k_o\xi_o}}(w_{k_o\xi_o:j_o}:\Lambda_{j_o}:?)$$

have exponent of growth $-r$. This done, it then follows that the relevant inner products at least exist, thus putting us in a position to make explicit computations.

We need a lemma.

LEMMA 7.1 *Let* (P,S) *be a* Γ-*cuspidal split parabolic subgroup of* G *with special split component* A. *Let* $\Phi \in \mathcal{E}_{cus}(\delta,\mathcal{O})$ -- *then*

$$\forall c > 1, \quad \exists q_c >> 0 \quad \text{st}$$

$$\sum_{\gamma \in \Gamma/\Gamma \cap P} \exp(-q_c \cdot ||H_{P|A}(x\gamma)||) \cdot |\Phi(x\gamma)|$$

has exponent of growth c.

PROOF Let

$$\{(P_i', S_i') : 1 \leq i \leq r_{oM}\}$$

be a set of Γ-percuspidal split parabolic subgroups of G which are dominated predecessors of (P,S) and with the property that

$$\{('P_i, 'S_i) : 1 \leq i \leq r_{oM}\}$$

is a set of representatives for the Γ_M-conjugacy classes of Γ_M-percuspidal split parabolic subgroups of M. Since

$$\Phi | K \times M/\{1\} \times \Gamma_M$$

is rapidly decreasing, estimation principles developed in [2] imply that

$$\forall q >> 0,\ \exists C_q > 0 \text{ st } \forall x \in G,\ \forall \gamma \in \Gamma$$

$$|\Phi(x\gamma)| \leq C_q \cdot \sum_{i=1}^{r_{oM}} \sum_{\gamma_i' \in \Gamma \cap P/\Gamma \cap P_i'} \exp(-q \cdot \| H_{'P_i|'A_i}(x\gamma\gamma_i') \|).$$

Thus

$$\sum_{\gamma \in \Gamma/\Gamma \cap P} \exp(-q \cdot \| H_{P|A}(x\gamma) \|) \cdot |\Phi(x\gamma)|$$

cannot exceed

$$C_q \cdot \sum_{i=1}^{r_{oM}} \sum_{\gamma \in \Gamma/\Gamma \cap P} \sum_{\gamma_i' \in \Gamma \cap P/\Gamma \cap P_i'}$$

$$\times \exp(-q \cdot \| H_{P|A}(x\gamma) \|) \exp(-q \cdot \| H_{'P_i|'A_i}(x\gamma\gamma_i') \|)$$

which, in view of the fact that

$$\mathfrak{a}_i' = {}'\mathfrak{a}_i \oplus \mathfrak{a},$$

is itself majorized by

$$C_q \cdot \sum_{i=1}^{r_{oM}} \sum_{\gamma_i \in \Gamma/\Gamma \cap P_i'} \exp(-qC \cdot \| H_{P_i'|A_i'}(x\gamma_i) \|),$$

C a positive absolute constant. However, in the notation of [2],

$$\sum_{\gamma_i \in \Gamma/\Gamma \cap P_i'} \exp(-qC \cdot \| H_{P_i'|A_i'}(x\gamma_i) \|),$$

qua a function on G, is precisely

$$\zeta_{qC}(P_i' : x),$$

the latter having exponent of growth c provided q is chosen large enough. //

Fix indices k_o and ξ_o -- then the preceding lemma may be used to control the growth of

$$\Theta_{\Phi_{k_o\xi_o}}(w_{k_o\xi_o:j_o}:\Lambda_{j_o}:?) \ .$$

Indeed,

$$c_{cus}(P_{k_o\xi_o}|A_{k_o\xi_o}:P_{j_o}|A_{j_o}:w_{k_o\xi_o:j_o}:\Lambda_{j_o})\Phi_{j_o}$$

is in $\mathcal{E}_{cus}(\delta,\mathcal{O}_{k_o\xi_o})$ so, since we are summing over $\Gamma/\Gamma\cap P_{k_o\xi_o}$, we need only show that given any $q >> 0$, subject to

$$\tau_{*,*}(F(w_{k_o\xi_o:j_o}):I_{P_{k_o\xi_o}}(\underset{\sim}{H}) - H_{P_{k_o\xi_o}|A_{k_o\xi_o}}(x)) \neq 0,$$

the entity

$$\exp(\langle H_{P_{k_o\xi_o}|A_{k_o\xi_o}}(x), w_{k_o\xi_o:j_o}\Lambda_{j_o} - \rho_{k_o\xi_o}\rangle)$$

admits the bound

$$C_q\cdot\exp(-q\cdot \|H_{P_{k_o\xi_o}|A_{k_o\xi_o}}(x)\|) \qquad (C_q > 0)$$

if Λ_{j_o} is sufficiently negative.

Set, for simplicity,

$$(P,S;A) = (P_{k_o\xi_o},S_{k_o\xi_o};A_{k_o\xi_o})$$

and then make the following temporary changes in the data:

$$\begin{cases} w_{k_o \xi_o : j_o} \to w_{j_o} \\ \rho_{k_o \xi_o} \to \rho . \end{cases}$$

In this notation we have:

LEMMA 7.2 *There exist positive constants* C_1, C_2, C_3, C_4 *with the property that*

$$\begin{cases} \forall H \in \mathfrak{a} \text{ st } \tau_{*,*}(F(w_{j_o}) : I_{\underset{\sim}{P}}(H) - H) \neq 0 \\ \forall \Lambda_{j_o} \in -\mathcal{C}_{P_{j_o}}(\check{\mathfrak{a}}_{j_o}) \end{cases}$$

$$\langle H, w_{j_o} \Lambda_{j_o} - \rho \rangle \leq - C_1 \cdot \|H\| \cdot \min_{\lambda_i} |(w_{j_o} \Lambda_{j_o}, \lambda_i)|$$

$$+ C_2 \cdot \|H\| + C_3 \cdot \|\Lambda_{j_o}\| + C_4,$$

λ_i *running through the simple roots of* $(P,S;A)$.

PROOF Since

$$\langle H, w_{j_o} \Lambda_{j_o} - \rho \rangle = \langle H - I_{\underset{\sim}{P}}(H), w_{j_o} \Lambda_{j_o} - \rho \rangle$$

$$+ \langle I_{\underset{\sim}{P}}(H), w_{j_o} \Lambda_{j_o} - \rho \rangle$$

$$\leq \langle H - I_{\underset{\sim}{P}}(H), w_{j_o} \Lambda_{j_o} \rangle - \langle H - I_{\underset{\sim}{P}}(H), \rho \rangle$$

$$+ C_3' \cdot \|\Lambda_{j_o}\| + C_4'$$

$$\leq \langle H - I_{\underset{\sim}{P}}(H), w_{j_o} \Lambda_{j_o} \rangle + C_2 \cdot \|H\|$$

$$+ C_3' \cdot \|\Lambda_{j_o}\| + C_4,$$

it will be enough to prove that

$$\langle H - I_P(\underset{\sim}{H}), w_{j_o}\Lambda_{j_o}\rangle \le -C_1 \cdot \|H\| \cdot \min_{\lambda_i} |(w_{j_o}\Lambda_{j_o}, \lambda_i)|$$

$$+ C_3'' \cdot \|\Lambda_{j_o}\|,$$

C_3 then being the sum of C_3' and C_3''. Given a simple root λ_i, determine H_i by the requirement

$$\langle H, \lambda_i \rangle = (H, H_i) \qquad (H \in \mathfrak{a}).$$

Now pass to coordinates via the map $T: R^{\ell} \to \mathfrak{a}$ defined according to the rule

$$T(t_1, \ldots, t_\ell) = \sum_{i=1}^{\ell} t_i H_i .$$

Then

$$\langle T(t_1, \ldots t_\ell), w_{j_o}\Lambda_{j_o}\rangle$$

$$= \sum_{i=1}^{\ell} t_i (w_{j_o}\Lambda_{j_o}, \lambda_i).$$

Furthermore, by definition (cf. [2]),

$$\tau_{*,*}(F(w_{j_o}) : T(t_1, \ldots, t_\ell)) = 1$$

iff

$$\begin{cases} \lambda_i \in F(w_{j_o}) \Longrightarrow \langle T(t_1, \ldots, t_\ell), \lambda^i \rangle > 0 \\ \lambda_i \notin F(w_{j_o}) \Longrightarrow \langle T(t_1, \ldots, t_\ell), \lambda^i \rangle \le 0, \end{cases}$$

that is, $\forall \lambda_i$,

$$t_i \text{ is } \begin{cases} > 0 & \text{if } \lambda_i \in F(w_{j_o}) \\ \leq 0 & \text{if } \lambda_i \notin F(w_{j_o}). \end{cases}$$

In this connection, recall that

$$\begin{cases} \lambda_i \in F(w_{j_o}) \Leftrightarrow w_{j_o}^{-1} \lambda_i < 0 \\ \lambda_i \notin F(w_{j_o}) \Leftrightarrow w_{j_o}^{-1} \lambda_i > 0. \end{cases}$$

It therefore follows that in these coordinates

$$\Lambda_{j_o} \in -\mathcal{C}_{P_{j_o}}(\check{\mathfrak{a}}_{j_o}) \Longrightarrow t_i(w_{j_o}\Lambda_{j_o}, \lambda_i) \geq 0 \qquad (\forall i).$$

Hence

$$\begin{aligned} &\langle H - I_P(\underset{\sim}{H}), w_{j_o}\Lambda_{j_o} \rangle \\ &= -\langle T(t_1, \ldots, t_\ell), w_{j_o}\Lambda_{j_o} \rangle \\ &= -\sum_{i=1}^{\ell} t_i(w_{j_o}\Lambda_{j_o}, \lambda_i) \\ &= -\sum_{i=1}^{\ell} |t_i(w_{j_o}\Lambda_{j_o}, \lambda_i)| \\ &\leq -(\sum_{i=1}^{\ell} |t_i|) \cdot \min_{\lambda_i} |(w_{j_o}\Lambda_{j_o}, \lambda_i)| \\ &\leq -C_1 \cdot \|H - I_P(\underset{\sim}{H})\| \cdot \min_{\lambda_i} |(w_{j_o}\Lambda_{j_o}, \lambda_i)| \\ &\leq -C_1 \cdot \|H\| \cdot \min_{\lambda_i} |(w_{j_o}\Lambda_{j_o}, \lambda_i)| + C_1 \cdot \|I_P(\underset{\sim}{H})\| \cdot \min_{\lambda_i} |(w_{j_o}\Lambda_{j_o}, \lambda_i)| \\ &\leq -C_1 \cdot \|H\| \cdot \min_{\lambda_i} |(w_{j_o}\Lambda_{j_o}, \lambda_i)| + C_3'' \cdot \|\Lambda_{j_o}\|, \end{aligned}$$

as desired. //

As Λ_{j_o} becomes ever more negative,

$$(w_{j_o}\Lambda_{j_o},\lambda_i) \to \begin{cases} +\infty & \text{if } \lambda_i \in F(w_{j_o}) \\ -\infty & \text{if } \lambda_i \notin F(w_{j_o}). \end{cases}$$

Thanks to the lemma, then, the sought for bound is manifest.

Having assured ourselves that the inner product

$$(E(\mathfrak{X}:G|\{1\}:P_{i_o}|A_{i_o}:T_{i_o}:\Lambda_{i_o}:?),\ \Theta_{\Phi_{k_o\xi_o}(w_{k_o\xi_o}:j_o}:\Lambda_{j_o}:?)})$$

exists when the parameters are conveniently restricted, let us proceed to its actual calculation. To keep the notation manageable, put

$$f = E(\mathfrak{X}:G|\{1\}:P_{i_o}|A_{i_o}:T_{i_o}:\Lambda_{i_o}:?).$$

Agreeing to suppress the k_o and ξ_o as above, we have

$$(f,\Theta_{\Phi(w_{j_o}:\Lambda_{j_o}:?)})$$

$$= \int_{G/\Gamma} f(x)\ (\sum_{\gamma\in\Gamma/\Gamma\cap P} \overline{\varphi(w_{j_o}:\Lambda_{j_o}:x\gamma)}\ )d_G(x)$$

or still

$$\int_{G/\Gamma\cap P} f(x)\overline{\varphi(w_{j_o}:\Lambda_{j_o}:x)}d_G(x)$$

or still

$$\int_K\int_A\int_{S/S\cap\Gamma} f(kas)\overline{\varphi(w_{j_o}:\Lambda_{j_o}:kas)}a^{2\rho}d_K(k)d_A(a)d_S(s)$$

or still

$$\int_K \int_{M/\Gamma_M} \int_A \int_{N/N\cap\Gamma} f(kman)\overline{\varphi(w_{j_o}:\Lambda_{j_o}:kman)}a^{2\rho}d_K(k)d_M(m)d_A(a)d_N(n)$$

or still

$$\int_K \int_{M/\Gamma_M} \int_A \int_{N/N\cap\Gamma} f(kman)\overline{\varphi(w_{j_o}:\Lambda_{j_o}:kma)}a^{2\rho}d_K(k)d_M(m)d_A(a)d_N(n)$$

or still

$$\int_K \int_{M/\Gamma_M} \int_A f^P(kma)\overline{\varphi(w_{j_o}:\Lambda_{j_o}:kma)}a^{2\rho}d_K(k)d_M(m)d_A(a)$$

or still

$$(-1)^{\#(F(w_{j_o}))} \cdot \int_K \int_{M/\Gamma_M} \int_A f_P(kma)[\ldots]a^{w_{j_o}\bar{\Lambda}_{j_o}}d_K(k)d_M(m)d_A(a),$$

where $[\ldots]$ is equal to

$$\tau_{*,*}(F(w_{j_o}):I_P(\underset{\sim}{H}) - \log a)$$

$$\times \overline{(c_{cus}(P|A:P_{j_o}|A_{j_o}:w_{j_o}:\Lambda_{j_o})\Phi_{j_o})(km)}.$$

On the basis of our negligibility assumption (cf. §3), there exist finitely

many elements $\Lambda_h \in \check{\mathfrak{a}} + \sqrt{-1}\,\check{\mathfrak{a}}$ and

$$u_h \otimes \Phi_h \in \mathbf{C}[\mathfrak{a}] \otimes \mathcal{E}_{cus}(\delta,\mathcal{O}),$$

which need not be explicated just yet, with the property that

$$f_P(=f_{K\times P}):$$

$$f_P(kma) = \sum_h a^{\Lambda_h}p_{u_h}(a)\Phi_h(km).$$

Inserting this information into the last relation, our inner product formula becomes

$$(-1)^{\#(F(w_{j_o}))} \cdot \sum_h (\Phi_h, c_{cus}(P|A:P_{j_o}|A_{j_o}:w_{j_o}:\Lambda_{j_o})\Phi_{j_o})$$

$$\times \int_{\mathfrak{a}} p_{u_h}(H) e^{(\Lambda_h + w_{j_o}\bar{\Lambda}_{j_o})(H)} \tau_{*,*}(F(w_{j_o}):I_P(\underset{\sim}{H}) - H) d_{\mathfrak{a}}(H)$$

or still

$$(-1)^{\#(F(w_{j_o}))} \cdot \sum_h (\Phi_h, c_{cus}(P|A:P_{j_o}|A_{j_o}:w_{j_o}:\Lambda_{j_o})\Phi_{j_o})$$

$$\times \int_{\mathfrak{a}} (\partial_{u_h} e^{\Lambda_h(H)}) e^{w_{j_o}\bar{\Lambda}_{j_o}(H)} \tau_{*,*}(F(w_{j_o}):I_P(\underset{\sim}{H}) - H) d_{\mathfrak{a}}(H)$$

or still

$$(-1)^{\#(F(w_{j_o}))} \cdot \sum_h (\Phi_h, c_{cus}(P|A:P_{j_o}|A_{j_o}:w_{j_o}:\Lambda_{j_o})\Phi_{j_o})$$

$$\times \partial_{u_h:\Lambda_h} \left[\int_{\mathfrak{a}} e^{(\Lambda_h + w_{j_o}\bar{\Lambda}_{j_o})(H)} \tau_{*,*}(F(w_{j_o}):I_P(\underset{\sim}{H}) - H) d_{\mathfrak{a}}(H)\right].$$

But (see Lemma 9.4 in [2])

$$\int_{\mathfrak{a}} e^{(\Lambda_h + w_{j_o}\bar{\Lambda}_{j_o})(H)} \tau_{*,*}(F(w_{j_o}):I_P(\underset{\sim}{H}) - H) d_{\mathfrak{a}}(H)$$

is equal to

$$(-1)^{\ell_o - \#(F(w_{j_o}))} \cdot \mathrm{vol}(\mathcal{C}_o)$$

$$\times \exp(\langle I_P(\underset{\sim}{H}), w_{j_o}\bar{\Lambda}_{j_o} + \Lambda_h\rangle)\cdot(1/\prod_{\lambda_i}(w_{j_o}\bar{\Lambda}_{j_o} + \Lambda_h, \lambda_i)).$$

Now explicate f_P:

$$f_P(kma) = \sum_{w_{i_o}\in W_{\{G\}}(\mathfrak{X};A,A_{i_o})} a^{w_{i_o}\Lambda_{i_o}}$$

$$\times\ (\nabla(\mathfrak{X}:G|(\{1\},\{1\}):P|A:P_{i_o}|A_{i_o}:w_{i_o}:\Lambda_{i_o})T_{i_o})(kma).$$

There is an identification

$$\mathrm{Hom}(S_{\mathfrak{X}(P,A)_{w_{i_o}}}, \mathcal{E}_{cus}(\delta,\mathcal{O})) \simeq \mathrm{Hom}(S_{\mathfrak{X}(P,A)_{w_{i_o}}}, C) \otimes \mathcal{E}_{cus}(\delta,\mathcal{O}),$$

hence a natural pairing (cf. §2)

$$S_{\mathfrak{X}(P,A)_{w_{i_o}}} \otimes \mathcal{E}_{cus}(\delta,\mathcal{O}) \times \mathrm{Hom}(S_{\mathfrak{X}(P,A)_{w_{i_o}}}, C) \otimes \mathcal{E}_{cus}(\delta,\mathcal{O}) \to C.$$

Accordingly,

$$(f, \Theta_{\Phi(w_{j_o}:\Lambda_{j_o}:?)})$$

$$= (-1)^{\ell_o} \cdot \mathrm{vol}(\mathcal{C}_o)\cdot\sum_{w_{i_o}\in W_{\{G\}}(\mathfrak{X};A,A_{i_o})} (?,?)_{w_{i_o}},$$

where $(?,?)_{w_{i_o}}$ is the result of pairing

$$\nabla(\mathfrak{X}:G|(\{1\},\{1\}):P|A:P_{i_o}|A_{i_o}:w_{i_o}:\Lambda_{i_o})T_{i_o}$$

with

$$d(\exp(\langle I_P(\underset{\sim}{H}),w_{j_o}\Lambda_{j_o}-?\rangle)\cdot(1/\textstyle\prod_{\lambda_i}(w_{j_o}\Lambda_{j_o}-?,\lambda_i)))(-w_{i_o}\bar{\Lambda}_{i_o})$$

$$\otimes\ c_{cus}(P|A:P_{j_o}|A_{j_o}:w_{j_o}:\Lambda_{j_o})\Phi_{j_o}.$$

REMARK In this connection, it should be observed that the expression for

$$(f,\Theta_{\Phi(w_{j_o}:\Lambda_{j_o}:?)})$$

thus derived is <u>formally</u> an immediate consequence of Lemma 5.7 from TES, as can be seen by explicating $\Phi(w_{j_o}:\Lambda_{j_o}:?)$. Indeed (cf. §9 in [2]),

$$\Phi(w_{j_o}:\Lambda_{j_o}:x)(?)$$

is equal to

$$(-1)^{\ell_o}\cdot\mathrm{vol}(\mathcal{C}_o)$$

times

$$\exp(\langle I_P(\underset{\sim}{H}),w_{j_o}\Lambda_{j_o}-?\rangle)\cdot(1/\textstyle\prod_{\lambda_i}(w_{j_o}\Lambda_{j_o}-?,\lambda_i))$$

$$\times\ (c_{cus}(P|A:P_{j_o}|A_{j_o}:w_{j_o}:\Lambda_{j_o})\Phi_{j_o}(x).$$

However, as was stressed at the beginning, the point is that the growth conditions needed to justify a direct application of Lemma 5.7 are, in the case at hand, simply not present.

There is another way to calculate the complex number $(?,?)_{w_{i_o}}$ supra which, in practice, is useful to know. The differential

$$d(\exp(\langle I_P(\underset{\sim}{H}),w_{j_o}\bar{\Lambda}_{j_o} + ?\rangle)\cdot(1/\prod_{i=1}^{\ell_o}(w_{j_o}\bar{\Lambda}_{j_o} + ?,\lambda_i)))(w_{i_o}\Lambda_{i_o})$$

is an element of

$$\mathrm{Hom}(S_{\mathfrak{X}(P,A)_{w_{i_o}}},C).$$

Tensor it on the right with the identity map on $E_{cus}(\delta,O)$ -- then

$$\begin{array}{c} d(\ldots)\otimes \mathrm{ID}:S_{\mathfrak{X}(P,A)_{w_{i_o}}}\otimes E_{cus}(\delta,O)\rightarrow C\otimes E_{cus}(\delta,O) \\ \qquad\qquad\qquad\qquad\qquad\qquad\qquad\qquad\wr\wr \\ \qquad\qquad\qquad\qquad\qquad\qquad\qquad\qquad E_{cus}(\delta,O). \end{array}$$

Applying $d(\ldots)\otimes \mathrm{ID}$ to $\nabla(\ldots)$ produces an element in $E_{cus}(\delta,O)$. Take its inner product with

$$c_{cus}(P|A:P_{j_o}|A_{j_o}:w_{j_o}:\Lambda_{j_o})\Phi_{j_o} .$$

We claim that the result is again $(?,?)_{w_{i_o}}$. Basically, it is a question here of keeping track of certain sign changes. To facilitate the verification, let $u\otimes\Phi$ be a generic term in

$$\nabla(\mathfrak{X}:G|(\{1\},\{1\}):P|A:P_{i_0}|A_{i_0}:w_{i_0}:\Lambda_{i_0})T_{i_0} \quad .$$

Put

$$\Psi = c_{cus}(P|A:P_{j_0}|A_{j_0}:w_{j_0}:\Lambda_{j_0})\Phi_{j_0} \quad .$$

Agreeing to write

$$\Delta(?)$$

for

$$\frac{\exp(\langle I_P(\underset{\sim}{H}),w_{j_0}\Lambda_{j_0} - ?\rangle)}{\prod_{\lambda_i}(w_{j_0}\Lambda_{j_0} - ?,\lambda_i)} \quad ,$$

it then need only be shown that

$$(u \otimes \Phi,\ d\Delta(-w_{i_0}\bar{\Lambda}_{i_0}) \otimes \Psi)$$

$$= (\Phi,\Psi) \cdot d\Delta^*(w_{i_0}\Lambda_{i_0})(u),$$

where, as usual,

$$\Delta^*(?) = \overline{\Delta(-\bar{?})}.$$

But

$$(u \otimes \Phi,\ d\Delta(-w_{i_0}\bar{\Lambda}_{i_0}) \otimes \Psi)$$

$$= (u \otimes \Phi,\ d[\Delta\cdot\Psi](-w_{i_0}\bar{\Lambda}_{i_0}))$$

$$= (\Phi,\ d[\Delta\cdot\Psi](-w_{i_0}\bar{\Lambda}_{i_0})(u^*))$$

$$= (\Phi, \partial_{u*}[\Delta\cdot\Psi](-w_{i_o}\bar{\Lambda}_{i_o}))$$

$$= (\Phi, [\partial_{u*}\Delta(-w_{i_o}\bar{\Lambda}_{i_o})]\Psi)$$

$$= (\Phi,\Psi)\cdot\overline{\partial_{u*}\Delta(-w_{i_o}\bar{\Lambda}_{i_o})}\quad .$$

Furthermore,

$$\overline{\partial_{u*}\Delta(-w_{i_o}\bar{\Lambda}_{i_o})}$$

$$= (1, \partial_{u*}\Delta(-w_{i_o}\bar{\Lambda}_{i_o}))$$

$$= (u, d\Delta(-w_{i_o}\bar{\Lambda}_{i_o}))$$

$$= (u, d_{\mathrm{Hom}}\Delta(-w_{i_o}\bar{\Lambda}_{i_o})\circ d1)$$

$$= (d_{\otimes}\Delta^*(w_{i_o}\Lambda_{i_o})u, d1)$$

$$= d\Delta^*(w_{i_o}\Lambda_{i_o})(u),$$

the constant term of $d_{\otimes}\Delta^*(w_{i_o}\Lambda_{i_o})u$. Hence the claim.

Coming back to our original problem, we thus arrive at the following conclusion, to wit: The inner product

$$(Q^{H}_{\sim}E(\mathfrak{X}:G|\{1\}:P_{i_o}|A_{i_o}:T_{i_o}:\Lambda_{i_o}:?),\ Q^{H}_{\sim}E(P_{j_o}|A_{j_o}:\Phi_{j_o}:\Lambda_{j_o}:?))$$

is equal to the sum

$$(-1)^{\ell_o}\cdot\mathrm{vol}(\mathcal{C}_o)\cdot\sum_{k_o=1}^{r_o}\sum_{\xi_o=1}^{r_{k_o}}\sum_{w_{k_o\xi_o:j_o}\in W(A_{k_o\xi_o},A_{j_o})}\sum_{w_{k_o\xi_o:i_o}\in W_{\{G\}}(\mathfrak{X};A_{k_o\xi_o},A_{i_o})}$$

of the

$$\nabla(\mathfrak{X}:G|(\{1\},\{1\}):P_{k_o\xi_o}|A_{k_o\xi_o}:P_{i_o}|A_{i_o}:w_{k_o\xi_o:i_o}:\Lambda_{i_o})T_{i_o}$$

paired with the

$$d(\exp(\langle I_{P_{k_o\xi_o}}(\underset{\sim}{H}), w_{k_o\xi_o:j_o}\Lambda_{j_o} - ?\rangle)\cdot(1/\prod_{\lambda_{k_o\xi_o}}(w_{k_o\xi_o:j_o}\Lambda_{j_o} - ?, \lambda_{k_o\xi_o})))(-w_{k_o\xi_o:i_o}\bar{\Lambda}$$

$$\otimes\ c_{cus}(P_{k_o\xi_o}|A_{k_o\xi_o}:P_{j_o}|A_{j_o}:w_{k_o\xi_o:j_o}:\Lambda_{j_o})\Phi_{j_o}.$$

While this result has been established with Λ_{i_o} and Λ_{j_o} subject to certain restrictions, it is clear that both sides of the equation are meromorphic in $(\Lambda_{i_o}, \bar{\Lambda}_{j_o})$. Equality thus obtains everywhere (in the sense of meromorphic functions, of course).

To summarize:

THEOREM 7.3 *Fix* $\underset{\sim}{H} \in \underset{\sim}{\mathfrak{a}}_Q$. *Let* $\mathcal{C}_o$ *be an association class of* Γ-*cuspidal split parabolic subgroups of* G. *Let* $\mathcal{C}_{i_o}$ *be a* G-*conjugacy class in* $\mathcal{C}_o$, $\mathfrak{X}$ *an equivariant system of admissible affine subspaces attached to* $\mathcal{C}_{i_o}$, $\{E,\nabla\}$ *an Eisenstein system belonging to* $\mathfrak{X}$; *let* $\mathcal{C}_{j_o}$ *be another* G-*conjugacy class in* $\mathcal{C}_o$ -- *then*

$$\forall \begin{cases} (P_{i_o}, S_{i_o}; A_{i_o}) \in \mathcal{C}_{i_o} \\ (P_{j_o}, S_{j_o}; A_{j_o}) \in \mathcal{C}_{j_o} \end{cases} \quad \forall\ \mathcal{O}_{i_o}, \mathcal{O}_{j_o} \in \underset{\sim}{\mathcal{O}}_o,$$

the inner product

$$(Q_{\sim}^{H}E(\mathfrak{X}:G|\{1\}:P_{i_o}|A_{i_o}:T_{i_o}:\Lambda_{i_o}:?),\ Q_{\sim}^{H}E(P_{j_o}|A_{j_o}:\Phi_{j_o}:\Lambda_{j_o}:?))$$

is equal to the sum

$$(-1)^{\ell_0}\cdot \mathrm{vol}(\mathcal{C}_0)\cdot \sum_{k_0=1}^{r_0}\sum_{\xi_0=1}^{r_{k_0}} \sum_{w_{k_0\xi_0:j_0}\in W(A_{k_0\xi_0},A_{j_0})} \sum_{w_{k_0\xi_0:i_0}\in W_{\{G\}}(\mathfrak{X};A_{k_0\xi_0},A_{i_0})}$$

of the

$$\nabla(\mathfrak{X}:G|(\{1\},\{1\}):P_{k_0\xi_0}|A_{k_0\xi_0}:P_{i_0}|A_{i_0}:w_{k_0\xi_0:i_0}:\Lambda_{i_0})T_{i_0}$$

paired with the

$$d(\exp(\langle I_{P_{k_0\xi_0}}(\underset{\sim}{H}),w_{k_0\xi_0:j_0}\Lambda_{j_0} - ?\rangle)\,(1/\prod_{\lambda_{k_0\xi_0}}(w_{k_0\xi_0:j_0}\Lambda_{j_0} - ?,\lambda_{k_0\xi_0})))(-w_{k_0\xi_0:i_0}\bar{\Lambda}_{i_0})$$

$$\otimes\ c_{cus}(P_{k_0\xi_0}|A_{k_0\xi_0}:P_{j_0}|A_{j_0}:w_{k_0\xi_0:j_0}:\Lambda_{j_0})\Phi_{j_0}\ .$$

We remark that if $\{E,\nabla\}$ is the canonical Eisenstein system, then the foregoing theorem reduces to Theorem 9.6 in [2] (cf. Theorem 5.1).

Needless to say, there is also a daggered version of the inner product formula which we had best set down explicitly.

Let again $\mathcal{C}_0$ be an association class of Γ-cuspidal split parabolic subgroups of G. Let $\mathcal{C}_{i_0}$ be a G-conjugacy class in $\mathcal{C}_0$, $\mathfrak{X}$ an equivariant system of admissible affine subspaces attached to $\mathcal{C}_{i_0}$. Fix an association class $\mathcal{C}$, dominating $\mathcal{C}_{i_0}$, such that $\mathcal{C}(i_0;\mathfrak{X}) \neq \emptyset$. Supposing that $\{E,\nabla\}$ is an Eisenstein system belonging to $\mathfrak{X}$, we shall then employ without comment ubiquitous notation (cf. §3). Let, therefore,

$$(P,S;A) \in \mathcal{C}_k \subset \mathcal{C}$$

with

$$(P,S;A) \succeq \begin{cases} (P_{i_o},S_{i_o};A_{i_o}) \\ (P_{j_o},S_{j_o};A_{j_o}), \end{cases}$$

the subscript 'i' on A having been suppressed, the split components being special. As in [2], denote by $\mathrm{Dom}_\Gamma(P)$ the set of all Γ-cuspidal split parabolic subgroups of G which are dominated predecessors of (P,S). Let P_{k_o} $(1 \leq k_o \leq r_o^\dagger)$ be a set of representatives for

$$P\backslash \mathcal{C}_o \cap \mathrm{Dom}_\Gamma(P).$$

Then

$$\mathcal{C}_o \cap \mathrm{Dom}_\Gamma(P) = \coprod_{k_o} \mathcal{C}_{k_o}(P)$$

where

$$\mathcal{C}_{k_o}(P) = P\cdot\{P_{k_o}\} \cap \mathcal{C}_o \cap \mathrm{Dom}_\Gamma(P).$$

Let $P_{k_o\xi_o}$ $(1 \leq \xi_o \leq r_{k_o}^\dagger)$ be a set of representatives for $\Gamma \cap P\backslash \mathcal{C}_{k_o}(P)$ -- then

$$\{P_{k_o\xi_o} : 1 \leq k_o \leq r_o^\dagger,\ 1 \leq \xi_o \leq r_{k_o}^\dagger\}$$

is a set of representatives for

$$\Gamma \cap P\backslash \mathcal{C}_o \cap \mathrm{Dom}_\Gamma(P).$$

Now introduce the partial truncation operator $Q_P^{\underset{\sim}{H}}$ (cf. [2]) -- then, under the supposition that $\underset{\sim}{H} \in \underset{\sim}{\mathfrak{a}}_Q$, the inner product

$$(Q_P^{\underset{\sim}{H}} E(\mathfrak{X}:P|A:P_{i_o}|A_{i_o}:T_{i_o}:\Lambda_{i_o}^{\dagger}:?),\ Q_P^{\underset{\sim}{H}} E(P_{j_o}^{\dagger}|A_{j_o}^{\dagger}:\Phi_{j_o}:\Lambda_{j_o}^{\dagger}:?))$$

is equal to the sum

$$(-1)^{\ell_o^{\dagger}} \cdot \mathrm{vol}(\mathcal{C}_o^{\dagger}) \cdot \sum_{k_o=1}^{r_o^{\dagger}} \sum_{\xi_o=1}^{r_{k_o}^{\dagger}} \sum_{w_{k_o\xi_o:j_o}^{\dagger} \in W(A_{k_o\xi_o}^{\dagger}, A_{j_o}^{\dagger})} \sum_{w_{k_o\xi_o:i_o}^{\dagger} \in W_{\mathcal{C}_k}^{\dagger}(\mathfrak{X}; A_{k_o\xi_o}, A_{i_o})}$$

of the

$$\nabla(\mathfrak{X}:P|(A,A):P_{k_o\xi_o}|A_{i_o\xi_o}:P_{i_o}|A_{i_o}:w_{k_o\xi_o:i_o}^{\dagger}:\Lambda_{i_o}^{\dagger})T_{i_o}$$

paired with the

$$d(\exp(\langle I_{P_{k_o\xi_o}^{\dagger}}(\underset{\sim}{H}), w_{k_o\xi_o:j_o}^{\dagger}\Lambda_{j_o}^{\dagger} - ?\rangle) \cdot (1/\prod_{\lambda_{k_o\xi_o}^{\dagger}} (w_{k_o\xi_o:j_o}^{\dagger}\Lambda_{j_o}^{\dagger} - ?, \lambda_{k_o\xi_o}^{\dagger})))(-w_{k_o\xi_o:i_o}^{\dagger}\bar{\Lambda}_{i_o}^{\dagger})$$

$$\otimes\ c_{cus}(P_{k_o\xi_o}^{\dagger}|A_{k_o\xi_o}^{\dagger}:P_{j_o}^{\dagger}|A_{j_o}^{\dagger}:w_{k_o\xi_o:j_o}^{\dagger}:\Lambda_{j_o}^{\dagger})\Phi_{j_o}\ .$$

§8. CONCLUDING REMARKS

The purpose of this § is to pinpoint the position of the present paper as regards the developments which are to follow in due course. We must stress that it is not just a question of increasing the generality of the results themselves but also one of the methods used in so doing. Indeed, as has been mentioned earlier, the situation is quite analogous to that encountered in the elementary theory of Eisenstein series where, via methods akin to those used by us here, one determines the 'cuspidal contribution' to the wave-packet spaces (cf. Chap. 4 in TES), the requisite generalization leading to the theory of Eisenstein systems. A similar theory should, therefore, be expected vis-a-vis the inner product problem.

To begin with, let us consider a special case, viz.: $\text{rank}(\Gamma) = 2$. Then our conclusions are definitive in so far as the inner product formula for two truncated Eisenstein series is concerned, be they cuspidal or residual. Accordingly, it would not be difficult to write down in explicit terms that part of the trace formula coming from the continuous spectrum. For an easy exercise, one could, e.g., work out the situation when

$$\begin{cases} G = SL(3,\mathbf{R}) \\ \Gamma = SL(3,\mathbf{Z}). \end{cases}$$

The general case cannot, in our opinion anyway, be treated by the methods employed in the work at hand. Our current success is in fact a direct reflection of the very fortunate circumstance that at the first level all the relevant poles are simple, hence all the relevant residues are limits. This, of course, will not be true in general, a major conundrum. Nevertheless, the theorems that have been obtained represent a partial justification for a philosophy which we have repeatedly emphasized: The theory of Eisenstein systems is a powerful tool for handling certain problems centering on the development of the trace formula.

This philosophy will be even more apparent in the next paper in this series. There, it will be necessary to shift our point of view, residues being regarded as integrals rather than limits. This means that drastically different methods will be needed, these being by comparison of substantial intricacy in that the analytical aspects of the theory of Eisenstein systems then comes to the foreground, the latter, it is to be observed, having played but an indirect role here.

There is one final comment which should be made. Limiting techniques were used in §6 to pass to the imaginary axis. Of course, the type of limit being evaluated is quite different from that encountered in §5. In contradistinction to the results of §5, however, the results of §6 per the leading term in the inner product formula (i.e., that part involving the c-functions), are evidently valid, as they stand, *in general*. Consequently, nothing further in this direction will be required in the future.

REFERENCES

[1] M.S. Osborne and G. Warner: *The Selberg trace formula I: Γ-rank one lattices*, Crelle's J., **324**(1981), pp. 1-113.

[2] M.S. Osborne and G. Warner: *The Selberg Trace Formula II: Partition, reduction, truncation*, Pacific J., To Appear.

NOTE

Sometime after this paper was completed, we received a preprint from J. Arthur with the title

On the inner product of truncated Eisenstein series.

In this work, Arthur treats the general case, i.e., makes no assumption on the level, but is only able to obtain a weaker result, viz. an asymptotic rather than an exact inner product formula. Furthermore, in that paper, there are no results comparable to those to be found here in §6 and §7, which, of course, are true in all generality.

University of Washington

Seattle, Washington 98195

General instructions to authors for PREPARING REPRODUCTION COPY FOR MEMOIRS

For more detailed instructions send for AMS booklet, "A Guide for Authors of Memoirs." Write to Editorial Offices, American Mathematical Society, P. O. Box 6248, Providence, R. I. 02940.

MEMOIRS are printed by photo-offset from camera copy fully prepared by the author. This means that, except for a reduction in size of 20 to 30%, the finished book will look exactly like the copy submitted. Thus the author will want to use a good quality typewriter with a new, medium-inked black ribbon, and submit clean copy on the appropriate model paper.

Model Paper, provided at no cost by the AMS, is paper marked with blue lines that confine the copy to the appropriate size. Author should specify, when ordering, whether typewriter to be used has PICA-size (10 characters to the inch) or ELITE-size type (12 characters to the inch).

Line Spacing – For best appearance, and economy, a typewriter equipped with a half-space ratchet – 12 notches to the inch – should be used. (This may be purchased and attached at small cost.) Three notches make the desired spacing, which is equivalent to 1-1/2 ordinary single spaces. Where copy has a great many subscripts and superscripts, however, double spacing should be used.

Special Characters may be filled in carefully freehand, using dense black ink, or INSTANT ("rub-on") LETTERING may be used. AMS has a sheet of several hundred most-used symbols and letters which may be purchased for $5.

Diagrams may be drawn in black ink either directly on the model sheet, or on a separate sheet and pasted with rubber cement into spaces left for them in the text. Ballpoint pen is *not* acceptable.

Page Headings (Running Heads) should be centered, in CAPITAL LETTERS (preferably), at the top of the page – just above the blue line and touching it.

LEFT-hand, EVEN-numbered pages should be headed with the AUTHOR'S NAME;

RIGHT-hand, ODD-numbered pages should be headed with the TITLE of the paper (in shortened form if necessary).

Exceptions: PAGE 1 and any other page that carries a display title require NO RUNNING HEADS.

Page Numbers should be at the top of the page, on the same line with the running heads.

LEFT-hand, EVEN numbers – flush with left margin;

RIGHT-hand, ODD numbers – flush with right margin.

Exceptions: PAGE 1 and any other page that carries a display title should have page number, centered below the text, on blue line provided.

FRONT MATTER PAGES should be numbered with Roman numerals (lower case), positioned below text in same manner as described above.

MEMOIRS FORMAT

It is suggested that the material be arranged in pages as indicated below.
Note: Starred items (*) are requirements of publication.

Front Matter (first pages in book, preceding main body of text).

Page i – *Title, *Author's name.

Page iii – Table of contents.

Page iv – *Abstract (at least 1 sentence and at most 300 words).

*1980 Mathematics Subject Classifications represent the primary and secondary subjects of the paper. For the classification scheme, see Annual Subject Indexes of MATHEMATICAL REVIEWS beginning in December 1978.

Key words and phrases, if desired. (A list which covers the content of the paper adequately enough to be useful for an information retrieval system.)

Page v, etc. – Preface, introduction, or any other matter not belonging in body of text.

Page 1 – Chapter Title (dropped 1 inch from top line, and centered).

Beginning of Text.

Footnotes: *Received by the editor date.

Support information – grants, credits, etc.

Last Page (at bottom) – Author's affiliation.

ABCDEFGHIJ–AMS–89876543

Memoirs of the American Mathematical Society

Number 284

Ethan Akin

Hopf bifurcation in the two locus genetic model

Published by the

AMERICAN MATHEMATICAL SOCIETY

Providence, Rhode Island, USA

July 1983 · Volume 44 · Number 284 (second of three numbers)

MEMOIRS of the American Mathematical Society (ISSN 0065-9266) is published bimonthly (each volume consisting usually of more than one number) by the American Mathematical Society at 201 Charles Street, Providence, Rhode Island 02904. Second Class postage paid at Providence, Rhode Island 02940. Postmaster: Send address changes to Memoirs of the American Mathematical Society, American Mathematical Society, P. O. Box 6248, Providence, RI 02940.

Table of Contents

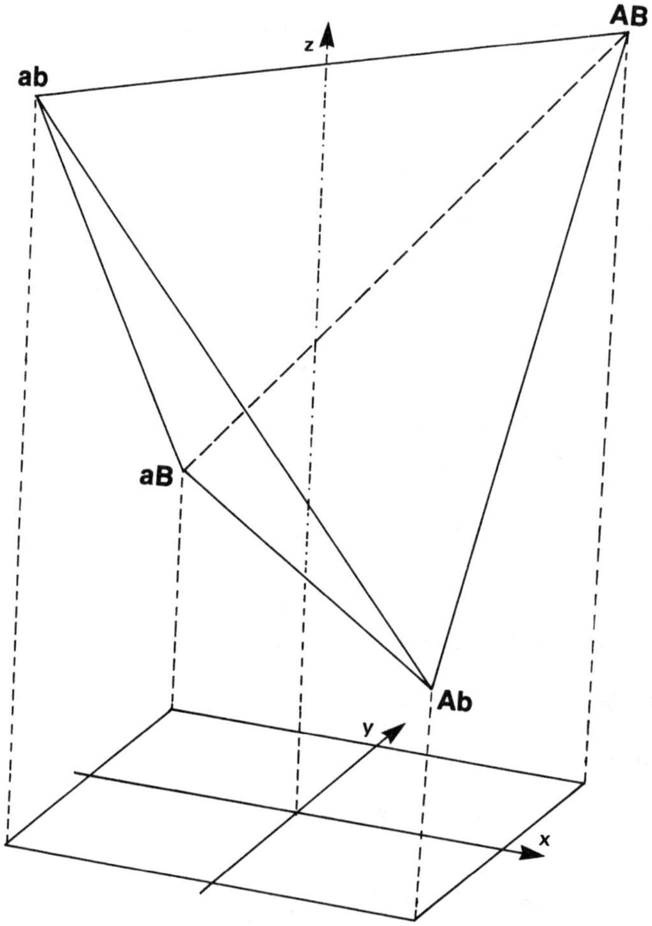
AB
z
ab
aB
Ab
y
x

ABSTRACT: Hopf bifurcations occur in the class of simple genetic models for the combined effect of selection and recombination. The demonstration of cycling in such models is biologically unexpected. To study this phenomenon we describe the locus of positions at which Hopf bifurcation occurs in the two-locus-two-allele model. The description is given by an explicit, computable parametrization which can be used to generate all possible examples. Asymptotic estimates show that limit cycles can occur.

AMS(MOS) subject classifications (1980): 34C25, 92A10.

Key words and phrases: Fisher's fundamental theorem, fitness, Hopf bifurcation, limit cycles, recombination, selection, two-locus-two-allele system.

Library of Congress Cataloging in Publication Data

Akin, Ethan, 1946-
Hopf bifurcation in the two locus genetic model.

(Memoirs of the American Mathematical Society, ISSN 0065-9266 ; no. 284)
Bibliography: p.
Includes index.
1. Genetics--Mathematical models. 2. Bifurcation theory. I. Title. II. Series.
QA3.A57 no. 284 [QH438.4.M3] 510s [575.1'0724] 83-6438
ISBN 0-8218-2284-5

For Joyce, still and again.

Introduction

In The Geometry of Population Genetics [GPG], I proved that cycles could occur as solutions of the differential equations modelling selection and recombination in multi-locus systems. With constant fitness parameters m_{ij} and so no frequency dependence this result is rather surprising. The proof in GPG did not provide any examples and left open the question of the stability of the cycles. The purpose of this monograph is to remedy these defects. We focus on the simplest case, the two-locus-two-allele model, and investigate for what values of fitness and recombination rate the cycles arise.

The book is designed for the use of several different classes of readers.

The biologist who is familiar with the two locus model and who merely wishes to know what the examples are can proceed directly to Chapter III. There is presented a program which can be used to compute essentially all examples which lead to limit cycles. A few numerical examples are also given there. If he has a computer handy he can generate examples himself and perhaps embark on some simulation studies. This approach allows the reader to avoid all the fancy mathematics that Shahshahani and I have introduced into the study of these models.

While the program of Chapter III is rather simple as programs go,

---

Received by the editor November 17, 1981.
Research supported by NSF Grant MC579-04713 and by RF-CUNY.

it is presented with no explanation. The reader who desires some understanding of the mathematics underlying the program, eg. the Hopf bifurcation phenomenon, is advised to look at Chapter I. This chapter is a survey of the mathematics behind and an outline of the constructi of the latter program. I also recommend this chapter to anyone doing serious mathematical work on the two locus model. It is my hope that such tools as the cylindrical frame and the square group will have a wider usefulness than just this work. The mathematical background is about at the level of Hirsch and Smale's book [5]. Some acquaintance with Chapter I of GPG is also necessary although intimacy is not required.

Of course, with Chapter I I hope to seduce the reader into the heart of the work which is Chapter II. The mathematics is not all that deep, there is just a lot of it. The background consists of Chapters I and IV of GPG. This work illustrates the rewards available to the reader who is willing to develop an understanding of the Shahshahani geometry.

I would like to thank my colleague Stanley Ocken for building a small span across the gap where my knowledge of computer programming is supposed to be. Also, I am grateful to his family Evelyn, Rachel and Dov for graciously enduring my presence while he worked with me at his home. Again I am grateful for the typing of Ms. Kate March. Finally, I would like to thank Joyce Akin for drawing the frontispiece.

Chapter I: Cycles in Simple Genetic Systems

In the two-locus-two-allele (TLTA) model the change in the proportions of the four gamete genotypes is described by a differential equation representing the sum of the effects of selection and recombination.

If recombination is ignored and selection is frequency independent then the genotype distribution moves so as to increase mean fitness. This is Fisher's "Fundamental Theorem". In fact, the selection vectorfield is exactly the gradient of mean fitness (times one-half) when the concept of gradient is interpreted properly (cf. [11]).

On the other hand, recombination alone moves the distribution towards linkage equilibrium along a line preserving the gene frequencies at the two loci. In the process entropy is increased.

For the combined field Ewens proved that mean fitness still increases if there is no epistasis [3], but Moran showed that this need no longer be true when epistasis occurs [8]. Wright's adaptive surface view of evolution suggests that while the combined effect of selection plus recombination need not act to maximize mean fitness it might be maximizing some general "fitness function". Perhaps some mix of mean fitness and entropy or perhaps mean fitness adjusted to allow for the epistatic effects would do.

In GPG I showed that this conjecture is false by proving that cycles can sometimes occur. For initial values on the cycle the system doesn't approach an equilibrium at all much less an equilibrium which is the local maximum of some function.

The theorem was proved by showing that in the family of dynamical systems coming from selection plus recombination a mathematical phenomenon called a Hopf bifurcation occurs. The proof indicates how to construct examples in principle but there are some practical difficulties. The design of the proof and the lack of a stock of examples leave open the possibility that these cycles are merely mathematical curiousities without any real significance. This could occur in a number of ways:

(a) Position Effects: It could be that the cycles only happen when the selection matrices have positive effects, i.e. the cis and trans heterozygotes have different fitnesses.

(b) Structural Instability: It could happen that the cycles occur only for very special choices of selection matrices and that any perturbation of the fitness numbers would cause the cycles to disappear. Mathematically, this would be the case if the Hopf bifurcations were degenerate.

(c) Orbital Instability: It could be that the cycles are never stable and that while initial values on the cycles don't go to equilibrium any initial value except these special ones does approach an equilibrium.

Either of the latter two possibilities would mean that the cycles would never be observed because the noise in a real system would jog the fitness numbers or initial value away from the special position that the cycles would need.

The purpose of this monograph is to resolve these questions and to build up a stock of examples. What we end up with is a description of all the selection matrices without position effects which produce these cycles. In the process we construct examples which eliminate the above possibilities. That is, we give examples such that the system need not approach equilibrium but an open set of initial values are attracted to a stable limit cycle. Furthermore, these examples are robust in that a slight change in selection or recombination values merely distorts the original cycle.

What then is the biological "feel" of these examples, the biological meaning of this counterintuitive behavior? To answer this requires a detailed study either by computer similation or qualitative analysis of these stable examples. What we provide here is a description of essentially all such examples.

In this introductory chapter we outline the ideas behind and the construction of the description of the entire class of examples. The detailed work is carried out in the next chapter. We then conclude with a chapter simply presenting the description as a program.

1. The Space of Selection Matrices.

Let the two alleles at the first locus be A and a and the alleles at the second be B and b. Following the usual convention we number the gamete types 1 through 4 via the ordering: AB, Ab, aB, ab. p_1 is the fraction of AB gametes in the gene pool, p_2 the fraction of Ab gametes, etc. Thus, the gamete distribution is a vector lying in:

$$\Delta = \{p \in \mathbb{R}^4 : p_i \geq 0 \ i = 1,\dots,4 \text{ and } \Sigma_i \ p_i = 1\}.$$

A distribution is called interior if all four gamete types occur. So the set of interior distributions is:

$$\mathring{\Delta} = \{p \in \Delta: p_i > 0 \ i = 1,\dots,4\}.$$

The marginal distributions. i.e. the gene distributions at each locus are defined by:

$$p_A = p_1 + p_2 \qquad p_B = p_1 + p_3$$
(1.1)
$$p_a = p_3 + p_4 \qquad p_b = p_2 + p_4$$

Define the quantity d by:

(1.2) $$d = p_1p_4 - p_2p_3.$$

It is easy to check that

(1.3)
$$\begin{aligned} p_1 &= p_Ap_B + d \\ p_2 &= p_Ap_b - d \\ p_3 &= p_ap_B - d \\ p_4 &= p_ap_b + d \end{aligned}$$

So, d is a measure of the degree of linkage disequilibrium. $d = 0$ if and only if the genotype distribution is the product of the marginal distributions or, equivalently, the states of the two loci are independent.

A selection matrix is a 4×4 matrix (m_{ij}) of real numbers where m_{ij} is the fitness of the zygote obtained by the union of gametes of type i and j. Here fitness is measured as a net relative reproductive rate. Since the zygote genotype doesn't depend on the order of the gametes, m_{ij} is symmetric:

(1.4) $$m_{ij} = m_{ji} \qquad i,j = 1,\ldots,4.$$

The absence of position effects corresponds to the assumption that the two double heterozygote types (cis and trans) have the same fitness:

(1.5) $$m_{14} = m_{23}.$$

From now on the term <u>selection matrix</u> is defined to include (1.5) as well as (1.4).

In that case the fitness numbers can be displayed in the usual 3×3 diagram:

Table

(1.6)

	aa	Aa	AA
BB	m_{33}	m_{13}	m_{11}
Bb	m_{34}	$m_{14}=m_{23}$	m_{12}
bb	m_{44}	m_{24}	m_{22}

The vector space of symmetric 4×4 matrices is ten dimensional. The subspace of selection matrices is nine dimensional. parametrized by the nine entries of the above table.

We now relate to this vector space a space of vectorfields or associated differential equations.

The differential equations for selection and recombination are given by (see [2. p. 197]):

$$\frac{dp_i}{dt} = p_i(m_i - \bar{m}) - rbd\xi_i \qquad (i = 1....,4)$$

(1.7) with

$$\xi_i = \begin{cases} +1 & i = 1,4 \\ -1 & i = 2,3. \end{cases}$$

Here $m_i = \Sigma_j\, p_j m_{ij}$ and $\bar{m} = \Sigma_i\, p_i m_i$ are the mean fitness of gamete type i and of the entire population respectively. d is given by (1.2). rb is a nonnegative constant with r the recombination rate in units of crossovers per birth and $b = b_{14} = b_{23}$ the birth rate for the double heterozygotes.

Notice that the behavior of (1.7) depends only on the ratio of the entries m_{ij} to the constant rb provided the latter is not zero. For if we multiply the matrix (m_{ij}) and the constant rb by a positive number k the resulting equation is the same as that obtained from (1.7) by the time change of replacing the variable t by t/k. So multiplication by k does not affect what happens but only the rate at which it happens. Thus, we can normalize the equations by assuming

$$\text{(1.8) (Normalization)} \qquad rb = 1.$$

To each selection matrix (m_{ij}) there is associated a vectorfield X^m on Δ whose associated differential equation is given by (1.7) with (1.8) assumed:

$$\text{(1.9)} \qquad (X^m_p)_i = p_i(m_i - \bar{m}) - d\xi_i \qquad i = 1,\dots,4.$$

Next, note that adding a common constant to the entries of (m_{ij}) doesn't affect X^m or the equation (1.7) at all. The constant is added to the m_i's and is then subtracted with $\bar{m}$. There is no convenient normalization for this condition so we simply define two selection matrices to be <u>equivalent</u> if they differ by a common constant in each entry, i.e. (m_{ij}) is equivalent to $(\tilde{m}_{ij})$ when $m_{ij} - \tilde{m}_{ij}$ doesn't depend on i or j. The vectorfield X^m depends only on the equivalence class of (m_{ij}), so we can assume the central entry of (1.6) is 0. We denote by M the space of selection matrices so normalized. M is the eight dimensional space of symmetric 4×4 matrices satisfying:

$$\text{(1.10)} \qquad m_{14} = m_{23} = 0.$$

Thus, the examples we will construct can be altered by addition of an arbitrary constant. Furthermore, the selection matrix entries really represent not the fitness values themselves but the ratio of the selection rates to the recombination rate. Compare the discussion in GPG page 72.

2. Eigenvalues, Eigenvectors and Linear Differential Equations.

Suppose V is a vector space and $L: V \to V$ is a linear mapping. L can be regarded as a vectorfield on V to which is associated the linear differential equation:

$$\frac{dx}{dt} = L(x) \qquad x \in V. \tag{2.1}$$

If L has a complex eigenvalue, then there exist solutions of (2.1) which go around the origin. They may spiral in towards 0 or spiral out or they may cycle around 0. In any case it is the imaginary part of the eigenvalue which is responsible for the angular motion around 0. This angular motion is the ultimate source of the cycles that we want to describe.

In this section we review the linear algebra of maps with complex eigenvalues and the associated theory of linear differential equations. In the following section we will apply all this to study nonlinear equations near equilibrium and then describe what a Hopf bifurcation is in some detail.

Throughout these two background sections we pay special attention to the case when the dimension of V is 3 because the TLTA systems are three-dimensional.

By choosing a basis for V we can represent the linear map L

by a matrix and compute its determinant. While the matrix representation depends on the choice of basis the determinant does not and so we can write it as det L.

The most important question about the determinant is whether it is zero or not. If det L is nonzero then L is called <u>nonsingular</u>. It is then a one-to-one and onto mapping and the inverse linear map can be defined. If det L vanishes then L called <u>singular</u> and is neither one-to-one nor onto. In fact, L is singular if and only if some nonzero vector is mapped to 0 just as the zero vector is. This means there exist vectors $x \in V$ such that

$$L(x) = 0 \quad \text{and} \quad x \neq 0. \tag{2.2}$$

In general, if I is the identity linear map, represented by the identity matrix with respect to any basis, then $sI - L$ is a linear map and we can compute its determinant. Setting the determinant equal to zero we get the <u>characteristic equation</u> for L in the variable s:

$$0 = \det(sI - L). \tag{2.3}$$

The characteristic equation is a polynomial of degree equal to the dimension of V. In particular, if the dimension of V is three then (2.3) can be written:

$$0 = s^3 - c_1 s^2 + c_2 s - c_3, \tag{2.4}$$

where the coefficients c_1, c_2 and c_3 can be computed from the matrix representation of L. For example, by setting $s = 0$ in (2.3) and (2.4) and pulling a factor of (-1) from each of the three columns of the matrix for $-L$ we see that:

(2.5) $$c_3 = \det L.$$

The roots of the characteristic equation are called the <u>eigenvalues of</u> L. σ is a real root of (2.3) if and only if $\sigma I - L$ is singular. From (2.2), σ is a real eigenvalue if and only if there is a vector $x \in V$, called an <u>eigenvector</u> of L associated with σ, such that:

(2.6) $$L(x) = \sigma x \quad \text{and} \quad x \neq 0.$$

However, a polynomial like (2.3) or (2.4) might have complex roots. At first glance $sI - L$ doesn't make sense for complex s since we can only multiply vectors by real scalars.

In order to interpret complex eigenvalues we define complex vectors using the same trick by which complex numbers are defined.

Remember that a complex number is written $s + it$ where s and t are real numbers and $i = \sqrt{-1}$. Similarly, we write a complex vector as $x + iy$ using a pair of vectors x and y in V. You add and subtract complex vectors by adding and subtracting the corresponding real and imaginary parts. Multiplication by complex scalars is defined by clearing parentheses and using $i \cdot i = -1$, just like multiplication among complex numbers themselves:

(2.7) $$(s + it)(x + iy) = (sx - ty) + i(tx + sy).$$

Define $L(x + iy) = L(x) + iL(y)$, extending the definition of the linear map to complex vectors.

Linear algebra works for complex vector spaces, too. Technically, this is because the set of complex numbers satisfy a list of algebraic properties called the axioms of a <u>field</u>, meaning you can add, subtract and multiply complex numbers and divide by them (except for 0).

If $\sigma = \varepsilon + i\lambda$ is a complex root of (2.3), i.e. a complex eigenvalue, then it has an associated complex eigenvector $x = u + iv$, i.e.,

(2.8) $L(u + iv) = (\varepsilon + i\lambda)\cdot(u + iv)$ and $u \neq 0$ or $v \neq 0$.

Expanding this out using (2.7) and equating real and imaginary parts we see that (2.8) is completely equivalent to two real vector equations:

$$L(u) = \varepsilon u - \lambda v$$
(2.9)
$$L(v) = \lambda u + \varepsilon v.$$

Complex roots of a real polynomial equation always occur in conjugate pairs, i.e. $\varepsilon - i\lambda$ is also a root. In fact, it is easy to check that the equations of (2.9) are also equivalent to (2.8)with all the pluses replaced by minuses. So the conjugate vector $u - iv$ is the eigenvector for the conjugate eigenvalue $\varepsilon - i\lambda$.

Now we can forget about complex vectors and define the nonzero pair of real vectors $\{u,v\}$ to be an eigenvector pair, or more sloppily a pair of eigenvectors, associated to the complex conjugate pair of eigenvalues $\varepsilon \pm i\lambda$ if u and v satisfy (2.9). Note that since λ is nonzero neither u nor v can be the zero vector unless both are.

The real part of the two members of the eigenvalue pair, ε, is the same for both. The imaginary parts have opposite signs so we can always assume λ is positive (if not, replace λ by $-\lambda$ and v by $-v$)

While a polynomial equation of degree n has at most n real roots it has exactly n real or complex roots (counting repeated roots). In particular. for equation (2.4) there are two possibilities:

either it has three real roots (counting possible repeated roots as in $0 = (s - 1)^3$) or it has one real root and one complex conjugate pair (and so no repeated roots). Calling these roots σ_1, σ_2 and σ_3 we can write the right side of (2.4) as the product of the factors $(s - \sigma_\alpha)$ with $\alpha = 1,2,3$. Expanding out and equating powers of s we get:

$$
\begin{aligned}
c_1 &= \sigma_1 + \sigma_2 + \sigma_3 \\
c_2 &= \sigma_1\sigma_2 + \sigma_2\sigma_3 + \sigma_3\sigma_1 \qquad (2.10) \\
c_3 &= \sigma_1\sigma_2\sigma_3
\end{aligned}
$$

In other words, $c_3 = \det L$ is the product of the eigenvalues. So $c_3 = 0$ if and only if one of the eigenvalues is 0, which is exactly what (2.2) says.

c_1, called the <u>trace</u> of L, is the sum of the eigenvalues.

An important special case occurs when there is a pure imaginary pair of roots, i.e. $\epsilon = 0$. This happens when the coefficients c_α $(\alpha = 1,2,3)$ are related to one another in a special way.

1 <u>Lemma</u>: Assume V is three dimensional. $L: V \to V$ has a pure imaginary pair of roots if and only if:

$$c_3 = c_1 \cdot c_2 \qquad (2.11)$$

and

$$c_2 > 0. \qquad (2.12)$$

<u>Proof:</u> If $\sigma_2 = i\lambda$ and $\sigma_3 = -i\lambda$ then $c_1 = \sigma_1 + \sigma_2 + \sigma_3 = \sigma_1$ and the trace c_1 is itself a root. This is exactly what (2.11) says since $s = c_1$ satisfies (2.4) when

$$0 = c_1^3 - c_1^3 + c_1c_2 - c_3$$

which is the same as (2.11).

Now $c_1 = \sigma_1$ is a root if and only if the other two roots σ_2 and σ_3 satisfy $\sigma_2 + \sigma_3 = 0$ or $\sigma_2 = -\sigma_3$. Then from (2.10) we get:

$$c_2 = \sigma_1\sigma_2 - \sigma_2^2 - \sigma_1\sigma_2 = -\sigma_2^2$$

So if σ_2 and σ_3 are real c_2 is negative or zero. If they are complex conjugates then $2\epsilon = \sigma_2 + \sigma_3 = 0$ and so $-\sigma_2^2 = -(i\lambda)^2 = \lambda^2$ which is positive. QED

Now suppose that V is a <u>Euclidean vector space</u>. This means that V comes equipped with an inner product or dot product (,) which can be used to define the length of a vector x in V by $\|x\|^2 = (x,x)$. For details on Euclidean vector spaces see [GPG; Sec. I.3].

If x is an eigenvector for a real eigenvalue, σ, then any non-zero multiple of x is also an eigenvector In particular, we can divide by $\|x\|$ to get an eigenvector of unit length. $-x/\|x\|$ is also a unit eigenvector, pointing in the opposite direction. If σ is not a repeated root of the characteristic equation then these two are the only unit eigenvectors. This device for picking out a special eigenvector from the infinite family of eigenvectors of different lengths is a useful mathematical tool called <u>normalization</u>.

For a conjugate pair of complex eigenvalues there is a normalization which is related in conception:

2 <u>Proposition</u>: Assume V is a Euclidean vector space and that $\epsilon \pm i\lambda$ with $\lambda > 0$ are eigenvalues

for $L: V \to V$. There is a pair of vector $\{e_u, e_v\}$ and a number $k > 0$ which satisfy the following properties:

(a) e_u and e_v have unit length and are orthogonal to one another, i.e.

$$\|e_u\| = \|e_v\| = 1 \quad \text{and} \quad (e_u, e_v) = 0.$$

(b) k is called the <u>skewness</u> of the pair and L relates e_u, e_v and k by:

$$L(e_u) = \epsilon e_u - \lambda k^{-1} e_v$$

(2.13)

$$L(e_v) = \lambda k e_u + \epsilon e_v.$$

We will call a pair $\{e_u, e_v\}$ a <u>normalized pair</u> for the pair of eigenvalues $\epsilon \pm i\lambda$ if it satisfies these conditions.

We will prove this result when we neet it, in Sec. II.4. For now we just note the following facts which will also be proved there.

<u>$k = 1$ case:</u> If $k = 1$ then comparing equations (2.13) with (2.9) we see that $\{e_u, e_v\}$ is an eigenvector pair for $\epsilon \pm i\lambda$. Property (a) says that it is an especially nice pair but we will show that any rotation of the pair still satisfies (a) and (b). That is, we can replace e_u by $\cos\theta\, e_u - \sin\theta\, e_v$ and e_v by $\sin\theta\, e_u + \cos\theta\, e_v$ for any angle θ.

<u>$k \neq 1$ case:</u> Now $\{e_u, e_v\}$ is not an eigenvector pair because (2.13) and (2.9) don't agree. Multiplying the first equation of (2.13) by k, we see that $\{u, v\}$ is an eigenvector pair where $u = ke_u$ and $v = e_v$. However, in this case the pair $\{e_u, e_v\}$ is uniquely determined except for possible sign changes and also the exchange replacing e_u by e_v and e_v by $-e_u$ which has the effect of replacing k by k^{-1}.

<u>dim V = 3 case</u>: When V is three dimensional there is a uniqu line through 0 perpendicular to the plane spanned by e_u and e_v (= the plane spanned by u and v). So up to multiplication by -1 there is a unique unit vector e_w perpendicular to e_u and e_v. Thus, $\{e_u, e_v, e_w\}$ is an orthonormal basis for V. e_w is usually not an eigenvector for L but we will use this basis to express the eigenvector for the real eigenvalue of L. Recall that there is a third, real eigenvalue in addition to the conjugate pair $\varepsilon \pm i\lambda$.

Some additional linear algebra is needed to carry out the details in Chap. II but we defer it to Sec. II.1 and turn now to linear differential equations.

If σ is a real eigenvalue for L with eigenvector x satisfying (2.6) then the function of t:

$$x(t) = e^{\sigma t}x \tag{2.14}$$

is the solution of equation (2.1) starting at x, i.e. with $x(0) = x$.

If $\sigma < 0$ then $x(t)$ approaches 0 along the x-ray as t goes to ∞. If $\sigma > 0$ then $x(t)$ moves out the x-ray roward ∞ as t goes to ∞. If $\sigma = 0$, i.e. equation (2.2) holds, then $x(t)$ remains fixed at x. This is called the <u>degenerate</u> case because every multiple of x is an equilibrium and so we have a line of equilibria instead of an isolated equilibrium.

Suppose $\sigma = \varepsilon + i\lambda$ is a complex eigenvalue with complex eigenvector $x = u + iv$. $x(t)$ defined by (2.14) is still a solution of (2.1) extended to apply to complex vectors. The real and imaginary parts of $x(t)$ are solutions of the original equation:

$$u(t) = e^{\varepsilon t}[(\cos \lambda t)u - (\sin \lambda t)v]$$

(2.15)

$$v(t) = e^{\varepsilon t}[(\sin \lambda t)u + (\cos \lambda t)v].$$

Note that $u(0) = u$ and $v(0) = v$. Here we have applied to (2.7) the classical Euler formula:

$$e^{s \pm it} = e^{s}[\cos t \pm i \sin t].$$

As mentioned in the introductory remarks to this section, the imaginary part, λ, of the eigenvalue imparts a rotational motion about zero in the uv plane to each of these solutions. The stability properties are determined by the real part, ε.

If $\varepsilon < 0$ then these solutions spiral inward approaching 0 as t goes to ∞. If $\varepsilon > 0$ then these solutions spiral outward for positive t. Even if $\varepsilon = 0$, the equilibrium at 0 is isolated in the uv plane but then every nonequilibrium solution in the uv plane cycles around 0. This is called the <u>harmonic case</u> because it occurs in physics in the simple harmonic motion of an ideal spring. A nonlinear example of this filling of a region by cycles around an equilibrium is given by the classical Lotka-Volterra predator-prey equations.

If all of the eigenvalues of L are negative real or have negative real parts then all of the solutions of (2.1) approach 0 as t goes to ∞. 0 is then called an <u>attractor</u>. If all the real parts are positive then the solutions move away from 0 which is then called a <u>repellor</u>.

We now turn to the study of equilibria of nonlinear equations and to the definition of Hopf bifurcation.

3. Nonlinear Equilibria and Hopf Bifurcations.

We will look at vectorfields X on the three dimensional space $\mathring{\Delta}$. The vectors tangent to $\mathring{\Delta}$ lie in:

$$(3.1) \qquad \mathbb{R}^4_0 = \{x \in \mathbb{R}^4 : \Sigma^4_{i=1} x_i = 0\}.$$

In applying the work of the previous section the subspace $\mathbb{R}^4_0$ of $\mathbb{R}^4$ will be the vector space V.

A vectorfield X is a function from $\mathring{\Delta}$ to $\mathbb{R}^4_0$ and is associated with the differential equation:

$$(3.2) \qquad \frac{dp}{dt} = X(p) \qquad p \in \mathring{\Delta},$$

where we will write X_p or $X(p)$ interchangeably for the vector at $p \in \mathring{\Delta}$

An <u>equilibrium point</u> is a point p_0 at which the vectorfield vanishes, i.e. $X(p_0) = 0$. In that case, the motion consisting of rest at p_0 is a solution of (3.2).

At an equilibrium p_0 there is defined a linear map $L: \mathbb{R}^4_0 \to \mathbb{R}^4_0$ approximating X near p_0. With respect to a coordinate system the matrix for L is the Jacobian matrix of first partial derivatives of the components of X at p_0. Introduce the change of variables $p = p_0 + x$ $(x \in \mathbb{R}^4_0)$ to replace p in $\mathring{\Delta}$ by x in $\mathbb{R}^4_0$. The equilibri[um] $p = p_0$ corresponds to $x = 0$. The <u>linearization</u> of (3.2) near p_0 is the linear differential equation:

$$(3.3) \qquad \frac{dx}{dt} = L(x) \qquad x \in \mathbb{R}^4_0.$$

Is (3.3) a good approximation to equation (3.2) for x near 0, i.e. p near p_0? The answer to this question depends on what kind of equilibrium p_0 is. Happily, there are conditions which depend only on the linearization L which allow us to give an affirmative answer.

p_0 is called a hyperbolic equilibrium if all of the eigenvalues of L have nonzero real parts. In particular, none of the eigenvalues are zero. For a hyperbolic equilibrium there is a theorem of Hartman which says in essence that the behavior of solutions of (3.2) and (3.3) near equilibrium are similar. A weak special case of this result that is in constant use goes back to Lyapunov. We call p_0 a linear attractor if all of the eigenvalues of L have negative real parts. We mentioned at the end of the previous section that in this case all of the solutions of (3.3) approach 0 as t goes to ∞ i.e. 0 is an attractor for (3.3). Similarly, Lyapunov's theorem says that in the linearly stable case all of the solutions of (3.2) beginning near the equilibrium p_0 approach p_0 as t goes to ∞, i.e. p_0 is an attractor for (3.2).

By contrast, if some of the eigenvalues of L are zero or pure imaginary then the "higher order terms" in (3.2) may induce orbital behavior which can't be deduced from the linearization (3.3).

p_0 is called a nondegenerate equilibrium if none of the eigenvalues of L is equal to zero or, equivalently, if the determinant of L does not vanish. This simply says that L is a nonsingular linear map.

If p_0 is nondegenerate then it varies smoothly as the parameters of the equation vary. This means that if X_1 is a new vectorfield which is a small perturbation of X, eg. X_1 is obtained by changing some coefficients slightly or by adding some small extra term, then X_1 has a unique equilibrium p_1 near p_0 and the linearization L_1 of X_1 at p_1 is a linear map near L. The condition of nonsingularity of L is needed because this result is based on the inverse function theorem (see [GPG, Thm. I.3.3, p. 30]).

A hyperbolic equilibrium is nondegenerate but the converse is not true. In fact, the nondegenerate equilibria which are not hyperbolic are exactly what we wish to study. We call such an equilibrium a <u>Hopf equilibrium</u>.

If p_0 is a Hopf equilibrium then some eigenvalue must have real part equal to zero. But since p_0 is nondegenerate none of the eigenvalues equal zero. Consequently, p_0 must have a pure imaginary eigenvalue. As complex eigenvalues occur in conjugate pairs, two of the eigenvalues must be of the form $\pm\, i\lambda_0$ with $\lambda_0 > 0$. Since $\mathbb{R}^4_0$ is three dimensional there is one remaining eigenvalue, μ_0, which is real and nonzero. In sum, p_0 is a Hopf equilibrium if the eigenvalues of L are of the form $\pm\, i\lambda_0$, μ_0 with $\lambda_0 > 0$ and $\mu_0 \neq 0$.

We have seen that associated with a conjugate pair of complex eigenvalues there exists an eigenvalue pair which span a plane lying, in this case, in $\mathbb{R}^4_0$. If p_0 is a Hopf equilibrium for X there is a nonlinear version of this plane called the <u>center manifold</u>. The center manifold is a surface in $\mathring{\Delta}$ passing through p_0. At p_0 the plane of vectors tangent to the surface is the plane associated with the complex eigenvalues.

The center manifold is important because it is a <u>local invariant manifold</u> for equation (3.2). This means that a solution path with initial value in the center manifold remains in that surface at least until it leaves a region of points near the equilibrium. Consequently, (3.2) restricts to define a dynamical system on the surface itself. This is now a two dimensional system with equilibrium at p_0.

Actually, the center manifold is defined not only for a Hopf equilibrium but also for the nearby equilibrium p_1 of a perturbation vectorfield X_1. The perturbation has to be small enough that the eigenvalues of L_1 are of the form $\varepsilon \pm i\lambda_1, \mu_1$ with $\lambda_1 > 0$ and $|\mu_1| > |\varepsilon|$. In other words, to define the center manifold the real part of the complex pair doesn't have to be zero it merely has to be smaller in absolute value than the third, real, eigenvalue.

The nonzero real μ_1 provides information about the local stability of the center manifold itself. If $\mu_1 < 0$ then the center manifold is attracting. Solution paths beginning near the center manifold move toward the manifold though not necessarily toward the equilibrium. If $\mu_1 > 0$ then the center manifold is repelling and nearby solution paths move away from the center manifold unless they begin exactly on it.

Restricting to the two dimensional system on the center manifold itself, the equilibrium is hyperbolic if $\varepsilon \neq 0$. The behavior is then like that of the linearization. If $\varepsilon < 0$ solutions spiral in toward equilibrium and if $\varepsilon > 0$ solutions spiral out.

Thus, if $\mu_1 < 0$ and $\varepsilon < 0$ the equilibrium p_1 of X_1 is an attractor in $\mathring{\Delta}$. Because we are assuming $|\mu_1| > |\varepsilon|$ solutions beginning near p_1 approach the center manifold relatively quickly and as they get near it they are observed to spiral in toward p_1. If $\mu_1 < 0$ and $\varepsilon > 0$ then solutions approach the center manifold and spiral outward more or less parallel to it. Technically, in each of these cases there are two solution paths which do not spiral. They constitute the (one dimensional) stable manifold of p_1 associated with μ_1. This is an arc crossing at p_1 the surface of the center manifold with tangent direction the eigenvector associated to μ_1.

Rather than try to understand the verbal description, the reader should digest the hyperbolic case ($\varepsilon \neq 0$, $\mu \neq 0$) by ruminating on a coordinate version of the linearization. This is a system of three real equations:

$$\frac{dx}{dt} = \varepsilon x + \lambda y$$

$$(3.4) \qquad \frac{dy}{dt} = -\lambda x + \varepsilon y$$

$$\frac{dz}{dt} = \mu z.$$

Here the center manifold is the xy plane and the manifold associated to μ (the stable manifold when $\mu < 0$) is the z axis.

By replacing the rectangular coordinates xy by the usual polar coordinates $r\theta$ we obtain (3.4) in cylindrical coordinates $r\theta z$:

$$\frac{dr}{dt} = \varepsilon r$$

$$(3.5) \qquad \frac{d\theta}{dt} = -\lambda$$

$$\frac{dz}{dt} = \mu z.$$

The obvious advantage of cylindrical coordinates is a theme to which we will return.

Return now to the case where p_0 is a Hopf equilibrium. The real eigenvalue, μ_0, is still nonzero and its sign determines whether the center manifold is attracting or not. Now $\varepsilon = 0$. So on the center manifold the linearization has a harmonic equilibrium not an attracting or repelling one. The solutions of (3.2) can and usually do behave quite differently from those of the linearization. In short, the stability properties of p_0 in the center manifold can't be determined from the linearization. However, there is a higher derivative test which comes to the rescue. To understand this condition we go back to elementary calculus and the related problem of looking for a local maximum of a real function of a single real variable $y = f(x)$.

If $f'(x_0) = 0$ then x_0 is called a critical point of f. If $f''(x_0) < 0$ then x_0 is a local maximum, like $y = -x^2$ with $x_0 = 0$, and if $f''(x_0) > 0$ then x_0 is a local minimum, like $y = x^2$ at $x_0 = 0$. This is the so-called second derivative test. If $f''(x_0) = 0$ then the second derivative test yields no information. In that case, $f'''(x_0) \neq 0$ means x_0 is a point of inflection which is neither a local maximum nor a local minimum, like $y = \pm x^3$. If $f'''(x_0) = 0$, then x_0 is a local maximum if $f^{iv}(x_0)$ is negative (like $y = -x^4$) and a local minimum if $f^{iv}(x_0)$ is positive (like $y = x^4$). If $f^{iv}(x_0)$ also equals zero then this fourth derivative test also fails.

Recall that the linearization at p_0 is computed via the first derivatives at p_0 of the components of X. Using the second and third derivatives at p_0 of the components of X there is defined a real number whose sign determines the stability of a Hopf equilibrium p_0 in its center manifold.

The direct computation of this number from its definition is a truly hideous task. This Herculean labor has been accomplished by Marsden and McCracken and they present a formula for this number, which we will denote MARMC in their honor ([7, Sec. 4A]). Mortals need only apply the formula.

If MARMC < 0 then p_0 is stable in the center manifold, i.e. solutions spiral in approaching p_0, and p_0 is called a <u>vague attractor</u>. If MARMC > 0 then solutions spiral out and p_0 is called a <u>vague repellor</u>.

If MARMC = 0 then this test yields no information and stability analysis would require a test in still higher derivatives. This labor they leave to the gods.

Thus, the pair of numbers μ_0 and MARMC play the same roles for stability analysis of a Hopf equilibrium which μ_1 and ε played for a hyperbolic equilibrium. In particular, if $\mu_0 < 0$ and MARMC < 0 then we call p_0 a Hopf attractor. As the name suggests, p_0 is attracting and nearby solutions look similar to those near a hyperbolic equilibrium or its linearization with $\mu < 0$ and $\varepsilon < 0$.

However, there is an important difference between hyperbolic equilibria and Hopf equilibria. A hyperbolic equilibrium is robust. This means that a perturbation of the equation moves the equilibrium to a nearby equilibrium with similar stability characteristics. By contrast for a Hopf equilibrium a small perturbation moves the equilibrium slightly (because a Hopf equilibrium is non-degenerate) but the stability characteristics may change radically. This is the phenomenon of Hopf bifurcation to which we now turn.

Suppose that p_0 is a Hopf equilibrium for X and that there is a family of vectorfields depending on a real parameter ρ with X the vectorfield corresponding to $\rho = 0$. Because p_0 is nondegenerate the perturbed equilibrium, their linearizations and their eigenvalues are differentiable functions of ρ. In particular, the real part ε of the complex eigenvalue pair is a function of ρ with $\varepsilon = 0$ when $\rho = 0$. We assume the condition:

$$\frac{d\varepsilon(\rho)}{d\rho} \neq 0 \qquad \text{at } \rho = 0. \tag{3.6}$$

This implies that as the parameter ρ changes sign so does ε and the conjugate pair of eigenvalues crosses the imaginary axis.

Furthermore, (3.6) implies that ε is a monotone function of ρ and so near 0 we can regard ρ as a function of ε with $\rho = 0$ when $\varepsilon = 0$. Substituting this function $\rho(\varepsilon)$ has the effect of changing parameters to replace the parameter ρ by ε itself.

Thus, we have a family of vectorfields X^{ε} on $\mathring{\Delta}$ depending smoothly on the parameter ε with $X^0 = X$. The equilibrium p_{ε} in $\mathring{\Delta}$, the linearization L_{ε}, and its eigenvalues, which are $\varepsilon \pm i\lambda_{\varepsilon}, \mu_{\varepsilon}$ with $\lambda_{\varepsilon} > 0$ and $\mu_{\varepsilon} \neq 0$, are all functions of ε for ε near 0.

See what happens as ε changes sign from negative to positive: μ_{ε} does not change sign so the stability of the center manifolds (which undulate slightly as ε varies) is not affected. However, since ε changes sign p_{ε} is attracting in its center manifold for $\varepsilon < 0$ and repelling for $\varepsilon > 0$. The orbits no longer spiral in, they spiral out. The behavior at $\varepsilon = 0$ itself is determined by the sign of MARMC.

The Hopf Bifurcation Theorem due to Poincare, Andronov and, finally, Hopf describes the occurence in the center manifolds of a one parameter family of cycles around the equilibria [7; Thm. 3.1 and Thm. 3.15]. Their location and stability characteristics depend on the sign of MARMC.

(1) <u>MARMC ≤ 0:</u> For every $\varepsilon > 0$, small enough, there is in the center manifold for the equilibrium p_{ε} of X^{ε} a unique cycle of the dynamical system X^{ε}. Furthermore, this cycle is attracting in the center manifold. Note that since $\varepsilon > 0$, the equilibrium p_{ε} is repelling. In fact, solutions beginning near enough to p_{ε} spiral outward to approach the cycle.

There is a somewhat anthropomorphic description of what is happening in this case.

p_0 is a vague attractor. Hence, for $\varepsilon < 0$ and for $\varepsilon = 0$ solutions spiral in toward p_0. Now let ε become slightly positive. The outlying points haven't heard the news yet of the parameter change so they continue to spiral inward. But they can't go all the way because now p_ε is unstable and, in fact, points near p_ε are trying to spiral outward. The inward spiraling regime and the outward spiraling one are separated by a single closed orbit which they all approach. As ε becomes increasingly positive the push outward from p_ε becomes stronger and the closed orbit increases in radius.

If, in addition, $\mu_0 < 0$, i.e. p_0 is a Hopf attractor, then not only the solutions on the center manifold but all the solutions near p_ε, except those on the stable manifold arc for p_ε, approach the center manifold and then spiral toward the cycle. So a Hopf bifurcation at a Hopf attractor leads to attracting limit cycles around an unstable equilibrium where $\varepsilon > 0$.

In this case cycles do not occur near p_ε for $\varepsilon \leq 0$.

(2) <u>MARMC > 0</u>: Here a unique cycle occurs for $\varepsilon < 0$ and none for $\varepsilon \geq 0$. The cycles are repelling in the center manifold, separating a family of solutions spiraling outward from a family of solutions spiraling inward to equilibrium.

(3) <u>MARMC = 0</u>: Here the cycles can occur on either side or both of $\varepsilon = 0$. In fact, they can all occur at $\varepsilon = 0$. This is called a degenerate Hopf bifurcation and is illustrated by the family (3.4) or (3.5). For $\varepsilon < 0$ everyone in the xy-plane spirals in toward 0. When $\varepsilon = 0$ everyone gets the news of the parameter change and for this critical value the plane is filled by the one parameter family of cycles. Then when $\varepsilon > 0$ everyone spirals out.

If one regards only stable gadgets as observable then it is clear that the Hopf attractor with $\mu_0 < 0$, MARMC < 0 is the most interesting case. Then limit cycles occur when X is perturbed to X^{ϵ} with $\epsilon > 0$. Furthermore, these cycles are robust in that they stay around after X^{ϵ} is further perturbed. By constrast the cycles of a degenerate Hopf bifurcation are not robust. A slight change in the parameter causes them all to disappear. Finally, when $\mu_0 < 0$ and MARMC > 0 the cycles, while robust will not be observed because they are orbitally unstable. A slight change in initial conditions allows the solution path to spiral away.

In summary, the most important examples of Hopf bifurcation for our purposes are those occuring at Hopf attractors where $\mu_0 < 0$ and MARMC < 0.

4. The Hopf Bifurcation Locus.

We now return to the TLTA model.

Recall that we introduced in Sec. 1 the eight dimensional vector-space M of equivalent classes of selection matrices and for each $m \in M$ defined the vectorfield X^m representing the combined effect of selection and recombination normalized by $rb = 1$. X^m is a vectorfield on Δ and so maps to $\mathbb{R}^4_0$, defined by (3.1). This simply says that $\Sigma_i\, p_i$ remains constant at 1 and so its derivative is zero. The reader can check directly from (1.9) that for each m and p: $\Sigma_i (X^m_p)_i = 0$.

The entire parametrized family is a function:

$$X: M \times \Delta \longrightarrow \mathbb{R}^4_0$$

$$X(m,p) \equiv X_p^m.$$

We now introduce the equilibrium locus Σ. Σ is the subset, technically a submanifold, of $M \times \Delta$ consisting of pairs (m,p) such that p is an equilibrium point for X^m. Thus, Σ is just the preimage of the zero vector under the map X:

$$\Sigma = X^{-1}(0) = \{(m,p) \in M \times \Delta : X_p^m = 0\}. \tag{4.1}$$

The dimension of $M \times \Delta = \dim M + \dim \Delta = 8 + 3 = 11$. On the other hand, $\mathbb{R}_0^4$, like Δ, is three dimensional and so the subset Σ is defined by three independent equations. Therefore, the dimension of $\Sigma = \dim M \times \Delta - \dim \mathbb{R}_0^4 = 11 - 3 = 8$, or

$$\dim \Sigma = \dim M = 8.$$

Note, however, that M is a vectorspace while Σ is a manifold, a curved subset of $M \times \Delta$ which looks locally like an eight dimensional vectorspace (cf. [GPG; Sec. I. 3]).

If $(m,p) \in \Sigma$ then the linearization of X^m at p is a linear mapping:

$$L^{(m,p)} : \mathbb{R}_0^4 \longrightarrow \mathbb{R}_0^4 \qquad (m,p) \in \Sigma.$$

As (m,p) varies in Σ the linear map varies smoothly.

The coefficients of the characteristic equation:

$$0 = \det(sI - L^{(m,p)}) = s^3 - c_1(m,p)s^2 + c_2(m,p)s - c_3(m,p)$$

are differentiable real valued functions of (m,p) in Σ.

Our goal in this work is to describe all of the equilibria at which a Hopf bifurcation can occur. So we want to focus attention on the pairs (m,p) where p is a Hopf equilibrium for X^m. We know from the previous sections that if p is a Hopf equilibrium then the linearization $L^{(m,p)}$ has a pure imaginary eigenvalue. So we define the Hopf bifurcation locus:

$$\Sigma_H = \{(m,p) \in \Sigma : L^{(m,p)} \text{ has an imaginary eigenvalue}\}. \tag{4.2}$$

Not all of the points of Σ_H are Hopf equilibria. Recall, in addition, the real eigenvalue μ must be nonzero.

Σ_H and the subset of Hopf equilibria can be described using Lemma 2.1:

1 Proposition: Let $(m,p) \in \Sigma$. $(m,p) \in \Sigma_H$ if and only if:

$$c_3(m,p) = c_1(m,p) \cdot c_2(m,p) \tag{4.3}$$

and

$$c_2(m,p) > 0. \tag{4.4}$$

(m,p) is a Hopf equilibrium if and only if these conditions hold and, in addition,

$$c_3(m,p) \neq 0. \tag{4.5}$$

In fact, the sign of the real eigenvalue μ of $L^{(m,p)}$ is the same as the sign of $c_3(m,p)$.

Proof: By Lemma 2.1, $L^{(m,p)}$ has an imaginary eigenvalue if and only if (4.3) and (4.4) hold. To be a Hopf equilibrium it must also be true that the determinant of $L^{(m,p)}$ is nonzero. From equation (2.5) we get condition (4.5). Directly from (2.10) instead we see that

$c_3 = \sigma_1\sigma_2\sigma_3 = \mu(i\lambda)(-i\lambda) = \lambda^2\mu$. So the sign of c_3 and μ are the same.

QED

From this result we get the following program constituting an implicit description of Σ_H:

Step A: Choose a fixed basis $\mathfrak{B} = \{v_1, v_2, v_3\}$ for $\mathbb{R}_0^4$ and write $X_p^m = X(m,p)$ in $\mathfrak{B}$ coordinates. That is, compute $X(m,p)_\alpha$ = the v_α coordinate of X_p^m for $\alpha = 1,2,3$. In general, for any x in $\mathbb{R}_0^4$ we can write $x = \Sigma_{\alpha=1}^3 x_\alpha v_\alpha$.

Step B: Compute the 3×3 Jacobian matrix:

(4.6) $$_\alpha J(m,p)_\beta \equiv \frac{\partial X(m,p+x)_\alpha}{\partial x_\beta}\bigg|_{x=0}.$$

When $(m,p) \in \Sigma$, the matrix of the linear map $L^{(m,p)}$ with respect to the basis $\mathfrak{B}$ is precisely $J(m,p)$.

Step C: Compute the coefficients $c_\alpha(m,p)$ $(\alpha = 1,2,3)$ by the formula:

$$s^3 - c_1 s^2 + c_2 s - c_3 = \det[sI - J(m,p)].$$

These computations yield explicit formulae for $X(m,p)_\alpha$ and $c_\alpha(m,p)$ $(\alpha = 1,2,3)$ in terms of the entries of (m_{ij}) and the coordinat[e]s of p. The eight dimensional subset Σ, of $M \times P$ is defined by the three equations $X(m,p)_\alpha = 0$ $(p = 1,2,3)$ The additional equation (4.3[)] defines a seven dimensional subset of Σ. Removing the set where $c_2 = 0$ cuts this subset into two open pieces and the positive piece is Σ_H. Finally, removing the set $c_3 = 0$ cuts Σ_H into two open pieces corresponding to the two possible signs for μ associated with a Hopf equilibrium.

There are two large problems with this implicit description which make it difficult to apply. We use instead a parametric description of Σ_H This consists of a space S_H of seven real parameters and a mapping σ_H of S_H onto Σ_H.

First we look at the difficulties with the implicit description. These are the problems that the parametric description solves.

Constructing examples of elements of Σ_H requires solving equation (4.3) which is a nonlinear equation. It is also rather messy, but no worse than some of the expressions that arise in the parametric approach. However, the nonlinearity means that it may not be possible to solve the equation explicitly. Because we are looking for points on a seven dimensional subset of the eight dimensional space Σ numerical methods leading to approximate solutions won't do. Among other things it isn't clear from this approach that Σ_H is nonempty. Theorem IV.2.3 of GPG implies that Hopf bifurcations occur but doesn't exclude the possibility that they all involve position effects and we have ruled out position effects in the definition of M.

The other difficulty concerns the use we want to make of an element (m_0,p_0) of Σ_H once we have constructed it. If $c_3 \neq 0$ so that (m_0,p_0) is a Hopf equilibrium then a Hopf bifurcation occurs at m_0 in the family of vectorfields X^m by varying m. But to locate and analyze the cycles associated with such a bifurcation we must be able to compute the number MARMC for X^{m_0} at p_0 and we must know how to vary m so that the real part of the eigenvalue pair becomes positive, if MARMC < 0, or negative, if MARMC > 0, because that is where the cycles live.

While Marsden and McCracken have a formula for MARMC it requires a special coordinate system with respect to which the linearization $L^{(m_0,p_0)}$ has a special form. In essence, it requires that the basis of $\mathbb{R}_0^4$ which is used consist of the eigenvectors of $L^{(m_0,p_0)}$.

So what we want from the parametrization is a space S_H of seven parameters such that each of the following are given by explicit formulae in the parameters:

(1) The entries (m_{ij}) and the coordinates of p, i.e. the components of the map $\sigma_H: S_H \to \Sigma_H$.

(2) The eigenvalues and eigenvectors of the linearization. These are what is needed to compute MARMC.

Finally, so that we explicitly describe Hopf bifurcations emanating from Hopf equilibria we parametrize not just Σ_H but the open subset Σ_C of Σ where:

$$\Sigma_C = \{(m,p) \in \Sigma: L^{(m,p)} \text{ has a complex eigenvalue}\}. \tag{4.5}$$

One of the eight parameters for Σ_C will be the real part, ε, of the complex eigenvalue pair. We then get the parametrization of Σ_H by setting $\varepsilon = 0$. We then get Hopf bifurcations by fixing the other parameters and varying ε near 0.

5. Parametrization of Σ_H.

We turn now to the parametric description. The first step, as in much applied mathematics, is the proper choice of three coordinates for Δ.

From biological considerations it is clear that two of the coordinates should measure the gene distributions at each locus, eg. $p_A = p_1 + p_2$ and $p_B = p_1 + p_3$. As it happens the natural coordinates are:

$$(5.1)\qquad \begin{aligned} x &= p_1 + p_2 - p_3 - p_4 = p_A - p_a = 2p_A - 1 \\ y &= p_1 - p_2 + p_3 - p_4 = p_B - p_b = 2p_B - 1. \end{aligned}$$

x and y vary between ± 1 and the origin $x = y = 0$ occurs when $p_A = p_a = p_B = p_b = 1/2$. This point, or rather the line segment in Δ lying over it, plays a special role in the theory as we will see below.

For the third coordinate there are three natural candidates:

$$(5.2)\qquad \begin{aligned} z &= p_1 - p_2 - p_3 + p_4 \\ L &= \ln p_1 - \ln p_2 - \ln p_3 + \ln p_4 \\ &= \ln p_1p_4 - \ln p_2p_3 = \ln(p_1p_4/p_2p_3). \\ d &= p_1p_4 - p_2p_3. \end{aligned}$$

Each has a special role to play and we use all three. z like x and y is linear in p_1, p_2, p_3, p_4 and so $\{x,y,z\}$ is a set of linear coordinates on Δ. However, $z = 0$ does not correspond to linkage equilibrium. On the other hand, d is a measure of linkage disequilibrium and arises naturally because it occurs in the definition of the vectorfields X^m (cf. (1.9)). From (5.2) L can be regarded as a logarithmic version of d or z. Like d, L is a measure of linkage disequilibrium. This means that p is in linkage equilibrium if and only if $d = 0$, or equivalently, if and only if $L = 0$. d is usually regarded as the difference measure of linkage while p_1p_4/p_2p_3, usually denoted Z is the ratio measure. L is just the log of Z. L and Z have the disadvantage that they are only defined for interior distributions, i.e. $p \in \mathring{\Delta}$. However, all our work will be in $\mathring{\Delta}$ as explained below. Finally L is in a technical sense dual to z and has a special role in the theory. To understand this we need a brief review of some ideas about the geometry of the Shahshahani metric from [GPG; Chap. I].

The Shahshahani metric associates to each point p of $\mathring{\Delta}$ an inner product $(\ ,\ )_p$ on $\mathbb{R}_0^4$ which varies from point to point. If $f: \mathring{\Delta} \to \mathbb{R}$ is a smooth function then the gradient of f, denoted $\bar{\nabla}f$, can be defined with respect to the Shahshahani metric. Like the ordinary gradient it is a vectorfield on $\mathring{\Delta}$ but because the concept of gradient depends on the metric it is a different vectorfield from the usual one (see [GPG, Sec. I.3]). It is the geometry of the Shahshahani metric rather than the geometry of the usual constant inner product which is appropriate for the study of linkage and selection. For example, Shahshahani proved that the selection portion of X^m is just the Shahshahani gradient of mean fitness $\bar{m}$ (times 1/2) [11; p. 5] and also [GPG; p. 51]. This result implies and interprets Fisher's Fundamental Theorem of Natural Selection and Kimura's Maximum Principle.

By direct computation using formulae of [GPG; p. 43] the gradient $\bar{\nabla}L$ is everywhere orthogonal to the gradients $\bar{\nabla}x$ and $\bar{\nabla}y$ when we regard x, y and L as functions on $\mathring{\Delta}$. Here orthogonality means the Shahshahani inner products vanish.

This orthogonality means that as one moves along a path in $\mathring{\Delta}$ always parallel to $\bar{\nabla}L$ x and y, or equivalently the gene distribution at each locus, remain constant. So flowing along $\bar{\nabla}L$ means moving along a segment in $\mathring{\Delta}$ lying over a fixed pair x and y, i.e. the gene distributions don't change only linkage does. This is the motion used to compute the partial derivative with respect to z in the xyz coordinate system (i.e. vary z, keeping x and y constant, and differentiate). This is the duality between L and z.

Alternatively, the orthogonality condition means that as we flow along any vectorfield which is a linear combination of $\bar{\nabla}x$ and $\bar{\nabla}y$, L remains constant. The subset of $\mathring{\Delta}$ defined by L = constant is a curved copy of the open xy-square. This means that for each real constant the subset of $\mathring{\Delta}$ defined by so fixing L maps in a one-to-one fashion onto the square $\{(x,y): -1 < x < 1 \text{ and } -1 < y < 1\}$. Thus, $\mathring{\Delta}$ is sliced up, "foliated" is the technical expression, into copies of the square parametrized by L (cf. [GPG; Thm. I.5.3, p. 46]).

If we think of xyz, xyL and xyd as alternate possible versions of a rectangular coordinate system on $\mathring{\Delta}$, equations (3.4) and (3.5) suggest that we look for the analogue of a cylindrical coordinate system.

Begin by defining θ as the angular coordinate $\tan^{-1}(y/x)$. As usual θ is undefined when x = y = 0 and is defined up to multiples of 2π. By the chain rule the gradient of θ satisfies:

(5.3) $$(x^2 + y^2)\bar{\nabla}\theta = x\bar{\nabla}y - y\bar{\nabla}x.$$

This suggests that we look at the vectorfield:

(5.4) $$\mathrm{Ang}_0 \equiv x\bar{\nabla}y - y\bar{\nabla}x.$$

Since Ang_0 is a linear combination of $\bar{\nabla}x$ and $\bar{\nabla}y$ it is everywhere orthogonal to $\bar{\nabla}L$. As it happens it is also everywhere orthogonal to $\bar{\nabla}d$. This is a significant extra condition. While L and d are both measures of linkage disequilibrium they are different functions. In fact the gradients $\bar{\nabla}L$ and $\bar{\nabla}d$ are linearly independent except when

$x = y = 0$. Thus, Ang_0 is really defined up to a multiple by this condition.

Ang_0 vanishes only when $x = y = 0$ and so we can define a unit vectorfield by dividing by its length:

$$Ang \equiv Ang_0/\|Ang_0\| \qquad (x \neq 0 \text{ or } y \neq 0). \tag{5.5}$$

We can also normalize $\bar{\nabla}L$ to get a vectorfield representing "vertical motion" where we are thinking of $\mathring{\Delta}$ as a three dimensional object hanging over and projecting to the xy-square:

$$Ver \equiv \bar{\nabla}L/\|\bar{\nabla}L\|. \tag{5.6}$$

In (5.5) and (5.6) the concept of length is that of the Shahshahani metric.

Since $\mathring{\Delta}$ is three dimensional there is one orthogonal direction left. It should represent radial motion. By direct computation in the next chapter we will see that

$$Rad_0 \equiv [x(1 - y^2) - 4dy]\bar{\nabla}x + [y(1 - x^2) - 4dx]\bar{\nabla}y \tag{5.7}$$

is orthogonal to Ang_0 and hence to Ang. As it is a linear combination of $\bar{\nabla}x$ and $\bar{\nabla}y$ it is also orthogonal to $\bar{\nabla}L$ and hence to Ver. Rad_0 also vanishes only at $x = y = 0$ and so we can define:

$$Rad \equiv Rad_0/\|Rad_0\| \qquad (x \neq 0 \text{ or } y \neq 0). \tag{5.8}$$

So we have a trio {Ver, Rad, Ang} of mutually orthogonal unit vectorfields defined at each point of $\mathring{\Delta}$ except on the line where $x = y = 0$. These three vectors form a basis for $\mathbb{R}^4_0$ which is an orthonormal basis with respect to the Shahshahani inner product at the point. Another word for an orthonormal basis is a frame and since the frame varies with the point this trio is a moving frame. The concept of a moving frame was introduced into differential geometry by Elie Cartan to do just the job we want: to generalize the concept of a coordinate system. We call this trio the <u>cylindrical frame</u>.

To see why these vectorfields don't quite define a coordinate system and to see how they do behave we look back at Rad_0 and compute:

$$
\begin{aligned}
\mathrm{Rad}_0 &= [x - xy^2 - 4dy]\bar{\nabla}x + [y - x^2y - 4dx]\bar{\nabla}y \\
&= \bar{\nabla}\tfrac{1}{2}[x^2 + y^2 - x^2y^2 - 8dxy] + 4xy\bar{\nabla}d.
\end{aligned} \tag{5.9}
$$

This suggests defining the radial coordinate R by:

$$
R^2 = x^2 + y^2 - x^2y^2 - 8dxy. \tag{5.10}
$$

Of course, it must be shown that the expression on the right is non-negative. We will show later that it is and that it vanishes exactly when $x = y = 0$.

For each fixed value of d the equation R = constant defines a closed curve around the origin in the xy square. If $d = 0$ these ovals are symmetric with respect to the x and y axes. For $d > 0$ they are stretched more along the line $x = y$ and compressed along $x = -y$. For $d < 0$ the reverse is true.

Recall that Ang is orthogonal to $\bar{\nabla}d$ and to Rad_0. So by (5.9) it is orthogonal to $\bar{\nabla}\frac{1}{2}R^2$. This means that as you flow along Ang, R remains constant. L and d also remain constant. Begin at any point in $\mathring{\Delta}$ with not both x and y equal zero. Flow along Ang, i.e. solve the differential equation $dp/dt = \mathrm{Ang}(p)$. L is constant and so you remain in a sheet L = constant mapping down onto the xy square.

Projecting to the square you proceed around an oval with R and d fixed moving counterclockwise (because Ang is a positive multiple of $\bar{\nabla}\theta$). Thus the motion of Ang consists of closed orbits around the origin in the L = constant sheets.

Since Rad is orthogonal to Ang it is orthogonal to $\bar{\nabla}\theta$ and so motion along Rad keeps L and θ constant. It is just radial motion outward from the origin in the sheets L = constant.

Finally, the motion of Ver is vertical moving up piercing each sheet, keep x and y, and hence θ, constant.

The reason that the framing doesn't come from a coordinate system has to do with the extra $\bar{\nabla}d$ term in (5.9). $\bar{\nabla}d$ is orthogonal to Ang but not to $\bar{\nabla}L$ and so not to Ver. As you move vertically x and y remain constant but d changes and so does R. The ovals in different L = constant sheets don't map down to the same thing and so don't fit together to form cylinders.

The cylindrical frame is important and it is time to describe its real virtues.

In looking for points of Σ_H, or more generally Σ_C, we are looking for complex eigenvalues of the linearization of X^m. Now the linearization of X^m is the sum of two terms corresponding to the two terms of the definition (1.9). The first term of X^m is the selection vectorfield and selection is a gradient. The linearization of a gradient field can always be represented by a symmetric matrix. (This is the content of [GPG; Thm. IV. 1.2, p. 175]). A symmetric matrix has only real eigenvalues. So if the linearization of the recombination term at a point p is also symmetric then p can't be paired with any m in M to lie in Σ_C. This condition and the condition that p be an equilibrium exclude all of the boundary points of Δ. So we look only at interior distributions, i.e. p in $\mathring{\Delta}$. The linearization of recombination also turns out to be symmetric if p is in linkage equilibrium ($d = L = 0$) or if p lies over $x = y = 0$. It is this that accounts for the special role played by this segment.

At all other points the antisymmetric part of the linearized recombination field is nonzero. For an antisymmetric matrix the eigenvalues are pure imaginary or zero. The importance of the cylindrical frame comes from the fact that for this antisymmetric portion of the linearization of recombination Ang is the vector associated with the 0 eigenvalue. It follows that the associated matrix with respect to the cylindrical frame has a simple form. In fact, {Ver,Rad} is a normalized pair (sensu Prop. 2.2) for the imaginary eigenvalues.

It is useful to contrast the cylindrical frame with the coordinate choice of Karlin and Feldman in [6]. They are studying the symmetric viability model and they exploit its symmetry to choose a linear system of coordinates with respect to which selection has a simple form. It is simple enough that even after recombination is added in they are able to study qualitatively the behavior of the vectorfield on the entire domain. We choose the cylindrical frame with an eye toward the recombination field because that is where the imaginary part of the eigenvalues comes from. We make no restriction on the type of selection matrix. So we range over a more general class of examples, but at a cost. We can only describe the behavior of each example near the equilibrium which defines it.

We now describe the parameters associated to a point $(m,p) \in \Sigma_C$.

Thus, p is an equilibrium for X^m and the linearization $L^{(m,p)}$ has a pair of complex conjugate eigenvalues $\varepsilon \pm i\lambda$ ($\lambda > 0$) and a real eigenvalue μ. With respect to the Shahshahani inner product at p, $(\, , \,)_p$, $\mathbb{R}^4_0$ is a Euclidean vectorspace with orthonormal basis given by the cylindrical frame at p. Using Prop. 2.2 we choose a normalized pair $\{e_u, e_v\}$ for the linearization and also get the real number $k > 0$, the skewness. As remarked after Prop. 2.2, $\dim \mathbb{R}^4_0 = 3$ implies there are exactly two unit vectors $\pm e_w$ orthogonal to the uv plane. If we insist that $\{e_u, e_v, e_w\}$ have the same orientation or handedness as the cylindrical frame then e_w is uniquely determined by e_u and e_v. Relating the two bases {Ver,Rad,Ang} and $\{e_u, e_v, e_w\}$ is a matrix O expressing the change of coordinates with respect to the two bases. Since the cylindrical frame is fixed this matrix determines $\{e_u, e_v, e_w\}$. Because both bases are orthonormal O is a special sort of matrix called an <u>orthogonal matrix</u>. Because the bases share the same orientation the determinant of O must be positive. In fact, because it is orthogonal det O must equal 1. In other words, the matrix O is an element of the set of orthogonal matrices of determinant one. This set is called the <u>special orthogonal group</u>, denoted SO(3). The special orthogonal group is a three dimensional submanifold of the nine dimensional vectorspace of all 3×3 matrices. In other words, only three of the nine entries of the matrix can be choosen independently the rest are determined by the condition that the matrix be orthogonal. A description of SO(3) using three angular variables is given in Chap. III.

The parametrization is obtained by going backwards in this construction. Define

$$(5.11) \qquad \ddot{\Delta} = \{p \in \mathring{\Delta}: d \neq 0 \ \text{ and } \ (x,y) \neq (0,0)\}.$$

So $\ddot{\Delta}$ is the open subset of the set of interior distributions obtained by removing the surface of linkage equilibrium points and the segment over the origin.

$$(5.12) \qquad \begin{gathered} S_C \subset \ddot{\Delta} \times SO(3) \times (0,\infty) \times \mathbf{R} \\ (p,O,k,\varepsilon) \in S_C \ \text{ if } \ dO_{33} > 0. \end{gathered}$$

So S_C is an open subset of an eight (= 3 + 3 + 1 + 1) dimensional space defined by the condition that the difference measure of linkage disequilibrium and the 33 entry of the matrix O have the same sign. This condition is related to the choice of sign in the definition of λ, i.e. $\lambda > 0$.

Now given any quadruple (p,O,k,ε), O determines a $(\ ,\ )_p$ orthonormal basis $\{e_u,e_v,e_w\}$ by transforming $\{Ver_p,Rad_p,Ang_p\}$. The conditions that p be an equilibrium for X^m, k be the skewness $\{e_u,e_w\}$ a normalized pair for and ε the real part of the complex eigenvalue pair turn out to determine a unique equivalence class of selection matrices m such that (m,p) is in Σ_C. This means that explicit formulae can be given for the entries m_{ij} and for the matrix of the linearization $L^{(m,p)}$ with respect to the basis $\{e_u,e_v,e_w\}$. From these formulae MARMC can be computed. Finally, Σ_H is parametrized by setting $\varepsilon = 0$ to define $S_H \subset \ddot{\Delta} \times SO(3) \times (0,\infty)$.

Actually, we have simplified slightly in that to use the parametrization in this form there is another expression relating p and O, like dO_{33} but more complicated, which may not vanish. We will ignore this technicality here.

The parametrized description yields a smooth function $\sigma_C: S_C \to \Sigma_C$. While the map is onto it is not one-to-one indicating that the choice used in defining the parameters for a pair (m,p) are not unique. If the skewness $k \neq 1$ then to each (m,p) there correspond four points of S_C, two with skewness k and two with skewness k^{-1}. This comes from the sign change possibilities described as the $k \neq 1$ case after Prop. 2.2. On the other hand when $k = 1$ there is a circle in S_C mapped to the point (m,p) because in the $k = 1$ case we were free to rotate the normalized pair through an arbitrary angle. This is not an accident. S_H is a seven dimensional manifold but Σ_H is a seven dimensional algebraic set defined by equation (4.3). The points where $k = 1$ correspond to the so-called singularity set where it is not a manifold, i.e. doesn't look like a smoothly bent piece of a vectorspace.

The parametrized description means that by choosing $p \in \overset{\shortparallel}{\Delta}$, a suitable $O \in SO(3)$ (eg. $dO_{33} > 0$) and a positive number k we can explicitly write down a selection matrix m_{ij} such that $(m,p) \in \Sigma_H$. We can compute μ and MARMC and if both are negative then p is a Hopf attractor for X^m. So we can use the parametrization with small positive values of ϵ to obtain a nearby selection matrix m^{ϵ}_{ij} such that $(m^{\epsilon},p) \in \Sigma_C$ and such that $X^{m^{\epsilon}}$ has a limit cycle near p. The computational program and such examples are described in Chapter III.

6. Symmetries.

As a preliminary to the study of the symmetries of the TLTA model we look at symmetries of the general model of selection on a set I consisting of n gametic genotypes. Recall from GPG that the state space is

$$\Delta = \{p \in \mathbb{R}^I : p_i \geq 0 \text{ and } \Sigma\, p_i = 1\}.$$

Selection coefficients are given by a symmetric matrix $(m_{ij}: i,j \in I)$.

A permutation of I is a one-to-one and onto function $g: I \to I$. So g is just a rearrangement of the list of genotypes replacing i by ig. Such a permutation induces a one-to-one, onto linear map of Δ to itself, defined by rearranging the coordinates the same way:

$$(6.1) \qquad g(p)_i \equiv p_{ig} \qquad \text{where} \qquad g: I \longrightarrow I \text{ is a permutation.}$$

Notice that g defined by (6.1) is the restriction to Δ of a linear mapping of $\mathbb{R}^I$ to itself. So at any point p, the derivative of g is this same linear map g itself, i.e.

$$(6.2) \qquad d_p g = g, \quad \text{i.e.} \quad d_p g(X)_i = X_{ig}$$

for any vector X.

Notice also that we regard a permutation g as acting on the right on indices. This means that $g_1 g_2$ on I is defined by

$$(6.3) \qquad i(g_1 g_2) = (ig_1)g_2.$$

This is done so that the permutations acting on points of Δ satisfy the usual composition rule:

$$(6.4) \qquad g_1(g_2(p)) = (g_1 g_2)(p)$$

when g_1 and g_2 are permutations. To see this

$$[g_1(g_2(p))]_i = [g_2(p)]_{ig_1} = p_{(ig_1)g_2} = p_{i(g_1g_2)} = [g_1g_2(p)]_i.$$

The permutations also act on matrices m_{ij} by:

(6.5) $$g(m)_{ij} \equiv m_{ig\,jg}.$$

Clearly, $g(m)_{ij}$ is symmetric if m_{ij} is. An argument analogous to the one above shows:

(6.6) $$g_1(g_2(m)) = (g_1g_2)(m).$$

We describe several equivalent forms of the condition that m_{ij} be g-invariant, i.e. $g(m) = m$.

<u>1 Theorem</u>: Let g be a permutation of I.

If $(m_{ij}: i,j \in I)$ is a symmetric matrix of selection coefficients then the following conditions are equivalent and define the condition (m_{ij}) is g-<u>invariant</u>.

(a) $m_{ij} = m_{ig\,jg}$ for all $i,j \in I$, i.e. $g(m) = m$.

(b) The mean fitness function $\bar{m}(p) = \Sigma\, p_i p_j m_{ij}$ satisfies:

(6.7) $$\bar{m}(p) = \bar{m}(g(p)) \quad \text{for all } p \in \Delta.$$

(c) Let $\bar{\nabla}(\frac{1}{2}\bar{m})$ denote the selection vectorfield on Δ. (Strictly speaking the gradient is only defined on the interior distribution $\mathring{\Delta}$, but the selection field is defined on all of Δ.) The selection field satisfies:

(6.8) $$d_p g(\bar{\nabla}_p(\tfrac{1}{2}\bar{m})) = \bar{\nabla}_{g(p)}(\tfrac{1}{2}\bar{m}) \quad \text{for all } p \in \Delta.$$

(d) If p_t is a solution path of the selection differential equation:

$$\frac{dp_t}{dt} = \bar{\nabla}_{p_t}(\frac{1}{2}\bar{m}), \qquad \text{i.e. } \frac{d(p_t)_i}{dt} = (p_t)_i(m_i(p_t) - \bar{m}(p_t))$$

then $g(p_t)$ is also a solution path.

The fixed point set of g is defined by:

(6.9) $\mathrm{Fix}(g) = \{p \in \Delta: g(p) = p\} = \{p \in \Delta: p_{ig} = p_i \text{ for all } i \in I\}$.

If (m_{ij}) is g-invariant then Fix(g) is an invariant manifold for the selection field, i.e. any solution of the selection equation which begins in Fix(g) remains in Fix(g).

Proof: We repeatedly use the idea that summing any function $f(i)$ over all $i \in I$ yields the same result as summing $f(ig)$ over all $i \in I$ because g is just a rearrangement of I. In the language of first year calculus, i is a "dummy variable". For example, for the mean fitness function $\bar{m}(p)$ we have:

$$\bar{m}(p) = \Sigma_{ij}\, p_i p_j m_{ij} = \Sigma_{ij}\, p_{ig} p_{jg} m_{ig\,jg}$$

$$= \Sigma_{ij}\, g(p)_i g(p)_j g(m)_{ij} = \overline{g(m)}(g(p)).$$

So we have proved:

(6.10) $$\overline{g(m)}(g(p)) = \bar{m}(p).$$

A similar argument shows the mean fitness of gamete type i, defined by $m_i(p) = \Sigma\, p_j m_{ij}$, satisfies

$$g(m)_i(g(p)) = m_{ig}(p).$$

This equation and (6.10) imply that the selection vectorfields associated with m and $g(m)$ are related by the function g on Δ:

(6.11) $$d_p g(\bar{\nabla}_p(\tfrac{1}{2}\bar{m})) = \bar{\nabla}_{g(p)}(\tfrac{1}{2}\overline{g(m)}).$$

To see this, look at the associated differential equation:

$$\frac{dp_i}{dt} = p_i(m_i(p) - \bar{m}(p)) = (\bar{\nabla}_p(\tfrac{1}{2}\bar{m}))_i$$

and apply g to get:

$$(d_p g(\tfrac{dp}{dt}))_i = \frac{dg(p)_i}{dt} = \frac{dp_{ig}}{dt}$$

$$= p_{ig}[m_{ig}(p) - \bar{m}(p)]$$

$$= g(p)_i[g(m)_i(g(p)) - \overline{g(m)}(g(p))]$$

$$= (\bar{\nabla}_{g(p)}(\tfrac{1}{2}\overline{g(m)}))_i.$$

Incidentally, this shows that if p_t is a solution path for the m-selection equation then $g(p_t)$ is a solution path for the $g(m)$-selection equation.

Using (6.10) and (6.11) we can rewrite (b) and (c) of the Theorem to describe the different formulae for g invariance by:

(a) $m_{ij} = g(m)_{ij}$ for all i,j

(b) $\bar{m}(p) = \overline{g(m)}(p)$ for all p

(c) $\bar{\nabla}_p(\tfrac{1}{2}\bar{m}) = \bar{\nabla}_p(\tfrac{1}{2}\overline{g(m)})$ for all p.

Clearly (a) implies (b) since equality between two selection matrices implies that they have the same fitness functions. (b) implies (c) because equality between two functions implies equality between their gradients.

On the other hand, equality of the gradients, which is (c), implies that the functions $\bar{m}$ and $\overline{g(m)}$ differ by a constant. However, this constant is zero because when $p = b$ is the center of the simplex defined by $b_i = 1/n$, then $g(b) = b$ and so $\bar{m}(b) = \overline{g(m)}(b)$ by (6.10) even without invariance. So $\bar{m} = \overline{g(m)}$ everywhere, which is (b).

If $\bar{m} = \overline{g(m)}$ everywhere then letting p be the ith vertex we see $m_{ii} = g(m)_{ii}$. Then look at p half-way between the i and j vertices. There:

$$\frac{1}{4}(m_{ii} + 2m_{ij} + m_{jj}) = \bar{m}(p) = \overline{g(m)}(p)$$
$$= \frac{1}{4}(g(m)_{ii} + 2g(m)_{ij} + g(m)_{jj}).$$

So $m_{ij} = g(m)_{ij}$, and we see that (b) implies (a).

(c) is equivalent to (d) in the theorem because $g(p_t)$ is a solution path for $\bar{\nabla}(\frac{1}{2}\,\overline{g(m)})$ and two vectorfields have the same solution paths if and only if they are equal.

Finally, if m is g invariant then when $p \in \mathrm{Fix}(g)$, $g(p) = p$ together with (6.11) imply:

$$(6.12) \qquad g(\bar{\nabla}_p(\tfrac{1}{2}\,\bar{m})) = \bar{\nabla}_p(\tfrac{1}{2}\,\bar{m}) \qquad \text{if } p \in \mathrm{Fix}(g).$$

This says that the vector $\bar{\nabla}_p(\frac{1}{2}\,\bar{m})$ is tangent to $\mathrm{Fix}(g)$ at p. So the selection field restricts to define a vectorfield on $\mathrm{Fix}(g)$. Solving the differential equation in $\mathrm{Fix}(g)$ we see that if p_t begins in $\mathrm{Fix}(g)$ then it stays there, i.e. $\mathrm{Fix}\,(g)$ is an invariant manifold. QED

<u>Remarks:</u> (a) $\mathrm{Fix}(g)$ is the intersection of Δ with a linear subspace of $\mathbb{R}^I$ and so is a particularly simple sort of submanifold.

(b) While we proved (6.11) by direct computation, it actually follows from (6.10) and the fact that $g\colon \mathring{\Delta} \to \mathring{\Delta}$ is an isometry with respect to the Shahshahani metric (cf. GPG p. 37), i.e.

$$(6.13) \qquad (X,Y)_p = (d_p g(X), d_p g(Y))_{g(p)}$$

for X and Y tangent vectors at p. This follows from the definition of the Shahshahani metric (cf. GPG p. 38) and another dummy variable summation argument.

(c) In general, if anything, eg. matrix, function or vectorfield is invariant under a permutation g, then it is invariant under the iterates of g: $g^2 = gg$, $g^3 = ggg$ etc. For example, $m = g(m)$ implies $m = g(m) = g(g(m)) = g^2(m)$.

In the special case of the TLTA model there are four gametic genotypes. Since the number of permutations of four things is $4! = 24$ it would seem there are 24 kinds of symmetry of the model. However, the four genotypes come from two loci of two alleles each. This additional structure in the model is destroyed by most of these permutations. There are only eight which are biologically sensible to think about. We can arrive at these eight possibilities different ways.

First, recall the x and y coordinates defined on Δ by (5.1). The pair (x,y) defines a map of the three dimensional simplex Δ onto the square $\{(x,y) \in \mathbb{R}^2: |x| \leq 1 \text{ and } |y| \leq 1\}$. The maps $g: \Delta \to \Delta$ that are of interest are the ones which are related to symmetries of this square. There are eight rigid motions of the square and these are related to the eight permutations of the TLTA model

Second, recall $d = p_1p_4 - p_2p_3$ and $L = \ell n\ p_1p_4 - \ell n\ p_2p_3$, measures of linkage defined by (5.2). In these definitions p_1, p_4 and p_2, p_3 are paired together. In order that d and L not be destroyed by the map g this pairing must be preserved. The eight possibilities come from three independent binary choices made in constructing the permutation: (1) Interchange the pair (p_1, p_4) with the pair (p_2, p_3) or not; (2) within the pair (p_1, p_4) interchange p_1 and

p_4 or not; (3) within the pair (p_2, p_3) interchange p_2 and p_3 or not.

Third, a permutation can be thought of as a way of renaming the genotypes. If (A,a) and (B,b) are the pairs of alleles at the first and second locus this pairing of names must be preserved. Again there are three binary choices: the choice of interchange between the pairs and with each pair.

Fourth, since we are assuming no position effects, the selection matrix can be displayed in a square table like (1.6). In order that the permutation not introduce position effects it must induce a symmetry of this square table. Again there are eight symmetries of the square.

We display these eight permutations in a table. The set of these eight mappings from Δ to Δ will be called the square group.

(6.14) Table

g	$g\begin{pmatrix} p_1 \\ p_2 \\ p_3 \\ p_4 \end{pmatrix}$	$g\begin{pmatrix} x \\ y \\ z \end{pmatrix}$	Fix(g) $gp = p$	Gene Symbol	Square Symbol
1	$\begin{pmatrix} p_1 \\ p_2 \\ p_3 \\ p_4 \end{pmatrix}$	$\begin{pmatrix} x \\ y \\ z \end{pmatrix}$	all p	a A b B	
π_+	$\begin{pmatrix} p_1 \\ p_3 \\ p_2 \\ p_4 \end{pmatrix}$	$\begin{pmatrix} y \\ x \\ z \end{pmatrix}$	$p_2 = p_3$ or $x = y$	a A ↕ ↕ b B	

continued

π_-	$\begin{pmatrix} p_4 \\ p_2 \\ p_3 \\ p_1 \end{pmatrix}$	$\begin{pmatrix} -y \\ -x \\ z \end{pmatrix}$	$p_1 = p_4$ or $x = -y$	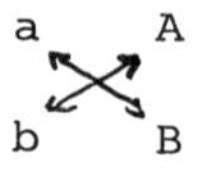	
π_0	$\begin{pmatrix} p_4 \\ p_3 \\ p_2 \\ p_1 \end{pmatrix}$	$\begin{pmatrix} -x \\ -y \\ z \end{pmatrix}$	$(p_1,p_2)=(p_4,p_3)$ or $(x,y) = (0,0)$	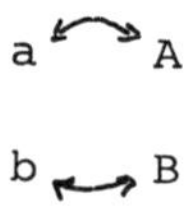	
π_x	$\begin{pmatrix} p_2 \\ p_1 \\ p_4 \\ p_3 \end{pmatrix}$	$\begin{pmatrix} x \\ -y \\ -z \end{pmatrix}$	$(p_1,p_3)=(p_2,p_3)$ or $(y,z) = (0,0)$	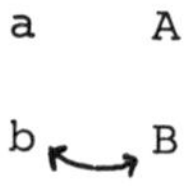	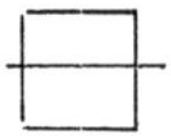
π_y	$\begin{pmatrix} p_3 \\ p_4 \\ p_1 \\ p_2 \end{pmatrix}$	$\begin{pmatrix} -x \\ y \\ -z \end{pmatrix}$	$(p_1,p_2)=(p_3,p_4)$ or $(x,z) = (0,0)$	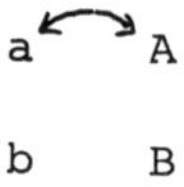	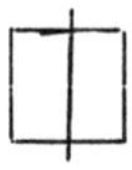
ρ_+	$\begin{pmatrix} p_2 \\ p_4 \\ p_1 \\ p_3 \end{pmatrix}$	$\begin{pmatrix} -y \\ x \\ -z \end{pmatrix}$	$p_1=p_2=p_3=p_4=\frac{1}{4}$ or $(x,y,z)=(0,0,0)$	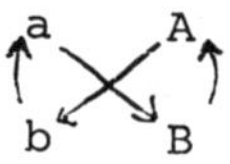	
ρ_-	$\begin{pmatrix} p_3 \\ p_1 \\ p_4 \\ p_2 \end{pmatrix}$	$\begin{pmatrix} y \\ -x \\ -z \end{pmatrix}$	$p_1=p_2=p_3=p_4=\frac{1}{4}$ or $(x,y,z)=(0,0,0)$	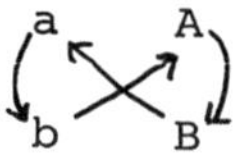	

In addition to the identity map 1 which leaves everything unchanged we have seven symmetries. The table says, for example, that $\pi_+(p_1,p_2,p_3,p_4) = (p_1,p_3,p_2,p_4)$ or equivalently $\pi_+(x,y,z) = (y,x,z)$. The fixed point set $\mathrm{Fix}(\pi_+)$ consists of the set of all p such that $p_2 = p_3$ or equivalently $x = y$. π_+ corresponds to changing the names of the alleles by just interchanging the two loci so A becomes B, a becomes b and vice-versa. On the xy square, π_+ induces the map $(x,y) \to (y,x)$ which is reflection across the diagonal $x = y$. Similarly, π_- induces reflection across the diagonal $x = -y$ while π_x and π_y induce reflections across the x and y axes respectively. π_0 induces the origin symmetry $(x,y) \to (-x,-y)$ which can also be regarded as a 180° rotation (in either direction). ρ_+ induces a positive 90° rotation while its inverse map ρ_- induces a negative 90° rotation.

We now relate these maps to selection and recombination. In particular, this result explains why in table (1.6) we put A on the upper right instead of the upper left as is usually done.

2 Theorem: Define the recombination field Rec: $\Delta \to \mathbf{R}^4_0$ by $\mathrm{Rec}_i(p) = -d\xi_i$, i.e. Rec is the recombination portion of the combined field X^m of (1.9).

If g is any of the symmetries of Table (6.14) then Rec is g-invariant, i.e.

$$d_p g(\mathrm{Rec}(p)) = \mathrm{Rec}(g(p)). \tag{6.15}$$

Consequently, the combined field X^m is g-invariant if and only if (m_{ij}) is g-invariant.

A selection matrix (m_{ij}) is g-invariant if and only if the 3×3 genetic table for (m_{ij}) (cf. (1.6)) is invariant under the symmetry of the square corresponding to g.

Proof: First note that for each g, d at g(p) equals $\pm$ d at p and $d_p g(\xi) = g(\xi)$ equals $\pm\xi$. In both cases the plus sign holds for $g = 1, \pi_+, \pi_-, \pi_0$ and the minus sign for the remaining g's. (6.15) follows from (6.2).

Since the Rec term of X^m is always g invariant, X^m is g-invariant if and only if the selection term $\bar{\nabla}(\frac{1}{2}\bar{m})$ is g-invariant. By Thm. 1 this occurs precisely when m_{ij} is g-invariant.

That g-invariance of m_{ij} corresponds to invariance of table (1.6) under the corresponding symmetry follows by looking at each case. π_+, for example, interchanges 2 and 3 leaving 1 and 4 alone. So $m_{ig\,jg} = m_{ij}$ for all i and j if and only if $m_{22} = m_{33}$, $m_{13} = m_{12}$ and $m_{24} = m_{34}$. This means table (1.6) is invariant under the reflection through the x = y diagonal, i.e. the diagonal from m_{44} to m_{11}.

QED

ρ_+ invariance is equivalent to ρ_- invariance and implies π_0 invariance. This follows from Remark (c) after Thm. 1 because $\rho_+^3 = \rho_-$, $\rho_-^3 = \rho_+$ and $\rho_+^2 = \rho_-^2 = \pi_0$. For example, ρ_+ invariance of m_{ij} is equivalent to $m_{11} = m_{22} = m_{33} = m_{44}$ and $m_{12} = m_{13} = m_{23} = m_{34}$, whereas π_0 invariance represents only half as many conditions. Since ρ_+ invariance is a stricter condition we will not bother with it, and we will focus only on the five π's. Notice that each of these is an involution, meaning $\pi^2 = 1$, the identity.

However ρ_+ is useful in relating different invariant examples.

3 <u>Lemma</u>: Let g and g_1 be two maps of the square group. The map $g: \Delta \to \Delta$ relates the vectorfields X^m and $X^{g(m)}$, i.e.

(6.16) $$d_p g(X^m(p)) = X^{g(m)}(g(p)).$$

Furthermore if m is g_1 invariant, then $g(m)$ is gg_1g^{-1} invariant.

<u>Proof</u>: (6.16) comes from adding (6.11) and (6.15). For the invariance result:

$$gg_1g^{-1}(g(m)) = gg_1(m) = g(m).$$

QED

4 Corollary: (a) The map π_0 relates the vectorfields X^m and $X^{\pi_0(m)}$. If m is g invariant for any g in the square group then $\pi_0(m)$ is also g invariant.

(b) The map ρ_+ relates the vectorfields X^m and $X^{\rho_+(m)}$. Furthermore,

$$\text{if } m \text{ is } \begin{Bmatrix} \pi_+ \\ \pi_- \\ \pi_0 \\ \pi_x \\ \pi_y \end{Bmatrix} \text{ invariant then } \rho_+(m) \text{ is } \begin{Bmatrix} \pi_- \\ \pi_+ \\ \pi_0 \\ \pi_y \\ \pi_x \end{Bmatrix} \text{ invariant.}$$

Proof: $\pi_0 g \pi_0 = g$ for all g, while $\rho_+\pi_+\rho_- = \pi_-$, $\rho_+\pi_-\rho_- = \pi_+$, etc.

QED

The result indicates that there are really three types of symmetric examples in the TLTA model. The origin symmetric or π_0 examples, the axis symmetric or π_x/π_y examples and the diagonal symmetric or π_+/π_- examples. We pair π_x with π_y and π_+ with π_- because, eg. if m is π_+ invariant then $\rho_+(m)$ is π_- invariant and the dynamical behavior of $X^{\rho_+(m)}$ is the same as that of X^m because the map ρ_+ relates the two vectorfields and so maps solutions of one onto solutions of the other.

The π_0-invariant case has been extensively studied in the literature under the name "two locus symmetric viability model". This work, culminating in Karlin and Feldman's treatment [6], is phrased in terms of the discrete time, difference equation model. But the results easily translate for the differential equation or vectorfield model.

How does all this relate to our previous concerns? We are looking for Hopf equilibria and the hope is that in some g-invariant cases these equilibria might be easier to find and describe than in the general model. The key is to look at the fixed point set Fix(g) and restrict the dynamical system to this lower dimensional, and hence simpler to study, submanifold.

In the case of π_0, π_x and π_y the fixed point set is an interval. A one dimensional system is particularly simple and it is easy to describe the equilibria in Fix(g) explicitly for each of these cases. These are the so-called symmetric equilibria. However, none of these can be Hopf equilibria because in each of these cases Fix(g) lies in a region where we know Hopf equilibria don't occur. On Fix(π_x) and Fix(π_y) d is identically zero. So all of these points are in linkage equilibrium. On the other hand, Fix(π_0) is the segment in Δ lying over the origin $(x,y) = (0,0)$.

Now there may be other equilibria in a g-invariant model other than those in Fix(g). In fact, the great advance of Karlin and Feldman in [6] was to discover, describe and analyze these non-symmetric equilibria in the π_0 invariant case. As far as I know some of these non-symmetric equilibria might be of Hopf type in some origin or axis symmetric case. However, away from Fix(g) the leverage which g-invariance gives to the local study of equilibria is considerably diminished. After all, if these equilibria were easy to deal with they would have been described before Karlin and Feldman got around to them. So instead of pursuing the origin and axis cases more deeply we turn instead to diagonal symmetry.

$Fix(\pi_+)$ and $Fix(\pi_-)$ are two dimensional convex cells in Δ which lie over the diagonals $x = y$ and $x = -y$ respectively in the xy square. Direct computation using the parametric description of Σ_H reveals that symmetric Hopf equilibria (i.e. Hopf equilibria in Fix) occur in the families of π_+ and π_- invariant systems. In these cases the two dimensional invariant manifold Fix is the center manifold. This explicit description of the center manifold reduces the computation of MARMC from a three-dimensional to a two-dimensional problem. The formulae are then much easier and for a numerical example can be carried out on a hand calculator. In Chap. III we describe symmetric π_+ invariant examples with Hopf attractors. Furthermore, the Hopf bifurcation itself can be accomplished within the family of π_+ invariant systems so that $Fix(\pi_+)$ remains the center manifold for each ϵ value. This implies that the cycles lie in $Fix(\pi_+)$ and so we can describe what they look like.

Throughout the cycle $x = y$ and so $p_A = p_B$. However, these gene frequencies oscillate back and forth between extreme values. Meanwhile, d and L oscillate back and forth between extreme values. The cycle is attracting in that if one begins at a point near the cycle one observes a spiral in $\mathring{\Delta}$ asymptotically approaching the cycle. The cycle is robust in that if one perturbs the coefficients m_{ij} slightly one observes a new cycle near the old one. Of course, if π_+-invariance is broken by the perturbation then the new cycle need not lie over the diagonal $x = y$ and so p_A need not remain equal to p_B.

7. Concluding Remarks.

In summary, the results of this work are as follows: First,

stable Hopf attractors exist in the family of TLTA models. Second, there is a program which describes all such examples and which shows how to obtain cycling attractors from each such example.

As far as understanding the meaning of this unexpected behavior, we are now on the threshold of enlightenment. There are two ways to proceed.

The way of the computer: Starting with any Hopf attractor with parameters (p,O,k,0) ($\mu < 0$ and MARMC < 0) we let ε be a small positive number and solve numerically the equation with parameters (p,O,k,ε). We then find a small cycle around p. Now by varying the selection numbers (m_{ij}) we increase the size of the cycle to get--we hope--an example with a cycle of reasonable size. Then for this example we use computer simulation to map out the domain of attraction of the cycle and of all other attractors. This requires solving the differential equation numerically with a variety of initial conditions. Contemplation of the resulting portrait should yield an understanding of why the example behaves as it does.

The way of symmetry: Restrict to $\mathrm{Fix}(\pi_+)$ and look at π_+ invariant examples. We know that Hopf attractors and the resulting cycles occur in this family of two-dimensional systems. Furthermore, two dimensional systems are not too difficult to study qualitatively without recourse to numerical simulations. So it should be possible to describe the behavior of all such systems. With a little luck it should then be possible to leave $\mathrm{Fix}(\pi_+)$ and study the behavior on the rest of Δ, following the lead of Karlin and Feldman's analysis of the π_0 invariant case.

Our work has shown that none of the three possibilities mentioned in the introduction to this chapter: position effects, structural instability and orbital instability, suffice to save the intuition that TLTA models always go to equilibrium. We have shown that robust cycles can occur as attractors even without position effects. There remains one possible way that only equilibria would be important as attractors. To describe this possibility we digress to introduce mutation.

The theorem on Hopf bifurcations in GPG is quite general. If you have any vectorfield which is not a gradient with respect to the Shahshahani metric then by varying the selection field added on, Hopf bifurcations occur. In particular, in a one locus model with three or more alleles mutation is usually not a gradient. So, for example, in the family of one-locus-three-allele models cycling occurs. These are two-dimensional systems and so are much easier than the TLTA models. However, I haven't studied them because I expect them to be without biological interest. Why?

First, there are the three problems of the introduction. I would not expect that position effects or orbital instability would be problems here either. However, it is a very real possibility that the Hopf bifurcations are all degenerate, particularly because mutation is a linear vectorfield (cf. GPG p. 64). In a study of related equations which occur in game dynamics [14] Zeeman showed that all the two dimensional Hopf bifurcations are degenerate.

Second, and more important, the cycles probably occur only when the selection numbers m_{ij} are tiny. Correcting an error in GPG on p. 181, we note that the mutation field is a gradient if all of the mutation rates are the same. In the notation of GPG p. 160, if $n_{ij} = n$ whenever $i \neq j$ then it is not hard to show that the mutation field is the gradient of $n \, \Sigma_i \, \ln p_i = n \, \ln \Pi_i \, p_i$. So the nongradient character of the mutation field depends on the difference between mutation rates being nonzero. Now so that selection does not swamp mutation and so the interaction product cycles, selection rates of the same order of magnitude as the mutation rates and possibly the difference between mutation rates would be required. But with such extremely weak selection the cycles would probably not be observable because they would be swamped by the noise of genetic drift.

In the TLTA examples, selection numbers are about the same order of magnitude as recombination. In the normalized model this means the m_{ij} values are around 1. This is weak selection but not negligible. However, it is possible that the examples are peculiar enough that the cycles while robust and attracting in the pure model are no longer observable when the noise of other factors like genetic drift occurs. It could be that the attraction is too weak to overcome real random perturbations. Alternatively, the cycle could be so small that it would be indistinguishable from the equilibrium that it is going around. This latter case would still be interesting, by the way, because the equilibrium would compute to be unstable but would appear to be stable.

All this says is that detailed study of the examples is needed. We now suggest some questions deserving further study.

How does this work relate to other models? I think the results should carry over directly for the discrete time difference equation version of the TLTA model. In fact a numerical solution of the differential equation is really a computation of the difference equation. The results should also carry over for the "correct" model of linkage and selection which does not have the Hardy-Weinberg condition assumed in. Nagylaki and Crow have studied such differential equations [9]. They have shown that if selection and recombination are fairly weak then there is an invariant manifold of quasi-Hardy-Weinberg equilibrium near the manifold of true Hardy-Weinberg equilibrium which is not in fact invariant. Because the attracting cycles we have found are robust they occur on the quasi Hardy-Weinberg manifold. All that is required is that the perturbing effects of selection and recombination be weak enough. This means that for any normalized example, we need only multiply the m_{ij}'s and rb by a small enough constant (cf. Sec. 1). Simulation studies would reveal how small the constant would have to be.

When do the cycles arise with the recombination rate as control parameter? Through any Hopf attractor there are many different Hopf bifurcations. The one we have used (i.e. fix p,O,k and vary ε) is convenient for computation. The biologically interesting family through a point (m,p) of Σ_H arises by varying the recombination rate r. Since we are looking at normalized models this corresponds to varying the m_{ij}'s by a common factor a. This means vary $m \in M$ along the ray $\{am: a > 0\}$. If (p,m) is a Hopf equilibrium (or more generally any element of Σ with p a nondegenerate equilibrium for X^m) then there is a unique function p(a) defined for a near 1 such that (p(a),am) remains in Σ. If (p,m) is Hopf then (p(a),am) is in Σ_C which is open in Σ. Now if (p,O,k,0) are parameters for (p,m) and if, in addition, $k \neq 1$ then there is a unique lifting to a function $(p(a),O(a),k(a),\varepsilon(a))$ of parameters for (p(a),am). Here $\varepsilon(1) = 0$. To check that X^{am} as a varies in a Hopf bifurcation it is necessary to compute the derivative $d\varepsilon(a)/da$ at $a = 1$ and make sure that it is not zero. Furthermore, if this derivative is positive (negative) then limit cycles occur for X^{am} with $a > 1$ (respectively with $a < 1$) provided (p,m) was a Hopf attractor. This means that cycles are caused by decreasing (respectively increasing) strength of recombination since a is essentially $1/r$.

What other kinds of attractors occur? As I remarked in GPG (p. 78) I think of these results as analogous to a theorem of Smale in [12]. There he showed that any kind of dynamical behavior can occur in a large class of ecological-type equations. In the TLTA model we are dealing with a much more restricted class of equations and so we discover a much more restricted class of anamolous dynamical behavior, namely cycles. Just as equilibria can bifurcate to yield cycles so can cycles bifurcate to yield invariant tori (a torus is the Cartesian product of circles), and these yield higher dimensiona tori. At least the torus is possible in the TLTA model though it may not occur. However, I would be willing to bet that invariant tori do occur in multilocus models with recombination. The next level of complication concerns the so-called strange attractors. About these I have no guess. It would be interesting to decide whether strange attractors do or do not occur in multilocus models.

Chapter II: Parametric Description of the Hopf Locus

1. Linear Maps, Bilinear Forms and Matrices.

In this introductory section we describe the matrices associated with linear maps and with bilinear forms. Our purpose is partly to review the linear algebra we will need and partly to set up some notation we will use.

A linear map between vectorspaces V_1 and V_2 is a function $L: V_1 \to V_2$ which relates the linear operations. That is, it satisfies:

$$L(v_1 + v_2) = L(v_1) + L(v_2)$$

$$L(sv) = sL(v)$$

For example, the identity map on a vectorspace V denoted $I_V: V \to V$ is clearly linear where $I_V(v) \equiv v$. We will omit the subscript when the domain is understood.

The class of linear maps is closed under various algebraic and set theory constructions for obtaining new maps from old. For example, if L and T are linear maps from V_1 to V_2 then so are $L + T$ and sL ($s \in \mathbb{R}$) defined by

$$(1.1) \qquad \begin{aligned} (L + T)(v) &\equiv L(v) + T(v) \\ (sL)(v) &\equiv sL(v). \end{aligned}$$

If $L_1: V_1 \to V_2$ and $L_2: V_2 \to V_3$ then the composition $L_2 \circ L_1: V_1 \to V_3$ defined by

$$(1.2) \qquad L_2 \circ L_1(v) \equiv L_2(L_1(v))$$

is a linear map. If $L: V_1 \to V_2$ is one-to-one and onto then the inverse function $L^{-1}: V_2 \to V_1$ is the unique function satisfying:

$$(1.3) \qquad L \circ L^{-1} = I_{V_2} \quad \text{and} \quad L^{-1} \circ L = I_{V_1}.$$

When L is linear so is L^{-1}.

When $V_2 = \mathbb{R}$, L is called a linear form. So a linear form on V is a real valued linear function of a vector variable in V.

A bilinear form H on V is a real valued function of two vector variables which is linear in each variable. That is, $H: V \times V \to \mathbb{R}$ satisfies

$$H(sv_1 + v_2, v_3) = sH(v_1, v_3) + H(v_2, v_3)$$

and

$$H(v_1, sv_2 + v_3) = sH(v_1, v_2) + H(v_1, v_3).$$

Addition and scalar multiplication of bilinear forms are defined by analogy with (1.1). If H and K are bilinear forms on V then $H + K$ and sH are where:

$$(1.4) \qquad \begin{aligned} (H + K)(v_1, v_2) &\equiv H(v_1, v_2) + K(v_1, v_2) \\ (sH)(v_1, v_2) &\equiv sH(v_1, v_2). \end{aligned}$$

Reversing the order of the vectors defines a new bilinear form called the adjoint H^*:

$$(1.5) \qquad H^*(v_1, v_2) \equiv H(v_2, v_1).$$

Note that $H^{**} = H$.

The bilinear form H is called self-adjoint or symmetric if $H^* = H$, i.e.

(1.6) $$H(v_1,v_2) = H(v_2,v_1)$$

H is called skew-adjoint or <u>anti-symmetric</u> if $H^* = -H$, i.e.

(1.7) $$H(v_1,v_2) = -H(v_2,v_1).$$

Every bilinear form can be decomposed uniquely as the sum of a symmetric form and an anti-symmetric form denoted SH and AH respectively:

(1.8) $$H = SH + AH \quad \text{where}$$
$$SH \equiv \frac{1}{2}(H + H^*)$$
$$AH \equiv \frac{1}{2}(H - H^*).$$

Recall from [GPG, Sec. I.3] that V is called a Euclidean vectorspace if it is equipped with a special bilinear form called an inner product, denoted (,). The inner product is symmetric and is also positive definite meaning:

$$(v,v) \geq 0 \quad \text{and} = 0 \text{ only if } v = 0.$$

Geometrical quantities such as the length or norm of a vector and the angle between two vectors are defined via the inner product.

On a Euclidean vectorspace the inner product allows us to set up a correspondence between bilinear forms and linear maps from the space to itself. If $L: V \to V$ then the associated bilinear form, denoted $H(L)$, is defined by:

(1.9) $$H(L)(v_1,v_2) \equiv (L(v_1), v_2).$$

On the other hand, if H is a bilinear form on V and $v_1 \in V$ then fixing the first variable at v_1 we define a linear form H_{v_1} :

$$H_{v_1}(v) \equiv H(v_1,v).$$

But the Riesz representation theorem [GPG; Thm. I.3.1, p. 25] says that every linear form on the Euclidean vectorspace V is obtained as the inner product with a unique vector called the gradient of the linear form. If we denote the gradient of H_{v_1} by $L(v_1)$--it depends on v_1 because H_{v_1} does--then this says

$$H_{v_1}(v) = (L(v_1),v).$$

In other words, the H we started with equals $H(L)$ for the function L which we have just constructed. You can check that the function L is linear because H is linear in the first variable.

This correspondence between bilinear forms and linear maps is itself a linear correspondence:

$$\begin{aligned} H(L+T) &= H(L) + H(T) \\ H(sL) &= sH(L). \end{aligned} \tag{1.10}$$

Using this correspondence we can define the adjoint of a linear map $L: V \to V$. Since $H(L)^*$ is a bilinear form it is associated to a linear map which we will call L^*, i.e. $H(L^*) = H(L)^*$. By symmetry of the inner product the adjoint L^* is characterized by the equation:

$$(L(v_1),v_2) = (v_1,L^*(v_2)). \tag{1.11}$$

We call L self adjoint or <u>symmetric</u> (respectively skew-adjoint or <u>antisymmetric</u>) if $H(L)$ is. So L is symmetric if $L = L^*$ and anti-symmetric if $L = -L^*$.

The tool which allows us to convert equations between vectors or linear maps into arrays of scalar equation is the concept of a basis for a vectorspace. A basis, $\mathfrak{B}$, of V is a list of vectors in V: $\{\xi_1,\ldots,\xi\}$ which is linearly independent and spans V. Note that the list is assumed to be ordered. By definition a reordering yields a new basis.

While a vectorspace has many different bases they all have the same number of elements. This number is called the dimension of V.

The basis $\mathfrak{B}$ associates to each vector v in V an element of $\mathbb{R}^n$, i.e. a list of n scalars called the coordinates with respect to $\mathfrak{B}$ (cf. [GPG; p. 21]). We will denote by $^{\mathfrak{B}}[v]$ the list of coordinates written as a column vector:

(1.12) $$^{\mathfrak{B}}_{j}[v] \equiv \text{the } \xi_j \text{ coordinate of } v.$$

By definition this means:

(1.13) $$v = \Sigma_{j=1}^{n} {}^{\mathfrak{B}}_{j}[v]\xi_j.$$

We will use subscripts and superscripts occurring on the left to refer to row variables and on the right to column variables. For example, $^{\mathfrak{B}}[v]$ is a column vector so j is a row variable.

In a Euclidean vectorspace $\mathfrak{B}$ is called an <u>orthonormal basis</u> or a <u>frame</u> if it satisfies

(1.14) $$(\xi_i,\xi_j) = \delta_{ij} \equiv \begin{cases} 1 & i = j \\ 0 & i \neq j \end{cases}.$$

This says that the vectors have unit length and are mutually orthogonal.

If $\mathfrak{B}$ is not an orthonormal basis then there is another basis called the <u>dual basis</u> associated to $\mathfrak{B}$. Denoted $\mathfrak{B}^* = \{\xi_1^*,\ldots,\xi_n^*\}$ this basis is related to $\mathfrak{B}$ by the equation:

(1.15) $$(\xi_i, \xi_j^*) = \delta_{ij}.$$

$\mathfrak{B}^*$ is defined using the Riesz representation theorem again. The idea is that the map from V to $\mathbb{R}$ which associates to $v \in V$ its jth $\mathfrak{B}$ coordinate, $\overset{\mathfrak{B}}{j}[v]$ is a linear form. So it can be represented as the inner product of v with a unique vector ξ_j^*:

(1.16) $$\overset{\mathfrak{B}}{j}[v] = (v, \xi_j^*).$$

Applying (1.16) with $v = \xi_i$ yields (1.15) because the ξ_j coordinate of ξ_i is δ_{ij}.

$\mathfrak{B}$ is an orthonormal basis iff it is self-dual, i.e. iff $\mathfrak{B}^* = \mathfrak{B}$. In any case, $\mathfrak{B}^{**} = \mathfrak{B}$, i.e. $\mathfrak{B}$ is the dual of $\mathfrak{B}^*$.

Dual bases arise naturally in the study of Riemannian manifolds. We digress to review some concepts from the latter part of [GPG; Sec. I.3].

A k-dimensional manifold M in $\mathbb{R}^n$ is defined locally by a coordinate system. This is a map $h: U \to M$ defined on a region U in $\mathbb{R}^k$. The inverse function of h defines on a piece of M a list of real valued functions $\{x_1, \ldots, x_k\}$ called the coordinates of M. On this piece of M are defined a list of vectorfields denoted $\{\partial_1, \ldots, \partial_k\}$. The notation comes from the fact that ∂_i corresponds to partial differentiation with respect to x_i. To be precise if f is a smooth real valued function on M then the directional derivative of f at p in the direction ∂_i is the partial with respect to x_i:

(1.17) $$\langle d_p f, \partial_i \rangle = \frac{\partial f}{\partial x_i} \quad \text{evaluated at } p.$$

See [GPG; p. 34]. ∂_i is obtained by transporting the standard basis vector e_i from $\mathbb{R}^k$ to M via h. Its motion corresponds to increasing

x_i at unit speed while leaving the other coordinates fixed.

Now suppose that M is a Riemannian manifold, i.e. to each point is associated an inner product which may vary with the point. As explained by [GPG; p. 35] the gradient of a smooth function is defined with respect to a Riemannian metric. The gradient of f is a vectorfield f satisfying:

(1.18) $$\langle d_p f, \xi\rangle = (\nabla_p f, \xi)_p$$

(cf [GPG; eq. I. (3.26), p. 35]).

Apply (1.18) with f the coordinate function x_j and ξ the vectorfield ∂_i. Note that $\partial x_j/\partial x_i = \delta_{ij}$:

(1.19) $$(\nabla_p x_j, \partial_i)_p = \langle d_p x_j, \partial_i\rangle = \delta_{ij}.$$

(1.19) says that the basis of vectors at p $\{\nabla_p x_1, \ldots, \nabla_p x_k\}$ is dual to the basis $\{\partial_1, \ldots, \partial_k\}$ at p. This illustrates that the concept of dual basis like the concept of gradient depends on the choice of inner product.

A Euclidean vectorspace is a special example of a Riemannian manifold with the inner product constant. If we identify each vector ξ_i of a basis $\mathfrak{B}$ with the vectorfield ∂_i which is constantly ξ_i then (1.16) says that the gradient of the coordinate function $v_j = {}_j^{\mathfrak{B}}[v]$ is the vectorfield which is constantly ξ_j^*.

Returning to the simpler category of linear algebra we describe the matrix associated to a linear map $L: V_1 \to V_2$.

Suppose that $\mathfrak{B}_1 = \{\xi_1, \ldots, \xi_n\}$ and $\mathfrak{B}_2 = \{\eta_1, \ldots, \eta_m\}$ are bases for V_1 and V_2 respectively. So dim $V_1 = n$ and dim $V_2 = m$. The associated matrix, denoted ${}^{\mathfrak{B}_2}[L]^{\mathfrak{B}_1}$, is an $m \times n$ matrix whose j^{th} column consists of the $\mathfrak{B}_2$ coordinates of $L(\xi_j)$, of the image of the jth vector in $\mathfrak{B}_1$:

(1.20) $$ {}^{\mathfrak{B}_2}_{i}[L]^{\mathfrak{B}_1}_{j} = {}^{\mathfrak{B}_2}_{i}[L(\xi_j)] = (\eta^*_i, L(\xi_j)) $$

where the second equation comes from (1.16). It assumes that the vectorspaces are Eudlidean and computes the inner product in V_2 where η^*_i and $L(\xi_j)$ live.

The properties of the correspondence between linear maps and matrices are collected in the following:

1 Proposition: (a) The matrix of L relates the coordinates of v and $L(v)$ using matrix multiplication:

(1.21) $$ {}^{\mathfrak{B}_2}[L(v)] = {}^{\mathfrak{B}_2}[L]^{\mathfrak{B}_1}\, {}^{\mathfrak{B}_1}[v]. $$

(b) The correspondence between matrices and linear maps is itself linear. If $L, T\colon V_1 \to V_2$ are linear maps, then:

(1.22) $$ {}^{\mathfrak{B}_2}[L + T]^{\mathfrak{B}_1} = {}^{\mathfrak{B}_2}[L]^{\mathfrak{B}_1} + {}^{\mathfrak{B}_2}[T]^{\mathfrak{B}_1}. $$
$$ {}^{\mathfrak{B}_2}[sL]^{\mathfrak{B}_1} = s\; {}^{\mathfrak{B}_2}[L]^{\mathfrak{B}_1}. $$

(c) Composition of linear maps corresponds to multiplication of matrices. If $L_1\colon V_1 \to V_2$, $L_2\colon V_2 \to V_3$ are linear maps and $\mathfrak{B}_\alpha$ is a basis for V_α $(\alpha = 1,2,3)$ then

(1.23) $$ {}^{\mathfrak{B}_3}[L_2 \circ L_1]^{\mathfrak{B}_1} = {}^{\mathfrak{B}_3}[L_2]^{\mathfrak{B}_2} \cdot {}^{\mathfrak{B}_2}[L_1]^{\mathfrak{B}_1}. $$

(d) The inverse linear map corresponds to the inverse matrix. If $L\colon V_1 \to V_2$ is one-to-one and onto then

(1.24) $$ {}^{\mathfrak{B}_1}[L^{-1}]^{\mathfrak{B}_2} = ({}^{\mathfrak{B}_2}[L]^{\mathfrak{B}_1})^{-1}. $$

(d) The matrix of the adjoint map with respect to the dual bases is the transpose of the original matrix If $L: V \to V$ with $\mathfrak{B}_1$ and $\mathfrak{B}_2$ bases for a Euclidean vectorspace V then

(1.25) $$ {}^{\mathfrak{B}_1^*}[L^*]^{\mathfrak{B}_2^*} = ({}^{\mathfrak{B}_2}[L]^{\mathfrak{B}_1})^t. $$

Here the superscript t refers to the matrix transpose.

Proof: These results are standard so we will just sketch some of the proofs. Usually the distinction between linear map and matrix and the dependence on the basis choice are not made quite so explicit. However, we will be changing bases quite a bit so this cumbersome notation will be useful.

To prove (1.21) we start with (1.13), apply L to both sides and use (1.20):

$$ \Sigma_i \, {}^{\mathfrak{B}_2}_{i}[L(v)]\,\eta_i = L(v) = L(\Sigma_j \, {}^{\mathfrak{B}_1}_{j}[v]\,\xi_j) $$
$$ = \Sigma_j \, {}^{\mathfrak{B}_1}_{j}[v]\,L(\xi_j) = \Sigma_{ij} \, {}^{\mathfrak{B}_1}_{j}[v] \, {}^{\mathfrak{B}_2}_{i}[L]_j^{\mathfrak{B}_1}\,\eta_i. $$

Equating coefficients of η_i gives the subscript version of (1.21).

Equation (1.21) characterizes the matrix ${}^{\mathfrak{B}_2}[L]^{\mathfrak{B}_1}$ meaning that the latter is the only matrix which satisfies the equation.

A similar little dance proves (1.23) while (1.22) comes from the linearity of the correspondence between vectors and coordinates.

(1.24) follows from (1.23) once we note that if $I: V \to V$ is the identity map of a vector space of dimension n and the same basis $\mathfrak{B}$ is chosen as domain and range basis then the associated matrix is the

$n \times n$ identity matrix, denoted I_n:

$$ {}^{\mathfrak{B}}[I]^{\mathfrak{B}} = I_n. \tag{1.26} $$

Here the ij entry of I_n is δ_{ij}. With the identity matrix we will omit the subscript n if the dimension is understood.

To prove (1.25) we apply (1.20) and (1.11):

$$ {}^{\mathfrak{B}_2}_{\ i}[L]^{\mathfrak{B}_1}_{j} = (\eta_i^*, L(\xi_j)) $$
$$ = (\xi_j, L^*(\eta_i^*)) = {}^{\mathfrak{B}_1^*}_{\ j}[L^*]^{\mathfrak{B}_2^*}_{i}. $$

QED

Now let H be a bilinear form on V. With respect to the basis $\mathfrak{B} = \{\xi_1,\ldots,\xi_n\}$ of V the matrix of H, denoted $[H]^{\mathfrak{B}}$, is the $n \times n$ matrix defined by:

$$ {}_i[H]^{\mathfrak{B}}_j = H(\xi_i, \xi_j). \tag{1.27} $$

The properties of the correspondence between bilinear forms and their matrices are collected in:

2 Proposition: (a) The matrix of H reduces the computation of the value of H to matrix multiplication:

$$ H(v_1,v_2) = ({}^{\mathfrak{B}}[v])^t \cdot [H]^{\mathfrak{B}} \cdot ({}^{\mathfrak{B}}[v]). \tag{1.28} $$

(b) The correspondence is linear:

$$ \begin{aligned} [H + K]^{\mathfrak{B}} &= [H]^{\mathfrak{B}} + [K]^{\mathfrak{B}}. \\ [sH]^{\mathfrak{B}} &= s[H]^{\mathfrak{B}}. \end{aligned} \tag{1.29} $$

(c) The adjoint of H corresponds to the transpose matrix:

(1.30) $$[H^*]^{\mathfrak{B}} = ([H]^{\mathfrak{B}})^t.$$

(d) If $L: V \to V$ is a linear map on a Euclidean vectorspace V then

(1.31) $$[H(L)]^{\mathfrak{B}} = ({}^{\mathfrak{B}^*}[L]^{\mathfrak{B}})^t.$$

In particular, if the basis $\mathfrak{B}$ is orthonormal then

(1.32) $$[H(L)]^{\mathfrak{B}} = ({}^{\mathfrak{B}}[L]^{\mathfrak{B}})^t.$$

Proof: (1.28) follows by substituting into $H(v_1,v_2)$ the expressions for v_1 and v_2 in terms of the basis á la (1.13), and expanding using bilinearity. (1.29) and (1.30) are easy.

For (1.31), use (1.27), (1.9) and (1.20):

$$ {}_i[H(L)]^{\mathfrak{B}}_j = H(L)(\xi_i,\xi_j) = (L(\xi_i),\xi_j)$$
$$= (\xi_j, L(\xi_i)) = {}^{\mathfrak{B}^*}_j[L]^{\mathfrak{B}}_i.$$

(1.32) follows because $\mathfrak{B}^* = \mathfrak{B}$ for an orthonormal basis. QED

Remarks: (a) Usually for a linear map $L: V \to V$ we choose the same basis $\mathfrak{B}$ for domain and range and refer to ${}^{\mathfrak{B}}[L]^{\mathfrak{B}}$ as the matrix of L with respect to $\mathfrak{B}$. (1.32) says that the matrix of L with respect to $\mathfrak{B}$ is the transpose of the matrix of $H(L)$ with respect to $\mathfrak{B}$. However, this only holds if the basis is orthonormal. Otherwise (1.31) says that the matrix of $H(L)$ is the transpose of a matrix for L with dual bases chosen as domain and range bases.

(b) If H is symmetric (or antisymmetric) then by (1.30) the matrix $[H]^{\mathfrak{B}}$ will be a symmetric matrix (resp. an antisymmetric matrix). So if L is symmetric (or antisymmetric) ${}^{\mathfrak{B}}[L]^{\mathfrak{B}}$ will be symmetric (resp. antisymmetric) if $\mathfrak{B}$ is an orthonormal basis. However if $\mathfrak{B}$ is not an orthonormal basis the symmetry or antisymmetry of L will not be apparent in the matrix representation ${}^{\mathfrak{B}}[L]^{\mathfrak{B}}$.

The above remarks raise the question of what happens to these matrices when the choice of bases is changed. In the final part of this section we examine this question.

Suppose that $\mathfrak{B}_1 = \{\xi_1,\dots,\xi_n\}$ and $\mathfrak{B}_2 = \{\eta_1,\dots,\eta_n\}$ are two bases on the vectorspace V. A change of basis often arises from a matrix ${}_jA_i$ which transforms $\mathfrak{B}_2$ into $\mathfrak{B}_1$:

$$\xi_j = \Sigma_{i=1}^{n}\ {}_jA_i\eta_i \tag{1.33}$$

This approach is useful because it says that if we write a formal column vector whose entries are themselves vectors, namely the η_i's, and multiply on the left by the matrix A then we get the column vector whose entries are the ξ_j's.

On the other hand, (1.33) simply tells the $\mathfrak{B}_2$ coordinates of the $\mathfrak{B}_1$ vectors:

$${}_jA_i = {}^{\mathfrak{B}_2}_{\ i}[\xi_j] = {}^{\mathfrak{B}_2}_{\ i}[I]^{\mathfrak{B}_1}_{\ j} = (\eta_i^*,\xi_j) \tag{1.34}$$

where the last equation comes from (1.20) in the Euclidean case.

So (1.34) says:

$$A^t = {}^{\mathfrak{B}_2}[I]^{\mathfrak{B}_1}. \tag{1.35}$$

In other words, the matrix which transforms $\mathfrak{B}_2$ vectors to $\mathfrak{B}_1$ vectors in the transpose of the matrix which transforms $\mathfrak{B}_1$ coordinates to $\mathfrak{B}_2$ coordinates since by (1.21):

$${}^{\mathfrak{B}_2}[v] = {}^{\mathfrak{B}_2}[I]^{\mathfrak{B}_1}\cdot{}^{\mathfrak{B}_1}[v]. \tag{1.36}$$

Of course, if the same basis is chosen as domain and range basis then the matrix of the identity map is the identity matrix. That is (1.26). If different bases are chosen then ${}^{\mathfrak{B}_2}[I]^{\mathfrak{B}_1}$ will not be the identity matrix.

The fundamental properties of the coordinate change matrix are summarized in:

3 Proposition: (a) The inverse basis change corresponds to the inverse matrix:

(1.37) $$ {}^{\mathfrak{B}_1}[I]^{\mathfrak{B}_2} = ({}^{\mathfrak{B}_2}[I]^{\mathfrak{B}_1})^{-1}. $$

(b) If V is Euclidean, then

(1.38) $$ {}^{\mathfrak{B}_1^*}[I]^{\mathfrak{B}_2^*} = ({}^{\mathfrak{B}_2}[I]^{\mathfrak{B}_1})^{t}. $$

(c) If $L: V \to V$ is a linear map then

(1.39) $$ \begin{aligned} {}^{\mathfrak{B}_2}[L]^{\mathfrak{B}_2} &= {}^{\mathfrak{B}_2}[I]^{\mathfrak{B}_1}\ {}^{\mathfrak{B}_1}[L]^{\mathfrak{B}_1}\ {}^{\mathfrak{B}_1}[I]^{\mathfrak{B}_2} \\ &= ({}^{\mathfrak{B}_1}[I]^{\mathfrak{B}_2})^{-1}\ {}^{\mathfrak{B}_1}[L]^{\mathfrak{B}_1}\ {}^{\mathfrak{B}_1}[I]^{\mathfrak{B}_2}. \end{aligned} $$

(d) If $H: V \times V \to \mathbb{R}$ is a bilinear form then

(1.40) $$ [H]^{\mathfrak{B}_2} = ({}^{\mathfrak{B}_1}[I]^{\mathfrak{B}_2})^{t} \cdot [H]^{\mathfrak{B}_1} \cdot {}^{\mathfrak{B}_1}[I]^{\mathfrak{B}_2}. $$

Proof: (1.37) and (1.38) follow from (1.24) and (1.25) respectively because $I_V^{-1} = I_V$ and $I_V^* = I_V$. (1.39) follows from (1.23). (1.40) follows from (1.28) and (1.36). QED

4 Corollary: If $\mathfrak{B}_1$ and $\mathfrak{B}_2$ are orthonormal bases then the coordinate change matrix is an orthogonal matrix, i.e.

$$(1.41) \qquad ({}^{\mathfrak{B}_2}[I]^{\mathfrak{B}_1})^t = ({}^{\mathfrak{B}_2}[I]^{\mathfrak{B}_1})^{-1}.$$

Proof: This follows from (1.37) and (1.38) since $\mathfrak{B}_1^* = \mathfrak{B}_1$ and $\mathfrak{B}_2^* = \mathfrak{B}_2$.

QED

5 Corollary: Let $\mathfrak{B}$ be the basis $\{\xi_1,\dots,\xi_n\}$ on a Euclidean vectorspace V and let $\mathfrak{B}^*$ be the dual basis. ${}^{\mathfrak{B}^*}[I]^{\mathfrak{B}}$ is the symmetric matrix of inner products:

$$(1.42) \qquad {}^{\mathfrak{B}^*}_{i}[I]^{\mathfrak{B}}_{j} = (\xi_i,\xi_j).$$

The inverse matrix is ${}^{\mathfrak{B}}[I]^{\mathfrak{B}^*}$ given by:

$$(1.43) \qquad {}^{\mathfrak{B}}_{i}[I]^{\mathfrak{B}^*}_{j} = (\xi_i^*,\xi_j^*).$$

Proof: ${}^{\mathfrak{B}^*}[I]^{\mathfrak{B}}$ is symmetric by (1.38). (1.42) follows from (1.34). The inverse matrix is ${}^{\mathfrak{B}}[I]^{\mathfrak{B}^*}$ by (1.37) and (1.43) follows from (1.34) again or from (1.42) itself applied to $\mathfrak{B}^*$. QED

2. Coordinates and the Cylindrical Frame.

$\mathbb{R}^4$ is the vectorspace of four-tuples of real numbers, i.e. $p \in \mathbb{R}^4$ if $p = (p_1,p_2,p_3,p_4)$. It is a Euclidean vectorspace with respect to the usual inner product (,):

$$(2.1) \qquad (p,q) = \Sigma_{i=1}^{4} p_i q_i.$$

The standard basis $\mathfrak{S}$ on $\mathbb{R}^4$ consists of the vectors $\{(1,0,0,0),(0,1,0,0),(0,0,1,0),(0,0,0,1)\}$. $\mathfrak{S}$ is an orthonormal basis. The name comes from the fact that the coordinates of the vector p with respect to $\mathfrak{S}$ are just the four numbers $p_1,\dots,p_4$. In the notation of Sec. 1:

(2.2) $$\mathfrak{S}_i[p] = p_i.$$

We will also use the notation $\mathfrak{S} = \{\partial_1, \partial_2, \partial_3, \partial_4\}$ indicating that we identify the vector (1,0,0,0) for example with the vectorfield ∂_1 which is constantly (1,0,0,0). (Cf. [GPG; p. 34].)

Define four vectors ξ^x, ξ^y, ξ^z and ξ^n in $\mathbb{R}^4$ by:

(2.3)
$$\begin{aligned} \xi^x &= (1,1,-1,-1) \\ \xi^y &= (1,-1,1,-1) \\ \xi^z &= (1,-1,-1,1) \\ \xi^n &= (1,1,1,1) \end{aligned}$$

Define new coordinate functions

(2.4)
$$\begin{aligned} x &= (\xi^x, p) = p_1 + p_2 - p_3 - p_4 \\ y &= (\xi^y, p) = p_1 - p_2 + p_3 - p_4 \\ z &= (\xi^z, p) = p_1 - p_2 - p_3 + p_4 \\ n &= (\xi^n, p) = p_1 + p_2 + p_3 + p_4. \end{aligned}$$

In matrix notation this says

(2.5)
$$\begin{pmatrix} x \\ y \\ z \\ n \end{pmatrix} = \begin{pmatrix} 1 & 1 & -1 & -1 \\ 1 & -1 & 1 & -1 \\ 1 & -1 & -1 & 1 \\ 1 & 1 & 1 & 1 \end{pmatrix} \begin{pmatrix} p_1 \\ p_2 \\ p_3 \\ p_4 \end{pmatrix}.$$

By equation (1.16), (2.4) says that x,y,z and n are the coordinates of p with respect to the dual basis to $\{\xi^x, \xi^y, \xi^z, \xi^n\}$. Since the ξ's are pairwise orthogonal and have length 2 this dual basis which we will denote by $\tilde{\mathfrak{B}}$ is given by

$$(2.6) \qquad \tilde{\mathfrak{B}} = \{\xi^x/4, \xi^y/4, \xi^z/4, \xi^n/4\} = \{\partial_x, \partial_y, \partial_z, \partial_n\}.$$

Here again we identify for example, the vector $\xi^x/4$ with the constant vectorfield and name it ∂_x.

The matrix of (2.5) whose rows are the ξ's is ${}^{\tilde{\mathfrak{B}}}[I]^{\mathfrak{S}}$. Multiplying this matrix by its transpose we get the identity matrix times 4. In other words

$$(2.7) \qquad {}^{\mathfrak{S}}[I]^{\tilde{\mathfrak{B}}} = ({}^{\tilde{\mathfrak{B}}}[I]^{\mathfrak{S}})^{-1} = \frac{1}{4}\cdot({}^{\tilde{\mathfrak{B}}}[I]^{\mathfrak{S}})^{t}.$$

Inverting equation (2.5) we get:

$$(2.8) \qquad 4\begin{pmatrix} p_1 \\ p_2 \\ p_3 \\ p_4 \end{pmatrix} = \begin{pmatrix} 1 & 1 & 1 & 1 \\ 1 & -1 & -1 & 1 \\ -1 & 1 & -1 & 1 \\ -1 & -1 & 1 & 1 \end{pmatrix}\begin{pmatrix} x \\ y \\ z \\ n \end{pmatrix}$$

The set of distributions Δ is a convex subset of the hypersurface defined by $\Sigma_i\, p_i = 1$, i.e. by $n = 1$. The tangent space of each point of $\mathring{\Delta}$ is $\mathbb{R}_0^4$ defined by $n = 0$ (cf. I.(3.1)). So on $\mathbb{R}_0^4$ we define the basis:

$$(2.9) \qquad \mathfrak{B} = \{\xi^x/4, \xi^y/4, \xi^z/4\} = \{\partial_x, \partial_y, \partial_z\}.$$

Here we are thinking of ∂_x, ∂_y and ∂_z as constant vectorfields in $\mathring{\Delta}$.

Recall from Chap. I, Sec. 1 that for the TLTA model the elements p of Δ as the distribution vectors on the four gamete genotypes AB, Ab, aB, ab. On Δ, $x = 2p_A - 1$ and $y = 2p_B - 1$ (cf. I. (5.1)), or

$$p_A = \frac{1+x}{2} \qquad p_a = \frac{1-x}{2}$$

$$p_B = \frac{1+y}{2} \qquad p_b = \frac{1-y}{2} .$$

Define $\tilde{d} = 4d$, where d is the difference measure of linkage disequilibrium, i.e.

(2.10) $$\tilde{d} = 4(p_1p_4 - p_2p_3).$$

From I.(1.3) we have:

(2.11) $$\begin{aligned} 4p_i &= (1+x)(1+y) + \tilde{d} \\ 4p_i &= (1+x)(1-y) - \tilde{d} \\ 4p_i &= (1-x)(1+y) - \tilde{d} \\ 4p_i &= (1-x)(1-y) + \tilde{d}. \end{aligned}$$

Substituting $4p_i = x + y + z + n = x + y + z + 1$ from the first row of (2.8) we get:

(2.12) $$\tilde{d} = z - xy.$$

Recall that the set of interior distributions $\mathring{\Delta}$ is a Riemannian manifold with respect to the Shahshahani metric. The reader should recall that such concepts as length, orthogonality and dual basis depend on the inner product. For $\mathbb{R}^4$ we always use the usual inner product (2.1). However, for $\mathbb{R}^4_0$ we will always use the Shahshahani metric $(\ ,\ )_p$ which assumes we are looking at tangent vectors of $\mathring{\Delta}$ based at some point p. This is why we regard the elements of $\mathfrak{B}$ as constant vectorfields on $\mathring{\Delta}$ rather than as vectors.

We now use the geometry of Shahshahani's Riemannian metric on

$\mathring{\Delta}$ to give another description of the elements of $\mathfrak{B}$.

In the notation of [GPG; p. 43] the coordinates x, y and z regarded as functions on $\mathring{\Delta}$ are among the class of special functions $E^a: \mathring{\Delta} \to \mathbb{R}$ defined by $E^a(p) = (a,p)$. The dual class of functions $L^b: \mathring{\Delta} \to \mathbb{R}$ are defined by $L^b(p) = (b, \ell n\ p)$. So we define:

$$
\begin{aligned}
L^x(p) &= (\xi^x, \ell n\ p) = \ell n\ p_1 + \ell n\ p_2 - \ell n\ p_3 - \ell n\ p_4 \\
L^y(p) &= (\xi^y, \ell n\ p) = \ell n\ p_1 - \ell_n\ p_2 + \ell_n\ p_4 - \ell n\ p_4 \\
L^z(p) &= (\xi^z, \ell n\ p) = \ell n\ p_1 - \ell n\ p_2 - \ell n\ p_3 + \ell n\ p_4 \\
L^n(p) &= (\xi^n, \ell n\ p) = \ell n\ p_1 + \ell n\ p_2 + \ell n\ p_3 + \ell n\ p_4.
\end{aligned} \tag{2.13}
$$

Notice that L^z is the log-ratio measure of linkage disequilibrium denoted by L in I.(5.2).

Computing the gradient of the first three with respect to the Shahshahani metric, it follows that:

$$\mathfrak{B} = \{\partial_x, \partial_y, \partial_z\} = \{\bar{\nabla} L^x/4, \bar{\nabla} L^y/4, \bar{\nabla} L^z/4\}. \tag{2.14}$$

For the formulae on [GPG;p.43] say, for example, that $\bar{\nabla} L^x = \Sigma_{i=1}^4 \xi_i^x \partial_i$ while (2.6) says that the coordinates of ∂_x with respect to $\mathfrak{S}$ are $\xi_i^x/4$ $(i = 1,\ldots,4)$.

We now see that the dual basis for $\mathfrak{B}$ consists of the gradients of the coordinate functions:

$$\mathfrak{B}^* = \{\bar{\nabla} x, \bar{\nabla} y, \bar{\nabla} z\}. \tag{2.15}$$

While this result follows from I. (1.19) it is easy to check I. (1.15) directly using [GPG; eq. I.(4.13), p. 43] which says that $(\bar{\nabla}_p E^a, \bar{\nabla}_p L^b)_p = (a,b)$ provided that $(\xi^n, b) = 0$. In general, if $(\xi^n, b) \neq 0$ the formula generalizes to

$$(\bar{\nabla}_p E^a, \bar{\nabla}_p L^b)_p = (a,b) - E^a(p)(\xi^n,b).$$

From this we compute:

$$(2.16)\qquad \begin{aligned}(\bar{\nabla}_p L^n, \bar{\nabla}_p x)_p &= -4x\\ (\bar{\nabla}_p L^n, \bar{\nabla}_p y)_p &= -4y\\ (\bar{\nabla}_p L^n, \bar{\nabla}_p z)_p &= -4z.\end{aligned}$$

Also by [GPG; I.(4.13)] the inner product between $\bar{\nabla}_p E^a$ and $\bar{\nabla}_p E^b$ is the covariance of a and b. We can write this:

$$(\bar{\nabla}_p E^a, \bar{\nabla}_p E^b)_p = E^{a*b}(p) - E^a(p)E^b(p)$$

where a*b is the vector defined by $(a*b)_i = a_i b_i$.

It is easy to check that

$$(2.17)\qquad \begin{aligned}&\xi^x * \xi^y = \xi^z \qquad \xi^y * \xi^z = \xi^x \qquad \xi^z * \xi^x = \xi^y\\ &\xi^x * \xi^x = \xi^y * \xi^y = \xi^z * \xi^z = \xi^n.\end{aligned}$$

This allows us to compute the matrix of Cor. 1.5:

$$(2.18)\qquad \begin{aligned}\mathfrak{B}_{[I]}\mathfrak{B}^* &= \begin{pmatrix} (\bar{\nabla}_p x, \bar{\nabla}_p x)_p & (\bar{\nabla}_p x, \bar{\nabla}_p y)_p & (\bar{\nabla}_p x, \bar{\nabla}_p z)_p \\ (\bar{\nabla}_p y, \bar{\nabla}_p x)_p & (\bar{\nabla}_p y, \bar{\nabla}_p y)_p & (\bar{\nabla}_p y, \bar{\nabla}_p z)_p \\ (\bar{\nabla}_p z, \bar{\nabla}_p x)_p & (\bar{\nabla}_p z, \bar{\nabla}_p y)_p & (\bar{\nabla}_p z, \bar{\nabla}_p z)_p \end{pmatrix}\\ &= \begin{pmatrix} 1 - x^2 & z - xy & y - xz \\ z - xy & 1 - y^2 & x - yz \\ y - xz & x - yz & 1 - z^2 \end{pmatrix}.\end{aligned}$$

In particular, we note from (2.12):

(2.19) $$(\bar\nabla_p x, \bar\nabla_p y)_p = z - xy = \tilde d$$

The inner products of (2.18) allow us to compute at each point p the lengths of the vectors of $\mathfrak{B}^*$ and the angles between them. In particular, for the angle between $\bar\nabla_p x$ and $\bar\nabla_p y$:

(2.20)
$$\text{cosine} = \frac{\tilde d}{\sqrt{(1-x^2)(1-y^2)}} = \frac{z - xy}{\sqrt{(1-x^2)(1-y^2)}}$$
$$\text{sine} = \sqrt{\frac{(1-x^2)(1-y^2)-\tilde d^2}{(1-x^2)(1-y^2)}} = \sqrt{\frac{1-x^2-y^2-z^2+2xyz}{(1-x^2)(1-y^2)}}$$

The numerator in the expression for the sine comes up repeatedly and so we give it a name:

(2.21) $$\mathrm{Sn} \equiv \sqrt{(1-x^2)(1-y^2)-\tilde d^2} = \sqrt{1-x^2-y^2-z^2+2xyz}$$

In particular , the expressions inside the radicals are nonnegative because $\text{cosine}^2 \leq 1$ (Schwartz inequality). They are strictly positive on $\mathring\Delta$ because the sine is never zero since $\bar\nabla x$ and $\bar\nabla y$ are everywhere linearly independent.

To compute the inverse matrix for ${}^{\mathfrak{B}}[I]^{\mathfrak{B}^*}$ we note that if $(\xi^n, a) = (\xi^n, b) = 0$ then

$$(\bar\nabla_p L^a, \bar\nabla_p L^b)_p = (a*b, p^{-1}) \equiv \Sigma_{i=1}^4 a_i b_i/p_i .$$

Define on $\mathring\Delta$ the functions:

(2.22)
$$\ell = \sqrt{(\xi^n, p^{-1})/16} = (p_1^{-1} + p_2^{-1} + p_3^{-1} + p_4^{-1})^{\frac{1}{2}}/4$$
$$\ell_x = (\xi^x, p^{-1})/16 = (p_1^{-1} + p_2^{-1} - p_3^{-1} - p_4^{-1})/16$$

(2.22) continued

$$\ell_y = (\mathfrak{s}^y, p^{-1})/16 = (p_1^{-1} - p_2^{-1} + p_3^{-1} - p_4^{-1})/16$$

$$\ell_z = (\mathfrak{s}^z, p^{-1})/16 = (p_1^{-1} - p_2^{-1} - p_3^{-1} + p_4^{-1})/16.$$

By (2.14), $(\partial_x, \partial_x)_p$, for example, equals $(\bar\nabla_p L^x, \bar\nabla_p L^x)_p/16$ which by the above formula and (2.17) is $(\mathfrak{s}^n, p^{-1}) = \ell^2$. Similarly, for the other inner products we get:

(2.23)

$$({}^{\mathfrak{B}}[I]^{\mathfrak{B}*})^{-1} = {}^{\mathfrak{B}*}[I]^{\mathfrak{B}} = \begin{pmatrix} (\partial_x,\partial_x)_p & (\partial_x,\partial_y)_p & (\partial_x,\partial_z)_p \\ (\partial_y,\partial_x)_p & (\partial_y,\partial_y)_p & (\partial_y,\partial_z)_p \\ (\partial_z,\partial_x)_p & (\partial_z,\partial_y)_p & (\partial_z,\partial_z)_p \end{pmatrix}$$

$$= \begin{pmatrix} \ell^2 & \ell_z & \ell_y \\ \ell_z & \ell^2 & \ell_x \\ \ell_y & \ell_x & \ell^2 \end{pmatrix}$$

We now do some algebra which will be useful for later calculations.

1 <u>Proposition</u>: We denote by Dt the determinant of the matrix ${}^{\mathfrak{B}}[I]^{\mathfrak{B}*}$. It satisfies

(2.24)
$$Dt = 2^8 p_1 p_2 p_3 p_4.$$

Dt is positive and ≤ 1 on $\mathring{\Delta}$.

The following equations are true:

(2.25)
$$\ell^2 = \frac{Sn^2}{Dt}$$

(2.26)
$$(x-yz)\ell_x + (y-xz)\ell_y = 1 - (1-z^2)\ell^2.$$

(2.27)
$$\ell_z + x\ell_y + y\ell_x + z\ell^2 = 0.$$

$$\frac{\ell_x \ell_y}{\ell^2} + x\ell_y + y\ell_x + z\ell^2 = \frac{\tilde{d}}{Sn^2} = \frac{z-xy}{Sn^2} \tag{2.28}$$

Proof: By direct computation the determinant of the matrix in (2.18) is given by

$$\begin{aligned} Dt &= (1-x^2)(1-y^2)(1-z^2) + 2(z-xy)(x-yz)(y-xz) \\ &\quad -(1-x^2)(x-yz)^2 - (1-y^2)(y-xz)^2 - (1-z^2)(z-xy)^2 \\ &= 1 - 2x^2 - 2y^2 - 2z^2 - 2x^2y^2 - 2x^2z^2 - 2y^2z^2 \\ &\quad + 8xyz + x^4 + y^4 + z^4. \end{aligned} \tag{2.29}$$

On the other hand, from (2.8) (with n = 1) we get

$$16p_1p_4 = (1+x+y+z)(1-x-y+z) = (1-x^2-y^2+z^2) + (2z-2xy)$$

$$16p_2p_3 = (1+x-y-z)(1-x+y-z) = (1-x^2-y^2+z^2) - (2z-2xy)$$

therefore

$$2^8 p_1p_2p_3p_4 = (1-x^2-y^2+z^2)^2 - (2z-2xy)^2$$

Expanding out this last expression you get the same formula as the latter one in (2.29).

Clearly, $Dt > 0$ on $\mathring{\Delta}$. To see that $\ell n\, Dt = 4\, \ell n\, 4 + L^n(p) \leq 0$ we note that by convexity of the log function:

$$4^{-1}(4\, \ell n\, 4 + \Sigma_i\, \ell n\, p_i) = 4^{-1}\, \Sigma_i\, \ell n\, 4p_i \leq \ell n(4^{-1}\, \Sigma_i\, 4p_i) = 0.$$

Now that we know the determinant of ${}^{\mathfrak{B}}[I]^{\mathfrak{B}*}$ we can compute its inverse using minors. Use the 33 minor for ℓ^2:

$$\begin{aligned} \ell^2 &= [(1-x^2)(1-y^2) - (z-xy)^2]/Dt \\ \ell_x &= [(y-xz)(z-xy) - (1-x^2)(x-yz)]/Dt \end{aligned} \tag{2.30}$$

(2.30) continued

$$\ell_y = [(x-yz)(z-xy) - (1-y^2)(y-xz)]/Dt$$

$$\ell_z = [(x-yz)(y-xz) - (1-z^2)(z-xy)]/Dt.$$

The first equation is (2.25).

To prove (2.26) note that since the matrices of (2.18) and (2.23) are inverses the product of the third row of the former with the third column of the latter is the 33 entry of I_3 which is 1. This says

$$(y-xz)\ell_y + (x-yz)\ell_x + (1-z^2)\ell^2 = 1.$$

To prove (2.27) we compute

$$c = x\ell_y + y\ell_x + z\ell^2$$

where we define the column vector $(a\ b\ c)^t$ to be ${}^{\mathfrak{B}*}[I]^{\mathfrak{B}}(x\ y\ z)^t$. By inverting the matrix we see that (a,b,c) is the solution of the equations

$$\begin{pmatrix} 1-x^2 & z-xy & y-xz \\ z-xy & 1-y^2 & x-yz \\ y-xz & x-yz & 1-z^2 \end{pmatrix} \begin{pmatrix} a \\ b \\ c \end{pmatrix} = \begin{pmatrix} x \\ y \\ z \end{pmatrix}.$$

By Cramer's rule, (Dt)c is given by the determinant

$$\begin{vmatrix} 1-x^2 & z-xy & x \\ z-xy & 1-y^2 & y \\ y-xz & x-yz & z \end{vmatrix} = \begin{vmatrix} 1 & z & x \\ z & 1 & y \\ y & x & z \end{vmatrix}.$$

The equality follows from two elementary column operations using the last column times x and y to add to the first and second columns, respectively. So

$$(Dt)c = z + x^2z + y^2z - 2xy - z^3.$$

Expanding out the expression for ℓ_z in (2.30) we see that $c = -\ell_z$.

To prove (2.28) we apply (2.27) which reduces the left side to

$$\frac{\ell_x\ell_y}{\ell^2} - \ell_z = \ell^{-2}[\ell_x\ell_y - \ell_z\ell^2].$$

By substituting the formulae of (2.30) one gets

$$Dt^2[\ell_x\ell_y - \ell_z\ell^2] = (z-xy)Dt.$$

To see this one multiplies out and notes that the only terms which don't have (z-xy) as a factor are two of the form $(1-x^2)(1-y^2)(x-yz)(y-xz)$ which occur with opposite sign. Cancelling these and factoring out $z - xy$ what remains is the first expression in (2.29) for Dt.

So from (2.25) the left side of (2.28) has been calculated to be

$$\frac{z-xy}{Dt} \div \frac{Sn^2}{Dt} = \frac{z-xy}{Sn^2}.$$

QED

<u>Remark</u>: Notice that ℓ^2 is the reciprocal of the harmonic mean of $4p_1,\ldots,4p_4$. Since the arithmetic mean which is one is always greater than or equal to the harmonic mean, $\ell^2 \geq 1$.

Now we turn to the cylindrical frame. We begin by defining the vectorfield Ang_0 on $\mathring{\Delta}$:

(2.31) $$Ang_0 = x\bar{\nabla}y - y\bar{\nabla}x.$$

We denote by An the expression which is the length of Ang_0.

$An^2 = \|Ang_0\|^2 = (Ang_0, Ang_0)_p = x^2(\bar{\nabla}_p y, \bar{\nabla}_p y)_p + y^2(\bar{\nabla}_p x, \bar{\nabla}_p x)_p$ $-2xy(\bar{\nabla}_p x, \bar{\nabla}_p y)$. So by (2.18):

$$An^2 = \|Ang_0\|^2 = x^2(1-y^2) + y^2(1-x^2) - 2xy\,\tilde{d}$$

(2.32)

$$= x^2 + y^2 - 2xyz.$$

Since $\bar{\nabla}_p x$ and $\bar{\nabla}_p y$ are linearly independent at all p we have $An^2 > 0$ except where $x = y = 0$. So we define the unit vectorfield:

$$(2.33) \qquad Ang = \frac{Ang_0}{An} = \frac{x\bar{\nabla}y - y\bar{\nabla}x}{\sqrt{x^2+y^2-2xyz}} \qquad \text{(undefined if } x = y = 0).$$

Now $4\partial_z = \bar{\nabla}L^z = \bar{\nabla}\,\ell n\, p_1p_4 - \bar{\nabla}\,\ell n\, p_2p_3$ is orthogonal to both $\bar{\nabla}x$ and $\bar{\nabla}y$ and so to Ang_0 and Ang.

On the other hand applying (2.16) we get that

$$(\nabla_p L^n, Ang_0)_p = -4(xy-yx) = 0.$$

Consequently, Ang_0 and Ang are also orthogonal to $\bar{\nabla}_p\,\ell n\, p_1p_4 + \bar{\nabla}_p\,\ell n\, p_2p_3$.

Hence, Ang is perpendicular to both $\bar{\nabla}_p\,\ell n\, p_1p_4$ and to $\bar{\nabla}_p\,\ell n\, p_2p_4$ and so to both $\bar{\nabla}_p\, p_1p_4$ and $\bar{\nabla}_p\, p_2p_3$. Subtracting we have proved:

$$(2.34) \qquad (Ang_0, \bar{\nabla}_p\tilde{d})_p = (Ang, \bar{\nabla}_p\tilde{d})_p = 0.$$

In other words, $\tilde{d}$ as well as L^z are constants of motion for the vectorfield Ang.

Now define the vectorfield

$$Rad_0 = (x(1-y^2)-\tilde{d}y)\bar{\nabla}x + (y(1-x^2)-\tilde{d}x)\bar{\nabla}y$$

$$= (x-yz)\bar{\nabla}x + (y-xz)\bar{\nabla}y.$$

Using (2.18) again

$$(\text{Ang}_0, \text{Rad}_0)_p = x(y(1-x^2)-\tilde{d}x)(1-y^2) - y(x(1-y^2)-\tilde{d}y)(1-x^2)$$

$$+ [x(x(1-y^2)-\tilde{d}y) - y(y(1-x^2)-\tilde{d}x)]\tilde{d}$$

$$= xy(1-x^2)(1-y^2)-\tilde{d}x^2(1-y^2) - yx(1-y^2)1-x^2)$$

$$+ \tilde{d}y^2(1-x^2) + [x^2(1-y^2)-y^2(1-x^2)]\tilde{d} = 0.$$

So Rad_0 is everywhere orthogonal to Ang_0. As for its length:

$$\|\text{Rad}_0\|^2 = (\text{Rad}_0, \text{Rad}_0)_p$$

$$= [(1-x^2)(1-y^2)-\tilde{d}^2][x^2(1-y^2)+y^2(1-x^2)-2xy\tilde{d}]$$

Proving this is a messy but direct computation. Expand $(\text{Rad}_0, \text{Rad}_0)_p$ using the first formula for Rad_0 and $(\bar{\nabla}_p x, \bar{\nabla}_p x)_p = 1-x^2$, $(\bar{\nabla}_p y, \bar{\nabla}_p y) = 1-y$ $(\bar{\nabla}_p x, \bar{\nabla}_p y)_p = \tilde{d}$. Collect powers of $\tilde{d}$. Then do the same for the produc of the above brackets.

We have already met, and named, each of the two bracketed expressions. From (2.21) and (2.32) we get:

$$\|\text{Rad}_0\| = \text{Sn}\cdot\text{An}. \tag{2.35}$$

Now Sn is everywhere positive on $\mathring{\Delta}$ and An is positive unless $x = y = 0$. So we can define the unit vectorfield

$$\text{Rad} \equiv \frac{\text{Rad}_0}{\text{Sn}\cdot\text{An}} \qquad \text{(undefined if } x = y = 0). \tag{2.36}$$

We digress for a moment to interpret the motion of Ang. First note:

$$\begin{aligned}\mathrm{Rad}_0 &= (x-xy^2-\tilde{d}y)\bar{\nabla}x + (y-yx^2-\tilde{d}x)\bar{\nabla}y \\ &= \bar{\nabla}\,\frac{1}{2}[x^2+y^2-x^2y^2-2xy\tilde{d}] + xy\bar{\nabla}\tilde{d}.\end{aligned}$$

We define

$$\begin{aligned}R^2 &= x^2 + y^2 - x^2y^2 - 2xy\tilde{d} \\ &= 1 - (1-x^2)(1-y^2) - 2xy\tilde{d} \\ &= x^2(1-y^2) + y^2(1-x^2) - 2xy\tilde{d} + x^2y^2 \\ &= \mathrm{An}^2 + x^2y^2.\end{aligned} \tag{2.37}$$

R^2 is positive unless $x = y = 0$ when $R^2 = 0$. Furthermore,

$$\mathrm{Rad}_0 = \bar{\nabla}\,\frac{1}{2}\,R^2 + xy\bar{\nabla}\tilde{d}. \tag{2.38}$$

Since Rad_0 and $\bar{\nabla}\tilde{d}$ are orthogonal to Ang it follows that R as well as $\tilde{d}$ are constants of motion for Ang. As described in Sec. I.5 this implies that the motion of Ang consists of closed cycles.

Since both Rad and Ang are linear combinations of $\bar{\nabla}x$ and $\bar{\nabla}y$ they are both perpendicular to $\partial_z = \bar{\nabla}L^z/4$. We complete the frame by normalizing this vector using the fact that $(\bar{\nabla}_p L^z, \bar{\nabla}_p L^z)_p = 16\ell^2$ (cf. (2.23)).

$$\mathrm{Ver} \equiv \frac{\partial_z}{\|\partial_z\|} = \frac{\bar{\nabla}L^z}{\|\bar{\nabla}L^z\|} = \frac{\partial_z}{\ell}\,. \tag{2.39}$$

So at each point p with $(x,y) \neq (0,0)$ we have defined an orthonormal basis, the <u>cylindrical frame</u>:

$$\mathfrak{B}_c \equiv \{\mathrm{Ver}, \mathrm{Rad}, \mathrm{Ang}\}. \tag{2.40}$$

Since $\mathfrak{B}_c$ is an orthonormal basis $\mathfrak{B}_c^* = \mathfrak{B}_c$.

We collect some calculations about the cylindrical frame:

2 Proposition: The following equations hold at all points p where $(x,y) \neq (0,0)$:

$$\bar{\nabla}_p \tilde{d} = (\frac{1}{\ell})\text{Ver} - (\frac{2\tilde{d}\,\text{An}}{\text{Sn}})\text{Rad}. \tag{2.41}$$

$$^{\mathfrak{B}}c_{[I]}{}^{\mathfrak{B}} = ({}^{\mathfrak{B}*}{}_{[I]}{}^{\mathfrak{B}}c)^t = \begin{pmatrix} \frac{\ell_y}{\ell} & \frac{\ell_x}{\ell} & \ell \\ \frac{x-yz}{\text{An Sn}} & \frac{y-xz}{\text{An Sn}} & 0 \\ \frac{-y}{\text{An}} & \frac{x}{\text{An}} & 0 \end{pmatrix} \tag{2.42}$$

$$^{\mathfrak{B}}{}_{[I]}{}^{\mathfrak{B}}c = ({}^{\mathfrak{B}}c_{[I]}{}^{\mathfrak{B}*})^t = \begin{pmatrix} 0 & \frac{x\ \text{Sn}}{\text{An}} & -\frac{(y-xz)}{\text{An}} \\ 0 & \frac{y\ \text{Sn}}{\text{An}} & \frac{(x-yz)}{\text{An}} \\ \frac{1}{\ell} & \frac{2(x-yz)(y-xz)}{\text{An Sn}} & \frac{x^2-y^2}{\text{An}} \end{pmatrix} \tag{2.43}$$

The matrix in (2.43) is the inverse of the matrix in (2.42).

Proof: We do the matrix computations first. Recall that $\mathfrak{B} = \{\partial_x, \partial_y, \partial_z$ and $\mathfrak{B}^*$ is the dual basis $\{\bar{\nabla}x, \bar{\nabla}y, \bar{\nabla}z\}$. We can write the definitions of Rad and Ang in matrix form:

$$\begin{pmatrix} \text{Rad} \\ \text{Ang} \end{pmatrix} = \begin{pmatrix} \frac{x-yz}{\text{An Sn}} & \frac{y-xz}{\text{An Sn}} \\ \frac{-y}{\text{An}} & \frac{x}{\text{An}} \end{pmatrix} \begin{pmatrix} \bar{\nabla}x \\ \bar{\nabla}y \end{pmatrix}$$

The determinant of the 2×2 matrix is:

$$\frac{x^2 - xyz + y^2 - xyz}{\text{An}^2\ \text{Sn}} = \frac{\text{An}^2}{\text{An}^2\text{Sn}} = \frac{1}{\text{Sn}}.$$

Inverting the 2×2 matrix we get

$$\begin{pmatrix} \bar{\nabla}x \\ \\ \bar{\nabla}y \end{pmatrix} = \begin{pmatrix} \frac{x\ Sn}{An} & \frac{-(y-xz)}{An} \\ \\ \frac{y\ Sn}{An} & \frac{(x-yz)}{An} \end{pmatrix} \begin{pmatrix} Rad \\ \\ Ang \end{pmatrix}$$

Since $\mathfrak{B}_c$ is an orthonormal basis we compute the components of $\bar{\nabla}z$ with respect to $\mathfrak{B}_c$ via the inner product:

$$(\bar{\nabla}_p z, Ver)_p = \frac{(\bar{\nabla}_p z, \partial_z)_p}{\|\partial_z\|} = \frac{1}{\|\partial_z\|} = \frac{1}{\ell}$$

$$(\bar{\nabla}_p z, Rad)_p = \frac{(\nabla_p z, Rad_0)_p}{An\ Sn} = \frac{2(x-yz)(y-xz)}{An\ Sn}$$

$$(\bar{\nabla}_p z, Ang)_p = \frac{(\bar{\nabla}_p z, Ang_0)_p}{An} = \frac{x^2-y^2}{An}$$

So we get the matrix equation of vectorfields:

$$(2.44)\qquad \begin{pmatrix} \bar{\nabla}x \\ \\ \bar{\nabla}y \\ \\ \bar{\nabla}z \end{pmatrix} = \begin{pmatrix} 0 & \frac{x\ Sn}{An} & -\frac{(y-xz)}{An} \\ \\ 0 & \frac{y\ Sn}{An} & \frac{(x-yz)}{An} \\ \\ \frac{1}{\ell} & \frac{2(x-yz)(y-xz)}{An\ Sn} & \frac{x^2-y^2}{An} \end{pmatrix} \begin{pmatrix} Ver \\ \\ Rad \\ \\ Ang \end{pmatrix} .$$

By (1.35) this 3×3 matrix is $({}^{\mathfrak{B}_c}[I]^{\mathfrak{B}*})^t$ which, in turn, equals ${}^{\mathfrak{B}}[I]^{\mathfrak{B}_c}$ by (1.38).

To get the components of Ver with respect to $\mathfrak{B}* = \{\bar{\nabla}x, \bar{\nabla}y, \bar{\nabla}z\}$ we must compute the inner products with respect to its dual basis $\mathfrak{B}$ (cf. eg (1.16)). Since Ver is ∂_z divided by its length ($= \ell$) we apply (2.23) to get

$$(Ver, \partial_x)_p = \frac{\ell_y}{\ell} \qquad (Ver, \partial_y)_p = \frac{\ell_x}{\ell}$$

$$(\mathrm{Ver},\partial_z)_p = \frac{\ell^2}{\ell} = \ell.$$

Hence

$$(2.45)\qquad \begin{pmatrix} \mathrm{Ver} \\ \mathrm{Rad} \\ \mathrm{Ang} \end{pmatrix} = \begin{pmatrix} \frac{\ell_y}{\ell} & \frac{\ell_x}{\ell} & \ell \\ \frac{x-yz}{\mathrm{An\ Sn}} & \frac{y-xz}{\mathrm{An\ Sn}} & 0 \\ \frac{-y}{\mathrm{An}} & \frac{x}{\mathrm{An}} & 0 \end{pmatrix} \begin{pmatrix} \bar{\nabla} x \\ \bar{\nabla} y \\ \bar{\nabla} z \end{pmatrix}$$

Again by (1.35) the matrix is $({}^{\mathfrak{B}*}[I]^{\mathfrak{B}}{}_{C})^t$ which equals ${}^{\mathfrak{B}}{}_{C}[I]^{\mathfrak{B}}$ by (1.38)

The matrices are inverses to one another by (1.37).

As for (2.41), since $\tilde{d} = z - xy$

$$\bar{\nabla}\tilde{d} = -y\nabla x - x\nabla y + \nabla z.$$

So we can compute $\bar{\nabla}\tilde{d}$ by multiplying (2.44) on the left by the row vector $(-y\ -x\ 1)$. The Ver component is thus equal to $1/\ell$. Without any computation we know that the Ang component is 0 since $\bar{\nabla}\tilde{d}$ is orthogonal to Ang. Finally, the Rad component is

$$\frac{-2xy\mathrm{Sn}^2 + 2(x-yz)(y-xz)}{\mathrm{An\ Sn}}.$$

The numerator (times 1/2) is:

$$-xy(1-x^2-y^2-z^2+2xyz) + (x-yz)(y-xz)$$

which is equal to

$$-(z-xy)(x^2+y^2-2xyz) = -\tilde{d}\ \mathrm{An}^2.$$

So the Rad component is

$$\frac{-2\tilde{d}\mathrm{An}^2}{\mathrm{An\ Sn}} = \frac{-2\tilde{d}\ \mathrm{An}}{\mathrm{Sn}}.$$

QED

3. Linearizations and Hessians.

A vectorfield on $\mathring{\Delta}$ is a smooth function $X: \mathring{\Delta} \to \mathbb{R}^4_0$. At any point $p \in \mathring{\Delta}$ we can differentiate this function to obtain the linearization of X at p. This linearization is a linear map

$$d_pX: \mathbb{R}^4_0 \longrightarrow \mathbb{R}^4_0.$$

The bilinear form associated to this linear map by the Shahshahani metric is called the Hessian at p, denoted $H_p(X)$. So for vectors $v_1, v_2 \in \mathbb{R}^4_0$

$$(3.1) \qquad H_p(X)(v_1, v_2) = (d_pX(v_1), v_2)_p.$$

The Hessian was introduced and studied in [GPG; Sec. IV.1]. For example, equation (1.32) occurred there as [GPG; Lemma IV.1.1, p. 174].

The purpose of this section is to describe the three terms which add up to the Hessian of X^m at an equilibrium point p.

X^m is the sum of two vectorfields. The first, with components $p_i(m_i - \bar{m})$ at p, is the selection vectorfield which is the gradient of $\bar{m}/2$. The second, with components $-d\xi^z_i$ at p, is the normalized recombination field which we will call Rec (see I. (1.6) and (1.7)). So

$$\mathrm{Rec}_i = -d\,\xi^z_i = -\tilde{d}\xi^z_i/4.$$

Recall that $\xi^z_i/4$ are the components of the constant vectorfield we have called ∂_z, the third element of the basis $\mathfrak{B}$ (cf. (2.8)). Hence

$$(3.3) \qquad \mathrm{Rec} = -\tilde{d}\,\partial_z.$$

Since ∂_z is a constant vectorfield its derivative is identically 0. So $d_p\mathrm{Rec}(v)$ is the derivative $d_p(-\tilde{d})(v)$ multiplied by ∂_z. In terms of the Hessian this says

$$H_p(\mathrm{Rec})(v_1,v_2) = (d_p\mathrm{Rec}(v_1),v_2)_p$$

$$= -(\bar\nabla_p\tilde d, v_1)_p\cdot(\partial_z,v_2)_p$$

or since $\partial_z = \|\partial_z\|\mathrm{Ver} = \ell\cdot\mathrm{Ver}$

(3.4) $$H_p(\mathrm{Rec})(v_1,v_2) = -\ell\cdot(\bar\nabla_p\tilde d,v_1)_p\cdot(\mathrm{Ver},v_2)_p.$$

Recall that Ver, unlike Rad and Ang, is defined everywhere in $\mathring{\Delta}$.

1 <u>Proposition</u>: (a) If $p \in \mathring{\Delta}$ at which $x = y = 0$, then

(3.5) $$H_p(\mathrm{Rec})(v_1,v_2) = -(\mathrm{Ver},v_1)_p(\mathrm{Ver},v_2)_p.$$

In particular, $H_p(\mathrm{Rec})$ is symmetric at such a point.

(b) If $p \in \mathring{\Delta}$ at which $(x,y) \neq (0,0)$ then

$$[H_p(\mathrm{Rec})]^{\mathfrak{B}}_{\;c} = ({}^{\mathfrak{B}}_{\;c}[d_p(\mathrm{Rec})]^{\mathfrak{B}}_{\;c})^t =$$

(3.6) $$\begin{pmatrix} -1 & 0 & 0 \\ \frac{2\ell\tilde d\ \mathrm{An}}{\mathrm{Sn}} & 0 & 0 \\ 0 & 0 & 0 \end{pmatrix}$$

In particular, if $\tilde d = 0$ then $H_p(\mathrm{Rec})$ is symmetric.

<u>Proof</u>: (a) If $x = y = 0$ then $\bar\nabla_p\tilde d = \bar\nabla_p z - x\bar\nabla_p y - y\bar\nabla_p x = \bar\nabla_p z$. Furthermore, from (2.18) $\bar\nabla_p z$ is orthogonal to $\bar\nabla_p x$ and $\bar\nabla_p y$ at such a point. So $\bar\nabla_p z$ is a multiple of ∂_z and hence of Ver. Suppose $\bar\nabla_p z = k$ Ver. Then:

$$1 = (\bar\nabla_p z,\partial_z)_p = (k\ \mathrm{Ver},\ell\ \mathrm{Ver})_p = k\ell.$$

So $k = 1/\ell$ and $\bar\nabla_p\tilde d = \bar\nabla_p z = (1/\ell)\mathrm{Ver}$. Substitute in (3.4) to get (3.5).

(b) The matrix is computed directly from the definition (1.27) and formula (3.4).

If v_2 = Rad or Ang then $H(v_1, v_2) = 0$ since these are perpendicular to Ver. So the second and third columns of the matrix vanish. If v_1 = Ang then $H(v_1, v_2) = 0$ since Ang is perpendicular to $\bar{\nabla}\tilde{d}$. So the third row vanishes.

For the remaining two entries we use equation (2.41). The 11 entry is $H_p(\text{Ver},\text{Ver}) =$

$$-\ell\cdot(\bar{\nabla}_p \tilde{d}, \text{Ver})_p (\text{Ver},\text{Ver})_p = -1.$$

The 21 entry is $H_p(\text{Rad},\text{Ver}) =$

$$-\ell\cdot(\bar{\nabla}_p \tilde{d}, \text{Rad})_p (\text{Ver},\text{Ver})_p = +\frac{\ell\, 2\tilde{d}\ \text{An}}{\text{Sn}}.$$

$[H]^{\mathfrak{B}_c} = ({}^{\mathfrak{B}_c}[L]^{\mathfrak{B}_c})^t$ by (1.32) since $\mathfrak{B}_c$ is an orthonormal basis.

QED

2 Corollary: If $p \in \mathring{\Delta}$ at which $(x,y) \neq (0,0)$ then $[H_p(\text{Rec})]^{\mathfrak{B}} =$

$$\begin{pmatrix} -(\frac{\ell_y}{\ell})^2 + \frac{2\ell_y(x-yz)\tilde{d}}{\text{Sn}^2} & -\frac{\ell_x\ell_y}{\ell^2} + \frac{2\ell_x(x-yz)\tilde{d}}{\text{Sn}^2} & -\ell_y + \frac{2\ell^2(x-yz)\tilde{d}}{\text{Sn}^2} \\ -\frac{\ell_x\ell_y}{\ell^2} + \frac{2\ell_y(y-xz)\tilde{d}}{\text{Sn}^2} & -(\frac{\ell_x}{\ell})^2 + \frac{2\ell_x(y-xz)\tilde{d}}{\text{Sn}^2} & -\ell_x + \frac{2\ell^2(y-xz)\tilde{d}}{\text{Sn}^2} \\ -\ell_y & -\ell_x & -\ell^2 \end{pmatrix}$$

(3.7)

Proof: By (1.40) the matrix $[H]^{\mathfrak{B}} = ({}^{\mathfrak{B}_c}[I]^{\mathfrak{B}})^t [H]^{\mathfrak{B}_c}\, {}^{\mathfrak{B}_c}[I]^{\mathfrak{B}}$ where ${}^{\mathfrak{B}_c}[I]^{\mathfrak{B}}$ is the matrix of (2.42). Multiply out. QED

The Hessian of the selection field $\bar{\nabla}\frac{1}{2}\bar{m}$ was computed in [GPG, IV, (1.17), p. 178]. It is the sum of two symmetric terms. The first is:

(3.8) $$H_1(v_1,v_2) = \Sigma_{i,j=1}^{4} m_{ij}(v_1)_i(v_2)_j.$$

We will call this bilinear form H(M). The second term is:

(3.9) $$H_2(v_1,v_2) = \Sigma_{i=1}^{4} p_i^{-1}(m_i-\bar{m})(v_1)_i(v_2)_i.$$

We will call this bilinear form $H(\Delta)$.

These formulae are defined for vectors $v_1,v_2 \in \mathbb{R}_0^4$. They can obviously be extended to define bilinear forms on $\mathbb{R}^4$ and they indicate directly the matrices with respect to the standard basis $\mathfrak{S}$:

(3.10) $${}_i[H_1]_j^{\mathfrak{S}} = m_{ij}$$

and

(3.11) $${}_i[H_2]_j^{\mathfrak{S}} = p_i^{-1}(m_i-\bar{m})\delta_{ij}.$$

As (m_{ij}) is assumed to be a selection matrix, i.e. to satisfy the CIS = TRANS condition I. (1.5) as well as symmetry I.(1.4), there are nine independent parameters which can be represented in the 3×3 genetic table I.(1.6). These nine entries are the standard selection parameters, which we will call the <u>$\mathfrak{S}$-parameters of selection</u>.

We can compute $[H_1]^{\tilde{\mathfrak{B}}} = ({}^{\mathfrak{S}}[I]^{\tilde{\mathfrak{B}}})^t [H_1]^{\mathfrak{S}\mathfrak{S}}[I]^{\tilde{\mathfrak{B}}}$ where ${}^{\mathfrak{S}}[I]^{\tilde{\mathfrak{B}}}$ is 1/4 times the matrix whose columns are $\xi^x, \xi^y, \xi^z, \xi^n$ (cf. (2.7)). Instead we begin with $[H_1]^{\tilde{\mathfrak{B}}}$ itself and use it to define nine parameters which we will call the <u>$\tilde{\mathfrak{B}}$-parameters of selection</u> .

3 <u>Proposition</u>: Define the 4×4 matrix

(3.12)
$$M = \begin{pmatrix} D_A & I & L & A \\ I & D_B & K & B \\ L & K & J & I \\ A & B & I & Y \end{pmatrix}.$$

$({}^{\widetilde{\mathfrak{B}}}[I]^{\mathfrak{S}})^t \; M \; {}^{\widetilde{\mathfrak{B}}}[I]^{\mathfrak{S}} = (m_{ij})$ is a selection matrix whose entries are described by the following 3×3 genetic table:

(3.13)

	aa	Aa	AA
BB	$Y-2A+2B+D_A+D_B$ $+2L-2K-4I+J$	$Y+2B-D_A+D_B$ $-2L-J$	$Y+2A+2B+D_A+D_B$ $+2L+2K+4I+J$
Bb	$Y-2A+D_A-D_B$ $+2K-J$	$Y-D_A-D_B$ $+J$	$Y+2A+D_A-D_B$ $-2K-J$
bb	$Y-2A-2B+D_A+D_B$ $-2L-2K+4I+J$	$Y-2B-D_A+D_B$ $+2L-J$	$Y+2A-2B+D_A+D_B$ $-2L+2K-4I+J$

Conversely, if (m_{ij}) is any selection matrix then the equation $M = ({}^{\mathfrak{S}}[I]^{\widetilde{\mathfrak{B}}})^t (m_{ij}) {}^{\mathfrak{S}}[I]^{\widetilde{\mathfrak{B}}}$ defines the matrix of $\widetilde{\mathfrak{B}}$ parameters of selection.

If M is the matrix of $\widetilde{\mathfrak{B}}$ parameters of selection then the $H(M)$ portion of the Hessian of the selection field has matrix with respect to $\mathfrak{B}$ given by the upper 3×3 submatrix of M, i.e.

(3.14)
$$[H(M)]^{\mathfrak{B}} = \begin{pmatrix} D_A & I & L \\ I & D_B & K \\ L & K & J \end{pmatrix}.$$

Proof: The correspondence between M and (m_{ij}) is an invertible linear map from the nine-dimensional space of symmetric 4×4 matrices to itself. It is only necessary to show that the CIS = TRANS condition that $m_{14} = m_{23}$ follows from the fact that $M_{12} = M_{34} = I$. Compute (m_{ij})

using formula (2.5) for $^{\widetilde{\mathfrak{B}}}[I]^{\mathfrak{S}}$. You get a matrix (m_{ij}) with $m_{14} = m_{23}$. From the results of the computation you fill in table (3.13) following the directions of the table I(1.6). For the converse one can either show that $m_{14} = m_{23}$ implies $M_{12} = M_{34}$ by direct computation or note that the space of matrices M of the form (3.12) is nine dimensional as is the space of selection matrices. Since the correspondence of M to (m_{ij}) is a one-to-one linear map into the space of selection matrices it must be an onto map by the dimension equality.

Finally, if $[H_1]^{\widetilde{\mathfrak{B}}} = M$ then to get $[H(M)]^{\mathfrak{B}}$ one must restrict the bilinear form to $\mathbb{R}^4_0$ from $\mathbb{R}^4$. In $\widetilde{\mathfrak{B}}$ coordinates $\mathbb{R}^4_0$ is the subspace defined by the vanishing of the last coordinate n. QED

The letters chosen for the entries of M are intended to correspond--rather roughly--to a decomposition of the 3×3 genetic table into additive, dominance and epistasis effects. Compare table (3.13) with the table in Crow and Kimura's book [2; p. 79].

It will be useful to have mean fitness and the selection field described in terms of the $\widetilde{\mathfrak{B}}$ parameters of selection.

4 <u>Proposition</u>: If M given by (3.12) is the matrix of $\widetilde{\mathfrak{B}}$ parameters of selection then at a point p in Δ the mean fitness is given by:

$$(3.15) \qquad \bar{m} = \begin{pmatrix} x \\ y \\ z \\ 1 \end{pmatrix}^t \begin{pmatrix} D_A & I & L & A \\ I & D_B & K & B \\ L & K & J & I \\ A & B & I & Y \end{pmatrix} \begin{pmatrix} x \\ y \\ z \\ 1 \end{pmatrix} .$$

At any interior point $p \in \mathring{\Delta}$ the selection field $\bar{\nabla}(\frac{1}{2}\,\bar{m}) =$

$$(3.16)\qquad \begin{aligned} &(A + xD_A + yI + zL)\bar{\nabla}x + \\ &(B + yD_B + xI + zK)\bar{\nabla}y + \\ &(I + xL + yK + zJ)\bar{\nabla}z. \end{aligned}$$

Proof: Regarding p as a vector in $\mathbb{R}^4$:

$$\begin{aligned} \bar{m} = H_1(p,p) &= ([p]^{\mathfrak{S}})^t (m_{ij})[p]^{\mathfrak{S}} \\ &= ([p]^{\tilde{\mathfrak{B}}})^t M[p]^{\tilde{\mathfrak{B}}}. \end{aligned}$$

But the $\tilde{\mathfrak{B}}$ coordinates of a point p in Δ are x,y,z and $n = 1$. This proves (3.15). To prove (3.16) we apply the gradient operator to both sides of (3.15). Since M is a symmetric matrix and since the gradient of 1 is 0 we have:

$$\bar{\nabla}(\tfrac{1}{2}\bar{m}) = \begin{pmatrix} x \\ y \\ z \\ 1 \end{pmatrix}^t M \begin{pmatrix} \bar{\nabla}x \\ \bar{\nabla}y \\ \bar{\nabla}z \\ 0 \end{pmatrix}.$$

Multiplying out we get (3.16). QED

5 Corollary: If M is the matrix of $\tilde{\mathfrak{B}}$ selection parameters then $p \in \mathring{\Delta}$ is equilibrium for the vectorfield X^m of selection and normalized recombination if and only if the following three equations are true at p:

$$(3.17)\qquad \begin{aligned} A &= \tilde{d}\,\ell_y - xD_A - yI - zL \\ B &= \tilde{d}\,\ell_x - yD_B - xI - zK \end{aligned}$$

(3.18) $$I + xL + yK + zJ = \tilde{d}\,\ell^2.$$

We refer to (3.18) as the CIS = TRANS equilibrium condition.

Proof: The combined field $X^m = \bar{\nabla}(\frac{1}{2}\bar{m}) + \text{Rec} = \bar{\nabla}(\frac{1}{2}\bar{m}) - \tilde{d}\partial_z$. $p \in \mathring{\Delta}$ is an equilibrium iff this vector vanishes at p. This occurs iff the inner product with ∂_x, ∂_y and ∂_z vanish at p. The three inner products yield the three equations (cf. (2.23)). QED

Remark: A and B do not occur in $[H(M)]^{\mathfrak{B}}$ given by (3.14). In the parametrized description of Σ_C which is our goal we will describe the entries of $[H(M)]^{\mathfrak{B}}$ in terms of the parameters. Then, at the end, A and B are defined by the equations of (3.17). On the other hand, (3.18) represents a linear relation on I, J, K and L which must be satisfied by the parametrized description. While this equation appears to be an equilibrium equation it actually comes from the absence of position effects. If position effects were admitted then this equation would be used to define the 34 entry of M which would not appear in $[H(M)]^{\mathfrak{B}}$ because it would not be assumed equal to the 12 entry.

Finally, we return to the bilinear form $H(\Delta)$ computed by (3.9).

If p were an equilibrium point for the selection field alone then $m_i - \bar{m}$ would vanish for $i = 1,\ldots,4$ and so this term would not occur in the Hessian of $\bar{\nabla}_p(\frac{1}{2}\bar{m})$. However, the equilibria we are interested in are those of the combined field X^m. At such an equilibrium we have:

$$p_i(m_i - \bar{m}) = \tilde{d}\,\xi_i^z/4 \qquad i = 1,\ldots,4.$$

So the formula (3.9) becomes:

$$H_2(v_1,v_2) = \frac{\tilde{d}}{4}\Sigma_{i=1}^{4} p_i^{-2}\, \xi_i^z (v_1)_i (v_2)_i$$

or

$$(3.19) \qquad [H_2]^{\mathfrak{S}} = \frac{\tilde{d}}{4} p_i^{-2} \xi_i^z \delta_{ij} \qquad i,j = 1,\ldots,4.$$

6 <u>Proposition</u>: Define the functions on $\mathring{\Delta}$:

$$(3.20) \qquad \tilde{\ell}_a = (\xi^a, p^{-2}) = \Sigma_{i=1}^{4} \xi_i^a p_i^{-2}/64 \qquad a = x,y,z,n.$$

Suppose $p \in \mathring{\Delta}$ is an equilibrium for the combined field X^m. Then at p the $H(\Delta)$ portion of the Hessian of the selection field has matrix given by:

$$(3.21) \qquad [H(\Delta)]^{\mathfrak{B}} = \tilde{d}\begin{pmatrix} \tilde{\ell}_z & \tilde{\ell}_n & \tilde{\ell}_x \\ \tilde{\ell}_n & \tilde{\ell}_z & \tilde{\ell}_y \\ \tilde{\ell}_x & \tilde{\ell}_y & \tilde{\ell}_z \end{pmatrix}.$$

<u>Proof</u>: If $[H_2]^{\mathfrak{S}}$ is given by (3.19), namely $\tilde{d}/4$ times the diagonal matrix with diagonal entries $p_1^{-2}, -p_2^{-2}, -p_3^{-2}, p_4^{-2}$, then $[H_2]^{\tilde{\mathfrak{B}}}$ is the 4×4 matrix $({}^{\mathfrak{S}}[I]^{\tilde{\mathfrak{B}}})^t [H_2]^{\mathfrak{S}\mathfrak{S}}[I]^{\tilde{\mathfrak{B}}}$. Computing with the matrix of (2.7) one gets:

$$\tilde{d}\begin{pmatrix} \tilde{\ell}_z & \tilde{\ell}_n & \tilde{\ell}_x & \tilde{\ell}_y \\ \tilde{\ell}_n & \tilde{\ell}_z & \tilde{\ell}_y & \tilde{\ell}_x \\ \tilde{\ell}_x & \tilde{\ell}_y & \tilde{\ell}_z & \tilde{\ell}_n \\ \tilde{\ell}_y & \tilde{\ell}_x & \tilde{\ell}_n & \tilde{\ell}_z \end{pmatrix}.$$

As in the proof of Prop. 3 you get $[H(\Delta)]^{\mathfrak{B}}$ by restricting to $\mathbb{R}_0^4$ where the last coordinate, n, vanishes. So the matrix we want is the upper left 3×3 submatrix. QED

We conclude by deriving a relation among the $\tilde{\ell}$'s.

7 Proposition: The following equation is true:

$$\tilde{\ell}_n + x\tilde{\ell}_x + y\tilde{\ell}_y + z\tilde{\ell}_z = \ell^2. \tag{3.22}$$

Proof: The left side of (3.22) is the product of the row vector $(x\ y\ z\ 1)$ times the column vector $(\tilde{\ell}_x\ \tilde{\ell}_y\ \tilde{\ell}_z\ \tilde{\ell}_n)^t$. Recall that $\tilde{\mathfrak{B}}_{[I]}{}^{\mathfrak{S}}$ is the matrix whose rows are $\xi^x, \xi^y, \xi^z, \xi^n$. So

$$\begin{pmatrix} x \\ y \\ z \\ 1 \end{pmatrix} = \tilde{\mathfrak{B}}_{[I]}{}^{\mathfrak{S}} \begin{pmatrix} p_1 \\ p_2 \\ p_3 \\ p_4 \end{pmatrix} = \tilde{\mathfrak{B}}_{[I]}{}^{\mathfrak{S}}(p)$$

$$64\begin{pmatrix} \tilde{\ell}_x \\ \tilde{\ell}_y \\ \tilde{\ell}_z \\ \tilde{\ell}_n \end{pmatrix} = \tilde{\mathfrak{B}}_{[I]}{}^{\mathfrak{S}} \begin{pmatrix} p_1^{-2} \\ p_2^{-2} \\ p_3^{-2} \\ p_4^{-2} \end{pmatrix} = \tilde{\mathfrak{B}}_{[I]}{}^{\mathfrak{S}}(p^{-2}).$$

Recalling that $\tilde{\mathfrak{B}}_{[I]}{}^{\mathfrak{S}}$ times its transpose is the identity matrix times 4 we get that the left side of (3.22) (times 64) is:

$$(\tilde{\mathfrak{B}}_{[I]}{}^{\mathfrak{S}}(p))^t\, \tilde{\mathfrak{B}}_{[I]}{}^{\mathfrak{S}}(p^{-2}) = (p)^t (\tilde{\mathfrak{B}}_{[I]}{}^{\mathfrak{S}})^t\, \tilde{\mathfrak{B}}_{[I]}{}^{\mathfrak{S}}(p^{-2})$$

$$= 4(p)^t(p^{-2}) = 4\ \Sigma\ p_i p_i^{-2} = 4\Sigma p_i^{-1} = 64\ \ell^2.$$

QED

Remark: Recall that $\ell^2 \geq 1$(cf. Remark after Prop. 2.1). In general, $\tilde{\ell}_n \geq \ell^4 \geq 1$ because by convexity of the function $y = x^2$:

$$\tilde{\ell}_n = \frac{1}{4}\ \Sigma(4p_i)^{-2} \geq (\frac{1}{4}\ \Sigma(4p_i)^{-1})^2 = \ell^4.$$

4. The Parametrization.

As indicated in Sec. I.5 the parametrization is based on the choice of a pair of vectors associated with a complex eigenvalue.

1 Proposition: Assume V is a Euclidean vectorspace and that $\varepsilon \pm i\lambda$ ($\lambda > 0$) are eigenvalues for $L: V \to V$. There exists an orthogonal pair of unit vectors $\{e_u, e_v\}$ called a normalized pair for the eigenvalue pair, and a positive real number k, called the skewness of the pair, such that

$$L(e_u) = \varepsilon e_u - \lambda k^{-1} e_v$$

(4.1)

$$L(e_v) = \lambda k e_u + \varepsilon e_v.$$

Now suppose $\varepsilon + i\lambda$ is a simple eigenvalue of L, i.e. it is not a repeated root of the characteristic equation. If $k \neq 1$ then the only other normalized pairs for L are $\{-e_u, -e_v\}$ with skewness k, $\{e_v, -e_u\}$ and $\{-e_v . e_u\}$ with skewness k^{-1}. If $k = 1$ then the skewness is 1 for all normalized pairs of which there is a one parameter family obtained by rotating the original pair, i.e. the set of normalized pairs is:

$$\{(\cos\theta\, e_u - \sin\theta\, e_v, \sin\theta\, e_u + \cos\theta\, e_v) : 0 \leq \theta < 2\pi\}.$$

Proof: If the complex vector $u + iv$ is an eigenvector for L with respect to the eigenvalue $\varepsilon + i\lambda$ then multiplying by any nonzero complex number $a + ib$ we get a new eigenvector $u_1 + iv_1$:

$$u_1 + iv_1 = (a+ib)(u+iv)$$

$$= (au-bv) + i(bu+av).$$

Furthermore, if $\epsilon + i\lambda$ is a simple root these are the only associated eigenvectors.

We want to choose a and b so that u_1 and v_1 are orthogonal. Computing the inner product

$$(4.2) \qquad (u_1,v_1) = (a^2-b^2)(u,v) + ab[(u,u)-(v,v)]$$

we set $(u_1,v_1) = 0$ and solve for the ratio $a/b = r$.

If $(u,v) \neq 0$, i.e. if the original pair was not orthogonal to begin with, then

$$(4.3) \qquad r - r^{-1} = [(v,v)-(u,u)]/(u,v).$$

As a function of $r > 0$, $r - r^{-1}$ is strictly increasing (its derivative is $1 + r^{-2} > 0$) and approachs $\pm\infty$ as r approaches ∞ and 0. Consequently, (4.3) has a unique positive solution, which could be obtained explicitly by multiplying by r and solving the quadratic equation.

This means that we can choose $a + ib$ so that $(a+ib)(u+iv)$ is an eigenvector with orthogonal parts. Assume that the original vector $u + iv$ was so chosen, i.e. that $(u,v) = 0$. In order that (u_1,v_1) also equal zero (4.2) implies two possibilities;

$k = 1$ <u>case</u>: $(u,u) = (v,v)$. Any $a + ib$ solves.

$k \neq 1$ <u>case</u>: $(u,u) \neq (v,v)$. $a + ib$ solves iff $a = 0$ or $b = 0$.

The names for the two cases comes from the definition:

$$k \equiv \|u\|/\|v\|.$$

$L(u+iv) = (\epsilon + i\lambda)(u + iv)$ can be rewritten (cf. I. (2.9)):

$$(4.4) \qquad \begin{aligned} L(u) &= \varepsilon u - \lambda v \\ L(v) &= \lambda u + \varepsilon v. \end{aligned}$$

Define $e_u = u/\|u\|$ and $e_v = v/\|v\|$. Divide the first equation of (4.4) by $\|u\|$ and the second by $\|v\|$. Substituting the definition of k you get (4.1).

Conversely, if $\{e_u, e_v\}$ satisfies (4.1) then $u = ke_u$ and $v = e_v$ satisfy (4.4) i.e. $ke_u + ie_v$ is an eigenvector for $\varepsilon + i\lambda$. In particular, if $k = 1$, $u + iv = e_u + ie_v$ is an eigenvector with real and imaginary parts orthogonal and of unit length. The $k = 1$ case above shows that multiplying by any nonzero $a + ib$ preserves the orthogonality condition. In order to preserve the lengths $a + ib$ must be a unit complex vector i.e. of the form $\cos\theta + i \sin\theta$. If $\varepsilon + i\lambda$ is a simple root this gives all of the normalized pairs.

If $k \neq 1$ then multiplying $ke_u + ie_v$ preserves orthogonality of the parts iff $ab = 0$. So we can multiply only by ± 1 and by $\pm ik^{-1}$ and preserve orthogonality together with unit length of the imaginary parts. This yields the four $k \neq 1$ possibilities where $\varepsilon + i\lambda$ is a simple root.

QED

Notice that in the three dimensional case complex eigenvalues are always simple as is the third eigenvalue which is real.

For the next result we need the concept of <u>orientation</u> or handedness. The property of orientation divides the set of bases of V into two disjoint classes. Bases $\mathfrak{B}_1$ and $\mathfrak{B}_2$ lie in the same class if the transition matrix ${}^{\mathfrak{B}_2}[I]^{\mathfrak{B}_1}$ has positive determinant and in opposite classes if the determinant is negative. A vector space is <u>oriented</u> if one of the classes is chosen to be labelled "positive" and the other "negative".

2 Proposition: Assume that V is a Euclidean vectorspace with $\dim V = 3$. Assume that the eigenvalues of $L: V \to V$ are $\varepsilon \pm i\lambda$ $(\lambda > 0)$ and $\varepsilon + \mu$. The following two conditions on an orthonormal basis $\mathfrak{B}_e = \{e_u, e_v, e_w\}$ and positive real number k are equivalent:

(a) $\{e_u, e_v\}$ is a normalized pair with skewness k for the eigenvalue pair $\varepsilon \pm i\lambda$.

(b) There exist real numbers γ_1, γ_2 such that the matrix of L with respect to $\mathfrak{B}_e$ is given by:

$$ {}^{\mathfrak{B}_e}[L]^{\mathfrak{B}_e} = \begin{pmatrix} \varepsilon & \lambda k & 2\gamma_1 \\ -\lambda k^{-1} & \varepsilon & 2\gamma_2 \\ 0 & 0 & \varepsilon+\mu \end{pmatrix}. \tag{4.5} $$

Furthermore, if V is assumed to be oriented then the choice of the pair $\{e_u, e_v\}$ uniquely determines e_w subject to the condition that $\mathfrak{B}_e$ have positive orientation.

Proof: The matrix equation (4.5) implies (4.1) by definition. So (b) implies (a).

If $\{e_u, e_v\}$ is a normalized pair then there is a unique line through 0 perpendicular to the plane they span. On this line there are exactly two unit vectors, say $\pm e_w$. Note that $\{e_u, e_v, e_w\}$ and $\{e_u, e_v, -e_w\}$ have opposite orientation because they are related by the matrix:

$$ \begin{pmatrix} 1 & 0 & 0 \\ 0 & 1 & 0 \\ 0 & 0 & -1 \end{pmatrix}. $$

Equation (4.1) implies that the first two columns of ${}^{\mathfrak{B}_e}[L]^{\mathfrak{B}_e}$ are as shown in (4.5). Since γ_1 and γ_2 are arbitrary we need only show that $\varepsilon + \mu$ is an eigenvalue of the matrix on the right. To see this you check directly that w =

$$(2(\gamma_1\mu+\lambda k\gamma_2), 2(\gamma_2\mu-\lambda k^{-1}\gamma_1), \mu^2+\lambda^2)$$

is an eigenvector with eigenvalue $\varepsilon + \mu$. In doing so it is a bit easier to check that w is an eigenvector with eigenvalue μ of the matrix obtained by subtracting the identity matrix times ε. QED

From this proof we can expand upon Prop. 2:

3 <u>Addendum</u>: Define $\mathfrak{B}_u = \{u, v, w\}$

$$u = ke_u$$

$$v = e_v \tag{4.6}$$

$$w = 2(\gamma_1\mu+\lambda k\gamma_2)e_u + 2(\gamma_2\mu-\lambda k^{-1}\gamma_1)e_v + (\mu^2+\lambda^2)e_w.$$

The matrix of L with respect to $\mathfrak{B}_u$ is:

$$ {}^{\mathfrak{B}_u}[L]^{\mathfrak{B}_u} = \begin{pmatrix} \varepsilon & \lambda & 0 \\ -\lambda & \varepsilon & 0 \\ 0 & 0 & \varepsilon+\mu \end{pmatrix} \tag{4.7}$$

The change of basis matrices are:

$$ {}^{\mathfrak{B}_e}[I]^{\mathfrak{B}_u} = \begin{pmatrix} k & 0 & 2(\gamma_1\mu+\lambda k\gamma_2) \\ 0 & 1 & 2(\gamma_2\mu-\lambda k^{-1}\gamma_1) \\ 0 & 0 & \mu^2+\lambda^2 \end{pmatrix} \tag{4.8}$$

$$(4.9)\qquad \mathfrak{B}_{u[I]}\mathfrak{B}_e = \begin{pmatrix} k^{-1} & 0 & -2(\gamma_1 k^{-1}\mu+\lambda\gamma_2)/(\mu^2+\lambda^2) \\ 0 & 1 & -2(\gamma_2\mu-\lambda k^{-1}\gamma_1)/(\mu^2+\lambda^2) \\ 0 & 0 & 1/(\mu^2+\lambda^2) \end{pmatrix}.$$

Proof: Matrix (4.7) comes from equation (4.4) for u and v and the fact that w is an eigenvector with eigenvalue $\varepsilon + \mu$ for L, as proved above. (4.8) follows directly from equations (4.6). (4.9) follows by inverting the matrix of (4.8), or by inverting the equation (4.6). QED

Remark: $\mathfrak{B}_e$ is an orthonormal basis while $\mathfrak{B}_u$ is not. On the other hand, the matrix form of (4.7) will be the one needed to compute MARMC. Note also that $\mathfrak{B}_e$ and $\mathfrak{B}_u$ have the same orientation since the determinant of the matrix of (4.8) is $k(\mu^2 + \lambda^2) > 0$.

We are now ready to build the parametric description of Σ_C and the subset Σ_H. We do it backwards by beginning with a point (m,p) of Σ_C and describing parameters for the point. These parameters are not unique. They will depend on a choice of normalized pair for the linearization of X^m at p. This means that the function from parameters to points (m,p) will not be one-to-one though it will be onto. We will be interested in which different sets of parameters correspond to the same point (m,p). This means we will want to know how much we could vary the choices in the construction we now undertake. It follows that we get different answers according to whether $k = 1$ or not.

To commence, suppose $(m,p) \in \Sigma_C$. Let $L^{(m,p)}: \mathbb{R}^4_0 \to \mathbb{R}^4_0$ be the linearization of X^m at p. By definition of Σ_C (cf. I.(4.5)) $L^{(m,p)}$ has a complex eigenvalue and hence a conjugate pair of complex eigenvalues $\varepsilon \pm i\lambda$ with $\lambda > 0$. The third eigenvalue, which is real, we label $\varepsilon + \mu$. Choose a normalized pair $\{e_u, e_v\}$ with skewness $k > 0$ for the eigenvalue pair $\varepsilon \pm i\lambda$. Note that the inner product we are using is the Shahshahani metric $(\ ,\ )_p$ at the equilibrium point p. The basis $\mathfrak{B}_e = \{e_u, e_v, e_w\}$ of Prop. 2 is uniquely defined from the choice of $\{e_u, e_v\}$ by the condition that $\mathfrak{B}_e$ have the same orientation as the cylindrical frame $\mathfrak{B}_c = \{\text{Ver}, \text{Rad}, \text{Ang}\}$ at p. Now define the matrix

(4.10) $$^{\mathfrak{B}_e}[I]^{\mathfrak{B}_c} = O = \begin{pmatrix} O_{11} & O_{12} & O_{13} \\ O_{21} & O_{22} & O_{23} \\ O_{31} & O_{32} & O_{33} \end{pmatrix}.$$

4 Lemma: O is a special orthogonal matrix, i.e., $O^t = O^{-1}$ and $\det O = 1$.

Proof: $O^t = O^{-1}$ by (1.41). Hence, $1 = \det(OO^t) = \text{Det}(O)^2$. So $\det(O) = \pm 1$. Since $\mathfrak{B}_e$ and $\mathfrak{B}_c$ have the same orientation $\det(O) = +1$.

QED

The set of special orthogonal 3×3 matrices, or the special orthogonal group as it is called, is a three dimensional subset of the nine dimensional vector space of 3×3 matrices. SO(3), as it is denoted, is actually a three dimensional submanifold so we can regard it as described by three real parameters. In Chapter III we give a description of SO(3) in terms of three angular variables.

The original parameters of Σ_C will be $O \in SO(3)$, $k \in (0,\infty)$, $p \in \mathring{\Delta}$ and $\mu, \varepsilon \in \mathbb{R}$. Now by looking at the Hessian matrices in various ways we will see how these parameters are related to the remaining variables of the situation. First, we will look at the antisymmetric parts and derive formulae for λ, γ_1 and γ_2 in terms of these parameters. Then, we will look at the symmetric parts and derive formulae for D_A, D_B, I, J, K, L among the $\tilde{\mathfrak{B}}$-parameters of selection. The remaining $\tilde{\mathfrak{B}}$-parameters A and B are then given by equations (3.17). Finally, the CIS = TRANS equilibrium condition (3.18) will yield a formula among the parameters which will usually allow us to solve for μ in terms of O, k and p. In Sec. 6 we will show how to compute MARMC.

The matrix of $L^{(m,p)}$ with respect to $\mathfrak{B}_e$ is given by (4.5). The Hessian $H_p(X^m)$ is the associated bilinear form. Since $\mathfrak{B}_e$ is orthonormal (1.32) implies that the matrix of $H_p(X^m)$ is given by the transpose:

$$(4.11) \qquad [H_p(X^m)]^{\mathfrak{B}_e} = \begin{pmatrix} \epsilon & -\lambda k^{-1} & 0 \\ \lambda k & \epsilon & 0 \\ 2\gamma_1 & 2\gamma_2 & \epsilon+\mu \end{pmatrix}.$$

5 <u>Theorem (Parametrization Condition</u> A): Let $(m,p) \in \Sigma_C$ and O be defined by (4.10). Then:

(a) $p \in \mathring{\Delta}$, $(x,y) \neq (0,0)$ and $\tilde{d} \neq 0$.

(b) In (4.11) γ_1, γ_2 and λ are given by:

$$(4.12) \qquad \begin{aligned} \gamma_1 &= -\frac{\ell\, \tilde{d}\, An}{Sn} \cdot O_{23} \\ \gamma_2 &= \frac{\ell\, \tilde{d}\, An}{Sn} \cdot O_{13} \\ \lambda &= \frac{\ell\, \tilde{d}\, An}{Sn(k+k^{-1})/2} \cdot O_{33} \end{aligned}$$

In particular, since $\lambda > 0$

$$(4.13) \qquad \tilde{d}\, O_{33} > 0.$$

Proof: Begin at an arbitrary point $p \in \mathring{\Delta}$. Choose an orthonormal basis $\mathfrak{B}_1$ at p. The Hessian $H_p(X^m)$ is the bilinear form associated to the linearization $L^{(m,p)}$. So if $H_p(X^m)$ were a symmetric form, the matrix ${}^{\mathfrak{B}_1}[L^{(m,p)}]^{\mathfrak{B}_1} = ([H_p(X^m)]^{\mathfrak{B}_1})^t$ (by (1.32)) would be a symmetric matrix by Remark (b) after Prop. 1.2. A symmetric matrix has only real eigenvalues and so at such a point $L^{(m,p)}$ could not have a complex eigenvalue, i.e. $(m,p) \notin \Sigma_C$.

On the other hand, $H_p(X^m)$ is the sum of the Hessian of the selection field $H_p(\nabla\overline{m}/2)$ which is always symmetric and $H_p(\text{Rec})$ which is symmetric when $(x,y) = (0,0)$ or $\tilde{d} = 0$ by Prop. 3.1. So if $p \in \mathring{\Delta}$ and $(x,y) = (0,0)$ or $\tilde{d} = 0$ then $L^{(m,p)}$ has only real eigenvalues. Since $L^{(m,p)}$ and its eigenvalues vary continuously with p, at a boundary point with $\tilde{d} = 0$ $L^{(m,p)}$ has only real eigenvalues. Finally, if p is a boundary point and $\tilde{d} \neq 0$ then p can't be an equilibrium for X^m, because the selection field points along the boundary and while the recombination field points into the interior of $\mathring{\Delta}$ (and is not zero since $\tilde{d} \neq 0$). Thus, if $(x,y) = (0,0)$ or $\tilde{d} = 0$ or p is on the boundary (m,p) can't lie in Σ_C, proving (a).

Recall from (1.8) that any bilinear form H can be decomposed uniquely into the sum of a symmetric form SH and an antisymmetric form AH where $AH = \frac{1}{2}(H - H^*)$. Since the matrix of H^* is the transpose of the matrix of H (cf. (1.30)) the matrix $[AH] = \frac{1}{2}([H] - [H]^t)$. So from (4.11):

$$(4.14)\quad [AH_p(X^m)]^{\mathfrak{B}_e} = \begin{pmatrix} 0 & -\lambda(k+k^{-1})/2 & -\gamma_1 \\ \lambda(k+k^{-1})/2 & 0 & -\gamma_2 \\ \gamma_1 & \gamma_2 & 0 \end{pmatrix} \equiv A$$

On the other hand, the Hessian of the selection field is symmetric so its antisymmetric part is zero. So from (3.6):

$$(4.15)\quad [AH_p(X^m)]^{\mathfrak{B}_c} = [AH_p(\mathrm{Rec})]^{\mathfrak{B}_c} = \begin{pmatrix} 0 & -\frac{\ell\tilde{d}\,An}{Sn} & 0 \\ \frac{\ell\tilde{d}\,An}{Sn} & 0 & 0 \\ 0 & 0 & 0 \end{pmatrix} \equiv B$$

But these two matrices are related by ${}^{\mathfrak{B}_e}[I]^{\mathfrak{B}_c} = O$ (cf. (1.40)):

$$(4.16)\qquad [AH_p(X^m)]^{\mathfrak{B}_c} = O^t[AH_p(X^m)]^{\mathfrak{B}_e}O, \qquad \text{i.e. } B = O^tAO.$$

In particular, since $O^t = O^{-1}$ the matrices of (4.14) and (4.15) have the same characteristic polynomial because

$$\det(sI-B) = \det(sI-O^tAO) = \det(sI-O^{-1}AO) = \det(O^{-1}(sI-A)O)$$

$$= \det(O^{-1})\det(sI-A)\det(O) = \det(sI-A).$$

The characteristic polynomials are $s(s^2 + \gamma_1^2 + \gamma_2^2 + (\frac{\lambda(k+k^{-1})}{2})^2$ and $s(s+(\ell\tilde{d}\,An/Sn)^2)$ respectively. So we have

$$(4.17\qquad \gamma_1^2 + \gamma_2^2 + (\frac{\lambda(k+k^{-1})}{2})^2 = (\frac{\ell\tilde{d}\,An}{Sn})^2.$$

and the eigenvalues of these matrices are $\pm i\ \ell\tilde{d}\,An/Sn$ and 0. Now look at the eigenvectors associated to 0.

The matrix B annihilates Ang or equally the multiple $(\ell\tilde{d}\,An/Sn)$Ang whose $\mathfrak{B}_c$ coordinates are given by:

$${}^{\mathfrak{B}_c}[(\ell\tilde{d}\,An/Sn)Ang] = \begin{pmatrix} 0 \\ 0 \\ \frac{\ell\tilde{d}\,An}{Sn} \end{pmatrix}$$

Consequently, from (4.16) the matrix A annihilates the column vector:

$$(4.18)\qquad {}^{\mathfrak{B}_e}[(\ell\tilde{d}\ \mathrm{An}/\mathrm{Sn})\mathrm{Ang}] = O\ {}^{\mathfrak{B}_c}[(\ell\tilde{d}\ \mathrm{An}/\mathrm{Sn})\mathrm{Ang}] = \frac{\ell\tilde{d}\ \mathrm{An}}{\mathrm{Sn}}\begin{pmatrix} o_{13} \\ o_{23} \\ o_{33} \end{pmatrix}$$

On the other hand, by inspection A annihilates

$$(4.19)\qquad \begin{pmatrix} \gamma_2 \\ -\gamma_1 \\ \frac{\lambda(k+k^{-1})}{2} \end{pmatrix}$$

and since 0 is a single root it only annihilates multiples of this vector. But by (4.17) the vectors of (4.18) and (4.19) have the same length (since $\mathfrak{B}_e$ is an orthonormal basis). So

$$\frac{\ell\tilde{d}\ \mathrm{An}}{\mathrm{Sn}}\begin{pmatrix} o_{13} \\ o_{23} \\ o_{33} \end{pmatrix} = \pm\begin{pmatrix} \gamma_2 \\ -\gamma_1 \\ \frac{\lambda(k+k^{-1})}{2} \end{pmatrix}$$

To complete the proof of (4.12) we must show that the sign is positive. This follows from the fact that $\mathfrak{B}_e$ and $\mathfrak{B}_c$ induce the same orientation and a technical argument which we will now sketch.

Consider the path of antisymmetric linear maps $L_t\colon \mathbb{R}_0^4 \to \mathbb{R}_0^4$ with ${}^{\mathfrak{B}_e}[L_t]^{\mathfrak{B}_e} = (A_t)^t$ where

$$A_t = \begin{pmatrix} 0 & -\lambda(k+k^{-1})/2 & -t\gamma_1 \\ \lambda(k+k^{-1})/2 & 0 & -t\gamma_2 \\ t\gamma_1 & t\gamma_2 & 0 \end{pmatrix}$$

0 is a simple eigenvalue for each L_t with eigenvector V_t satisfying:

$$\mathfrak{B}_c[v_t] = \begin{pmatrix} t\gamma_2 \\ -t\gamma_1 \\ \lambda(k+k^{-1})/2 \end{pmatrix}.$$

Since $\lambda(k+k^{-1})/2 > 0$ v_t pierces the e_u e_v plane in the same direction as e_w for all t between 0 and 1.

When $t = 0$, L_0 is a negative 90° rotation in the e_u e_v plane times the positive scalar multiple $\lambda(k+k^{-1})/2$:

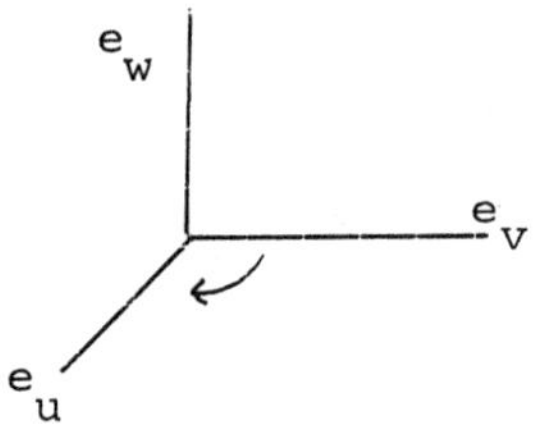

$t = 0$

When $t = 1$ the matrix of L_1 with respect to $\mathfrak{B}_c$ is B and so induces a negative (or positive) 90° rotation on the Ver-Rad plane if $\tilde{d}$ is positive (or negative).

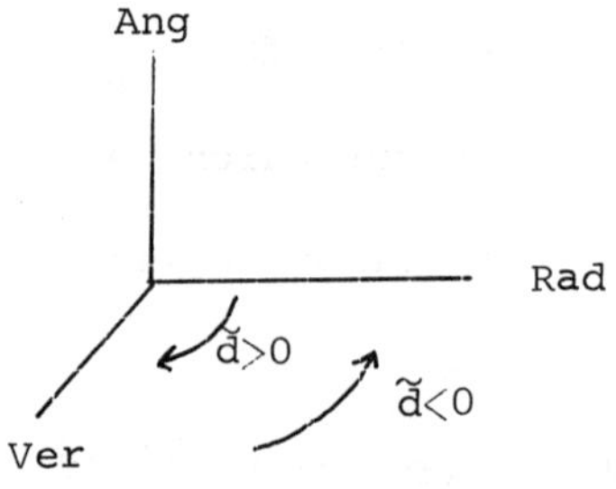

$t = 1.$

So v_1 induces the same orientation on the Ver-Rad plane as Ang if $\tilde{d} > 0$ and the opposite orientation if $\tilde{d} < 0$. In other words, v_1 points in th same direction as $\tilde{d}$ Ang. QED

6 <u>Theorem (Parametrization Condition</u> S): Let $(m,p) \in \Sigma_C$ and O be defined by (4.10) and $[H_p(X^m)]^{\mathfrak{A}_e}$ by (4.11). Then the matrix of $\tilde{\mathfrak{B}}$-parameters of selection is given by the matrix equation:

$$(4.20)\quad [H(M)]^{\mathfrak{B}} = -[H(\Delta)]^{\mathfrak{B}} - [SH(Rec)]^{\mathfrak{A}} + \varepsilon\, {}^{\mathfrak{B}*}[I]^{\mathfrak{B}} + \mu A_0(O) + A_1(O)$$

where the matrices:

$$[H(M)]^{\mathfrak{B}} = \begin{pmatrix} D_A & I & L \\ I & D_B & K \\ L & K & J \end{pmatrix} \quad \text{(cf. (3.14))}$$

$$[H(\Delta)]^{\mathfrak{B}} = \tilde{d}\begin{pmatrix} \tilde{\ell}_z & \tilde{\ell}_n & \tilde{\ell}_x \\ \tilde{\ell}_n & \tilde{\ell}_z & \tilde{\ell}_y \\ \tilde{\ell}_x & \tilde{\ell}_y & \tilde{\ell}_z \end{pmatrix} \quad \text{(cf. (3.21))}$$

$${}^{\mathfrak{B}*}[I]^{\mathfrak{B}} = \begin{pmatrix} \ell^2 & \ell_z & \ell_y \\ \ell_z & \ell^2 & \ell_x \\ \ell_y & \ell_x & \ell^2 \end{pmatrix} \quad \text{(cf. (2.23))}$$

$$(4.21)\, A_1(O) = (O\, {}^{\mathfrak{B}}c[I]^{\mathfrak{B}})^t \begin{pmatrix} 0 & \lambda(k-k^{-1})/2 & \gamma_1 \\ \lambda(k-k^{-1})/2 & 0 & \gamma_2 \\ \gamma_1 & \gamma_2 & 0 \end{pmatrix} O\, {}^{\mathfrak{B}}c[I]^{\mathfrak{B}}$$

$$A_0(O) = (O\, {}^{\mathfrak{B}}c[I]^{\mathfrak{B}})^t \begin{pmatrix} 0 & 0 & 0 \\ 0 & 0 & 0 \\ 0 & 0 & 1 \end{pmatrix} O\, {}^{\mathfrak{B}}c[I]^{\mathfrak{B}}.$$

$$ {}^{\mathfrak{B}}c_{[I]}{}^{\mathfrak{B}} = \begin{pmatrix} \frac{\ell_y}{\ell} & \frac{\ell_x}{\ell} & \ell \\ \frac{x-yz}{An\ Sn} & \frac{y-xz}{An\ Sn} & 0 \\ -\frac{y}{An} & \frac{x}{An} & 0 \end{pmatrix} \quad \text{(cf. (2.42))} $$

$[SH(Rec)]^{\mathfrak{B}} =$ (cf. (3.7))

$$ \begin{pmatrix} -(\frac{\ell_y}{\ell})^2 + \frac{2\ell_y(x-yz)\tilde{d}}{Sn^2} & -\frac{\ell_x\ell_y}{\ell^2} + \frac{[\ell_x(x-yz)+\ell_y(y-xz)]\tilde{d}}{Sn^2} & -\ell_y + \frac{\ell^2(x-yz)\tilde{d}}{Sn^2} \\ -\frac{\ell_x\ell_y}{\ell^2} + \frac{[\ell_x(x-yz)+\ell_y(y-xz)]\tilde{d}}{Sn^2} & -(\frac{\ell_x}{\ell})^2 + \frac{2\ell_x(y-xz)\tilde{d}}{Sn^2} & -\ell_x + \frac{\ell^2(y-xz)\tilde{d}}{Sn^2} \\ -\ell_y + \frac{\ell^2(x-yz)\tilde{d}}{Sn^2} & -\ell_x + \frac{\ell^2(y-xz)\tilde{d}}{Sn^2} & -\ell^2 \end{pmatrix} $$

=

[SH(Rec)] second version

$$ \begin{pmatrix} -(\frac{\ell_y}{\ell})^2 + \frac{2\ell_y(x-yz)\tilde{d}}{Sn^2} & -\frac{\ell_x\ell_y}{\ell^2} + \frac{\tilde{d}}{Sn^2} - \frac{\ell^2(1-z^2)\tilde{d}}{Sn^2} & -\ell_y + \frac{\ell^2(x-yz)\tilde{d}}{Sn^2} \\ -\frac{\ell_x\ell_y}{\ell^2} + \frac{\tilde{d}}{Sn^2} - \frac{\ell^2(1-z^2)\tilde{d}}{Sn^2} & -(\frac{\ell_x}{\ell})^2 + \frac{2\ell_x(y-xz)\tilde{d}}{Sn^2} & -\ell_x + \frac{\ell^2(y-xz)\tilde{d}}{Sn^2} \\ -\ell_y + \frac{\ell^2(x-yz)\tilde{d}}{Sn^2} & -\ell_x + \frac{\ell^2(y-xz)\tilde{d}}{Sn^2} & -\ell^2 \end{pmatrix} $$

Proof: This formula comes from the symmetric part of the Hessian $SH = \frac{1}{2}(H + H^*)$ and so $[SH] = \frac{1}{2}([H] + [H]^t)$.

From (4.11), $[SH_p(X^m)]^{\mathfrak{B}}e$ can be written as the sum of three terms:

$$\epsilon I_3 + \begin{pmatrix} 0 & \lambda(k-k^{-1})/2 & \gamma_1 \\ \lambda(k-k^{-1})/2 & 0 & \gamma_2 \\ \gamma_1 & \gamma_2 & 0 \end{pmatrix} + \mu \begin{pmatrix} 0 & 0 & 0 \\ 0 & 0 & 0 \\ 0 & 0 & 1 \end{pmatrix}.$$

To get $[SH_p(X^m)]^{\mathfrak{B}}$ we apply (1.40) with ${}^{\mathfrak{B}}e_{[I]}{}^{\mathfrak{B}} = {}^{\mathfrak{B}}e_{[I]}{}^{c}\,{}^{\mathfrak{B}}c_{[I]}{}^{\mathfrak{B}} = O\,{}^{\mathfrak{B}}c_{[I]}{}^{\mathfrak{B}}$. ${}^{\mathfrak{B}}c_{[I]}{}^{\mathfrak{B}}$ is given by (2.42). Applying (1.40) to the ϵI_3 term we note that

$$(O\,{}^{\mathfrak{B}}c_{[I]}{}^{\mathfrak{B}})^t I_3 O\,{}^{\mathfrak{B}}c_{[I]}{}^{\mathfrak{B}} = ({}^{\mathfrak{B}}c_{[I]}{}^{\mathfrak{B}})^t O^t O\,{}^{\mathfrak{B}}c_{[I]}{}^{\mathfrak{B}}.$$

But O^tO is the identity and $({}^{\mathfrak{B}}c_{[I]}{}^{\mathfrak{B}})^t = {}^{\mathfrak{B}*}[I]^{\mathfrak{B}c}$ by (1.38) so the chain of equalities continues

$$= {}^{\mathfrak{B}*}[I]^{\mathfrak{B}c}\,{}^{\mathfrak{B}c}c_{[I]}{}^{\mathfrak{B}} = {}^{\mathfrak{B}*}[I]^{\mathfrak{B}}.$$

The last is given by (2.23). So we have

$$[SH_p(X^m)]^{\mathfrak{B}} = \epsilon\,{}^{\mathfrak{B}*}[I]^{\mathfrak{B}} + A_1(O) + \mu(A_0(O).$$

On the other hand, $[SH_p(X^m)]^{\mathfrak{B}}$ consists of three terms which we have already computed namely the symmetrization of $[H_p(Rec)]^{\mathfrak{B}}$ given by (3.7) and two selection terms $[H(M)]^{\mathfrak{B}}$ given by (3.14) and $[H(\Delta)]^{\mathfrak{B}}$ given by (3.21), because p is an equilibrium point. So

$$[SH_p(X^m)]^{\mathfrak{B}} = [H(M)]^{\mathfrak{B}} + [H(\Delta)]^{\mathfrak{B}} + [SH(Rec)]^{\mathfrak{B}}.$$

Setting these two terms equal and solving for $[H(M)]^{\mathfrak{B}}$ gives (4.20).

The second formula for $[SH(Rec)]^{\mathfrak{B}}$ comes from substitutions using (2.26) in the 12 and 21 positions. QED

Since λ, γ_1 and γ_2 are given by (4.12) this equation gives explicit, though messy, formulae for the $\tilde{\mathfrak{B}}$ parameters D_A, D_B, I, J K, L in terms of O, k, p, ε and μ. The remaining $\tilde{\mathfrak{B}}$ parameters of selection, namely A and B, are obtained from (3.17). Then the selection matrix m_{ij} comes from the table in Prop. 3.3 with arbitrary Y.

To actually write out the entries of $A_1(O)$ for general O leads to expressions complicated enough that it is difficult to interpret them. So we won't bother. However $A_0(O)$ is not too bad and we will need it.

First, if $B = O^{\mathfrak{B}_C}[I]^{\mathfrak{B}}$ and $U_{ij} = 1$ when ij = 33 and 0 otherwise then

$$(4.22) \qquad A_0(O) = B^tUB = \begin{pmatrix} (b_{31})^2 & b_{31}b_{32} & b_{31}b_{33} \\ b_{31}b_{32} & (b_{32})^2 & b_{32}b_{33} \\ b_{31}b_{33} & b_{32}b_{33} & (b_{33})^2 \end{pmatrix}$$

So we only need to know the third row of B:

$$(4.23) \qquad \begin{aligned} b_{31} &= O_{31}\frac{\ell_y}{\ell} + O_{32}\frac{(x-yz)}{An} - O_{33}\frac{y}{An} \\ b_{32} &= O_{31}\frac{\ell_x}{\ell} + O_{32}\frac{(y-xz)}{An} + O_{33}\frac{x}{An} \\ b_{33} &= O_{31}\ell. \end{aligned}$$

Finally, we have to deal with the CIS = TRANS equation (3.18).

7 Theorem (Parametrization Condition CT): Let $(m,p) \in \Sigma_C$ and O be defined by (4.10). Using the notation of Thm. 6, we define the expressions

(4.24) $Dn = \tilde{d}(\ell O_{31})^2$

$$+ (O_{31}(x\ell + \frac{\ell_x}{\ell}) + O_{32}\frac{y-xz}{An\ Sn} + O_{33}\frac{x}{An})(O_{31}(y\ell + \frac{\ell_y}{\ell}) + O_{32}\frac{x-yz}{An\ Sn} - O_{33}\frac{y}{An}).$$

(4.25) $Nm = \tilde{d}\ell^2 - [A_1(O)_{12} + xA_1(O)_{13} + yA_1(O)_{23} + zA_1(O)_{33}].$

Then the following equation is true.

(4.26) $$Nm - \mu\ Dn = 0.$$

In particular, if $Dn \neq 0$ then

(4.27) $$\mu = Nm/Dn.$$

Proof: For the CIS = TRANS equation we must compute I + xL + yK + zJ. By (4.20) it is the sum of six terms:

$$-\tilde{d}(\tilde{\ell}_n + x\tilde{\ell}_x + y\tilde{\ell}_y + z\tilde{\ell}_z) \qquad (1)$$

$$+(\frac{\ell_x \ell_y}{\ell^2} + x\ell_y + y\ell_x + z\ell^2) \qquad (2)$$

$$+(-\frac{\tilde{d}}{Sn^2} + \frac{(1-z^2)\tilde{d}\ell^2}{Sn^2} - \frac{x(x-yz)\tilde{d}\ell^2}{Sn^2} - \frac{y(y-xz)\tilde{d}\ell^2}{Sn^2}) \qquad (3)$$

$$+\varepsilon(\ell_z + x\ell_y + y\ell_x + z\ell^2) \qquad (4)$$

$$+(A_1(O)_{12} + xA_1(O)_{13} + yA_1(O)_{23} + zA_1(O)_{33}) \qquad (5)$$

$$+\mu(A_0(O)_{12} + xA_0(O)_{13} + yA_0(O)_{23} + zA_0(O)_{33}) \qquad (6)$$

Term (1) comes from $-[H(\Delta)]^{\mathfrak{B}}$ and by (3.22) it equals $-\tilde{d}\ell^2$.

Terms (2) and (3) come from $[SH(Rec)]^{\mathfrak{B}}$ in the second form. By (2.28) Term (2) equals $\tilde{d}/Sn^2$. Since $(1-z^2)-x(x-yz)-y(y-xz) = Sn^2$, Term (3) equals

$$- \frac{\tilde{d}}{Sn^2} + \tilde{d}\ell^2.$$

Term (4) comes from $\varepsilon^{\mathfrak{B}*}[I]^{\mathfrak{B}}$ and by (2.27) it equals 0.

So the sum of Terms (1) through (4) is 0.

Term (5) comes from $A_1(O)$ and by (4.25) it equals $\tilde{d}\ell^2$ - Nm.

Term (6) comes from $\mu A_0(O)$ and by (4.22) the coefficient of μ is:

$$b_{31}b_{32} + xb_{31}b_{33} + yb_{32}b_{33} + z(b_{33})^2$$

$$= (b_{32} + xb_{33})(b_{31} + yb_{33}) + (z-xy)(b_{33})^2.$$

Substituting from (4.23) and recalling that $z - xy = \tilde{d}$, we see that the coefficient of μ is Dn.

Equation (3.18) says that $I + xL + yK + zJ = \tilde{d}\ell^2$ or by the above computations $\tilde{d}\ell^2 - Nm + \mu Dn = \tilde{d}\ell^2$, which is (4.26). QED

The Parametrization Theorem amounts to the observation that all the reasoning used to deduce these conditions is reversible.

8 Parametrization Theorem: Define the open subset of the eight-dimensional manifold $\mathring{\Delta} \times SO(3) \times \mathbb{R}^2$:

(4.28) $$S_C = \{(p,O,k,\varepsilon): (x,y) \neq (0,0),\ k > 0,\ \tilde{d}\,O_{33} > 0\}.$$

On S_C define real valued functions λ, γ_1, γ_2 by (4.12), noting that $\lambda > 0$ since $\tilde{d}O_{33} > 0$. Define the matrix valued function $A_1(O)$ by (4.21). Then define the real valued functions Dn and Nm by (4.24) and (4.25). Note that none of these functions depend on ε.

Now define the eight-dimensional submanifold of $S_C \times \mathbb{R}$:

(4.29) $$\tilde{S}_C = \{(p,0,k,\varepsilon,\mu): Nm - \mu Dn = 0\}.$$

On $\tilde{S}_C$ define the real valued functions D_A, D_B, I, J, K, L by the matrix equation (4.20). Define the real valued functions A, B by (3.17). Finally, define the selection matrix m_{ij} (the $\mathfrak{S}$-parameters of selection) by the table (3.13). Define $Y = D_A + D_B - J$ to satisfy the normalization condition $m_{14} = m_{23} = 0$ (see table (3.13)). Different choices of Y yield equivalent selection matrices.

In sum we have defined a map from $\tilde{S}_C$ to the space M of normalized selection matrices. The function sending $(p\ 0,k,\varepsilon,\mu)$ to the pair $(m.p)$ defines a smooth map

$$\sigma_C: \tilde{S}_C \longrightarrow \mathring{\Delta} \times M$$

satisfying the following properties:

(a) If $(m,p) = \sigma_C(p,0,k,\varepsilon,\mu)$ then $(m,p) \in \Sigma_C$ i.e. p is an equilibrium point for X^m and the linearization $L^{(m,p)}$ of X^m at p has a complex eigenvalue. In fact, $\varepsilon \pm i\lambda$ is the conjugate pair of complex eigenvalues and $\varepsilon + \mu$ is the real eigenvalue. Moreover, if the orthonormal basis $\mathfrak{B}_e = \{e_u . e_v . e_w\}$ is defined by the condition that $0 = {}^{\mathfrak{A}}e_{[I]}{}^{\mathfrak{B}}c$ then the matrix of $L^{(m.p)}$ with respect to $\mathfrak{B}_e$ is given by (4.5).

(b) σ_C maps $\tilde{S}_C$ onto Σ_C. Furthermore, if we define $S_H = \{(p,0,k\ \varepsilon\} \in S_C: \varepsilon = 0\}$ and $\tilde{S}_H = \{(p,0,k,\varepsilon,\mu) \in \tilde{S}_C: \varepsilon = 0\}$ then $S_C = S_H \times \mathbb{R}$, $\tilde{S}_C = \tilde{S}_H \times \mathbb{R}$. Defining σ_H to be the restriction of σ_C to $\tilde{S}_H$, σ_H maps $\tilde{S}_H$ onto Σ_H. Furthermore, σ_H maps $\{(p.0,k,\mu): \mu \neq 0\}$ onto the set of Hopf equilibria.

<u>Proof</u>: By definition of A and B equations (3.17) are true. Since

I,J,K,L have been defined by (4.20) the proof of Theorem 7 implies that $I + xL + yK + zJ = \tilde{d}\ell^2$, i.e. (3.18) holds. Consequently, p is an equilibrium point for X^m by Cor. 3.5. This means that by Prop. 3.6, $[H(\Delta)]^{\mathfrak{B}}$ is given by equation (3.21), with $[SH_p(X^m)]^{\mathfrak{B}} = [H(M)]^{\mathfrak{B}} + [H(\Delta)]^{\mathfrak{B}} + [SH(Rec)]^{\mathfrak{B}}$. So by (4.20) $[SH_p(X^m)]^{\mathfrak{B}} = \epsilon^{\mathfrak{B}*}[I]^{\mathfrak{B}} + \mu A_0(O) + A_1(O)$. Defining the basis $\mathfrak{B}_e$ by $O = {}^{\mathfrak{B}_e}[I]^{\mathfrak{B}_c}$ or ${}^{\mathfrak{B}_c}[I]^{\mathfrak{B}_e} = O^{-1} = O^t$ so that the columns of O^t given the $\mathfrak{B}_c$ coordinates of the elements of $\mathfrak{B}_e$ it follows that

$$[SH_p(X^m)]^{\mathfrak{B}_e} = \begin{pmatrix} \epsilon & \lambda(k-k^{-1})/2 & \gamma_1 \\ \lambda(k-k^{-1})/2 & \epsilon & \gamma_2 \\ \gamma_1 & \gamma_2 & \epsilon+\mu \end{pmatrix}.$$

On the other hand since the equations (4.12) are true the proof of Theorem 5 gives

$$[AH_p(X^m)]^{\mathfrak{B}_e} = \begin{pmatrix} 0 & \lambda(k+k^{-1})/2 & -\gamma_1 \\ -\lambda(k+k^{-1})/2 & 0 & -\gamma_2 \\ \gamma_1 & \gamma_2 & 0 \end{pmatrix}.$$

Adding these two equations we get (4.11), which is equivalent to (4.5) with $L = L^{(m,p)}$.

So $(m,p) \in \Sigma_C$ and σ_C maps into Σ_C. That it is onto follows from Theorem 5, 6 and 7 themselves which show how to obtain a set of parameters for any (m,p) in Σ_C. Since $(m,p) \in \Sigma_H$ precisely when $\epsilon = 0$ i.e. the eigenvalues are purely imaginary the rest of (b) is clear.

QED

Remark: In practice we will use not $\tilde{S}_C$ and $\tilde{S}_H$ but

$$\mathring{S}_C = \{(p,O,k,\varepsilon) \in S_C: Dn \neq 0\}$$

(4.30)

$$\mathring{S}_H = \{(p,O,k,\varepsilon) \in \mathring{S}_H: \varepsilon = 0\}$$

and define μ to be the function Nm/Dn. With this definition of μ, σ_C maps $\mathring{S}_C$ into Σ_C and $\mathring{S}_H$ into Σ_H and hits all but an exceptional set of one dimension less than the image.

Notice that on $\mathring{S}_C$ all of the functions are rational functions i.e. ratios of polynomials in the variables p_i, O_{ij}, k and ε, except for An, Sn, $\sqrt{Dt}$ (and $\ell = Sn/\sqrt{Dt}$) which are square roots of simple polynomials in x, y and z or the p_i's. This is important in constructing examples because we can't round off our computations and hope to remain in the submanifold Σ_C or Σ_H. We will see how this works when we construct numerical examples in Chap. III.

We digress to describe the extent to which σ_C fails to be one-to-one. Essentially, if the skewness $k \neq 1$ then $\sigma_C(p,O,k,\varepsilon,\mu)$ is also the image of exactly three other points which can be described by the language of group actions. This rather technical result is included for mathematical completeness and will not be needed later. We will merely sketch the proof.

9 <u>Theorem</u>: Let Z_4 be the cyclic group of order 4 consisting of the multiplicative group of square roots of -1, i.e.

$$Z_4 = \{i, i^2 = -1, i^3 = -i, i^4 = 1\}.$$

Define the 3 × 3 matrix by:

$$J = \begin{pmatrix} 0 & -1 & 0 \\ 1 & 0 & 0 \\ 0 & 0 & 1 \end{pmatrix}$$

Note that $J^3 = J^{-1} = J^t$ and so $J \in SO(3)$ and $J^4 = I_3$. The map $i \to J$ defines a representation of Z_4 in $SO(3)$.

Define the action of Z_4 on

$$i*(p,O,k,\varepsilon) = (p,JO,k^{-1},\varepsilon).$$

The action extends to an action on $\tilde{S}_C$ and restricts to give actions on $S_H, \tilde{S}_H, \mathring{S}_C, \mathring{S}_H$. All of these are free actions.

$$\sigma_C(i*(p,O,k,\varepsilon,\mu)) = \sigma_C(p,O,k,\varepsilon,\mu).$$

Consequently, σ_C can be defined on the quotient space $\tilde{S}_C/\mathbb{Z}_4$ and restricts to define σ_H on $\tilde{S}_H/\mathbb{Z}_4$.

The equation $k = 1$ defines a $\mathbb{Z}_4$ invariant closed codimension 1 submanifold of $\tilde{S}_C$ which we will denote $\tilde{S}_C^1$. So $\tilde{S}_H^1/\mathbb{Z}_4$ is a codimension 1 submanifold of $\tilde{S}_H/\mathbb{Z}_4$. On the complement of this submanifold, i.e. on $(\tilde{S}_H - \tilde{S}_H^1)/\mathbb{Z}_4$, σ_C is one-to-one. On $\tilde{S}_H^1/\mathbb{Z}_4$ σ_C is a circle bundle over its image.

Proof: The idea behind all this is best seen by looking at the basis $\mathfrak{B}_e = \{e_u,e_v,e_w\}$ where $\{e_u,e_v\}$ is a normalized pair for $L^{(m,p)}$ with skewness k. Acting by i replaces the pair $\{e_u,e_v\}$ by $\{-e_v,e_u\}$ with skewness k^{-1}. i^2 yields $\{-e_u,-e_v\}$ with skewness k and i^3 yields $\{e_v,-e_u\}$ with skewness k^{-1}. In each case, e_w is the third vector in the basis. If $k \neq 1$ these are the only choices. If $k = 1$ on the other hand we can extend the action to one of the entire circle

$\{e^{i\theta}: 0 \le \theta < 2\pi\}$ in the complex plane. The only details which should be checked are that the formulae for $A_0(O)$ and $A_1(O)$ are left invariant by the action and hence so are Dn and Nm. This is clear for $A_0(O)$ because JO has the same third row as O (see (4.22) and (4.23)). For $A_1(O)$ one notes that in (4.12) action by i leaves λ alone and replaces (γ_1,γ_2) by $(-\gamma_2,+\gamma_1)$. Since k is replaced by k^{-1}, we replace

$$\begin{pmatrix} 0 & \lambda(k-k^{-1})/2 & \gamma_1 \\ \lambda(k-k^{-1})/2 & 0 & \gamma_2 \\ \gamma_1 & \gamma_2 & 0 \end{pmatrix} \text{ by } \begin{pmatrix} 0 & -\lambda(k-k^{-1})/2 & -\gamma_2 \\ -\lambda(k-k^{-1})/2 & 0 & +\gamma_1 \\ -\gamma_2 & +\gamma_1 & 0 \end{pmatrix}$$

$$= \begin{pmatrix} 0 & -1 & 0 \\ +1 & 0 & 0 \\ 0 & 0 & 1 \end{pmatrix} \begin{pmatrix} 0 & \lambda(k-k^{-1})/2 & \gamma_1 \\ \lambda(k-k^{-1})/2 & 0 & \gamma_2 \\ \gamma_1 & \gamma_2 & 0 \end{pmatrix} \begin{pmatrix} 0 & 1 & 0 \\ -1 & 0 & 0 \\ 0 & 0 & 1 \end{pmatrix}$$

Substituting in formula (4.21) for $A_1(O)$ with O replaced by JO we get the old $A_1(O)$ back since $J^tJ = I_3$. QED

We conclude this section by illustrating the computations for the simplest special case, namely $O = I_3$. This requires that p satisfy the condition $\tilde{d} > 0$.

10 <u>Example</u>: Let $p \in \mathring{\Delta}$ with $(x,y) \neq (0,0)$ and $\tilde{d} > 0$. Let $O = I_3$. For the point (p,I_3,k,ε):

$$(4.30) \qquad \gamma_1 = \gamma_2 = 0, \ \lambda = 2\ell\tilde{d}\ \mathrm{An}/[\mathrm{Sn}(k+k^{-1})], \quad \mathrm{Dn} = -xy/\mathrm{An}^2.$$

$$\mathrm{Nm} = \tilde{d}\ell^2 - \frac{\tilde{d}(k-k^{-1})}{\mathrm{Sn}^2(k+k^{-1})}[1 - \mathrm{Sn}^2\ell^2].$$

Now with μ such that $Nm - \mu\, Dn = 0$, the $\widetilde{\mathfrak{B}}$ parameters of selection are given by:

$$(4.31)\quad D_A = -\tilde{d}\,\tilde{\ell}_z + \left(\frac{\ell_y}{\ell}\right)^2 - \frac{4\tilde{d}(x-yz)\ell_y}{(k^2+1)Sn^2} + \mu\frac{y^2}{An^2} + \varepsilon\ell^2$$

$$D_B = -\tilde{d}\,\tilde{\ell}_z + \left(\frac{\ell_x}{\ell}\right)^2 - \frac{4\tilde{d}(y-xz)\ell_x}{(k^2+1)Sn^2} + \mu\frac{x^2}{An^2} + \varepsilon\ell^2$$

$$I = -\tilde{d}\,\tilde{\ell}_n + \frac{\ell_x\ell_y}{\ell^2} - \frac{2\tilde{d}[(x-yz)\ell_x + (y-xz)\ell_y]}{(k^2+1)Sn^2} - \mu\frac{xy}{An^2} + \varepsilon\ell_z$$

$$K = -\tilde{d}\,\tilde{\ell}_y + \ell_x - \frac{2\tilde{d}(y-xz)\ell^2}{(k^2+1)Sn^2} + \varepsilon\ell_x$$

$$L = -\tilde{d}\,\tilde{\ell}_x + \ell_y - \frac{2\tilde{d}(x-yz)\ell^2}{(k^2+1)Sn^2} + \varepsilon\ell_y$$

$$J = -\tilde{d}\,\tilde{\ell}_z + \ell^2 + \varepsilon\ell^2$$

$$A = \tilde{d}\,\ell_y - xD_A - yI - zL$$

$$Y = D_A + D_B - J$$

$$B = \tilde{d}\,\ell_x - yD_B - xI - zK.$$

<u>Proof</u>: The formulae for γ_1, γ_2 and λ come from (4.12). For the rest we must compute $A_1(0)$ and $A_0(0)$. In (4.23) $b_{31} = -y/An$, $b_{32} = x/An$ and $b_{33} = 0$. So from (4.22):

$$A_0(0) = \begin{pmatrix} \frac{y^2}{An^2} & \frac{-xy}{An^2} & 0 \\ \frac{-xy}{An^2} & \frac{x^2}{An^2} & 0 \\ 0 & 0 & 0 \end{pmatrix}.$$

By (4.21), $A_1(0)$ is the product of $\ell(k-k^{-1})/2$ times

$({}^{\mathfrak{B}}C_{[I]}{}^{\mathfrak{B}})^t V({}^{\mathfrak{B}}C_{[I]}{}^{\mathfrak{B}})$ where

$$V = \begin{pmatrix} 0 & 1 & 0 \\ 1 & 0 & 0 \\ 0 & 0 & 0 \end{pmatrix}.$$

Multiply out to get

$$A_1(0) = \frac{\lambda(k-k^{-1})}{2}\begin{pmatrix} \frac{2\ell_y(x-yz)}{\ell \text{ An Sn}} & \frac{\ell_y(y-xz)+\ell_x(x-yz)}{\ell \text{ An Sn}} & \frac{\ell(x-yz)}{\text{An Sn}} \\ \frac{\ell_y(y-xz)+\ell_x(x-yz)}{\ell \text{ An Sn}} & \frac{2\ell_x(y-xz)}{\ell \text{ An Sn}} & \frac{\ell(y-xz)}{\text{An Sn}} \\ \frac{\ell(x-yz)}{\text{An Sn}} & \frac{\ell(y-xz)}{\text{An Sn}} & 0 \end{pmatrix}$$

Note that $\lambda/\ell\text{An Sn} = \tilde{d}/\text{Sn}^2[\frac{1}{2}(k+k^{-1})]$ and $\lambda\ell/\text{An Sn} = \ell^2\tilde{d}/\text{Sn}^2[\frac{1}{2}(k+k^{-1})]$. So

$$A_1(0) = \frac{k-k^{-1}}{k+k^{-1}}\begin{pmatrix} \frac{2\tilde{d}\ell_y(x-yz)}{\text{Sn}^2} & \frac{\tilde{d}[\ell_y(y-xz)+\ell_x(x-yz)]}{\text{Sn}^2} & \frac{\ell^2\tilde{d}(x-yz)}{\text{Sn}^2} \\ \frac{\tilde{d}[\ell_y(y-xz)+\ell_x(x-yz)]}{\text{Sn}^2} & \frac{2\tilde{d}\ell_x(y-xz)}{\text{Sn}^2} & \frac{\ell^2\tilde{d}(y-xz)}{\text{Sn}^2} \\ \frac{\ell^2\tilde{d}(x-yz)}{\text{Sn}^2} & \frac{\ell^2\tilde{d}(y-xz)}{\text{Sn}^2} & 0 \end{pmatrix}$$

Now the formulae for D_A, B_B, I, K, L, J follow directly from (4.20). Note in combining corresponding terms of $A_1(0)$ and $-[SH(\text{Rec})]^{\mathfrak{B}}$ (in the first form) that

$$\frac{k-k^{-1}}{k+k^{-1}} - 1 = \frac{-2k^{-1}}{k+k^{-1}} = \frac{-2}{k^2+1}.$$

Finally, to compute Nm we note that by (2.26):

$$A_1(0)_{12} = \frac{k-k^{-1}}{k+k^{-1}} \frac{\tilde{d}}{Sn^2}[1 - \ell^2(1 - z^2)].$$

Since $1 - z^2 - x(x-yz) - y(y-xz) = Sn^2$

$$A_1(0)_{12} + A_1(0)_{13} + yA_1(0)_{23} + zA_1(0)_{33} = \frac{k-k^{-1}}{k+k^{-1}}[\frac{\tilde{d}}{Sn^2} - \ell^2\tilde{d}]$$

So the formula for Nm follows directly from (4.25). QED

11 Addendum: With p fixed we can regard Nm as a function of k.

If $1 \leq 2\, Sn^2\, \ell^2$ then $Nm > 0$ for all $k > 0$.

If $1 > 2\, Sn^2\, \ell^2$ then Nm is a decreasing function of k which is negative for k near ∞ and is positive for $k \leq 1$.

Proof: $(k-k^{-1})/(k+k^{-1})$ is a strictly increasing function approaching 1 as $k \to \infty$ and -1 as $k \to 0$. It equals 0 when $k = 1$. So we have from (4

$$\lim_{k\to\infty} Nm = \frac{\tilde{d}}{Sn^2}[2\, Sn^2\, \ell^2 - 1]$$

$$\lim_{k\to 0} Nm = \frac{\tilde{d}}{Sn^2} > 0$$

$$Nm = \tilde{d}\lambda^2 > 0 \text{ when } k = 1.$$

If $1 \leq Sn^2\ell^2$ then Nm is a nondecreasing function of k since the coefficient of $(k-k^{-1})/(k+k^{-1})$ is positive. The first term in Nm is positive and the second term in Nm is nonnegative so $Nm > 0$ for all

If $1 > Sn^2\ell^2$ then the coefficient of $(k-k^{-1})/(k+k^{-1})$ is negative and so Nm is a decreasing function of k. Hence,

$$Nm > \lim_{k\to\infty} Nm = \frac{\tilde{d}}{Sn^2}[2Sn^2\ell^2 - 1]$$

for all k. So if $1 \leq 2\, Sn^2 \ell^2$ Nm is again positive for all k.

Finally, if $1 > 2Sn^2 \ell^2$ then Nm approaches a negative limit as $k \to \infty$ and so Nm takes on negative values for large k. Since Nm is decreasing there is exactly one change of sign which occurs at a value of $k > 1$ since $Nm > 0$ when $k = 1$ and hence for all $k \leq 1$. QED

We are interested in controlling the sign of Nm in constructing examples because for a Hopf attractor $\mu = Nm/Dn < 0$.

Notice that since $\ell^2 = Sn^2/Dt$ and $Dt \leq 1$, $\ell^2 \geq Sn^2$. Also from (2.20) $(1-x^2)(1-y^2) \geq Sn^2$.

5. Symmetric Cases.

We now apply the discussion of Sec. I.6 to describe the π_+ and π_- invariant examples in Σ_C. We begin by describing g-invariance of the selection matrix in terms of the $\widetilde{\mathfrak{B}}$ parameters of selection.

1 Proposition: With $g = \pi_+, \pi_-, \pi_0, \pi_x, \pi_y$, the involutions in the square group defined by Table I. (6.14) the selection matrix m_{ij} is g invariant if and only if the elements of M defined by (3.12) satisfy the equations:

π_+ invariance: $B = A$, $D_B = D_A$ and $L = K$.

π_- invariance: $B = -A$, $D_B = D_A$ and $L = -K$.

π_0 invariance: $A = B = L = K = 0$.

π_x invariance: $B = L = I = 0$.

π_y invariance: $A = K = I = 0$.

Proof: By Thm. I.6.2 (m_{ij}) is g invariant if and only if the 3×3 genetic table, which in this case is (3.13), is invariant under the associated symmetry of the square. For each case this leads to a system of equations reducing to the ones above.

For π_+, for example, reflecting across the $x = y$ diagonal we get: $2B - D_A + D_B - 2L = 2A + D_A - D_B - 2K$, $-2B - D_A + D_B + 2L = -2A + D_A - D_B + 2K$, and $-2A + 2B + 2L - 2K = 2A - 2B - 2L + 2K$. Adding the first two equations we get $D_A = D_B$ which reduces the first two equations to $2B - 2L = 2A - 2K$ or $B - A = L - K$. The third equation says $B - A = K - L$. So $B - A = L - K = 0$ and $B = A$, $L = K$. QED

2 Corollary: Let $(p,O,k,\varepsilon,\mu) \in \tilde{S}_C$ with $(m,p) = \sigma_C(p,O,k,\varepsilon,\mu)$.

(a) If $x = y$ then m is π_+ invariant if and only if $D_B = D_A$ and $L = K$.

(b) If $x = -y$ then m is π_- invariant if and only if $D_B = D_A$ and $L = -K$.

Proof: For (a), $A = B$ follows from $x = y$ and the other two conditions by equations (3.17). (b) is similar. QED

Remark: In this proof we have used the fact that $x = y$ iff $p_2 = p_3$ and this implies $\ell_x = \ell_y$. Also, $x - yz = y - xz$ and $\tilde{\ell}_x = \tilde{\ell}_y$. On the other hand, $x = -y$ iff $p_1 = p_4$ and this implies $x - yz = -(y-xz)$, $\ell_x = -\ell_y$ and $\tilde{\ell}_x = -\tilde{\ell}_y$.

We now describe the invariant examples.

3 Theorem: Let $(p,O,k,\varepsilon,\mu) \in \tilde{S}_C$ with $(m,p) = \sigma_C(p,O,k,\varepsilon,\mu)$. Suppose O has the form:

$$(5.1)\qquad O = \begin{pmatrix} \cos\theta & -\sin\theta & 0 \\ \pm\sin\theta & \pm\cos\theta & 0 \\ 0 & 0 & \pm 1 \end{pmatrix} \quad \text{(signs the same as that of } \tilde{d}).$$

If $x = y$ then m is π_+ invariant.

If $x = -y$ then m is π_- invariant.

<u>Proof</u>: For the purposes of this proof we will say that a 3×3 symmetric matrix A is of π_+ type if $A_{11} = A_{22}$ and $A_{31} = A_{32}$ and is of π_- type if $A_{11} = A_{22}$ and $A_{31} = -A_{32}$. By Cor. 2 we must prove that $x = y$ implies M is of π_+ type and $x = -y$ implies M is of π_- type where $M \equiv [H(M)]^{\mathfrak{B}}$ is the matrix of (3.14).

Suppose $x = y$. We prove that each term on the right of equation (4.20) is a matrix of π_+ type. It then follows that the sum $= M$ is. For $[H(\Delta)]^{\mathfrak{B}}$, $[SH(Rec)]^{\mathfrak{B}}$ and ${}^{\mathfrak{B}*}[I]^{\mathfrak{B}}$ this follows by inspection using the Remark after Cor. 2, i.e. $\ell_x = \ell_y$, $(x-yz) = (y-xz)$ and $\tilde{\ell}_x = \tilde{\ell}_y$. Furthermore, from (4.21) and (4.22) we have:

$$(5.2)\qquad A_0(O) = \frac{x^2}{An^2}\begin{pmatrix} 1 & -1 & 0 \\ -1 & 1 & 0 \\ 0 & 0 & 0 \end{pmatrix}$$

because $O_{31} = O_{32} = 0$, $O_{33}^2 = 1$ and $x = y$. So $A_0(O)$ is of π_+ type.

Finally, we compute $A_1(O)$ using (4.21). Note that by (4.12) $\gamma_1 = \gamma_2 = 0$. Next define

$$(5.3)\qquad U(\theta) = \begin{pmatrix} \sin 2\theta & \cos 2\theta & 0 \\ \cos 2\theta & -\sin 2\theta & 0 \\ 0 & 0 & 0 \end{pmatrix}$$

Using the identities $2 \sin\theta \cos\theta = \sin 2\theta$ and $\cos^2\theta - \sin^2\theta = \cos 2\theta$

it is easy to check that $\pm U(\theta)$ (same sign as $\tilde{d}$) =

$$O^t \begin{pmatrix} 0 & 1 & 0 \\ 1 & 0 & 0 \\ 0 & 0 & 0 \end{pmatrix} O .$$

Consequently from (4.21):

(5.4) $$A_1(O) = \pm \frac{\lambda(k-k^{-1})}{2} ({}^{\mathfrak{B}}c_{[I]}{}^{\mathfrak{B}})^t U(\theta)\, {}^{\mathfrak{B}}c_{[I]}{}^{\mathfrak{B}}$$

with the same sign as d.

Since we are in the x = y case:

(5.5) $${}^{\mathfrak{B}}c_{[I]}{}^{\mathfrak{B}} = \begin{pmatrix} \frac{\ell_x}{\ell} & \frac{\ell_x}{\ell} & \ell \\ \frac{x(1-z)}{An\ Sn} & \frac{x(1-z)}{An\ Sn} & 0 \\ -\frac{x}{An} & \frac{x}{An} & 0 \end{pmatrix}.$$

Direct computation shows that $A_1(O)$ is of π_+ type.

When x = -y the proof is similar. (5.2) is replaced by

(5.6) $$A_0(O) = \frac{x^2}{An^2} \begin{pmatrix} 1 & 1 & 0 \\ 1 & 1 & 0 \\ 0 & 0 & 0 \end{pmatrix}$$

and (5.4) still holds but with (5.5) replaced by:

(5.7) $${}^{\mathfrak{B}}c_{[I]}{}^{\mathfrak{B}} = \begin{pmatrix} -\frac{\ell_x}{\ell} & \frac{\ell_x}{\ell} & \ell \\ \frac{x(1+z)}{An\ Sn} & -\frac{x(1+z)}{An\ Sn} & 0 \\ \frac{x}{An} & \frac{x}{An} & 0 \end{pmatrix}.$$ QED

The importance of the invariant examples comes from the fact that we can explicitly describe the center manifolds. This is contained in the following result which includes a converse to Thm. 3.

4 <u>Theorem</u>: Let $(p,O,k,\varepsilon,\mu) \in \tilde{S}_C$ with $\mu \neq 0$ and let $(m,p) = \sigma_C(p,O,k,\varepsilon,\mu)$.

If $x = y$ (at p) and m is π_+ invariant then $\mathrm{Fix}(\pi_+) = \{q \in \mathring{\Delta}: x = y \text{ at } q\}$ is the center manifold for X^m with respect to the equilibrium p.

If $x = -y$ (at p) and m is π_- invariant then $\mathrm{Fix}(\pi_-) = \{q \in \mathring{\Delta}: x = -y \text{ at } q\}$ is the center manifold for X^m with respect to the equilibrium p.

In each of these cases O has the form (5.1) for some $0 \leq \theta < 2\pi$. In fact, this is true even if $\mu = 0$.

<u>Proof</u>: Let $\mathfrak{B}_e = \{e_u, e_v, e_w\}$ be the orthonormal basis determined by O as in Thm. 4.8. Then the matrix of the linearization $L^{(m,p)}$ with respect to the basis is given by (4.5). Performing the further change of basis to $\mathfrak{B}_u$ given by (4.6) the matrix becomes that of (4.7). Because $\lambda > 0$ formula (4.7) implies that the only subspaces of $\mathbb{R}^4_0$ left invariant by $L^{(m,p)}$ i.e. mapped into themselves are, in addition to the trivial extremes of $\mathbb{R}^4_0$ and the zero subspaces, the plane spanned by u and v = the plane spanned by e_u and e_v and the line spanned by w.

If $x = \pm y$ and m is $\pi_\pm$ invariant then $\mathrm{Fix}(\pi_\pm)$ is an invariant manifold for the vectorfield X^m. This implies that its tangent plane at the equilibrium p is invariant for $L^{(m,p)}$. Since this tangent plane is two dimensional it must be the e_u e_v plane. If $\mu \neq 0$ then the center manifold is defined and is characterized by being an invariant manifold with this tangent plane at p. So $Fix(\pi_{\pm})$ is the center manifold when $\mu \neq 0$.

Whether $\mu = 0$ or not the fact that the tangent plane to $Fix(\pi_{\pm})$ is the e_u e_v plane implies (5.1). This is because the diagonals of the square are composed of four rays from the origin and the origin itself. Since $(x,y) \neq (0,0)$ at p it follows that the tangent plane of $Fix(\pi_{\pm})$ at p is spanned by Ver and Rad. So the orthonormal pairs $\{e_u, e_v\}$ and {Ver, Rad} span the same plane. It follows that the perpendicular vector e_w is perpendicular to the Ver-Rad plane, i.e. $e_w = \pm$ Ang and $\{e_u, e_v\}$ is obtained by a rotation and possibly a reflection from the pair (Ver, Rad). This implies (5.1) with the signs determined by the conditions that $\tilde{d}O_{33} > 0$ and $\det(O) = 1$. QED

So if $p \in \mathring{\Delta}$ ($(x,y) \neq (0,0)$ and $\tilde{d} \neq 0$) and $x = \pm y$ then with a given k and ε we are free to choose O anywhere in the open subset of the three dimensional space SO(3) determined by the conditions $\tilde{d}O_{33} > 0$ and $Dn \neq 0$. However, if we want an invariant example we are cut down from three to one degree of freedom namely the choice of θ in (5.1), with the sign determined by $\tilde{d}$ ($Dn \neq 0$ is automatic).

Cor I.6.4 showed that if m is π_+ invariant then $\rho_+(m)$ is π_- invariant and vice-versa. This suggests that we need only look at π_+ invariant examples. We verify this by showing that the action of the square group on Δ "lifts" to an action of the parameter space $\tilde{S}_C$.

5 Lemma: Let V be a vector space and $g: V \to V$ be a linear isomorphism. If $\mathfrak{B} = \{\xi_1, \ldots, \xi_n\}$ is a basis for V then $g\mathfrak{B}$ denotes the basis $\{g\xi_1, \ldots, g\xi_n\}$. Now let $L: V \to V$ be a linear map and $\mathfrak{B}_1$, $\mathfrak{B}_2$ be bases

for V. The following matrix identity holds:

$$(5.8)\qquad {}^{g\mathfrak{B}_2}[gLg^{-1}]^{g\mathfrak{B}_1} = {}^{\mathfrak{B}_2}[L]^{\mathfrak{B}_1}.$$

In particular,

$$(5.9)\qquad {}^{g\mathfrak{B}_2}[I]^{g\mathfrak{B}_1} = {}^{\mathfrak{B}_2}[I]^{\mathfrak{B}_1}.$$

<u>Proof</u>: Suppose $\mathfrak{B}_1 = \{\xi_1,\dots,\xi_n\}$, $\mathfrak{B}_2 = \{\eta_1,\dots,\eta_n\}$. The matrix $a_{ij} = {}^{\mathfrak{B}_2}_{\ i}[L]^{\mathfrak{B}_1}_{\ j}$ is characterized by the equation

$$L(\xi_i) = \Sigma_j\, a_{ij}\eta_j.$$

But this equation is equivalent to

$$gLg^{-1}(g\xi_i) = \Sigma_j\, a_{ij}g(\eta_j).$$ QED

<u>Remark</u>: Notice that if V is Euclidean and g is an isometry then $(g\mathfrak{B})^* = g(\mathfrak{B}^*)$. In particular, if $\mathfrak{B}$ is orthonormal then $g\mathfrak{B}$ is.

6 <u>Theorem:</u> (a) The square group acts on $M \times \Delta$ by $g(m,p) = (g(m),g(p))$. Σ, Σ_C, Σ_H and the set of Hopf equilibria are invariant subsets of $M \times \Delta$, i.e. if (m,p) lies in one of these sets then $g(m,p)$ does.

(b) Define the group homomorphism (a,v) on the square group to $\mathbb{Z}_2 \times \mathbb{Z}_2$ where $\mathbb{Z}_2$ is the additive group of integers mod 2, i.e. $\mathbb{Z}_2 = \{0,1\}$, by:

$$(5.10)\qquad (a(g),v(g)) = \begin{cases} (0,0) & g = 1,\pi_0 \\ (1,0) & g = \pi_+,\pi_- \\ (0,1) & g = \rho_+,\rho_- \\ (1,1) & g = \pi_x,\pi_y \end{cases}$$

Let $\Delta(a,b,c)$ stand for the 3×3 diagonal matrix with entries a,b,c on the diagonal.

The square group acts on SO(3) by:

$$(5.11)\qquad g(O) = \Delta(1,1,(-1)^{a(g)+v(g)}) O \Delta((-1)^{v(g)},1,(-1)^{a(g)}).$$

(c) The square group acts on $\tilde{S}_C$ by $g(p,O,k,\varepsilon,\mu) = (g(p),g(O),k,\varepsilon,\mu)$. The subsets $\tilde{S}_H$, $\mathring{S}_C$, $\mathring{S}_H$ are invariant subsets. Furthermore, the map $\sigma_C\colon \tilde{S}_C \to \Sigma_C$ is equivariant, i.e. $\sigma_C(g(p,O,k,\varepsilon,\mu)) = g\sigma_C(p,O,k,\varepsilon,\mu)$.

Proof: (a) It is clear that if m_1 and m_2 are equivalent selection matrices, i.e. they differ by a constant independent of i and j, then $g(m_1)$ and $g(m_2)$ defined by I.(6.5) are also equivalent. In fact, they differ by the same constant. So the square group acts on M. By Lemma I.6.3 the map g relates the vector fields X^m and $X^{g(m)}$. In particular, if p is an equilibrium for X^m then $g(p)$ is an equilibrium for $X^{g(m)}$ (cf. I.(6.16)). This says $(m,p) \in \Sigma$ implies $g(m,p) \in \Sigma$. Furthermore, since g is linear, $d_p g = g$ (cf. I.(6.2)) and so if we linearize we get

$$(5.12)\qquad L^{(g(m),g(p))} = gL^{(m,p)}g^{-1}.$$

So the two linearizations have the same eigenvalues. The sets Σ_C, Σ_H, etc. are defined by eigenvalue conditions and so they are invariant.

(b) It can be checked directly that a and v defined by (5.10) are group homomorphisms to $\mathbb{Z}_2$ by noting that the kernels are normal subgroups. We will instead look at the cylindrical frame $\mathfrak{B}_C = \{\text{Ver},\text{Rad},\text{Ang}\}$ at any point $p \in \mathring{\Delta}$ with $(x,y) \neq (0,0)$.

By Remark (b) after Thm. I.6.1, $g\colon \mathring{\Delta} \to \mathring{\Delta}$ is an isometry of the Shahshahani metric for each g in the square group. As remarked on

GPG, p. 37 this implies that for f a real-valued function on $\mathring{\Delta}$:

$$g(\bar{\nabla}_p(f \circ g)) = d_p g(\bar{\nabla}_p(f \circ g)) = \bar{\nabla}_{g(p)} f. \tag{5.13}$$

For example, notice that $L^z(g(p)) = (-1)^{v(g)} L^z(p)$, i.e. $L^z \circ g = \pm L^z$ with the minus sign for $q = \pi_x, \pi_y, \rho_+, \rho_-$ and the plus sign for the rest. (5.13) then implies

$$g(\bar{\nabla} L^z) = (-1)^{v(g)} \bar{\nabla} L^z.$$

Since g is an isometry we can divide by the length at each point and get the first of the following equations:

$$g(\mathrm{Ver}_p) = (-1)^{v(g)} \mathrm{Ver}_{g(p)}$$

$$g(\mathrm{Rad}_p) = \mathrm{Rad}_{g(p)} \tag{5.14}$$

$$g(\mathrm{Ang}_p) = (-1)^{a(g)} \mathrm{Ang}_{g(p)}.$$

The other equations are proved by looking at cases and using Table I. (6.14) together with (5.13) above. In each case prove the correspondence for Rad_0 and Ang_0 by direct computation and then divide by the length.

For example, consider Ang_0 at $g(p)$ which is $x(g(p))\bar{\nabla}_{g(p)}y - y(g(p))\bar{\nabla}_{g(p)}x$. Suppose $g = \rho_+$. Then $x(g(p)) = -y(p)$ and $y(g(p)) = x(p)$ by Table I. (6.14) and so (5.13) implies then $\bar{\nabla}_{g(p)}y = g\bar{\nabla}_p(y \circ g) = g(\bar{\nabla}_p x)$ while $\bar{\nabla}_{g(p)}x = -g(\bar{\nabla}_p y)$. So Ang_0 at $g(p)$ is $g(-y(p)\bar{\nabla}_p x + x(p)\bar{\nabla}_p y) = g(\mathrm{Ang}_0(p))$. So (5.14) is true for Ang with $g = \rho_+$ and $a(\rho_+) = 0$. A similar argument through the eight cases proves (5.14).

There is a fancier argument which avoids the case computations. Notice that any symmetry of the square preserves the origin and maps rays to rays. So any member of the square group maps the origin segment in Δ, defined by $(x,y) = (0,0)$, to itself andit maps each two dimensional slice defined by y/x = constant into another slice defined by y/x = some other constant. Since we already know $g(\mathrm{Ver}) = \pm \mathrm{Ver}$ and since g preserves orthogonality and length it follows that $g(\mathrm{Rad}) = \pm\mathrm{Rad}$ because the tangent space of each slice is spanned by Ver and Rad. This latter sign must always be positive because g preserves the origin segment and so maps the outward pointing vector Rad to an outward pointing vector. Finally, orthogonality then implies $g(\mathrm{Ang}) = \pm \mathrm{Ang}$ and the sign is determined by the signs on Ver, Rad and the condition that g preserve orientation $(g = 1,\pi_x,\pi_0,\pi_y)$ or reverse orientation $(g = \pi_+,\pi_-,\rho_+,\rho_-)$ on $\mathring{\Delta}$.

It follows that v and a are group homomorphisms and so (5.11) defines an action of the square group on SO(3). Notice that by (5.14):

$$(5.15) \qquad {}^{g\mathfrak{B}_c}{}_{[I]}{}^{\mathfrak{B}_c} = \Delta((-1)^{v(g)},1,(-1)^{a(g)}) = {}^{\mathfrak{B}_c}{}_{[I]}{}^{g\mathfrak{B}_c}.$$

The factor $\Delta(1,1,(-1)^{v(g)+a(g)})$ is put in to make $\det(gO) = +1$.

(c): What we will prove is that $(m,p) = \sigma_C(p,O,k,\varepsilon,\mu)$ implies that the parameters $(g(p),g(O),k,\varepsilon,\mu)$ can be chosen for the equilibrium $g(p)$ of $X^{g(m)}$ in the construction of Sec. 4. This implies that $(g(p),g(O),k,\varepsilon,\mu) \in \tilde{S}_C$ and its image under σ_C is $(g(m),g(p))$.

For (p,O,k,ε,μ) the basis $\mathfrak{B}_e = \{e_u,e_v,e_w\}$ is orthonormal at p, $O = {}^{\mathfrak{B}_e}{}_{[I]}{}^{\mathfrak{B}_c}$ and for $L = L^{(m,p)}$ equation (4.5) holds. Since g is an isometry, $g\mathfrak{B}_e = \{ge_u,ge_v,ge_w\}$ and $g\mathfrak{B}_c$ are orthonormal at $g(p)$. Lemma 5 implies that $O = {}^{g\mathfrak{B}_e}{}_{[I]}{}^{g\mathfrak{B}_c}$ while ${}^{g\mathfrak{B}_e}{}_{[gLg^{-1}]}{}^{g\mathfrak{B}_e}$ is given by the matrix of (4.5). Since $L^{(gm,gp)} = gLg^{-1}$ by (5.12) it follows that $\{ge_u,ge_v\}$

is a normalized pair for the linearization of $X^{g(m)}$ at $g(p)$ with the same skewness k, the same ε and the same μ. Now the basis which must be chosen is not necessarily $g\mathfrak{B}_e$ but what we will denote by $(\pm)g\mathfrak{B}_e = \{ge_u, ge_v, \pm ge_w\}$ where the positive sign is chosen if $g\mathfrak{B}_e$ has the same orientation as $\mathfrak{B}_c$, i.e. if g is orientation preserving, and the minus sign if g is orientation reversing. So

$$\Delta(1,1,(-1)^{v(g)+a(g)}) = {}^{(\pm)g\mathfrak{B}_e}[I]^{g\mathfrak{B}_e}.$$

The parameters we have constructed for $X^{g(m)}$ at $g(p)$ are $(g(p), {}^{(\pm)g\mathfrak{B}_e}[I]^{\mathfrak{B}_c}, k, \varepsilon, \mu)$. However:

$$\begin{aligned} {}^{(\pm)g\mathfrak{B}_e}[I]^{\mathfrak{B}_c} &= {}^{(\pm)g\mathfrak{B}_e}[I]^{g\mathfrak{B}_e}\; {}^{g\mathfrak{B}_e}[I]^{g\mathfrak{B}_c}\; {}^{g\mathfrak{B}_c}[I]^{\mathfrak{B}_c} \\ &= \Delta(1,1,(-1)^{v(g)+a(g)}) O \Delta((-1)^{v(g)},1,(-1)^{a(g)}) \\ &= gO. \end{aligned}$$

Since $\tilde{S}_H$ is defined by $\varepsilon = 0$ it is invariant. Finally to see that $\mathring{S}_C$ and $\mathring{S}_H$ are invariant it suffices to see then $Dn \circ g = \pm Dn$. This can be shown directly from (4.24) by computing cases. QED

7 <u>Corollary</u>: Let $(m,p) = \sigma_C(p,O,k,\varepsilon,\mu)$. Suppose that $x = y$ at p and O is of the form (5.1) so that (m,p) is a π_+ invariant example. Then $(\rho_+(m),\rho_+(p)) = \sigma_C(\rho_+(p),\rho_+(O),k,\varepsilon,\mu)$ is a π_- invariant example with $\rho_+(O) = \Delta(1,1,-1) O \Delta(-1,1,1)$. Furthermore, the map $\rho_+ : \mathring{\Delta} \to \mathring{\Delta}$ relates the vectorfields X^m and $X^{\rho_+(m)}$.

This construction is a one-to-one, onto correspondence between π_+ and π_- invariant examples and associated examples exhibit the same dynamical behavior.

Proof: That $X^{\rho_+(m)}$ is π_- invariant and is related to X^m by ρ_+ follow from Cor. I.6.4. The parameter correspondence comes from the previous theorem and the fact that $a(\rho_+) = 0$, $v(\rho_+) = 1$. Since $\tilde{d}$ at $\rho_+(p)$ is the negative of $\tilde{d}$ at p it follows that $\rho_+ O = \Delta(1,1,-1) O \Delta(-1,1,1)$ has the form (5.1), too, with θ replaced by $\theta + \frac{\pi}{2}$.

Since the vectorfields X^m and $X^{\rho_+(m)}$ are related by ρ_+, all of the dynamical behavior of X^m, eg. equilibria, cycles etc. are mapped by ρ_+ to corresponding behavior of $X^{\rho_+(m)}$. QED

6. MARMC

We begin with the formula of Marsden and McCracken.

Suppose that (x_1,x_2,x_3) are a coordinate system near a point p and that (X^1,X^2,X^3) are the components of a smooth vectorfield with equilibrium at p. So

(6.1)
$$\frac{dx_1}{dt} = X^1(x_1,x_2,x_3)$$
$$\frac{dx_2}{dt} = X^2(x_1,x_2,x_3)$$
$$\frac{dx_3}{dt} = X^3(x_1,x_2,x_3)$$

is the associated differential equation and since p is an equilibrium:

(6.2)
$$X^1 = X^2 = X^3 = 0 \quad \text{at} \quad p.$$

We will use subscripts to represent partial derivatives at p so:

$$(6.3)\qquad X^i_j \equiv \frac{\partial X^i}{\partial x_j} \qquad X^i_{jk} = \frac{\partial^2 X^i}{\partial x_j \partial x_k}$$

$$X^i_{jk\ell} = \frac{\partial^3 X^i}{\partial x_j \partial x_k \partial x_\ell} \qquad (i,j,k,\ell = 1,2,3)$$

with all partial derivatives evaluated at p.

We assume that the linearization of the vectorfield at p is given by (4.7) (with $\varepsilon = 0$) with respect to the coordinates (x_1, x_2, x_3). This says:

$$(6.4)\qquad \begin{pmatrix} X^1_1 & X^1_2 & X^1_3 \\ X^2_1 & X^2_2 & X^2_3 \\ X^3_1 & X^3_2 & X^3_3 \end{pmatrix} = \begin{pmatrix} 0 & \lambda & 0 \\ -\lambda & 0 & 0 \\ 0 & 0 & \mu \end{pmatrix} \qquad (\lambda > 0 \text{ and } \mu \neq 0).$$

The formula for MARMC is given in [7; Sec. 4A]. I am defining MARMC to be v'''(0), computed there, times the positive factor $4\lambda/3\pi$. It is given in three pieces:

$$(6.5)\ \text{MARMC} = [\hat{X}^1_{111} + \hat{X}^1_{122} + \hat{X}^2_{112} + \hat{X}^2_{222}] + \frac{1}{\lambda}[-X^1_{11}\cdot X^1_{12} + X^2_{22}\cdot X^2_{12} + X^2_{11}\cdot X^2_{12} - X^1_{22}\cdot X^1_{12} + X^1_{11}\cdot X^2_{11} - X^1_{22}\cdot X^2_{22}],$$

where

$$(6.6)\qquad \hat{X}^i_{jk\ell} = X^i_{jk\ell} + X^i_{j3}\cdot f_{k\ell} + X^i_{k3}\cdot f_{j\ell} + X^i_{\ell 3}\cdot f_{jk}$$

$$(i,j,k,\ell = 1,2)$$

and finally, the f_{ij} are given by the matrix formulae:

(6.7) $$\begin{pmatrix} f_{11} \\ f_{12}=f_{21} \\ f_{22} \end{pmatrix} = \frac{-1}{\Delta}\begin{pmatrix} 2\lambda^2+\mu^2 & -\lambda\mu & 2\lambda^2 \\ \lambda\mu & \mu^2 & -\lambda\mu \\ 2\lambda^2 & 2\lambda\mu & 2\lambda^2+\mu^2 \end{pmatrix}\begin{pmatrix} X^3_{11} \\ X^3_{12} \\ X^3_{22} \end{pmatrix}$$

(6.8) $$\Delta = \mu(\lambda^2 + \mu^2).$$

What is going on here is that the center manifold is the graph of a function $x_3 = f(x_1,x_2)$ and MARMC (6.5) is computed from the two-dimensional system which is the restriction to the center manifold. (6.6) to (6.8) are formulae to relate the computation of the partials of the restricted vectorfield back to the partials of the original one.

If the original system is two dimensional then this extra step is not needed, i.e. if (X^1,X^2) is a vectorfield with respect to coordinates (x_1,x_2), so that

(6.9) $$\frac{dx_1}{dt} = X^1(x_1,x_2)$$
$$\frac{dx_2}{dt} = X^2(x_1,x_2)$$
$$X^1 = X^2 = 0 \quad \text{at} \quad p$$

and the linearization satisfies:

(6.10) $$\begin{pmatrix} X^1_1 & X^1_2 \\ X^2_1 & X^2_2 \end{pmatrix} = \begin{pmatrix} 0 & \lambda \\ -\lambda & 0 \end{pmatrix}$$

Then we have:

(6.11) MARMC =

$$[X^1_{111} + X^1_{122} + X^2_{112} + X^2_{222}] +$$

$$\frac{1}{\lambda}[-X^1_{11}\cdot X^1_{12} + X^2_{22}\cdot X^2_{12}$$
$$+ X^2_{11}\cdot X^2_{12} - X^1_{22}\cdot X^1_{12}$$
$$+ X^1_{11}\cdot X^2_{11} - X^1_{22}\cdot X^2_{22}]$$

This computation of MARMC requires not only that p be a Hopf equilibrium but also that the coordinate system (x_1,x_2,x_3) be such that the linearization matrix X^i_j be of the form given by (6.4). In general to apply the formulae we must first make a linear change of coordinates to get this special form.

1 Lemma: Let $(Y^1,Y^2,\ldots,Y^n)$ be the components of a vectorfield with respect to a coordinate system $(y_1,\ldots,y_n)$, i.e. the associated differential equation is:

(6.12) $$\frac{dy_\alpha}{dt} = Y^\alpha(y_1,\ldots,y_n) \qquad \alpha = 1,\ldots,n.$$

Let $(x_1,\ldots,x_n)$ be obtained by a linear change of coordinates:

(6.13)
$$y_\alpha = \Sigma^n_{i=1}\, {}_\alpha A_i x_i$$
$$x_i = \Sigma^n_{\alpha=1}\, {}_i B_\alpha y_\alpha$$

so that A and B are inverse $n \times n$ matrices. Then at any point p we have:

(6.14)
$$X^i = \Sigma^n_{\alpha=1}\, {}_iB_\alpha\, Y^\alpha$$
$$X^i_j = \Sigma^n_{\alpha,\beta=1}\, {}_iB_\alpha\, Y^\alpha_\beta\, {}_\beta A_j$$
$$X^i_{jk} = \Sigma^n_{\alpha,\beta,\gamma=1}\, {}_iB_\alpha Y^\alpha_{\beta\gamma}\, {}_\beta A_j\, {}_\gamma A_k$$
$$X^i_{jk\ell} = \Sigma^n_{\alpha,\beta,\gamma,\delta=1}\, {}_iB_\alpha\, Y^\alpha_{\beta\gamma\delta}\, {}_\beta A_j\, {}_\gamma A_k\, {}_\delta A_\ell \, (i,j,k,\ell = 1,\ldots,n).$$

Proof: Substituting $y = Ax$ in (6.12) and multiplying by B we get:

$$X^i(x_1,\ldots,x_n) = \Sigma_\alpha\, {}_iB_\alpha Y^\alpha(\Sigma_{\beta_1}\, {}_1A_{\beta_1} x_{\beta_1},\ldots,\Sigma_{\beta_n}\, {}_nA_{\beta_n} x_{\beta_n}).$$

Now just differentiate to get the formulae of (6.14).

Notice that the "tensorial" transformations of (6.14) only work because the coordinate change is linear. QED

The second equation in (6.14) says that the two matrix representations of the linearization: (Y^α_β) and (X^i_j) are related by the equation $(X^i_j) = A^{-1}(Y^\alpha_\beta)A$. This is just a variation in notation from equation (1.39). In particular if $n = 3$ and the x coordinate system is chosen so that (6.4) holds then we can compute MARMC from the $Y^\alpha_{\beta\gamma}$'s, $Y^\alpha_{\beta\gamma\delta}$'s, the ${}_iB_\alpha$'s and the ${}_\alpha A_j$'s using (6.5) through (6.8) and (6.14). If $n = 2$ and (6.10) holds then we use (6.11) and (6.14).

Now suppose that $(p,0,k,0,\mu) \in \tilde{S}_H$ with $\mu \neq 0$. We apply these computations with (y_1,y_2,y_3) defined to be the coordinates (x,y,z), i.e. coordinates with respect to the $\mathfrak{B}$ basis for $\mathbb{R}^4_0$, while (x_1,x_2,x_3) will be the coordinates with respect to the $\mathfrak{B}_u$ basis of Addendum 4.3.

To find Y^1, Y^2, Y^3 for X^m we want to write

$$X^m = Y^1\partial_x + Y^2\partial_y + Y^3\partial_z,$$

i.e. we write X^m in terms of the basis $\mathfrak{B}$.

From (3.3) we know that the recombination term $\text{Rec} = -\tilde{d}\partial_z$. On the other hand, equation (3.16) expresses the selection term $\bar{\nabla}(\frac{1}{2}\bar{m})$ in terms of the basis $\mathfrak{B}^*$, i.e. we read off ${}^{\mathfrak{B}^*}[\bar{\nabla}(\frac{1}{2}\bar{m})]$. Using ${}^{\mathfrak{B}}[I]^{\mathfrak{B}^*}$ from (2.18) we get:

(6.15)
$$Y^1 = (A + xD_A + yI + zL)(1 - x^2) + (B + yD_B + xI + zK)(z - xy) + (I + xL + yK + zJ)(y - xz)$$

$$Y^2 = (A + xD_A + yI + zL)(z - xy) + (B + yD_B + xI + zK)(1 - y^2) + (I + xL + yK + zJ)(x - yz)$$

$$Y^3 = (A + xD_A + yI + zL)(y - xz) + (B + yD_B + xI + zK)(x - yz) + (I + xL + yK + zJ)(1 - z^2) - (z - xy).$$

Notice that in these formulae we are thinking of (x,y,z) as varying in $\mathring{\Delta}$. Once we take partial derivatives and evaluate at the equilibrium point p then (x,y,z) will be the values associated with that point p just as in all the formulae for the selection parameters D_A, D_B etc. themselves.

The second partials are given by:

(6.16)
$$Y^1_{11} = -2A - 6D_Ax - 4Iy - 4Lz$$
$$Y^1_{12} = -B + L - 4Ix - 2D_By - 2Kz$$
$$Y^1_{22} = 2K - 2D_Bx$$
$$Y^1_{13} = -4Lx - 2Ky - 2Jz$$
$$Y^1_{23} = D_B + J - 2Kx$$
$$Y^1_{33} = 2K - 2Jx$$

$$Y^2_{11} = 2L - 2D_A y$$

$$Y^2_{12} = -A + K - 2D_A x - 4I_y - 2Lz$$

$$Y^2_{22} = -2B - 4Ix - 6D_B y - 4Kz$$

$$Y^2_{13} = D_A + J - 2Ly$$

$$Y^2_{23} = -2Lx - 4Ky - 2Jz$$

$$Y^2_{33} = 2L - 2Jy$$

$$Y^3_{11} = 2I - 2D_A z$$

$$Y^3_{12} = 1 + D_A + D_B - 2Iz$$

$$Y^3_{22} = 2I - 2D_B z$$

$$Y^3_{12} = -A + K - 2D_A x - 2Iy - 4Lz$$

$$Y^3_{23} = -B + L - 2Ix - 2D_B y - 4Kz$$

$$Y^3_{33} = -2I - 4Lx - 4Ky - 6Jz$$

From these formulae it is easy to write down the formulae for the third partials $Y^{\alpha}_{\beta\gamma\delta}$ $(\alpha,\beta,\gamma,\delta) = 1,2,3)$.

We have not bothered to write down the formulae for the first partials Y^{α}_{β}. This is because the $\tilde{\mathfrak{B}}$ parameters of selection are defined by the given element $(p,0,k,0,\mu)$ of $\tilde{S}_H$. Letting $L^{(m,p)}$ denote the linearization of X^m at p we know that

$$_{\alpha}^{\mathfrak{B}}[L^{(m,p)}]_{\beta}^{\mathfrak{B}} = Y^{\alpha}_{\beta}$$

But by definition of the parametrization ${}^{\mathfrak{B}}{}_{e}[L^{(m,p)}]^{\mathfrak{B}}{}_{e}$ is given by (4.5) with $\varepsilon = 0$ where $0 = {}^{\mathfrak{B}}{}_{e}[I]^{\mathfrak{B}}{}_{c}$. Now define $\mathfrak{B}_u = \{u,v,w\}$ by (4.6). We know from (4.7) that we can use (x_1,x_2,x_3) coordinates of the $\mathfrak{B}_u$ basis where $(y_1,y_2,y_3) = (x,y,z)$ are coordinates of the $\mathfrak{B}$ basis. Technically, to get coordinates on $\mathring{\Delta}$ we translate from $\mathbb{R}_0^4$ to Δ by the vector $\xi^n/4 = (1,1,1,1)/4$. So Lemma 1 applies with

$$A = {}^{\mathfrak{B}}[I]^{\mathfrak{B}}{}_u = {}^{\mathfrak{B}}[I]^{\mathfrak{B}}{}_c\, {}^{\mathfrak{B}}{}_c[I]^{\mathfrak{B}}{}_e\, {}^{\mathfrak{B}}{}_e[I]^{\mathfrak{B}}{}_u \tag{6.17}$$

$$B = {}^{\mathfrak{B}}{}_u[I]^{\mathfrak{B}} = {}^{\mathfrak{B}}{}_u[I]^{\mathfrak{B}}{}_e\, {}^{\mathfrak{B}}{}_e[I]^{\mathfrak{B}}{}_c\, {}^{\mathfrak{B}}{}_c[I]^{\mathfrak{B}}.$$

where ${}^{\mathfrak{B}}{}_e[I]^{\mathfrak{B}}{}_u$ and its inverse ${}^{\mathfrak{B}}{}_u[I]^{\mathfrak{B}}{}_e$ are given by (4.8) and (4.9); ${}^{\mathfrak{B}}{}_e[I]^{\mathfrak{B}}{}_c$ and its inverse ${}^{\mathfrak{B}}{}_c[I]^{\mathfrak{B}}{}_e$ are O and O^t; ${}^{\mathfrak{B}}{}_c[I]^{\mathfrak{B}}$ and its inverse ${}^{\mathfrak{B}}[I]^{\mathfrak{B}}{}_c$ are given by (2.42) and (2.43).

Putting these computations all together we can compute MARMC as a function of the parameter p,O,k,μ. The actual formula is clearly too messy to write down explicitly and so MARMC is best described by a program of putting together the various pieces. We will do this in the next chapter.

We now discribe MARMC for the π_+ invariant case of the previous section. Suppose that $(p,O,k,O,\mu) \in \tilde{S}_H$ with $x = y$ and O of the form (5.1). We restrict to the invariant manifold $\mathrm{Fix}(\pi_+)$. Notice that we need not assume $\mu \neq 0$.

On $\mathrm{Fix}(\pi_+)$ we can use $(y_1,y_2) = (x,z)$ as ccordinates as y always equals x there. At each point the tangent space of $\mathrm{Fix}(\pi_+)$ is

$$\{v \in \mathbb{R}_0^4 : v_2 = v_3\} \tag{6.18}$$

and the basis corresponding to the coordinates (x,z) is $\hat{\mathfrak{B}}_+ = \{(\xi^x + \xi^y)/4, \xi^z/4\} = \{\partial_x + \partial_y, \partial_z\}$. Notice that as one moves **according** to $\partial_x + \partial_y$, z remains constat, while x and y each increase at unit speed and so remain equal. On $\mathrm{Fix}(\pi_+)$ we want to write

$$X^m = Y^1(\partial_x + \partial_y) + Y^2\partial_z.$$

Since m is π_+ invariant by Thm. 5.3, it follows from Prop. 5.1 that $A = B$, $D_A = D_B$ and $K = L$ as well as $x = y$ on $\mathrm{Fix}(\pi_+)$. So Y^1 is just Y^1 of (6.15) and Y^2 is Y^3 of (6.15).

(6.19)
$$\begin{aligned}
Y^1 &= (A + xD_A + xI + zK)(1 - x^2 + z - x^2) \\
&\quad + (I + xK + xK + zJ)(x - xz) \\
&= A(1 + z - 2x^2) + D_A(x + xz - 2x^3) \\
&\quad + K(z + z^2 + 2x^2 - 4x^2z) + I(2x - 2x^3) + J(xz - xz^2)
\end{aligned}$$

$$\begin{aligned}
Y^2 &= 2(A + xD_A + xI + zK)(x - xz) \\
&\quad + (I + xK + xK + zJ)(1 - z^2) - (z - x^2) \\
&= A(2x - 2xz) + D_A(2x^2 - 2x^2z) \\
&\quad + K(2x + 2xz - 4xz^2) + I(1 + 2x^2 - z^2 - 2x^2z) \\
&\quad + J(z - z^3) - (z - x^2).
\end{aligned}$$

(6.20)
$$\begin{aligned}
Y^1_{11} &= -4A + 4K - 12D_Ax - 12Ix - 8Kz \\
Y^1_{12} &= D_A + J - 8Kx - 2Jz \\
Y^1_{22} &= 2K - 2Jx
\end{aligned}$$

$$\begin{aligned}
Y^2_{11} &= 2 + 4D_A + 4I - 4D_Az - 4Iz \\
Y^2_{12} &= -2A + 2K - 4D_Ax - 4Ix - 8Kz \\
Y^2_{22} &= -2I - 8Kx - 6Jz.
\end{aligned}$$

Again from these formulae it is easy to write down the partials $Y^{\alpha}_{\beta\gamma\delta}$ $(\alpha,\beta,\gamma,\delta = 1,2)$.

To change variables to (x_1,x_2) so that (6.10) holds we define the reduced cylindrical frame $\hat{\mathfrak{B}}_c = \{\mathrm{Ver},\mathrm{Rad}\}$ and define $\hat{O}$ to be the upper left 2×2 minor of O from (5.1), i.e.

$$\hat{O} = \begin{pmatrix} \cos\theta & -\sin\theta \\ \pm\sin\theta & \pm\cos\theta \end{pmatrix} \quad \text{(signs the same as that of } \tilde{d}\text{).} \tag{6.21}$$

Then by (5.1) if $\hat{\mathfrak{B}}_e = \{e_u,e_v\}$ i.e. the normalized pair itself, then

$$ {}^{\hat{\mathfrak{B}}_e}[I]^{\hat{\mathfrak{B}}_c} = \hat{O} \quad \text{and} \quad {}^{\hat{\mathfrak{B}}_c}[I]^{\hat{\mathfrak{B}}_e} = \hat{O}^t. \tag{6.22}$$

Notice that $\hat{O}$ is an orthogonal 2×2 matrix with $\det \hat{O} = \pm 1$ (sign the same as $\tilde{d}$).

Define $\hat{\mathfrak{B}}_u = \{u,v\} = \{k\, e_u, e_v\}$ via (4.6) so that

$$ {}^{\hat{\mathfrak{B}}_e}[I]^{\hat{\mathfrak{B}}_u} = \begin{pmatrix} k & 0 \\ 0 & 1 \end{pmatrix} \quad \text{and} \quad {}^{\hat{\mathfrak{B}}_u}[I]^{\hat{\mathfrak{B}}_e} = \begin{pmatrix} k^{-1} & 0 \\ 0 & 1 \end{pmatrix}. \tag{6.23}$$

With respect to $\hat{\mathfrak{B}}_u$, the linearization restricted to $\mathrm{Fix}(\pi_+)$ has the form (6.1). So Lemma 1 applies with

$$\begin{aligned} A &= {}^{\hat{\mathfrak{B}}_+}[I]^{\hat{\mathfrak{B}}_u} = {}^{\hat{\mathfrak{B}}_+}[I]^{\hat{\mathfrak{B}}_c}\; {}^{\hat{\mathfrak{B}}_c}[I]^{\hat{\mathfrak{B}}_e}\; {}^{\hat{\mathfrak{B}}_e}[I]^{\hat{\mathfrak{B}}_u} \\ B &= {}^{\hat{\mathfrak{B}}_u}[I]^{\hat{\mathfrak{B}}_+} = {}^{\hat{\mathfrak{B}}_u}[I]^{\hat{\mathfrak{B}}_e}\; {}^{\hat{\mathfrak{B}}_e}[I]^{\hat{\mathfrak{B}}_c}\; {}^{\hat{\mathfrak{B}}_c}[I]^{\hat{\mathfrak{B}}_+} \end{aligned} \tag{6.24}$$

where ${}^{\hat{\mathfrak{B}}_e}[I]^{\hat{\mathfrak{B}}_u}$ and ${}^{\hat{\mathfrak{B}}_u}[I]^{\hat{\mathfrak{B}}_e}$ are given by (6.23) and ${}^{\hat{\mathfrak{B}}_e}[I]^{\hat{\mathfrak{B}}_c}$ and ${}^{\hat{\mathfrak{B}}_c}[I]^{\hat{\mathfrak{B}}_e}$ are given by (6.22). Finally from (2.42) and (2.43) we get:

(6.25)
$$\hat{\mathfrak{B}}_{C[I]}\hat{\mathfrak{B}}_{+} = \begin{pmatrix} \frac{2\ell_x}{\ell} & \ell \\ \frac{2x(1-z)}{An\ Sn} & 0 \end{pmatrix}$$

(6.26)
$$\hat{\mathfrak{B}}_{+[I]}\hat{\mathfrak{B}}_{C} = \begin{pmatrix} 0 & \frac{x\ Sn}{An} \\ \frac{1}{\ell} & \frac{2x^2(1-z)^2}{An\ Sn} \end{pmatrix} .$$

We conclude by stating the cycling result which the computation of MARMC gives us.

2 <u>Theorem</u>: Let $(p,0,k,0,\mu) \in \tilde{S}_H$ with $\mu \neq 0$. Then for all ε, $(p,0,k,\varepsilon,\mu) \in \tilde{S}_C$. Let $(m^\varepsilon,p) = \sigma_C(p,0,k,\varepsilon,\mu)$ so that (m^0,p) is associated with the original list of parameters.

If MARMC > 0 then p is a vague repellor in its center manifold for X^{m^0}. For $\varepsilon < 0$ and sufficiently close to 0, cycles occur in the center manifold of p for X^{m^ε}. The cycles are robust and are repelling in the center manifold.

If MARMC < 0 the p is a vague attractor in its center manifold for X^{m^0}. For $\varepsilon > 0$ and sufficiently close to 0, cycles occur in the center manifold of p for X^{m^ε}. The cycles are robust and are attracting in the center manifold. If, in addition, $\mu < 0$ then the cycles are attractors in $\mathring{\Delta}$, i.e. are limit cycles.

<u>Proof</u>: That $(p,0,k,\varepsilon,\mu) \in \tilde{S}_C$ for all ε follows from (4.29) because the equation $Nm - \mu\ D_u = 0$ is independent of ε. Then X^{m^ε} is a Hopf bifurcation as the parameter ε varies through 0. The results now follow from the discussion in Chap. I, Sec. 3. QED

7. Asymptotic Estimates.

Even for the symmetric examples, the formula for MARMC is too complicated to study directly. In the search for Hopf attractors, one procedure is just to use the rather simple computer program implicit in the previous description of MARMC. We then hunt around plugging in values for the parameters. In the next chapter we will give a list of some examples found this way. Another approach is to fix p, O and $\varepsilon = 0$ and let k vary. Everything then becomes a function of k. In particular, μ and MARMC become rational functions of k and as k approaches ∞ the sign of these functions is the sign of the coefficient the highest power of k. In the symmetric case, at least, these coefficients are not too bad.

We will use the notation $\emptyset(k^n)$ to represent terms in powers of k at most n. In other words, terms collected in $\emptyset(k^n)$ bounded by some constant times k^n. This is usually written $O(k^n)$. We are using $\emptyset$ to avoid confusion with the orthogonal matrix O.

Restricting to the case where $x = y$, $\tilde{d} > 0$ and $O = I_3$ we can apply the formulae of Example 4.10:

$$(7.1) \quad \gamma_1 = \gamma_2 = 0, \qquad \lambda = 2\ell\tilde{d}\,An/[Sn(k+k^{-1})], \qquad Dn = -x^2/An^2.$$

$$(7.2) \qquad Nm = \frac{\tilde{d}}{Sn^2}[2Sn^2\ell^2 - 1] + \frac{2\tilde{d}}{Sn^2(k^2+1)}[Sn^2\ell^2 - 1]$$

$$= \frac{\tilde{d}}{Sn^2}[2Sn^2\ell^2 - 1] + \emptyset(k^{-2})$$

$$\mu = Nm/Dn$$

where we have used $(k-k^{-1})/(k+k^{-1}) = 1 - 2/(k^2 + 1)$ to get (7.2) from (4.30).

(7.3) $$D_A = D_B = -\tilde{d}\tilde{\ell}_z + (\frac{\ell_x}{\ell})^2 - Nm + \varepsilon\ell^2 - \frac{4\tilde{d}x(1-z)\ell_x}{(k^2+1)Sn^2}$$

$$= -\tilde{d}\tilde{\ell}_z + (\frac{\ell_x}{\ell})^2 - Nm + \emptyset(k^{-2}) \qquad \text{(when } \varepsilon = 0).$$

$$I = -\tilde{d}\tilde{\ell}_n + (\frac{\ell_x}{\ell})^2 + Nm + \varepsilon\ell_z - \frac{4\tilde{d}x(1-z)\ell_x}{(k^2+1)Sn^2}$$

$$= -\tilde{d}\tilde{\ell}_n + (\frac{\ell_x}{\ell})^2 + Nm + \emptyset(k^{-2}) \qquad \text{(when } \varepsilon = 0).$$

$$K = L = -\tilde{d}\tilde{\ell}_x + \ell_x + \varepsilon\ell_x - \frac{2\tilde{d}x(1-z)\ell^2}{(k^2+1)Sn^2}$$

$$= -\tilde{d}\tilde{\ell}_x + \ell_x + \emptyset(k^{-2}) \qquad \text{(when } \varepsilon = 0).$$

$$J = -\tilde{d}\tilde{\ell}_z + \ell^2 + \varepsilon\ell^2$$

$$= -\tilde{d}\tilde{\ell}_z + \ell^2 \qquad \text{(when } \varepsilon = 0).$$

$$A = B = \tilde{d}\ell_x - x(D_A + I) - zK. \qquad Y = 2D_A - J.$$

Here we have used the fact that $\mu(x^2/An^2) = \mu(-Dn) = -Nm$.

Since x, z, ℓ_x, etc. are fixed all of the above $\tilde{\mathfrak{B}}$-parameters of selection are of the form constant + $\emptyset(k^{-2})$. Consequently, so are the partial derivatives Y^i_{jk} (6.20) and the third partials $Y^i_{jk\ell}$.

Now we give the asymptotic estimates for μ and MARMC.

1 <u>Theorem:</u> Choose $p \in \mathring{\Delta}$ with $x = y \neq 0$ and $\tilde{d} > 0$, i.e. $z > x^2$. Choose $O = I_3$ and $\varepsilon = 0$. Let $k > 1$ vary in $\mathbb{R}$. Then as functions of k:

(7.4) $$\mu = -\frac{\tilde{d}\,An^2}{x^2 Sn^2}[2\,Sn^2\ell^2 - 1] + \emptyset(k^{-2})$$

(7.5) $$\text{MARMC} = k^4 \frac{x(1-z)}{\tilde{d}An^2 \ell^4} Y^1_{22} \cdot [Y^1_{12} + Y^2_{22} + (\frac{2x(1+x)(1-z)(x-z)}{Sn^2}) Y^1_{22}] + \emptyset(k^3),$$

where the partial derivatives Y^i_{jk} are given by

$$Y^1_{22} = 2K - 2xJ$$

$$Y^1_{12} = D_A - 8xK + (1-2z)J$$

$$Y^2_{22} = -2I - 8xK - 6zJ.$$

Proof: The estimate for μ follows immediately from (7.1) and (7.2).

For MARMC we begin by noting that

(7.6) $$\lambda^{-1} = \frac{Sn(k+k^{-1})}{2\tilde{d}\ An\ \ell} = k \cdot \frac{Sn}{2\tilde{d}\ An\ \ell} + \emptyset(k^{-1}).$$

Now the partial derivatives X^i_{jk}'s and $X^i_{jk\ell}$'s that occur in (6.11) are obtained from the Y^i_{jk}'s and $Y^i_{jk\ell}$'s by transforming as in Lemma 6.1 using the matrices:

$$A = \begin{pmatrix} 0 & \frac{x\ Sn}{An} \\ \frac{1}{\ell} & \frac{2x^2(1-z)^2}{An\ Sn} \end{pmatrix} \begin{pmatrix} k & 0 \\ 0 & 1 \end{pmatrix}$$

$$B = \begin{pmatrix} k^{-1} & 0 \\ 0 & 1 \end{pmatrix} \begin{pmatrix} \frac{2\ell_x}{\ell} & \ell \\ \frac{2x(1-z)}{An\ Sn} & 0 \end{pmatrix}$$

(cf. equations (6.23 -26) and $\hat{O} = I_2$).

Writing out the elements of A and B we get

(7.7) $${}_2A_1 = k \cdot \frac{1}{\ell} \qquad {}_2B_1 = \frac{2x(1-z)}{An\ Sn}$$

$$ {}_1A_2 = \frac{x\ Sn}{An} \qquad {}_2B_1 = \frac{2x(1-z)}{An\ Sn} $$

$$ {}_2A_2 = \frac{2x^2(1-z)^2}{An\ Sn} \qquad {}_1B_1 = k^{-1}\cdot\frac{2\ell_x}{\ell} $$

$$ {}_1A_1 = 0 \qquad {}_2B_2 = 0. $$

Now we apply (6.14) noting that k occurs with power at most 1 in the ${}_iA_j$'s and at most 0 in the ${}_iB_j$'s. Consequently

$$ X^i_{jk\ell} = \Sigma\ {}_iB_\alpha Y^\alpha_{\beta\gamma\delta}\ {}_\beta A_j\ {}_\gamma A_k\ {}_\delta A_\ell = \emptyset(k^3) $$

for $i,j,k,\ell = 1,2$.

$$ X^i_{jk} = \Sigma\ {}_iB_\alpha\ Y^\alpha_{\beta\gamma}\ {}_\beta A_j\ {}_\gamma A_k = \emptyset(k^2) $$

for $i,j,k = 1,2$.

Now we look at the X^i_{jk}'s more closely.

The only way we can actually get a k^2 term in X^i_{jk} is with ${}_iB_\alpha = {}_2B_1$, ${}_\beta A_j = {}_\gamma A_k = {}_2A_1$, i.e. $(i,j,k) = (2,1,1)$ and $(\alpha,\beta,\gamma) = (1,2,2)$ So

(7.8)
$$ \begin{aligned} X^2_{11} &= {}_2B_1 Y^1_{22}({}_2A_1)^2 + \emptyset(k) \\ &= k^2\cdot\frac{2x(1-z)}{An\ Sn\ \ell^2}\ Y^1_{22} + \emptyset(k). \end{aligned} $$

For the remaining X^i_{jk}'s getting even a k^1 term requires special conditions. Among the choices for ${}_iB_\alpha$, ${}_\beta A_j$, ${}_\gamma A_k$, two of the three have to be the high power choices ${}_2B_1$ and ${}_2A_1$ and the remaining choice is one of the two other nonzero entries of the A or B matrix. This leads to six k^1 terms:

(7.9) $$X^2_{12} = {}_2B_1Y^1_{21}\ {}_2A_1\ {}_1A_2 + {}_2B_1Y^1_{22}\ {}_2A_1\ {}_2A_2 + \emptyset(k^0)$$

$$= k\cdot\left[\frac{2x^2(1-z)}{An^2\ \ell}\,Y^1_{12} + \frac{4x^3(1-z)^3}{An^2 Sn^2\ \ell}\,Y^1_{22}\right] + \emptyset(k^0)$$

$$X^1_{11} = {}_1B_2Y^2_{22}(A_{21})^2 + {}_1B_1Y^1_{22}(A_{21})^2 + \emptyset(k^0)$$

$$= k\cdot\left[\frac{1}{\ell}\,Y^2_{22} + \frac{2\ell_x}{\ell^3}\,Y^1_{22}\right] + \emptyset(k^0).$$

The other two terms lead to X^2_{21} which is the same as X^2_{12}. $X^i_{jk} = \emptyset(k^0)$ for the remaining triples (i,j,k).

Now we substitute in formula (6.11) for MARMC. Note that the $X^i_{jk\ell}$ terms are $\emptyset(k^3)$ while inside the second bracket we can only get a k^3 power in terms in which X^2_{11} occur. So

(7.10) $$\text{MARMC} = [\emptyset(k^3)] + \lambda^{-1}[X^2_{11}(X^2_{12} + X^1_{11}) + \emptyset(k^2)]$$

$$= k\cdot\frac{Sn}{2\tilde{d}\ An\ \ell}[X^2_{11}(X^2_{12} + X^1_{11})] + \emptyset(k^3),$$

by (7.6).

Completing the proof of (7.5) is now just a little algebra. From (2.32) we have $An^2 = 2x^2(1-z)$ since $x = y$. Hence from (7.9)

$$X^2_{12} + X^1_{11} = k\cdot\frac{1}{\ell}\Big[Y^1_{12} + Y^2_{22} + \Big(\frac{2x(1-z)^2}{Sn^2} + \frac{2\ell_x}{\ell^2}\Big)Y^1_{22}\Big] + \emptyset(k^0).$$

Now by (2.25), $\ell^2 = Sn^2/Dt$ and so the coefficient of Y^1_{22} above is $2/Sn^2$ times

$$x(1-z)^2 + Dt\ \ell_x = x(1-z)^2 + [x^2(1-z)^2 - x(1-x^2)(1-z)]$$

by (2.30) since $x = y$. Factoring out $x(1-z)$ and simplifying this

$$= x(1-z)(1+x)(x-z).$$

$$X^2_{12} + X^1_{11} = k\cdot \frac{1}{\ell}[Y^1_{12} + Y^2_{22} + (\frac{2x(1+x)(1-z)(x-z)}{Sn^2})Y^1_{22}] + \emptyset(k^0).$$

Substituting this equation and (7.8) in (7.10) yields (7.5). QED

2 Example: Choose $x = y = z$ between 0 and 1 and $O = I_3$. So (cf. (2.8) and (2.12)):

(7.11) $$p_1 = (1+3x)/4 \qquad p_2 = p_3 = p_4 = (1-x)/4$$

$$\tilde{d} = x(1-x) > 0.$$

(a) If $0 < x < .7$ then $\mu < 0$ for all k.

(b) MARMC < 0 for k sufficiently large provided

(7.12) $$(K - xJ)(D_A - 2I - 16xK + (1 - 8x)J) < 0.$$

Proof: Since $x = z$ the coefficient of Y^1_{22} in the brackets of (7.5) is zero. Since $x > 0$ the sign of the coefficient of k^4 in MARMC is the sign of $Y^1_{22}\cdot(Y^1_{12} + Y^2_{22})$. This proves (b).

To prove (a) we note that by Addendum 4.11 after Example 4.10 it suffices to show that $2\,Sn^2\ell^2 - 1 > 0$ when $0 < x < .7$. By (2.25) $\ell^2 = Sn^2/Dt$ so it suffices to show

$$2\,Sn^4 - Dt > 0.$$

By (7.11) and (2.24) $Dt = (1-x)^3(1+3x)$. By (2.21) $Sn^2 = 1 - 3x^2 + 2x^3 = (1-x)^2(1+2x)$.

$$2Sn^4 - Dt = (1-x)^3[2(1-x)(1+2x)^2 - (1+3x)]$$

$$= (1-x)^3[1 + 3x - 8x^3].$$

The cubic in brackets is positive when $x = 0$ and negative when $x = 1$. Since its derivative is $3 - 24x^2$ it has one root between 0 and 1. So $Nm > 0$ and $\mu = Nm/Dn < 0$ provided x is smaller than this root. As the cubic is positive for $x = .7$ the root is larger than .7. Consequently, $\mu < 0$ for $0 < x < .7$. QED

Finally, to get a numerical value of x which satisfies (7.12)—at least for large enough k--we notice that playing with various examples suggest that J usually dominates the other $\tilde{\mathfrak{B}}$-parameters of selection. The coefficient of J^2 in (7.12) is $-x(1-8x)$ which is negative if $x < 1/8$. So we use $x = .1$ and the asymptotic estimates of (7.3):

$x = y = z = .1 \qquad \tilde{d} = x - x^2 = .09$

$p_1 = (1+3x)/4 = .325 \qquad p_2 = p_3 = p_4 = (1-x)/4 = .225$

$\ell_x = \ell_y = \ell_x = (p_1^{-1} - p_2^{-1})/16 \sim -.085$

$\ell^2 = (p_1^{-1} + 3p_2^{-1})/16 \sim 1.026$

$\tilde{\ell}_x = \tilde{\ell}_y = \tilde{\ell}_z = (p_1^{-2} - p_2^{-2})/64 \sim -.16$

$\tilde{\ell}_n = (p_1^{-2} + 3p_2^{-2})/64 \sim 1.074$

$Sn^2 = (1-x)^2(1+2x) = .972.$

$$Nm = \frac{\tilde{d}}{Sn^2}[2\, Sn^2 \ell^2 - 1] + \emptyset(k^{-2}) \sim .092 + \emptyset(k^{-2})$$

$D_A \sim -.07 + \emptyset(k^{-2})$

$I \sim .0025 + \emptyset(k^{-2})$

$K \sim -.07 + \emptyset(k^{-2})$

$J \sim 1.04$

$A = B \sim .006 + \emptyset(k^{-2})$

$(K - xJ)(D_A - 2I - 16xK + (1-8x)J) \sim -.043.$

So we have shown that with $p = (.325,.225,.225,.225)$, $O = I_3$, $\varepsilon = 0$ and k sufficiently large, p is a Hopf attractor for the combined field X^m. Perturbing the field by letting ε be positive but small, we obtain limit cycles.

Precise numerical examples are given in the next chapter.

For the general case where symmetry is not assumed we can get asymptotic results by varying $\tilde{d}$ as well as k. Fix x and y and O. Then everything becomes a function of $\tilde{d}$ and k and we see what happens as $\tilde{d}$ approaches 0 and k approaches ∞.

Now $z = xy + \tilde{d} = xy + \emptyset(\tilde{d})$ and so substituting in the formulae Sec 2:

(7.13)
$$Sn^2 = (1 - x^2)(1 - y^2)$$
$$Dt = (1 - x^2 - y^2 + x^2y^2)^2 = Sn^4$$
$$\ell^2 = 1/Sn^2$$
$$\ell_x = -x/Sn^2$$
$$\ell_y = -y/Sn^2$$

$$An^2 = x^2(1-y^2) + y^2(1-x^2),$$

where all of these equations are true up to addition of $\emptyset(\tilde{d})$.

Now fix $O = I_3$ so that we must regard $\tilde{d}$ as approaching 0 from the positive direction. From Example 4.10:

(7.14) $$\gamma_1 = \gamma_2 = 0, \qquad \lambda = \tilde{d}[2An/Sn^2(k+k^{-1})] + \emptyset(\tilde{d}^2),$$

$$Dn = -xy/An^2, \qquad Nm = \tilde{d}[1/Sn^2] + \emptyset(\tilde{d}^2),$$

$$\mu = \tilde{d}[-An^2/xy\ Sn^2] + \emptyset(\tilde{d}^2)$$

$$D_A = [y^2 + \varepsilon]/Sn^2 + \emptyset(\tilde{d})$$

$$D_B = [x^2 + \varepsilon]/Sn^2 + \emptyset(\tilde{d})$$

$$I = [xy + \varepsilon xy]/Sn^2 + \emptyset(\tilde{d})$$

$$K = [-x - \varepsilon x]/Sn^2 + \emptyset(\tilde{d})$$

$$L = [-y - \varepsilon y]/Sn^2 + \emptyset(\tilde{d})$$

$$J = [1 + \varepsilon]/Sn^2 + \emptyset(\tilde{d})$$

$$A = [-xy^2 - \varepsilon x]/Sn^2 + \emptyset(\tilde{d})$$

$$B = [-x^2y - \varepsilon y]/Sn^2 + \emptyset(\tilde{d}).$$

Furthermore, all of these estimates are uniform in k, i.e. the $\emptyset(\tilde{d})$ terms can be bounded by some constant times $\tilde{d}$ where the constant depends on x and y but is independent of k.

3 _Theorem_: Choose x and y with $|x|,|y| < 1$ and $xy > 0$. Let p in $\mathring{\Delta}$ vary over this value of (x,y) with $\tilde{d} > 0$ and approaching 0. Choose $O = I_3$ and $\varepsilon = 0$. Let $k > 1$ vary in $\mathbb{R}$. Then

$$\mu = \tilde{d}[-An^2/xy\ Sn^2] + \emptyset(\tilde{d}^2)$$

uniformly in k. So for $\tilde{d}$ sufficiently small $\mu < 0$ for all k.

(7.15) $$\text{MARMC} = \tilde{d}^{-1}k^4(-2xy)\cdot\left[\frac{(1-y^2)+(1-x^2)}{An^2} + 15 - 8xy\right] + \emptyset(\tilde{d}^{-1}k^3, \tilde{d}^0k^4).$$

So for $\tilde{d}$ sufficiently small and k sufficiently large MARMC < 0 and p is a Hopf attractor for X^m where $(m,p) = \sigma_H(p, I_3, k, 0)$

Proof: The estimating procedure is similar to that of Thm. 1 with additional complications coming from (6.6) and (6.7).

Each of the X^i_{jk}'s and $X^i_{jk\ell}$'s are of the form constant plus $\emptyset(\tilde{d})$. To get the X^i_{jk}'s and $X^i_{jk\ell}$'s we must use (6.14) and the matrices (cf. (6.17)):

$$A = \begin{pmatrix} 0 & \frac{x\ Sn}{An} & \frac{(\mu^2+\lambda^2)(-y(1-x^2))}{An} \\ 0 & \frac{y\ Sn}{An} & \frac{(\mu^2+\lambda^2)(-x(1-y^2))}{An} \\ kSn & \frac{2x^2y^2\,Sn}{An} & \frac{(\mu^2+\lambda^2)(x^2-y^2)}{An} \end{pmatrix}$$

$$B = \begin{pmatrix} \frac{k^{-1}(-y)}{Sn} & \frac{k^{-1}(-x)}{Sn} & \frac{k^{-1}}{Sn} \\ \frac{x(1-y^2)}{An\ Sn} & \frac{y(1-x^2)}{An\ Sn} & 0 \\ \frac{-y}{An(\mu^2+\lambda^2)} & \frac{x}{An(\mu^2+\lambda^2)} & 0 \end{pmatrix}$$

Here we have substituted using (7.13) and $z = xy$ in (2.42) and (2.43) to get ${}^{\mathfrak{B}}[I]^{\mathfrak{B}}_{c}$ and ${}^{\mathfrak{B}}_{c}[I]^{\mathfrak{B}}$ up to terms of the form $\emptyset(\tilde{d})$.

To compute MARMC using (6.5) we begin by looking at the third partials and prove:

$$\hat{X}^i_{jk\ell} = \emptyset(\tilde{d}^{-1}k^3) \qquad (k,j,k,\ell = 1,2). \tag{7.16}$$

While this is similar to the corresponding result in Thm. 1, more work is needed to get it.

Since the third partials themselves $X^i_{jk\ell}$ involve neither the third column of A nor the third row of B we have

$$X^i_{jk\ell} = \emptyset(k^3) \qquad (i,j,k,\ell = 1,2).$$

The remaining terms are of the form:

$$-\frac{M}{\Delta}\cdot X^3_{ij}\cdot X^k_{\ell 3} \qquad (i,j,k,\ell = 1,2),$$

where M is an entry of the matrix in (6.7). To get $X^k_{\ell 3}$ one third column entry of A is used and to get X^3_{ij} one third row entry of B is used. So the factors of $\mu^2 + \lambda^2$ cancel. Furthermore, the first column of A is used at most once to get $X^k_{\ell 3}$ and so k occurs with power at most 3 in $X^3_{ij}\cdot X^k_{\ell 3}$. Finally, the M/Δ factors are bounded by $\tilde{d}^{-1}$. (7.16) follows.

From (7.14):

$$\frac{1}{\lambda} = \tilde{d}^{-1}\, k[Sn^2/2\ An] + \emptyset(\tilde{d}^{-1}k^0).$$

The rest of the argument is very close to that of Thm. 1. We want to estimate X^i_{jk} $(i,j,k = 1,2)$ and so we only need the first and second rows of B and the first and second columns of A.

The only k^2 terms we can get come from

$$X^2_{11} = {}_2B_1Y^1_{33}({}_3A_1)^2 + {}_2B_2Y^2_{33}({}_3A_1)^2 + \emptyset(k) \tag{7.17}$$

$$= k^2(Sn/An)(Y^1_{33}x(1-y^2) + Y^2_{33}y(1-x^2)) + \emptyset(k).$$

The k^1 terms we can get are:

$$(7.18)\qquad X^1_{11} = {}_1B_1Y^1_{33}({}_3A_1)^2 + {}_1B_2Y^2_{33}({}_3A_1)^2 + {}_1B_3Y^3_{33}({}_3A_1)^2 + \emptyset(k^0)$$

$$= k(Sn)[Y^1_{33}(-y) + Y^2_{33}(-x) + Y^3_{33}] + \emptyset(k^0).$$

and

$$(7.19)\qquad X^3_{12} = {}_2B_1Y^1_{31}\,{}_3A_1\,{}_1A_2 + {}_2B_1Y^1_{32}\,{}_3A_1\,{}_2A_2 + {}_2B_1Y^1_{33}\,{}_3A_1\,{}_3A_2 + {}_2B_2Y^2_{31}\,{}_3A_1\,{}_1A_2 + {}_2B_2Y^2_{32}\,{}_3A_1\,{}_2A_2 + {}_2B_2Y^2_{33}\,{}_3A_1\,{}_3A_2 + \emptyset(k^0$$

$$= k(Sn/An^2)[Y^1_{13}\,x^2(1-y^2) + Y^1_{23}\,xy(1-y^2) + Y^1_{33}\,2x^3y^2(1-y^2) + Y^2_{13}xy(1-x^2) + Y^2_{23}\,y^2(1-x^2) + Y^2_{33}\,2x^2y^3(1-x^2)] + \emptyset(k^0).$$

The remaining terms are $\emptyset(k^0)$.

So from (6.5) we have

$$\text{MARMC} = (\tfrac{1}{\lambda})X^2_{11}(X^2_{12} + X^1_{11}) + \emptyset(\tilde{d}^{-1}k^3).$$

To complete the computation we use (6.16), $z = xy$, and (7.14) (with $\varepsilon = 0$) to get (ignoring $\emptyset(\tilde{d})$ terms):

$$Y^1_{33} = -4x/Sn^2 \qquad Y^2_{33} = -4y/Sn^2 \qquad Y^3_{33} = 0$$

$$Y^1_{13} = 4xy/Sn^2 \qquad Y^1_{23} = (1 + 3x^2)Sn^2$$

$$Y^2_{13} = (1 + 3y^2)/Sn^2 \qquad Y^2_{23} = 4xy/Sn^2$$

$$X^2_{11} = k^2(1/\text{An Sn})(-4x^2(1-y^2)-4y^2(1-x^2)) + \emptyset(k')$$

$$= k^2(-4\ \text{An}/\text{Sn}) + \emptyset(k^1).$$

$$X^1_{11} = k(8xy/\text{Sn}) + \emptyset(k^0)$$

$$X^2_{12} = k(1/\text{An}^2\text{Sn})[4x^3y(1-y^2) + (1+3x^2)xy(1-y^2)$$

$$-8x^4y^2(1-y^2) + (1+3y^2)xy(1-x^2)+4xy^3(1-x^2)-8x^2y^4(1-x^2)]$$

$$+ \emptyset(k^0).$$

Since $\text{An}^2 = x^2(1-y^2) + y^2(1-x^2)$ we have

$$X^1_{11} + X^2_{12} = k(xy/\text{An}^2\text{Sn})[(1-y^2)(1+15x^2-8x^3y)$$

$$+ (1-x^2)(1+15y^2-8xy^3)] + \emptyset(k^0).$$

From this the estimate for MARMC follows by substitution.

If $xy > 0$ then the coefficient of $\tilde{d}^{-1}k^4$ is negative because $xy < 1$ and so

$$15 - 8xy > 7.$$

QED

Chapter III: Programs and Examples

1. The General Program.

We recall, in outline, the nature of the two-locus-two-allele model with alleles A, a at the first locus and B, b at the second.

The four haploid genotypes AB, Ab, aB and ab are numbered 1,...,4. The gene pool is described by a vector $p \in \mathbb{R}^4$ with $p_i \geq 0$ and $\Sigma_i\, p_i = 1$ $(i = 1,\ldots,4)$.

The action of selection on the gene pool is determined by the choice of a 4 by 4 matrix whose entry m_{ij} is the fitness of the ij diploid type $(i,j = 1,\ldots,4)$. Since the ij and ji diploid types are the same, $m_{ij} = m_{ji}$. In the absence of position effects the two double heterozygotes have the same fitness, i.e. $m_{14} = m_{23}$. A symmetric matrix satisfying this additional condition is called a selection matrix.

Every selection matrix determines a vectorfield X^m on the states of the gene pool whose associated differential equation describes the combined effect of selection and recombination:

$$\frac{dp_i}{dt} = p_i(m_i - \bar{m}) - d\xi_i \qquad (i = 1,\ldots,4)$$

(1.1) where

$$\xi_i = \begin{cases} +1 & i = 1,4 \\ -1 & i = 2,3. \end{cases}$$

Here $m_i = \Sigma\, p_j m_{ij}$, $\bar{m} = \Sigma\, p_i m_i$ and $d = p_1 p_4 - p_2 p_3$.

In this section we summarize the work of the preceding chapter by presenting a program for determining examples of selection matrices m_{ij} whose associated dynamical system given by (1.1) exhibits limit cycles arising from a Hopf bifurcation. In other words there exists a nonequilibrium periodic solution of (1.1) such that for initial values near enough to the cycle the solution path approaches the cycle asymptotically as t approaches ∞. Furthermore, these cycles are robust in that a slight change in the selection matrix merely distorts the cycle.

Notice that equation (1.1) has been normalized by the assumption of a recombination rate equal to unity. In other words, the usual equations of selection and recombination are of the form:

$$\frac{dp_i}{dT} = p_i(M_i - \bar{M}) - Rd\,\xi_i \qquad (i = 1,\dots,4) \tag{1.2}$$

where M_{ij} is the selection matrix and R is the recombination rate per unit time T. So $R = r \cdot b > 0$ with r the average number of crossovers per birth and b is the birth rate for the double heter-zygotes.

We can use the normalization because equations (1.1) and (1.2) are related by the time change:

$$t = T/R \quad \text{and} \quad T = Rt, \tag{1.3}$$

provided that:

$$m_{ij} = M_{ij}/R \quad \text{and} \quad M_{ij} = R\, m_{ij}. \tag{1.4}$$

In other words, the choice of the positive number R is completely arbitrary and the m_{ij}'s of our examples represent the ratio between the true fitness values and the recombination rate. A solution of (1.1) with period $2\pi/\lambda$ in t is a solution of (1.2) with period $2\pi R/\lambda$ in T.

The program consists of two parts.

In the first part the input consists of the choice of certain parameters. These are (1) a distribution vector p, (2) a special orthogonal matrix O, i.e. O is a 3×3 matrix with determinant 1 whose transpose equals its inverse and (3) a positive number k.

The program then computes a number of algebraic expressions and imposes certain inequality conditions. These are of two kinds.

Certain inequalities have to hold in order that the selection matrix associated with the parameters be defined. These are $dO_{33} > 0$ (and so $d \neq 0$, i.e. p cannot be in linkage equilibrium) and either $p_A = p_1 + p_2 \neq 1/2$ or $p_B = p_1 + p_3 \neq 1/2$. In addition, polynomials labeled Nm and Dn may not vanish.

If these inequalities hold then the output consists of numbers $\lambda > 0$, $\mu \neq 0$ and a selection matrix $m_{ij}(\varepsilon)$ which is a linear function of a new real variable ε.

The second part of the program uses the $\varepsilon = 0$ output of the first part to compute a number labelled MARMC.

The second kind of inequality--labelled cycling conditions--are required in order to get stable limit cycles. These conditions are $\mu < 0$ and MARMC < 0.

When an example is computed satisfying all of these conditions then with selection matrix $m_{ij}(\varepsilon)$ for $\varepsilon > 0$ and small enough limit cycles occur in (1.1). The cycles are roughly cantered about p at a distance some constant times ε as ε varies near zero. The period is approximately $2\pi/\lambda$ in t. (These estimates come from [7; p. 65 and p. 79].)

Before elaborating the program itself we make some general remarks about it.

The program per se doesn't guarantee that any examples exist. A priori it could be, for example, that MARMC is always positive. However, the results of Chap. II. Sec 7 do guarantee that some solutions exist and points out where to look for them. For others it is a matter of hunting.

Since only the sign of MARMC is important it can be computed as a decimal (i.e. using floating point numbers) provided that the result is large enough that the round off error not affect the sign. However, for the $m_{ij}(\epsilon)$ coefficients themselves exact answers are required. This is because, while the cycles are robust it is not clear how robust, i.e. how small the allowable variation in parameter values can be, until you have it. For this purpose it is useful to use rational values for the parameters p_i, O_{ij} and k. Luckily, all of the computed expressions are rational functions of the parameter values except for three: An, Sn and ℓ. These are square roots of simple polynomials. So the output can be computed as fractions with radicals of whole numbers. This will be illustrated by the list of examples in Sec. 3.

This rationality problem brings us to how to get orthogonal matrices.

A special orthogonal matrix is a general rotation in $\mathbb{R}^3$ and so can be written as the product of three rotations about the coordinate axes, eg. O can be represented:

$$O = O_x(\varphi)O_y(\psi)O_z(\theta) \tag{1.5}$$

$$O_x(\varphi) = \begin{pmatrix} 1 & 0 & 0 \\ 0 & \cos\varphi & -\sin\varphi \\ 0 & \sin\varphi & \cos\varphi \end{pmatrix}$$

$$O_y(\psi) = \begin{pmatrix} \cos\psi & 0 & -\sin\psi \\ 0 & 1 & 0 \\ \sin\psi & 0 & \cos\psi \end{pmatrix}$$

$$O_z(\theta) = \begin{pmatrix} \cos\theta & -\sin\theta & 0 \\ \sin\theta & \cos\theta & 0 \\ 0 & 0 & 1 \end{pmatrix}$$

So choosing an element O amounts to choosing three angular variables θ, ψ, φ. To get rational entries one chooses the angles to be defined by so-called Pythagorean triples:

(1.6) $$\text{sine} = \pm \frac{a}{c} \qquad \text{cosine} = \pm\frac{b}{c}$$

where a,b and c are nonnegative integers such that $a^2 + b^2 = c^2$. For example: (a,b,c) or (b,a,c) =

$$(1,0,1), \quad (3,4,5) \quad \text{or} \quad (5,12,13).$$

The first choices lead to angles which are multiples of 90°.

If you want more triples, there is a classic result of elementary number theory which tells how to get all of them (eg. [10, p. 99]). Choose whole numbers r and s with $r > s$ and let (a,b,c) or (b,a,c) = $(r^2 - s^2, 2rs, r^2 + s^2)$.

Part I.

Step A: Choose $p \in \mathring{\Delta}$, i.e. $p \in \mathbb{R}^4$ with $p_i > 0$ and $\Sigma_i\, p_i = 1$.

Compute:

$$x = p_1 + p_2 - p_3 - p_4$$

$$y = p_1 - p_2 + p_3 - p_4 \qquad \text{(II. (2.4))}$$

$$z = p_1 - p_2 - p_3 + p_4$$

$$\tilde{d} = z - xy \qquad \text{(II. (2.12))}$$

<u>Conditions:</u> $\tilde{d} \neq 0$ and $(x,y) \neq (0,0)$.

<u>Compute</u>:

$$\ell_x = (p_1^{-1} + p_2^{-1} - p_3^{-1} - p_4^{-1})/16$$

$$\ell_y = (p_1^{-1} - p_2^{-1} + p_3^{-1} - p_4^{-1})/16 \qquad \text{(II. (2.22))}$$

$$\ell_z = (p_1^{-1} - p_2^{-1} - p_3^{-1} + p_4^{-1})/16$$

$$\ell = \sqrt{(p_1^{-1} + p_2^{-1} + p_3^{-1} + p_4^{-1})/16}$$

$$\tilde{\ell}_x = (p_1^{-2} + p_2^{-2} - p_3^{-2} - p_4^{-2})/64$$

$$\tilde{\ell}_y = (p_1^{-2} - p_2^{-2} + p_3^{-2} - p_4^{-2})/64 \qquad \text{(II. (3.20))}$$

$$\tilde{\ell}_z = (p_1^{-2} - p_2^{-2} - p_3^{-2} + p_4^{-2})/64$$

$$\tilde{\ell}_n = (p_1^{-2} + p_2^{-2} + p_3^{-2} + p_4^{-2})/64$$

$$Sn = \sqrt{1 - x^2 - y^2 - z^2 + 2xyz} \qquad \text{(II. (2.21))}$$

$$An = \sqrt{x^2 + y^2 - 2xyz} \qquad \text{(II. (2.32))}$$

$$C = \begin{pmatrix} \frac{\ell_y}{\ell} & \frac{\ell_x}{\ell} & \ell \\ \frac{x-yz}{An\ Sn} & \frac{y-xz}{An\ Sn} & 0 \\ \frac{-y}{An} & \frac{x}{An} & 0 \end{pmatrix} \qquad \text{(II. (2.42))}$$

$$C^{-1} = \begin{pmatrix} 0 & \frac{x\ Sn}{An} & -\frac{(y-xz)}{An} \\ 0 & \frac{y\ Sn}{An} & \frac{(x-yz)}{An} \\ \ell^{-1} & \frac{2(x-yz)(y-xz)}{An\ Sn} & \frac{x^2-y^2}{An} \end{pmatrix} \qquad \text{(II. (2.43))}$$

(In the notation of Chap. II, $C = {}^{\mathfrak{B}}{}_{C}[I]^{\mathfrak{B}}$ and C^{-1} is its inverse = ${}^{\mathfrak{B}}[I]^{\mathfrak{B}}{}_{C}$.)

<u>Choose</u>: $O \in SO(3)$, i.e. O is a 3×3 matrix with $O^{-1} = O^t$ and $\det O = 1$.

<u>Compute</u>:

$$Dn = \tilde{d}(O_{31}\ell)^2 + \qquad \text{(II. (4.24))}$$

$$[O_{31}(x\ell+\frac{\ell_x}{\ell}) + O_{32}(\frac{y-xz}{An\ Sn}) + O_{33}(\frac{x}{An})][O_{31}(y\ell+\frac{\ell_y}{\ell}) + O_{32}(\frac{x-yz}{An\ Sn}) - O_{33}(\frac{y}{An})].$$

<u>Conditions</u>: $\tilde{d}O_{33} > 0$ and $Dn \neq 0$.

<u>Compute</u>:

$$A_0(O) = (OC)^t \begin{pmatrix} 0 & 0 & 0 \\ 0 & 0 & 0 \\ 0 & 0 & 1 \end{pmatrix} (OC). \qquad \text{(II. (4.21))}$$

<u>Step C</u>: <u>Choose</u>: $k > 0$.

Compute:

$$\gamma_1 = -\left(\frac{\tilde{d}\ \ell\ An}{Sn}\right)O_{23}$$

$$\gamma_2 = \left(\frac{\tilde{d}\ \ell\ An}{Sn}\right)O_{13} \qquad \text{(II.(4.12))}$$

$$\lambda = \left(\frac{2\tilde{d}\ \ell\ An}{Sn(k+k^{-1})}\right)O_{33}$$

$$A_1(O) = (OC)^t \begin{pmatrix} 0 & \frac{\lambda(k-k^{-1})}{2} & \gamma_1 \\ \frac{\lambda(k-k^{-1})}{2} & 0 & \gamma_2 \\ \gamma_1 & \gamma_2 & 0 \end{pmatrix} (OC) \qquad \text{(II.(4.21))}$$

$$Nm = \tilde{d}\ell^2 - [A_1(O)_{12} + xA_1(O)_{13} + yA_1(O)_{23} + zA_1(O)_{33}] \qquad \text{(II.(4.25))}$$

$$\mu = Nm/Dn \qquad \text{(II.(4.27))}.$$

Condition (cycling): $\mu < 0$.

Compute: (With ε left as a free variable) (II.(4.20))

$$D_A = \varepsilon(\ell^2) - \tilde{d}\ \tilde{\ell}_z + A_1(O)_{11} + \mu A_0(O)_{11} + \left(\frac{\ell_y}{\ell}\right)^2 - \frac{2\tilde{d}(x-yz)\ell_y}{Sn^2}$$

$$D_B = \varepsilon(\ell^2) - \tilde{d}\tilde{\ell}_z + A_1(O)_{22} + \mu A_0(O)_{22} + \left(\frac{\ell_x}{\ell}\right)^2 - \frac{2\tilde{d}(y-xz)\ell_x}{Sn^2}$$

$$I = \varepsilon(\ell_z) - \tilde{d}\tilde{\ell}_n + A_1(O)_{12} + \mu A_0(O)_{12} + \frac{\ell_x \ell_y}{\ell^2} - \frac{\tilde{d}[(x-yz)\ell_x + (y-xz)\ell_y]}{Sn^2}$$

$$K = \varepsilon(\ell_x) - \widetilde{d}\widetilde{\ell}_y + A_1(0)_{23} + \mu A_0(0)_{23}$$
$$+ \ell_x - \frac{\widetilde{d}(y-xz)\ell^2}{Sn^2}$$

$$L = \varepsilon(\ell_y) - \widetilde{d}\widetilde{\ell}_x + A_1(0)_{13} + \mu A_0(0)_{13}$$
$$+ \ell_y - \frac{\widetilde{d}(x-yz)\ell^2}{Sn^2}$$

$$J = \varepsilon(\ell^2) - \widetilde{d}\widetilde{\ell}_z + A_1(0)_{33} + \mu A_0(0)_{33}$$
$$+ \ell^2.$$

$$A = \widetilde{d}\,\ell_y - xD_A - yI - zL$$

$$B = \widetilde{d}\,\ell_x - yD_B - xI - zK.$$

Step D: Choose: Y a real number. This choice does not affect the dynamics. To normalize, compute $Y = D_A + D_B - J$.

Compute: (The table of selection parameters (II.(3.13)).)

	aa	Aa	AA
BB	m_{33}	$m_{13} = m_{31}$	m_{11}
Bb	$m_{34} = m_{43}$	$m_{14} = m_{41} =$ $m_{23} = m_{32}$	$m_{12} = m_{21}$
bb	m_{44}	$m_{24} = m_{42}$	m_{22}

is given by

$Y - 2A + 2B + D_A + D_B$ $+ 2L - 2K - 4I + J$	$Y + 2B - D_A + D_B$ $- 2L - J$	$Y + 2A + 2B + D_A + D_B$ $+ 2L + 2K + 4I + J$
$Y - 2A + D_A - D_B$ $+ 2K - J$	$Y - D_A - D_B$ $+ J$	$Y + 2A + D_A - D_B$ $- 2K - J$
$Y - 2A - 2B + D_A + D_B$ $- 2L - 2K + 4I + J$	$Y - 2B - D_A + D_B$ $+ 2L - J$	$Y + 2A - 2B + D_A + D_B$ $- 2L + 2K - 4I + J$

Part II.

Step A: In Part I results set $\varepsilon = 0$. Compute: (II. (6.16))

$$Y^1_{11} = -2A - 6xD_A - 4yI - 4zL$$

$$Y^1_{22} = 2K - 2xD_B$$

$$Y^1_{33} = 2K - 2xJ$$

$$Y^1_{12} = Y^1_{21} = -B + L - 4xI - 2yD_B - 2zK$$

$$Y^1_{13} = Y^1_{31} = -4xL - 2yK - 2zJ$$

$$Y^1_{23} = Y^1_{32} = D_B + J - 2xK$$

$$Y^2_{11} = 2L - 2yD_A$$

$$Y^2_{22} = -2B - 4xI - 6yD_B - 4zK$$

$$Y^2_{33} = 2L - 2yJ$$

$$Y^2_{12} = Y^2_{21} = -A + K - 2xD_A - 4yI - 2zL$$

$$Y^2_{13} = Y^2_{31} = D_A + J - 2yL$$

$$Y^2_{23} = Y^2_{32} = -2xL - 4yK - 2zJ$$

$$Y^3_{11} = 2I - 2zD_A$$

$$Y^3_{22} = 2I - 2zD_B$$

$$Y^3_{33} = -2I - 4xL - 4yK - 6zJ$$

$$Y^3_{12} = Y^3_{21} = 1 + D_A + D_B - 2zI$$

$$Y^3_{13} = Y^3_{31} = -A + K - 2xD_A - 2yI - 4zL$$

$$Y^3_{23} = Y^3_{32} = -B + L - 2xI - 2yD_B - 4zK.$$

$$Y^1_{111} = -6D_A \qquad Y^1_{222} = 0 \qquad Y^1_{333} = 0$$

$$Y^1_{112} = Y^1_{121} = Y^1_{211} = -4I \qquad Y^1_{122} = Y^1_{212} = Y^1_{221} = -2D_B$$

$$Y^1_{113} = Y^1_{131} = Y^1_{311} = -4L \qquad Y^1_{133} = Y^1_{313} = Y^1_{331} = -2J$$

$$Y^1_{223} = Y^1_{232} = Y^1_{322} = 0 \qquad Y^1_{233} = Y^1_{323} = Y^1_{332} = 0$$

$$Y^1_{123} = Y^1_{132} = Y^1_{213} = Y^1_{231} = Y^1_{312} = Y^1_{321} = -2K$$

$$Y^2_{111} = 0 \qquad Y^2_{222} = -6D_B \qquad Y^2_{333} = 0$$

$$Y^2_{112} = Y^2_{121} = Y^2_{211} = -2D_A \qquad Y^2_{122} = Y^2_{212} = Y^2_{221} = -4I$$

$$Y^2_{113} = Y^2_{131} = Y^2_{311} = 0 \qquad Y^2_{133} = Y^2_{313} = Y^2_{331} = 0$$

$$Y^2_{223} = Y^2_{232} = Y^2_{322} = -4K \qquad Y^2_{233} = Y^2_{323} = Y^2_{332} = -2J$$

$$Y^2_{123} = Y^2_{132} = Y^2_{213} = Y^2_{231} = Y^2_{312} = Y^2_{321} = -2L$$

$$Y^3_{111} = 0 \qquad Y^3_{222} = 0 \qquad Y^3_{333} = -6J$$

$$Y^3_{112} = Y^3_{121} = Y^3_{211} = 0 \qquad Y^3_{122} = Y^3_{212} = Y^3_{221} = 0$$

$$Y^3_{113} = Y^3_{131} = Y^3_{311} = -2D_A \qquad Y^3_{133} = Y^3_{313} = Y^3_{331} = -4L$$

$$Y^3_{223} = Y^3_{232} = Y^3_{322} = -2D_B \qquad Y^3_{233} = Y^3_{323} = Y^3_{332} = -4K$$

$$Y^3_{123} = Y^3_{132} = Y^3_{213} = Y^3_{231} = Y^3_{312} = Y^3_{321} = -2I.$$

Step B: Compute:

$$Q = C^{-1}O^t \begin{pmatrix} k & 0 & 2(\mu\gamma_1 + \lambda k\gamma_2) \\ 0 & 1 & 2(\mu\gamma_2 - \lambda k^{-1}\gamma_1) \\ 0 & 0 & \mu^2 + \lambda^2 \end{pmatrix}$$

(II.(6.17))

$$Q^{-1} = \begin{pmatrix} k^{-1} & 0 & -2(k^{-1}\mu\gamma_1 + \lambda\gamma_2)/(\mu^2 + \lambda^2) \\ 0 & 1 & -2(\mu\gamma_2 - \lambda k^{-1}\gamma_1)/(\mu^2 + \lambda^2) \\ 0 & 0 & 1/(\mu^2 + \lambda^2) \end{pmatrix} OC.$$

$$X^i_{jk} = \Sigma^3_{\alpha,\beta,\gamma=1} (Q^{-1})_{i\alpha} Y^\alpha_{\beta\gamma} Q_{\beta j} Q_{\gamma k}$$

with $i,j,k = 1,2,3.$ (II.(6.14))

$$X^i_{jk\ell} = \Sigma^3_{\alpha,\beta,\gamma,\delta=1} (Q^{-1})_{i\alpha} Y^\alpha_{\beta\gamma\delta} Q_{\beta j} Q_{\gamma k} Q_{\delta\ell}$$

with $i = 1,2,3$ and $j,k,\ell = 1,2,$

$$f_{11} \quad \frac{-[(2\lambda^2 + \mu^2)X^3_{11} - \lambda\mu X^3_{12} + 2\lambda^2 X^3_{22}]}{\mu(\lambda^2 + \mu^2)}$$

$$f_{12} = f_{21} = \frac{-[\lambda\mu X^3_{11} + \mu^2 X^3_{12} - \lambda\mu X^3_{22}]}{\mu(\lambda^2 + \mu^2)} \qquad \text{(II.(6.8))}$$

$$f_{22} = \frac{-[2\lambda^2 X^3_{11} + 2\lambda\mu X^3_{12} + (2\lambda^2 + \mu^2)X^3_{22}]}{\mu(\lambda^2 + \mu^2)}$$

$$\hat{X}^i_{jk\ell} = X^i_{jk\ell} + X^i_{j3} f_{k\ell} + X^i_{k3} f_{j\ell} + X^i_{\ell 3} f_{jk}$$

$$\text{with } i,j,k,\ell = 1,2. \qquad \text{(II.(6.7))}.$$

$$\begin{aligned} \text{MARMC} = {} & [\hat{X}^1_{111} + \hat{X}^1_{122} + \hat{X}^2_{112} + \hat{X}^2_{222}] + \\ & \lambda^{-1}[-X^1_{11}\cdot X^1_{12} + X^2_{22}\cdot X^2_{12} \\ & \quad + X^2_{11}\cdot X^2_{12} - X^1_{22}\cdot X^1_{12} \\ & \quad + X^1_{11}\cdot X^2_{11} - X^1_{22}\cdot X^2_{22}]. \end{aligned} \qquad \text{(II.(6.6))}.$$

Condition (Cycling): MARMC < 0.

2. The Symmetric Program:

The symmetric or π_+ case demands special input and yields special examples.

The choices for the general program have 7 degrees of freedom: (1) 3 from p because $\Sigma\, p_i = 1$, (2) 3 from O corresponding to the three angular variables and (3) 1 from k.

For the symmetric program $p_2 = p_3$ is required so the p choice has only 2 degrees of freedom. Only one angular variable is free for

0. So the choices for the symmetric program have 4 degrees of freedom.

The output is an example whose vectorfield X^m is invariant under the symmetry π_+ of the state space of the gene pool defined by

$$\pi_+(p_1,p_2,p_3,p_4) = (p_1,p_3,p_2,p_4). \tag{2.1}$$

Note that the fixed point set of π_+ is:

$$\text{Fix}(\pi_+) = \{p: p_2 = p_3\}. \tag{2.2}$$

So our p choice in the symmetric program is required to lie in $\text{Fix}(\pi_+)$.

When the program produces a symmetric example with $\mu < 0$ and MARMC < 0 then the cycles for X^m with $m_{ij} = m_{ij}(\varepsilon)$ ($\varepsilon > 0$ and small) also lie in $\text{Fix}(\pi_+)$.

Part I.

Step A: Choose: $p \in \mathring{\Delta}$ with $p_2 = p_3$.

Compute:

$$x = p_1 - p_4$$

$$z = p_1 - 2p_2 + p_4$$

$$\tilde{d} = z - x^2$$

Conditions: $\tilde{d} \neq 0$ and $x \neq 0$, i.e. $p_1 \neq p_4$.

Compute:

$$\ell_x = (p_1^{-1} - p_4^{-1})/16$$

$$\ell_z = (p_1^{-1} - 2p_2^{-1} + p_4^{-1})/16$$

$$\ell = \sqrt{(p_1^{-1} + 2p_2^{-1} + p_4^{-1})/16}$$

$$\tilde{\ell}_x = (p_1^{-2} - p_4^{-2})/64$$

$$\tilde{\ell}_z = (p_1^{-2} - 2p_2^{-2} + p_4^{-2})/64$$

$$\tilde{\ell}_n = (p_1^{-2} + 2p_2^{-2} + p_4^{-2})/64$$

$$\text{Sn} = \sqrt{(1 - z)(1 - 2x^2 + z)}$$

$$\text{An} = \sqrt{2x^2(1 - z)}$$

$$C = \begin{pmatrix} \frac{\ell_x}{\ell} & \frac{\ell_x}{\ell} & \ell \\ \frac{x(1-z)}{\text{An Sn}} & \frac{x(1-z)}{\text{An Sn}} & 0 \\ \frac{-x}{\text{An}} & \frac{x}{\text{An}} & 0 \end{pmatrix}$$

$$\hat{C} = \begin{pmatrix} \frac{2\ell_x}{\ell} & \ell \\ \frac{2x(1-z)}{\text{An Sn}} & 0 \end{pmatrix} \qquad \text{(II.(6.25))}$$

$$\hat{C}^{-1} = \begin{pmatrix} 0 & \frac{x\ \text{Sn}}{\text{An}} \\ \ell^{-1} & \frac{2x^2(1-z)^2}{\text{An Sn}} \end{pmatrix} \qquad \text{(II.(6.26))}$$

Step B: Choose: $0 \leq \theta < 2\pi$.

Compute: (II.(5.1))

$$O = \begin{pmatrix} \cos\theta & -\sin\theta & 0 \\ \pm\sin\theta & \pm\cos\theta & 0 \\ 0 & 0 & \pm 1 \end{pmatrix}$$

$$\hat{O} = \begin{pmatrix} \cos\theta & -\sin\theta \\ \pm\sin\theta & \pm\cos\theta \end{pmatrix}$$

with the sign choices made the same as the sign of $\tilde{d}$, i.e. $\tilde{d} = \pm|d|$.

$$Dn = -\frac{x^2}{An^2}.$$

<u>Step C</u>: <u>Choose</u>: $k > 0$.

<u>Compute</u>:

$$\lambda = \frac{2|\tilde{d}|\ell\, An}{Sn(k+k^{-1})}$$

$$A_1(O) = \frac{\tilde{d}\ell\, An(k-k^{-1})}{Sn(k+k^{-1})} C^t \begin{pmatrix} \sin 2\theta & \cos 2\theta & 0 \\ \cos 2\theta & -\sin 2\theta & 0 \\ 0 & 0 & 0 \end{pmatrix} C$$

(II.(5.4)).

$$Nm = \tilde{d}\ell^2 - [A_1(O)_{12} + 2xA_1(O)_{13} + zA_1(O)_{33}]$$

$$\mu = Nm/Dn.$$

<u>Condition (Cycling)</u>: $\mu < 0$.

<u>Compute</u>: (With ε left as a free variable): (II.(5.2))

$$D_A = \varepsilon(\ell^2) - \tilde{d}\tilde{\ell}_z - Nm + A_1(O)_{11}$$

$$+ \left(\frac{\ell_x}{\ell}\right)^2 - \frac{2\tilde{d}x(1-z)\ell_x}{Sn^2}$$

$$I = \varepsilon(\ell_z) - \tilde{d}\tilde{\ell}_n + Nm + A_1(O)_{12} + (\frac{\ell_x}{\ell})^2 - \frac{2\tilde{d}x(1-z)\ell_x}{Sn^2}$$

$$K = \varepsilon(\ell_x) - d\tilde{\ell}_x + A_1(O)_{13} + \ell_x - \frac{\tilde{d}x(1-z)\ell^2}{Sn^2}$$

$$J = \varepsilon(\ell^2) - \tilde{d}\tilde{\ell}_z + A_1(O)_{33} + \ell^2$$

$$A = \tilde{d}\ell_x - x(D_A + I) - zK.$$

Step D: Choose: Y a real number. This choice does not affect the dynamics. To normalize, compute $Y = 2D_A - J$.

Compute: (Normalized version on the right)

$$m_{11} = Y + 4A + 2D_A + 4K + 4I + J = 4(A + D_A + K + I)$$

$$m_{22} = m_{33} = Y + 2D_A - 4I + J = 4(D_A - I)$$

$$m_{44} = Y - 4A + 2D_A - 4K + 4I + J = 4(-A + D_A - K + I)$$

$$m_{12} = m_{21} = m_{13} = m_{31} = Y + 2A - 2K - J = 2(A + D_A - K - J)$$

$$m_{24} = m_{42} = m_{34} = m_{43} = Y - 2A + 2K - J = 2(-A + D_A + K - J)$$

$$m_{14} = m_{41} = m_{23} = m_{32} = Y - 2D_A + J = 0.$$

Part II.

Step A: In Part I results set $\varepsilon = 0$.

Compute: (II.(6.20))

$$Y^1_{11} = -4A + 4K - 12x(D_A + I) - 8zK$$

$$Y^1_{22} = 2K - 2xJ$$

$$Y^1_{12} = Y^1_{21} = D_A + J - 8xK - 2zJ$$

$$Y^2_{11} = 2 + 4D_A + 4I - 4z(D_A + I)$$

$$Y^2_{22} = -2I - 8xK - 6zJ$$

$$Y^2_{12} = Y^2_{21} = -2A + 2K - 4x(D_A + I) - 8zK$$

$$Y^1_{111} = -12(D_A + I) \qquad Y^1_{222} = 0$$

$$Y^1_{112} = Y^1_{121} = Y^1_{211} = -8K \qquad Y^1_{122} = Y^1_{212} = Y^1_{221} = -2J$$

$$Y^2_{111} = 0 \qquad Y^2_{222} = -6J$$

$$Y^2_{112} = Y^2_{121} = Y^2_{211} = -4(D_A+I) \qquad Y^2_{122} = Y^2_{212} = Y^2_{221} = -8K$$

Step B: Compute:

$$\hat{Q} = \hat{C}^{-1}\hat{O}^t \begin{pmatrix} k & 0 \\ 0 & 1 \end{pmatrix}$$

$$\hat{Q}^{-1} = \begin{pmatrix} k^{-1} & 0 \\ 0 & 1 \end{pmatrix} \hat{O}\hat{C}. \qquad \text{(II.(6.24))}$$

$$X^i_{jk} = \Sigma^2_{\alpha,\beta,\gamma=1} (\hat{Q}^{-1})_{i\alpha} Y^\alpha_{\beta\gamma} Q_{\beta j} Q_{\gamma k}$$

$$X^i_{jk\ell} = \Sigma^2_{\alpha,\beta,\gamma,\delta=1} (\hat{Q}^{-1})_{i\alpha} Y^\alpha_{\beta\gamma\delta} Q_{\beta j} Q_{\gamma k} Q_{\delta\ell}$$

with $i,j,k,\ell = 1,2.$

$$\text{MARMC} = [X^1_{111} + X^1_{122} + X^2_{112} + X^2_{222}] +$$

$$\lambda^{-1}[-X^1_{11}\cdot X^1_{12} + X^2_{22}\cdot X^2_{12}$$

$$+ X^2_{11}\cdot X^2_{12} - X^1_{22}\cdot X^1_{12}$$

$$+ X^1_{11}\cdot X^2_{11} - X^1_{22}\cdot X^2_{22}].$$

Condition (Cycling): MARMC < 0.

3. Sample Computations.

A sample version of the symmetric program was executed with the help of my colleague Stanley Ocken. In particular, he introduced me to the MACSYMA package at MIT which did the needed algebra. The two examples described below come from the two theorems of Sec. II.7.

1 Example: According to Thm. II.7.3 Hopf attractors occur when $x = y >$ provided $\tilde{d} > 0$ is small and $k > 0$ is large. So we choose:

$$x = y = \frac{3}{4} \qquad \tilde{d} = \frac{1}{32}$$

$$z = \tilde{d} + xy = \frac{19}{32}$$

$$p = (\frac{99}{128},\frac{13}{128},\frac{13}{128},\frac{3}{128}) \sim (.773,.102,.102,.023).$$

We then compute:

$$\ell_x = \ell_y = -\frac{3328}{1287} \sim -2.59 \qquad \ell_z = \frac{1952}{1287} \sim 1.52$$

$$\ell^2 = \frac{5120}{1287} \sim 3.99$$

$$\text{Sn} = \frac{\sqrt{195}}{32} \sim .436 \qquad \text{An} = \frac{3\sqrt{13}}{16} \sim .676$$

We choose $\theta = 0$, i.e. the orthogonal matrix O is the identity and allow k as well as ε to remain as free variables.

$$\lambda = \frac{4}{\sqrt{429}}\left(\frac{k}{k^2+1}\right) \sim .193\left(\frac{k}{k^2+1}\right), \qquad \gamma_1 = \gamma_2 = 0$$

$$\mu = -\frac{10}{99} + \frac{\sqrt{143}}{330\sqrt{5}}\left(\frac{k^2-1}{k^2+1}\right) \sim -.101 + .016\left(\frac{k^2-1}{k^2+1}\right)$$

$$D_A = D_B = \frac{1{,}689{,}568}{1{,}656{,}369} - \frac{44}{15\sqrt{715}}\left(\frac{k^2-1}{k^2+1}\right) + \frac{5120}{1287}\varepsilon$$

$$\sim 1.02 - .110\left(\frac{k^2-1}{k^2+1}\right) + 3.99\varepsilon$$

$$I = \frac{178{,}776}{1{,}656{,}369} - \frac{4}{\sqrt{715}}\left(\frac{k^2-1}{k^2+1}\right) + \frac{1952}{1287}\varepsilon$$

$$\sim .108 - .150\left(\frac{k^2-1}{k^2+1}\right) + 1.52\varepsilon$$

$$K = L = -\frac{3{,}141{,}632}{1{,}656{,}369} + \frac{8}{3\sqrt{715}}\left(\frac{k^2-1}{k^2+1}\right) - \frac{3328}{1287}\varepsilon$$

$$\sim -1.90 + .100\left(\frac{k^2-1}{k^2+1}\right) - 2.59\varepsilon$$

$$J = \frac{5{,}272{,}576}{1{,}656{,}369} + \frac{5120}{1287}\varepsilon$$

$$\sim 3.18 + 3.99\varepsilon$$

$$A = B = -\frac{876{,}512}{1{,}656{,}369} + \frac{217}{60\sqrt{715}}\left(\frac{k^2-1}{k^2+1}\right) - \frac{3328}{1287}\varepsilon$$

$$\sim -.529 + .135\left(\frac{k^2-1}{k^2+1}\right) - 2.59\varepsilon .$$

MARMC is negative for all positive k.

Notice that since $(k^2-1)/(k^2+1)$ is at most 1, μ is also negative for all positive k. Consequently, if we fix k and look at the one parameter family of selection matrices depending on ε, p is a Hopf attractor for the corresponding equation (1.1) when $\varepsilon = 0$. When $\varepsilon > 0$ and sufficiently small, p is a saddle point equilibrium surrounded by an attracting, limit cycle lying in the plane $x = y$.

Choosing $k = 1$ and $Y = 2D_A - J$ to normalize with $m_{14} = m_{23} = 0$ we get the following selection table (accurate to three significant digits):

k = 1	aa	Aa	AA
BB	$3.65 + 9.88\varepsilon$	-1.58	$-5.20 + 1.32\varepsilon$
Bb	-7.06	0	-1.58
bb	$14.2 + 42.8\varepsilon$	-7.06	$3.65 + 9.88\varepsilon$

2 Example: According to Thm. II.7.1 and the Example which follows it Hopf attractors occur when $k > 0$ is large and

$$x = y = z = \frac{1}{10} \qquad \tilde{d} = \frac{9}{100}$$

$$p = (\frac{13}{40},\frac{9}{40},\frac{9}{40},\frac{9}{40}) = (.325,.225,.225,.225).$$

$$\ell_x = \ell_y = \ell_z = -\frac{10}{117} \sim -.0855 \qquad \ell^2 = \frac{120}{117} \sim 1.03$$

$$\mathrm{Sn} = \frac{9\sqrt{3}}{5\sqrt{10}} \sim .986 \qquad \mathrm{An} = \frac{3}{10\sqrt{5}} \sim .134$$

Again we choose $\theta = 0$ and allow k as well as ε to vary:

$$\lambda = \frac{3}{5\sqrt{585}}(\frac{k}{k^2+1}) \sim .0248(\frac{k}{k^2+1}), \qquad \gamma_1 = \gamma_2 = 0.$$

$$\mu = -\frac{54}{325} + \frac{1}{100\sqrt{390}}\left(\frac{k^2-1}{k^2+1}\right) \sim -.166$$

$$D_A = D_B = -\frac{527}{7605} - \frac{1}{45\sqrt{390}}\left(\frac{k^2-1}{k^2+1}\right) + \frac{120}{117}\varepsilon$$

$$\sim -.0693 - .0011\left(\frac{k^2-1}{k^2+1}\right) + 1.03\varepsilon$$

$$I = \frac{32}{7605} - \frac{1}{30\sqrt{390}}\left(\frac{k^2-1}{k^2+1}\right) - \frac{10}{117}\varepsilon$$

$$\sim .00421 - .00169\left(\frac{k^2-1}{k^2+1}\right) - .0855\varepsilon$$

$$K = L = -\frac{605}{7605} + \frac{1}{6\sqrt{390}}\left(\frac{k^2-1}{k^2+1}\right) - \frac{10}{117}\varepsilon$$

$$\sim -.0745 + .0084\left(\frac{k^2-1}{k^2+1}\right) - .0855\varepsilon$$

$$J = \frac{7910}{7605} + \frac{120}{117}\varepsilon$$

$$\sim 1.04 + 1.03\varepsilon$$

$$A = B = \frac{51.5}{7605} - \frac{1}{90\sqrt{390}}\left(\frac{k^2-1}{k^2+1}\right) - \frac{10}{117}\varepsilon$$

$$\sim .00677 - .00056\left(\frac{k^2-1}{k^2+1}\right) - .0855\varepsilon.$$

MARMC is negative for all $k > 1.02$.

Notice that μ is again negative for all positive k.

Choosing $k = 1.1$ and $k = 7$ we get the following selection tables (accurate to three significant digits). Again we choose $Y = 2D_A - J$ to get $m_{14} = m_{23} = 0$.

k = 1.1	aa	Aa	AA
BB	-.294 + 4.44ε	-2.05	-.550 + 3.08ε
Bb	-2.39	0	-2.05
bb	.027 + 4.44ε	-2.39	-.294 + 4.44ε

k = 7	aa	Aa	AA
BB	-.292 + 4.44ε	-2.07	-.532 + 3.08ε
Bb	-2.38	0	-2.07
bb	-.010 + 4.44ε	-2.38	-.292 + 4.44ε

I would like to conclude with some speculations suggested by these examples.

In example 1 and example 2 (k = 1.1) the maximum of mean fitness occurs at a vertex, i.e. fixation at each locus. In fact, in example 1 three of the four vertices are local maxima of mean fitness and so are attracting under selection alone. The introduction of recombination makes the attraction of such a vertex equilibrium even stronger. However there is in each case an interior saddle point equilibrium for selection alone. I presume that recombination acts to stabilize this equilibrium and so that the Hopf attractor comes from it. However, before it becomes stable it is surrounded by a stable cycle. This may be how the cycles arise in all of these models.

In example 2 with k = 7 the maximum of mean fitness occurs on the edge between fixation at AB and fixation at ab. The effect of recombination is to move this equilibrium into the interior. However, I suspect that the Hopf attractor is related still to the effect of recombination on the interior saddle point equilibrium of selection alone.

Mathematics Department
The City College
137 St. and Convent Ave.
New York, N. Y. 10031

Bibliography

[1] = [GPG] Akin, E. The Geometry of Population Genetics, Lecture Notes in Biomathematics #31 Springer-Verlag (1979).

[2] Crow, J.F. and Kimura, M. An Introduction to Population Genetics Theory, Harper and Row Publishers, Inc. (1970).

[3] Ewens, W.J. "With Additive Fitness, the Mean Fitness Increases" (1969) Nature 221: 1076.

[4] Ewens, W.J. Population Genetics, Metheun and Co., Ltd. (1969).

[5] Hirsch, M. and Smale, S. Differential Equations, Dynamical Systems, and Linear Algebra, Academic Press, Inc. (1974).

[6] Karlin, S. and Feldman, M.W. "Linkage and Selection: Two Locus Symmetric Viability Model" (1970) Theo. Pop. Biol. 1: 39-71.

[7] Marsden, J.E. and McCracken, M. The Hopf Bifurcation and its Applications, Springer-Verlag (1976).

[8] Moran, P.A.P. "On the Nonexistence of Adaptive Topographies" (1964) Ann. Human Genet. 27: 383-393.

[9] Nagylaki, T. and Crow, J. "Continuous Selection Models" (1974) Theo. Pop. Biol. 5: 257-283.

[10] Niven, I. and Zuckerman, H.S. An Introduction to the Theory of Numbers, John Wiley and Sons, Inc. (1960).

[11] Shahshahani, S. A New Mathematical Framework for the Study of Linkage and Selection, Memoirs AMS #211 (1979).

[12] Smale, S. "On the Differential Equations of Populations in Competition" (1976) J. Math. Bio. 3: 5-7.

[13] Wright, S. "'Surfaces' of Selective Value" (1967) PNAS 58: 165-172.

[14] Zeeman, E.C. "Population Dynamics from Game Theory" in Proc. Int. Conf. Global Theory of Dynamical Systems, Lecture Notes in Mathematics # 819 Springer-Verlag (1979).

Index

General instructions to authors for

PREPARING REPRODUCTION COPY FOR MEMOIRS

For more detailed instructions send for AMS booklet, "A Guide for Authors of Memoirs." Write to Editorial Offices, American Mathematical Society, P. O. Box 6248, Providence, R. I. 02940.

MEMOIRS are printed by photo-offset from camera copy fully prepared by the author. This means that, except for a reduction in size of 20 to 30%, the finished book will look exactly like the copy submitted. Thus the author will want to use a good quality typewriter with a new, medium-inked black ribbon, and submit clean copy on the appropriate model paper.

Model Paper, provided at no cost by the AMS, is paper marked with blue lines that confine the copy to the appropriate size. Author should specify, when ordering, whether typewriter to be used has PICA-size (10 characters to the inch) or ELITE-size type (12 characters to the inch).

Line Spacing – For best appearance, and economy, a typewriter equipped with a half-space ratchet – 12 notches to the inch – should be used. (This may be purchased and attached at small cost.) Three notches make the desired spacing, which is equivalent to 1-1/2 ordinary single spaces. Where copy has a great many subscripts and superscripts, however, double spacing should be used.

Special Characters may be filled in carefully freehand, using dense black ink, or INSTANT ("rub-on") LETTERING may be used. AMS has a sheet of several hundred most-used symbols and letters which may be purchased for $5.

Diagrams may be drawn in black ink either directly on the model sheet, or on a separate sheet and pasted with rubber cement into spaces left for them in the text. Ballpoint pen is *not* acceptable.

Page Headings (Running Heads) should be centered, in CAPITAL LETTERS (preferably), at the top of the page – just above the blue line and touching it.

LEFT-hand, EVEN-numbered pages should be headed with the AUTHOR'S NAME;

RIGHT-hand, ODD-numbered pages should be headed with the TITLE of the paper (in shortened form if necessary).

Exceptions: PAGE 1 and any other page that carries a display title require NO RUNNING HEADS.

Page Numbers should be at the top of the page, on the same line with the running heads.

LEFT-hand, EVEN numbers – flush with left margin;

RIGHT-hand, ODD numbers – flush with right margin.

Exceptions: PAGE 1 and any other page that carries a display title should have page number, centered below the text, on blue line provided.

FRONT MATTER PAGES should be numbered with Roman numerals (lower case), positioned below text in same manner as described above.

MEMOIRS FORMAT

It is suggested that the material be arranged in pages as indicated below.
Note: Starred items (*) are requirements of publication.

Front Matter (first pages in book, preceding main body of text).

Page i – *Title, *Author's name.

Page iii – Table of contents.

Page iv – *Abstract (at least 1 sentence and at most 300 words).

*1980 Mathematics Subject Classifications represent the primary and secondary subjects of the paper. For the classification scheme, see Annual Subject Indexes of MATHEMATICAL REVIEWS beginning in December 1978.

Key words and phrases, if desired. (A list which covers the content of the paper adequately enough to be useful for an information retrieval system.)

Page v, etc. – Preface, introduction, or any other matter not belonging in body of text.

Page 1 – Chapter Title (dropped 1 inch from top line, and centered).

Beginning of Text.

Footnotes: *Received by the editor date.
Support information – grants, credits, etc.

Last Page (at bottom) – Author's affiliation.

ABCDEFGHIJ–AMS–89876543

Memoirs of the American Mathematical Society
Number 285

Maury Bramson

Convergence of solutions of the Kolmogorov equation to travelling waves

Published by the
AMERICAN MATHEMATICAL SOCIETY
Providence, Rhode Island, USA

July 1983 · Volume 44 · Number 285 (end of volume)

MEMOIRS of the American Mathematical Society

This journal is designed particularly for long research papers (and groups of cognate papers) in pure and applied mathematics. It includes, in general, longer papers than those in the TRANSACTIONS.

Mathematical papers intended for publication in the Memoirs should be addressed to one of the editors. Subjects, and the editors associated with them, follow:

MEMOIRS are printed by photo-offset from camera-ready copy fully prepared by the authors. Prospective authors are encouraged to request booklet giving detailed instructions regarding reproduction copy. Write to Editorial Office, American Mathematical Society, P.O. Box 6248, Providence, Rhode Island 02940. For general instructions, see last page of Memoir.

SUBSCRIPTION INFORMATION. The 1983 subscription begins with Number 272 and consists of six mailings, each containing one or more numbers. Subscription prices for 1983 are $104.00 list; $52.00 member. Each number may be ordered separately; *please specify number* when ordering an individual paper. For prices and titles of recently released numbers, refer to the New Publications sections of the **NOTICES** of the American Mathematical Society.

BACK NUMBER INFORMATION. For back issues see the AMS Catalogue of Publications.

TRANSACTIONS of the American Mathematical Society

This journal consists of shorter tracts which are of the same general character as the papers published in the MEMOIRS. The editorial committee is identical with that for the MEMOIRS so that papers intended for publication in this series should be addressed to one of the editors listed above.

Subscriptions and orders for publications of the American Mathematical Society should be addressed to American Mathematical Society, P. O. Box 1571, Annex Station, Providence, R. I. 02901. *All orders must be accompanied by payment.* Other correspondence should be addressed to P. O. Box 6248, Providence, R. I. 02940.

MEMOIRS of the American Mathematical Society (ISSN 0065-9266) is published bimonthly (each volume consisting usually of more than one number) by the American Mathematical Society at 201 Charles Street, Providence, Rhode Island 02904. Second Class postage paid at Providence, Rhode Island 02940. Postmaster: Send address changes to Memoirs of the American Mathematical Society, American Mathematical Society, P. O. Box 6248, Providence, RI 02940.

TABLE OF CONTENTS

ABSTRACT

The classic Kolmogorov equation $u_t = \frac{1}{2}u_{xx} + f(u)$ is investigated under general initial data. A necessary and sufficient condition on the initial data is given for convergence to a travelling wave as $t \to \infty$. In the case of convergence, a formula is given for computing the position of the wave. The methodology involves use of the Feynman-Kac integral and sample path estimates for Brownian motion.

1980 Mathematics Subject Classification. 60J60, 35K55.

Library of Congress Cataloging in Publication Data

Bramson, Maury, 1951-
Convergence of solutions of the Kolmogorov equation to travelling waves.

(Memoirs of the American Mathematical Society, ISSN 0065-9266 ; no. 285)
1. Diffusion processes. 2. Brownian motion processes. 3. Differential equations, Parabolic--Numerical solutions. I. Title. II. Title: Kolmo equation to travelling waves. III. Series.
QA3.A57 no. 285 [QA274.75] 510s [519.2'33] 83-6
ISBN 0-8218-2285-3

1. INTRODUCTION

Background. In their famous 1937 paper, "Étude de l'équation de la diffusion avec croissance de la quantité de matière et son application à un problème biologique", Kolmogorov, Petrovsky, and Piscounov analyzed the asymptotic behavior as $t \to \infty$ for a class of nonlinear diffusion equations of the form

$$(1.1) \qquad u_t = \frac{1}{2} u_{xx} + f(u) \; .$$

This equation arose in the context of a genetics model for the spread of an advantageous gene through a population. The solution $u(t,x)$ to (1.1) measured the proportion of the population possessing this gene as the biological system evolved. The forcing term f was assumed to be in $C^1[0,1]$ and satisfy the conditions

$$(1.2a) \qquad f(0) = f(1) = 0 \; , \; f(u) > 0 \quad \text{for} \quad 0 < u < 1 \; ,$$

and

$$(1.2b) \qquad f'(0) = 1 \; , \; f'(u) \leq 1 \quad \text{for} \quad 0 < u \leq 1 \; .$$

Under Heaviside initial data,

$$(1.3) \qquad \begin{aligned} u(0,x) &= 0 \quad \text{for} \quad x > 0 \\ &= 1 \quad \text{for} \quad x \leq 0 \; , \end{aligned}$$

Kolmogorov et al. showed that $u(t,x)$ approached a travelling wave $w(x)$ as $t \to \infty$:

Received by the editors August 21, 1981.

The research in this paper was supported in part by the National Science Foundation Grant NSF-MCS-78-00168.

(1.4) $u(t,x+m(t)) \to w(x)$ uniformly in x ,

for appropriate choice of $m(t)$. They identified $w(x)$ as the solution (unique up to translation) of the equation

(1.5) $$0 = \frac{1}{2}w_{xx} + 2^{1/2}w_x + f(w)$$

satisfying

(1.6a) $0 < w(x) < 1$ for all x

and

(1.6b) $$w(x) \to 0 \text{ as } x \to \infty$$ $$\to 1 \text{ as } x \to -\infty ,$$

and showed that the asymptotic rate of propagation of the wave is $2^{1/2}$:

(1.7) $\dot{m}(t) \to 2^{1/2}$ as $t \to \infty$.

Fisher [6] also investigated in 1937 an equation satisfying (1.1) and (1.2) which arose in the context of the same genetics model. While also conjecturing that (1.4) held, he however only investigated certain tail properties of $w(x)$. The equation (1.1) presently goes under several aliases: the Kolmogorov equation, K-P-P equation, and the Fisher equation. We shall employ the first appellation.

In addition to arising in the context of genetics, the Kolmogorov equation has also appeared in the context of chemical combustion theory and flame propagation, where $u(t,x)$ denotes the temperature of the reaction. For references, see Aronson-Weinberger [2] and Fife-McLeod [5]. The equation moreover occurs in the context of one-dimensional branching diffusions. One can show quite simply that if $\mathfrak{X}_1(t),\ldots,\mathfrak{X}_{n(t)}(t)$ denote the positions of the particles existing at time t for branching Brownian motion (where $n(t)$ is itself random), then

$$(1.8) \qquad u(t,x) \equiv P[\max_{1 \leq k \leq n(t)} \mathfrak{X}_k(t) > x]$$

satisfies (1.1). In particular if the branching mechanism is binary, then $f(u) = u - u^2$. (See, e.g., McKean [15].)

Starting with Kanel [13] in the early 1960's, the Kolmogorov equation has become the object of renewed interest. Over the past decade, there has been considerable work done on extensions and generalizations of (1.1)-(1.4). In particular, interest has centered on extending (1.1) to $f(u)$ more general than in (1.2), and to higher spatial dimensions. Papers worthy of note include Aronson-Weinberger [2,3], Fife-McLeod [5], Jones [11], and McKean [15]. (See [5] for a more comprehensive bibliography.)

<u>The problem</u>. Until recently, little progress had been made on two particular questions involving the Kolmogorov equation. Under (1.2), it was not known for which initial data other than Heaviside data convergence to a travelling wave as in (1.4) continues to hold. Also, the higher order behavior of the centering term $m(t)$ was unknown, even under Heaviside initial data. The latter question was partially answered by Bramson [4], who showed that under Heaviside initial data (and assuming that $-\infty < f''(0) < 0$),

$$(1.9) \qquad m(t) = 2^{1/2}t - 3 \cdot 2^{-3/2} \log t + O(1) .$$

Uchiyama [17] was able to make substantial headway into the former question, and demonstrated convergence to a travelling wave for a wide range of initial data. Under the assumptions that $u(0,x)$ is differentiable, that

$$(1.10a) \qquad 0 \leq u(0,x) \leq 1 \quad \text{for all } x ,$$

$$(1.10b) \qquad \liminf_{x \to -\infty} u(0,x) > 0 ,$$

and

$$(1.10c) \qquad \liminf_{x \to \infty} \frac{-u_x(0,x)}{u(0,x)} \geq 2^{1/2} ,$$

he demonstrated that (1.4) and (1.5) remain valid. His argument relied heavily on the differentiability of $u(0,x)$ and made use of certain well-behaved comparison functions in the time-dependent phase plane of $u(t,x)$.

Under inequality (1.10c), the initial data has a right tail which drops off at least exponentially at rate $2^{1/2}$. This behavior is critical to the large time behavior of $u(t,x)$, as an investigation of the travelling wave solutions of (1.1) reveals. Substitution into (1.1) shows that travelling waves

$$(1.11) \qquad w^{\lambda}(x) = u(t,x+\lambda t)$$

satisfy

$$(1.12) \qquad 0 = \frac{1}{2} w^{\lambda}_{xx} + \lambda w^{\lambda}_{x} + f(w^{\lambda}) .$$

Phase plane analysis shows that for $\lambda \geqq 2^{1/2}$, there is a unique (up to translation) solution (1.11) which satisfies (1.6), whereas for $\lambda < 2^{1/2}$, there is no such solution. Moreover,

$$(1.13) \qquad w^{2^{1/2}}(x) \sim Cxe^{-2^{1/2}x} \quad \text{for large } x ,$$

and for $\lambda > 2^{1/2}$,

$$(1.14) \qquad w^{\lambda}(x) \sim C_{\lambda} e^{-bx} \quad \text{for large } x ,$$

where $b = \lambda - \sqrt{\lambda^2 - 2} < 2^{1/2}$, and C , C_{λ} are constants. Thus initial data with relatively fat right tails as in (1.14) ought not be expected to obey (1.4), (1.5), and (1.7). Instead, such data should in some sense be in the "domain of attraction" of $w^{\lambda}(x)$, $\lambda > 2^{1/2}$. Proofs to that effect may be found in [15] and [17]. It is the purpose of this article to analyze the solution to the Kolmogorov equation under arbitrary initial data. Criteria will be given for convergence to the travelling waves specified in (1.12), as well as a formula for the value of the centering term $m(t)$.

Results. We summarize our results in the form of two theorems, Theorems A and B, and in several examples. Theorem A gives necessary and sufficient conditions on the initial data for convergence to the travelling waves $w^{\lambda}(x)$, $\lambda \geq 2^{1/2}$. We require here the additional assumption (1.2') on the forcing term f . No assumptions are made regarding the monotonicity and differentiability of the initial data.

Theorem A. Let $u(t,x)$ be a solution of

(1.1) $$u_t = \frac{1}{2}u_{xx} + f(u) ,$$

with $0 \leq u(0,x) \leq 1$, and where

(1.2a) $$f \in C^1[0,1] , f(0) = f(1) = 0 , f(u) > 0 \text{ for } 0 < u < 1 ,$$

and

(1.2b) $$f'(0) = 1 , f'(u) \leq 1 \text{ for } 0 < u \leq 1 .$$

In addition, assume that

(1.2') $$1 - f'(u) = O(u^{\rho}) \text{ for some } \rho > 0 .$$

Then for the case $\lambda > 2^{1/2}$,

(1.15) $$u(t,x+m(t)) \to w^{\lambda}(x) \text{ uniformly in } x \text{ as } t \to \infty$$

for some choice of $m(t)$ iff for some (all) $h > 0$,

(1.16) $$\lim_{t \to \infty} \frac{1}{t} \log \left[\int_t^{t(1+h)} u(0,y)\, dy\right] = -b$$

where $b = \lambda - \sqrt{\lambda^2 - 2} < 2^{1/2}$, and for some $\eta > 0$, $M > 0$, $N > 0$,

(1.17) $$\int_x^{x+N} u(0,y)\, dy > \eta \text{ for } x \leq -M .$$

For the case $\lambda = 2^{1/2}$,

$$(1.18)\qquad u(t,x+m(t)) \to w^{2^{1/2}}(x) \text{ uniformly in } x \text{ as } t\to\infty$$

iff for some (all) $h > 0$,

$$(1.19)\qquad \limsup_{t\to\infty} \frac{1}{t}\log\left[\int_t^{t(1+h)} u(0,y)\,dy\right] \leqq -2^{1/2} ,$$

and (1.17) is satisfied.

For initial data for which either (1.15) or (1.18) holds, it is simple to show that

$$(1.20)\qquad m(t)/t \to \lambda \quad \text{as} \quad t\to\infty .$$

More accurate estimation of $m(t)$ requires more work. Theorem B gives a precise formula for $m(t)$ in the case where $\lambda > 2^{1/2}$.

<u>Theorem B</u>. Let $u(t,x)$ be a solution of (1.1), (1.2), and (1.2') as in Theorem A. If (1.16) and (1.17) are satisfied, then $m(t)$ may be chosen so that

$$(1.21)\qquad m(t) = \sup\{x: \varphi(t,x) \geqq 1\} ,$$

where

$$(1.22)\qquad \varphi(t,x) = e^t\int_{-\infty}^{\infty} u(0,y)\,\frac{e^{-(x-y)^2/2t}}{\sqrt{2\pi t}}\,dy .$$

(Note that since $w^\lambda(x)$ is unique only up to translation, translates of (1.21) may also be chosen.)

As an application of Theorem B, one may calculate $m(t)$ explicitly as in the following example.

<u>Example 1</u>. Let $g(x) = e^{bx}u(0,x)$ for some $b < 2^{1/2}$. If for each fixed $t_0 > 1$ and large enough x ,

(1.23) $$t_0^{-\gamma} \leq g(tx)/g(x) \leq t_0^{\gamma}$$

for $1 \leq t \leq t_0$, where $\gamma > 0$ is fixed, then

(1.15) $$u(t,x+m(t)) \to w^{\lambda}(x) .$$

One may choose

(1.24) $$m(t) = \lambda t + \frac{1}{b} \log g((\lambda - b)t) .$$

For initial data chosen so that (1.18) is satisfied, determination of $m(t)$ is more difficult. In Section 9, we will give an implicit formula for $m(t)$; in the following examples, we can evaluate $m(t)$ explicitly.

Example 2. If

(1.25) $$\lim_{x \to \infty} e^{bx} u(0,x) = 0$$

for some $b > 2^{1/2}$, then

(1.18) $$u(t,x+m(t)) \to w^{2^{1/2}}(x) .$$

One may choose

(1.26) $$m(t) = 2^{1/2} t - 3 \cdot 2^{-3/2} \log t .$$

The estimate (1.26) is an improvement of (1.9).

The behavior of $m(t)$ for initial data with right tails in the range between those for Examples 1 and 2 is of particular interest, and is elucidated by the following examples.

Example 3. Let $h(x) = e^{2^{1/2}x} u(0,x)$, where $h(x) = O(x^{\gamma})$, $\gamma > 0$. Then (1.18) holds, and

(1.27a) $$m(t) = 2^{1/2} t - 3 \cdot 2^{-3/2} \log t + b(t) ,$$

where

$$\text{(1.27b)} \qquad b(t) = 2^{-1/2}\log\left[\int_0^\infty x\,h(x)e^{-x^2/2t}\,dx\right] \vee 0 \quad .$$

Example 4. Suppose in particular that

$$\text{(1.28)} \qquad h(x) = x^{\alpha} \quad \text{for} \quad x \geq 1 \ ,$$

where $\alpha > -2$. Then $m(t)$ may be chosen so that

$$\text{(1.29)} \qquad m(t) = 2^{1/2}t + (\alpha-1)2^{-3/2}\log t \quad .$$

If $h(x) = x^{-2}$ for $x \geq 1$, then

$$\text{(1.30)} \qquad m(t) = 2^{1/2}t - 3\cdot 2^{-3/2}\log t + 2^{-1/2}\log\log t \quad .$$

Heuristics. The basic technique employed in this article is the use of the Feynman-Kac integral formula in conjunction with sample path estimates for Brownian motion. The estimates thus obtained for the right tail of $u(t,x)$ are precise enough to allow comparison with the right tail of $w^{\lambda}(x)$, $\lambda \geq 2^{1/2}$. Such comparison is sufficient to imply convergence to $w^{\lambda}(x)$ as in Theorem A and enable location of the wave as in Theorem B and the above examples.

The Feynman-Kac formula states that for $u(t,x)$ satisfying the linear equation

$$\text{(1.31)} \qquad u_t = \frac{1}{2}u_{xx} + k(t,x)u \ ,$$

$$\text{(1.32)} \qquad u(t,x) = E\left[\exp\left\{\int_0^t k(t-s,\mathfrak{X}_x(s))\,ds\right\}u(0,\mathfrak{X}_x(t))\right] \ ,$$

where $\mathfrak{X}_x(t)$ is Brownian motion started at x . (For references, see Friedman [7], page 148, or Kac [12].) Equality (1.32) states that $u(t,x)$ is the weighted average of the different sample paths of Brownian motion. In our case, we set

$$\text{(1.33)} \qquad k(s,y) = f(u(s,y))/u(s,y) \quad .$$

$k(s,y)$ is here not explicitly known, and will vary along different paths. Since for $s > 0$, $0 < u(s,y) < 1$ for all y , we do however know from (1.2) that

(1.34) $$0 < k(s,y) < 1 \quad .$$

Moreover, under conditions (1.16) or (1.19), and (1.17), for $s \geq s(\epsilon)$, $0 < \epsilon < 1$,

(1.35a) $$\lim_{y \to \infty} u(s,y) = 0$$

and

(1.35b) $$\liminf_{y \to -\infty} u(s,y) \geq \epsilon \quad .$$

Therefore for $s \geq \hat{s}(\epsilon)$,

(1.36a) $$\lim_{y \to \infty} k(s,y) = 1$$

and

(1.36b) $$\limsup_{y \to -\infty} k(s,y) \leq \epsilon \quad .$$

We interpret (1.32) and (1.36) as stating that the weighting of a path will be nearly maximal at large values of y and not significant at small values of y , with the transition occurring in the vicinity of the wave (near where $u(s,y) = 1/2$) . Consequently, paths which remain substantially above the wave "most" of the time are weighted heavily and need to be considered when calculating (1.32), whereas paths which remain substantially below the wave a "significant" portion of the time are insignificantly weighted, and may be neglected. In practice, "heavily weighted" paths and "insignificant" paths will be defined in terms of paths not crossing / crossing certain curves in space-time. Both bounds for the weight of such paths and their probability must be computed. The fact that the position of the wave is not a priori known compounds the difficulties.

The detailed behavior of the right side of (1.32) is quite different depending on whether (1.16) or (1.19) (and (1.17)) is assumed to hold. The two cases need to be handled separately, with convergence to $w^{\lambda}(x)$, $\lambda > 2^{1/2}$, being much easier to treat.

If (1.16) is satisfied, one is able to compare (1.32) with the more elementary quantity

$$(1.37) \qquad e^{t} E[u(0,\mathfrak{X}_x(t))] \ .$$

The major contribution to (1.37) comes from paths $x(s)$, $0 \leq s \leq t$, with

$$(1.38) \qquad x(t) \sim x - bt \ ,$$

where $b = \lambda - \sqrt{\lambda^2 - 2}$ and $\lambda - b > 0$. This behavior is a compromise between two conflicting effects: 1) for $y \leq x$, $P[\mathfrak{X}_x(t) \in dy]$ is increasing in y , and 2) as $y \to \infty$, $u(0,y)$ is on the average decreasing rapidly in y ; this behavior is easy to verify in the special case where $u(0,y) \sim e^{-by}$. But under (1.16), (1.32) (with $k(s,y)$ chosen as in (1.33)) is of the same order of magnitude as (1.37) at large values of x . This occurs because, setting

$$(1.39) \qquad m_{1/2}(t) = \sup\{x\colon\ u(t,x) \geq 1/2\} \ ,$$

for $x = m_{1/2}(t) + z$ where $z > 0$ is reasonably large, Brownian motion satisfying (1.38) will with a positive probability (which is not time dependent) always remain in the region of the (s,y) plane where $u(s,y) \sim 0$ and hence $k(s,y) \sim 1$. Calculation of (1.32) therefore reduces to the simpler problem of computing (1.37), from which one obtains for large z and t , that

$$(1.40) \qquad u(t,m_{1/2}(t) + z) \sim Ce^{-bz}$$

where C is some constant. Comparing (1.40) with (1.14), we see that $u(t,x)$ has essentially the same tail as $w^{\lambda}(x)$. A comparison argument will then show that (1.15) must in fact hold.

In order to investigate the case where (1.19) is satisfied, we again focus our attention on paths satisfying (1.38), which in this case reduces to $x(t) \sim 0$. Therefore, Brownian motion paths $\mathfrak{X}_x(s)$ which satisfy (1.38) do not typically tend to climb away from $m_{1/2}(t-s)$ and into the region $k \sim 1$ for large s. Rather, the typical path may cross $m_{1/2}(t-s)$ frequently, and be insufficiently weighted to affect (1.32). Instead, $u(t,x)$ is determined by those atypical paths which bend up quickly away from $m_{1/2}(t-s)$ and remain substantially above $m_{1/2}(t-s)$ except at values of s close to 0 and t (the end points).

A path $x(s)$ will be "heavily weighted" if for small r,

$$\begin{aligned} x(s) &> m_{1/2}(t-s) + s^{\delta} \qquad \text{for } r \leq s \leq t/2 \\ &> m_{1/2}(t-s) + (t-s)^{\delta} \quad \text{for } t/2 \leq s \leq t-r , \end{aligned} \tag{1.41}$$

where $0 < \delta < 1/2$ is fixed. It will be "insignificant" if for large r,

$$x(s) < m_{1/2}(t-s) - s^{\delta} \qquad \text{for some } r \leq s \leq t/2 \tag{1.42a}$$

or

$$x(s) < m_{1/2}(t-s) - (t-s)^{\delta} \quad \text{for some } t/2 \leq s \leq t-r . \tag{1.42b}$$

(The definitions of "heavily weighted" and "insignificant" paths employed in the text will actually be a bit different than those given here.) For paths satisfying (1.41),

$$\exp\{\int_0^t k(t-s,x(s))\,ds\} \sim e^t , \tag{1.43}$$

whereas for paths satisfying (1.42),

$$\exp\{\int_0^t k(t-s,x(s))\,ds\} \ll e^t . \tag{1.44}$$

Cumbersome estimates for Brownian bridge enable one to compute bounds on the probability that a path satisfies (1.41). In conjunction with (1.43) and (1.44), these bounds show that for reasonably large values of z, $u(t,m_{1/2}(t)+z)$ is essentially determined by the "heavily weighted" paths of

(1.41) and only marginally by the more numerous "insignificant" paths of (1.42). (The paths which are neither of type (1.41) nor type (1.42) may also be eliminated.) One obtains for large z and t, that

$$(1.45) \qquad u(t,m_{1/2}(t)+z) \sim C'ze^{-2^{1/2}z} ,$$

where C' is some constant. Comparison of (1.45) with (1.13) shows that $u(t,x)$ has essentially the same tail as $w^{2^{1/2}}(x)$. The same comparison argument as cited before implies (1.18).

The essential work in calculating $m(t)$ is in each case performed en route to obtaining (1.15) and (1.18). In the case $\lambda > 2^{1/2}$, comparison with (1.38) is sufficient. In the case $\lambda = 2^{1/2}$, one needs to compute the probability of the "heavily weighted" paths of (1.41).

Outline. Section 2 will introduce basic notation, and state certain elementary properties of Brownian motion and Brownian bridge which will be useful later on. In Section 3, certain basic facts about the Kolmogorov equation will be stated, and the proof of the original Kolmogorov theorem with Heaviside initial data will be outlined. Section 4 is devoted to obtaining estimates for (1.31) in the case where $k(t,x) \equiv 1$. These estimates are then applied in Section 5, where Theorems A and B for the case $\lambda > 2^{1/2}$ are demonstrated. The converse direction of Theorem A in the case $\lambda = 2^{1/2}$ is also demonstrated in this section, where it is shown that the condition (1.19) is necessary for (1.18) to hold. The remaining four sections of this article are devoted to the asymptotic behavior of $u(t,x)$ under (1.19). In Section 6, estimates are obtained on the probability that Brownian bridge lies above certain curves. In Section 7, these estimates are applied to obtain certain bounds involving the Feynman-Kac integral. These bounds are then applied in Section 8 and 9 to demonstrate convergence to the travelling wave moving at rate $\lambda = 2^{1/2}$, and to obtain formulas for the centering term $m(t)$.

2. SOME PROPERTIES OF BROWNIAN MOTION AND BROWNIAN BRIDGE

In this section we introduce basic notation and state certain elementary properties of Brownian motion and Brownian bridge which we will find useful later on. If we denote standard Brownian motion by $\mathfrak{X}$, then

$$(2.1) \qquad \mathfrak{z}^t(s) = \mathfrak{X}(s) - \frac{s}{t}\mathfrak{X}(t) \ , \ 0 \leq s \leq t \ ,$$

defines a new Gaussian process called Brownian bridge (ending at time t). For the sake of concreteness, it is assumed in this section that $\mathfrak{X}$ and $\mathfrak{z}^t$ are defined on $(C[0,t], \mathfrak{B}^t)$, where $C[0,t]$ denotes the continuous functions on $[0,t]$, and $\mathfrak{B}^t$ is the Borel σ-algebra. We will sometimes also be dealing with the completions of $\mathfrak{B}^t$, which will be denoted by $\bar{\mathfrak{B}}^t$. When confusion is not likely, we will omit the superscripts from $\mathfrak{z}^t$, $\mathfrak{B}^t$, and $\bar{\mathfrak{B}}^t$.

$p^1(s_1,x_1;s_2,x_2;\ldots;s_k,x_k)$ will denote the joint density of Brownian motion at points $x_1,x_2,\ldots,x_k$ for times $s_1,s_2,\ldots,s_k$, and

$$(2.2) \qquad p^1(s_1,x_1;\ldots;s_k,x_k \mid s_{k+1},x_{k+1};\ldots;s_{k+j},x_{k+j}) = \frac{p^1(s_1,x_1;\ldots;s_k,x_k;s_{k+1},x_{k+1};\ldots;s_{k+j},x_{k+j})}{p^1(s_{k+1},x_{k+1};\ldots;s_{k+j},x_{k+j})}$$

will denote the conditional density. Similarly, $p^0(s_1,x_1;\ldots;s_k,x_k)$ and $p^0(s_1,x_1;\ldots;s_k,x_k \mid s_{k+1},x_{k+1};\ldots;s_{k+j},x_{k+j})$ will stand for the density and joint density of Brownian bridge. We will sometimes find it convenient to deal with Brownian motion starting at x and Brownian bridge starting at x and ending at y. Employ $\mathfrak{X}_x$ to denote such a Brownian motion and $\mathfrak{z}_{x,y}$ to denote such a Brownian bridge. Note that $\mathfrak{z}_{x,y}(s)$, $0 \leq s \leq t$, is stochastically equivalent to $\mathfrak{z}(s) + sy/t + (t-s)x/t$, $0 \leq s \leq t$. The above notation may be applied as well to the case where $x,y \neq 0$. In (2.2), $s_{k+1} = 0$ and $x_{k+1} = x$ is taken to mean that Brownian motion starts at x. Analogous notation is used for $\delta_{x,y}$. If no mention is made of the time index 0 (and t), then it is assumed that $x = 0$ (and $y = 0$).

Since (2.2) defines a family of regular conditional probabilities for Brownian motion (with the analog holding for Brownian bridge),

$$p^1(s_1,x_1;\ldots;s_k,x_k;A \mid s_{k+1},x_{k+1};\ldots;s_{k+j},x_{k+j})$$

and

$$P^1[A \mid s_{k+1},x_{k+1};\ldots;s_{k+j},x_{k+j}]$$

(respectively, $p^0(s_1,x_1;\ldots;s_k,x_k;A \mid s_{k+1},x_{k+1};\ldots;s_{k+j},x_{k+j})$ and

$$P^0[A \mid s_{k+1},x_{k+1};\ldots;s_{k+j},x_{k+j}])$$

define canonical versions of conditional densities and conditional probabilities when interpreted in the obvious manner, where $A \in \mathcal{B}$. For $A_1,A_2 \in \mathcal{B}$ and $P^i[A_1 \mid s_{k+1},x_{k+1};\ldots;s_{k+j},x_{k+j}] > 0$, we define

$$p^i(s_1,x_1;\ldots;s_k,x_k;A_2 \mid s_{k+1},x_{k+1};\ldots;s_{k+j},x_{k+j};A_1)$$

$$= \frac{p^i(s_1,x_1;\ldots;s_k,x_k;A_1 \cap A_2 \mid s_{k+1},x_{k+1};\ldots;s_{k+j},x_{k+j})}{P^i[A_1 \mid s_{k+1},x_{k+1};\ldots;s_{k+j},x_{k+j}]}$$

for $i = 0,1$. ($P^i[A_2 \mid s_{k+1},x_{k+1};\ldots;s_{k+j},x_{k+j};A_1]$ is defined similarly.) We use the notation $P^i[s_1,x_1;\ldots;s_k,x_k \mid s_{k+1},x_{k+1};\ldots;s_{k+j},x_{k+j};A_2]$ for $P^i[A_1 \mid s_{k+1},x_{k+1};\ldots;s_{k+j},x_{k+j};A_2]$ if $A_1 = \{g \in C[0,t]: g(s_1) \leqq x_1,\ldots, g(s_k) \leqq x_k\}$.

The following notation will be convenient. If $\ell(s)$ is defined on $[s_1,s_2]$ (perhaps taking the value $-\infty$), then B_ℓ (or $B_\ell[s_1,s_2]$ in case of ambiguity) will denote the set of paths lying strictly above ℓ on $[s_1,s_2]$. If $\ell(s) \equiv x$, then we will write B_x; $B_\ell(S)$ will mean the set of paths lying strictly above $\ell(s)$ at the times $s_1,\ldots,s_n$, where $S = \{s_k: k = 1,\ldots,n\}$.

Similarly, B^{ℓ} will denote the set of paths lying strictly below ℓ on $[s_1,s_2]$; $B_{\ell_1}^{\ell_2}$ will denote the set of paths lying strictly between ℓ_1 and ℓ_2 on $[s_1,s_2]$. A (with or without subscripts) will be used for an unspecified element of $\mathcal{B}$. We will sometimes have occasion to use the notation $\lfloor x \rfloor$ (respectively $\lceil x \rceil$), which means the greatest integer less than or equal to x (respectively, the least integer greater than or equal to x).

In the remainder of this section, we state some lemmas concerning Brownian motion and Brownian bridge which will be useful later on. We begin by listing (without proof) some well-known properties of Brownian bridge.

<u>Lemma 2.1</u>. Brownian bridge (ending at time t) possesses the following properties:

(a) $$p^0(s_1,x_1;\ldots;s_k,x_k) = p^1(s_1,x_1;\ldots;s_k,x_k \mid t,0)$$
$$= p^1(s_1,x_1+\frac{s_1}{t}y;\ldots;s_k,x_k+\frac{s_k}{t}y \mid t,y) .$$

(b) $\mathfrak{z}(s)$ is stochastically equivalent to $\frac{(t-s)}{t}\,\mathfrak{x}\left(\frac{st}{t-s}\right)$, and $\mathfrak{x}(s)$ is stochastically equivalent to $\frac{(s+t)}{t}\,\mathfrak{z}\left(\frac{st}{t-s}\right)$.

(c) The Brownian bridge is strong Markov.

(d) $\mathfrak{z}(s)$ for $0 \leq s \leq t$ is independent of $\mathfrak{x}(t)$.

(e) $\mathfrak{z}(s)$ for $0 \leq s \leq t$ is stochastically equivalent to $\hat{\mathfrak{z}}(s) \overset{\text{def}}{=} \mathfrak{z}(t-s)$ for $0 \leq s \leq t$.

(f) $\mathfrak{z}^{s'}(s) = \mathfrak{z}^{t}(s) - \frac{s}{s'}\,\mathfrak{z}^{t}(s')$ for $0 \leq s \leq s' \leq t$.

(g) $\mathfrak{z}^{s'}(s)$ for $0 \leq s \leq s'$ is independent of $\mathfrak{z}^{t}(s')$.

The next lemma gives several applications of the reflection principle.

<u>Lemma 2.2</u>. (a) For $x_1,x_2 > 0$,

(2.3) $$P[\mathfrak{z}^t(s) > -\frac{s}{t}x_1 - \frac{(t-s)}{t}x_2 \text{ for } 0 \leq s \leq t] = 1-\exp[-2x_1x_2/t] .$$

(b) For $x > 0$ and $0 < s_o < t$,

$$(2.4) \qquad P[\mathfrak{z}^t(s) > -x \text{ for } 0 \leq s \leq s_o] > 1 - \frac{2\sqrt{s_o}}{x} e^{-x^2/2s_o} .$$

(c) For $x_1, x_2 > 0$ and $0 \leq x \leq x_2/2$,

$$(2.5) \qquad p^1(t,x; B^{x_2}_{-x_1}) \geq \sqrt{\frac{2}{\pi t^3}}\, x_1 e^{-x^2/2t}[x - (x_1 + 2x_2)e^{-x_2^2/2t} - \frac{x_1}{t}(x + x_1)^2] .$$

(We will be interested in the case where $x_1 \ll x \ll x_2, t$.)

(d) For $x_1, x_2 > 0$ and fixed c_o ,

$$(2.6) \qquad P[\mathfrak{z}^t_{x_1,x_2}(s) > c_o(s \wedge (t-s)) \text{ for } 0 \leq s \leq t] =$$
$$\sqrt{\frac{2}{\pi t}} \int_0^\infty (1-e^{-4x_1 x/t})(1-e^{-4x_2 x/t}) \exp[-2(x + \frac{c_o t}{2} - \frac{x_1 + x_2}{2})^2/t]\, dx .$$

<u>Proof</u>. (a) Application of Lemma 2.1 (a) shows that

$$P[\mathfrak{z}^t(s) > -\frac{s}{t}x_1 - \frac{(t-s)}{t}x_2 \text{ for } 0 \leq s \leq t] = P^1[B_{-x_2} \mid t, x_1 - x_2] .$$

By the reflection principle, this equals

$$\frac{p^1(t, x_1 - x_2; B_{-x_2})}{p^1(t, x_1 - x_2)} = 1 - \frac{p^1(t, -x_1 - x_2)}{p^1(t, x_1 - x_2)}$$

$$= 1 - \exp[-2x_1 x_2/t] .$$

(b) Applying Lemma 2.1 (a), it is easy to check that

$$P[\mathfrak{z}^t(s) > -x \text{ for } 0 \leq s \leq s_o] = 1 - P[\mathfrak{z}^t(s) \leq -x \text{ for some } s \in [0, s_o]]$$
$$\geq 1 - 2P[\mathfrak{x}(s) \leq -x \text{ for some } s \in [0, s_o]] .$$

By the reflection principle, this is at least

$$1 - 4\int_{-\infty}^{-x} \frac{e^{-y^2/2s_o}}{\sqrt{2\pi s_o}}\, dy > 1 - \frac{2\sqrt{s_o}}{x} e^{-x^2/2s_o} .$$

(c) Repeated application of the reflection principle shows that

$$p^1(t,x\,;\,B_{-x_1}^{x_2}) \geqq p^1(t,x) - p^1(t,x+2x_1) - p^1(t,x-2x_2) + p^1(t,x-2x_1-2x_2) \;.$$

This equals

$$\frac{1}{\sqrt{2\pi t}}\,[e^{-x^2/2t}(1-e^{-2x_1(x+x_1)/t}) - e^{-(x-2x_2)^2/2t}(1-e^{-2x_1(x_1+2x_2-x)/t})] \;.$$

It is easy to compute that for $0 \leqq x \leqq x_2/2$, this is at least

$$\sqrt{\frac{2}{\pi t^3}}\, x_1 e^{-x^2/2t}[x-(x_1+2x_2)e^{-x_2^2/2t} - \frac{x_1}{t}(x+x_1)^2] \;.$$

(d) The assertion follows by applying parts (f) and (g) of Lemma 2.1 to subdivide $\mathfrak{s}^t_{x_1,x_2}$ into Brownian bridges over the intervals $[0,t/2]$ and $[t/2,t]$, and then applying part (a) to each of the pieces. //

For the remaining results of this section, we will use the following lemma on the regularity of hitting probabilities of curves by Brownian bridge (or equivalently, by Brownian motion). Let $\ell(s)$, $0 \leqq s \leqq t$, denote an arbitrary function which is bounded above (but not necessarily measurable or finite valued). Choose $S_k = \{s_{k,1},\dots,s_{k,2^k}\}$, $s_{k,j} \in [0,t]$, so that

(2.7a) $\quad s_{k,j} \in \mathcal{J}_{k,j} \equiv [(j-1)t/2^k, jt/2^k]$

and

(2.7b) $\quad \ell(s_{k,j}) \geqq \sup\{\ell(s)\colon s \in \mathcal{J}_{k,j}\} - 2^{-k}$.

Also, set $S^n = \bigcup_{k=1}^{n} S_k$ and define $\ell^{S^n}(s)$, $0 \leqq s \leqq t$, as

$$(2.8)\qquad \begin{aligned} \ell^{S^n}(s) &= \ell(s) \quad \text{for } s \in S^n \\ &= -\infty \quad \text{otherwise .} \end{aligned}$$

Then, the following result holds:

<u>Lemma 2.3</u>. Let ℓ , S^n , and ℓ^{S^n} , $n=1,2,\dots,$ be defined as above.

Then, $A \equiv \{\mathfrak{z} : \mathfrak{z}(s) > \ell(s)$ for $0 < s < t\}$ is measurable with respect to $\bar{\mathfrak{B}}$, and

$$(2.9) \qquad P[\mathfrak{z}(s) > \ell^{S^n}(s) \text{ for } 0 < s < t] \downarrow P[A] .$$

<u>Proof</u>. Since $\mathfrak{z}(s)$ has continuous sample paths, it follows without difficulty from the definition of S^n that

$$\bar{A} \equiv \{\mathfrak{z} : \mathfrak{z}(s) \geq \ell(s) \text{ for } 0 < s < t\} = \bigcap_{n=1}^{\infty} \{\mathfrak{z} : \mathfrak{z}(s) \geq \ell^{S^n}(s) \text{ for } 0 < s < t\} .$$

Thus, (2.9) is valid if ">" is replaced with "$\geq$". Also,

$$\{\mathfrak{z} : \mathfrak{z}(s) > \ell^{S^n}(s) \text{ for } 0 < s < t\} \subset \{\mathfrak{z} : \mathfrak{z}(s) \geq \ell^{S^n}(s) \text{ for } 0 < s < t\}$$

for all n; to demonstrate the lemma it therefore suffices to show that $A \in \bar{\mathfrak{B}}$ and $P[A] = P[\bar{A}]$. We will find it simpler to instead think in terms of

$$A_0(\bar{A}_0) \equiv \{\mathfrak{z} : \mathfrak{z}(s) > (\geq) \ell(s) \text{ for } 0 < s \leq t/2\}$$

$$A_1(\bar{A}_1) \equiv \{\mathfrak{z} : \mathfrak{z}(s) > (\geq) \ell(s) \text{ for } t/2 \leq s < t\} ,$$

where $A = A_0 \cap A_1$ and $\bar{A} = \bar{A}_0 \cap \bar{A}_1$.

First of all, introduce the processes ${}_\epsilon\mathfrak{z}(s)$, $0 \leq s \leq t/2$, where

$${}_\epsilon\mathfrak{z}(s) \equiv \mathfrak{z}(s) - \epsilon s .$$

It is easy to check that the measures induced by $\mathfrak{z}(t/2)$ and ${}_\epsilon\mathfrak{z}(t/2)$, μ and ${}_\epsilon\mu$, are mutually absolutely continuous. Moreover, for each $M > 0$,

$$\frac{d\,{}_\epsilon\mu}{d\mu} \to 1 \text{ uniformly in } [-M,M] \text{ as } \epsilon \to 0 .$$

It therefore follows from Lemma 2.1 (a) that

$$P[\mathfrak{z}(s) \geq \ell(s) + \epsilon s \text{ for } 0 < s \leq t/2] = P[{}_\epsilon\mathfrak{z}(s) \geq \ell(s) \text{ for } 0 < s \leq t/2]$$
$$\uparrow P[\mathfrak{z}(s) \geq \ell(s) \text{ for } 0 < s \leq t/2]$$

as $\epsilon \downarrow 0$, which equals $P[\bar{A}_0]$. But note that

$$\bigcup_{\epsilon>0} \{\mathfrak{z} : \mathfrak{z}(s) \geq \ell(s) + \epsilon s \quad \text{for} \quad 0 < s \leq t/2\} \subset A_0 \subset \bar{A}_0 .$$

Thus, A_0 is squeezed in between two sets having the same measure, which implies that $A_{1/2} \in \bar{\mathfrak{B}}$ with $P[A_0] = P[\bar{A}_0]$. Similar reasoning may be applied to A_1 and $\bar{A}_1$ to show that $A_1 \in \bar{\mathfrak{B}}$ with $P[A_1] = P[\bar{A}_1]$. Since $A = A_0 \cap A_1$ and $\bar{A} = \bar{A}_0 \cap \bar{A}_1$, it is obvious that $A \in \bar{\mathfrak{B}}$ and $P[A] = P[\bar{A}]$, which concludes the argument. //

The remainder of this section consists of four further lemmas, each of which should become intuitively clear after a little thought. The next lemma states that Brownian motion restricted to lying above a fixed function retains certain of its regularity properties.

Lemma 2.4. (a) Define $\hat{\ell}(s) = \ell(t-s)$ for $0 \leq s \leq t$. Then for all x_1 and x_2,

$$(2.10) \qquad p^1(t,x_2 ; B_\ell \mid 0,x_1) = p^1(t,x_1 ; B_{\hat{\ell}} \mid 0,x_2) .$$

(b) For all $r \in (0,t)$ and x,

$$(2.11) \quad p^1(t,x ; B_\ell[0,t]) = \int_{-\infty}^{\infty} p^1(r,y ; B_\ell[0,r]) p^1(t,x ; B_\ell[r,t] \mid r,y)\, dy .$$

(c) $p^1(t,x ; B_\ell^L[0,t])$ is jointly continuous in t and x for (t,x) in the interior of $G = \{(s,y) : \ell(s) < y < L(s)\}$.

Proof. (a) Let $S = \{s_k : k=1,\ldots,n\}$, where $0 = s_1 < s_2 < \ldots < s_n = t$, and $\hat{S} = \{s_k' : k=1,\ldots,n\}$, where $s_k' = t - s_k$. Then, since

$$\begin{aligned} p^1(s_{k+1},y_{k+1} \mid s_k,y_k) &= p^1(s_{k+1}-s_k,y_{k+1}-y_k) \\ &= p^1(s_{k+1}-s_k,y_k-y_{k+1}) , \end{aligned}$$

it is an elementary observation that

$$p^1(t,x_2;B_\ell[S] \mid 0,x_1)$$

$$= \int_{-\infty}^{\ell(s_2)} dy_2 \cdots \int_{-\infty}^{\ell(s_{n-1})} dy_{n-1}\, p^1(s_2,y_2 \mid 0,x_1) \cdots$$

$$\cdot\, p^1(s_{n-1},y_{n-1} \mid s_{n-2},y_{n-2})\, p^1(t,x_2 \mid s_{n-1},y_{n-1})$$

(2.12)
$$= \int_{-\infty}^{\hat{\ell}(s_2')} dy_{n-1} \cdots \int_{-\infty}^{\ell(s_{n-1}')} dy_2\, p^1(s_{n-1}',y_{n-1} \mid 0,x_2) \cdots$$

$$\cdot\, p^1(s_2',y_2 \mid s_3',y_3)\, p^1(t,x_1 \mid s_2',y_2)$$

$$= p^1(t,x_1;B_{\hat{\ell}}[\hat{S}] \mid 0,x_2) \quad .$$

Choose S^n as in Lemma 3. Then

$$\lim_{n\to\infty} p^1(t,x_2;B_\ell[S^n] \mid 0,x_1) = p^1(t,x_2;B_\ell \mid 0,x_1)$$

and

$$\lim_{n\to\infty} p^1(t,x_1;B_\ell[\hat{S}^n] \mid 0,x_2) = p^1(t,x_1;B_\ell \mid 0,x_2) \quad ,$$

and part (a) follows from (2.12).

(b) One may reason as in part (a). The discrete analogue of (2.11) is easy to obtain. Then apply monotone convergence and Lemma 2.3 to obtain (2.11).

(c) The statement can be shown by using the decomposition

$$p^1(t,x;B_\ell^L[0,t])$$

$$= \int_{-\infty}^{\infty} dy\, p^1(t-\delta,y;B_\ell^L[0,t-\delta]) \cdot p^1(t,x;B_\ell^L[t-\delta,t] \mid t-\delta,y) \quad ,$$

noting that for $(t,x) \in \overset{o}{G}$ and small $x-y$,

$$p^1(t,x;B_\ell^L[t-\delta,t] \mid t-\delta,y)/p^1(\delta,x-y) \to 1 \quad \text{as} \quad \delta \to 0 \ ,$$

and then applying the smoothness of $p^1(\delta,x-y)$ in both arguments. //

In the next lemma, we make use of coupling. If $Z_i(s)$, $i=1,\ldots,n$, denote processes, then the joint process $Z(s) = (\tilde{Z}_1(s),\ldots,\tilde{Z}_n(s))$ is said to couple $Z_i(s)$ if for each i , $Z_i(s)$ and $\tilde{Z}_i(s)$ are stochastically equivalent. A simple example is the case where $Z_{x_1}(s;\ell)$ and $Z_{x_2}(s;\ell)$ denote Brownian bridges on $[0,t]$ with $Z_{x_i}(0;\ell) = x_i$, $Z_{x_i}(t;\ell) = 0$, and $x_1 < x_2$, and which are conditioned to lie above the function $\ell(s)$. (Assume that $P^0[B_\ell \mid 0,x_i] > 0$.) Then it is simple to find a coupling $Z(s) = (\tilde{Z}_{x_1}(s;\ell), \tilde{Z}_{x_2}(s;\ell))$ so that

$$(2.13) \qquad \tilde{Z}_{x_1}(s;\ell) \leq \tilde{Z}_{x_2}(s;\ell) \quad \text{for} \quad 0 \leq s \leq t .$$

For, assuming that $Z_{x_i}(s;\ell)$ are defined on the same probability space, we may set $T = \inf\{s \in [0,t]: Z_{x_1}(s;\ell) \geq Z_{x_2}(s;\ell)\}$, and define

$$(2.14a) \qquad \tilde{Z}_{x_1}(s;\ell) = Z_{x_1}(s;\ell) \quad \text{for} \quad 0 \leq s \leq t$$

and

$$(2.14b) \qquad \begin{aligned} \tilde{Z}_{x_2}(s;\ell) &= Z_{x_2}(s;\ell) \quad \text{for} \quad 0 \leq s \leq T \\ &= Z_{x_1}(s;\ell) \quad \text{for} \quad T \leq s \leq t . \end{aligned}$$

Since $Z_{x_i}(s;\ell)$ are strong Markov and have the same transition probabilities, and since T is a stopping time, it follows from (2.14) that $\tilde{Z}_{x_i}(s;\ell)$ are equivalent to $Z_{x_i}(s;\ell)$. From construction, (2.13) is satisfied, and so $Z(s)$ is as desired. We now apply an extension of this reasoning to demonstrate Lemma 2.5. (In practice, coupling is done more systematically by directly constructing a process with the prescribed joint generators. For more detail, see Griffeath [10].)

<u>Lemma 2.5</u>. Let S denote a finite set in $[0,t]$, and let $\ell_i^S(s)$ be defined as in (2.8) for functions $\ell_1(s) \leq \ell_2(s)$ if $s \in [0,t]$. Then there is a coupling $Z(s) = (\tilde{Z}_0(s;\ell_1^S), \tilde{Z}_0(s;\ell_2^S))$ of $Z_0(s;\ell_i^S)$ for which

$$(2.15) \qquad \tilde{Z}_0(s;\ell_1^S) \leq \tilde{Z}_0(s;\ell_2^S) \quad \text{for} \quad 0 \leq s \leq t .$$

<u>Proof</u>. Since one may proceed inductively in the construction of ℓ_2^S from

ℓ_1^S by altering a single point at a time, it is no loss of generality to assume that

$$\ell_1^S(s) = \ell_2^S(s) \quad \text{if} \quad s \neq s_o ,$$

for some fixed point $s_o \in [0,t]$. It is easy to see that in this case,

$$P[Z_0(s_o;\ell_1^S) \leq x] \geq P[Z_0(s_o;\ell_2^S) \leq x] \quad \text{for all} \quad x .$$

In particular, we may assume that the processes are already coupled so that $Z_0(s_o;\ell_1^S) \leq Z_0(s_o;\ell_2^S)$. As in (2.14a), we set $\tilde{Z}_0(s;\ell_1^S)$ equal to $Z_0(s;\ell_1^S)$ for $s \in [0,t]$. To construct $\tilde{Z}_0(s;\ell_2^S)$, we consider the intervals $[0,s_o)$ and $(s_o,t]$ separately. Mimicking (2.14b), define

$$\begin{aligned} \bar{Z}_0(s;\ell_2^S) &= Z_0(s;\ell_2^S) \quad \text{for} \quad 0 \leq s \leq T \\ &= Z_0(s;\ell_1^S) \quad \text{for} \quad T \leq s \leq t , \end{aligned} \tag{2.16}$$

where $T = \inf\{s \in [s_o,t] : Z_0(s;\ell_1^S) = Z_0(s;\ell_2^S)\}$. Since $Z_0(s;\ell_i^S)$ are strong Markov and T is a stopping time, $Z_0(s;\ell_2^S)$ and $\bar{Z}_0(s;\ell_2^S)$ are equivalent. Also, (2.15) clearly holds for $s \in [s_o,t]$. The reasoning in the interval $[0,s_o]$ follows analogously once one realizes that one may apply the equivalent of (2.16) to $\hat{Z}_0(s;\ell_2^S) \equiv \bar{Z}_0(t-s;\ell_2^S)$. Therefore, applying Lemma 2.4(a) to make the switch back to forward time, one may conclude that

$$\begin{aligned} \tilde{Z}_0(s;\ell_2^S) &= Z_0(s;\ell_1^S) \quad \text{for} \quad 0 \leq s \leq t-U \\ &= \bar{Z}_0(s;\ell_2^S) \quad \text{for} \quad t-U \leq s \leq t , \end{aligned}$$

where $t-U = \sup\{s \in [0,s_o] : Z_0(s;\ell_1^S) = Z_0(s;\ell_2^S)\}$, is also equivalent to $Z_0(s;\ell_2^S)$. (U is a stopping time for $\hat{Z}_0$.) Clearly, (2.15) is now satisfied over $[0,t]$, which completes the proof. //

The following lemma is a consequence of Lemma 2.5, and will prove to be quite useful in Section 6.

Lemma 2.6. Assume that $\ell_1(s)$, $\ell_2(s)$, and $\Lambda(s)$ satisfy $\ell_1(s) \leq \ell_2(s) < \Lambda(s)$ for $s \in [0,t]$, and that $P^0[B_{\ell_2}[0,t]] > 0$. Then,

$$P^0[B^\Lambda \mid B_{\ell_1}] \geq P^0[B^\Lambda \mid B_{\ell_2}] \tag{2.17}$$

and

$$P^0[B_\Lambda \mid B_{\ell_1}] \leq P^0[B_\Lambda \mid B_{\ell_2}] . \tag{2.18}$$

Proof. Let S_1^n and S_2^n , $n=1,2,\ldots,$ denote finite subsets of $[0,t]$ chosen so that (2.17) is satisfied for ℓ_1 and ℓ_2 , respectively. Also, set $S^n = S_1^n \cup S_2^n$. By Lemma 2.3, $P^0[B_{\ell_i}[S^n]] \downarrow P^0[B_{\ell_i}]$ as $n \uparrow \infty$. Analogous reasoning shows that $P^0[B^\Lambda[S^n]] \downarrow P^0[B^\Lambda]$, and consequently $P^0[B^\Lambda[S^n] \mid B_{\ell_i}[S^n]] \to P^0[B^\Lambda \mid B_{\ell_i}]$ as $n \to \infty$. Therefore, to demonstrate (2.17) it is sufficient to show that

$$P^0[B^\Lambda[S^n] \mid B_{\ell_1}[S^n]] \geq P^0[B^\Lambda[S^n] \mid B_{\ell_2}[S^n]] \tag{2.19}$$

for each n . Similarly, to demonstrate (2.18), it is sufficient to show that

$$P^0[B_\Lambda[S^n] \mid B_{\ell_1}[S^n]] \leq P^0[B_\Lambda[S^n] \mid B_{\ell_2}[S^n]] . \tag{2.20}$$

But (2.19) and (2.20) are both immediate consequences of (2.15), with $S = S^n$. //

We conclude this section with a computation which will be used in Proposition 6.1.

Lemma 2.7. Set

$$\begin{aligned} \Lambda_t(s) &= Cs^\epsilon \quad \text{for } 0 \leq s \leq t/2 \\ &= C(t-s)^\epsilon \quad \text{for } t/2 \leq s \leq t , \end{aligned}$$

where $\epsilon > 1/2$ and $C > 0$. Then,

$$P^0[B^{\wedge_t}[r,t-r] \mid B_0[r,t-r]] \to 1 \text{ uniformly in } t \text{ as } r \to \infty .$$

Proof. We will find it more convenient to first investigate $1 - P^0[B^{\wedge_t}[r,t-r] \mid B_{-c}[r,t-r]]$, where $0 < c < 1$. By (2.17), this is at most

$$(2.21)\quad \begin{aligned} 1 - P^0[B^{\wedge_t}[r,t-r] \mid B_{-c}[0,t]] &\leq \sum_{k=\lceil r \rceil}^{\lceil t-r \rceil} (1 - P^0[B^{\wedge_t}[k-1,k] \mid B_{-c}[0,t]]) \\ &\leq 2 \sum_{k=\lceil r \rceil}^{\lceil t/2 \rceil} (1 - P^0[B^{\wedge_t}[0,k] \mid B_{-c}[0,t]]) , \end{aligned}$$

where the last inequality follows from the symmetry of $\wedge_t$ around $t/2$. We now obtain upper bounds for the individual summands. It can be checked that

$$(2.22)\quad \begin{aligned} &P^0[B^{\wedge_t}[0,k] \mid B_{-c}[0,t]] \\ &= \int_{-c}^{\wedge_t(k)} dx\, p^0(k,x) \cdot P^0[B_{-c}^{\wedge_t(k)}[0,k] \mid k,x] \cdot P^0[B_{-c}[k,t] \mid k,x] / P^0[B_{-c}[0,t]] \\ &\geq \int_0^{\wedge_t(k)/2} . \end{aligned}$$

Applying Lemmas 2.1 and 2.2, we obtain the following bounds for the individual components of (2.21):

$$\begin{aligned} P^0[B_{-c}^{\wedge_t(k)}[0,k] \mid k,x] &= P^1[B_{-c}^{\wedge_t(k)}[0,k] \mid k,x] \\ &\geq \frac{2c}{k}[x - 3\wedge_t(k) e^{-\wedge_t^2(k)/k} - \frac{c}{k}(x+1)^2] , \end{aligned}$$

$$P^0[B_{-c}[k,t] \mid k,x] \geq 2cx(1 - \frac{cx}{t})/t ,$$

and

$$P^0[B_{-c}[0,t]] \leq 2c^2/t$$

for $0 \leq x \leq \wedge_t(k)/2$. Therefore, for $k \leq t/2$, (2.22) is at least

$$\int_0^{\Lambda_t(k)/2} \frac{4x}{u_k}\cdot\frac{e^{-x^2/2u_k}}{\sqrt{2\pi u_k}}[x-3\Lambda_t(k)e^{-\Lambda_t^2(k)/k}-c(x+1)^2/k][1-cx/t]\,dx ,$$

where $u_k = k(t-k)/t$. After some computation, it can be shown that this is at least

$$(2.23)\qquad (1-c\Lambda_t(k)/t)(1-a_1Cke^{-\Lambda_t^2(k)/k}-a_2c) ,$$

where a_1 and a_2 are fixed constants.

Now, note that by (2.17) of Lemma 2.6, $P^0[B^{\Lambda_t}[0,k] \mid \underline{B}_c[0,t]]$ is an increasing function of $c > 0$. Therefore, we may set $c=0$ in (2.23) to obtain

$$P^0[B^{\Lambda_t}[0,k] \mid \underline{B}_c[0,t]] \geqq 1-a_1Cke^{-\Lambda_t^2(k)/k}$$

$$= 1-a_1Cke^{-Ck^\delta} ,$$

where $\delta = 2\varepsilon-1 > 0$. Going back to (2.21), we see that

$$(2.24)\qquad 1-P^0[B^{\Lambda_t}[r,t-r] \mid \underline{B}_c[r,t-r]] \leqq 2a_1C\sum_{k=\lceil r\rceil}^{\lceil t/2\rceil} ke^{-Ck^\delta}$$

for all $0<c<1$. Since the left side of (2.24) is clearly continuous in c , it follows that

$$(2.25)\qquad 1-P^0[B^{\Lambda_t}[r,t-r] \mid B_0[r,t-r]] \leqq 2a_1C\sum_{k=\lceil r\rceil}^{\infty} ke^{-Ck^\delta} .$$

But the right side of (2.25) is finite and does not depend on t . Consequently,

$$1-P^0[B^{\Lambda_t}[r,t-r] \mid B_0[r,t-r]] \to 0$$

uniformly in t as $r\to\infty$, which demonstrates the lemma. //

3. BASIC PROPERTIES OF THE KOLMOGOROV EQUATION

In Section 3 we discuss the basic properties of the Kolmogorov equation

$$(1.1) \qquad u_t = \frac{1}{2}u_{xx} + f(u) \ .$$

We begin by outlining the construction of solutions of (1.1), and giving certain properties of the associated travelling waves $w^{\lambda}(x) = u(t,x+\lambda t)$. We then state the maximum principle and an important extension (Proposition 3.2), which will lead up to a proof of the Kolmogorov theorem. The remainder of the section is devoted to obtaining various regularity properties for the solution of (1.1), which we will use in later sections.

The Kolmogorov equation is sufficiently well-behaved under (1.2) so that there is no difficulty in establishing either existence or uniqueness of the solution under measurable initial data $u(0,x)$ satisfying $0 \leq u(0,x) \leq 1$ for all x . These assumptions on the initial data will be made implicitly throughout the remainder of the article. Existence may be done by Picard iteration, where $u^n(t,x)$, $n=1,2,\ldots,$ are constructed recursively so that

$$(3.1a) \qquad u_t^1 = \frac{1}{2}u_{xx}^1$$

and

$$(3.1b) \qquad u_t^n = \frac{1}{2}u_{xx}^n + f(u^{n-1})$$

for $n \geq 2$. By inverting (3.1b), it is not difficult to show that $u^n(t,x)$ converges uniformly on compact time intervals to a solution of (1.1). Uniqueness is the consequence of a simple Gronwall inequality.

The travelling wave solutions of (1.1), i.e., solutions $u(t,x)$ which satisfy

$$(1.11) \qquad w^{\lambda}(x) = u(t,x+\lambda t)$$

where

(1.6a) $0 < w(x) < 1$ for all x

and

$$w^{\lambda}(x) \to 0 \text{ as } x \to \infty$$
(1.6b)
$$\to 1 \text{ as } x \to -\infty ,$$

are also not difficult to investigate. It is immediate from (1.1) that $w^{\lambda}(x)$ satisfies

(1.12) $$0 = \frac{1}{2} w^{\lambda}_{xx} + \lambda w^{\lambda}_{x} + f(w^{\lambda}) ,$$

or equivalently,

(3.2) $$\dot{w}^{\lambda} = p$$
$$\dot{p} = -2\lambda p - 2f(w^{\lambda}) .$$

Investigation of the singularities at $(0,0)$ and $(1,0)$ in the phase plane (w^{λ}, p) shows that

(3.3) $(0,0)$ is a node for $\lambda \geqq 2^{1/2}$
$(0,0)$ is a spiral point for $0 < \lambda < 2^{1/2}$

and that the other singularity

(3.4) $(1,0)$ is hyperbolic for all $\lambda > 0$.

A little thought shows that (1.6) rules out the existence of a travelling wave for $\lambda < 2^{1/2}$. One can on the other hand show that for $\lambda \geqq 2^{1/2}$, there is a unique solution (up to translation) of (1.12) which satisfies (1.6). Moreover, it is easy to demonstrate from (3.2) that for $\lambda > 2^{1/2}$,

(1.14) $$w^{\lambda}(x) \sim Ce^{-bx} \text{ as } x \to \infty ,$$

where $b = \lambda - \sqrt{\lambda^2 - 2} < 2^{1/2}$ and $C > 0$ depends on the choice of w^{λ} .

For $\lambda = 2^{1/2}$, (3.2) has a degenerate eigenvector, and one may show by means of a shooting argument that under (1.2'),

$$(1.13) \qquad w^{2^{1/2}}(x) \sim Cxe^{-2^{1/2}x} \quad \text{as} \quad x \to \infty$$

(Aronson [1]). The dissimilar behavior of the tails in (1.13) and (1.14) will turn out to be significant when we study the asymptotic behavior (in t) of $u(t,x)$ under general initial data. (The left tails of w^{λ} may also be computed, but are not relevant for our purposes.) For more detail up through (3.4), see [3].

The Kolmogorov equation, although nonlinear, still exhibits parabolic behavior. In particular, a maximum principle is valid. We quote here the version found in [2].

Proposition 3.1 (Maximum Principle). Let $u^1(t,x) \in [0,1]$ and $u^2(t,x) \in [0,1]$ satisfy the inequalities

$$(3.5) \qquad u^2_t - \tfrac{1}{2}u^2_{xx} - f(u^2) \geqq u^1_t - \tfrac{1}{2}u^1_{xx} - f(u^1) \quad \text{in} \quad (0,T] \times (a,b) \ ,$$

$$(3.6) \qquad 0 \leqq u^1(0,x) \leqq u^2(0,x) \leqq 1 \quad \text{in} \quad (a,b) \ ,$$

where $0 < T \leqq \infty$, $-\infty \leqq a < b \leqq \infty$, and f satisfies (1.2). Moreover, if $a > -\infty$, assume that

$$(3.7a) \qquad 0 \leqq u^1(t,a) \leqq u^2(t,a) \leqq 1 \quad \text{on} \quad [0,T]$$

and if $b < \infty$, assume that

$$(3.7b) \qquad 0 \leqq u^1(t,b) \leqq u^2(t,b) \leqq 1 \quad \text{on} \quad [0,T] \ .$$

Then $u^2 \geqq u^1$, and if $u^2(0,x) > u^1(0,x)$ in an open subinterval of (a,b) , then $u^2 > u^1$ in $(0,T] \times (a,b)$.

Proof. By the mean-value theorem,

$$(u^2 - u^1)_t - \frac{1}{2}(u^2 - u^1)_{xx} \geq f(u^2) - f(u^1)$$

$$= f'(u^1 + \theta(u^2 - u^1))(u^2 - u^1)$$

for some function $\theta \in (0,1)$. Let $\alpha = \max_{[0,1]} f'(u)$, and define

$$v(t,x) = e^{-\alpha t}(u^2(t,x) - u^1(t,x)) .$$

Then,

$$v_t - \frac{1}{2}v_{xx} \geq [f'(u^1 + \theta(u^2 - u^1)) - \alpha]v .$$

The initial and boundary data of v are nonnegative, whereas the coefficient of v is nonpositive. The proposition therefore follows from the usual maximum principle for linear parabolic inequalities. (See, e.g., Protter-Weinberger [16], page 172.) //

(One can also, if one wishes, give a probabilistic proof of Proposition 3.1 which uses the Feynman-Kac formula.)

The maximum principle will serve as an invaluable tool in obtaining estimates for $u(t,x)$. We mention in passing that since 0 and 1 are both zeroes of f , the maximum principle trivially implies that if

$$0 \leq u(0,x) \leq 1 \quad \text{for all} \quad x ,$$

then in fact

$$0 \leq u(t,x) \leq 1 \quad \text{for all} \quad t \quad \text{and} \quad x .$$

Also, if $u(0,x)$ is monotone, then by the maximum principle, so is $u(t,x)$. Typically, the maximum principle gives restrictions on growth. Later in the section, we will present various applications of the maximum principle in both its linear and nonlinear forms. For the moment, we restrict ourselves to the following simple result, which limits the rate of propagation of $u(t,x)$.

Corollary 1. Let $u^1(t,x)$ and $u^2(t,x)$ be solutions of (1.1) where $u^1(t,x)$

satisfies (1.2) and $u^2(t,x)$ satisfies the linear equation

$$(3.8) \qquad u_t = \frac{1}{2}u_{xx} + u .$$

If $u^1(0,x) = u^2(0,x)$ for all x, then

$$u^1(t,x) \leq u^2(t,x) \quad \text{for all } t \text{ and } x .$$

Consequently under Heaviside initial data, the centering term $m(t)$ of $u^1(t,x)$ satisfies

$$(3.9) \qquad m(t) < 2^{1/2}t - 2^{-3/2}\log t + C ,$$

for some constant C.

Proof. The first inequality follows immediately from the maximum principle and $f(u) \leq u$ for $0 \leq u \leq 1$. The solution $u^2(t,x)$ is easily computed to be

$$(3.10) \qquad u^2(t,x) = e^t \int_{-\infty}^{0} (2\pi t)^{-1/2} e^{-(x-y)^2/2t}\, dy ,$$

from which one obtains (3.9) without difficulty. //

The estimate (3.9) is only one-sided, and turns out to differ from the actual value of $m(t)$ by $2^{-1/2}\log t$. However, as will be seen shortly, this estimate does provide us a way of ruling out $w^\lambda(x)$, $\lambda > 2^{1/2}$, as a possible limit for $u(t,x+m(t))$ under Heaviside initial data.

En route to demonstrating the Kolmogorov theorem for Heaviside initial data, we next state a natural generalization of the maximum principle. The maximum principle says that if $u^1(t,x)$ and $u^2(t,x)$ satisfy (1.1) and (1.2), then the difference

$$(3.11) \qquad u^2(t,x) - u^1(t,x)$$

remains nonnegative for all t if it is nonnegative at $t = 0$. We show that it is also true that if (3.11) has at most one sign change at $t = 0$, then

(3.11) will continue to have at most one sign change for all t. This statement is the essential step in the reasoning leading up to the Kolmogorov theorem, and implies that the data $u(t,x)$ is "stretching" as t increases. We give here the elegant proof from McKean [15]. The proof is probabilistic and employs the Feynman-Kac integral, which was briefly introduced in Section 1. (See Friedman [7], page 148, or Kac [12] for a more detailed explanation.) Analytic proofs seem to be more cumbersome.

Proposition 3.2. Let $u^1(t,x)$ and $u^2(t,x)$ satisfy (1.1) and (1.2) with initial data chosen so that if $x_1 < x_2$,

$$u^2(0,x_1) > u^1(0,x_1) \text{ implies that } u^2(0,x_2) \geq u^1(0,x_2) . \tag{3.12}$$

Then for all t, if $x_1 < x_2$,

$$u^2(t,x_1) > (\geq)\, u^1(t,x_1) \text{ implies that } u^2(t,x_2) > (\geq)\, u^1(t,x_2) . \tag{3.13}$$

Proof. As in Proposition 3.1, if $v(t,x) \equiv u^2(t,x) - u^1(t,x)$, then

$$v_t = \frac{1}{2} v_{xx} + f'(u^1 + \theta(u^2 - u^1))v \tag{3.14}$$

for some function $\theta \in (0,1)$. To demonstrate (3.13), it suffices to show that if $v(t,x_1) > 0$ and $x_1 < x_2$, then $v(t,x_2) > 0$; the demonstration that $v(t,x_2) < 0$ implies $v(t,x_1) < 0$ is the same. We first rewrite (3.14) as

$$v_t = \frac{1}{2} v_{xx} + k(t,x)v ,$$

where $k(t,x) = f'(u^1 + \theta(u^2 - u^1))$. By the Feynman-Kac formula,

$$v(t,x) = E[\exp\{\int_0^T k(t-s,\mathfrak{X}_x(s))\, ds\} v(t-T,\mathfrak{X}_x(T))] , \tag{3.15}$$

where $0 \leq T \leq t$. ($\mathfrak{X}_x$ denotes Brownian motion started at x.) It is not difficult to see that

$$M(T) = \exp\{\int_0^T k(t-s,\mathfrak{X}_x(s))\, ds\} v(t-T,\mathfrak{X}_x(T))$$

is a martingale. Therefore, (3.15) is also valid when T is a non-constant stopping time. In particular,

$$T = \inf_{0 \leq s \leq t} \{s: M(s) = 0\} \wedge t$$

is a stopping time. Consequently, if we set $x = x_1$,

$$0 < v(t,x_1) = E[M(T)] ,$$

from which it follows that $P[M(T) > 0] > 0$. In particular, there is some continuous path $\gamma(s)$ with $\gamma(0) = x_1$ so that

$$(3.16) \qquad v(t-s,\gamma(s)) > 0 \quad \text{for} \quad 0 \leq s \leq t .$$

On the other hand, one may again reason as above, but this time setting

$$T' = \inf_{0 \leq s \leq t} \{s: \mathfrak{X}_x(s) = \gamma(s)\} \wedge t$$

and $x = x_2$. By (3.12), $M(T') \geq 0$, where strict inequality holds if $T' < t$. It is easy to see that $P[T' < t] > 0$. We therefore obtain that

$$v(t,x_2) = E[M(T')] > 0 ,$$

which completes the proof. //

We now state two corollaries to Proposition 3.2 which will be employed in the proof of the Kolmogorov theorem. Here and later on, for $0 < \epsilon < 1$, we set

$$(3.17) \qquad m_\epsilon(t) = \sup\{x: u(t,x) \geq \epsilon\} .$$

(Set $m_\epsilon(t) = -\infty$ if the set is void.)

Corollary 1. Let $u(t,x)$ be the solution of (1.1) and (1.2) with Heaviside initial data, i.e., data satisfying (1.3). Then for fixed ϵ, $0 < \epsilon < 1$,

$$(3.18) \qquad \begin{aligned} u(t,x+m_\epsilon(t)) &\uparrow w(x) \quad \text{for} \quad x > 0 \\ &\downarrow w(x) \quad \text{for} \quad x \leq 0 \end{aligned}$$

for some function $w(x)$ as $t \uparrow \infty$.

Proof. For each $t_0, a \in \mathbb{R}^+$, we may set

$$u^1(t,x) = u(t,x+m_\epsilon(t_0))$$
$$u^2(t,x) = u(t+a,x+m_\epsilon(t_0+a)) .$$

Since $u^1(t,x)$ has Heaviside initial data, it is obvious that (3.12) is satisfied in this context. For $t=t_0$ and $t=t_0+a$, $u(t,x+m_\epsilon(t))=\epsilon$; it therefore follows from Proposition 3.2 that

$$\begin{aligned} u(t_0+a,x+m_\epsilon(t_0+a)) &\geqq u(t_0,x+m_\epsilon(t_0)) \quad \text{for } x \geqq 0 \\ &\leqq u(t_0,x+m_\epsilon(t_0)) \quad \text{for } x \leqq 0 . \end{aligned}$$

Since t_0 and a were chosen arbitrarily, (3.18) follows. //

For the next corollary, we formalize the concept of "stretching" anticipated in (3.18). If g and h are functions on $\mathbb{R}$ with h being monotonically decreasing, then g is said to be <u>more stretched</u> than h, written $g \overset{s}{\geqq} h$, if for each c and $x_1 < x_2$, $g(x_1) > h(x_1+c)$ implies that $g(x_2) > h(x_2+c)$ and $g(x_1) \geqq h(x_1+c)$ implies that $g(x_2) \geqq h(x_2+c)$. If g and h are differentiable, $g \overset{s}{\geqq} h$ means that g lies above h in their mutual phase plane, that is, if $g(x_1) = h(x_2)$ for some x_1 and x_2, then $g'(x_1) \geqq h'(x_2)$.

Corollary 2. Let $u^1(t,x)$ and $u^2(t,x)$ be solutions of (1.1) and (1.2) with

$$(3.19) \qquad u^1(0,x) \overset{s}{\leqq} u^2(0,x) .$$

Then for all t,

$$(3.20) \qquad u^1(t,x) \overset{s}{\leqq} u^2(t,x) .$$

In particular, if $u^1(t,x)$ has Heaviside initial data, then

(3.21) $u^1(t,x) \overset{s}{\leq} w^{2^{1/2}}(x)$ for all t .

By the maximum principle, the solution $u(t,x)$ to the Kolmogorov equation with Heaviside initial data remains monotone for all time. Inequality (3.21) shows that the family of distribution functions $u(t,\cdot)$, $t \in \mathbb{R}^+$, is tight after appropriate centering.

<u>Proof of Corollary 2</u>. Inequality (3.20) is a direct application of the definition of "$\overset{s}{<}$" and Proposition 3.2. (3.21) follows upon setting $u^2(0,x) = w^{2^{1/2}}(x)$, and observing that (3.19) is in this case trivial and that $w^{2^{1/2}}(x)$ is a travelling wave. //

We proceed on to the classic Kolmogorov result.

<u>Theorem (Kolmogorov-Petrovsky-Piscounov)</u>. Let $u(t,x)$ be the solution to

(1.1) $$u_t = \frac{1}{2}u_{xx} + f(u)$$

with Heaviside initial data, where f satisfies (1.2). Then

(1.4) $$u(t,x+m_{1/2}(t)) \to w^{2^{1/2}}(x)$$

uniformly in x as $t \to \infty$, where $w^{2^{1/2}}(x)$ is the solution of (1.5) and (1.6) with $w^{2^{1/2}}(0) = 1/2$. The centering term $m_{1/2}(t)$ satisfies

(3.22) $$m_{1/2}(t)/t \to 2^{1/2} \quad \text{as} \quad t \to \infty .$$

(Kolmogorov et al. actually showed the slightly stronger $\dot{m}_{1/2}(t) \to 2^{1/2}$.)

<u>Proof</u>. By Corollary 1 to Proposition 3.2,

(3.18) $$u(t,x+m_{1/2}(t)) \begin{cases} \uparrow w(x) & \text{for } x > 0 \\ \downarrow w(x) & \text{for } x \leq 0 \end{cases}$$

for some $w(x)$. Since $u(t,x)$ is monotonically decreasing, so is $w(x)$. Also, Corollary 2 states that

$$u(t,x) \overset{s}{\leq} w^{2^{1/2}}(x) \tag{3.23}$$

for all t ; therefore

$$\begin{aligned} w(x) &\leq w^{2^{1/2}}(x) \quad \text{for } x>0 \\ &\geq w^{2^{1/2}}(x) \quad \text{for } x\leq 0 . \end{aligned} \tag{3.24}$$

In particular, $w(x)$ is nondegenerate with

$$0 < w(x) < 1 \quad \text{for } -\infty < x < \infty \tag{1.6a}$$

and

$$\begin{aligned} w(x) &\to 0 \quad \text{as } x\to\infty \\ &\to 1 \quad \text{as } x\to-\infty . \end{aligned} \tag{1.6b}$$

The next step is to verify that $w(x)$ is in fact a travelling wave, that is, satisfies (1.12). This can be established in any one of a number of ways, where typically one rewrites (1.1) as an integral operator for u , applies (3.18) to pass to the limit as $t\to\infty$, and then differentiates to obtain (1.12). [14] inverted the heat operator; we summarize here the argument in [15], in which one integrates twice with respect to the space variable and once with respect to time. (If one wishes, one may instead with a little more work here avoid referring back to (1.1) to exhibit (1.12).) Set $v(t,x) = u(t,x+m_{1/2}(t))$. Integration of (1.1) and application of (3.18) shows for each x that as $t\to\infty$,

$$\begin{aligned} o(1) &= \int_t^{t+1} ds \int_0^x dy \int_0^y dz\, [\, \tfrac{1}{2} v_{zz} + \dot{m}_{1/2} v_z + f(v)\,] \\ &= \tfrac{1}{2} w(x) - \tfrac{1}{4} - \tfrac{a}{2} x \\ &\quad + [m_{1/2}(t+1) - m_{1/2}(t)]\cdot[\int_0^x (w - \tfrac{1}{2})\, dy + o(1)] \end{aligned} \tag{3.25}$$

$$+\int_0^x dy \int_0^y f(w)\,dz + o(1) \quad ,$$

where $a = \lim_{t\to\infty} v_y(t,0)$. The third line of (3.25) is justified by the mean value theorem. (It follows from (3.23) without too much difficulty that $\dot{m}$ is bounded below.) Note that $m_{1/2}(t+1) - m_{1/2}(t)$ is the only quantity on the right side of (3.25) which depends on t . Therefore, fixing x so that $\int_0^x (w - \frac{1}{2})\,dy \neq 0$, one sees that

(3.26) $\quad m_{1/2}(t+1) - m_{1/2}(t) \to \lambda$

for some λ . Differentiation of (3.25) after passing to the limit yields (1.12).

It follows from (3.9) that $\lambda \leq 2^{1/2}$. (One may also exclude $w^{\lambda}(x)$, $\lambda > 2^{1/2}$, on the grounds that $w^{\lambda}(x)$ lies above $w^{2^{1/2}}(x)$ in the phase plane, which contradicts (3.24).) But, as has already been discussed, (1.6) excludes the possibility $\lambda < 2^{1/2}$. Therefore, $\lambda = 2^{1/2}$ and $w(x) = w^{2^{1/2}}(x)$. Since $u(t,x)$ is monotone in x and $w^{2^{1/2}}(x)$ is continuous, convergence in (1.4) is in fact uniform. Finally note that the limit in (3.22) follows from (3.26). //

In the remainder of the section, we devote our attention to obtaining various asymptotic estimates for solutions of (1.1) and (1.2), which will be applied in later sections. Typically, we will be interested in examining the right tail behavior and obtaining bounds for the growth of $u(t,x)$ under different initial data. The maximum principle will serve as the principle tool. We will also be putting to use our knowledge of the asymptotic nature of $w^{\lambda}(x)$ for $\lambda > 2^{1/2}$, and the two last propositions stated here will employ the Kolmogorov theorem.

For the questions we consider, we will typically not be assuming monotonicity of initial data. The next lemma states, however, that in the case where the initial data is restricted to lying below some cut-off point, the data will remain monotone above that point for all t. This result will be important in proving Proposition 7.1. The elliptic analogue of the lemma may be found in Gidas-Ni-Nirenberg [8].

<u>Lemma 3.1</u>. Assume that $u(t,x)$ satisfies (1.1) and (1.2) with

$$u(0,x) = 0 \quad \text{if} \quad x \geq M, \tag{3.27}$$

for some M. Then for $y \geq x \geq M$,

$$u(t,y) \leq u(t,2x-y) \tag{3.28}$$

for each t, and hence

$$u_x(t,x) \leq 0. \tag{3.29}$$

<u>Proof</u>. Let $u^1(t,y)$ and $u^2(t,y)$ be the solutions of (1.1) on $\mathbb{R}^+ \times [x,\infty)$ with initial data

$$u^1(0,y) = 0 \quad \text{for} \quad y \geq x$$

and

$$u^2(0,y) = u(0,2x-y) \quad \text{for} \quad y \geq x,$$

and with boundary conditions

$$u^1(t,x) = u^2(t,x) = u(t,x)$$

for $t \geq 0$. By the maximum principle,

$$u^1(t,y) \leq u^2(t,y)$$

for all t and $y \geq x$. Since $x \geq M$,

$$u^1(t,y) = u(t,y)$$

for $y \geq x$, whereas since the heat operator is invariant under reflection,

$$u^2(t,y) = u(t,2x-y)$$

for all y. Hence

$$u(t,y) \leq u(t,2x-y)$$

for $y \geq x$, which demonstrates the lemma. //

(Under the assumption $\int_{-\infty}^{\infty} u(0,x)dx > 0$, one can with a little more work demonstrate the strict inequality analogues of (3.28) and (3.29).)

Lemma 3.2 is a simple application of the maximum principle, and tells us that the semigroup operator associated with the Kolmogorov equation is continuous for bounded time in the uniform topology.

<u>Lemma 3.2</u>. Let $u^1(t,x)$ and $u^2(t,x)$ be solutions of (1.1) with $f'(u) \leq 1$ and which are bounded over finite time. If

$$u^2(0,x) - u^1(0,x) \leq \epsilon \quad \text{for all } x, \tag{3.30}$$

then

$$u^2(t,x) - u^1(t,x) \leq \epsilon e^t \quad \text{for all } t \text{ and } x. \tag{3.31}$$

Consequently, if

$$|u^2(0,x) - u^1(0,x)| \leq \epsilon \quad \text{for all } x,$$

then

$$|u^2(t,x) - u^1(t,x)| \leq \epsilon e^t \quad \text{for all } t \text{ and } x. \tag{3.32}$$

<u>Proof</u>. The function $v(t,x) \equiv u^2(t,x) - u^1(t,x)$ satisfies

$$v_t = \frac{1}{2}v_{xx} + \frac{f(u^2) - f(u^1)}{u^2 - u^1}v, \tag{3.33}$$

where the coefficient of v is always at most 1. If $v^+(t,x)$ satisfies (3.33) with $v^+(0,x) = v(0,x) \vee 0$, then by the maximum principle,

(3.34) $v^+(t,x) \geq v(t,x) \vee 0$ for all t and x .

Let $\bar{v}(t,x)$ be the solution of

$$\bar{v}_t = \frac{1}{2}\bar{v}_{xx} + \bar{v}$$

with initial data $\bar{v}(0,x) \equiv \epsilon$. Then by the maximum principle and (3.30),

(3.35) $v^+(t,x) \leq \bar{v}(t,x)$ for all t and x .

Using the substitution $\hat{v}(t,x) = e^{-t}\bar{v}(t,x)$ it is easy to see that

$$\bar{v}(t,x) = \epsilon e^t$$

and therefore by (3.34) and (3.35),

$$v(t,x) \leq \epsilon e^t \; .$$ //

Lemma 3.2 will frequently be useful. As a consequence of the lemma, we show that data which converges to a travelling wave will propagate at an asymptotic linear rate.

Corollary 1. Let $u(t,x)$ be a solution of (1.1) and (1.2) with

(3.36) $u(t,x+m(t)) \to w^\lambda(x)$ uniformly in x as $t \to \infty$

for some $m(t)$ and $\lambda \geq 2^{1/2}$. For fixed s_o ,

$$m(t+s) - m(t) \to \lambda s \text{ uniformly in } s \in [0,s_o] \text{ as } t \to \infty \; .$$

Consequently, $m(t)/t \to \lambda$ as $t \to \infty$.

Proof. By Lemma 3.2 and (3.36),

$$\sup_x | u(t+s,x+\lambda s+m(t)) - w^\lambda(x)|$$

(3.37) $$\leq e^{s_o} \sup_x |u(t,x+m(t)) - w^\lambda(x)|$$

$$\to 0 \text{ uniformly in } s \in [0,s_o] \text{ as } t \to \infty \; .$$

Comparison of (3.37) with (3.36) shows that

$$m(t+s) - m(t) \to \lambda s \quad \text{uniformly in} \quad s \in [0,s_0] \quad \text{as} \quad t \to \infty \,. \qquad //$$

The next lemma will be needed in Lemma 3.4, which in turn leads to Proposition 3.3.

Lemma 3.3. Let $u^1(t,x)$ and $u^2(t,x)$ be solutions of (1.1) with $f'(u) \leq 1$. If $0 \leq u^i(0,x) \leq 1$ for all x, and

$$(3.38) \qquad u^1(0,x) = u^2(0,x) \quad \text{for} \quad x > 0 \,,$$

then for some constant C,

$$(3.39) \qquad |u^2(t,x) - u^1(t,x)| \leq Ct^{-1/4} \quad \text{for} \quad x \geq b(t) \,,$$

where $b(t) = 2^{1/2}t - 2^{-5/2}\log t$.

Proof. Define v, v^+, $\bar{v}$, and $\hat{v}$ as in the previous lemma, but where $\bar{v}$ and $\hat{v}$ have the same initial data as v^+. By (3.8)

$$\hat{v}(0,x) = 0 \quad \text{for} \quad x > 0 \,.$$

Since $0 \leq \hat{v}(0,x) \leq 1$ for all x and $\hat{v}(t,x)$ satisfies the heat equation, it follows that

$$\hat{v}(t,x) \leq \int_{-\infty}^{0} \frac{1}{\sqrt{2\pi t}} e^{-(x-y)^2/2t} dy \,,$$

which by direct computation is at most $Ct^{-1/4}e^{-t}$ for $x \geq b(t)$ and an appropriate constant C. Therefore,

$$v^+(t,x) \leq Ct^{-1/4} \quad \text{for} \quad x \geq b(t) \,,$$

and (3.39) follows from (3.34) and (3.35). //

We now employ Lemma 3.3 and our knowledge of the behavior of the right tail of $w^\lambda(x)$, $\lambda > 2^{1/2}$, to demonstrate Lemma 3.4. Lemma 3.4 states that if the initial data $u(0,x)$ has a right tail which is asymptotically the same as that of $w^\lambda(x)$, then in fact $u(t,x+\lambda t) \sim w^\lambda(x)$ for values of x which are not decreasing too rapidly in t.

Lemma 3.4. Let $u(t,x)$ satisfy (1.1), (1.2), and if $\lambda = 2^{1/2}$, (1.2'). Assume that

$$u(0,x) = \gamma(x)w^\lambda(x) , \tag{3.40}$$

where $\lambda \geqq 2^{1/2}$ and $\gamma(x) \to 1$ as $x \to \infty$. Then

$$\sup_{x \geqq b_\lambda(t)} |u(t,x+\lambda t) - w^\lambda(x)| \to 0 \quad \text{as} \quad t \to \infty , \tag{3.41}$$

where $b_\lambda(t) = (2^{1/2} - \lambda)t - \frac{1}{8}\log t$.

(The term $\frac{1}{8}\log t$ is only of concern in the case where $\lambda = 2^{1/2}$, and may otherwise be dropped.)

Proof. For each N, set

$$\begin{aligned} u^N(0,x) &= u(0,x) \quad \text{for} \quad x > N \\ &= w^\lambda(x) \quad \text{for} \quad x \leqq N . \end{aligned}$$

By (1.13) and (1.14), $w^\lambda(x)$, $\lambda \geqq 2^{1/2}$, has an exponential tail. It therefore follows from (3.40), that for each $\delta > 0$ we may select N so that

$$w^\lambda(x+\delta) \leqq u^N(0,x) \leqq w^\lambda(x-\delta) \quad \text{for all} \quad x . \tag{3.42}$$

By the maximum principle,

$$w^\lambda(x+\delta) \leqq u^N(t,x+\lambda t) \leqq w^\lambda(x-\delta) \quad \text{for all} \quad t \quad \text{and} \quad x .$$

But from Lemma 3.3,

$$|u^N(t,x) - u(t,x)| \leqq Ct^{-1/4}$$

for $x \geqq b(t) + N$. Therefore,

$$w^{\lambda}(x+\delta) - Ct^{-1/4} \leqq u(t,x+\lambda t) \leqq w^{\lambda}(x-\delta) + Ct^{-1/4}$$

for $x \geqq b_{\lambda}(t) + N(t)$, where $N(t) \to -\infty$ as $t \to \infty$. Letting $t \to \infty$ and then $\delta \to 0$, we obtain (3.41), where convergence is uniform in $x \geqq b_{\lambda}(t)$ because $w^{\lambda}(x)$ is uniformly continuous. //

Proposition 3.3 states that if the right tail of $u(t,x+m(t))$, for some $m(t)$, approaches that of $w^{\lambda}(x)$ as $t \to \infty$, then $u(t,x+m(t)) \sim w^{\lambda}(x)$ for x which are not decreasing too rapidly in t . Since Propositions 5.1 and 7.1 will show (for reasonable initial data) that $u(t,x) \sim 1$ outside this range, this relation will in fact extend to all x . We therefore have a useful tool which enables us to restrict our attention to the tail of $u(t,x+m(t))$ while demonstrating convergence to a travelling wave. The proof of Proposition 3.3 employs Lemmas 3.2 and 3.4.

<u>Proposition 3.3</u>. Let $u(t,x)$ satisfy (1.1), (1.2), and if $\lambda = 2^{1/2}$, (1.2'). Suppose that for some $m(t)$ and N ,

(3.43) $$\gamma_1^{-1}(x)w^{\lambda}(x) - \gamma_2(t) \leqq u(t,x+m(t)) \leqq \gamma_1(x)w^{\lambda}(x) + \gamma_2(t) \quad \text{for} \quad x \geqq N ,$$

where $\lambda \geqq 2^{1/2}$ and

(3.44) $$\begin{aligned} &\gamma_1(x) \to 1 \quad \text{as} \quad x \to \infty \\ &\gamma_2(t) \to 0 \quad \text{as} \quad t \to \infty . \end{aligned}$$

Then

(3.45) $$\sup_{x \geqq c(t)} |u(t,x+m(t)) - w^{\lambda}(x)| \to 0 \quad \text{as} \quad t \to \infty ,$$

where $c(t) \to -\infty$ as $t \to \infty$.

<u>Proof</u>. Let $v^{\pm}(t,x)$ denote the solutions of (1.1) with

$$\begin{aligned} v^{+}(0,x) &= \gamma_1(x)w^{\lambda}(x) \quad &&\text{for} \quad x \geqq N \\ &= 1 &&\text{for} \quad x < N \end{aligned}$$

and

$$v^-(0,x) = \gamma_1^{-1}(x) w^\lambda(x) \quad \text{for} \quad x \geqq N$$
$$= 0 \qquad \text{for} \quad x < N \ .$$

By Lemma 3.4, $\forall \epsilon > 0 \ \exists\, s_0$ so that for $s \geqq s_0$,

(3.46) $$|v^\pm(s, x+\lambda s) - w^\lambda(x)| < \epsilon \quad \text{for} \quad x \geqq b_\lambda(s) \ .$$

But,

$$u(t-s, x+m(t-s)) - v^+(0,x) \leqq \gamma_2(t-s)$$
$$u(t-s, x+m(t-s)) - v^-(0,x) \geqq -\gamma_2(t-s) \ .$$

Therefore by Lemma 3.2, for t chosen large enough so that $\gamma_2(t-s) < \epsilon e^{-s}$,

(3.47) $$u(t, x+m(t-s)) - v^+(s,x) < \epsilon$$
$$u(t, x+m(t-s)) - v^-(s,x) > -\epsilon \ .$$

By (3.46) and (3.47), for large enough t ,

(3.48) $$|u(t, x+\lambda s + m(t-s)) - w^\lambda(x)| < 2\epsilon \quad \text{for} \quad x \geqq b_\lambda(s) \ .$$

Comparison of (3.48) with (3.43) shows that for fixed s ,

(3.49) $$m(t) - m(t-s) \to \lambda s \quad \text{as} \quad t \to \infty \ .$$

Since $b_\lambda(s) \to -\infty$ as $s \to \infty$, by letting $s_0 \to \infty$, we see that (3.45) follows from (3.48), (3.49), and the uniform continuity of $w^\lambda(x)$. //

We introduce at this point a little terminology. We will frequently be dealing with forcing terms f as in (1.2) which are <u>concave</u>, that is for which $f(u)/u$ is decreasing in u . In particular, this implies that $f(u^1+u^2) \leqq f(u^1) + f(u^2)$ for $u^1 \geqq 0$, $u^2 \geqq 0$, and $u^1 + u^2 \leqq 1$. Two cases of future interest are $\bar{f}$ and $\underline{f}$, where

$$\begin{aligned} \bar{f}(u) &= u \cdot \max\{f(u')/u' : u' \geqq u\} \\ \underline{f}(u) &= u \cdot \min\{f(u')/u' : u' \leqq u\} \quad , \end{aligned} \tag{3.50}$$

and f satisfies (1.2). Clearly $\underline{f}(u) \leqq f(u) \leqq \bar{f}(u)$. Therefore, if $\underline{u}(t,x)$ and $\bar{u}(t,x)$ are respectively the solutions of

$$u_t = \frac{1}{2} u_{xx} + \underline{f}(u) \tag{1.1'}$$

and

$$u_t = \frac{1}{2} u_{xx} + \bar{f}(u) \tag{1.1''}$$

with $\underline{u}(0,x) = u(0,x) = \bar{u}(0,x)$, it follows that $\underline{u}(t,x) \leqq u(t,x) \leqq \bar{u}(t,x)$. Computations involving $u(t,x)$ may therefore sometimes be reduced to the special case where f is concave, in which context the following simple lemma holds. In this article (as was also the case in [17]), Lemma 3.5 will find considerable use.

<u>Lemma 3.5</u>. Let $u^1(t,x)$, $u^2(t,x)$, and $u^3(t,x)$ all be solutions of (1.1) and (1.2) with the same f , which is assumed to be concave. Then if

$$u^3(0,x) \leqq u^1(0,x) + u^2(0,x) \quad \text{for all } x , \tag{3.51}$$

it follows that

$$u^3(t,x) \leqq u^1(t,x) + u^2(t,x) \quad \text{for all } t \text{ and } x . \tag{3.52}$$

Also, for $c > 1$, if

$$u^2(0,x) \leqq c u^1(0,x) \quad \text{for all } x , \tag{3.53}$$

then

$$u^2(t,x) \leqq c u^1(t,x) \quad \text{for all } t \text{ and } x . \tag{3.54}$$

<u>Proof</u>. Set $v(t,x) = u^1(t,x) + u^2(t,x)$. By (1.1) and the concavity of f ,

$$\begin{aligned} v_t &= \frac{1}{2} v_{xx} + f(u^1) + f(u^2) \\ &\geqq \frac{1}{2} v_{xx} + f(v) \ , \end{aligned} \tag{3.55}$$

where we set $f(v) = 0$ if $v \geq 1$. It follows from (3.51), (3.55), and the maximum principle that $u^3(t,x) \leqq v(t,x)$ for all t and x. Inequality (3.54) follows from (3.53) by analogous reasoning. //

We will conclude this section with two consequences of the Kolmogorov theorem. The first, Proposition 3.4, states that general nonzero data will grow and spread at at least the same rate as does Heaviside initial data. In particular, isolated nonzero data cannot exist without expanding at a linear rate. This behavior, although intuitively quite clear, requires a little work because of the specialized nature of the Kolmogorov proof, which makes heavy use of the Heaviside initial data.

We first state two preparatory lemmas.

<u>Lemma 3.6</u>. Let $u(t,x)$ be a solution to (1.1) and (1.2) such that

$$u(t_0,0) \geqq \eta \tag{3.56}$$

for some t_0 and η, with $t_0 \geqq 1$ and $\eta > 0$. Then for either $\mathcal{J} = [0,1]$ or $\mathcal{J} = [-1,0]$,

$$u(t_0,x) \geqq \eta/10 \quad \text{for } x \in \mathcal{J} \ . \tag{3.57}$$

<u>Proof</u>. Since $u_t \leqq \frac{1}{2} u_{xx} + u$,

$$u(t_0,0) \leqq e \int_{-\infty}^{\infty} u(t_0-1,y) \frac{e^{-y^2/2}}{\sqrt{2\pi}} \, dy \ .$$

Therefore,

$$u(t_0,0) \leqq 2e \int_{\mathcal{J}} u(t_0-1,y) \frac{e^{-y^2/2}}{\sqrt{2\pi}} \, dy \tag{3.58}$$

for either $\mathcal{J} = \mathcal{J}^+$ or $\mathcal{J} = \mathcal{J}^-$, where $\mathcal{J}^+ = [0,\infty)$ and $\mathcal{J}^- = (-\infty,0]$. Note that

$$\min_{y\in\mathcal{J}^{\pm}}\{e^{-(y\mp1)^2/2}/e^{-y^2/2}\} = e^{-1/2} \quad ,$$

from which we obtain

$$(3.59) \qquad u(t_0,0) \leqq 2e^{3/2}\int_{\mathcal{J}} u(t_0-1,y)\,\frac{e^{-(x-y)^2/2}}{\sqrt{2\pi}}\,dy$$

for $x\in\mathcal{J}^+$ (if $\mathcal{J}=\mathcal{J}^+$) or $x\in\mathcal{J}^-$ (if $\mathcal{J}=\mathcal{J}^-$). On the other hand, $u_t \geqq \frac{1}{2}u_{xx}$. It follows from (3.59) and the maximum principle, that

$$u(t_0,0) \leqq 2e^{3/2}u(t_0,x) \quad .$$

Therefore, by (3.56),

$$u(t_0,x) \geqq \eta/2e^{3/2} > \eta/10 \quad \text{for} \quad x\in\mathcal{J} \quad .$$

//

Lemma 3.7 gives a lower bound on the growth rate of solutions of (1.1) with small initial data.

Lemma 3.7. Let $u(t,x)$ be a solution to (1.1) and (1.2) with

$$(3.60) \qquad \begin{aligned} u(0,x) &= \eta \quad \text{if} \quad x\in\mathcal{J} \\ &= 0 \quad \text{if} \quad x\notin\mathcal{J} \ , \end{aligned}$$

for some interval $\mathcal{J}$. For each choice of θ, $0<\theta<1$, $t \leqq -\frac{1}{\theta}\log\eta$, and all x,

$$(3.61) \qquad u(t,x) \geqq C_1^{\theta}\eta e^{\theta t}\int_{\mathcal{J}} \frac{e^{-(x-y)^2/2t}}{\sqrt{2\pi t}}\,dy \ ,$$

where $C_1^{\theta} > 0$ is some constant which does not depend on η. If (1.2') is also satisfied, then one may set $\theta = 1$.

Proof. Let $u^{\theta}(t,x)$ denote the solution to $u_t^{\theta} = \frac{1}{2}u_{xx}^{\theta} + f^{\theta}(u)$ with $u^{\theta}(0,x) \equiv u(0,x)$ and $f^{\theta}(u) = f(u)\wedge\theta u$. By the maximum principle,

$$u^{\theta}(t,x) \leqq u(t,x) \quad \text{for all} \quad t \quad \text{and} \quad x \ ;$$

it therefore suffices to demonstrate (3.61) for u^{θ} instead of for u.

Since $u^\theta(t,x)$ satisfies $u_t^\theta \leqq \frac{1}{2}u_{xx}^\theta + \theta u^\theta$ and (3.60),

(3.62) $\quad u^\theta(t,x) \leqq \eta e^{\theta t}$ for all t and x.

But by (1.2), if $0 < \theta < 1$, then $f^\theta(u^\theta) = \theta u^\theta$ for $0 \leqq u^\theta \leqq C_1^\theta$ and some $C_1^\theta > 0$. Applying (3.62), we thus have the bound $f^\theta(u^\theta) \geqq g(t)u^\theta$, where $g(t) = \theta$ for $t \leqq \frac{1}{\theta}\log(C_1^\theta/\eta)$. Consequently,

$$u_t^\theta \geqq \frac{1}{2}u_{xx}^\theta + g(t)u^\theta ,$$

and another application of the maximum principle shows that

$$(3.63) \qquad u^\theta(t,x) \geqq e^{\theta t}\cdot \exp\left[\int_0^t (g(s)-\theta)\,ds\right]\int_{-\infty}^{\infty} u(0,y)\,\frac{e^{-(x-y)^2/2t}}{\sqrt{2\pi t}}\,dy .$$

For $t < -\frac{1}{\theta}\log\eta$, it is easy to check that

$$\exp\left[\int_0^{-\frac{1}{\theta}\log\eta} (g(s)-\theta)ds\right] \geqq C_1^\theta ,$$

and so (3.63) is at least

$$C_1^\theta \eta e^{\theta t}\int_{\mathcal{J}} \frac{e^{-(x-y)^2/2t}}{\sqrt{2\pi t}}\,dy .$$

If (1.2') is also satisfied, we set $u^\theta = u$. Application of the same reasoning as above allows us to set $g(t) = 1 - C_2(\eta e^t)^\rho$ for $t \leqq -\log\eta$ and some $C_2 > 0$. In this case, it is easy to see that one may choose

$$C_1^1 = \exp\left[-\int_{-\infty}^{-\log\eta} (g(s)-1)\,ds\right] = e^{-C_2/\rho} > 0 . \qquad //$$

In Proposition 3.4, we show that the spread of the solution of the Kolmogorov equation under arbitrary initial data lags behind the spread under Heaviside initial data (in the positive direction) by at most a constant. This result and the weaker version (3.66) will play an important role in establishing estimates needed for Sections 7-9.

<u>Proposition 3.4</u>. Let $u(t,x)$ be a solution to (1.1) where f satisfies

(1.2) and is concave. Assume that

$$(3.64)\qquad u(t_0,0) \geqq \eta$$

for some $t_0 \geqq 1$ and $\eta > 0$. Then $\forall\theta$, $0 < \theta < 1$, and $\forall\epsilon > 0$, $\exists T$ (not depending on η) so that for $t \geqq T$,

$$(3.65)\qquad u(t+t_0-\frac{1}{\theta}\log\eta, x) > w^{2^{1/2}}(|x| - m_{1/2}(t) + c^\theta) - \epsilon ,$$

where c^θ is some constant and $m_{1/2}(t)$ is the median of the solution under Heaviside initial data. In particular, $\forall\delta > 0$,

$$(3.66)\qquad u(t+t_0-\frac{1}{\theta}\log\eta, x) \to 1 \quad \text{as} \quad t\to\infty$$

uniformly in $|x| \leqq (2^{1/2}-\delta)t$ and $\eta > 0$. If (1.2') is also satisfied, then one may set $\theta = 1$ in (3.65) and (3.66).

Note that because of the comparison $f \geqq \underline{f}$ (where $\underline{f}$ is as in (3.50)) and the maximum principle, (3.66) extends to nonconcave f.

Proof. The idea of the proof is to employ the maximum principle to compare the right tail of $u(t,x)$ with the solution of (1.1) under Heaviside initial data, whose asymptotic behavior we know by the Kolmogorov theorem. Proposition 3.3 then allows extension of the estimate to the entire wave, from which (3.65) will follow.

First of all, set $t_1 = -\frac{1}{\theta}\log\eta$ and $t_2 = t_0 + t_1$. By Lemmas 3.6 and 3.7, and the maximum principle,

$$u(t_2,x) \geqq C_3^\theta \int_{\mathcal{J}} \frac{e^{-(x-y)^2/2t}}{\sqrt{2\pi t}}\, dy \quad \text{for all } x ,$$

where $\mathcal{J}$ denotes either $[0,1]$ or $[-1,0]$ and $C_3^\theta > 0$ does not depend on η. (If (1.2') is valid, we may set $\theta = 1$.) For $|x| \leqq 1$, it therefore follows that

$$(3.67)\qquad u(t_2+1,x) \geqq C_4^\theta$$

for some $0 < C_4^\theta \leqq 1$. By the maximum principle, if $u^1(t,x)$ satisfies (1.1) with

$$u^1(0,x) = C_4^\theta \quad \text{for } x \in (-1,0] \qquad (3.68)$$
$$\qquad = 0 \quad \text{for } x \notin (-1,0] ,$$

then

$$u(t+t_2+1,x) \geqq u^1(t,x) \quad \text{for all } t \text{ and } x . \qquad (3.69)$$

We also introduce the second comparison function $u^2(t,x)$, where $u^2(t,x)$ is the solution of (1.1) with Heaviside initial data. By Lemma 3.5,

$$u^1(t,x) \geqq C_4^\theta[u^2(t,x) - u^2(t,x+1)] \qquad (3.70)$$

for all t and x .

Next, set $z = x - m_{1/2}(t)$. It follows from the Kolmogorov theorem that as $t \to \infty$, the right side of (3.70) converges uniformly in z to

$$C_4^\theta [w^{2^{1/2}}(z) - w^{2^{1/2}}(z+1)] . \qquad (3.71)$$

Because of (1.13), for large z and some constant c_1^θ , we may approximate (3.71) by $w^{2^{1/2}}(z+c_1^\theta)$. More specifically,

$$u^1(t,z+m_{1/2}(t) - c_1^\theta) \geqq \gamma_1^{-1}(z)w^{2^{1/2}}(z) - \gamma_2(t) \qquad (3.72)$$

for appropriate $\gamma_1(z)$ and $\gamma_2(t)$ satisfying $\gamma_1(z) \to 1$ as $z \to \infty$ and $\gamma_2(t) \to 0$ as $t \to \infty$. It therefore follows from Proposition 3.3, that $\forall \epsilon > 0 \ \exists T_1$ so that if $t \geqq T_1$,

$$u^1(t,z+m_{1/2}(t) - c_1^\theta) > w^{2^{1/2}}(z) - \epsilon/2 \qquad (3.73)$$

for $z \geqq c^\theta(t)$, where $c^\theta(t) \to -\infty$ as $t \to \infty$. Note that for large enough t , $u^1(t,m_{1/2}(t) + c^\theta(t) - c_1^\theta) > 1 - \epsilon$, and that by Lemma 3.1 and (3.68), $u^1(t,x)$ is decreasing in $x \geqq 0$. By (3.68) and (3.73), we thus obtain that for $x \geqq 0$,

$$u(t+t_0-\tfrac{1}{\theta}\log\eta+1,x) \geqq u^1(t,x) \tag{3.74}$$
$$> w^{2^{1/2}}(x-m_{1/2}(t)+c_1^\theta)-\epsilon\ .$$

Also, by (1.4) and the corollary to Lemma 3.2, $m_{1/2}(t+1)-m_{1/2}(t) \to 2^{1/2}$ as $t\to\infty$. A cosmetic change therefore yields

$$u(t+t_0-\tfrac{1}{\theta}\log\eta,x) > w^{2^{1/2}}(x-m_{1/2}(t)+c^\theta)-\epsilon \tag{3.75}$$

for $c^\theta = c_1^\theta + 2$. The inequality (3.65) finally follows from (3.75) and the symmetry of the problem. (3.66) follows from (3.65) upon recalling that $m_{1/2}(t)/t \to 2^{1/2}$ as $t\to\infty$ and $w^{2^{1/2}}(x)\to 1$ as $z\to-\infty$. //

Proposition 3.5 tells us that the position of the wave $u(t,x)$ under Heaviside initial data (and for concave f) is nearly convex in t. This result will be important in obtaining finer estimates for $m_{1/2}(t)$ in Sections 7-9. Proposition 3.5 is a cleaner version of Step 1 of Theorem 9.2 in [17].

Proposition 3.5. Let $u(t,x)$ satisfy (1.1), (1.2), and (1.3), and assume that f is concave. Then for some constant $M > 0$,

$$m_{1/2}(s) - \tfrac{s}{t}m_{1/2}(t) \leqq M \tag{3.76}$$

for all s and t, $s \leqq t$.

Proof. We first show that for some constant M,

$$m_{1/2}(t_1+s)-m_{1/2}(t_1) \leqq m_{1/2}(t_2+s)-m_{1/2}(t_2)+M \tag{3.77}$$

for all $0 \leqq t_1 \leqq t_2$ and $s \geqq 0$. By the first corollary to Proposition 3.2,

$$u(t_2,x+m_{1/2}(t_2)) \geqq u(t_1,x+m_{1/2}(t_1)) \quad \text{for } x > 0$$
$$\geqq 1/2 \quad \text{for } x \leqq 0\ ;$$

in either case,

$$u(t_2,x+m_{1/2}(t_2)) \geqq \tfrac{1}{2}u(t_1,x+m_{1/2}(t_1))\ .$$

Therefore by Lemma 3.5,

$$u(t_2+s,x+m_{1/2}(t_2)) \geqq \tfrac{1}{2}u(t_1+s,x+m_{1/2}(t_1)) \ ,$$

and so from the monotonicity of u,

$$m_{1/4}(t_2+s)-m_{1/2}(t_2) \geqq m_{1/2}(t_1+s)-m_{1/2}(t_1) \ . \tag{3.78}$$

By Corollary 2 of Proposition 3.2, $m_{1/2}(t)-m_{1/4}(t)$ is bounded for all t; (3.77) thus follows from (3.78).

We show now that if (3.76) is violated for some s_1 and t, then (3.77) fails, and hence obtain a contradiction. Let T denote the first point r in $[s_1,t]$ at which

$$\alpha = \inf\{m_{1/2}(r)/r : r \in [s_1,t]\}$$

is attained. Since $m_{1/2}(r)$ is continuous, T exists. Let t_1 denote the last point r in $[0,s_1]$ for which $m_{1/2}(r) = \alpha r$. Set $t_2 = s_1$ and $s = T-s_1$. Then it can be shown with a little work that

$$m_{1/2}(t_2) > \alpha t_2 + M$$

and

$$m_{1/2}(t_2+s) = \alpha(t_2+s) \ ,$$

and therefore

$$m_{1/2}(t_2+s)-m_{1/2}(t_2) < \alpha s - M \ . \tag{3.79}$$

Moreover,

$$m_{1/2}(t_1) = \alpha t_1$$

and

$$m_{1/2}(t_1+s) \geqq \alpha(t_1+s) \ ,$$

and therefore

$$(3.80) \qquad m_{1/2}(t_1 + s) - m_{1/2}(t_1) \geqq \alpha s \quad .$$

But (3.79) and (3.80) together contradict (3.77), and therefore (3.76) must hold. //

4. ASYMPTOTIC BEHAVIOR OF THE KOLMOGOROV EQUATION UNDER FULL HEATING

In Section 4 we investigate the asymptotic behavior as $t \to \infty$ of solutions of the linear heat equation

$$(3.8) \qquad u_t = \frac{1}{2}u_{xx} + u$$

under initial data with right tail behavior of the type specified in (1.16) and (1.19) of Theorem A. Since the kernel of (3.8) is

$$(4.1) \qquad (2\pi t)^{-1/2} e^t e^{-(x-y)^2/2t} ,$$

such asymptotic behavior may be calculated directly. It will be shown in Proposition 4.1 that such solutions of (3.8) have right tails which are, if $\lambda > 2^{1/2}$, almost exponential. In Section 5, we compare (3.8) with (1.1), and then apply Proposition 3.3 to the latter solution to obtain Theorems A and B.

We will find it convenient to introduce some terminology here before proceeding. Set

$$(4.2) \qquad \varphi(t,x) = (2\pi t)^{-1/2} e^t \int_{-\infty}^{\infty} u(0,y) e^{-(x-y)^2/2t} dy ,$$

which is the solution of (3.8) under initial data $u(0,y)$. As usual, $u(0,y)$ is assumed to be measurable with $0 \leq u(0,y) \leq 1$. To avoid having to worry about trivial behavior, we always assume that $\int_{-\infty}^{\infty} u(0,y)\,dy > 0$. Also, set

$$(4.3) \qquad m^{\varphi}(t) = \max\{x: \varphi(t,x) \geq 1\} ,$$

where $m^{\varphi}(t) \equiv -\infty$ if the set is void. For computations involving (4.2), we will be primarily interested in the regions $\mathcal{A}_T = A_1 \cup A_2$ in $R^+ \times R$, where

$$A_1 = \{(t,x): t \geq T, x \geq b_0 t\}$$

and

$$A_2 = \{(t,x):\ 1 \leqq t \leqq T\ ,\ x \geqq b_0 T\}\ .$$

Here, $T \geqq 1$, and b_0 is considered fixed with $b < b_0 < 2^{1/2}$, where b is chosen as in (1.16). Also, recall that as in previous sections, $0 < b \leqq 2^{1/2}$ and $\lambda \geqq 2^{1/2}$ are assumed to have the relationship $\lambda = \frac{1}{2} + \frac{b}{2}$, or equivalently, $b = \lambda - \sqrt{\lambda^2 - 2}$.

In our first two results of the section, Lemmas 4.1 and 4.2, we compute upper and lower bounds on $\varphi(t,x)$. These results are then applied and extended in Proposition 4.1 and its corollaries to obtain precise estimates for the rate of decrease of $\varphi(t,x)$ for large x . In Proposition 4.2, we compute an alternative characterization for the conditions (1.16) and (1.19). The basic approach taken in computing these results is to decompose the region of integration in (4.2) into two parts, the sets $\mathcal{J}$ and $\mathcal{J}^c$, where $\mathcal{J} = (x-(b+\delta)t, x-(b-\delta)t)$ and $\mathcal{J}^c$ is its complement. The effect of the integral over $\mathcal{J}^c$ on $\varphi(t,x)$ will be negligible for large t owing to the nature of the initial data (satisfying (1.16) or (1.19)), whereas the integral over $\mathcal{J}$ may be accurately calculated for δ chosen small enough so as to allow only a small change of (4.1) over the interval. This procedure allows us to deduce the same asymptotic behavior for initial data as in (1.16) as for the special case with initial data where the right tail is exponential and $\varphi(t,x)$ is easy to compute.

<u>Lemma 4.1</u>. Assume that for some $h > 0$ and t_0 ,

$$\log\left[\int_t^{t(1+h)} u(0,y)\,dy\right] \leqq -bt \quad \text{for} \quad t \geqq t_0\ , \tag{4.4}$$

where $0 < b \leqq 2^{1/2}$. Then for each $\delta \geqq 0$ and some constant C ,

$$\log\left[e^t \int_{\mathcal{J}^c} u(0,y) \frac{e^{-(x-y)^2/2t}}{\sqrt{2\pi t}}\,dy\right] \leqq -b(x-\lambda t) - \frac{\delta^2}{3} t - \log(t \wedge 1) + C \tag{4.5}$$

for all t and x , where $\mathcal{J} = (x-(b+\delta)t, x-(b-\delta)t)$.

Proof. Inequality (4.5) may be shown by direct estimation of the terms

$$\int_{x+n}^{x+n+1} u(0,y) \frac{e^{-(x-y)^2/2t}}{\sqrt{2\pi t}} dy .$$

From (4.4),

$$\log [\int_{x+n}^{x+n+1} u(0,y)\, dy] \leq C_1 - b(x+n)$$

for $n \in \mathbb{Z}$ and some constant C_1. It is therefore easy to check that

$$\log [e^t \int_{x+n}^{x+n+1} u(0,y) \frac{e^{-(x-y)^2/2t}}{\sqrt{2\pi t}} dy] \leq -b(x-\lambda t) - (bt+n_1)^2/2t - \frac{1}{2}\log t + C_2 ,$$

where $n_1 = n$ for $n \geq -bt$, $n_1 = n+1$ for $n < -bt$, and C_2 is some constant. Consequently,

$$\log [e^t \int_{\mathcal{J}^c} u(0,y) \frac{e^{-(x-y)^2/2t}}{\sqrt{2\pi t}} dy]$$

$$\leq -b(x-\lambda t) + C_2 + \log [2 \sum_{n \geq \delta t - 1} e^{-n^2/2t}/\sqrt{2\pi t}] .$$

With a little more computation, this can be shown to be at most

$$-b(x-\lambda t) - \frac{\delta^2}{3} t - \log (t \wedge 1) + C$$

for appropriate choice of C. //

We will find the following variation of Lemma 4.1 useful.

Corollary 1. Assume that for some $h > 0$

(4.6) $$\limsup_{t \to \infty} \frac{1}{t} \log [\int_t^{t(1+h)} u(0,y)\, dy] \leq -b ,$$

where $0 < b \leq 2^{1/2}$. Then for $\delta \geq 0$ and $0 < \epsilon < b$,

(4.7) $$\log [e^t \int_{\mathcal{J}^c} u(0,y) \frac{e^{-(x-y)^2/2t}}{\sqrt{2\pi t}} dy] \leq -b(x-\lambda t) + \epsilon x - \frac{\delta^2}{4} t$$

for all x and for large enough t. In particular,

$$(4.8)\qquad \log\varphi(t,x) \leqq -b(x-\lambda t)+\epsilon x \ .$$

(The bound $\epsilon < b$ is relevant only for negative values of x.)

Proof. Inequality (4.7) follows by applying Lemma 4.1 to slightly smaller values of b than given in (4.6), and noting that $b\lambda$ is increasing in b. Inequality (4.8) follows from (4.7) by setting $\delta = 0$. //

Lemma 4.2 gives a partial converse to (4.8) in the case where $b < 2^{1/2}$.

Lemma 4.2. Assume that for some $h > 0$,

$$\liminf_{t\to\infty} \frac{1}{t}\log\left[\int_t^{t(1+h)} u(0,y)\,dy\right] \geqq -b \ ,$$

where $0 < b < 2^{1/2}$. Also, assume that $M > \lambda$ is fixed. Then $\forall\epsilon > 0\ \exists T$ such that for $(t,x)\in\mathcal{A}_T$ and $x \leqq Mt$,

$$(4.9)\qquad \log\varphi(t,x) \geqq -b(x-\lambda t) - \epsilon x \ .$$

Proof. We may assume without loss of generality that $h \leqq b/(M-b)$. Then for $x \leqq Mt$,

$$(4.10)\qquad \begin{aligned} \log\varphi(t,x) &\geqq \log\left[e^t\int_{x-bt}^{(x-bt)(1+h)} u(0,y)\,\frac{e^{-(x-y)^2/2t}}{\sqrt{2\pi t}}\,dy\right] \\ &\geqq \frac{(2-b^2)}{2}\,t + \log\left[\frac{1}{\sqrt{2\pi t}}\int_{x-bt}^{(x-bt(1+h)} u(0,y)\,dy\right] , \end{aligned}$$

since $e^{-(x-y)^2/2t}$ is minimized at $x-bt$. By assumption, $\forall\epsilon > 0\ \exists T'$ such that for $t \geqq T'$,

$$(4.11)\qquad \log\left[\frac{1}{\sqrt{2\pi t}}\int_t^{t(1+h)} u(0,y)\,dy\right] \geqq -(b+\epsilon)t \ .$$

For $(t,x)\in\mathcal{A}_T$, $x-bt \geqq (b_0-b)t$. Therefore for $T = T'/(b_0-b)$,

$$\begin{aligned} \log\varphi(t,x) &\geqq \frac{(2-b^2)}{2}\,t - (b+\epsilon)(x-bt) \\ &\geqq -b(x-\lambda t) - \epsilon x \end{aligned}$$

follows from (4.10) and (4.11). //

The upper and lower bounds obtained for $\varphi(t,x)$ in Lemmas 2.1 and 2.2 will be applied in the following two corollaries. The first corollary gives the first-order asymptotic behavior of $m^{\varphi}(t)$.

Corollary 1. Assume that the initial data satisfies (1.16) with $0<b<2^{1/2}$. Then,

$$(4.12a) \qquad \lim_{t\to\infty} \frac{1}{t} m^{\varphi}(t) = \lambda .$$

If the initial data satisfies (1.19), then

$$(4.12b) \qquad \lim_{t\to\infty} \frac{1}{t} m^{\varphi}(t) = 2^{1/2} .$$

Proof. If (1.16) holds, then (4.12a) is a direct consequence of (4.8) and (4.9). If (1.19) holds, the upper bound given by (4.8) is still valid, whereas the lower bound is easy to compute directly from the definition of $\varphi(t)$. //

Corollary 2. Assume that the initial data satisfies (1.16) with $0<b<2^{1/2}$. Then $\forall \epsilon > 0 \ \exists T$ such that for $(t_1,x_1),(t_2,x_2) \in \mathcal{A}_T$ and $x_1 \leq Mt_1$ with $M>\lambda$ fixed,

$$(4.13) \qquad \log[\varphi(t_2,x_2)/\varphi(t_1,x_1)] \leq b\lambda(t_2-t_1) - b(x_2-x_1) + \epsilon(x_1+x_2) .$$

Proof. Apply (4.8) and (4.9). //

For Section 5 we will need to obtain estimates on the behavior of $\varphi(t_2,x_2)/\varphi(t_1,x_1)$ for large t_i and x_i. Corollary 2 is enough to give us one-sided bounds saying that the ratio decreases at least exponentially quickly in x_2 for x_2 enough greater than x_1. In Proposition 4.1 we give fairly precise two-sided estimates for the case $x_i \leq Mt_i$.

Proposition 4.1. Assume that the initial data satisfies (1.16) with

$0 < b < 2^{1/2}$. Then $\forall \epsilon > 0 \ \exists T$ such that for $(t_1,x_1),(t_2,x_2) \in \mathcal{A}_T$ and $x_i \leqq Mt_i$ with $M > \lambda$ fixed,

$$(4.14) \qquad \begin{aligned} &|\log[\varphi(t_2,x_2)/\varphi(t_1,x_1)] - b\lambda(t_2-t_1) + b(x_2-x_1)| \\ &\leqq \epsilon[|t_2-t_1| + |x_2-x_1| + 1] \ . \end{aligned}$$

In particular for fixed $N > 0$,

$$(4.15) \qquad \log[\varphi(t_2,x_2)/\varphi(t_1,x_1)] = b\lambda(t_2-t_1) - b(x_2-x_1) + o_N(1)$$

if

$$t_2 \in [t_1-N, t_1+N] \quad \text{and} \quad x_2 \in [x_1-N, x_1+N] \ ,$$

where $o_N(1)$ depends only on t_1 and N .

<u>Proof</u>. By (4.8) and (4.9), $\forall \epsilon_1 > 0 \ \exists T$ such that if $(t_i,x_i) \in \mathcal{A}_T$, $x_i \leqq Mt_i$, then

$$(4.16) \qquad |\log\varphi(t_i,x_i) + b(x_i-\lambda t_i)| \leqq \epsilon_1 x_i \ .$$

Therefore,

$$(4.17) \qquad |\log[\varphi(t_2,x_2)/\varphi(t_1,x_1)] - b\lambda(t_2-t_1) + b(x_2-x_1)| \leqq \epsilon_1(x_1+x_2) \ .$$

Set $t_3 = t_1 \vee t_2$ and $t_4 = t_1 \wedge t_2$. If (t_i,x_i) , $i = 1,2$, satisfy

$$(4.18) \qquad |t_2-t_1| + |x_2-x_1| \geqq \sqrt{\epsilon_1}\, t_3 \ ,$$

then it follows from (4.17) that

$$\begin{aligned} &|\log[\varphi(t_2,x_2)/\varphi(t_1,x_1)] - b\lambda(t_2-t_1) + b(x_2-x_1)| \\ &\leqq 2M\sqrt{\epsilon_1}\,[|t_2-t_1| + |x_2-x_1|] \ . \end{aligned}$$

Inequality (4.14) follows upon setting $\epsilon = 2M\sqrt{\epsilon_1}$.

We still need to show (4.14) in the case where (4.18) fails. For $i = 1,2$, let

$$\mathcal{J}_i = (x_i - (b+\delta)t_i \,,\, x_i - (b-\delta)t_i)$$

and

$$\mathcal{J} = (x_3 - (b+2\delta)t_3 \,,\, x_3 - (b-2\delta)t_3) \;,$$

where x_3 is the coordinate corresponding to t_3. Also, set $\delta = 3\epsilon = 6M\sqrt{\epsilon_1}$, which we may assume is at most $b/2$. Since (4.18) fails, it may be checked that $\mathcal{J}_1 \cup \mathcal{J}_2 \subset \mathcal{J}$. To demonstrate (4.14), we first note from (4.7) that for T chosen large enough,

$$\log [e^{t_i} \int_{\mathcal{J}^c} u(0,y) \frac{e^{-(x_i-y)^2/2t_i}}{\sqrt{2\pi t_i}} dy] \leq \log [e^{t_i} \int_{\mathcal{J}_i^c} u(0,y) \frac{e^{-(x_i-y)^2/2t_i}}{\sqrt{2\pi t_i}} dy]$$

$$\leq -b(x_i - \lambda t_i) + \epsilon_1 x_i - 2\epsilon^2 t_i \;.$$

Since $x_i \leq Mt_i$, this is at most

$$-b(x_i - \lambda t_i) - \epsilon_1 x_i - \epsilon^2 t_i \;. \tag{4.19}$$

On the other hand, (4.16) implies that

$$\log \varphi(t_i, x_i) \geq -b(x_i - \lambda t_i) - \epsilon_1 x_i \;. \tag{4.20}$$

For $t_i >> 1/\epsilon^2$, the quantity in (4.19) is substantially less than that on the right side of (4.20). A more quantitative statement is that for

$$\varphi^{\delta}(t_i, x_i) \equiv \int_{\mathcal{J}} u(0,y) \frac{e^{-(x_i-y)^2/2t_i}}{\sqrt{2\pi t_i}} dy \;,$$

then

$$\begin{aligned} &|\log [\varphi(t_2,x_2)/\varphi(t_1,x_1)] - \log [\varphi^{\delta}(t_2,x_2)/\varphi^{\delta}(t_1,x_1)]| \\ &\leq \log [1 + e^{-\epsilon^2 t_4}] \leq e^{-\epsilon^2 t_4} \;, \end{aligned} \tag{4.21}$$

which $\to 0$ as T (and hence t_4) $\to \infty$. Now, by maximizing and minimizing the normal density over $\mathcal{J}$, one may calculate directly that

$$(4.22)\quad \begin{aligned} &|\log[\varphi^\delta(t_2,x_2)/\varphi^\delta(t_1,x_1)] - b\lambda(t_2-t_1) + b(x_2-x_1)| \\ &\leq C\epsilon\,[|t_2-t_1| + |x_2-x_1| + 1] \end{aligned}$$

for some computable constant C. Together, the equations in (4.21) and (4.22) show that (4.14) is also valid when (4.18) fails. //

We now obtain two important corollaries of Proposition 4.1, which will be vital for estimates in Section 5 regarding the asymptotic behavior of $u(t,x)$ and $k(t,x)$. The first corollary follows from (4.15), and states that $\varphi(t,x+m^\varphi(t))$ has a right tail which approximates the tail of $w^\lambda(x)$, $\lambda > 2^{1/2}$, for large t. Once we show that $u(t,x+m^\varphi(t)) \sim \varphi(t,x+m^\varphi(t))$ for large t and x, Corollary 1 may be applied in conjunction with Proposition 3.3 to demonstrate that $u(t,x+m^\varphi(t)) \to w^\lambda(x)$ for all x as $t\to\infty$. In Corollary 1, $w^\lambda(x)$ will stand for the unique solution of (1.6) and (1.12) with

$$(4.23)\quad e^{bx}w^\lambda(x) \to 1 \quad \text{as} \quad x\to\infty .$$

<u>Corollary 1</u>. Assume that (1.16) holds with $0<b<2^{1/2}$. Then there exist $\gamma_1(x)$ and $\gamma_2(t)$ which satisfy

$$(3.44a)\quad \gamma_1(x) \to 1 \quad \text{as} \quad x\to\infty$$

and

$$(3.44b)\quad \gamma_2(t) \to 0 \quad \text{as} \quad t\to\infty$$

so that for $t,x \geq 0$,

$$(4.24)\quad \gamma_1(x)w^\lambda(x) - \gamma_2(t) \leq \varphi(t,x+m^\varphi(t)) \leq \gamma_1(x)w^\lambda(x) + \gamma_2(t) .$$

<u>Proof</u>. Corollary 1 of Lemma 4.2 states that $\lim_{t\to\infty} \frac{1}{t}m^\varphi(t) = \lambda$. Consequently $(t,m^\varphi(t)) \in \mathcal{A}_T$ for $t\geq T$ chosen large enough. It thus follows from (4.15) of the proposition upon setting $t_i = t$, $x_1 = m^\varphi(t)$, and $x_2 = x+m^\varphi(t)$, that

(4.25) $$e^{-bx} - \gamma(t,x) \leq \varphi(t,x+m^{\varphi}(t)) \leq e^{-bx} + \gamma(t,x)$$

for some $\gamma(t,x)$ which satisfies

(4.26) $$\Gamma_1(t) \equiv \sup\{\gamma(t,x): 0 \leq x \leq N(t)\} \to 0 \quad \text{as} \quad t \to \infty ,$$

where $N(t) \to \infty$ as $t \to \infty$. On the other hand, if $x_2 \leq Mt$, we may apply (4.14) of the proposition, and if $x_2 > Mt$, Corollary 2 of Lemma 4.2, to conclude that $\varphi(t,x)$ is decreasing to 0 at least exponentially quickly as $x \to \infty$, and therefore that

(4.27) $$\Gamma_2(t) \equiv \sup\{\varphi(t,x+m^{\varphi}(t)): x > N(t)\} \to 0 \quad \text{as} \quad t \to \infty .$$

Now, set $\gamma_2(t) = \Gamma_1(t) \vee \Gamma_2(t) \vee e^{-bN(t)}$. Then from above,

(4.28) $$e^{-bx} - \gamma_2(t) \leq \varphi(t,x+m^{\varphi}(t)) \leq e^{-bx} + \gamma_2(t)$$

for $x \geq 0$. If we let $\gamma_1(x) = 1/e^{bx}w^{\lambda}(x)$, then (4.28) becomes

$$\gamma_1(x)w^{\lambda}(x) - \gamma_2(t) \leq \varphi(t,x+m^{\varphi}(t)) \leq \gamma_1(x)w^{\lambda}(x) + \gamma_2(t) .$$

(4.23) implies (3.44a), whereas (3.44b) follows from (4.26) and (4.27). //

The second corollary is a consequence of (4.13) and (4.14), and states that $\varphi(t,x)$ is quite small for x sufficiently greater than λt. We will apply this result in Section 5 to show that $k(t,x) \sim 1$ in this region, and that, in particular, Lemma 5.1 follows.

<u>Corollary 2</u>. Assume that (1.16) is satisfied with $0 < b < 2^{1/2}$. Set $\alpha = \frac{b}{2}(\lambda - 2^{1/2})$ and $\beta = \frac{b}{2}$. For large enough t, if $0 \leq s \leq t - e^{-\alpha t/2}$ and $y \geq m^{\varphi}(t) - 2^{1/2}s + z$ with $z \geq 1$, then

(4.29) $$\log \varphi(t-s,y) \leq -(\alpha s + \beta z) .$$

<u>Proof</u>. First restrict s to $[0,t-1]$. Since $m^{\varphi}(t)/t \to \lambda$ with $\lambda > b_0$,

a little computation shows that for each T , t may be chosen large enough so that if $y \geq m^{\varphi}(t) - 2^{1/2}s + z$, then $(t-s,y) \in \mathcal{A}_T$. Therefore, for $m^{\varphi}(t) - 2^{1/2}s + z \leq y \leq M(t-s)$, we may apply (4.14) of the proposition to conclude that

$$
\begin{aligned}
\log\varphi(t-s,y) &\leq -b\lambda s - b(y - m^{\varphi}(t)) + \epsilon[s + |y - m^{\varphi}(t)| + 1] \\
&\leq -b\lambda s + b(2^{1/2}s - z) + \epsilon[(2^{1/2} + 1)s + z + 1] \qquad (4.30) \\
&\leq -\frac{b}{2}(\lambda - 2^{1/2})s - \frac{b}{2}z = -(\alpha s + bz)
\end{aligned}
$$

for ϵ chosen small enough. In a similar manner, one may apply Corollary 2 of Lemma 4.2 to the case where $y > M(t-s)$ to again obtain (4.29). For $s \in [t-1, t-e^{-\alpha t/2}]$, we may apply Lemma 4.1 with $\delta = 0$ to obtain

$$
\begin{aligned}
\log\varphi(t-s,y) &\leq -(b-\epsilon)(y - \lambda(t-s)) - \log(t-s) + C \\
&\leq -(b-\epsilon)(m^{\varphi}(t) - 2^{1/2}t + z) + \frac{\alpha t}{2} + C_1 \\
&\leq -(\alpha t + \beta z)
\end{aligned}
$$

for small enough ϵ and large enough t , since $m^{\varphi}(t)/t \to \lambda$ as $t \to \infty$. //

The conditions (1.16) and (1.19) on the initial data, which we have been using throughout this section, may be reformulated in terms of a simple condition on the asymptotic behavior of $\varphi(t,x)$. This equivalence will be used in Section 5 to demonstrate that conditions (1.16) and (1.19) are in fact necessary for convergence to travelling waves. Lemma 4.3 gives an estimate which is needed in one direction; the main result is stated in Proposition 4.2. We will be using here the abbreviated notation

$$\varphi^{\lambda}(t) \equiv \varphi(t,\lambda t) .$$

Lemma 4.3. Assume that

$$\frac{1}{t}\log\varphi^{\lambda}(t) \to 0 \quad \text{as} \quad t \to \infty \qquad (4.31)$$

for some $\lambda > 2^{1/2}$. Then for some choice of $c > 0$ and C ,

$$(4.32) \qquad \log\left[e^t \int_{\mathcal{J}^c} u(0,y) \frac{e^{-(\lambda t-y)^2/2t}}{\sqrt{2\pi t}}\, dy\right] \leqq -c\delta^2 t + C \quad \text{for all } t$$

if $\delta < \frac{1}{2}(\lambda-b) \wedge 1$ and $\mathcal{J} = ((\lambda - b - \delta)t\,,\, (\lambda - b + \delta)t)$.

Proof. We decompose the above integral into 3 pieces, which we proceed to estimate. First, rewrite

$$(4.33) \qquad e^t \int_{(\lambda-b+\delta)t}^{\infty} u(0,y) \frac{e^{-(\lambda t-y)^2/2t}}{\sqrt{2\pi t}}\, dy$$

as

$$(4.34) \qquad e^{t(1+\epsilon)} \int_{(\lambda-b+\delta)t}^{\infty} u(0,y) \frac{1}{\sqrt{2\pi t(1+\epsilon)}} \exp\left[-\frac{(\lambda t(1+\epsilon)-y)^2}{2t(1+\epsilon)}\right] A(y)\, dy \ ,$$

where $\epsilon = \delta/2(\lambda-b)$ and

$$(4.35) \qquad A(y) = \sqrt{1+\epsilon}\, \exp\left[-t\epsilon - \frac{(\lambda t-y)^2}{2t} + \frac{(\lambda t(1+\epsilon)-y)^2}{2t(1+\epsilon)}\right] \ .$$

It can be computed that

$$\sup\{A(y)\colon y \geqq (\lambda-b+\delta)t\} = A((\lambda-b+\delta)t) \leqq \sqrt{1+\epsilon}\, \exp[-\delta\epsilon t/4]$$
$$\leqq 2\exp[-c_1\delta^2 t] \ ,$$

for an appropriate constant $c_1 > 0$. Therefore, (4.34) is at most

$$2\exp[-c_1\delta^2 t]\,\varphi^{\lambda}(t(1+\epsilon)) \ ,$$

which by (4.31) is at most

$$(4.36) \qquad C_1 \exp[-c_2\delta^2 t]$$

for appropriate constants $c_2 > 0$ and C . We may estimate

$$(4.37) \qquad e^t \int_0^{(\lambda-b-\delta)t} u(0,y) \frac{e^{-(\lambda t-y)^2/2t}}{\sqrt{2\pi t}}\, dy$$

in the same manner as (4.33), but employing $\epsilon = -\delta/2(\lambda-b)$ instead. Then,

$$\sup\{A(y):\ 0 \leq y \leq (\lambda-b-\delta)t\} = A((\lambda-b+\delta)t)$$

$$\leq \sqrt{1-\epsilon}\exp[\delta\epsilon t/4] \leq \exp[-c_1\delta^2 t] ,$$

from which it follows that (4.37) is at most as great as (4.36). Also, since $u(0,y) \leq 1$ for all y, it is easy to compute that

$$(4.38) \qquad e^t\int_{-\infty}^{0} u(0,y)\,\frac{e^{-(\lambda t-y)^2/2t}}{\sqrt{2\pi t}}\,dy \leq C_2\exp[-t(\lambda^2-2)/2]$$

for some C_2. Inequality (4.32) then follows from (4.36) and (4.38). //

Proposition 4.2. The conditions

$$(1.16) \qquad \lim_{t\to\infty}\frac{1}{t}\log[\int_t^{t(1+h)} u(0,y)\,dy] = -b \quad \text{for all (some)} \quad h > 0$$

and

$$(4.31) \qquad \lim_{t\to\infty}\frac{1}{t}\log\varphi^\lambda(t) = 0$$

are equivalent for $\lambda = \frac{1}{b}+\frac{b}{2} > 2^{1/2}$. The conditions

$$(1.19) \qquad \limsup_{t\to\infty}\frac{1}{t}\log[\int_t^{t(1+h)} u(0,y)\,dy] \leq -2^{1/2} \quad \text{for all (some)} \quad h > 0$$

and

$$(4.39) \qquad \lim_{t\to\infty}\frac{1}{t}\log\varphi^{2^{1/2}}(t) = 0$$

are also equivalent.

Proof. The case $\lambda > 2^{1/2}$. With the aid of (4.8) and (4.9), (4.31) follows directly from (1.16). We now show that (4.31) implies (1.16). Set

$$\varphi_\delta^\lambda(t) = e^t\int_{(\lambda-b-\delta)t}^{(\lambda-b+\delta)t} u(0,y)\,\frac{e^{-(\lambda t-y)^2/2t}}{\sqrt{2\pi t}}\,dy .$$

Minimizing and maximizing the normal density over $[(\lambda-b-\delta)t,(\lambda-b+\delta)t]$, one obtains

$$e^t \cdot e^{-(b+\delta)^2/2t} \int_{t'}^{t'(1+h_\delta)} u(0,y)\,dy \leq \sqrt{2\pi t}\, \varphi_\delta^\lambda(t)$$

(4.40)

$$\leq e^t \cdot e^{-(b-\delta)^2/2t} \int_{t'}^{t'(1+h_\delta)} u(0,y)\,dy \;,$$

where $t' = (\lambda-b-\delta)t$ and $h_\delta = 2\delta/(\lambda-b-\delta)$. But (4.31) in conjunction with Lemma 4.3 implies that $\lim_{t\to\infty} \frac{1}{t} \log \varphi_\delta^\lambda(t) = 0$. It therefore follows from (4.40) that

$$\limsup_{t\to\infty} \frac{1}{t} \log \Big[\int_t^{t(1+h_\delta)} u(0,y)\,dy\Big] \leq \frac{(b+\delta)^2-2}{2(\lambda-b-\delta)} \tag{4.41}$$

and

$$\liminf_{t\to\infty} \frac{1}{t} \log \Big[\int_t^{t(1+h_\delta)} u(0,y)\,dy\Big] \geq \frac{(b-\delta)^2-2}{2(\lambda-b-\delta)} \;. \tag{4.42}$$

Note that (4.41) and (4.42) are also valid for all $h > h_\delta$. Letting $\delta \to 0$ in both cases, we obtain

$$\lim_{t\to\infty} \frac{1}{t} \log \Big[\int_t^{t(1+h)} u(0,y)\,dy\Big] = -b \;.$$

<u>The case $\lambda = 2^{1/2}$</u>. It is easy to compute from (1.19) that

$$\liminf_{t\to\infty} \frac{1}{t} \log \varphi^{2^{1/2}}(t) \geq 0 \;.$$

On the other hand, application of (4.8) to (1.19) shows that

$$\limsup_{t\to\infty} \frac{1}{t} \log \varphi^{2^{1/2}}(t) \leq 0 \;.$$

(4.39) thus follows from (1.19). The other direction is also simple. Minimize the normal density over $[\delta t, \delta t(1+h)]$ to obtain

$$\varphi^{2^{1/2}}(t) \geq (2\pi t)^{-1/2} e^t \int_{\delta t}^{\delta t(1+h)} u(0,y)\, e^{-(2^{1/2}t-y)^2/2t}\,dy$$

(4.43)

$$\geq \exp\big[\delta t(2^{1/2} - \delta/2)\big] \int_{\delta t}^{\delta t(1+h)} u(0,y)\,dy \;.$$

If we assume (4.39), it follows from (4.43) that

$$\limsup_{t\to\infty} \frac{1}{t} \log \left[\int_t^{t(1+h)} u(0,y)\, dy \right] \leq \delta/2 - 2^{1/2} \ .$$

Letting $\delta \to 0$, we obtain (1.19). //

We conclude this section with a simple computation, which ensures that $\varphi(t,x)$ does not rise or fall too sharply.

Lemma 4.4. For each $\alpha, M > 0$, there is a $\beta > 0$ such that for $x_1 \leq x_2 \leq x_1 + Mt$ and $\varphi(t,x_1) \geq e^{-\alpha t}$,

$$\varphi(t,x_2)/\varphi(t,x_1) \geq \frac{1}{2} e^{-\beta(x_2-x_1)} \ . \tag{4.44}$$

Proof. Define $a(t)$ so that

$$(2\pi t)^{-1/2} e^t \int_{-\infty}^{-a(t)} e^{-y^2/2t}\, dy = \frac{1}{2} e^{-\alpha t} \ .$$

It is easy to check that

$$a(t) \leq \gamma t \quad \text{for some} \quad \gamma \ .$$

It is also easy to see that

$$\begin{aligned} e^t \int_{-\infty}^{x_1-a(t)} u(0,y) \frac{e^{-(x_1-y)^2/2t}}{\sqrt{2\pi t}}\, dy &\leq e^t \int_{-\infty}^{x_1-a(t)} \frac{1}{\sqrt{2\pi t}} e^{-(x_1-y)^2/2t}\, dy \\ &= \frac{1}{2} e^{-\alpha t} \leq \frac{1}{2} \varphi(t,x_1) \ . \end{aligned} \tag{4.45}$$

On the other hand, since

$$e^{-(x_2-y)^2/2t} / e^{-(x_1-y)^2/2t} \ , \quad y \in [x_1 - a(t), \infty) \ ,$$

achieves its minimum at $y = x_1 - a(t)$,

$$\varphi(t,x_2) \geqq e^t \int_{x_1-a(t)}^{\infty} u(0,y) \frac{e^{-(x_2-y)^2/2t}}{\sqrt{2\pi t}}\, dy$$

$$\geqq e^t \cdot \exp\{-(x_2-x_1)[x_2-x_1+2a(t)]/2t\} \int_{x_1-a(t)}^{\infty} u(0,y) \frac{e^{-(x_1-y)^2/2t}}{\sqrt{2\pi t}}\, dy$$

$$\geqq e^{-\beta(x_2-x_1)}\, e^t \int_{x_1-a(t)}^{\infty} u(0,y) \frac{e^{-(x_1-y)^2/2t}}{\sqrt{2\pi t}}\, dy \ ,$$

where $\beta = \frac{1}{2}(M+2\gamma)$. By (4.45), this is at most

$$\frac{1}{2} e^{-\beta(x_2-x_1)} \varphi(t,x_1) \ .$$

Therefore, (4.44) follows. //

5. CONVERGENCE OF THE KOLMOGOROV EQUATION IN THE CASE $\lambda > 2^{1/2}$

We now apply the estimates obtained in Section 4 to derive Theorems A and B for the case $\lambda > 2^{1/2}$. We will show that the conditions

$$(1.16) \qquad \lim_{t\to\infty} \frac{1}{t}\log\left[\int_t^{t(1+h)} u(0,y)\,dy\right] = -b$$

for some $h > 0$, and

$$(1.17) \qquad \int_x^{x+N} u(0,y)\,dy > \eta \quad \text{if} \quad x \leqq -M ,$$

for some $\eta, M, N > 0$, are sufficient to imply convergence to a travelling wave, i.e.,

$$(1.15) \qquad u(t,x+m(t)) \to w^{\lambda}(x) \quad \text{uniformly in } x \text{ as } t\to\infty ,$$

for some $m(t)$. As before, λ and b are related by $b = \lambda - \sqrt{\lambda^2 - 2} < 2^{1/2}$. The methodology employed in the proof also enables one to compute the centering term $m(t)$ explicitly in terms of the initial data $u(0,x)$. We will also demonstrate the converse direction, which states that if (1.15) holds, then (1.16) and (1.17) must follow. The argument employed for this direction does not rely on λ being strictly greater than $2^{1/2}$. We will therefore also be able to show here that if $\lambda = 2^{1/2}$ is assumed, then (1.17) and

$$(1.19) \qquad \limsup_{t\to\infty} \frac{1}{t}\log\left[\int_t^{t(1+h)} u(0,y)\,dy\right] \leqq -2^{1/2}$$

both follow from (1.15). The other direction, that (1.17) and (1.19) together imply (1.15) for $\lambda = 2^{1/2}$, is considerably more complicated, and will be demonstrated in Sections 6-9.

Convergence to a travelling wave and the converse direction of the statement will be proved in Theorems 1 and 2, respectively. We first demonstrate the following lemma, which establishes lower bounds on the amount of heating occuring in the Feynman-Kac integral along a class of paths in $[0,t]\times R$. These bounds will be applied in Theorem 1 for comparison of $u(t,x)$ with $\varphi(t,x)$.

We will continue here to use the notation of previous sections, especially that of Section 4. Recall, in particular, the quantities $\varphi(t,x)$ and $m^{\varphi}(t)$. Also, as before, $k(t,x) \equiv f(u(t,x))/u(t,x)$ where f satisfies (1.2).

<u>Lemma 5.1</u>. Assume that $u(t,x)$ satisfies (1.1) under conditions (1.2) and (1.2'), and that its initial data $u(0,x)$ satisfies (1.16) with $0<b<2^{1/2}$. Let $x(s)$, $0 \le s \le t$, be continuous with

$$x(s) \geq m^{\varphi}(t) - 2^{1/2}s + z \tag{5.1}$$

and $z \geq 1$. Then for some T, $t \geq T$ implies that

$$t - e^{-\alpha_1 t} - Ce^{-\beta_1 z} \leq \int_0^t k(t-s,x(s))\,ds \leq t \tag{5.2}$$

where α_1, β_1, and C are positive constants depending on f.

<u>Proof</u>. Since $k(s,y) \leq 1$ for all s and y, the right inequality of (5.2) is trivial. In particular, $u(s,y) \leq \varphi(s,y)$ for all s and y. Therefore by (1.2'),

$$k(s,y) = \frac{f(u(s,y))}{u(s,y)} \geq 1 - C_1(u(s,y))^{\rho}$$

$$\geq 1 - C_1(\varphi(s,y))^{\rho}$$

for some C_1 depending only on f. Corollary 2 of Proposition 4.1 states that for $0 \leq s \leq t - e^{-\alpha t/2}$ and $y > m^{\varphi}(\varphi) - 2^{1/2}s + z$ with $z \geq 1$,

$$\log\varphi(t-s,x) \leq -(\alpha s + \beta z)$$

for specified constants $\alpha,\beta > 0$. Consequently,

$$k(t-s,y) \geqq 1 - C_1 \exp[-\rho(\alpha s + \beta z)]$$

for $0 \leqq s \leqq t - e^{-\alpha t/2}$. Since $k(t-s,y)$ is nonnegative, it follows that

$$\int_0^t k(t-s,x(s))\,ds \geqq t - e^{-\alpha_1 t} - Ce^{-\beta_1 z}$$

for $x(s)$ satisfying (5.1), and positive constants α_1 , β_1 , and C . //

In Theorem 1, $w^\lambda(x)$ will denote the specific translate of the travelling wave which satisfies

$$e^{bx} w^\lambda(x) \to 1 \quad \text{as} \quad x \to \infty \; . \tag{4.23}$$

Also, recall from Section 4 that

$$m^\varphi(t) = \max\{x\colon\ e^t \int_{-\infty}^{\infty} u(0,y) \frac{e^{-(x-y)^2/2t}}{\sqrt{2\pi t}}\,dy = 1\} \; .$$

<u>Theorem 1</u>. Let $u(t,x)$ be a solution of

$$u_t = \frac{1}{2}u_{xx} + f(u) \; , \tag{1.1}$$

where

$$\begin{aligned} &f \in C^1[0,1] \; , \; f(0) = f(1) = 0 \; , \; f(u) > 0 \quad \text{for} \quad 0 < u < 1 \; , \\ &f'(0) = 1 \; , \; f'(u) \leqq 1 \quad \text{for} \quad 0 < u \leqq 1 \; , \end{aligned} \tag{1.2}$$

and

$$1 - f'(u) = O(u^\rho) \quad \text{for some} \quad \rho > 0 \; . \tag{1.2'}$$

Assume that for some $h > 0$,

$$\lim_{t\to\infty} \frac{1}{t} \log\left[\int_t^{t(1+h)} u(0,y)\,dy\right] = -b \tag{1.16}$$

with $0 < b < 2^{1/2}$, and for some $\eta, M, N > 0$,

(1.17) $\int_{x}^{x+N} u(0,y)\,dy > \eta$ if $x \leq -M$.

Then for $m^{\varphi}(t)$ as defined above,

(5.3) $u(t,x+m^{\varphi}(t)) \to w^{\lambda}(x)$ uniformly in x as $t\to\infty$,

where $\lambda = \frac{1}{b}+\frac{b}{2}$.

<u>Proof</u>. By the Feynman-Kac formula,

$$u(t,x) = E[\exp\{\int_0^t k(t-s,\mathfrak{X}_x(s))\,ds\}\,u(0,\mathfrak{X}_x(t))]$$

(5.4)

$$= \int_{-\infty}^{\infty} u(0,y)\,\frac{e^{-(x-y)^2/2t}}{\sqrt{2\pi t}}\,E[\exp\{\int_0^t k(t-s,\mathfrak{z}_{x,y}(s))\,ds\}]\,dy \ .$$

$\mathfrak{X}_x$ and $\mathfrak{z}_{x,y}$ denote Brownian motion and Brownian bridge starting at x (and ending at y for the bridge) and are as defined in Section 2. The basic argument of the proof provided here will be to show that if x is somewhat greater than $m^{\varphi}(t)$, then for "most" paths $\mathfrak{X}_x$, $\int_0^t k(t-s,\mathfrak{X}_x(s))\,ds \sim t$. It will follow that $u(t,x) \sim \varphi(t,x)$, and therefore by Corollary 1 of Proposition 4.1, that $u(t,z+m^{\varphi}(t)) \sim w^{\lambda}(z)$ for large z , where $z = x - m^{\varphi}(t)$. Then by Proposition 3.3, this region extends as $t\to\infty$ so that $u(t,z+m^{\varphi}(t)) \sim w^{\lambda}(z)$ for $z \geq c(t)$, where $c(t) \to -\infty$. If $u(0,x)$ is monotone, this completes the proof. For nonmonotone $u(0,x)$, an additional proposition is included after Theorem 1 to ensure that $u(t,z+m^{\varphi}(t)) \to 1$ as $t \to \infty$ for $z < c(t)$.

We proceed with the argument. First, set

$$r(t,x) = x - b_1 t$$

and

$$\ell^{z}_{x,y}(s) = \frac{s}{t}(y - b_2 t) + \frac{t-s}{t}(x - \frac{z}{2})$$

for $0 \leq s \leq t$ and $z \geq 2$, where we choose $b < b_1 < 2^{1/2}$ and $b_2 = 2^{1/2} - b_1$. Clearly, (5.4) is at least

$$(5.5)\quad \int_{r(t,x)}^{\infty} u(0,y)\,\frac{e^{-(x-y)^2/2t}}{\sqrt{2\pi t}}\,E[\exp\{\int_0^t k(t-s,\mathfrak{z}_{x,y}(s))\,ds\};\ B_{\ell^z_{x,y}}]\,dy\ .$$

Note that B_ℓ is the set of paths lying above the line segment with endpoints $y-b_2t$ and $x-z/2$. By Lemma 2.2(a), for each x and y,

$$(5.6)\qquad P[B_{\ell^z_{x,y}}] = 1-e^{-b_2z}\ .$$

Also, set $z = x - m^\varphi(t)$ (consequently, $x > m^\varphi(t)$). It is easy to check that for $\mathfrak{z}_{x,y} \in B_\ell$ and $y \geq r(t,x)$,

$$(5.7)\qquad \begin{aligned} \mathfrak{z}_{x,y}(s) &> x - 2^{1/2}s - \frac{z}{2} \\ &= m^\varphi(t) - 2^{1/2}s + \frac{z}{2} \end{aligned}$$

for all $0 \leq s \leq t$. Invoking Lemma 5.1, it follows from (5.7) that for $\mathfrak{z}_{x,y} \in B_\ell$,

$$(5.8)\qquad \begin{aligned} \exp\{\int_0^t k(t-s,\mathfrak{z}_{x,y}(s))\,ds\} &\geq \exp\{t-e^{-\alpha t}-Ce^{-\beta z}\} \\ &\geq e^t(1-e^{-\alpha t}-Ce^{-\beta z})\ , \end{aligned}$$

where α, β, and C are positive constants. With the aid of (5.6) and (5.8), we see that (5.5) is at least

$$(5.9)\qquad (1-e^{-b_2z})(1-e^{-\alpha t}-Ce^{-\beta z})\,e^t\int_{r(t,x)}^{\infty} u(0,y)\,\frac{e^{-(x-y)^2/2t}}{\sqrt{2\pi t}}\,dy\ .$$

On the other hand, if $0 < \epsilon < (b_1-b)^2/12\lambda$, then

$$(5.10)\qquad \begin{aligned} e^t\int_{-\infty}^{r(t,x)} u(0,y)\,\frac{e^{-(x-y)^2/2t}}{\sqrt{2\pi t}}\,dy &\leq \exp[-b(x-\lambda t)+\epsilon x-(b_1-b)^2t/4] \\ &\leq e^{-b_3t} \end{aligned}$$

for large enough t, where $b_3 = (b_1-b)^2/12$. The first inequality in (5.10) follows from (4.7), and the second inequality from $x > m^\varphi(t) = \lambda t + o(t)$ ((4.12a)). Consequently, (5.9), and thus $u(t,z+m^\varphi(t))$, is at least

(5.11) $$(1-e^{-b_2 z})(1-e^{-\alpha t}-Ce^{-\beta z})(\varphi(t,z+m^{\varphi}(t))-e^{-b_3 t})$$

for large enough t and $z \geqq 2$.

By employing (5.11), it is not difficult to choose functions $\hat{\gamma}_1(z)$ and $\hat{\gamma}_2(t)$ with

(3.44a) $$\hat{\gamma}_1(z) \to 1 \quad \text{as} \quad z \to \infty$$

and

(3.44b) $$\hat{\gamma}_2(t) \to 0 \quad \text{as} \quad t \to \infty ,$$

so that

(5.12) $$\hat{\gamma}_1(z)\varphi(t,z+m^{\varphi}(t)) - \hat{\gamma}_2(t) \leqq u(t,z+m^{\varphi}(t))$$

for $z \geqq 2$. On the other hand, the inequality

(5.13) $$u(t,z+m^{\varphi}(t)) \leqq \varphi(t,z+m^{\varphi}(t))$$

is immediate. We have thus shown that for large z and t , u is accurately approximated by φ . But we also know from Corollary 1 of Proposition 4.1, that

(4.24) $$\gamma_1(z)w^{\lambda}(z) - \gamma_2(t) \leqq \varphi(t,z+m^{\varphi}(t)) \leqq \gamma_1(z)w^{\lambda}(z) + \gamma_2(t)$$

for $\gamma_1(z)$ and $\gamma_2(t)$ satisfying (3.44a) and (3.44b), respectively. It is a simple matter to construct from (4.24), (5.12), and (5.13) functions $\tilde{\gamma}_1(z)$ and $\tilde{\gamma}_2(t)$ such that

$$\tilde{\gamma}_1^{-1}(z)w^{\lambda}(z) - \tilde{\gamma}_2(t) \leqq u(t,z+m^{\varphi}(t)) \leqq \tilde{\gamma}_1(z)w^{\lambda}(z) + \tilde{\gamma}_2(t)$$

for $z \geqq 2$, and which also satisfy (3.44a) and (3.44b).

We are now finally in a position to apply Proposition 3.3 to conclude

$$(5.14)\qquad \sup_{z \geq c(t)} |u(t,z+m^{\varphi}(t)) - w^{\lambda}(z)| \to 0 \quad \text{as} \quad t\to\infty$$

for some $c(t) \to -\infty$ as $t\to\infty$. If the initial data $u(0,x)$ is monotone, then $u(t,x)$ will remain monotone for all time, and so (5.3) follows from the fact that $w^{\lambda}(z) \to 1$ as $z \to -\infty$. For nonmonotone initial data, it still needs to be demonstrated that $u(t,z+m^{\varphi}(t)) \to 1$ uniformly in $z < c(t)$. We postpone the proof until Proposition 5.1, which follows immediately. //

Proposition 5.1 (Conclusion to Theorem 1). Let $u(t,x)$ satisfy the same conditions as in Theorem 1. If $c(t)$ is chosen so that $c(t) \to -\infty$ as $t \to \infty$, then

$$(5.15)\qquad \liminf_{t\to\infty}\ \inf_{z< c(t)} u(t,z+m^{\varphi}(t)) = 1 .$$

Proof. It follows easily from (1.17) that $u(1,y) > \eta'$ for some $\eta' > 0$ and all $y \leq -M$. We may therefore apply (3.66) of Proposition 3.4 to conclude that $\forall\delta > 0$,

$$(5.16)\qquad \liminf_{t\to\infty} \{u(t,x)\colon x< (2^{1/2}-\delta)t\} = 1 .$$

It remains to be shown, that

$$(5.17)\qquad \liminf_{t\to\infty} \{u(t,x)\colon b_0 t < x < m^{\varphi}(t) - c(t)\} = 1$$

for some b_0 , $b < b_0 < 2^{1/2}$.

For fixed ϵ_1 , $0 < \epsilon_1 < b/3\lambda$, set

$$r_1(t,x) = t - \frac{1+\epsilon_1}{b\lambda} \log\varphi(t,x) .$$

Also, choose T large enough so that Proposition 4.1 holds with $\epsilon = \epsilon_1/8$ and $M = 4\lambda^2/b$, and

$$|m^{\varphi}(t) - \lambda t| < \epsilon t \quad \text{for} \quad t \geq T .$$

We will first show that for $t \geqq \frac{2\lambda}{b} T$ and $b_0 t < x < m^{\varphi}(t) - c(t)$,

$$(5.18) \qquad -\frac{3\epsilon_1}{2} \log\varphi(t,x) \leqq \log\varphi(r_1(t,x),x) \leqq -\frac{\epsilon_1}{2} \log\varphi(t,x) .$$

Since $c(t) \to -\infty$, it follows from (4.14) of Proposition 4.1 that $\varphi(t,x) \to \infty$ uniformly in $x \in (b_0 t, m^{\varphi}(t) - c(t))$ as $t \to \infty$. We are thus stating in (5.18) that $\varphi(r_1(t,x),x) \to 0$ uniformly as $t \to \infty$, but since ϵ_1 is small, this convergence is not "too quick". Moreover, for x in this range, it will be demonstrated that $u(r_1(t,x),x) \sim \varphi(r_1(t,x),x)$. After verifying (5.18), we will then apply Proposition 3.4 (with $t_0 = r(t,x)$) to conclude that $u(s,x)$ grows rapidly enough from time $r(t,x)$ to t to ensure (5.17) as $t \to \infty$. The idea is thus to demonstrate (5.17) by backtracking in time from t to slightly before when $m^{\varphi}(s)$ "crosses" the level x , and then to show that the intervening time up to t is great enough so that $u(t,x) \sim 1$.

To demonstrate (5.18), note that from Proposition 4.1 and the first corollary to Lemma 4.2,

$$(5.19) \qquad \log\varphi(t,x) \leqq [b(\lambda - b) + \epsilon_1/4]t$$

if $t \geqq \frac{2\lambda}{b} T$ and $b_0 t < x < m^{\varphi}(t) - c(t)$. It follows that

$$(5.20) \qquad r_1(t,x) \geqq [\frac{b}{\lambda} - \frac{3\epsilon_1}{2}]t \geqq \frac{b}{2\lambda} t .$$

Therefore, $r_1(t,x) \geqq T$ and $x \leqq M r_1(t,x)$, and so we may apply (4.14) of Proposition 4.1 to conclude that

$$|\log[\varphi(r_1(t,x),x)/\varphi(t,x)] + (1+\epsilon_1)\log\varphi(t,x)| \leqq \epsilon[\frac{1+\epsilon_1}{b\lambda}\log\varphi(t,x) + 1] .$$

Hence,

$$|\log\varphi(r_1(t,x),x) + \epsilon_1 \log\varphi(t,x)| \leqq \frac{\epsilon_1}{2}\log\varphi(t,x) ,$$

which is (5.18).

Now, refer back to (5.11) of the proof of Theorem 1, which states that for $z \geqq 2$,

$$(5.21) \qquad u(s,x) \geqq (1-e^{-b_2 z})(1-e^{-\alpha s}-Ce^{-\beta z})(\varphi(s,x)-e^{-b_3 s}) \ ,$$

where, as before, $z = x - m^{\varphi}(s)$ and α , β , b_2 , and C are all positive. Since $\varphi(t,x) \to \infty$ uniformly in $b_0 t < x < m^{\varphi}(t) - c(t)$, the right hand inequality of (5.18) in conjunction with (4.14) implies that

$$x - m^{\varphi}(r_1(t,x)) \to \infty \text{ uniformly for } x \in [b_0 t, m^{\varphi}(t) - c(t)] \ .$$

Thus, (5.21) remains valid if one substitutes $r_1(t,x)$ for s . We are of course free to choose ϵ_1 so as to satisfy

$$\epsilon_1 < b(\lambda - b) \wedge b_3/12\lambda(\lambda - b) \ .$$

Applying (5.19) to the left side of (5.18), one may compute that for such ϵ_1 ,

$$\varphi(r_1(t,x),x) > e^{-b_3 r_1(t,x)/2} \ .$$

It therefore follows from (5.21) (and (5.20)) that for $b_0 t < x < m^{\varphi}(t) - c(t)$,

$$(5.22) \qquad u(r_1(t,x),x) \geqq C_2(t)\,\varphi(r_1(t,x),x) \ ,$$

where $C_2(t) \to 1$ as $t \to \infty$. Consequently, one final application of (5.18) shows that

$$(5.23) \qquad \log u(r_1(t,x),x) \geqq -3\epsilon_1 \log\varphi(t,x)$$

for large enough t . We are finally in a position to apply Proposition 3.4. One may write

$$u(t,x) = u(s(t,x) + r_1(t,x) + 3\epsilon_1 \log\varphi(t,x), x) \ ,$$

where if $\epsilon < 1/6$,

$$s(t,x) \equiv \left(\frac{1+\epsilon_1}{b\lambda} - 3\epsilon_1\right)\log\varphi(t,x)$$

$$\to \infty$$

uniformly for $b_0 t < x < m^{\varphi}(t) - c(t)$ as $t \to \infty$. (5.17) then follows from (3.66) (with $\theta = 1$) and (5.23). //

Having obtained sufficient conditions on the initial data of $u(t,x)$ for convergence to a travelling wave, we pause to comment on the translation term $m^{\varphi}(t)$. We already computed in (4.12a) that

$$(5.24) \qquad m^{\varphi}(t)/t \to \lambda \quad \text{as} \quad t \to \infty$$

under the condition (1.16). In fact, the corollary to Lemma 3.2 states that under convergence to a travelling wave, (5.24) must indeed occur. Higher order behavior of $m^{\varphi}(t)$ depends, on the other hand, on the more precise nature of the right tail of $u(0,x)$. The definition of $m^{\varphi}(t)$ is, however, explicit enough to frequently permit calculation, which we now do in a specific case.

Example. Let $g(x) = e^{bx}u(0,x)$ for some $b < 2^{1/2}$. Assume that for each fixed $t_0 > 1$ and large enough x,

$$(1.23) \qquad t_0^{-\gamma} \leq g(tx)/g(x) \leq t_0^{\gamma}$$

for $1 \leq t \leq t_0$, where $\gamma > 0$ is fixed. Under (1.17), it then follows that

$$(5.3) \qquad u(t,x+m^{\varphi}(t)) \to w^{\lambda}(x) \quad \text{uniformly in } x \text{ as } t \to \infty .$$

Here,

$$(5.25) \qquad m^{\varphi}(t) = \lambda t + \frac{1}{b} \log g((\lambda - b)t) + o(1) .$$

Proof. It is easy to show that under (1.23), (1.16) is satisfied. (5.3) therefore follows from Theorem 1. To demonstrate (5.25), we first note that by (5.24) and the corollary to Lemma 4.1,

$$1 = e^t \int_{-\infty}^{\infty} u(0,y) \frac{1}{\sqrt{2\pi t}} \exp[-(m^\varphi(t) - y)^2/2t]\, dy$$

(5.26)

$$= (1+\epsilon(t))\, e^t \int_{(\lambda-b-\delta)t}^{(\lambda-b+\delta)t} u(0,y) \frac{1}{\sqrt{2\pi t}} \exp[-(m^\varphi(t) - y)^2/2t]\, dy$$

for fixed $\delta > 0$, where $\epsilon(t) \to 0$ as $t \to \infty$. Because of (1.23), this is at most

$$(5.27) \quad C_\delta(t) g((\lambda-b)t) e^t \int_{(\lambda-b-\delta)t}^{(\lambda-b+\delta)t} e^{-by} \frac{1}{\sqrt{2\pi t}} \exp[-(m^\varphi(t) - y)^2/2t]\, dy \ ,$$

for large enough t, where

$$C_\delta(t) = \left(\frac{\lambda-b+\delta}{\lambda-b-\delta}\right)^{2\gamma} (1+\epsilon(t)) \ .$$

(5.27) is at most

$$C_\delta(t) g((\lambda-b)t) e^t \int_{-\infty}^{\infty} e^{-by} \frac{1}{\sqrt{2\pi t}} \exp[-(m^\varphi(t) - y)^2/2t]\, dy \ ;$$

the substitution $y' = y - m^\varphi(t)$ shows that this equals

$$(5.28) \qquad C_\delta(t) g((\lambda-b)t) \exp[(1+\frac{b^2}{2})t - b\, m^\varphi(t)] \ .$$

Since $b\lambda = 1+\frac{b^2}{2}$, (5.26) - (5.28) show that

$$(5.29) \qquad 1 \leq C_\delta(t) g((\lambda-b)t) \exp[b(\lambda t - m^\varphi(t))] \ .$$

Similarly, one may obtain the upper bound

$$(5.30) \qquad 1 \geq C_\delta^{-1}(t) g((\lambda-b)t) \exp[b(\lambda t - m^\varphi(t))] \ .$$

Letting $\delta \to 0$ as $t \to \infty$, (5.29) and (5.30) together imply that

$$g((\lambda-b)t) \exp[b(\lambda t - m^\varphi(t))] \to 1$$

as $t \to \infty$. Consequently,

$$m^\varphi(t) = \lambda t + \frac{1}{b} \log g((\lambda-b)t) + o(1) \ . \qquad //$$

We now demonstrate the converse direction of Theorem 1. We will find it convenient to apply Proposition 4.2, which gives an alternate criterion for convergence of $u(t,x)$ to a travelling wave. The proof, as in Theorem 1, makes use of the Feynman-Kac formula, although the computations are simpler in this case. We include the case $\lambda = 2^{1/2}$, which requires no extra work in this direction.

<u>Theorem 2</u>. Let $u(t,x)$ satisfy conditions (1.1), (1.2), and (1.2') as in Theorem 1. Assume that for some $m(t)$,

(1.15) $$u(t,x+m(t)) \to w^{\lambda}(x) \text{ uniformly in } x \text{ as } t\to\infty ,$$

where $\lambda \geq 2^{1/2}$. Then, if $\lambda > 2^{1/2}$,

(1.16) $$\lim_{t\to\infty} \frac{1}{t} \log \Big[\int_t^{t(1+h)} u(0,y)\,dy\Big] = -b$$

for all $h > 0$, where $b = \lambda - \sqrt{\lambda^2 - 2}$, whereas if $\lambda = 2^{1/2}$,

(1.19) $$\limsup_{t\to\infty} \frac{1}{t} \log \Big[\int_t^{t(1+h)} u(0,y)\,dy\Big] \leq -2^{1/2} .$$

In either case, for some $\eta, M, N > 0$,

(1.17) $$\int_x^{x+N} u(0,y)\,dy > \eta \quad \text{for} \quad x \leq -M .$$

<u>Proof</u>. We apply Proposition 4.2, which states that both (1.16) and (1.19) are consequences of

(4.31) $$\lim_{t\to\infty} \frac{1}{t} \log \varphi(t,\lambda t) = 0$$

for $\lambda \geq 2^{1/2}$. One-half of (4.31) is easy to show. Since we are assuming (1.15), it follows from the corollary to Lemma 3.2 that

(5.31) $$m(t) = \lambda t - \varepsilon(t) ,$$

where $\varepsilon(t) = o(t)$. We may assume here without loss of generality that $m(t) = m_{1/2}(t)$. Since

$$\varphi(t,m(t)) \geqq u(t,m(t)) = 1/2 ,$$

it follows from Lemma 4.4 that

$$\varphi(t,\lambda t) \geqq \frac{1}{4} e^{-\beta|\varepsilon(t)|}$$

for some $\beta > 0$. Consequently,

$$\liminf_{t\to\infty} \frac{1}{t} \log \varphi(t,\lambda t) \geqq 0 .$$

The other one-half of (4.31), namely that

$$\limsup_{t\to\infty} \frac{1}{t} \log \varphi(t,\lambda t) \leqq 0 , \tag{5.32}$$

requires more work. If (5.32) is violated and $\limsup_{t\to\infty} \frac{1}{t} \log \varphi(t,\lambda t) \geqq \varepsilon > 0$, then again applying Lemma 4.4, we may find a $\delta > 0$ such that

$$\limsup_{t\to\infty} e^{-\delta t} \varphi(t,(\lambda+\delta)t) \geqq 1 . \tag{5.33}$$

This conclusion will eventually get us into trouble, as we shall see.

First, we introduce the family of lines

$$\ell_t(s) = \lambda s + \frac{\delta}{2} t , \quad -\infty < s < \infty .$$

By (5.31),

$$\ell_t(s) - m(s) \to \infty \text{ uniformly for } s \in [0,t] \text{ as } t \to \infty .$$

It therefore follows from (1.15) and the fact that $w^{\lambda}(x) \to 0$ as $x \to \infty$, that

$$u(s,x) \to 0 \text{ uniformly in } x \geqq \ell_t(s) \text{ as } s,t \to \infty$$

with $s \leqq t$. Consequently for each $\eta > 0$, we may choose s_0 and t_0

large enough so that for $s \geqq s_0$, $t \geqq t_0$, and $s \leqq t$,

(5.34) $\quad u(s,x) < \eta$ for $x \geqq \ell_t(s)$.

We proceed to obtain a lower bound on $u(t,(\lambda+\delta)t)$ by arguing in a manner reminiscent of the proof of Theorem 1. We apply the Feynman-Kac representation, from which we compute a lower bound by analyzing those paths lying above $\ell_t(s)$. First, observe that

$$u(t,x) = \int_{-\infty}^{\infty} u(0,y) \frac{e^{-(x-y)^2/2t}}{\sqrt{2\pi t}} E[\exp\{\int_0^t k(t-s,\mathfrak{z}_{x,y}(s))\, ds\}]\, dy ,$$

where as usual $\mathfrak{z}_{x,y}$ denotes Brownian bridge starting at x and ending at y. Here, we shall fix $x = (\lambda+\delta)t$. The above integral is at least

$$(5.35) \quad \int_{\delta t}^{\infty} u(0,y) \frac{e^{-(x-y)^2/2t}}{\sqrt{2\pi t}} E[\exp\{\int_0^t k(t-s,\mathfrak{z}_{x,y}(s))\, ds\};\ A_y(t)]\, dy ,$$

where

$$A_y(t) = \{\mathfrak{z}_{x,y}:\ \mathfrak{z}_{x,y}(s) \geqq \frac{s}{t}y + \frac{t-s}{t}x - \frac{\delta}{2}t \quad \text{for } 0 \leqq s \leqq t\} .$$

Note that since $A_y(t)$ is the set of paths lying above the line segment with endpoints $y - \frac{\delta}{2}t$ and $x - \frac{\delta}{2}t$, it follows from Lemma 2.2(a), that

$$(5.36) \quad P[A_y(t)] = 1 - e^{-\delta^2 t/2} .$$

For $\mathfrak{z}_{x,y} \in A_y(t)$ and $y \geqq \delta t$, one may also check that

$$(5.37) \quad \mathfrak{z}_{x,y}(s) > \ell_t(t-s) \quad \text{for } 0 \leqq s \leqq t .$$

Recall now that under (1.2'), $k(s,y') \geqq 1 - C(u(s,y'))^\rho$ for constants $C,\rho > 0$. Therefore by (5.34) and (5.37), for $\mathfrak{z}_{x,y} \in A_y(t)$ with $y \geqq \delta t$, $0 \leqq s \leqq t-s_0$, and $t \geqq t_0$,

$$k(t-s,\mathfrak{z}_{x,y}(s)) \geqq 1 - C\eta^\rho ,$$

and consequently

$$\int_0^t k(t-s,\mathfrak{z}_{x,y}(s))\,ds \geq \int_0^{t-s_0} k(t-s,\mathfrak{z}_{x,y}(s))\,ds$$

$$\geq (1-C\eta^\rho)(t-s_0)\ .$$

This inequality together with (5.36) shows that (5.35) is at least

$$(5.38)\qquad (1-\exp[-\delta^2 t/2])\cdot\exp[-C\eta^\rho t-s_0]\cdot e^t\int_{\delta t}^{\infty} u(0,y)\,\frac{e^{-(x-y)^2/2t}}{\sqrt{2\pi t}}\,dy$$

for large enough t.

On the other hand,

$$e^t\int_{-\infty}^{\delta t} u(0,y)\,\frac{e^{-(x-y)^2/2t}}{\sqrt{2\pi t}}\,dy \leq e^t\int_{2^{1/2}t}^{\infty}\frac{e^{-y^2/2t}}{\sqrt{2\pi t}}\,dy \leq C_1/\sqrt{t}$$

for appropriate C_1, since $x=(\lambda+\delta)t \geq (2^{1/2}+\delta)t$. But if (5.33) is to hold, then

$$e^{-\delta t}\varphi(t,(\lambda+\delta)t) \geq 1/2\ ,$$

and hence

$$\varphi(t,(\lambda+\delta)t) \geq 1$$

for certain arbitrarily large values of t. For these values of t, (5.38), and hence $u(t,x)$, is at least

$$(5.39)\qquad C_2(t)\exp[-C_3\eta^\rho t]\,\varphi(t,(\lambda+\delta)t)\ ,$$

where $C_2(t)\to\infty$ as $t\to\infty$ and $C_3 = 2C$. Finally, choose η so that $C_3\eta^\rho \leq \delta/2$. Since $u(t,x)\leq 1$, (5.39) implies that for the above values of t,

$$C_2(t)e^{-\delta t/2}\varphi(t,(\lambda+\delta)t) \leq 1\ .$$

This is a contradiction, and means that (5.33) cannot hold. Therefore (5.32) is valid, and hence (4.31) follows.

We still need to show (1.17). Assume that

$$(5.40) \qquad \int_x^{x+N} u(0,y)\,dy \leq \eta$$

for some choice of x, N, and η. Then for any x_0,

$$(5.41) \qquad e^t \int_{\mathcal{I}} u(0,y) \frac{e^{-(x_0-y)^2/2t}}{\sqrt{2\pi t}}\,dy \leq \eta e^t/\sqrt{2\pi t} ,$$

where $\mathcal{I} = (x,x+N)$. On the other hand, for x_0 chosen so that

$$(5.42) \qquad x_0 \in [x + N/3 + 2^{1/2}t \, , \, x + 2N/3 - 2^{1/2}t] ,$$

it is easy to show that

$$(5.43) \qquad \begin{aligned} e^t \int_{\mathcal{I}^c} u(0,y) \frac{e^{-(x_0-y)^2/2t}}{\sqrt{2\pi t}}\,dy &\leq e^t \int_{\mathcal{I}^c} \frac{e^{-(x_0-y)^2/2t}}{\sqrt{2\pi t}}\,dy \\ &\leq \frac{1}{\sqrt{2\pi t}}\, e^{-N^2/18t} . \end{aligned}$$

It follows from (5.41) and (5.43) that for x_0 chosen as in (5.42),

$$u(t,x_0) \leq \varphi(t,x_0) \leq \frac{1}{\sqrt{2\pi t}} [\eta e^t + e^{-N^2/18t}] .$$

Therefore, for each fixed $\epsilon > 0$ and $t > 0$,

$$(5.43) \qquad u(t,x_0) < \epsilon$$

if η is chosen small enough and N large enough in (5.40). Now, if (1.17) were not to hold, then arbitrarily negative x satisfying (5.40) could be found for fixed η and fixed N. It would thus follow from (5.43) that for each t,

$$\liminf_{x \to \infty} u(t,x) = 0 .$$

This contradicts (1.15), and so (1.17) must indeed hold. //

6. HITTING PROBABILITIES FOR BROWNIAN BRIDGE

In this section, we make comparative estimates on the probability that Brownian bridge will hit certain curves. Our main results will show that if two curves are "close enough" to one another, then the hitting probabilities will be nearly the same. In Proposition 6.1, we will consider a case where the two curves assume the same values at $s = 0$ and $s = t$, and where as s and $t-s$ increase, the curves only gradually separate; in Proposition 6.2, we will extend this to a case in which the curves may differ substantially at fixed times, but are close to translates of one another. These estimates will be applied in Sections 7 through 9, where we will define curves which divide $[0,\infty)\times R$ into regions of nearly full heating/uncertain heating and uncertain heating/insignificant heating with respect to the term $k(t,x)$ defined in (1.33). With the aid of these estimates, we will be able to compare the hitting probabilities of these two curves, and thereby apply the Feynman-Kac representation to obtain sharp upper and lower bounds for the Kolmogorov equation.

To state Proposition 6.1, we introduce the following notation. For a given family of functions L_t defined at least on $[0,t]$, define the functions $\underline{L}_t$ and $\overline{L}_t$ so that

$$(6.1a)\qquad \begin{aligned} \underline{L}_t(s) &= L_t(s) - Cs^{\delta} && \text{for } 0 \leqq s \leqq t/2 \\ &= L_t(s) - C(t-s)^{\delta} && \text{for } t/2 \leqq s \leqq t \end{aligned}$$

and

$$(6.1b)\qquad \begin{aligned} \overline{L}_t(s) &= L_t(s) + Cs^{\delta} && \text{for } 0 \leqq s \leqq t/2 \\ &= L_t(s) + C(t-s)^{\delta} && \text{for } t/2 \leqq s \leqq t , \end{aligned}$$

where C and δ are constants with $C > 0$ and $0 < \delta < 1/2$. Of course,

$\underline{L}_t(s) \leq L_t(s) \leq \bar{L}_t(s)$ for all t and $s \in [0,t]$. We will typically wish to consider L_t for which

(6.2) $\underline{L}_t(s) \leq 0$ if $s \in [r_o, t-r_o]$

for all t and some fixed r_o ; L_t need not be finite valued. We will not make any regularity assumptions on L_t , although it would be possible to paraphrase the results in Sections 7-9 so that consideration here of continuous L_t would suffice.

In Proposition 6.1, we will seek to compare the hitting probabilities of $\underline{L}_t$ and $\bar{L}_t$ on the interval $[r, t-r]$, where t and r are large. Since we will not know either function explicitly, direct computation does not make sense. Rather, we will obtain estimates by deforming one curve into the other. To do so, we will need to apply a special case of the Cameron-Martin formula. One may find a treatment of the general case in various places, e.g. in [7] or [9]. We will be interested here in the special case where the drift of the diffusion is only time-dependent. The formula states that if $\alpha(s)$ is a measurable L^2 function on $[0,t]$, and $\mathfrak{X}(s)$ denotes Brownian motion, then the probability measure $\mathbb{P}^1_\alpha$ defined on $C[0,t]$ by

(6.3) $$\mathbb{P}^1_\alpha(A) = E[\exp\{\int_0^t \alpha(s)d\mathfrak{X}(s) - \frac{1}{2}\int_0^t \alpha^2(s)ds\} ; A]$$

is the measure of the diffusion in $\mathbb{R}$ with drift $\alpha(s)$ at time s and unit variance. In this case, (6.3) may be proved without too much difficulty by discrete approximation of α . Here, there is also no difficulty in deriving the analogue of (6.3) for Brownian bridge. For $\mathfrak{z}^t(s)$ denoting Brownian bridge terminating at time t ,

(6.4) $$\mathbb{P}^0_\alpha(A) = E[\exp\{\int_0^t \alpha(s)d\mathfrak{z}^t(s) - \frac{1}{2}\int_0^t \alpha^2(s)ds + \frac{1}{2t}(\int_0^t \alpha(s)\,ds)^2\} ; A]$$

is the measure of the diffusion with drift $\alpha(s)$ which is conditioned to return to 0 at time t .

Proposition 6.1. Let $\underline{L}_t$ and $\bar{L}_t$ denote families of functions defined as in (6.1) so that (6.2) holds. Then,

$$(6.5)\qquad \frac{P[\mathfrak{z}^t(s) > \bar{L}_t(s) \quad \text{for} \quad r \le s \le t-r]}{P[\mathfrak{z}^t(s) > \underline{L}_t(s) \quad \text{for} \quad r \le s \le t-r]} \to 1 \quad \text{as} \quad r \to \infty$$

uniformly in t.

Proof. The basic idea of the proof is to deform the curve $\bar{L}_t$ into $\underline{L}_t$ by modifying Brownian bridge so that it becomes a diffusion with drift. The Cameron-Martin formula then enables us to express this new process in terms of a weighted average of the original Brownian bridge. We show that, as a rule, the weighting that is involved is insignificant, and therefore the probabilities that this diffusion lies above $\underline{L}_t$ and that Brownian bridge lies above $\underline{L}_t$ are almost equal for large r. Since the former is also the probability that the bridge lies above $\bar{L}_t$, (6.5) follows.

We first observe that

$$(6.6)\quad P[\mathfrak{z}(s) > \bar{L}_t(s) \quad \text{for} \quad r \le s \le t-r] = P[\mathfrak{z}(s) - \beta_{r,t}(s) > \underline{L}_t(s) \quad \text{for} \quad r \le s \le t-r],$$

where

$$(6.7)\qquad \beta_{r,t}(s) = \begin{cases} 2Cr^{\delta-1}s & \text{for} \quad 0 \le s \le r \\ 2Cs^{\delta} & \text{for} \quad r \le s \le t/2 \\ 2C(t-s)^{\delta} & \text{for} \quad t/2 \le s \le t-r \\ 2Cr^{\delta-1}(t-s) & \text{for} \quad t-r \le s \le t . \end{cases}$$

(We assume here that $t \ge 2r$.) By Lemma 2.3, these probabilities are well-defined. The process $\mathfrak{z}(s) - \beta_{r,t}(s)$, $0 \le s \le t$, is a diffusion with drift $\alpha(s) = -\dot{\beta}_{r,t}(s)$ and unit variance. Therefore by (6.4), we may rewrite the probabilities in (6.6) as

$$(6.8)\qquad E[\exp\{-\int_0^t \dot{\beta}_{r,t}(s)d\mathfrak{z}(s) - \frac{1}{2}\int_0^t \dot{\beta}^2_{r,t}(s)\,ds\}; A_{r,t}] ,$$

where

$$A_{r,t} = \{\mathfrak{z}: \mathfrak{z}(s) > \underline{L}_t(s) \quad \text{for} \quad r \leq s \leq t-r\} \ .$$

We wish to compare the quantity in (6.8) with $P[A_{r,t}]$.

To investigate (6.8), we restrict $A_{r,t}$ to the subset $A^1_{r,t}$, where

$$A^1_{r,t} = A_{r,t} \cap \{\mathfrak{z}: \mathfrak{z}(s) < \Lambda_t(s) \quad \text{for} \quad r \leq s \leq t-r\}$$

and

$$\begin{aligned} \Lambda_t(s) &= 2Cs^{\epsilon} \qquad \text{for} \quad 0 \leq s \leq t/2 \\ &= 2C(t-s)^{\epsilon} \quad \text{for} \quad t/2 \leq s \leq t \ , \end{aligned}$$

with $\frac{1}{2} < \epsilon < 1-\delta$. To analyze the behavior on $A^1_{r,t}$, note that $\dot{\beta}_{r,t}(s)$ is decreasing in s and constant on $[0,r]$ and $[t-r,t]$. Define $\ell_{r,t}(s)$ as in (6.7), but with δ replaced by ϵ . It is not difficult to see that for $\mathfrak{z} \in A^1_{r,t}$,

$$\int_0^t \dot{\beta}^2_{r,t}(s) d\mathfrak{z}(s) < \int_0^t \dot{\beta}_{r,t}(s) d\ell_{r,t}(s) \ .$$

A simple computation shows that this is at most $C_1 r^{\delta+\epsilon-1}$, where C_1 depends on C . Also,

$$\int_0^t \dot{\beta}^2_{r,t}(s)\, ds \leq C_2 r^{2\delta-1}$$

for appropriate choice of C_2 . Consequently, (6.8) is at least

$$(6.9) \qquad \exp\{-C_3 r^{\delta+\epsilon-1}\} P[A^1_{r,t}]$$

for some C_3 . The quantity $\delta+\epsilon-1$ is negative, and thus (6.9) approaches $P[A^1_{r,t}]$ for large r . We have thus reduced the problem to showing that

$$(6.10) \qquad P[A^1_{r,t}]/P[A_{r,t}] \to 1 \quad \text{as} \quad r \to \infty$$

uniformly in t . But by Lemma 2.6 and (6.2),

$$P[A^1_{r,t}]/P[A_{r,t}] = P^0[B^{\wedge t}[r,t-r] \mid B_{\underline{L}_t}[r,t-r]]$$
$$\geq P^0[B^{\wedge t}[r,t-r] \mid B_0[r,t-r]]$$

for $r \geq r_0$, which by Lemma 2.7,

$$\to 1 \quad \text{as} \quad r \to \infty$$

uniformly in t. The proof is thus complete. //

In order to state Proposition 6.2, we introduce the following additional notation. If $L(s)$ is a function on the interval $[0,t]$, then $\theta_{r,t} \circ L$ is defined so that

$$(6.11) \qquad \theta_{r,t} \circ L(s) = \begin{cases} L(s+s^\delta) + C_1 s^\delta & \text{for } r \leq s \leq t/2 \\ L(s+(t-s)^\delta) + C_1(t-s)^\delta & \text{for } t/2 \leq s \leq t-2r \\ L(s) & \text{otherwise,} \end{cases}$$

where $0 < \delta < 1/2$. We set $C_1 = C/2$, where $C > 0$ is chosen as in (6.1). $\theta_{r,t}$ shifts L up and to the left, and becomes more pronounced as s nears $t/2$. We also define $\theta^{-1}_{r,t}$ so that

$$(6.12) \qquad \theta^{-1}_{r,t} \circ L = \inf\{\ell : \theta_{r,t} \circ \ell \geq L\} .$$

Over $[r+r^\delta, t-2r]$, $\theta^{-1}_{r,t}$ is just the inverse of $\theta_{r,t}$, and shifts L back down and to the right. (6.12) also specifies $\theta^{-1}_{r,t}$ in the two regions where the inverse is ambiguous: over $[r, r+r^\delta)$, where $\theta^{-1}_{r,t} \circ L = -\infty$, and over $(t-2r, t-2r+(2r)^\delta)$, where $\theta^{-1}_{r,t} \circ L$ is the maximum of the shifted value of L and of L itself. We employ $\theta_{r,t}$ and $\theta^{-1}_{r,t}$ to define $\underline{\mathcal{L}}_{r,t}$ and $\bar{\mathcal{L}}_{r,t}$ so that

$$(6.13) \qquad \underline{\mathcal{L}}_{r,t} = \theta^{-1}_{r,t} \circ L$$

and

(6.14) $\bar{\mathcal{L}}_{r,t} = \theta_{r,t} \circ L \vee \theta^{-1}_{r,t} \circ L \vee L$.

$\bar{\mathcal{L}}_{r,t}$ and $\underline{\mathcal{L}}_{r,t}$ should be thought of as raised and lowered shifts of L with certain lateral movement included; $\bar{\mathcal{L}}_{r,t}$ is defined so as to ensure that $\bar{\mathcal{L}}_{r,t} \geq \underline{\mathcal{L}}_{r,t}$ and $\bar{\mathcal{L}}_{r,t} \geq L$. The manner in which $\theta_{r,t}$ is defined near the endpoints of $[r,t-r]$ is just a technicality, the point being to ensure that $\theta_{r,t} \circ L$ and $\theta^{-1}_{r,t} \circ L$ are only defined in terms of values of L on $[r,t-r]$, so that Proposition 6.2 below will hold. Proposition 6.2 states that the hitting probabilities of $\underline{\mathcal{L}}_{r,t}$, L_t , and $\bar{\mathcal{L}}_{r,t}$ over $[r,t-r]$ are close for large r ; this result will be applied extensively in Sections 7-9 to obtain upper and lower bounds for the Feynman-Kac integral associated with (1.1).

Proposition 6.2. Let $\underline{\mathcal{L}}_{r,t}$ and $\bar{\mathcal{L}}_{r,t}$ denote families of functions defined as in (6.13) and (6.14) in terms of L_t so that (6.2) holds. Then,

$$\frac{P[\mathfrak{z}^t(s) > \bar{\mathcal{L}}_{r,t}(s) \quad \text{for} \quad r \leq s \leq t-r]}{P[\mathfrak{z}^t(s) > L_t(s) \quad \text{for} \quad r \leq s \leq t-r]} \to 1 \quad \text{as} \quad r \to \infty \tag{6.15}$$

and

$$\frac{P[\mathfrak{z}^t(s) > L_t(s) \quad \text{for} \quad r \leq s \leq t-r]}{P[\mathfrak{z}^t(s) > \underline{\mathcal{L}}_{r,t}(s) \quad \text{for} \quad r \leq s \leq t-r]} \to 1 \quad \text{as} \quad r \to \infty , \tag{6.16}$$

with convergence in both cases being uniform in t .

Proof. Since $\bar{\mathcal{L}}_{r,t}(s) \geq L_t(s)$ for all s , it suffices to show that

$$P^0[B_{\bar{\mathcal{L}}}[r,t-r] \mid B_L[r,t-r]] \to 1 \quad \text{as} \quad r \to \infty \tag{6.17}$$

uniformly in t , in order to demonstrate (6.15). (We suppress subscripts when convenient.) We will show here that

$$P^0[B_{\theta \circ L}[r,t-r] \mid B_L[r,t-r]] \to 1 \quad \text{as} \quad r \to \infty \tag{6.18}$$

uniformly in t . The argument that $P^0[B_{\theta^{-1} \circ L} \mid B_L] \to 1$ is the same; together, these two limits yield (6.17).

It is a simple consequence of Proposition 6.1 that (6.18) is equivalent to

$$P^0[B_{\theta\circ L}[r,t-r] \mid B_{\bar{L}}[r,t-r]] \to 1 \quad \text{as} \quad r \to \infty \tag{6.19}$$

uniformly in t. To demonstrate (6.19), we show that $P^0[B^c_{\theta\circ L} \mid B_{\bar{L}}] \to 0$ as $r \to \infty$ uniformly in t. (B^c_L stands here for the complement of B_L.) The argument is computational, the point being that in order for a path which lies above $\bar{L}$ to hit $\theta \circ L$, it must fall a certain distance over a comparatively short period of time. Since except for a modest sideways shift, $\bar{L}$ lies considerably above $\theta \circ L$, this cannot occur with any sizable probability.

Proceeding now, we note that

$$\begin{aligned} &P^0[B^c_{\theta\circ L}[r,t-r] \mid B_{\bar{L}}[r,t-r]] \\ &\leq \sum_{k=\lfloor r\rfloor}^{\lfloor t-r\rfloor} P^0[B^c_{\theta\circ L}[k\vee r,(k+1)\wedge(t-r) \mid B_{\bar{L}}[r,t-r]] \ . \end{aligned} \tag{6.20}$$

For $s > t-2r$, $\theta\circ L(s) = L(s) < \bar{L}(s)$; consequently, for $k > t-2r$, the above summands are zero. To handle the case for $k \leq t-2r$, choose i_k so that

$$\begin{aligned} i_k \in \mathcal{J}_k &= [k+k^\delta, k+k^\delta+2] && \text{if } k \leq t/2 \\ &= [k+(t-k)^\delta, k+(t-k)^\delta+2] && \text{if } k \geq t/2 \end{aligned}$$

and

$$\bar{L}_t(i_k) \geq \sup\{\bar{L}_t(s):\ s\in\mathcal{J}_k\} - 1 \ .$$

It follows from (6.1), (6.11), and a little computation that

$$\begin{aligned} \sup\{\theta_{r,t}\circ L_t(s):\ s\in[k_1,k_2+1]\} &\leq \bar{L}_t(i_k) - C_1k_1^\delta + (1+2C_1) && \text{for } k\leq t/2 \\ &\leq \bar{L}_t(i_k) - C_1(t-k_1)^\delta + (1+2C_1) && \text{for } k\geq t/2 , \end{aligned} \tag{6.21}$$

where $k_1 = k\vee r$ and $k_2 = (k+1)\wedge(t-2r) - 1$. Also, by (2.18), the sum in

(6.20) is at most

$$\sum_{k=\lfloor r\rfloor}^{\lfloor t-2r\rfloor} P^0[B^c_{\theta o L}[k_1,k_2+1] \mid \mathfrak{z}^t(i_k) > \bar{L}_t(i_k)] \ ,$$

which, with the aid of Lemma 2.1, is easily seen to be at most

$$\sum_{k=\lfloor r\rfloor}^{\lfloor t-2r\rfloor} P^0[B^c_{\theta o L}[k_1,k_2+1] \mid i_k, \bar{L}_t(i_k)]$$

(6.22)
$$= \sum_{k=\lfloor r\rfloor}^{\lfloor t-2r\rfloor} P[\mathfrak{z}^{i_k}(s) \leq \theta_{r,t} \circ L_t(s) - \frac{s}{i_k}\bar{L}_t(i_k) \text{ for some } s \in [k_1,k_2+1]] \ .$$

Since $\bar{L}_t(i_k) \leq 4C_1 i_k^\delta$, it follows from (6.21) and a little more computation, that (6.22) is at most

(6.23)
$$\sum_{k=\lfloor r\rfloor}^{\lfloor t/2\rfloor} P[\mathfrak{z}^{i_k}(s) \leq -C_1k^\delta + C_2 \text{ for some } s \in [k,i_k]]$$
$$+ \sum_{k=\lceil t/2\rceil}^{\lfloor t-2r\rfloor} P[\mathfrak{z}^{i_k}(s) \leq -C_1(t-k)^\delta + C_2 \text{ for some } s \in [k,i_k]]$$

for some constant C_2 , which depends on C_1 . Because $i_k - k \leq k^\delta \wedge (t-k)^\delta + 2$, one may apply Lemma 2.2(b) to conclude that (6.23) is at most

(6.24)
$$C_3 \sum_{k=\lfloor r\rfloor}^{\infty} e^{-C_4 k^\delta} < \infty \ ,$$

for appropriate choice of $C_3, C_4 > 0$. As $r \to \infty$, the right side of (6.24) $\to 0$, where the rate of convergence does not depend on t . We have thus shown that $P^0[B^c_{\theta o L} \mid B_{\bar{L}}] \to 0$ uniformly in t as $r \to \infty$. (6.19) and consequently (6.15) follow.

Since $\underline{\mathcal{L}}_{r,t}(s) \leq \bar{\mathcal{L}}_{r,t}(s)$ for all s , (6.15) provides an upper bound for (6.16), and to demonstrate (6.16), we need only show that

(6.25)
$$P^0[B_L[r,t-r] \mid B_{\underline{\mathcal{L}}}[r,t-r]] \to 1 \text{ as } r \to \infty$$

uniformly in t . For $L'_{r,t} \equiv \underline{\mathcal{L}}_{r,t}$, one may check that $\underline{L}'_{r,t} \leq 0$ for r chosen large enough. Therefore, we may apply (6.18) with $L = \underline{\mathcal{L}}$, and obtain

$$(6.26)\qquad P^0[B_{\theta\circ\underline{\mathcal{L}}}[r,t-r] \mid B_{\underline{\mathcal{L}}}[r,t-r]] \to 1 \quad \text{as} \quad r \to \infty$$

uniformly in t. But $\underline{\mathcal{L}}$ was constructed so that $L_{r,t} \leqq \theta_{r,t} \circ \underline{\mathcal{L}}_{r,t}$ for all r and t. (6.25) is thus an immediate consequence of (6.26), which completes the proof. //

Since the processes $\mathfrak{z}^t_{x,y}(s) - \frac{s}{t}y - \frac{t-s}{t}x$, $0 \leqq s \leqq t$, are stochastically equivalent to $\mathfrak{z}^t(s)$, it is not difficult to see that the limits in Propositions 6.1 and 6.2 are in fact also uniform over $x,y > 0$ if $\mathfrak{z}^t$ is replaced by $\mathfrak{z}^t_{x,y}$. Such behavior is not too surprising, as the higher the Brownian bridge starts and finishes, the less dependent the hitting probabilities should be on the particular curves in question. In Lemma 6.1 and Proposition 6.3, we make this last statement more precise.

Lemma 6.1. Assume that the functions ℓ_1 and ℓ_2 are upper semi-continuous except at at most finitely many points, and that $\ell_1(s) \leqq \ell_2(s)$ for $s \in [0,t]$. Then, provided that the denominator is nonzero,

$$(6.27)\qquad p^1(t,x;B_{\ell_2})/p^1(t,x;B_{\ell_1}) \quad \text{is increasing in } x .$$

Proof. The proof is essentially the same as that of Lemma 5 in [4]. We first set

$$g(t,x;c) \equiv cp^1(t,x;B_{\ell_2}) - p^1(t,x;B_{\ell_1}) .$$

To demonstrate (6.27), it suffices to show that for all $c \geqq 1$ and $\ell_2(t) < x_1 \leqq x_2$,

$$(6.28)\qquad g(t,x_1;c) > 0 \quad \text{implies that} \quad g(t,x_2;c) \geqq 0 .$$

The fact that $g(s,x;c)$ is unbounded in s and x will prove unpleasant later on. We will therefore modify $g(s,x;c)$ without altering (6.28). With this in mind, set

$$g_y(s,x;c) = cp^1(s,x;B_{\ell_2} \mid 0,y) - p^1(s,x;B_{\ell_1} \mid 0,y)$$

for $0 \leq s \leq t$. Clearly $g_0 = g$. By reasoning as in Lemma 2.4(c), it is not difficult to show that $g_y(s,x;c) \to g_0(s,x;c)$ as $y \downarrow 0$. Consequently, if (6.28) is in fact violated for some c_0 , x_1 , and x_2 , we may then define an initial distribution μ with continuous density ν so that for

$$h(s,x) \equiv \int dy\, \nu(y)\, g_y(s,x;c_0)\ , \quad 0 \leq s \leq t\ , \tag{6.29}$$

$$h(t,x_1) > 0 \quad \text{and} \quad h(t,x_2) < 0\ . \tag{6.30}$$

The remainder of the proof is devoted to demonstrating that (6.30) cannot hold, which we show by means of a martingale argument. We proceed with the construction of the martingale. First note that by Lemma 2.4(b), for $0 \leq s_1 < s_2 \leq t$, $i = 1,2$,

$$\int dy\, \nu(y) p^1(t,y;B_{\hat{\ell}_i}[s_1,t] \mid s_1,y_1)$$
$$= \int dy_2\, p^1(s_2,y_2;B_{\hat{\ell}_i}[s_1,s_2] \mid s_1,y_1) \int dy\, \nu(y)\, p^1(t,y;B_{\hat{\ell}_i}[s_2,t] \mid s_2,y_2)\ ,$$

where $\hat{\ell}_i(s) = \ell_i(t-s)$ for $s \in [0,t]$. Therefore, if we set $S_x^i = \min\{s: \mathfrak{X}_x(s) \leq \hat{\ell}_i(s)\} \wedge t$ for $x > \ell_i(t)$, then

$$\begin{aligned} H_x^i(s) &= \int dy\, \nu(y)\, p^1(t,y;B_{\hat{\ell}_i}[s,t] \mid s,\mathfrak{X}_x(s)) && \text{if } s < S_x^i \\ &= 0 && \text{if } s \geq S_x^i \end{aligned}$$

defines a nonnegative martingale on $[0,t]$ (on $\{\mathcal{B}^s\}$). By Lemmas 2.4(a) and (2.4c), $H_x^i(s)$ is continuous where $(s,\mathfrak{X}_x(s)) \in \operatorname{int}\{(s,y): \ell_i(s) < y\}$. Since ℓ_i is upper semi-continuous except at at most a finite number of points, it follows without difficulty that $H_x^i(s)$ is almost surely right continuous on $[0,t]$. Consequently,

$$H_x(s) = c_0 H_x^2(s) - H_x^1(s)$$

is also (a.s.) a right continuous martingale. Moreover, for $s < S_x^2$ (which includes $\{H_x(s) > 0\}$), it follows from the definition of $H_x^i(s)$ and Lemma

2.4(a), that

$$H_x(s) = \int dy\, \nu(y)\, g_y(t-s, \mathfrak{X}_x(s)\,;\, c_0)$$
$$= h(t-s, \mathfrak{X}_x(s)) \quad .$$

We now argue as in Proposition 3.2, and introduce the stopping time

$$T = \inf\{s\colon H_{x_1}(s) \leqq 0\} \wedge t \quad ,$$

where x_1 is as in (6.30). By the Optional Sampling Theorem,

$$E[H_{x_1}(T)] = E[H_{x_1}(0)] = h(t,x_1) > 0 \quad ,$$

from which it follows that $P[H_{x_1}(T) > 0] > 0$. In particular, there is some continuous path $\gamma(s)$ with $\gamma(s) > \hat{\ell}_2(s)$ for $s \in [0,t]$, so that

$$(6.31) \qquad h(t-s,\gamma(s)) > 0 \quad \text{for} \quad 0 \leqq s \leqq t \quad .$$

On the other hand, one may reason similarly with the stopping time

$$T' = \inf\{s\colon \mathfrak{X}_{x_2}(s) = \gamma(s)\} \wedge t \quad .$$

Since $c_0 \geqq 1$, $H_{x_2}(T') \geqq 0$ for $T' = t$, whereas by (6.31), $H_{x_2}(T') > 0$ for $T' < t$. Therefore,

$$h(t,x_2) = E[H_{x_2}(T')] \geqq 0 \quad ,$$

which violates (6.30). Consequently, (6.27) must hold and the proof is complete. //

In Proposition 6.3, we restate Lemma 6.1 in a form which will be more amenable to application in Sections 7-9.

<u>Proposition 6.3</u>. Assume that the functions ℓ_1 and ℓ_2 are upper semi-continuous except at at most finitely many points, and that $\ell_1(s) \leqq \ell_2(s)$ for $s \in [0,t]$. Then, provided that the denominator is nonzero,

(6.32) $$\frac{P[\mathfrak{z}_{x,y}(s) > \ell_2(s) \quad \text{for} \quad 0 \leq s \leq t]}{P[\mathfrak{z}_{x,y}(s) > \ell_1(s) \quad \text{for} \quad 0 \leq s \leq t]} \quad \text{is increasing in } x \text{ and } y .$$

Proof. As before, we set $\hat{\ell}_i(s) = \ell_i(t-s)$. By Lemma 6.1, for $x \geq 0$,

$$p^1(t,x;B_{\hat{\ell}_2} \mid 0,y)/p^1(t,x;B_{\hat{\ell}_1} \mid 0,y)$$

is increasing in x . Consequently, by Lemma 2.4(a),

(6.33) $$p^1(t,y;B_{\ell_2} \mid 0,x)/p^1(t,y;B_{\ell_1} \mid 0,x)$$

is also increasing in x . Applying Lemma 6.1 again, we see that (6.33) is also increasing in y . Division of both terms by $p^1(t,y \mid 0,x)$ then reduces (6.33) to (6.32). //

We note here that $m_{1/2}(t)$, which was defined in (1.39), is upper semi-continuous. $L(t)$ will later be defined in terms of $m_{1/2}(t)$, and also be upper semi-continuous. The functions $\underline{L}$, $\bar{L}$, $\underline{\mathcal{L}}$, and $\bar{\mathcal{L}}$ defined in this section will therefore also be upper semi-continuous except at perhaps finitely many points, and will thus be subject to Proposition 6.3.

7. ESTIMATES FOR THE KOLMOGOROV EQUATION IN THE CASE $\lambda = 2^{1/2}$

In this section, we present several of the main estimates which are needed to exhibit convergence of $u(t,x)$ to the travelling wave $w^{2^{1/2}}(x)$. These estimates will be essential for proving the main results in Sections 8 and 9. The first estimate, Proposition 7.1, is a statement of regularity for solutions $u(t,x)$ of (1.1) and (1.2) with initial data satisfying (1.17) and (1.19), which it says are asymptotically monotone in x. For fixed t, $u(t,x)$ may therefore be conceptualized as being close to 0 for large values of x, close to 1 for small values of x, and having some (presumably narrow) region of intermediate values. Propositions 7.2 and 7.3 deal with lower and upper bounds for $u(t,x)$. Proposition 7.2 states that over paths with high enough trajectories, $k = f(u)/u$ is close enough to 1 to ensure essentially full heating in the corresponding Feynman-Kac integral. Proposition 7.3 states that the combined effect of paths whose trajectories are not always high, is small. The functions $\bar{\mathcal{L}}$ and $\underline{\mathcal{L}}$, which were introduced in Section 6, are used here to define functions $\bar{\mathcal{m}}$ and $\underline{\mathcal{m}}$ which specify the appropriate regions. Propositions 7.4 and 7.5 are reformulations of the results of Section 6 in a more convenient setting, and state that the probabilities of Brownian bridge lying above certain functions related to $\bar{\mathcal{m}}$ and $\underline{\mathcal{m}}$ are asymptotically equal. Propositions 7.2-7.5 together compose the key estimates for evaluating the asymptotic behavior of $u(t,x)$. The last result of the section, Proposition 7.6, states that only part of the initial data is relevant in determining the asymptotic behavior of $u(t,x)$. This result will be used for the case covered in Section 9 to show that the functions $\bar{\mathcal{m}}$ and $\underline{\mathcal{m}}$ behave as we need them to in order to be able to apply Propositions 7.2-7.5.

Proposition 7.1 is an application of Lemma 3.1. Since by Proposition 3.4,

$u(t,x) \to 1$ uniformly in $x \leq (2^{1/2}-\delta)t$ (with $\delta > 0$ fixed), one needs here only show that $u(t,x)$ is asymptotically monotone in the range $((2^{1/2}-\delta)t,\infty)$. The basic idea of the argument is to restrict the initial data to the interval $(-\infty,y(s)]$, for some function $y(s)$. Denote by $u^s(t,x)$ the accompanying solution. Then on account of Lemma 3.1, for $y(t) \leq x_1 \leq x_2$, $u^t(t,x_1) \geq u^t(t,x_2)$. On the other hand by Lemma 4.1, $u^t(t,x_i) \sim u(t,x_i)$ in this range. The desired monotonicity of $u(t,x)$ thus follows. In Proposition 7.1, as in the remainder of the section, we automatically exclude the possibility of trivial initial data, and assume that $\int_{-\infty}^{\infty} u(0,y)\,dy > 0$.

<u>Proposition 7.1</u>. Let $u(t,x)$ satisfy (1.1) and (1.2), and assume that for some $h > 0$,

(1.19) $$\limsup_{t\to\infty} \frac{1}{t}\log\left[\int_t^{t(1+h)} u(0,y)\,dy\right] \leq -2^{1/2} .$$

Then,

(7.1) $$\limsup_{t\to\infty} \{u(t,x_2) - u(t,x_1) : 0 \leq x_1 \leq x_2\} = 0 .$$

If it is also assumed that for some $\eta, M, N > 0$,

(1.17) $$\int_x^{x+N} u(0,y)\,dy > \eta \quad \text{for} \quad x < -M ,$$

then (7.1) extends to

(7.2) $$\limsup_{t\to\infty} \{u(t,x_2) - u(t,x_1) : x_1 \leq x_2\} = 0 .$$

<u>Proof</u>. It is a simple consequence of (3.66) that for each $\delta > 0$, $u(t,x) \to 1$ as $t \to \infty$ uniformly in $x \in [0,(2^{1/2}-\delta)t]$. If (1.17) is assumed to hold, then for appropriate $\eta' > 0$, $u(1,x) \geq \eta'$ for $x \in (-\infty,0]$, and therefore $u(t,x) \to 1$ as $t \to \infty$ uniformly in $x \in (-\infty,0]$ as well. To demonstrate (7.1) and (7.2), it therefore suffices to show that under (1.19),

$$\limsup_{t\to\infty} \{u(t,x_2) - u(t,x_1): (2^{1/2}-\delta)t \leqq x_1 \leqq x_2\} = 0 \tag{7.3}$$

for some $\delta > 0$.

To demonstrate (7.3), we let $u^s(t,x)$, $s \geqq 0$, denote the solution of (1.1) and (1.2) with initial data

$$\begin{aligned} u^s(0,x) &= u(0,x) \quad \text{for} \quad x \leqq \delta_1 s \\ &= 0 \qquad\quad\ \text{for} \quad x > \delta_1 s , \end{aligned}$$

for some fixed $0 < \delta_1 < 1$. On the one hand, note that with $s = t$, it follows from Lemma 3.1 that

$$u^t(t,x_2) \leqq u^t(t,x_1) \quad \text{for} \quad \delta_1 t \leqq x_1 \leqq x_2 . \tag{7.4}$$

On the other hand, if we set $v^s(t,x) = u(t,x) - u^s(t,x)$, then

$$\begin{aligned} v^s(0,x) &= u(0,x) \quad \text{for} \quad x > \delta_1 s \\ &= 0 \qquad\quad\ \text{for} \quad x \leqq \delta_1 s . \end{aligned}$$

Since by (1.2), $f'(u) \leqq 1$ for all u, v^s satisfies $v_t^s = \frac{1}{2} v_{xx}^s + k^s(t,x)v^s$, where $k^s(t,x) \leqq 1$ for all t and x. Consequently,

$$v^s(t,x) \leqq e^t \int_{\delta_1 s}^{\infty} u(0,y) \frac{e^{-(x-y)^2/2t}}{\sqrt{2\pi t}} dy .$$

Since (1.19) is assumed, it follows without difficulty from (4.7) that with $s = t$,

$$v^t(t,x) \to 0 \quad \text{uniformly in} \quad x \geqq (2^{1/2}-\delta_1^2/6)t \quad \text{as} \quad t \to \infty . \tag{7.5}$$

Applying both (7.4) and (7.5), we see that

$$\begin{aligned} &\sup\,[u(t,x_2) - u(t,x_1)] \\ &\leqq \sup\,[u(t,x_2) - u^t(t,x_2)] + \sup\,[u^t(t,x_2) - u^t(t,x_1)] \\ &\qquad + \sup\,[u^t(t,x_1) - u(t,x_1)] \end{aligned}$$

$\to 0$ uniformly in $(2^{1/2}-\delta)t \leq x_1 \leq x_2$ as $t \to \infty$,

if $0 < \delta < \delta_1^2/6 \wedge (2^{1/2}-\delta_1)$. Thus, (7.3) is valid, and the proof is complete. //

In order to state Propositions 7.2 through 7.5, we introduce some new terminology here. Recall that in Section 6 we introduced families of functions L_t , $\underline{L}_t$, $\bar{L}_t$, $\underline{\mathcal{L}}_{r,t}$, and $\bar{\mathcal{L}}_{r,t}$ such that (6.2) was assumed to hold. $\underline{L}$ was defined by lowering L slightly, and $\bar{L}$ by raising L slightly. By shifting L slightly to the right as well as lowering L , we obtained $\underline{\mathcal{L}}$, whereas by shifting to the left and raising L (and maximizing over L and $\underline{\mathcal{L}}$), we obtained $\bar{\mathcal{L}}$. It was shown in Propositions 6.1 and 6.2 that the hitting probabilities for all these functions were approximately equal for large r . We assume here that L is of the form

$$L_{r,t}(s) = m_{1/2}(s) - \frac{s}{t}m_{1/2}(t) - \frac{t-s}{t}\alpha(r,t) , \quad 0 \leq s \leq t , \tag{7.6}$$

where $m_{1/2}$ should be thought of as the (upper) median of some solution $u(t,x)$ of (1.1) and (1.2) (as defined in (1.39)). $\alpha(r,t)$ will be specified later on, but will be assumed to satisfy

$$\alpha(r,t) = o(t) , \tag{7.7}$$

where $o(t)$ is uniform in r . With $\underline{\mathcal{L}}$ and $\bar{\mathcal{L}}$ defined in terms of L as previously specified, we set

$$\begin{aligned} \underline{\mathcal{M}}_{r,t}(s) &= \underline{\mathcal{L}}_{r,t}(s) + \frac{s}{t}m_{1/2}(t) + \frac{t-s}{t}\alpha(r,t) \\ &= m_{1/2}(s) + \underline{\mathcal{L}}_{r,t}(s) - L_{r,t}(s) \end{aligned} \tag{7.8}$$

for $0 \leq s \leq t$. Similarly, set

$$\begin{aligned} \bar{\mathcal{M}}_{r,t}(s) &= \bar{\mathcal{L}}_{r,t}(s) + \frac{s}{t}m_{1/2}(t) + \frac{t-s}{t}\alpha(r,t) \\ &= m_{1/2}(s) + \bar{\mathcal{L}}_{r,t}(s) - L_{r,t}(s) . \end{aligned} \tag{7.9}$$

$m_{1/2}$ will be assumed to be upper semicontinuous, and therefore so will L. It follows that $\underline{\mathcal{L}}$, $\bar{\mathcal{L}}$, $\underline{\mathcal{M}}$, and $\bar{\mathcal{M}}$ are upper semicontinuous except at finitely many points. (If one wishes to do so for aesthetic reasons, one may modify $\underline{\mathcal{L}}$, $\bar{\mathcal{L}}$, $\underline{\mathcal{M}}$, and $\bar{\mathcal{M}}$ to be continuous.)

$\underline{\mathcal{M}}$ and $\bar{\mathcal{M}}$ have the same basic relationship to $m_{1/2}$ as $\underline{\mathcal{L}}$ and $\bar{\mathcal{L}}$ have to L. In particular, we will be interested in the hitting probabilities of $\underline{\mathcal{M}}$ and $\bar{\mathcal{M}}$ for Brownian bridge with endpoints at or above $m_{1/2}(t)$ and $\alpha(r,t)$; in this context Proposition 6.2 is easily reformulated as Proposition 7.4 in terms of $\underline{\mathcal{M}}$ and $\bar{\mathcal{M}}$. For the sake of concreteness, we set $C = 2C_1 = 8$ in the formulas (6.1) and (6.11) for $\underline{L}$, $\bar{L}$, $\underline{\mathcal{L}}$, and $\bar{\mathcal{L}}$. We will also frequently assume the following analogue of (6.2): for r chosen large enough and for all t,

$$(7.10)\qquad \begin{aligned} m_{1/2}(s) &\leq \tfrac{s}{t} m_{1/2}(t) + \tfrac{t-s}{t}\alpha(r,t) + 8s^{\delta} && \text{for } s \in [r,t/2] \\ &\leq \tfrac{s}{t} m_{1/2}(t) + \tfrac{t-s}{t}\alpha(r,t) + 8(t-s)^{\delta} && \text{for } s \in [t/2,t-r] . \end{aligned}$$

Moreover, we may assume that $r \leq t/2$ always holds.

As in previous sections $k(t,x) \equiv f(u(t,x))/u(t,x)$, where $u(t,x)$ is a solution of (1.1). Also, recall the quantities $\varphi(t,x)$ and $m^{\varphi}(t)$, which occur in Sections 4 and 5 in the context $k \equiv 1$. Although we will need to devote a considerable amount of effort to determine $m_{1/2}(t)$, it is at this point easy to obtain the useful first-order estimate

$$(3.22')\qquad m_{1/2}(t)/t \to 2^{1/2} \quad \text{as} \quad t \to \infty$$

under (1.19). For, from Corollary 1 of Lemma 4.2, $m^{\varphi}(t)/t \to 2^{1/2}$ as $t \to \infty$. By the maximum principle, $u(t,x) \leq \varphi(t,x)$ for all t and x, and so $\limsup_{t\to\infty} m_{1/2}(t)/t \leq 2^{1/2}$. On the other hand, $\liminf_{t\to\infty} m_{1/2}(t)/t \geq 2^{1/2}$ follows from Proposition 3.4.

We now demonstrate Proposition 7.2, which gives a lower bound on $k(s,y)$ for $y > \bar{\mathcal{M}}(s)$. The key point here is that, owing to the definition of $\bar{\mathcal{M}}$,

(7.14) (and (7.17)) holds for large enough r . Consequently, $u(s,y) \sim 0$ for $y > \bar{m}_{r,t}(s)$; otherwise u grows sufficiently rapidly so that by time $s+s^{\delta}$ (or $s+(t-s)^{\delta}$ if $s \geq t/2$), $u(s+s^{\delta},y)$ (respectively, $u(s+(t-s)^{\delta},y)$) is too large to satisfy (7.14). $u(s,y) \sim 0$ then implies that $k(s,y) \sim 1$.

Proposition 7.2. Let $u(t,x)$ be a solution to (1.1), (1.2), and (1.2'), with initial data satisfying (1.19) for some $h > 0$. Then for r chosen large enough,

$$
(7.11) \qquad \begin{aligned} k(s,y) &\geq 1 - C_2 e^{-\rho s^{\delta}} && \text{if } r \leq s \leq t/2 \\ &\geq 1 - C_2 e^{-\rho(t-s)^{\delta}} && \text{if } t/2 \leq s \leq t-2r \end{aligned}
$$

for $y > \bar{m}_{r,t}(s)$ and some constant C_2 . Consequently for continuous $x(s)$ with $x(s) > \bar{m}_{r,t}(t-s)$ in $[2r,t-r]$,

$$
(7.12) \qquad e^{3r-t} \exp\left[\int_{2r}^{t-r} k(t-s,x(s))\, ds\right] \to 1 \text{ as } r \to \infty
$$

uniformly in t .

Proof. It follows from the definition of $L_{r,t}$ and $\bar{m}_{r,t}$ that for $s \in [r,t/2]$,

$$
(7.13) \qquad \begin{aligned} \bar{m}_{r,t}(s) &\geq \theta_{r,t} \circ L_{r,t}(s) + \tfrac{s}{t} m_{1/2}(t) + \tfrac{t-s}{t}\alpha(r,t) \\ &= m_{1/2}(s+s^{\delta}) + [4 + (\alpha(r,t) - m_{1/2}(t))/t]s^{\delta} \ . \end{aligned}
$$

By (3.22') on page 100 and (7.7), this implies that

$$
(7.14) \qquad \bar{m}_{r,t}(s) \geq m_{1/2}(s+s^{\delta}) \quad \text{for } s \in [r,t/2]
$$

and r (and hence t) chosen large enough. On the other hand, $u(s+s^{\delta}, m_{1/2}(s+s^{\delta})) = 1/2$, and so from Proposition 7.1,

$$
u(s+s^{\delta}, m_{1/2}(s+s^{\delta}) + z) \leq 3/4 \quad \text{for } z \geq 0
$$

and large s . We may therefore apply (3.66) of Proposition 3.4 to conclude that for an appropriate constant C_3 and s chosen large enough,

$$u(s,m_{1/2}(s+s^{\delta})+z) \leqq C_3 e^{-s^{\delta}} \quad \text{for} \quad z \geqq 0 . \tag{7.15}$$

Together with (7.14), this shows that for $y > \bar{\mathcal{m}}_{r,t}(s)$,

$$u(s,y) \leqq C_3 e^{-s^{\delta}} \quad \text{for} \quad s \in [r,t/2] \tag{7.16}$$

and large r . Reasoning analogous to that leading up to (7.14) shows that

$$\bar{\mathcal{m}}_{r,t}(s) \geqq m_{1/2}(s+(t-s)^{\delta}) \quad \text{for} \quad s \in [t/2,t-2r] \tag{7.17}$$

and large r . Proceeding as in (7.15) and (7.16), one therefore also obtains

$$u(s,y) \leqq C_3' e^{-(t-s)^{\delta}} \quad \text{for} \quad s \in [t/2,t-2r] \tag{7.18}$$

and large r . Application of (1.2') immediately reduces (7.16) and (7.18) to (7.11). Since

$$\int_r^{\infty} e^{-\rho s^{\delta}} ds \to 0 \quad \text{as} \quad r \to \infty ,$$

(7.12) follows directly from (7.11). //

We will show in Proposition 7.3 that the contribution to the Feynman-Kac integral by paths hitting $\underline{\mathcal{m}}$ is quite small. To do so, it is necessary to take into account where individual paths hit $\underline{\mathcal{m}}$. This requires a fair amount of notation. First, we introduce the random variables

$$S^1(r,t) = \sup\{s\colon 2r \leqq s \leqq t/2 , \ \mathfrak{z}_{x,y}(s) \leqq \underline{\mathcal{m}}_{r,t}(t-s)\} \tag{7.19a}$$

and

$$S^2(r,t) = \inf\{s\colon t/2 \leqq s \leqq t-r , \ \mathfrak{z}_{x,y}(s) \leqq \underline{\mathcal{m}}_{r,t}(t-s)\} , \tag{7.19b}$$

and put

$$(7.20)\quad \begin{aligned} S(r,t) &= S^1(r,t) \quad \text{for} \quad S^1(r,t)+S^2(r,t) > t \\ &= S^2(r,t) \quad \text{for} \quad S^1(r,t)+S^2(r,t) \leqq t \ . \end{aligned}$$

Set $S^1 = 0$ if the first set is void and $S^2 = t$ if the second set is void. (In the above notation, the subscripts x and y are suppressed.) $S(r,t)$ is the time nearest $t/2$ at which $\mathfrak{z}_{x,y}$ hits $\underline{\mathcal{M}}_{r,t}$. We also introduce the process $\tilde{\mathfrak{z}}_{x,y}(s) = (\mathfrak{z}_{x,y}(s+t/2), \hat{\mathfrak{z}}_{y,x}(s+t/2))$, where $\hat{\mathfrak{z}}_{y,x}(s) = \mathfrak{z}_{x,y}(t-s)$. Since $S^2(r,t)$ is a stopping time for $\mathfrak{z}_{x,y}(s+t/2)$, and $t-S^1(r,t)$ is a stopping time for $\hat{\mathfrak{z}}_{y,x}(s+t/2)$, $\tilde{S}(r,t) = \min\{S^2(r,t), t-S^1(r,t)\}$ is a stopping time for $\tilde{\mathfrak{z}}_{x,y}(s)$. $\tilde{\mathfrak{z}}_{x,y}$ is strong Markov; the strong Markov property may therefore be applied at $\tilde{S}(r,t)$. We will use this fact in conjunction with $\mathfrak{z}_{x,y}$ and $S(r,t)$ for (7.36) in Lemma 7.2.

We next introduce

$$(7.21)\quad G_{x,y}(r_1;r,t) = \{\mathfrak{z}_{x,y}\colon r_1 \leqq S(r,t) \leqq t-r_1\}\ , \quad \text{where} \quad r_1 \in [r,t/2]\ .$$

These sets may also be written as

$$\{\mathfrak{z}_{x,y}\colon \mathfrak{z}_{x,y}(s) \leqq \underline{\mathcal{M}}_{r,t}(t-s) \quad \text{for some} \quad s \in [r_1 \vee 2r, t-r_1]\}\ .$$

In particular, we abbreviate and set

$$(7.22)\quad G_{x,y}(r,t) = \{\mathfrak{z}_{x,y}\colon \mathfrak{z}_{x,y}(s) \leqq \underline{\mathcal{M}}_{r,t}(t-s) \quad \text{for some} \quad s\in[2r,t-r]\}\ .$$

Also, introduce $I_j = [j,j+1) \cup (t-j-1,t-j]$, $j=0,1,\ldots,j_0-1$, where $j_0 < t/2 \leqq j_0+1$, and $I_{j_0} = [j_0, t-j_0]$, and define

$$(7.23)\quad A_j(r,t) = \{\mathfrak{z}_{x,y}\colon S(r,t)\in I_j\}\ ,\quad j=0,1,\ldots,j_0$$

and

$$(7.24)\quad \begin{aligned} A_j^1(r,t) &= \{\mathfrak{z}_{x,y}\in A_j(r,t)\colon \mathfrak{z}_{x,y}(s) > -(s\wedge(t-s)) + \tfrac{s}{t}y + \tfrac{t-s}{t}x \quad \text{for} \quad s\in I_j\} \\ A_j^2(r,t) &= A_j(r,t) - A_j^1(r,t)\ . \end{aligned}$$

The sets $A_j(r,t)$ partition $G_{x,y}(r,t)$; $A_j^1(r,t)$ will in each case be the main component of $A_j(r,t)$. When convenient, we shorten notation by dropping indices.

We do most of the work for Proposition 7.3 in the form of two lemmas. The first, Lemma 7.1, states that $P[G^c(j;r,t)]$ does not increase too rapidly in j . (G^c denotes the complement of G .) On the other hand, Lemma 7.2 states that the weight attached to paths in $A_{j-1}^1(r,t) \subset G^c(j;r,t)$ drops off comparatively quickly in j . $P[A_{j-1}^2(r,t)]$ is also easily shown to be quite small. Partitioning $G(r,t)$ into $\{A_j(r,t): j=1,2,\ldots\}$, we are therefore able to show in Proposition 7.3 that the contribution to $E[\exp\{\int_{2r}^{t-r}\ldots\}u(0,\mathfrak{x}_x(t))]$ by paths lying in $G(r,t)$ is small in comparison with $e^{t-3r}P[G^c(r,t)]$. Since we will be able to show later on with the aid of Proposition 7.2 that the contribution to the Feynman-Kac integral by paths lying in $G^c(r,t)$ is of the order of magnitude of this last quantity, it will follow from Proposition 7.3 that we may essentially disregard the sets $G(r,t)$ for large r when computing $u(t,x)$.

Lemma 7.1 below makes use of Lemma 2.6 in computing $P[G^c(r_1;r,t)]$.

<u>Lemma 7.1</u>. Let $u(t,x)$ satisfy (1.1) and (1.2). Assume that for r chosen large enough and all t , $m_{1/2}(t)$ satisfies (7.10). Then if $r_1 \in [r,t/2]$, $y \geq \alpha(r,t)$, and $x \geq m_{1/2}(t)$,

$$(7.25) \qquad P[G_{x,y}^c(r_1;r,t)] \leq C_3\frac{r_1}{r}P[G_{x,y}^c(r,t)]$$

for some constant C_3 , where $G_{x,y}$ is defined as in (7.22) and α satisfies (7.7).

<u>Proof</u>. By (2.18),

$$(7.26) \qquad P[G_{x,y}^c(r,t)\mid G_{x,y}^c(r_1;r,t)] \geq P[\mathfrak{z}_{x,y}(s) > \mathcal{M}_{r,t}(t-s) \text{ for } s\in\mathcal{J}] ,$$

where $\mathcal{J} = [2r,r_1)\cup(t-r_1,t-r]$. From (7.10) and a little work, it follows that the right side of (7.26) is at least

$$P[\mathfrak{z}(s) > 0 \quad \text{for} \quad s \in \mathcal{J}]$$

for large enough r. A second application of (2.18) shows that this is at least

$$(P[\mathfrak{z}(s) > 0 \quad \text{for} \quad s \in [r, r_1]])^2 .$$

Since $r_1 \leq t/2$, it is not difficult to show that

$$P[\mathfrak{z}(s) > 0 \quad \text{for} \quad s \in [r, r_1]] \geq C_4 \sqrt{r/r_1}$$

for some constant $C_4 > 0$. (Use, e.g., Lemma 2.1(b) and the reflection principle.) Therefore,

$$P[G^c_{x,y}(r,t) \mid G^c_{x,y}(r_1; r,t)] \geq C_4^2 r/r_1 ,$$

which demonstrates (7.25) with $C_3 = 1/C_4^2$. //

In Lemma 7.2, we show that the weight attached to the paths in $A^1_j(r,t)$ drops off quickly in j. The reasoning applied here is similar to that in Proposition 7.2, although we now obtain an upper bound. The key point is that the inequality (7.31) holds for large r. Since $u(s,y')$ must grow at fixed y' as s increases, it will then by the case (in (7.33)) that $u(s_2,y') \sim 1$ if $0 \leq y' \leq \underline{\mathfrak{m}}(s) + 2(t-s)^\delta$ and $s_2 \geq s$ (for $s \in [t/2, t-2r]$; analogous behavior holds for $s \in [r, t/2]$). Thus, after $\mathfrak{z}_{x,y}$ hits $\underline{\mathfrak{m}}$, it must pass through a large rectangular box in which $k \sim 0$. Since $\mathfrak{z}_{x,y}$ will typically linger inside this box for a while, during which time there is little heating, such paths will have comparatively small weight attached to them. As j increases, the dimensions of the box increase, and so the weight decreases.

<u>Lemma 7.2</u>. Let $u(t,x)$ be a solution of (1.1) satisfying (1.2) and (1.19). Then for r chosen large enough and $j \geq \lfloor r \rfloor$,

$$(7.27)\quad e^{3r-t}E[\exp\{\int_{2r}^{t-r}k(t-s,\mathfrak{z}_{x,y}(s))\,ds\};A_j^1(r,t)] \leq C_4 e^{-j^\delta/4}P[A_j^1(r,t)]$$

for $y \geq \alpha(r,t)$, $x \geq m_{1/2}(t)$, and some constant C_4 , where A_j^1 is defined in (7.24) and $\alpha(r,t)$ satisfies (7.7).

Proof. Since $k(s,y') \leq 1$ for all s and y' ,

$$(7.28)\quad \begin{aligned} &e^{3r-t}E[\exp\{\int_{2r}^{t-r}k(t-s,\mathfrak{z}_{x,y}(s))\,ds\};A_j^1(r,t)] \\ &\leq e^{-j^\delta/2}E[\exp\{\int_{j-j^\delta/2}^{j}k(t-s,\mathfrak{z}_{x,y}(s))\,ds\};A_j^1(r,t),\ S(r,t)=S^1] \\ &+ e^{-j^\delta/2}E[\exp\{\int_{t-j}^{t-j+j^\delta/2}k(t-s,\mathfrak{z}_{x,y}(s))\,ds\};A_j^1(r,t),\ S(r,t)=S^2]\ , \end{aligned}$$

where S^1 , S^2 , and S are defined in (7.19) and (7.20). We will show that the first term on the right is at most

$$(7.29)\quad 2e^{-j^\delta/4}P[A_j^1(r,t)]\ .$$

The same bound and derivation hold for the second term as well; together, these estimates demonstrate (7.27) for $C_4 = 4$.

It follows from the definition of $L_{r,t}$ and $\underline{m}_{r,t}$ that for $s \in [2r,t/2]$,

$$(7.30)\quad \begin{aligned} \underline{m}_{r,t}(t-s) &= \theta_{r,t}^{-1}\circ L_{r,t}(t-s) + \frac{(t-s)}{t}m_{1/2}(t) + \frac{s}{t}\alpha(r,t) \\ &= m_{1/2}(s_1) - [4 + (\alpha(r,t) - m_{1/2}(t))/t]s^\delta + o_1(1)\ , \end{aligned}$$

where $s_1 = t - s - s^\delta + o_2(1)$ and $o_1(1), o_2(1) \to 0$ as $r \to \infty$. Together with (3.22') on page 100 and (7.7), (7.30) implies that for r (and hence t) chosen large enough,

$$(7.31)\quad \underline{m}_{r,t}(t-s) \leq m_{1/2}(s_1) - 2s^\delta\ .$$

On the other hand, since $u(s_1, m_{1/2}(s_1)) = 1/2$, it follows from (3.66) of

Proposition 3.4 that for each $\epsilon > 0$ and r chosen large enough, if $s_2 \geq t-s$, then

(7.32) $$u(s_2, m_{1/2}(s_1)) \geq 1-\epsilon \ .$$

By Proposition 7.1, for appropriately large r, (7.32) also holds at $(s_2, m_{1/2}(s_1) - z)$ for $0 \leq z \leq m_{1/2}(s_1)$. Consequently by (7.31), for $0 \leq y' \leq \mathcal{m}_{r,t}(t-s) + 2s^{\delta}$ with $s \in [2r, t/2]$,

(7.33) $$u(s_2, y') \geq 1-\epsilon \ .$$

Now, set

$$D_j = \{\mathfrak{z}_{x,y} : \mathfrak{z}_{x,y}(s) - \mathfrak{z}_{x,y}(S^1) \leq 2j^{\delta} \quad \text{for} \quad s \in [j - j^{\delta}/2, S^1]\} \ .$$

Under $A_j^1(r,t)$ and $S(r,t) = S^1$, $S^1 \in [j, j+1)$. Therefore as a special case of (7.33), we have here that if $\mathfrak{z}_{x,y} \in D_j$, then

$$u(t-s, \mathfrak{z}_{x,y}(s)) \geq 1-\epsilon \quad \text{for} \quad s \in [j - j^{\delta}/2, j] \ .$$

If ϵ is chosen appropriately, this implies that $k(t-s, \mathfrak{z}_{x,y}(s)) \leq 1/2$. Consequently, under $A_j^1(r,t)$ and $S(r,t) = S^1$,

(7.34) $$\exp\Big[\int_{j-j^{\delta}/2}^{j} k(t-s, \mathfrak{z}_{x,y}(s))\, ds\Big] \leq e^{j^{\delta}/4} \quad \text{for} \quad \mathfrak{z}_{x,y} \in D_j \ .$$

By (2.4) of Lemma 2.2, for j chosen large enough,

(7.35) $$P[\mathfrak{z}^{t'}(s) < \tfrac{1}{2} j^{\delta} - 1 \quad \text{for} \quad 0 \leq s \leq \tfrac{1}{2}(j+1)^{\delta}] \geq 1 - e^{-j^{\delta}/4}$$

for arbitrary $t' \geq j+1$. Since we are assuming $y \geq \alpha(r,t)$ and $x \geq m_{1/2}(t)$, where $\alpha(r,t)$ and $m_{1/2}(t)$ satisfy (7.7) and (3.22'), it can be calculated that for $\mathfrak{z}_{x,y} \in A_j^1(r,t)$ and $S(r,t) = S^1$,

$$\mathfrak{z}_{x,y}(S^1) - x \geq -3S^1 \geq -3(j+1)$$

if r (and hence t) is chosen large enough. Application of (7.33) and a little computation therefore show that

$$(7.36)\qquad P[D_j^c \mid A_j^1(r,t)\,,\, S(r,t) = S^1] \leq e^{-j^\delta/4}\ .$$

Putting everything together, we see that (7.34) and (7.36) imply that

$$e^{-j^\delta/2} E[\exp\{\int_{j-j^\delta/2}^{j} k(t-s,\mathfrak{z}_{x,y}(s))\,ds\}\,;\, A_j^1(r,t)\,,\, S(r,t) = S^1]$$

$$\leq e^{-j^\delta/2} E[e^{j^\delta/4}\,;\, D_j \cap A_j^1(r,t)\,,\, S(r,t) = S^1]$$

$$+\, e^{-j^\delta/2} E[e^{j^\delta/2}\,;\, D_j^c \cap A_j^1(r,t)\,,\, S(r,t) = S^1]$$

$$\leq 2e^{-j^\delta/4} P[A_j^1(r,t)\,,\, S(r,t) = S^1] \leq 2e^{-j^\delta/4} P[A_j^1(r,t)]\ .$$

The last quantity in (7.37) is the bound given in (7.29). This is what we needed to show, and so the proof is complete. //

Having already done the main calculations in Lemmas 7.1 and 7.2, we are in a position to demonstrate Proposition 7.3 quickly.

<u>Proposition 7.3</u>. Let $u(t,x)$ be a solution of (1.1) satisfying (1.2) and (1.19). Assume that for large enough r and all t, (7.10) is satisfied. Then for $y \geq \alpha(r,t)$ and $x \geq m_{1/2}(t)$,

$$(7.38)\qquad e^{3r-t} E[\exp\{\int_{2r}^{t-r} k(t-s,\mathfrak{z}_{x,y}(s))\,ds\}\,;\, G_{x,y}(r,t)] \leq P[G_{x,y}^c(r,t)]/r^2$$

for large r, where $k(t,x) = f(u(t,x))/u(t,x)$, $G_{x,y}$ is defined in (7.22), and α satisfies (7.7).

<u>Proof</u>. Decompose the left side of (7.38) into

$$(7.39)\qquad e^{3r-t} \sum_{j=\lfloor r \rfloor}^{j_0} E[\exp\{\cdot\}\,;\, A_j^1(r,t)] + e^{3r-t} \sum_{j=\lfloor r \rfloor}^{j_0} E[\exp\{\cdot\}\,;\, A_j^2(r,t)]\ ,$$

where $A_j^i(r,t)$, $i = 1,2$, are defined in (7.24). By Lemma 7.2, the first sum is at most

$$C_4 \sum_{j=\lfloor r \rfloor}^{j_0} e^{-j^{\delta}/4} P[A_j^1(r,t)] \leq C_4 \sum_{j=\lfloor r \rfloor}^{j_0} e^{-j^{\delta}/4} P[G_{x,y}^c(j+1;r,t)] \ .$$

On the other hand,

$$P[A_j^2(r,t)]$$

$$\leq P[\mathfrak{z}_{x,y}(s) \leq -(s \vee (t-s)) + \tfrac{s}{t}y + \tfrac{t-s}{t}x \quad \text{for some} \quad s \in I_j \mid G_{x,y}^c(j+1;r,t)]$$

$$\cdot P[G_{x,y}^c(j+1;r,t)] \ ,$$

which by (2.18) is at most

$$P[\mathfrak{z}_{x,y}(s) \leq -(s \vee (t-s)) + \tfrac{s}{t}y + \tfrac{t-s}{t}x \quad \text{for some} \quad s \in I_j] \cdot P[G_{x,y}^c(j+1;r,t)]$$

$$\leq e^{-j/2} P[G_{x,y}^c(j+1;r,t)] \ .$$

Therefore, since $k \leq 1$, (7.39) is at most

$$(1 + C_4) \sum_{j=\lfloor r \rfloor}^{j_0} (e^{-j^{\delta}/4} + e^{-j/2}) P[G_{x,y}^c(j+1;r,t)] \ ,$$

which by Lemma 7.1, is at most

$$C_5 P[G_{x,y}^c(r,t)] \sum_{j=\lfloor r \rfloor}^{\infty} (j+1)(e^{-j^{\delta}/4} + e^{-j/2})$$

for appropriate C_5 . As $r \to \infty$, the above sum $\to 0$ more quickly than r^{-n} for each fixed n ; (7.38) follows with $n = 2$. //

In Proposition 7.2, we showed for large r that paths lying above $\bar{m}_{r,t}(s)$ over the interval $[r,t-2r]$ enjoy essentially full heating on $[r,t-2r]$. In Proposition 7.3, we showed that the combined effect of those paths hitting $\underline{m}_{r,t}(s)$ over $[r,t-2r]$ is in a certain sense small. To tie these two results together, we will need to know that the probabilities of Brownian bridge lying above $\bar{m}_{r,t}$ and $\underline{m}_{r,t}$ are essentially the same. This is done in Proposition 7.4 for the interval $[r,t-r]$ by paraphrasing Proposition 6.2. However, in order to obtain precise estimates for the associated

Feynman-Kac integral, we will also need to be able to investigate the effect of heating over the remaining intervals $[0,2r]$ and/or $[t-r,t]$. (Consider r as being fixed as $t\to\infty$.) For Brownian bridge with endpoints x and y close to $m_{1/2}(t)$ and $\alpha(r,t)$, such computations are in general not simple. However, if one or more of these endpoints is allowed to go to infinity with r fixed, then $\mathfrak{z}_{x,y}$ should typically remain large enough in $[0,r]$ and/or $[t-2r,t]$ so that $u\sim 0$ and hence $k\sim 1$ at these times. For the two classes of initial data considered in Sections 8 and 9, it will be the case that computation of these limits is sufficient for our purposes. In Section 8, we will be able to allow $x\to\infty$, whereas in Section 9 we will allow $x,y\to\infty$. In these contexts, we will wish to reformulate Proposition 7.4 in terms of the probabilities of lying above certain functions $\bar{\mathcal{M}}^{x}_{r,t}$, $\bar{\mathcal{M}}^{x,y}_{r,t}$, and $\underline{\mathcal{M}}'_{r,t}$ over the entire interval $[0,t]$. This is done in Proposition 7.5.

Proposition 7.4. Let $u(t,x)$ be a solution to (1.1) and (1.2). Also, assume that

$$(7.10)\qquad \begin{aligned} m_{1/2}(s) &\leqq \tfrac{s}{t}m_{1/2}(t) + \tfrac{t-s}{t}\alpha(r,t) + 8s^{\delta} && \text{for } s\in[r,t/2] \\ &\leqq \tfrac{s}{t}m_{1/2}(t) + \tfrac{t-s}{t}\alpha(r,t) + 8(t-s)^{\delta} && \text{for } s\in[t/2,t-r] \end{aligned}$$

for r chosen large enough, where $\alpha(r,t)$ satisfies (7.7). Then

$$(7.40)\qquad \frac{P[\mathfrak{z}_{x,y}(s) > \bar{\mathcal{M}}_{r,t}(t-s) \quad \text{for } s\in[r,t-r]]}{P[\mathfrak{z}_{x,y}(s) > m_{1/2}(t-s) \quad \text{for } s\in[r,t-r]]} \to 1$$

and

$$(7.41)\qquad \frac{P[\mathfrak{z}_{x,y}(s) > m_{1/2}(t-s) \quad \text{for } s\in[r,t-r]]}{P[\mathfrak{z}_{x,y}(s) > \underline{\mathcal{M}}_{r,t}(t-s) \quad \text{for } s\in[r,t-r]]} \to 1$$

uniformly in t, $x \geqq m_{1/2}(t) + M_1$, and $y \geqq \alpha(r,t) + M_1$ as $r\to\infty$, where M_1 is constant.

Proof. For $x' = x - m_{1/2}(t) - M_1$ and $y' = y - \alpha(r,t) - M_1$, the left side of (7.40) may be rewritten as

$$\frac{P[\mathfrak{z}_{x',y'}(s) > \mathcal{L}_{r,t}(t-s) - M_1 \quad \text{for} \quad s \in [r,t-r]]}{P[\mathfrak{z}_{x',y'}(s) > L_{r,t}(t-s) - M_1 \quad \text{for} \quad s \in [r,t-r]]} ,$$

where $L_{r,t}$ is defined as in (7.6). On account of (7.10),

$$\underline{L}_{r,t}(s) - M_1 \leq 0 \quad \text{for} \quad s \in [r,t-r]$$

and r chosen large enough. The limit in (7.40) with $x = m_{1/2}(t) + M_1$ and $y = \alpha(r,t) + M_1$ then follows from Proposition 6.2; uniformity over $x \geq m_{1/2}(t) + M_1$ and $y \geq \alpha(r,t) + M_1$ follows from (6.32). The demonstration of (7.41) is the same. //

In order to state Proposition 7.5, we first introduce the functions $\bar{\mathfrak{m}}^x_{r,t}$, $\bar{\mathfrak{m}}^{x,y}_{r,t}$, and $\underline{\mathfrak{m}}'_{r,t}$, where

$$(7.42) \qquad \begin{aligned} \bar{\mathfrak{m}}^x_{r,t}(s) &= \bar{\mathfrak{m}}_{r,t}(s) && \text{for} \quad s \in [0,t-2r] \\ &= (x + m_{1/2}(t))/2 && \text{for} \quad s \in (t-2r,t] , \end{aligned}$$

$$(7.43) \qquad \begin{aligned} \bar{\mathfrak{m}}^{x,y}_{r,t}(s) &= \bar{\mathfrak{m}}_{r,t}(s) && \text{for} \quad s \in [r,t-2r] \\ &= y/2 && \text{for} \quad s \in [0,r) \\ &= (x + m_{1/2}(t))/2 && \text{for} \quad s \in (t-2r,t] , \end{aligned}$$

and

$$(7.44) \qquad \begin{aligned} \underline{\mathfrak{m}}'_{r,t}(s) &= \underline{\mathfrak{m}}_{r,t}(s) && \text{for} \quad s \in [r,t-2r] \\ &= -\infty && \text{for} \quad s \in [0,r) \cup (t-2r,t] . \end{aligned}$$

Since $\underline{\mathfrak{m}}_{r,t} \leq \bar{\mathfrak{m}}_{r,t}$, it follows that $\underline{\mathfrak{m}}'_{r,t} \leq \bar{\mathfrak{m}}^{\cdot}_{r,t}$, where $\cdot$ denotes either x or x,y.

To demonstrate Proposition 7.5, we apply Proposition 7.4 together with Lemma 2.6, and note that since as $x \to \infty$, $(x - m_{1/2}(s))/2 \to \infty$ on $(t-2r,t]$, the probability of Brownian bridge hitting $\bar{\mathfrak{m}}^{\cdot}_{r,t}$ on $(t-2r,t]$ (and on $[0,r)$ if $y \to \infty$) decreases to zero if r is increased slowly enough with respect to

x (and y) and t.

Proposition 7.5. Let $u(t,x)$ be a solution to (1.1) and (1.2) with initial data satisfying (1.19) for some $h > 0$. Also, assume that (7.10) holds for r chosen large enough, where $\alpha(r,t)$ satisfies (7.7). Then

$$(7.45) \qquad \frac{P[\mathfrak{z}_{x,y}(s) > \bar{\mathcal{M}}^{x_0}_{r,t}(t-s) \text{ for } s \in [0,t-r]]}{P[\mathfrak{z}_{x,y}(s) > \mathcal{M}'_{-r,t}(t-s) \text{ for } s \in [0,t-r]]} \to 1$$

uniformly in $t \geq 8r$, $x \geq x_0 \geq m_{1/2}(t) + 8r$, and $y \geq \alpha(r,t)$ as $r \to \infty$. Similarly,

$$(7.46) \qquad \frac{P[\mathfrak{z}_{x,y}(s) > \bar{\mathcal{M}}^{x_0,y_0}_{r,t}(t-s) \text{ for } s \in [0,t]]}{P[\mathfrak{z}_{x,y}(s) > \mathcal{M}'_{-r,t}(t-s) \text{ for } s \in [0,t]]} \to 1$$

uniformly in $t \geq 8r$, $x \geq x_0 \geq m_{1/2}(t) + 8r$, and $y \geq y_0 \geq \alpha(r,t) \vee 8r$ as $r \to \infty$.

Proof. We will demonstrate only (7.45) since the proof of (7.46) is similar. First of all, one may calculate under $t \geq 8r$, $x \geq x_0 \geq m_{1/2}(t) + 8r$, and $y \geq \alpha(r,t)$, that because of (3.22') on page 100 and (7.7),

$$\frac{s}{t}y + \frac{t-s}{t}x \geq \frac{1}{2}(x_0 + m_{1/2}(t)) + r \quad \text{for} \quad s \in [0,2r]$$

and large enough r. Therefore,

$$(7.47) \qquad \begin{aligned} &P[\mathfrak{z}_{x,y}(s) > \tfrac{1}{2}(x_0 + m_{1/2}(t)) \quad \text{for} \quad s \in [0,2r]] \\ &\geq P[\mathfrak{z}^t(s) > -r \quad \text{for} \quad s \in [0,2r]] , \end{aligned}$$

which by (2.4)

$$\to 1 \text{ uniformly in } t \text{ as } r \to \infty .$$

Since $\bar{\mathcal{M}}^{x_0}_{r,t}(s) = \bar{\mathcal{M}}_{r,t}(s)$ on $[r,t-2r]$, we may rewrite

$$(7.48)\qquad \frac{P[\mathfrak{z}_{x,y}(s) > \bar{\mathfrak{m}}_{r,t}^{x_0}(t-s) \quad \text{for} \quad s \in [0,t-r]]}{P[\mathfrak{z}_{x,y}(s) > \bar{\mathfrak{m}}_{r,t}(t-s) \quad \text{for} \quad s \in [2r,t-r]]}$$

as

$$P^0[B_{\bar{\mathfrak{m}}^{x_0}}[t-2r,t] \mid B_{\bar{\mathfrak{m}}^{x_0}}[r,t-2r]\,;\,0,y\,;\,t,x] \;,$$

where we suppress the subscripts r and t. It follows from (2.18) that this is at least

$$(7.49)\qquad P^0[B_{\bar{\mathfrak{m}}^{x_0}}[t-2r,t] \mid 0,y\,;\,t,x] \;.$$

By (7.47), this quantity

$$(7.50)\qquad \to 1 \quad \text{as} \quad r \to \infty$$

uniformly in $t \geqq 8r$, $x \geqq x_0 \geqq m_{1/2}(t) + 8r$, and $y \geqq \alpha(r,t)$. By applying Proposition 3.4, one may also show that for large enough r, $\underline{\mathfrak{m}}_{r,t}(s) \leqq m_{1/2}(t)$ on $s \in [t-2r,t-r]$. Reasoning as in (7.48)-(7.50), one may also conclude that

$$(7.51)\qquad \frac{P[\mathfrak{z}_{x,y}(s) > \underline{\mathfrak{m}}_{r,t}(t-s) \quad \text{for} \quad s \in [r,t-r]]}{P[\mathfrak{z}_{x,y}(s) > \underline{\mathfrak{m}}'_{r,t}(t-s) \quad \text{for} \quad s \in [0,t-r]]} \to 1$$

as $r \to \infty$. On the other hand, by Proposition 7.4,

$$(7.52)\qquad \frac{P[\mathfrak{z}_{x,y}(s) > \bar{\mathfrak{m}}_{r,t}(t-s) \quad \text{for} \quad s \in [r,t-r]]}{P[\mathfrak{z}_{x,y}(s) > \underline{\mathfrak{m}}_{r,t}(t-s) \quad \text{for} \quad s \in [r,t-r]]} \to 1$$

as $r \to \infty$ uniformly in t, $x \geqq m_{1/2}(t)$, and $y \geqq \alpha(r,t)$. Together, (7.48)-(7.50), (7.51), and (7.52) imply (7.45). //

Before doing Proposition 7.6, we first state an elementary lemma, which says that if the initial data $u(0,x)$ is great enough for values of x above 0, then the nature of the initial data below 0 has little effect on the position of the median of $u(t,x)$. In order to avoid getting tangled up over notation involving solutions $u(t,x)$ of (1.1) associated with different initial data, we will find it convenient here and later on to employ the following terminology. Denote by $u^g(t,x)$ the solution of (1.1) with initial data $g(x)$; $u^H(t,x)$ will denote the solution under Heaviside initial data. Similarly, we will denote by $m^u_{1/2}(t)$ (or equivalently, $m^g_{1/2}(t)$ if $g(x) = u(0,x)$) and $m^H_{1/2}(t)$ the positions of the medians of $u(t,x)$ and $u^H(t,x)$. $k^u(t,x)$ and $k^H(t,x)$ will have analogous interpretations. In stating the lemma, we will also find it convenient to employ the notation $u \overset{s}{\leq} v$ introduced before Corollary 2 of Proposition 3.2.

Lemma 7.3. Let $u(t,x)$ be a solution of (1.1) and (1.2) for which f is concave. Assume $u(t,x)$ has initial data satisfying

$$(7.52) \qquad u(0,x) \overset{s}{\leq} w^{2^{1/2}}(x)$$

and

$$(7.53) \qquad u(0,x) \geq g(x) \ ,$$

where $\int_{-\infty}^{\infty} g(x)\,dx > 0$. Also, let $v(t,x)$ denote the solution of (1.1) with initial data

$$\begin{aligned} v(0,x) &= u(0,x) \quad \text{for} \quad x > 0 \\ &= 1 \qquad\quad \text{for} \quad x \leq 0 \ . \end{aligned}$$

Then,

$$(7.54) \qquad m^u_{1/2}(t) \geq m^v_{1/2}(t) + C_1$$

for some constant C_1 (depending on $g(x)$) and all t .

Proof. Since f is concave, it follows from Lemma 3.5 that

$$v(t,x) \leq u(t,x) + u^H(t,x) \quad \text{for all} \quad t \quad \text{and} \quad x \ .$$

Therefore,

(7.55) $m^{v}_{1/2}(t) \leq m^{u}_{1/4}(t) \vee m^{H}_{1/4}(t)$ for all t .

On the other hand, by Proposition 3.4 and (7.53),

(7.56) $m^{u}_{1/4}(t) \geqq m^{H}_{1/4}(t) + C_2$ if $t \geqq t_0$,

for appropriate t_0 and C_2 (both depending on $g(x)$). Also, by (7.52) and Corollary 2 of Proposition 3.2, $u(t,x) \overset{s}{\leqq} w^{2^{1/2}}(x)$ for all t , and therefore

(7.57) $m^{u}_{1/2}(t) \geqq m^{u}_{1/4}(t) + C_3$ for all t

and appropriate C_3 . For $t \geqq t_0$, (7.54) follows immediately from (7.55)-(7.57). For $t < t_0$, (7.54) follows from (7.52), (7.53), and a simple continuity argument. //

In order to state Proposition 7.6, we introduce the functions $\beta_r(t)$ and $b_r(t)$, where

(7.58) $\beta_r(t) = \inf\{x\colon\ x + 2^{1/2}s \geqq m_{1/2}(s)$ for $s \in [r,t]\}$

and

(7.59) $b_r(t) = \inf\{s \in [r,t]\colon\ \beta_r(t) + 2^{1/2}s = m_{1/2}(s)\}$.

Note that since $m_{1/2}(t)$ is upper semi-continuous,

$$m_{1/2}(b_r(t)) = \beta_r(t) + 2^{1/2}\, b_r(t) \quad .$$

To demonstrate in Proposition 7.6 that the initial data over $(-\infty, \beta_r(t)]$ does not affect $u(t,x)$ significantly for $x \geqq m_{1/2}(t)$, we shall reason as follows. Since f is assumed to be concave, it is not difficult to show that the "contribution" of $u(0,x)$ in the range $(-\infty, \beta_r(t)]$ to $u(t,x)$ is at most $u^{H}(t,x-\beta_r(t))$. But by the corollary to Proposition 3.1,

$2^{1/2}r - m^H_{1/2}(r) \to \infty$ as $r \to \infty$. It follows that

$$m^u_{1/2}(b_r(t)) - [m^H_{1/2}(b_r(t)) + \beta_r(t)] \to \infty$$

uniformly in t as $r \to \infty$. Once having fallen behind $u(s,x)$, $u^H(s,x-\beta_r(t))$ will not at later times catch up; it is therefore also true that

$$m^u_{1/2}(t) - [m^H_{1/2}(t) + \beta_r(t)] \to \infty \quad \text{as} \quad r \to \infty .$$

With a little more work, one may conclude that $u^H(t,x-\beta_r(t))/u(t,x) \to 0$ uniformly in t and $x \geqq m_{1/2}(t)$ as $r \to \infty$.

Proposition 7.6. Let $u(t,x)$ be a solution of (1.1) and (1.2) for which f is assumed concave. Then,

$$(7.60) \quad E[\exp\{\int_0^t k(t-s,\mathfrak{X}_x(s))\,ds\}u(0,\mathfrak{X}_x(t)) ; \mathfrak{X}_x(t) \leqq \beta_r(t)]/u(t,x) \to 0$$

uniformly in t and $x \geqq m_{1/2}(t)$ as $r \to \infty$, where β_r is as defined in (7.58).

Proof. Let $v(s,x)$ denote the solution of (1.1) and (1.2) with initial data

$$\begin{aligned} v(0,x) &= 0 && \text{for } x > \beta_r(t) \\ &= u(0,x) && \text{for } x \leqq \beta_r(t) . \end{aligned}$$

By the maximum principle, $v(s,y) \leqq u(s,y)$ for all s and y. Since f is concave, it follows that $k^v(s,y) \geqq k^u(s,y)$ for all s and y. Consequently,

$$\begin{aligned} E[\exp\{\cdot\}u(0,\mathfrak{X}_x(t)) ; \mathfrak{X}_x(t) \leqq \beta_r(t)] &\leqq v(t,x) \\ &\leqq u^H(t,x-\beta_r(t)) \end{aligned}$$

for all t and x. To demonstrate (7.60), it therefore suffices to show that

(7.61) $\quad u^H(t,x-\beta_r(t))\,/\,u(t,x) \to 0$

uniformly in t and $x \geqq m_{1/2}(t)$ as $r \to \infty$.

From the definition of $b_r(t)$, $m^u_{1/2}(b_r(t)) = 2^{1/2}\,b_r(t) + \beta_r(t)$. But from Corollary 1 of Proposition 3.1,

$$m^H_{1/2}(b_r(t)) \leqq 2^{1/2}\,b_r(t) - 2^{-3/2}\log b_r(t) + C$$

for some constant C. Therefore,

$$\begin{aligned} m^u_{1/2}(b_r(t)) - [m^H_{1/2}(b_r(t)) + \beta_r(t)] &\geqq 2^{-3/2}\log b_r(t) - C \\ &\geqq 2^{-3/2}\log r - C \;. \end{aligned} \tag{7.62}$$

To demonstrate (7.61), we need to show that the analog of (7.62) holds at time t. Employing the terminology of the second corollary to Proposition 3.2, we see that since $u(0,x) \overset{s}{\geqq} u^H(0,x)$,

(7.63) $\quad u(b_r(t),x) \overset{s}{\geqq} u^H(b_r(t),x)$.

So, by (7.62) and (7.63), if $u^G(t,x)$ denotes the solution of (1.1) and (1.2) with initial data

$$\begin{aligned} G(x) &= u^H(b_r(t),x) \quad \text{for} \quad x > m^H_{1/2}(b_r(t)) \\ &= 0 \qquad\qquad\quad\;\; \text{for} \quad x \leqq m^H_{1/2}(b_r(t)) \;, \end{aligned}$$

then for all x,

$$u(b_r(t),x) \geqq u^G(0,x+\beta_r(t) + 2^{-3/2}\log r - C) \;,$$

and therefore by the maximum principle,

(7.64) $\quad m^u_{1/2}(b_r(t)+s) \geqq m^G_{1/2}(s) + \beta_r(t) + 2^{-3/2}\log r - C$

for all s. Now, we may assume here that $r \geqq 1$, in which case

$$u^H(1,x) \overset{s}{\leqq} u^H(b_r(t),x) \overset{s}{\leqq} w^{2^{1/2}}(x) \;,$$

where $w^{2^{1/2}}$ is the particular solution of (1.5) with $w^{2^{1/2}}(0) = 1/2$. Application of Lemma 7.3 then gives

$$(7.65) \qquad m^G_{1/2}(s) \geq m^H_{1/2}(b_r(t) + s) + C_1$$

for all s and some constant C_1. Setting $s = t - b_r(t)$, we thus obtain from (7.64) and (7.65) that

$$(7.66) \qquad m^u_{1/2}(t) - [m^H_{1/2}(t) + \beta_r(t)] \geq 2^{-3/2} \log r + C_2$$

for some constant C_2.

We claim that (7.61) is a simple consequence of (7.66) and Corollary 2 of Proposition 2. If $y = x - m^u(t)$, then

$$(7.67) \qquad \frac{u^H(t,x-\beta_r(t))}{u(t,x)} = \frac{u^H(t,y+m^u_{1/2}(t)-\beta_r(t))}{u^H(t,y+m^H_{1/2}(t))} \cdot \frac{u^H(t,y+m^H_{1/2}(t))}{u(t,y+m^u_{1/2}(t))} .$$

By Corollary 2, $u^H(t,x) \overset{s}{\leq} w^{2^{1/2}}(x)$, and by (1.13), $w^{2^{1/2}}(x) \sim C_3 x e^{-2^{1/2}x}$ as $x \to \infty$. Therefore, if we choose $y_1 \geq y \geq 0$ to be the value where $w^{2^{1/2}}(y_1) = u^H(t,y+m^H_{1/2}(t))$, then

$$(7.68) \qquad \frac{u^H(t,y+m^u_{1/2}(t)-\beta_r(t))}{u^H(t,y+m^H_{1/2}(t))} \leq \frac{w^{2^{1/2}}(y_1+c_r(t))}{w^{2^{1/2}}(y_1)} \leq C_4\, c_r(t)\, e^{-2^{1/2}c_r(t)}$$

for some constant C_4, where $c_r(t)$ denotes the left side of (7.66). By (7.66), $c_r(t) \to \infty$ uniformly in t as $r \to \infty$; it follows that the left side of (7.68) $\to 0$ uniformly in t and $y \geq 0$ as $r \to \infty$. On the other hand, since $u^H(t,x) \overset{s}{\leq} u(t,x)$, another application of the corollary shows that

$$(7.69)\qquad \frac{u^H(t,y+m^H_{1/2}(t))}{u(t,y+m^u_{1/2}(t))} \leqq \frac{u^H(t,m^H_{1/2}(t))}{u(t,m^u_{1/2}(t))} = 1 .$$

(7.61) follows directly from (7.67)-(7.69). //

8. CONVERGENCE OF THE KOLMOGOROV EQUATION IN THE CASE $\lambda = 2^{1/2}$ FOR FINITE INITIAL MASS

In this section and in Section 9, we demonstrate convergence of $u(t,x)$ to the travelling wave moving at rate $2^{1/2}$ under initial data which is assumed to satisfy

$$(1.17) \qquad \int_x^{x+N} u(0,y)\,dy > \eta \quad \text{if} \quad x \leqq -M ,$$

for some $\eta, M, N > 0$, and

$$(1.19) \qquad \limsup_{t \to \infty} \frac{1}{t} \log \left[\int_t^{t(1+h)} u(0,y)\,dy\right] \leqq -2^{1/2}$$

for some $h > 0$. Although we will in both sections employ the estimates which were derived in Sections 6 and 7, the problem seems to break down into two different cases depending on whether or not

$$(8.1) \qquad \int_0^\infty y\, e^{2^{1/2} y}\, u(0,y)\,dy < \infty \quad .$$

If (8.1) is satisfied, then we will say that $u(t,x)$ has <u>finite initial mass</u>; otherwise, $u(t,x)$ is said to have <u>infinite initial mass</u>. In this section, we treat the case where (8.1) holds; in Section 9, we will treat the case of infinite initial mass.

The procedure which we apply in this section to demonstrate convergence to the travelling wave $w^{2^{1/2}}(x)$ is based on showing that under (8.1),

$$(8.2) \qquad m_{1/2}(t) = 2^{1/2} t - 3 \cdot 2^{-3/2} \log t + O(1) \; .$$

(We are, of course, excluding the trivial case where $\int_{-\infty}^{\infty} u(0,y)\,dy = 0$.) Knowledge of (8.2) gives us a great amount of information regarding the behavior of $k(s,y) = f(u(s,y))/u(s,y)$. The families of functions $\bar{\mathcal{m}}(s)$ and

$\underline{\mathcal{m}}(s)$, $0 \leq s \leq t$, (whose sub- and superscripts we omit) were constructed in terms of $m_{1/2}(t)$ in Section 7 so that $k(s,y) \sim 1$ for $y > \bar{\mathcal{m}}(s)$ and $k(s,y) \sim 0$ for $y < \underline{\mathcal{m}}(s)$ (Propositions 7.2 and 7.3), and so that their hitting probabilities for Brownian bridge are almost the same (Propositions 7.4 and 7.5). Both $\bar{\mathcal{m}}(s)$ and $\underline{\mathcal{m}}(s)$ should be thought of as being close to $m_{1/2}(s)$. One may therefore employ (8.2) to show that the probability of Brownian bridge lying above either function for all $s \in [0,t]$ is in each case approximately C/t , where C depends on the endpoints of the bridge, which are assumed to lie close to the endpoints of $m_{1/2}(s)$. We will therefore be able to obtain accurate estimates on relevant quantities involving the weights $\exp[\int_0^t k(t-s,\mathfrak{z}_{x,y}(s))\,ds]$ of the Feynman-Kac integral which corresponds to $u(t,x)$. These estimates will be sufficiently accurate for us to compute the tail behavior of $u(t,x)$, which by Proposition 3.3 will be enough to imply convergence of $u(t,x+m_{1/2}(t))$ to the travelling wave $w^{2^{1/2}}(x)$.

The section is structured as follows. First we demonstrate that $2^{1/2}t - 3\cdot 2^{-3/2}\log t$ is essentially a lower bound for $m_{1/2}(t)$ (Proposition 8.1). This follows from Proposition 3.5, which says that under Heaviside initial data, $m_{1/2}(t)$ is almost convex. We then employ Proposition 8.1 to show that $2^{1/2}t - 3\cdot 2^{-3/2}\log t$ is in fact essentially an upper bound as well (Proposition 8.2). We also obtain at this point accurate bounds on the right tail of $u(t,x)$ ((8.16) and (8.29)). These bounds will be used in conjunction with Corollary 2 of Proposition 8.2 and Lemma 8.1 to avoid the necessity of comparing hitting probabilities of $\bar{\mathcal{m}}(s)$ and $\underline{\mathcal{m}}(s)$ for values of s close to 0 . The estimates we then perform will be asymptotically precise, rather than only of bounded error, as would otherwise be the case. In Proposition 8.3, we demonstrate several of the main estimates which are needed in the proof of the theorem. Upper and lower bounds for $u(t,x)$ in terms of the hitting probabilities of $\underline{\mathcal{m}}$ and $\bar{\mathcal{m}}$ are obtained, which are then shown to be asymptotically equal. Theorem 3, the principal result of the section, states that under (8.1) and (1.17), $u(t,x+m(t))$ converges to the travelling wave $w^{2^{1/2}}(x)$, where the centering term $m(t)$ may be specified as $m(t) = 2^{1/2}t - 3\cdot 2^{-3/2}\log t$.

At various points in this section, it will be convenient for us to perform estimates on the initial data restricted to the interval $[y_0,\infty)$ for some $y_0 \leq 0$ which is chosen so that

$$(8.3) \qquad \int_{y_0}^{\infty} u(0,y)\,dy > 0 \ .$$

It will also be convenient for us to use the notation introduced in Section 7, where $u^g(t,x)$, $m^g_{1/2}(t)$, and $k^g(t,x)$ stand for the quantities $u(t,x)$, $m_{1/2}(t)$, and $k(t,x)$ corresponding to the solution of (1.1) with initial data $g(x)$. The superscript H will in each case denote the solution with Heaviside initial data. The term $\alpha(r,t)$ which was introduced in (7.6) in conjunction with $\underline{\mathcal{m}}$ and $\bar{\mathcal{m}}$ will in this section typically denote either y_0 or $-\log r$, where y_0 is assumed to satisfy (8.3). In either case (7.7) is clearly satisfied with $\alpha(r,t) = o(t)$, since we may assume that $r \leq t$.

Proposition 8.1 is our first result on the asymptotic behavior of $m_{1/2}(t)$ under general initial data beyond that $m_{1/2}(t)/t \to 2^{1/2}$ as $t \to \infty$. To demonstrate the proposition, we may employ Proposition 3.4 and replace $u(0,x)$ with Heaviside initial data. Proposition 3.5 states that under Heaviside initial data (and concave f), $m_{1/2}(t)$ is almost convex. This enables us to conclude that $\bar{\mathcal{m}}(s)$, $0 \leq s \leq t$, is almost convex, from which one obtains a lower bound on the probability of Brownian bridge lying above $\bar{\mathcal{m}}$ if one replaces $\bar{\mathcal{m}}$ by the straight line with the same endpoints. From Proposition 7.2, if $x(s) > \bar{\mathcal{m}}(t-s)$ for s not close to either 0 or t, then $k(t-s, x(s)) \sim 1$, and therefore the bound given in (8.8) holds. Proposition 8.1 then follows from these two bounds after a little computation.

<u>Proposition 8.1</u>. Let $u(t,x)$ be a solution to (1.1), (1.2), and (1.2'). Then,

(8.4) $\quad m_{1/2}(t) \geqq 2^{1/2}t - 3\cdot 2^{-3/2}\log t + C_0$ for large enough t ,

where C_0 depends on f and the initial data, but not on t .

Proof. Owing to the maximum principle, one may in (1.1) replace f by $\underline{f}$ (introduced in (3.50)). We may thus assume here that f is concave. Also, by Proposition 3.4,

$$m_{1/2}(t) \geqq m^H_{1/2}(t) + C_1$$

for large enough t . It therefore suffices for us to demonstrate (8.4) under the assumption of Heaviside initial data.

In order to demonstrate (8.4), we will next establish two inequalities. First of all, note that by Proposition 3.5,

$$(8.5) \qquad m_{1/2}(s) - \frac{s}{t}m_{1/2}(t) \leqq M$$

for some M and all $0 \leqq s \leqq t$, if t is chosen large enough. We recall $\underline{\mathcal{m}}_{r,t}(s)$ and $\bar{\mathcal{m}}_{r,t}(s)$ from (7.8) and (7.9), and set $\alpha(r,t) = y_0$, where y_0 is fixed. Owing to (8.5), r may be chosen large enough so that (7.10) holds. We may therefore apply Proposition 7.4 to obtain

$$P[\mathfrak{z}_{x,y}(s) > \bar{\mathcal{m}}_{r,t}(t-s) \quad \text{for} \quad r \leqq s \leqq t-r]$$

$$(8.6) \qquad \geqq C_1\, P[\mathfrak{z}_{x,y}(s) > \underline{\mathcal{m}}_{r,t}(t-s) \quad \text{for} \quad r \leqq s \leqq t-r]$$

$$\geqq C_1\, P[\mathfrak{z}(s) > 0 \quad \text{for} \quad r \leqq s \leqq t-r] \ ,$$

for $y \geqq y_0$, $x \geqq m_{1/2}(t)$, and r chosen large enough, where $C_1 > 0$. Application of the reflection principle and a little estimation then shows that the last term is at least

$$(8.7) \qquad C_2/t \ ,$$

where $C_2 > 0$ depends on just y_0 and r , both of which are considered fixed. On the other hand, the other inequality is just a direct application of Proposition 7.2, and states that if $\mathfrak{z}_{x,y}(s) > \bar{m}_{r,t}(t-s)$ for $2r \leq s \leq t-r$, then

$$\exp\left[\int_{2r}^{t-r} k(t-s,\mathfrak{z}_{x,y}(s))\,ds\right] \geq C_3\, e^t \tag{8.8}$$

for $C_3 > 0$ not depending on t.

To demonstrate (8.4), we now employ the Feynman-Kac representation

$$u(t,x) = \int_{-\infty}^{\infty} u(0,y) \frac{e^{-(x-y)^2/2t}}{\sqrt{2\pi t}} E\left[\exp\left\{\int_0^t k(t-s,\mathfrak{z}_{x,y}(s))\,ds\right\}\right] dy \, .$$

This is at least

$$\int_{-\infty}^{\infty} u(0,y) \frac{e^{-(x-y)^2/2t}}{\sqrt{2\pi t}} E\left[\exp\left\{\int_{2r}^{t-r} k(t-s,\mathfrak{z}_{x,y}(s))\,ds\right\}\right] dy \, .$$

Application of (8.8) and then (8.6)-(8.7) shows that if $x \geq m_{1/2}(t)$, this last quantity is at least

$$\begin{aligned} &C_3\, e^t \int_{-\infty}^{\infty} u(0,y) \frac{e^{-(x-y)^2/2t}}{\sqrt{2\pi t}} P[\mathfrak{z}_{x,y}(s) > \bar{m}_{r,t}(t-s) \text{ for } 2r \leq s \leq t-r]\,dy \\ &\geq C_4 \frac{e^t}{t} \int_{y_0}^{\infty} u(0,y) \frac{e^{-(x-y)^2/2t}}{\sqrt{2\pi t}}\, dy \, , \end{aligned} \tag{8.9}$$

for some $C_4 > 0$.

Now, choose y_0 small enough so that (8.3) is satisfied. It is an easy matter to compute that if $x = 2^{1/2}t - 3\cdot 2^{-3/2}\log t + z_1$, with $|z_1| \leq C_5 t^{1/2}$ for some C_5, then

$$\int_{y_0}^{\infty} u(0,y) \frac{e^{-(x-y)^2/2t}}{\sqrt{2\pi t}}\, dy \geq C_6\, t\, e^{-t} \cdot e^{-2^{1/2} z_1}$$

for some $C_6 > 0$. Together with (8.9), this last inequality shows that for $|z_1| \leq C_5 t^{1/2}$ and $x \geq m_{1/2}(t)$,

$$u(t,x) \geq C_7\, e^{-2^{1/2} z_1} \tag{8.10}$$

for some $C_7 > 0$. Observe that if (8.4) were violated, then for any fixed z_1 , t could be chosen so that $x \geqq m_{1/2}(t)$, and thus (8.10) would necessarily hold. But if $z_1 < 2^{-1/2} \log C_7$, then $u(t,x) > 1$, which is not possible. Therefore, (8.4) must indeed hold, which completes the proof. //

Here and later on in this section, we will find it useful to employ the following notation. We introduce $n(s)$, where

$$(8.11) \qquad n(s) = 2^{1/2}s - 3\cdot 2^{-3/2}\log s + C_0 \quad \text{for} \quad s \in [0,t] ,$$

and C_0 is chosen as in (8.4) of Proposition 8.1. (We may assume that $C_0 \leqq 0$.) By Proposition 8.1, $n(s) \leqq m_{1/2}(s)$ for large enough s . We also set

$$(8.12) \qquad \bar{y} = y - \alpha(r,t) \ , \ z = x - m_{1/2}(t) \ , \text{ and } \ \bar{z} = x - n(t) \ ,$$

where $\alpha(r,t) = o(t)$, and will be specified more precisely later on. In conjunction with (8.11), we introduce the functions $\underline{\mathcal{L}}^n$, $\bar{\mathcal{L}}^n$, $\underline{\mathcal{M}}^n$, and $\bar{\mathcal{M}}^n$, which are defined in the same manner as $\underline{\mathcal{L}}$, $\bar{\mathcal{L}}$, $\underline{\mathcal{M}}$, and $\bar{\mathcal{M}}$ were in Sections 6 and 7, but are constructed from $n(s)$ rather than $m_{1/2}(s)$. It is not difficult to show that $\underline{\mathcal{M}}^n$ lies slightly below n , and $\bar{\mathcal{M}}^n$ lies slightly above n .

The following corollary of Proposition 8.1 gives an upper bound on the probability that $\mathfrak{z}_{x,y}$ lies above $\underline{\mathcal{M}}$. It is shown by noting that Propositions 8.1 and 6.2 enable us to substitute for this probability the probability that Brownian bridge lies above a particular line, which is easily computed. The corollary will be applied in Propositions 8.2 and 8.3(b).

<u>Corollary 1</u>. Let $u(t,x)$ be a solution to (1.1), (1.2), and (1.2') with initial data satisfying (8.1). Assume that $\alpha(r,t) = O(\log r)$ in (7.6). Then for $x \geqq n(t) + 1$, $y \geqq \alpha(r,t) + 1$, and large r ,

$$(8.13)\quad \begin{aligned} &P[\mathfrak{z}_{x,y}(s) > \underline{\mathfrak{m}}_{r,t}(t-s) \quad \text{for} \quad s\in[2r,t-r]] \\ &\leq C_1 r(1-\exp[-2\bar{y}\bar{z}/t]) \end{aligned}$$

for some constant C_1 .

Proof. By Proposition 8.1, $n(s) \leq m_{1/2}(s)$ for $s \geq r$ and r chosen large enough, whereas by (3.22') on page 100, $m_{1/2}(t)/t \to 2^{1/2}$ as $t \to \infty$. It therefore follows from the construction of $\underline{\mathfrak{m}}$ in (7.6) and (7.8) that for large enough t ,

$$\underline{\mathfrak{m}}^n_{r,t}(s) - \underline{\mathfrak{m}}_{r,t}(s) \leq s^\delta \wedge (t-s)^\delta \quad \text{if} \quad s\in[r,t] ,$$

where $\delta < 1/2$. Consequently,

$$(8.14)\quad \begin{aligned} &P[\mathfrak{z}_{x,y}(s) > \underline{\mathfrak{m}}_{r,t}(t-s) , \ s\in[2r,t-r]] \\ &\leq P[\mathfrak{z}_{x,y}(s) > \underline{\mathfrak{m}}^n_{r,t}(t-s) - s^\delta \wedge (t-s)^\delta , \ s\in[2r,t-r]] . \end{aligned}$$

The right side of (8.14) equals

$$(8.15)\quad P[\mathfrak{z}_{\bar{z},\bar{y}}(s) > \underline{\mathcal{L}}^n_{r,t}(t-s) - s^\delta \wedge (t-s)^\delta , \ s\in[2r,t-r]] ,$$

where $\bar{y}$ and $\bar{z}$ are defined in (8.12). Owing to the explicit nature of $n(s)$, one may without difficulty check that for large r ,

$$|\mathcal{L}^n_{r,t}(t-s)| < 2(s^\delta \wedge (t-s)^\delta) \quad \text{if} \quad s\in[2r,t-2r] .$$

Therefore, by Propositions 6.1 and 6.3, (8.15) is at most

$$C_2 P[\mathfrak{z}_{\bar{z},\bar{y}}(s) > 0 \quad \text{for} \quad s\in[2r,t-2r]] .$$

Applying Lemma 2.2(a) and (2.18), one may conclude that this is at most

$$C_1 r(1-\exp[-2\bar{y}\bar{z}/t])$$

for $\bar{y} \geq 1$, $\bar{z} \geq 1$, and appropriate C_1 . //

(8.21),

$$E[\exp\{\int_0^t k^H(t-s,\mathfrak{z}_{x,y}(s))\,ds\}] \leqq C_5^H e^t(1-\exp[-2\hat{y}\bar{z}/t])$$

for $x \geqq m_{1/2}(t)+1$ and all y. But we are assuming that $u(0,x)=1$ on $(-\infty,0]$ and so $u(s,y') \geqq u^H(s,y')$ for all s and y'. Again by the concavity of f, it follows that $k(s,y') \leqq k^H(s,y')$. Therefore,

$$(8.22) \qquad E[\exp\{\int_0^t k(t-s,\mathfrak{z}_{x,y}(s))\,ds\}] \leqq C_5^H e^t(1-\exp[-2\hat{y}\bar{z}/t])$$

for $x \geqq m_{1/2}(t)+1$ and all y.

We immediately obtain from (8.22) and the Feynman-Kac formula that

$$(8.23) \qquad \begin{aligned} u(t,x) &= \int_{-\infty}^{\infty} u(0,y)\,\frac{e^{-(x-y)^2/2t}}{\sqrt{2\pi t}}\,E[\exp\{\int_0^t k(t-s,\mathfrak{z}_{x,y}(s))\,ds\}]\,dy \\ &\leqq C_5\, e^t \int_{-\infty}^{\infty} u(0,y)\,\frac{e^{-(x-y)^2/2t}}{\sqrt{2\pi t}}\,(1-e^{-2\hat{y}\bar{z}/t})\,dy \end{aligned}$$

for all $x \geqq m_{1/2}(t)+1$. Since we are assuming that (8.3) is satisfied with $y_0 = 0$, it is not difficult to check that for $x \geqq m_{1/2}(t)+1 \geqq t$, this is at most

$$(8.24) \qquad C_6\, e^t \int_0^{\infty} u(0,y)\,\frac{e^{-(x-y)^2/2t}}{\sqrt{2\pi t}}\,(1-e^{-2y\bar{z}/t})\,dy$$

for appropriate C_6. In order to obtain (8.16) from (8.24), we will need to replace $\bar{z}$ by z. To justify this substitution, we need to show that $\bar{z}-z = m_{1/2}(t)-n(t)$ is bounded. First of all, note that for $z_2 = x-2^{1/2}t$, (8.24) may be rewritten as

$$(8.25) \qquad \begin{aligned} &\frac{C_6\, e^{-2^{1/2}z_2}}{\sqrt{2\pi t}} \int_0^{\infty} e^{2^{1/2}y}\,u(0,y)\,e^{-(z_2-y)^2/2t}\,(1-e^{-2y\bar{z}/t})\,dy \\ &\leqq C_6\, t^{-3/2}\bar{z}\, e^{-2^{1/2}z_2} \int_0^{\infty} y\, e^{2^{1/2}y}\,u(0,y)\,dy \,. \end{aligned}$$

By (8.1) and (8.12), this is at most

$$
\begin{aligned}
C_7\, t^{-3/2}\, \bar{z}\, e^{-2^{1/2} z_2} &\leq C_8\, \bar{z}\, e^{-2^{1/2}\bar{z}} \\
&= C_8(x-n(t))\, e^{-2^{1/2}(x-n(t))}
\end{aligned}
\tag{8.26}
$$

for appropriate C_7 , C_8 , and $x \geqq m_{1/2}(t)+1$. Inequalities (8.23)-(8.26) show that for $x \geqq m_{1/2}(t)+1$ and large t ,

$$u(t,x) \leqq C_8(x-n(t))\, e^{-2^{1/2}(x-n(t))} \tag{8.27}$$

Since $n(t) \leqq m_{1/2}(t)$, (8.17) follows. On the other hand, comparison of $u(t,x)$ with $u^H(t,x)$ and (3.20) show that $u(t,m_{1/2}(t))$ is bounded below by a positive constant for large t . It therefore follows from (8.27) that

(8.28) $m_{1/2}(t) - n(t)$ is bounded for large t .

(8.18) thus follows from the definition of $n(t)$. To demonstrate (8.16), note that because of (8.28), $\bar{z} - z$ is bounded. Since $1 - \exp[-2yz/t]$ is concave in z , (8.16) follows from (8.24) for appropriate C_2 . //

We now derive two corollaries from Propositions 8.1 and 8.2. The first corollary gives an analogue of (8.16), but with the inequality going in the opposite direction. The proof is similar to that of Proposition 8.1, although we now require the estimate (8.2).

<u>Corollary 1</u>. Let $u(t,x)$ be a solution to (1.1), (1.2), and (1.2') with initial data satisfying (8.1). Then for $x \geqq m_{1/2}(t)$ and all t ,

$$u(t,x) \geqq C_5\, e^t \int_{\alpha(r,t)}^{\infty} u(0,y)\, \frac{e^{-(x-y)^2/2t}}{\sqrt{2\pi t}}\, (1 - e^{-2\bar{y}z/t})\, dy \tag{8.29}$$

if $\alpha(r,t) = O(\log r)$, where $C_5 > 0$ depends on α , f , and the initial data, but not on t or x .

<u>Proof</u>. By Proposition 8.1 and 8.2, for s chosen large enough,

$$C_0 \leqq m_{1/2}(s) - 2^{1/2}s + 3\cdot 2^{-3/2}\log s \leqq C_4 \quad . \tag{8.30}$$

Since $\alpha(r,t) = O(\log r)$, it is therefore an easy matter to check that $m_{1/2}(s)$ satisfies (7.10). Application of Proposition 7.4 thus shows that for r chosen large enough, $y \geqq \alpha(r,t)$, and $x \geqq m_{1/2}(t)$,

$$\begin{aligned} &P[\mathfrak{z}_{x,y}(s) > \bar{\mathcal{m}}_{r,t}(t-s) \quad \text{for} \quad s \in [2r,t-r]] \\ &\geqq C_6\, P[\mathfrak{z}_{x,y}(s) > \underline{\mathcal{m}}_{r,t}(t-s) \quad \text{for} \quad s \in [2r,t-r]] \end{aligned} \tag{8.31}$$

for appropriate $C_6 > 0$. It moreover follows from (8.30) that

$$\underline{\mathcal{m}}_{r,t}(t-s) \leqq \frac{s}{t}\alpha(r,t) + \frac{t-s}{t}m_{1/2}(t) \quad \text{for} \quad s \in [r,t-r]$$

for large r , and so the right side of (8.31) is at least

$$C_6\, P[\mathfrak{z}_{z,\bar{y}}(s) > 0 \quad \text{for} \quad s \in [r,t-r]] \ , \tag{8.32}$$

where z and $\bar{y}$ are as in (8.12). By Lemma 2.2(a), this last quantity is at least

$$C_7\,(1 - \exp[-2\bar{y}z/t]) \tag{8.33}$$

for appropriate $C_7 > 0$. Combining (8.31)-(8.33), we see that for $y \geqq \alpha(r,t)$ and $x \geqq m_{1/2}(t)$,

$$P[\mathfrak{z}_{x,y}(s) > \bar{\mathcal{m}}_{r,t}(t-s)\ ,\ s \in [2r,t-r]] \geqq C_7(1 - \exp[-2\bar{y}z/t]) \quad . \tag{8.34}$$

On the other hand, by Proposition 7.2,

$$\exp\left[\int_{2r}^{t-r} k(t-s,\mathfrak{z}_{x,y}(s))\,ds\right] \geqq C_8\, e^t \tag{8.35}$$

if $\mathfrak{z}_{x,y}(s) > \bar{\mathcal{m}}_{r,t}(t-s)$ for $2r \leqq s \leqq t-r$, where $C_8 > 0$ (and r is fixed).

Applying (8.34) and (8.35), we see that for $x \geqq m_{1/2}(t)$,

$$u(t,x) = \int_{-\infty}^{\infty} u(0,y) \frac{e^{-(x-y)^2/2t}}{\sqrt{2\pi t}} E[\exp\{\int_0^t k(t-s,\mathfrak{z}_{x,y}(s))\,ds\}]\,dy$$

$$\geqq C_5\, e^t \int_{\alpha(r,t)}^{\infty} u(0,y) \frac{e^{-(x-y)^2/2t}}{\sqrt{2\pi t}} (1 - e^{-2\bar{y}z/t})\,dy \;,$$

which gives us (8.29). //

So far in this section, all our estimates have been accurate only up to a bounded error. In order to be able to demonstrate convergence of $u(t,x+m_{1/2}(t))$ to $w^{2^{1/2}}(x)$, we will need to make precise estimates on the tail of u. Our previous estimates were dependent on Propositions 7.2 and 7.3, and computed upper and lower bounds for u and $m_{1/2}$ in terms of the probabilities of Brownian bridge lying above $\underline{\mathcal{m}}(s)$ and $\bar{\mathcal{m}}(s)$ on $[r,t-2r]$, where r was fixed. In order to improve these estimates, it is sufficient to reapply this reasoning, but we need to allow $r \to \infty$ and somehow also take into account the behavior of $k(s,y)$ on $[0,r]$ and $[t-2r,t]$. As $r \to \infty$, Propositions 7.2 and 7.3 give us sharp results on the behavior of $k(s,y)$ in $[r,t-2r]$. Also, for x chosen large enough relative to r, we will show in Proposition 8.3(a) that $k(t-s,\mathfrak{z}_{x,y}(s)) \sim 1$ for $s \in [0,2r]$ and $\mathfrak{z}_{x,y} > \bar{\mathcal{m}}^x$. The availability of Proposition 7.5 enables us moreover to make asymptotically precise comparisons on the probabilities of Brownian bridge lying above $\underline{\mathcal{m}}'$ and $\bar{\mathcal{m}}^x$ over $[r,t]$. Using these results, we are therefore able to obtain accurate estimates on the behavior of k over the extended interval $[r,t]$ as well. For the case of finite initial mass, the behavior of k over the time interval $[0,r]$ is more difficult to handle. We may however avoid having to make any direct estimates on this interval with the aid of the second corollary to Proposition 8.2 and the corollary to Lemma 8.1. As a consequence of these corollaries, it is shown in part (c) of Proposition 8.3 that the lower and upper bounds for $u(t,x)$ obtained in parts (a) and (b) in terms of $u(r,\cdot)$ are essentially as close as the bounds one obtains by totally ignoring the effect of k over $[0,r]$. (For infinite initial mass, it will be shown in Section 9 that one may on the other hand allow $y \to \infty$ as $t \to \infty$ for the endpoint y, and therefore analyze the interval $[0,r]$ in the same manner as the interval $[t-2r,t]$.)

Corollary 2 is an application of the two-sided bounds on the right tail of $u(t,x)$ given by (8.16) and (8.29), and states that the right tail of $u(t,x)$ is at least as fat as that of the normal distribution.

Corollary 2. Let $u(t,x)$ be a solution of (1.1), (1.2), and (1.2') with initial data satisfying (8.1). Then for $x_2 \geq x_1 \geq 0$ and large enough t,

$$\frac{u(t,x_2)}{u(t,x_1)} \geq C_6 \frac{p^1(t,x_2-y_0)}{p^1(t,x_1-y_0)}, \tag{8.36}$$

where y_0 satisfies (8.3) and C_6 does not depend on t or x.

Proof. Application of (8.16) of Proposition 8.2 together with (8.29) of Corollary 1 (with $\alpha(r,t) = y_0$) shows that for $x_2 \geq x_1 \geq m_{1/2}(t)+1$ and large enough t,

$$\frac{u(t,x_2)}{u(t,x_1)} \geq \frac{C_7 \int_{y_0}^{\infty} u(0,y)\, e^{-(x_2-y)^2/2t}(1-e^{-2\bar{y}z_2/t})\,dy}{\int_{y_0}^{\infty} u(0,y)\, e^{-(x_1-y)^2/2t}(1-e^{-2\bar{y}z_1/t})\,dy}, \tag{8.37}$$

where $\bar{y} = y - y_0$, $z_i = x_i - m_{1/2}(t)$, and $C_7 = C_5/C_2 > 0$. The inequality in (8.37) will continue to hold if z_2 is replaced by z_1 on the right side. Also,

$$e^{-(x_2-y)^2/2t}/e^{-(x_1-y)^2/2t} \quad \text{is increasing in } y .$$

It follows that the right side of (8.37) is at least

$$C_7 e^{-(x_2-y_0)^2/2t}/e^{-(x_1-y_0)^2/2t} = C_7 p^1(t,x_2-y_0)/p^1(t,x_1-y_0)\ ;$$

(8.36) is thus valid over $x_2 \geq x_1 \geq m_{1/2}(t)+1$. On the other hand, Proposition 7.1 and Corollary 2 of Proposition 3.2 imply that for

$0 \leq x \leq m_{1/2}(t)+1$ and large t, $u(t,x)$ is bounded below by some $C_8 > 0$. Also, since $y_0 \leq 0$, $p^1(t,x-y_0)$ is decreasing in $x \geq 0$. Therefore,

$$\frac{u(t,x_2)}{u(t,x_1)} \geq C_8 \geq C_8 \frac{p^1(t,x_2-y_0)}{p^1(t,x_1-y_0)}$$

for $0 \leq x_1 \leq x_2 \leq m_{1/2}(t)+1$. It follows that (8.36) is valid over the entire range $0 \leq x_1 \leq x_2$ with $C_6 = C_7 C_8$. //

The following simple lemma and its corollary will be crucial in allowing us to forego any calculation of $k(s,y)$ on $[0,r]$ when estimating $u(t,x)$.

<u>Lemma 8.1</u>. Assume that $g(x)$ and $h(x)$ are nonnegative and increasing in x. Then for any probability measure μ,

$$\int_{-\infty}^{\infty} g(x)h(x)\mu(dx) \geq \int_{-\infty}^{\infty} g(x)\mu(dx) \cdot \int_{-\infty}^{\infty} h(x)\mu(dx) \quad (8.38)$$

<u>Proof</u>. Since $h(x)$ is increasing in x, it is easy to check that

$$\int_{y}^{\infty} h(x)\mu(dx) \geq \int_{y}^{\infty} \mu(dx) \cdot \int_{-\infty}^{\infty} h(x)\mu(dx) .$$

Since $g(x)$ is also increasing, integration by parts then shows that

$$\int_{-\infty}^{\infty} g(x)h(x)\mu(dx) \geq \int_{-\infty}^{\infty} g(x)\mu(dx) \cdot \int_{-\infty}^{\infty} h(x)\mu(dx) .$$ //

<u>Corollary 1</u>. Assume that $h(x)$ is increasing in x with $0 \leq h(x) \leq 1$, and that $g(x)$ is nonnegative with $g(x_1) \leq C\,g(x_2)$ for $x_1 \leq x_2$ and some constant C. Then, if $\int_{-\infty}^{\infty} h(x)\mu(dx) \geq 1-\epsilon$ where μ is a probability measure,

$$\int_{-\infty}^{\infty} g(x)h(x)\mu(dx) \geq (1-C\epsilon)\int_{-\infty}^{\infty} g(x)\mu(dx) . \quad (8.39)$$

<u>Proof</u>. Introduce $g_1(x) \equiv \sup\{g(y) : y \leq x\}$. By (8.38),

$$\int_{-\infty}^{\infty} g_1(x)h(x)\mu(dx) \geq \int_{-\infty}^{\infty} g_1(x)\mu(dx) \cdot \int_{-\infty}^{\infty} h(x)\mu(dx) ,$$

which by assumption is at least

$$(1-\epsilon)\int_{-\infty}^{\infty} g_1(x)\mu(dx) \quad .$$

Consequently,

$$\int_{-\infty}^{\infty} g(x)(1-h(x))\mu(dx)/\int_{-\infty}^{\infty} g(x)\mu(dx)$$

$$(8.40) \quad \leq C\int_{-\infty}^{\infty} g_1(x)(1-h(x))\mu(dx)/\int_{-\infty}^{\infty} g_1(x)\mu(dx)$$

$$\leq C\epsilon \quad .$$

(8.39) and (8.40) are equivalent. //

In Proposition 8.3, we do much of the dirty work required for Theorem 3. In parts (a) and (b) we obtain lower and upper bounds on $u(t,x)$ which are analogous to (8.29) and (8.16), but are more precise. We proceed in the same manner as in the computation of these earlier bounds, but more carefully. For part (a), Proposition 7.2 is again applied over the time interval $[r,t-2r]$, whereas it is not difficult to show directly that $k(t-s,\mathfrak{z}_{x,y}(s)) \sim 1$ if $s \in [0,2r]$ and $\mathfrak{z}_{x,y}(s)$ is large enough. Together, these estimates show that essentially "full heating" occurs for paths remaining above $\bar{\mathcal{M}}^x$ on $[r,t]$, and thus (a) follows. For part (b), Proposition 7.3 is again applied to show that only paths remaining above $\underline{\mathcal{M}}'$ on $[r,t]$ have a substantial impact on the Feynman-Kac integral. In part (c), the last two corollaries are applied to show that the ratio of the lower and upper bounds for $u(t,x)$ just obtained in (a) and (b) is at least as great as the analogous ratio one obtains upon substituting the Gaussian kernel in for $u(r,\cdot)$. On account of Proposition 7.5, this last ratio $\to 1$ as $r \to \infty$. The bounds computed in parts (a) and (b) for $u(t,x)$ are thus increasingly sharp as $r \to \infty$; it remains to compute these bounds in Theorem 3.

<u>Proposition 8.3</u>. Let $u(t,x)$ be a solution to (1.1), (1.2), and (1.2') with initial data satisfying (8.1). Set $\alpha(r,t) = -\log r$ in (7.6). Then for large r,

(a) $u(t,x) \geq C_1(r)\, e^{t_1} \int_{-\infty}^{\infty} u(r,y) \frac{e^{-(x-y)^2/2t_1}}{\sqrt{2\pi t_1}} P[\mathfrak{z}_{x,y}^{t_1}(s) > \bar{\mathcal{m}}_{r,t}^{x}(t-s)\ , s\in[0,t_1]]\, dy$

if $x \geq m_{1/2}(t)+r$, and

(b) $u(t,x) \leq C_2(r)\, e^{t_1} \int_{-\infty}^{\infty} u(r,y) \frac{e^{-(x-y)^2/2t_1}}{\sqrt{2\pi t_1}} P[\mathfrak{z}_{x,y}^{t_1}(s) > \underline{\mathcal{m}}'_{r,t}(t-s)\ , s\in[0,t_1]]\, dy$

if $x \geq m_{1/2}(t)$, where $t_1 = t-r$, and $C_1(r)\uparrow 1$ and $C_2(r)\downarrow 1$ as $r\uparrow\infty$. Denote by $\psi_1(r;t,x)$ and $\psi_2(r;t,x)$ the right hand sides of (a) and (b). Then for $t \geq 8r$ and $x \geq m_{1/2}(t)+8r$,

(c) $1 \leq \psi_2(r;t,x)/\psi_1(r;t,x) \leq \gamma(r)$,

where $\gamma(r)\downarrow 1$ as $r\uparrow\infty$.

Proof. (a) From the Feynman-Kac formula,

$$u(t,x) = \int_{-\infty}^{\infty} u(r,y) \frac{e^{-(x-y)^2/2t_1}}{\sqrt{2\pi t_1}} E[\exp\{\int_0^{t_1} k(t-s,\mathfrak{z}_{x,y}^{t_1}(s))\, ds\}]\, dy$$

(8.41)

$$\geq \int_{-\infty}^{\infty} u(r,y) \frac{e^{-(x-y)^2/2t_1}}{\sqrt{2\pi t_1}} E[\exp\{\cdot\}\ ;\ \mathfrak{z}_{x,y}^{t_1}(s) > \bar{\mathcal{m}}_{r,t}^{x}(t-s)\ , s\in[0,t_1]]\, dy\ .$$

By Proposition 7.2, if $x(s) > \bar{\mathcal{m}}_{r,t}(t-s)$ on $[2r,t-r]$, then

(7.12) $$e^{3r-t} \exp[\int_{2r}^{t-r} k(t-s,x(s))\, ds] \to 1 \quad \text{as} \quad r\to\infty$$

uniformly in t . Note that from (7.42), $\bar{\mathcal{m}}_{r,t}^{x}(t-s) = \bar{\mathcal{m}}_{r,t}(t-s)$ for $s\in[2r,t-r]$. On the other hand, Propositions 8.1 and 8.2 show that $m_{1/2}(t) - 2^{1/2}t + 3\cdot 2^{-3/2}\log t$ is bounded for large t , and that $u(t,x) \leq C_3\, z\, e^{-2^{1/2}z}$ for $x \geq m_{1/2}(t)+1$ and $z = x - m_{1/2}(t)$. Therefore for z chosen large enough, if $x(s) > \bar{\mathcal{m}}_{r,t}^{x}(t-s) \equiv (x+m_{1/2}(t))/2$ on $[0,2r]$, then

$$u(t-s,x(s)) \leq C_7\, z\, e^{-z/2^{1/2}}$$

for appropriate C_7 . Consequently, by (1.2'), $k(t-s,x(s)) \geq 1 - C_8\, e^{-\rho' z}$ for appropriate $\rho' > 0$ and C_8 . One thus concludes that if $x(s) > \bar{m}^{x}_{r,t}(t-s)$ on $[0,2r]$, then

$$e^{-2r} \exp\left[\int_0^{2r} k(t-s,x(s))\, ds\right] \geq \exp\left[-2C_8\, r\, e^{-\rho' z}\right] \tag{8.42}$$
$$\to 1 \quad \text{as} \quad r \to \infty$$

uniformly in t and $z = x - m_{1/2}(t) \geq r$. Together, (7.12) and (8.42) imply that for $x(s) > \bar{m}^{x}_{r,t}(t-s)$ on $[0,t-r]$,

$$e^{r-t} \exp\left[\int_0^{t-r} k(t-s,x(s))\, ds\right] \to 1 \quad \text{as} \quad r \to \infty \tag{8.43}$$

uniformly in t and $x - m_{1/2}(t) \geq r$. Together with (8.41), (8.43) demonstrates part (a).

(b) In order to demonstrate (b), we employ the inequality

$$u(t,x) \leq C_2'(r) \int_{-\log r}^{\infty} u(0,y)\, \frac{e^{-(x-y)^2/2t}}{\sqrt{2\pi t}}\, E\left[\exp\left\{\int_0^t k(t-s,\mathfrak{z}_{x,y}(s))\, ds\right\}\right] dy \tag{8.44}$$

for $x \geq m_{1/2}(t)$, where $C_2'(r) \downarrow 1$ as $r \uparrow \infty$. To exhibit (8.44), one may show from the maximum principle and a little computation, that

$$\int_{-\infty}^{-\log r} u(0,y)\, \frac{e^{-(x-y)^2/2t}}{\sqrt{2\pi t}}\, E[\exp\{\cdot\}]\, dy \leq u^H(t,x+\log r) \quad , \tag{8.45}$$

where u^H is the solution of (1.1) with Heaviside initial data and forcing term $\bar{f}$ (as defined in (3.50)). From (8.16) and (8.29) with $\alpha(r,t) \equiv y_0$, it follows after a little estimation that

$$u^H(t,x+\log r)/u(t,x) \to 0 \quad \text{as} \quad r \to \infty \tag{8.46}$$

uniformly in t and in $x \geq m_{1/2}(t)$. (8.44) then follows from (8.45), (8.46), and the Feynman-Kac formula.

Proceeding with the demonstration of (b), we employ the Feynman-Kac formula and rewrite $u(t,x)$ as

$$\int_{-\infty}^{\infty} u(0,y) \frac{e^{-(x-y)^2/2t}}{\sqrt{2\pi t}} E[\exp\{\cdot\}; \mathfrak{z}^t_{x,y}(s) > \mathcal{M}'_{r,t}(t-s), s \in [0,t]] dy$$

(8.47)

$$+ \int_{-\infty}^{\infty} u(0,y) \frac{e^{-(x-y)^2/2t}}{\sqrt{2\pi t}} E[\exp\{\cdot\}; G_{x,y}(r,t)] dy ,$$

where $G_{x,y}$ is defined in (7.22). On account of (8.44), one may replace the integral $\int_{-\infty}^{\infty}$ by the upper bound $C'_2(r)\int_{-\log r}^{\infty}$ in both terms. It is easy to see that the first term is at most

$$C'_2(r)\int_{-\infty}^{\infty} u(r,y) \frac{e^{-(x-y)^2/2t_1}}{\sqrt{2\pi t_1}} E[\exp\{\cdot\}; \mathfrak{z}^{t_1}_{x,y}(s) > \mathcal{M}'_{r,t}(t-s), s \in [0,t_1]] dy$$

(8.48)

$$\leq C'_2(r)\, e^{t_1} \int_{-\infty}^{\infty} u(r,y) \frac{e^{-(x-y)^2/2t_1}}{\sqrt{2\pi t_1}} P[\mathfrak{z}^{t_1}_{x,y}(s) > \mathcal{M}'_{r,t}(t-s), s \in [0,t_1]] dy .$$

On the other hand, owing to (8.4) and (8.18), for $\alpha(r,t) = -\log r$, (7.10) holds for large enough r. By Proposition 7.3, the second term in (8.47) is therefore at most

(8.49) $$C'_2(r) \frac{e^t}{r^2} \int_{-\log r}^{\infty} u(0,y) \frac{e^{-(x-y)^2/2t}}{\sqrt{2\pi t}} P[\mathfrak{z}^t_{x,y}(s) > \mathcal{M}_{r,t}(t-s), s \in [2r,t-r]] dy$$

for $x \geq m_{1/2}(t)$ and large enough r. By Corollary 1 of Proposition 8.1, this is at most

(8.50) $$C_1 C'_2(r) \frac{e^t}{r} \int_{-\log r}^{\infty} u(0,y) \frac{e^{-(x-y)^2/2y}}{\sqrt{2\pi t}} (1 - e^{-2\bar{y}\bar{z}/t})\, dy ,$$

where $\bar{y}$ and $\bar{z}$ are defined in (8.12). Moreover, because of (8.4) and (8.18), $z - \bar{z}$ is bounded. Therefore by (8.29), (8.50) is at most

(8.51) $$C_9\, u(t,x)/r$$

for $x \geq m_{1/2}(t)$ and appropriate C_9. Combining the upper bound given by (8.48) for the first term in (8.47) together with the upper bound given by (8.51) for the second term in (8.47), one obtains the bound for $u(t,x)$ asserted in (b).

(c) First, choose $y_0 \leq 0$ so that $\int_{y_0}^{\infty} u(0,y)\,dy > 0$. As in (b), owing to (8.4) and (8.18), (7.10) is satisfied for large enough r for $\alpha(r,t) = -\log r$. Application of Proposition 7.5 therefore shows that

$$(8.52)\qquad \frac{P[\mathfrak{z}^t_{x,y_0}(s) > \bar{\mathcal{M}}^x_{r,t}(t-s)\,,\ s \in [0,t_1]]}{P[\mathfrak{z}^t_{x,y_0}(s) > \mathcal{M}'_{-r,t}(t-s)\,,\ s \in [0,t_1]]} \geq 1 - \epsilon(r)$$

for $t \geq 8r$, $x - m_{1/2}(t) \geq 8r$, and r large enough, where $\epsilon(r) \downarrow 0$ as $r \uparrow \infty$. Now set

$$h^x_{r,t}(y) = \frac{P[\mathfrak{z}^{t_1}_{x,y}(s) > \bar{\mathcal{M}}^x_{r,t}(t-s)\,,\ s \in [0,t_1]]}{P[\mathfrak{z}^{t_1}_{x,y}(s) > \mathcal{M}'_{-r,t}(t-s)\,,\ s \in [0,t_1]]}$$

and

$$\mu^x_{r,t}(dy) = \frac{p^1(r,y-y_0)p^1(t_1,x-y)}{p^1(t,x-y_0)} \cdot \frac{P[\mathfrak{z}^{t_1}_{x,y}(s) > \mathcal{M}'_{-r,t}(t-s)\,,\ s \in [0,t_1]]}{P[\mathfrak{z}^{t}_{x,y_0}(s) > \mathcal{M}'_{-r,t}(t-s)\,,\ s \in [0,t_1]]}\,dy\,.$$

Since $\mathcal{M}'_{-r,t} \leq \bar{\mathcal{M}}^x_{r,t}$, $0 \leq h^x_{r,t}(y) \leq 1$, and by Proposition 6.3, $h^x_{r,t}(y)$ is increasing in y. Moreover, $\mu^x_{r,t}$ is a probability measure, which by Lemma 2.1(f) and (8.52) satisfies

$$\int_{-\infty}^{\infty} h^x_{r,t}(y)\,\mu^x_{r,t}(dy) \geq 1 - \epsilon(r)$$

for $t \geq 8r$, $x - m_{1/2}(t) \geq 8r$, and r large enough. Also, set

$$\begin{aligned} g_r(y) &= u(r,y)/p^1(r,y-y_0) && \text{for } y \geq 0 \\ &= 0 && \text{for } y < 0\,. \end{aligned}$$

By Corollary 2 of Proposition 8.2, $g_r(y_1) \leq C_6^{-1} g_r(y_2)$ for $0 \leq y_1 \leq y_2$ and fixed $C_6 > 0$. Consequently, all the conditions needed in order to apply the corollary to Lemma 8.1 are satisfied. We may therefore conclude that

$$\int_0^\infty u(r,y)p^1(t_1,x-y)\, P[\mathfrak{z}_{x,y}^{t_1}(s) > \bar{\mathfrak{m}}_{r,t}^x(t-s)\,,\ s\in[0,t_1]]\,dy$$

(8.53)

$$\geq (1 - C_6^{-1}\epsilon(r))\int_0^\infty u(r,y)p^1(t_1,x-y)\, P[\mathfrak{z}_{x,y}^{t_1}(s) > \underline{\mathfrak{m}}'_{r,t}(t-s)\,,\ s\in[0,t_1]]\,dy$$

for $t \geq 8r$, $x - m_{1/2}(t) \geq 8r$, and r large enough. Moreover, since $m_{1/2}(t)/t \to 2^{1/2}$ as $t \to \infty$, $p^1(t_1,x-y)$ is decreasing rapidly as y decreases through negative values. Therefore, since $P[\mathfrak{z}_{x,y}^{t_1}(s) > \underline{\mathfrak{m}}'_{r,t}(t-s)]$ is also decreasing, the integrand on the right side of (8.53) is insignificant over the range $(-\infty,0)$ in comparison with over $[0,\infty)$. Reasoning along these lines, one can show without difficulty that the right side of (8.53) is at least

(8.54) $$(1 - \epsilon_1(r))\int_{-\infty}^\infty u(r,y)p^1(t_1,x-y)\, P[\mathfrak{z}_{x,y}^{t_1}(s) > \underline{\mathfrak{m}}'_{r,t}(t-s)\,,\ s\in[0,t_1]]\,dy$$

for $t \geq 8r$, $x - m_{1/2}(t) \geq 8r$, and r large enough, where $\epsilon_1(r) \downarrow 0$ as $r \uparrow \infty$. The right inequality in (c) follows then from (8.53) and (8.54), with $\gamma(r) = C_2(r)/C_1(r)(1-\epsilon_1(r))$. Since $\underline{\mathfrak{m}}' \leq \bar{\mathfrak{m}}^x$, the left inequality in (c) is obvious. //

We are now ready to prove the main result of this section, Theorem 3. Theorem 3 states that under initial data which has finite mass and which satisfies (1.17), the solution of the Kolmogorov equation converges uniformly to the travelling wave moving at rate $2^{1/2}$, where the centering term $m(t)$ may be chosen to be $m(t) = 2^{1/2}t - 3\cdot 2^{-3/2}\log t$. The basic tactic here is the same as was employed in Section 5 to demonstrate convergence to $w^\lambda(x)$, $\lambda > 2^{1/2}$, in Theorem 1. Specifically, we derive accurate enough estimates on u to show that $u(t,x+m(t)) \sim C x e^{-2^{1/2}x}$ for large x, which we know behaves asymptotically like $w^{2^{1/2}}(x)$. We then apply Proposition 3.3 to conclude that $u(t,x+m(t)) \to w^{2^{1/2}}(x)$ uniformly in $x \geq c(t)$, where $c(t) \to -\infty$. Since the initial data is assumed to satisfy (1.17), it then follows from Proposition 7.1 that $u(t,x)$ is nearly monotone for large t, and so this limit extends to all x. But, we have already shown in Proposition 8.3(c) that the lower and upper bounds for $u(t,x)$ given in parts (a) and (b) are both asymptotically sharp. To analyze $u(t,x+m(t))$, it therefore suffices to compute the analogue of the integrals in (a) and (b) for some convenient curve $n(s)$ which satisfies $\underline{\mathcal{m}}' \leq n \leq \bar{\mathcal{m}}^x$. In (8.60), we choose this curve to be linear, in which case the integral reduces down to (8.63). Estimation of this last quantity in (8.64)-(8.69) demonstrates that $u(t,x+m(t)) \sim C x e^{-2^{1/2}x}$ as desired.

<u>Theorem 3</u>. Let $u(t,x)$ be a solution of

$$u_t = \frac{1}{2}u_{xx} + f(u) , \tag{1.1}$$

where

$$\begin{aligned} &f \in C^1[0,1] , \ f(0) = f(1) = 0 , \ f(u) > 0 \ \text{ for } \ 0 < u < 1 , \\ &f'(0) = 1 , \ f'(u) \leq 1 \ \text{ for } \ 0 < u \leq 1 , \end{aligned} \tag{1.2}$$

and

$$1 - f'(u) = O(u^\rho) \ \text{ for some } \ \rho > 0 . \tag{1.2'}$$

Assume that

$$\int_0^\infty y e^{2^{1/2}y} u(0,y)\, dy < \infty , \tag{8.1}$$

and that for some $\eta, M, N > 0$,

$$\int_x^{x+N} u(0,y)\, dy > \eta \ \text{ for } \ x \leq -M . \tag{1.17}$$

Then,

(1.18) $u(t,x+m(t)) \to w^{2^{1/2}}(x)$ uniformly in x as $t \to \infty$,

where

(1.26) $m(t) = 2^{1/2}t - 3\cdot 2^{-3/2}\log t$,

and $w^{2^{1/2}}(x)$ satisfies (1.5) and (1.6).

Note that the particular choice of $w^{2^{1/2}}(x)$ here depends on the initial data $u(0,x)$.

Proof. In order to demonstrate (1.18) it will suffice for us to show that for each $i = i_0, i_0+1, i_0+2, \ldots,$ and $x \in [i,i+1)$,

(8.55a) $C x e^{-2^{1/2}x}\gamma_1^{-1}(x) - \gamma_2^i(t) \leq u(t,x+m(t)) \leq C x e^{-2^{1/2}x}\gamma_1(x) + \gamma_2^i(t)$,

where

(8.55b)
$$\gamma_1(x) \to 1 \text{ as } x \to \infty$$
$$\gamma_2^i(t) \to 0 \text{ as } t \to \infty ,$$

and $C > 0$ and $i_0 \in \mathbb{Z}^+$ are fixed. Owing to Proposition 7.1, (8.55) is equivalent to the apparently stronger assertion

$$C x e^{-2^{1/2}x}\gamma_1^{-1}(x) - \gamma_2(t) \leq u(t,x+m(t)) \leq C x e^{-2^{1/2}x}\gamma_1(x) + \gamma_2(t)$$

for all t and $x \geq i_0$, where γ_1 and γ_2 again satisfy (8.55b). Moreover, by (1.13), $w^{2^{1/2}}(x)$ may be chosen so that $e^{2^{1/2}x}w^{2^{1/2}}(x)/x \to C$ as $x \to \infty$; therefore,

$$\gamma_1^{-1}(x)\, w^{2^{1/2}}(x) - \gamma_2(t) \leq u(t,x+m(t)) \leq \gamma_1(x)\, w^{2^{1/2}}(x) + \gamma_2(t)$$

for $x \geq i_0$, and a possibly different choice of $\gamma_1(x)$ which still satisfies (8.55b). It thus follows from Proposition 3.3 that

$$(3.45)\qquad \sup_{x \geq c(t)} |u(t,x+m(t)) - w^{2^{1/2}}(x)| \to 0 \quad \text{as} \quad t \to \infty$$

for some $c(t)$ where $c(t) \to -\infty$. Applying Proposition 7.1 once again, we see that (3.45) in fact extends to all x, from which (1.18) follows.

In order to demonstrate (8.55), we will apply Propositions 7.2-7.5 together with the estimates on $u(t,x)$ and $m_{1/2}(t)$ derived in this section. We first recall that because of the lower bounds on $k(s,y)$ given by Propositions 7.2 and 8.2, we were able to show in Proposition 8.3(a) that for $t_1 = t-r$ and $x \geq m_{1/2}(t) + r$,

$$(8.56)\qquad \begin{aligned} u(t,x) &\geq C_1(r)\, e^{t_1} \\ &\cdot \int_{-\infty}^{\infty} u(r,y) \frac{e^{-(x-y)^2/2t_1}}{\sqrt{2\pi t_1}} P[\mathfrak{z}^{t_1}_{x,y}(s) > \bar{\mathcal{m}}^{x}_{r,t}(t-s),\ s \in [0,t_1]]\, dy , \end{aligned}$$

where $C_1(r) \uparrow 1$ as $r \uparrow \infty$. By applying Proposition 7.3, we also showed in Proposition 8.3(b) that for $x \geq m_{1/2}(t)$,

$$(8.57)\qquad \begin{aligned} u(t,x) &\leq C_2(r)\, e^{t_1} \\ &\cdot \int_{-\infty}^{\infty} u(r,y) \frac{e^{-(x-y)^2/2t_1}}{\sqrt{2\pi t_1}} P[\mathfrak{z}^{t_1}_{x,y}(s) > \underline{\mathcal{m}}'_{r,t}(t-s),\ s \in [0,t_1]]\, dy , \end{aligned}$$

where $C_2(r) \downarrow 1$ as $r \uparrow \infty$. Denote by $\psi_1(r;t,x)$ the right hand side of (8.56), and by $\psi_2(r;t,x)$ the right hand side of (8.57). With the aid of Proposition 7.5 and Lemma 8.1, we were moreover able to show in Proposition 8.3(c) that

$$(8.58)\qquad 1 \leq \psi_2(r;t,x)/\psi_1(r;t,x) \leq \gamma(r)$$

for r chosen large enough, $t \geq 8r$, and $x - m_{1/2}(t) \geq 8r$, where $\gamma(r) \downarrow 1$ as $r \uparrow \infty$.

$\psi_1(r;t,x)$ and $\psi_2(r;t,x)$ each provide us with asymptotically accurate estimates on $u(t,x)$. To exhibit (8.55), it therefore suffices to compute $\psi(r;t,x)$ for large values of r , t , and x for any ψ which is sandwiched in between ψ_1 and ψ_2 . Here, we choose $\psi(r;t,x)$ equal to

$$(8.59) \qquad e^{t_1}\int_{-\infty}^{\infty} u(r,y)\frac{e^{-(x-y)^2/2t_1}}{\sqrt{2\pi t_1}} P[\mathfrak{z}_{x,y}^{t_1}(s) > n_{r,t}(t-s)\ ,\ s\in[0,t_1]]\,dy\ ,$$

where

$$(8.60) \qquad n_{r,t}(s) = (2^{1/2}t - 3\cdot 2^{-3/2}\log t)(s-r)/(t-r) + 2^{1/2}r(t-s)/(t-r)$$

for $s\in[r,t]$. $n_{r,t}$ is the line segment running from $2^{1/2}r$ at time r to $2^{1/2}t - 3\cdot 2^{-3/2}\log t$ at time t . Since we already know that $m_{1/2}(s) - 2^{1/2}s + 3\cdot 2^{-3/2}\log s$ is bounded for large enough s , it is not difficult to check that

$$(8.61) \qquad \underline{\mathfrak{M}}'_{r,t}(t-s) \leqq n_{r,t}(t-s) \leqq \bar{\mathfrak{M}}^{x}_{r,t}(t-s) \quad \text{for} \quad s\in[0,t-r]\ ,$$

with $r \geqq r_0$ and $x - m_{1/2}(t) \geqq C$ for appropriate r_0 and C . Therefore

$$\psi_1(r;t,x)/C_1(r) \leqq \psi(r;t,x) \leqq \psi_2(r;t,x)/C_2(r)$$

over the same range of values, and so by (8.58),

$$(8.62) \qquad \gamma^{-1}(r) \leqq u(t,x)/\psi(r;t,x) \leqq \gamma(r)$$

for $r \geqq r_1$, $t \geqq 8r$, and $x - m_{1/2}(t) \geqq 8r$, where r_1 is fixed.

Now by Lemma 2.2(a), for $x_1 = x - 2^{1/2}t + 3\cdot 2^{-3/2}\log t > 0$ and $y_1 = y - 2^{1/2}r$, (8.59) equals

$$(8.63) \qquad e^{t_1}\int_{2^{1/2}r}^{\infty} u(r,y)\frac{e^{-(x-y)^2/2t_1}}{\sqrt{2\pi t_1}}(1 - e^{-2x_1y_1/t_1})\,dy\ .$$

Substituting in $x_2 = x - 2^{1/2}t$, it may be checked that (8.63), and hence $\psi(r;t,x)$, equals

$$(8.64)\qquad \sqrt{\frac{t^3}{2\pi t_1^3}}\, e^{-2^{1/2}x_1}$$

$$\cdot \int_0^\infty e^{2^{1/2}y_1} u(r,y_1+2^{1/2}r)\, e^{-(x_2-y_1)^2/2t_1} \cdot t_1(1-e^{-2x_1y_1/t_1})\, dy_1 \quad ,$$

which we write as $C(r;t,x_1)x_1 e^{-2^{1/2}x_1}$. Therefore by (8.62)-(8.64),

$$(8.65)\qquad \gamma^{-1}(r)\, C(r;t,x_1)x_1 e^{-2^{1/2}x_1} \leqq u(t,x) \leqq \gamma(r)\, C(r;t,x_1)x_1 e^{-2^{1/2}x_1}$$

for $r \geqq r_1$, $t \geqq 8r$, and $x - m_{1/2}(t) \geqq 8r$.

We proceed to reduce (8.65) to (8.55). Note that for fixed x_1 , y_1 , and r ,

$$t_1(1-\exp[-2x_1y_1/t_1]) \uparrow 2x_1y_1 \quad \text{as} \quad t \uparrow \infty \quad .$$

Also, by (8.1),

$$\int_0^\infty y\, e^{2^{1/2}y}\, u(r,y)\, dy < \infty$$

for all r . A dominated convergence argument therefore shows that for fixed x_1 and r , (8.64) converges to

$$(8.66)\qquad \sqrt{\frac{2}{\pi}}\, x_1 e^{-2^{1/2}x_1} \int_0^\infty y\, e^{2^{1/2}y}\, u(r,y+2^{1/2}r)\, dy$$

as $t \to \infty$; we rewrite (8.66) as $C(r)x_1 e^{-2^{1/2}x_1}$. It is also not difficult to show that in fact, over $x_1 \in [0,i]$ with i fixed,

$$(8.67)\qquad C(r;t,x_1) \to C(r) \quad \text{uniformly in} \quad x_1 \quad \text{as} \quad t \to \infty \quad .$$

Moreover, since $\gamma(r) \to 1$ as $r \to \infty$, it follows from (8.65) and (8.67) that

$$(8.68)\qquad C(r) \to C \quad \text{as} \quad r \to \infty \quad ,$$

where $0 < C < \infty$.

We are finally in a position to demonstrate (8.55). Set

$$\gamma_1(x_1) = \left(\frac{C(r)}{C} \vee \frac{C}{C(r)}\right) \gamma(r)$$

and

$$\gamma_2^i(t) = 2 \sup\{|C(r;t,x_1) - C(r)| : x_1 \in [i,i+1)\} ,$$

where $r = \lfloor x_1 \rfloor / 8$. If one chooses $i_0 \geqq 8r_1$ large enough so $\gamma(i_0) \leqq 2$, then one can check that by (8.65),

$$\text{(8.69)} \qquad C x_1 e^{-2^{1/2} x_1} \gamma_1^{-1}(x) - \gamma_2^i(t) \leqq u(t,x) \leqq C x_1 e^{-2^{1/2} x_1} \gamma_1(x) + \gamma_2^i(t)$$

for $i \geqq i_0$ and $x_1 \in [i,i+1)$. By (8.68), $\gamma_1(i) \to 1$ as $i \to \infty$, whereas by (8.67), $\gamma_2^i(t) \to 0$ as $t \to \infty$. Therefore, (8.55) holds, and the proof of the theorem is complete.

//

9. CONVERGENCE OF THE KOLMOGOROV EQUATION IN THE CASE $\lambda = 2^{1/2}$ FOR INFINITE INITIAL MASS

In Section 8, we demonstrated convergence of $u(t,x)$ to the travelling wave moving at rate $2^{1/2}$ under the assumption that $u(t,x)$ has finite initial mass, that is,

$$(8.1) \qquad \int_0^\infty y\, e^{2^{1/2}y}\, u(0,y)\, dy < \infty \quad .$$

On the other hand, we already know from Theorem 2 of Section 5 that if $u(t,x)$ is assumed to converge to the travelling wave moving at rate $2^{1/2}$, then

$$(1.19) \qquad \limsup_{t\to\infty} \frac{1}{t} \log \Big[\int_t^{t(1+h)} u(0,y)\, dy \Big] \leq -2^{1/2}$$

for each $h > 0$. In this section, we handle the complement of (8.1) under (1.19), and show that if (under 1.19)

$$(9.1) \qquad \int_0^\infty y\, e^{2^{1/2}y}\, u(0,y)\, dy = \infty \quad ,$$

then $u(t,x)$ also converges to the travelling wave moving at rate $2^{1/2}$. As before, we also assume here that the left tail of $u(0,y)$ satisfies

$$(1.17) \qquad \int_x^{x+N} u(0,y)\, dy > \eta \quad \text{for} \quad x \leq -M \; ,$$

for some choice of $\eta, M, N > 0$. The behavior of the solutions $u(t,x)$ under (8.1) and (9.1) is however not entirely analogous, the difference becoming discernible in the limiting behavior of the centering term $m(t)$. Whereas under (8.1), $m(t)$ may be chosen so that

$$m(t) = 2^{1/2}t - 3\cdot 2^{-3/2} \log t \quad ,$$

under (9.1), $m(t)$ may instead satisfy a variety of behavior, and in fact

$$(9.2) \qquad m(t) - 2^{1/2}t + 3\cdot 2^{-3/2}\log t \to \infty \quad \text{as} \quad t \to \infty$$

in all cases. Later on in this section after demonstrating convergence of $u(t,x)$, we will devote our attention to analyzing the asymptotic behavior of $m(t)$ under specific choices of initial data.

As has been the case throughout the paper, the methodology employed in this section makes heavy use of the Feynman-Kac integral. One would therefore like to be able to proceed much as before, and subdivide the half-plane $R^+ \times R$ more-or-less into regions where $u(s,y) \sim 0$ and $u(s,y) \sim 1$, the "boundary" being given by the median $m_{1/2}(s)$. The term $k(s,y) = f(u(s,y))/u(s,y)$ appearing in the integral will in these regions satisfy $k(s,y) \sim 1$ and $k(s,y) \sim 0$. In computing the Feynman-Kac integral, one is therefore able to ignore paths which proceed too far into the region where $u(s,y) \sim 1$, and instead restrict one's attention to the other paths. This approach thus hinges on being able to obtain sufficiently precise estimates on the behavior of $m_{1/2}(s)$ in order to obtain accurate bounds on the probability that a Brownian bridge always lies above $m_{1/2}(s)$ (or, more precisely, that Brownian bridge lies above the curves $\bar{\mathcal{M}}(s)$ and $\underline{\mathcal{M}}(s)$ which were defined in Section 7, and are close approximations of $m_{1/2}(s)$).

The preliminary results in this section are devoted to demonstrating (9.39) and (9.40) of Proposition 9.3. (9.40) gives an upper bound on the thickness of the right tail of $u(t,x)$ for large time, which is essentially the tail of the travelling wave $w^{2^{1/2}}(x)$. Corollary 2 of Proposition 3.2 on the other hand gives us a lower bound which is asymptotically the same. One may therefore employ Proposition 3.3 to conclude as in previous sections that not just the right tail, but the entire solution $u(t,x)$ must converge to $w^{2^{1/2}}(x)$ under appropriate translation. The details are carried out in Theorem 4. The other estimate (9.39) allows us to estimate $u(t,x)$ by approximating its Feynman-Kac integral so that only paths lying above $m_{1/2}(s)$ are counted, for which the heating term $k(s,y)$ is replaced by 1. (9.39) serves as an implicit formula for $m_{1/2}(t)$. Utilizing the estimates given in Lemmas 9.3 and 9.4, we show in Theorem 5 that $m_{1/2}(t)$ is essentially the only solution of (9.39). Employing (9.39), we then show that one is able to compute $m_{1/2}(t)$ explicitly (up to a constant dependent on $w^{2^{1/2}}$) for a large class of initial data.

In order to obtain the estimates leading up to Proposition 9.3, one first requires some basic knowledge on the behavior of $m_{1/2}(t)$. Information to the effect that $m_{1/2}(t)$ is almost convex, as was the case in Section 8, would be enough. Although such behavior for $m_{1/2}(t)$ is no longer generally the case under infinite initial mass, Proposition 7.6 and Lemma 9.1 furnish an alternative approach. According to Proposition 7.6, those Brownian motion paths whose terminal values lie below appropriately chosen $\beta_r(t)$ may be neglected from the Feynman-Kac integral for $u(t,x)$, as their contribution is insignificant. Lemma 9.1 states on the other hand that the line segment $\ell_{r,t}(s)$ with endpoints $\beta_r(t)$ and $m_{1/2}(t)$ at 0 and t lies essentially above $m_{1/2}(s)$ for intermediate values of s . Therefore $m_{1/2}(s)$, "as seen from" Brownian bridges with endpoints above $\beta_r(t)$ and $m_{1/2}(t)$, is small enough to enable one to apply the upper and lower estimates from Sections 6 and 7 involving the curves $\underline{\mathcal{m}}$ and $\bar{\mathcal{m}}$ to such Brownian bridges. Employing these estimates, one obtains Propositions 9.1 and 9.2, which give upper and lower bounds for $u(t,x)$ which are reminiscient of the inequalities of Propositions 8.3(a) and 8.3(b). Manipulation of these bounds then leads to Proposition 9.3.

In order to state Lemma 9.1, we recall the following terminology from Section 7:

$$(7.58) \qquad \beta_r(t) = \inf\{x:\ x+2^{1/2}s \geqq m_{1/2}(s) \quad \text{for} \quad s\in[r,t]\}$$

and

$$(7.59) \qquad b_r(t) = \inf\{s\in[r,t]:\ \beta_r(t)+2^{1/2}s = m_{1/2}(s)\} \quad .$$

We also introduce the family of line segments

$$(9.3)\qquad \ell_{r,t}(s) = \frac{s}{t}m_{1/2}(t) + \frac{t-s}{t}\beta_r(t) \quad \text{for} \quad 0 \leqq s \leqq t \ .$$

The assertion of the lemma is that $m_{1/2}(s)$ is never (for $s \in [r,t]$) much greater than $\ell_{r,t}(s)$. This follows from Theorem 3 and Proposition 3.4, which give a lower bound on the difference $m_{1/2}(t) - m_{1/2}(s)$.

<u>Lemma 9.1</u>. Let $u(t,x)$ be a solution to (1.1) and (1.2), and let $\ell_{r,t}(s)$ and $\beta_r(t)$ be defined as above. Then if $r \geqq 2$,

$$(9.4)\qquad \ell_{r,t}(s) - m_{1/2}(s) \geqq c_t(s) \quad \text{for} \quad r \leqq s \leqq t - C_1$$

where

$$(9.5)\qquad \begin{aligned} c_t(s) &= C_2 \log s && \text{for} \quad 0 \leqq s \leqq t/2 \\ &= C_2 \log(t-s) && \text{for} \quad t/2 \leqq s \leqq t \ , \end{aligned}$$

and C_1 and C_2 are constants.

<u>Proof</u>. From the definition of $\beta_r(t)$, it is not difficult to check that

$$(9.6)\qquad \ell_{r,t}(s) - m_{1/2}(s) \geqq \frac{s}{t}[(m_{1/2}(t) - m_{1/2}(b_r(t))) - 2^{1/2}(t - b_r(t))]$$

for $r \leqq s \leqq t$. Let $m^H_{1/2}(s)$ denote the median of the solution of (1.1) under Heaviside initial data with forcing term $\underline{f}$ defined as in (3.50). It follows from Proposition 3.4 with $t_0 = s$ that

$$(9.7)\qquad m_{1/2}(t) - m_{1/2}(s) \geqq m^H_{1/2}(t-s) + C_3 \quad \text{if} \quad 2 \leqq s \leqq t - C_1$$

for appropriate constants C_1 and C_3 . We also know from Theorem 3 that the right hand side is at least

$$(9.8)\qquad 2^{1/2}(t-s) - 3\cdot 2^{-3/2}\log(t-s) + C_4$$

for some constant C_4 . Plugging in $s = b_r(t)$ in (9.8) and applying (9.6) - (9.8), we thus obtain

$$\ell_{r,t}(s) - m_{1/2}(s) \geq -3\cdot 2^{-3/2} \tfrac{s}{t} \log(t - b_r(t)) + C_4 \quad \text{for} \quad r \leq s \leq t - C_1 \ .$$

By the concavity of log , this is at least

$$-3\cdot 2^{-3/2} \log(s+1) + C_4 \ ,$$

from which the first line of (9.5) follows.

To demonstrate the second line of (9.5), note that

$$\ell_{r,t}(s) - m_{1/2}(s) = \tfrac{t-s}{t}(\beta_r(t) - m_{1/2}(t)) + m_{1/2}(t) - m_{1/2}(s) \ .$$

By (9.7), (9.8), and the definition of $\beta_r(t)$, this is at least

$$-3\cdot 2^{-3/2} \log(t-s) + C_4 \quad \text{for} \quad r \leq s \leq t - C_1 \ . \qquad //$$

We recall from (3.22') on page 100, that $m_{1/2}(t)/t \to 2^{1/2}$ as $t \to \infty$ under initial data $u(0,x)$ which satisfies (1.19). It is therefore easy to check that $\beta_r(t) = o(t)$, where $o(t)$ is uniform in $r \leq t/2$. We may therefore set

$$\alpha(r,t) = \beta_r(t)$$

in this section, where $\alpha(r,t)$ is the term which was introduced in (7.6) in conjunction with $\underline{\mathfrak{m}}$ and $\bar{\mathfrak{m}}$. Owing to Lemma 9.1, $m_{1/2}(t)$ will satisfy (7.10) in this context. This enables us to employ Propositions 7.4 and 7.5 in this section to obtain bounds on the hitting probabilities of $\underline{\mathfrak{m}}$ and $\bar{\mathfrak{m}}$ by Brownian bridges with appropriate terminal points.

The next result, Lemma 9.2, will be employed to provide an extension of sorts to Proposition 7.6. By Proposition 7.6, one may omit all paths of Brownian motion whose endpoints lie below $\beta_r(t)$ in the computation of the Feynman-Kac integral of $u(t,x)$ for large t . Lemma 9.2 states that in a somewhat different setting - for those Brownian motion paths lying above any of a class of functions h - the relative effect of paths with endpoints bounded above by any fixed M becomes negligible for large t if (9.1) is satisfied. Lemma 9.2 will be applied later on to justify replacement of $\beta_r(t)$ by appropriate $N_r(t) \geq \beta_r(t)$ as the lower limit of integration within the Feynman-Kac integral, where $N_r(t) \to \infty$ as $t \to \infty$. In many instances, it will already be true that $\beta_r(t) \to \infty$ as $t \to \infty$, and so this step will frequently not be necessary. In the demonstration of Lemma 9.2, we are able to replace $h_t(s)$ by the explicit lower bound $2^{1/2}s - s^{\delta}$. Since this bound is nearly linear, application of Proposition 6.1 allows us to instead calculate the hitting probability of a straight line by Brownian bridge. The desired result then follows from some further estimation involving the normal kernel.

Lemma 9.2. Let $u(t,x)$ be a solution to (1.1) and (1.2), with initial data $u(0,x)$ satisfying (9.1). Also, let h_t denote a family of functions with

$$h_t(s) \geq 2^{1/2}s - s^{\delta} \quad \text{on} \quad [r_0, t-r_0] \quad ,$$

where r_0 is fixed. Then, provided that the denominator is nonzero, one may for fixed r_1 and r_2 choose $M(t) \to \infty$ as $t \to \infty$ so that

$$(9.9)\qquad \frac{E[u(0,\mathfrak{X}_x(t))\,;\,\mathfrak{X}_x(t) \leq M(t)\,,\,\mathfrak{X}_x(s) > h_t(t-s) \quad \text{for} \quad s \in [r_1, t-r_2]]}{E[u(0,\mathfrak{X}_x(t))\,;\,\mathfrak{X}_x(s) > h_t(t-s) \quad \text{for} \quad s \in [r_1, t-r_2]]}$$

$$\to 0 \quad \text{as} \quad t \to \infty \quad \text{uniformly in} \quad x \geq 2^{1/2}t - t^{\delta} + 1 \quad .$$

In particular, (9.9) is valid over $x \geq m_{1/2}(t)$ for $h_t(s) = m_{1/2}(s)$ and for $h_t(s) = \underline{\mathcal{m}}_{r,t}(s)$ if $r_2 > 0$.

Proof. In order to demonstrate (9.9), it suffices to show that for each fixed $M \geq 1$,

$$(9.10)\qquad \frac{\int_{-\infty}^{M} u(0,y)\, e^{-(x-y)^2/2t}\, P[\mathfrak{z}_{x,y}(s) > h_t(t-s)\,,\; s \in [r_1, t-r_2]]\, dy}{\int_{M}^{\infty} u(0,y)\, e^{-(x-y)^2/2t}\, P[\mathfrak{z}_{x,y}(s) > h_t(t-s)\,,\; s \in [r_1, t-r_2]]\, dy} \to 0$$

as $t \to \infty$ uniformly in $x \geq 2^{1/2}t - t^{\delta} + 1$. We may assume here that $r_0 \geq r_1$

and $r_0 \geqq r_2$. It then follows from Proposition 6.3 that

$$(9.11) \qquad \frac{P[\mathfrak{z}_{x,y}(s) > h_t(t-s) \quad \text{for} \quad s \in [r_1, t-r_2]]}{P[\mathfrak{z}_{x,y}(s) > 2^{1/2}(t-s) - (t-s)^\delta \quad \text{for} \quad s \in [r_0, t-r_0]]}$$

is increasing in y . Owing to the monotonicity of $P[\mathfrak{z}_{x,y}(s) > h_t(t-s)]$, it therefore suffices for us to instead demonstrate

$$(9.12) \qquad \frac{P[\mathfrak{z}_{x,M}(s) > 2^{1/2}(t-s) - (t-s)^\delta , s \in [r_0, t-r_0]] \int_{-\infty}^{M} u(0,y)\, e^{-(x-y)^2/2t}\, dy}{\int_M^\infty u(0,y)\, e^{-(x-y)^2/2t}\, P[\mathfrak{z}_{x,y}(s) > 2^{1/2}(t-s) - (t-s)^\delta , s \in [r_0, t-r_0]]\, dy}$$
$$\to 0 \quad \text{as} \quad t \to \infty \quad \text{uniformly in} \quad x \geqq 2^{1/2}t - t^\delta + 1$$

rather than (9.9) or (9.10). If we set $z = x - 2^{1/2}t + t^\delta$, then

$$(9.13) \qquad \begin{aligned} &P[\mathfrak{z}_{x,y}(s) > 2^{1/2}(t-s) - (t-s)^\delta \quad \text{for} \quad s \in [r_0, t-r_0]] \\ &= P[\mathfrak{z}_{z,y}(s) > (t-s)t^{\delta-1} - (t-s)^\delta \quad \text{for} \quad s \in [r_0, t-r_0]] \quad . \end{aligned}$$

Also, by Proposition 6.1, for r_0 chosen large enough,

$$(9.14) \qquad \begin{aligned} &P[\mathfrak{z}_{z,y}(s) > 0 \quad \text{for} \quad s \in [r_0, t-r_0]] \\ &\leqq P[\mathfrak{z}_{z,y}(s) > (t-s)t^{\delta-1} - (t-s)^\delta \quad \text{for} \quad s \in [r_0, t-r_0]] \\ &\leqq C_3\, P[\mathfrak{z}_{z,y}(s) > 0 \quad \text{for} \quad s \in [r_0, t-r_0]] \end{aligned}$$

for all t , $y, z \geqq 0$, and some constant C_3 . Lemma 2.6 may be employed to show that

$$(9.15) \qquad \begin{aligned} &P[\mathfrak{z}_{z,y}(s) > 0 \quad \text{for} \quad s \in [0,t]] / P[\mathfrak{z}_{z,y}(s) > 0 \quad \text{for} \quad s \in [r_0, t-r_0]] \\ &\geqq P[\mathfrak{z}_{z,y}(s) > 0 \quad \text{for} \quad s \in [0, r_0] \cup [t-r_0, t]] \quad , \end{aligned}$$

where the last probability is bounded below by a positive constant if $y \geqq 1$, $z \geqq 1$, and r_0 is fixed. Application of (9.13)-(9.15) shows that the left side of (9.12) is at most

$$(9.16)\qquad \frac{C_4\, P[\mathfrak{z}_{z,M}(s) > 0 \quad \text{for} \quad s\in[0,t]]\int_{-\infty}^{M} u(0,y)\, e^{-(x-y)^2/2t}\,dy}{\int_M^{\infty} u(0,y)\, e^{-(x-y)^2/2t}\, P[\mathfrak{z}_{z,y}(s) > 0 \quad \text{for} \quad s\in[0,t]]\,dy}$$

for appropriate C_4 since $M \geq 1$.

To complete the proof, we need to show that (9.16) $\to 0$. To see this, first note that by Lemma 2.2(a),

$$P[\mathfrak{z}_{z,y}(s) > 0 \quad \text{for} \quad 0 \leq s \leq t] = 1 - \exp[-2yz/t] \;.$$

If we restrict y to the range $2M \leq y \leq a(t)$, where $a(t) \to \infty$ as $t \to \infty$ but $a(t) = o(t^{1/2})$, then it may be checked that

$$\frac{P[\mathfrak{z}_{z,y}(s) > 0 \quad \text{for} \quad s\in[0,t]]}{P[\mathfrak{z}_{z,M}(s) > 0 \quad \text{for} \quad s\in[0,t]]}\, e^{-(x-y)^2/2t}$$

$$(9.17)\qquad \geq C_5\, e^{yz/2t} \cdot \frac{1-e^{-2yz/t}}{1-e^{-2yM/t}} \cdot e^{2^{1/2}y} \cdot e^{-(x-M)^2/2t}$$

$$\geq C_6\, y\, e^{2^{1/2}y} \cdot e^{-(x-M)^2/2t}$$

for some $C_5, C_6 > 0$ (which depend on M). Plugging in (9.17), it follows that (9.16) is at most

$$(9.18)\qquad \frac{C_7 \int_{-\infty}^{M} u(0,y)\, e^{-(x-y)^2/2t}\,dy}{e^{-(x-M)^2/2t}\int_{2M}^{a(t)} u(0,y)\, y\, e^{2^{1/2}y}\,dy}$$

for appropriate $C_7 > 0$. Since $x \geq 2^{1/2}t - t^{\delta} \geq t + M$ for large enough t,

$$\int_{-\infty}^{M} u(0,y)\, e^{(x-M)^2/2t}/e^{(x-y)^2/2t}\,dy \leq C_8 \int_{-\infty}^{0} e^{y}\,dy$$

$$< \infty$$

for large t. On the other hand, it follows from (9.1) that as $t \to \infty$ (and hence $a(t) \to \infty$),

$$\int_{2M}^{a(t)} y\, e^{2^{1/2} y}\, u(0,y)\, dy \to \infty \ .$$

We therefore see that (9.18) $\to 0$ uniformly in $x \geqq 2^{1/2}t - t^{\delta} + 1$ as $t \to \infty$. Consequently, (9.16) $\to 0$ as well, which demonstrates (9.9) of the lemma. On account of Theorem 3 and Proposition 3.4, it is not difficult to check that for $\delta > 0$, $m_{1/2}(s) \geqq 2^{1/2}s - s^{\delta}$ and $\underline{\mathcal{m}}_{r,t}(s) \geqq 2^{1/2}s - s^{\delta}$ if $s \in [2r, t-2r]$ and r is chosen large enough. Since $r_2 > 0$ will ensure that the denominator is nonzero, (9.9) is applicable to $m_{1/2}(s)$ and $\underline{\mathcal{m}}_{r,t}(s)$ with $r_0 = 2r$.

//

We now employ Proposition 7.6 and Lemmas 9.1 and 9.2 to compute an upper bound on $u(t,x)$. Because of Proposition 7.6, we know we may for large t omit all paths ending below $\beta_r(t)$ in the Feynman-Kac integral corresponding to $u(t,x)$. On the other hand, owing to Lemma 9.1, we may apply Proposition 7.3 to those paths ending above $\beta_r(t)$. The bound we thus obtain is (9.19), but with $\beta_r(t)$ instead of $N_r(t)$ serving as the lower limit of integration. Employment of Lemma 9.2 then assures us that this lower bound may be increased to $N_r(t)$ for appropriate $N_r(t)$, where $N_r(t) \to \infty$ as $t \to \infty$. Note that since Proposition 7.6 is only stated for concave forcing terms f, Proposition 9.1 and its consequences suffer the same restriction. (In Theorem 4 we will remove this restriction by means of a domination argument which compares $u(t,x)$ with solutions with concave f.)

Proposition 9.1. Let $u(t,x)$ be a solution to (1.1), where f satisfies (1.2) and is concave. Also, assume that the initial data $u(0,x)$ satisfy (1.19) and (9.1). Then for $x \geqq m_{1/2}(t)$ and large enough r,

$$(9.19)\quad \begin{aligned} u(t,x) &\leqq C_1(r)\, e^{t} \\ &\cdot \int_{N_r(t)}^{\infty} u(0,y)\, \frac{e^{-(x-y)^2/2t}}{\sqrt{2\pi t}}\, P[\mathfrak{z}_{x,y}(s) > \underline{\mathcal{m}}_{r,t}(t-s),\ 2r \leqq s \leqq t-r]\, dy \ , \end{aligned}$$

where $C_1(r) \to 1$ as $r \to \infty$ and $N_r(t) \geqq \beta_r(t)$ with $N_r(t) \to \infty$ as $t \to \infty$.

Proof. Since f is concave, it follows from Proposition 7.6 that

$$u(t,x) \leqq C_2(r)\, E[\exp\{\int_0^t k(t-s,\mathfrak{X}_x(s))\,ds\}\, u(0,\mathfrak{X}_x(t))\, ; \mathfrak{X}_x(t) > \beta_r(t)]$$

(9.20)
$$= C_2(r) \int_{\beta_r(t)}^{\infty} u(0,y) \frac{e^{-(x-y)^2/2t}}{\sqrt{2\pi t}} E[\exp\{\int_0^t k(t-s,\mathfrak{z}_{x,y}(s))\,ds\}]\,dy$$

for $x \geqq m_{1/2}(t)$, where $C_2(r) \to 1$ as $r \to \infty$. On the other hand by Lemma 9.1,

(9.4) $$\ell_{r,t}(s) - m_{1/2}(s) \geqq c_t(s) ,$$

where $c_t(s)$ satisfies (9.5). (7.10) is therefore satisfied with $\alpha(r,t) = \beta_r(t)$. Invoking the notation of Section 7, it thus follows from Proposition 7.3 that

(9.21) $$E[\exp\{\int_0^t k(t-s,\mathfrak{z}_{x,y}(s))\,ds\}; G_{x,y}(r,t)] \leqq e^t P[G^c_{x,y}(r,t)]/r^2$$

for $y \geqq \beta_r(t)$, $x \geqq m_{1/2}(t)$, and large enough r. (9.21) states that the combined effect on the Feynman-Kac integral of those paths crossing $\underline{\mathcal{M}}_{r,t}$ in the range $[r,t-2r]$ is small in relation to the other paths, and becomes negligible as $r \to \infty$. Applying (9.21), and keeping in mind that $k \leqq 1$, if follows that (9.20) is at most

(9.22) $$C_3(r)\, e^t \int_{\beta_r(t)}^{\infty} u(0,y) \frac{e^{-(x-y)^2/2t}}{\sqrt{2\pi t}} P[\mathfrak{z}_{x,y}(s) > \underline{\mathcal{M}}_{r,t}(t-s),\ 2r \leqq s \leqq t-r]\,dy$$

for $x \geqq m_{1/2}(t)$, where $C_3(r) \to 1$ as $r \to \infty$. We now apply Lemma 9.2 to (9.22), with $r_1 = 2r$, $r_2 = r$, and $M(t) = M_r(t)$. Setting $N_r(t) = \beta_r(t) \vee M_r(t)$, we see that (9.22) is at most

$$C_1(r)\, e^t \int_{N_r(t)}^{\infty} u(0,y) \frac{e^{-(x-y)^2/2t}}{\sqrt{2\pi t}} P[\mathfrak{z}_{x,y}(s) > \underline{\mathcal{M}}_{r,t}(t-s),\ 2r \leqq s \leqq t-r]\,dy$$

for $x \geqq m_{1/2}(t)$, where $C_1(r) \to 1$ as $r \to \infty$. //

In order to obtain the desired tail estimate given in (9.40) of Proposition 9.3, we will also need the following corollary, which gives an upper bound on $u(t,x_2)$ which decreases exponentially quickly in x_2 for $x_2 \geq m_{1/2}(t)$. Here and later on, z_i will denote $x_i - m_{1/2}(t)$.

Corollary 1. Let $u(t,x)$ be a solution to (1.1), where f satisfies (1.2) and is concave. Also, assume that the initial data $u(0,x)$ satisfy (1.19) and (9.1). Choose $\delta_1 > 0$. If t and r are chosen large enough, then under $m_{1/2}(t) < x_1 \leq x_2$,

$$u(t,x_2) \leq e^{-\delta_2 t} + C_1(r)(2\pi t)^{-1/2}\, e^t \frac{z_2}{z_1} \exp[-(2^{1/2}-\delta_1)(z_2-z_1)] \tag{9.23}$$
$$\cdot \int_{N_r(t)}^{\infty} u(0,y)\, e^{-(x-y)^2/2t}\, P[\mathfrak{z}_{x_1,y}(s) > \mathcal{m}_{r,t}(t-s),\ 2r \leq s \leq t-r]\, dy ,$$

where $C_1(r)$ and $N_r(t)$ are as in Proposition 9.1, and $\delta_2 > 0$.

Proof. By Proposition 9.1,

$$u(t,x_2) \leq C_1(r)\, e^t \tag{9.24}$$
$$\cdot \int_{N_r(t)}^{\infty} u(0,y) \frac{e^{-(x_2-y)^2/2t}}{\sqrt{2\pi t}} P[\mathfrak{z}_{x_2,y}(s) > \mathcal{m}_{r,t}(t-s),\ 2r \leq s \leq t-r]\, dy ,$$

where $C_1(r) \to 1$ as $r \to \infty$ and $\beta_r(t) \leq N_r(t) \to \infty$ as $t \to \infty$. In order to derive (9.23) from (9.24), we first decompose the above integral into the pieces $\int_{\delta t}^{\infty}$ and $\int_{N_r(t)}^{\delta t}$, with $\delta = \delta_1/2$.

Estimation of $\int_{\delta t}^{\infty}$ is not difficult. We note that on account of (1.19) and (3.22') on page 100, $m_{1/2}(t)/t \to 2^{1/2}$ as $t \to \infty$. It therefore follows from (4.7) and a little computation, that for appropriate $\delta_2 > 0$ and large enough t (depending on δ),

$$e^t \int_{\delta t}^{\infty} u(0,y) \frac{e^{-(x-y)^2/2t}}{\sqrt{2\pi t}}\, dy \leq e^{-\delta_2 t} \tag{9.25}$$

if $x_2 \geq m_{1/2}(t)$.

Reduction of $\int_{N_r(t)}^{\delta t}$ to the form given in (9.23) requires substitution of x_1 for x_2 in the kernel and in $P[\cdot]$. Owing to (9.4) and (9.5), it is not difficult to show that

$$\text{(9.26)} \qquad \mathcal{m}_{r,t}(t-s) - \frac{s}{t}\beta_r(t) - \frac{t-s}{t}m_{1/2}(t) \leq 0 \quad \text{for} \quad 2r \leq s \leq t-r$$

and r chosen large enough. Therefore by Proposition 6.3,

$$\frac{P[\mathfrak{z}_{x_2,y}(s) > \mathcal{m}_{r,t}(t-s)\,,\, 2r \leq s \leq t-r]}{P[\mathfrak{z}_{x_1,y}(s) > \mathcal{m}_{r,t}(t-s)\,,\, 2r \leq s \leq t-r]} \leq \frac{P[\mathfrak{z}_{z_2,\bar{y}}(s) > 0\,,\, 0 \leq s \leq t]}{P[\mathfrak{z}_{z_1,\bar{y}}(s) > 0\,,\, 0 \leq s \leq t]} ,$$

where $z_i = x_i - m_{1/2}(t)$ and $\bar{y} = y - \beta_r(t)$. Since $z_i, \bar{y} > 0$, it is a simple consequence of Lemma 2.2(a) that this ratio is at most z_2/z_1 . Substitution thus shows that

$$\text{(9.27)} \quad \begin{aligned} &\int_{N_r(t)}^{\delta t} u(0,y)\, \frac{e^{-(x_2-y)^2/2t}}{\sqrt{2\pi t}}\, P[\mathfrak{z}_{x_2,y}(s) > \mathcal{m}_{r,t}(t-s)\,,\, 2r \leq s \leq t-r]\, dy \\ &\leq \frac{z_2}{z_1} \int_{N_r(t)}^{\delta t} u(0,y)\, \frac{e^{-(x_2-y)^2/2t}}{\sqrt{2\pi t}}\, P[\mathfrak{z}_{x_1,y}(s) > \mathcal{m}_{r,t}(t-s)\,,\, 2r \leq s \leq t-r]\, dy \ . \end{aligned}$$

On the other hand, since $m_{1/2}(t)/t \to 2^{1/2}$, it is easy to check that for $y \leq \delta t$, $x_2 \geq x_1 \geq m_{1/2}(t)$, and t chosen large enough,

$$\text{(9.28)} \qquad e^{(x_1-y)^2/2t} / e^{(x_2-y)^2/2t} \leq \exp[-(2^{1/2} - \delta_1)(z_2 - z_1)] \ .$$

The right side of (9.27) is therefore at most

$$\text{(9.29)} \quad \begin{aligned} &\frac{z_2}{z_1} \exp[-(2^{1/2} - \delta_1)(z_2 - z_1)] \\ &\cdot \int_{N_r(t)}^{\delta t} u(0,y)\, \frac{e^{-(x_1-y)^2/2t}}{\sqrt{2\pi t}}\, P[\mathfrak{z}_{x_1,y}(s) > \mathcal{m}_{r,t}(t-s)\,,\, 2r \leq s \leq t-r]\, dy \ . \end{aligned}$$

(9.23) follows from (9.24)-(9.29). //

Proposition 9.2 gives a lower bound for $u(t,x)$ in terms of $P[A_{r,t}(x,y)]$, where

$$(9.30) \qquad A_{r,t}(x,y) = \{\mathfrak{z}_{x,y} : \mathfrak{z}_{x,y}(s) > \bar{\mathcal{m}}_{r,t}^{x,y}(t-s) \quad \text{for} \quad 0 \leq s \leq t\}$$

with $\bar{\mathcal{m}}_{r,t}^{x,y}(s)$ defined as in (7.43). Together with Proposition 9.1 and its corollary, Proposition 9.2 furnishes the main bounds required for Proposition 9.3. As does Proposition 9.1, Proposition 9.2 utilizes the results in Section 7; this time Proposition 7.2 furnishes the crucial estimate. The bound C_3 obtained here is unfortunately not completely satisfactory. To derive Proposition 9.3, we will need the stronger bound given by $C_3 = 1$. This improved estimate will be computed from the cruder version given here in the course of demonstrating Proposition 9.3.

Proposition 9.2. Let $u(t,x)$ be a solution to (1.1), (1.2), and (1.2'), with initial data satisfying (1.19) and (9.1). Then for $x \geq m_{1/2}(t)$,

$$(9.31) \qquad u(t,x) \geq C_3(r,t)\, e^t \int_{N_r(t)}^{\infty} u(0,y) \frac{e^{-(x-y)^2/2t}}{\sqrt{2\pi t}} P[A_{r,t}(x,y)]\, dy ,$$

where

$$(9.32) \qquad \lim_{r\to\infty} \lim_{t\to\infty} C_3(r,t) = C_3 ,$$

$N_r(t)$ is defined as in Proposition 9.1 with $N_r(t) \to \infty$ as $t \to \infty$, and $C_3 > 0$ is chosen so as to satisfy (9.38).

Proof. By the Feynman-Kac formula,

$$u(t,x) = \int_{-\infty}^{\infty} u(0,y) \frac{e^{-(x-y)^2/2t}}{\sqrt{2\pi t}} E[\exp\{\int_0^t k(t-s,\mathfrak{z}_{x,y}(s))\, ds\}]\, dy .$$

This is at least

$$\int_{N_r(t)}^{\infty} u(0,y)\,\frac{e^{-(x-y)^2/2t}}{\sqrt{2\pi t}}\,E[\exp\{\int_0^t \cdots\}\,;\,A_{r,t}(x,y)]\,dy$$

$$\geq \int_{N_r(t)}^{\infty} u(0,y)\,\frac{e^{-(x-y)^2/2t}}{\sqrt{2\pi t}}\,P[A_{r,t}(x,y)] \tag{8.61}$$

$$\cdot \inf\{\exp[\int_0^{2r} + \int_{2r}^{t-r} + \int_{t-r}^{t}]:\, \mathfrak{z}_{x,y}\in A_{r,t}(x,y)\}\,dy \quad .$$

We proceed to compute lower bounds over $A_{r,t}(x,y)$ for each of the above three integrals. We first see that from Proposition 7.2,

$$\inf\{\exp[\int_{2r}^{t-r} k(t-s,\mathfrak{z}_{x,y}(s))\,ds]:\, \mathfrak{z}_{x,y}\in A_{r,t}(x,y)\} \geq C_4(r)\,e^{t-3r} \quad , \tag{9.34}$$

where $C_4(r)\to 1$ as $r\to\infty$. (Note that $\bar{\mathcal{M}}^{x,y}_{r,t}(s) = \bar{\mathcal{M}}_{r,t}(s)$ on $[r,t-2r]$.) On the other hand on account of (1.19), it follows from Lemma 4.1 (with $\delta = 0$) that for fixed r and appropriate $\epsilon(N)\to 0$ as $N\to\infty$,

$$\sup\{u(s,y') : \epsilon(N) \leq s \leq r \ ,\ y' \geq N\} \to 0 \quad \text{as} \quad N\to\infty \ .$$

Therefore,

$$\inf\{k(s,y') : \epsilon(N) \leq s \leq r \ ,\ y' \geq N\} \to 1 \quad \text{as} \quad N\to\infty \ . \tag{9.35}$$

Since $N_r(t)\to\infty$ as $t\to\infty$, $\bar{\mathcal{M}}^{x,y}_{r,t}(s)\to\infty$ uniformly over $0\leq s\leq r$ and $y\geq N_r(t)$ as $t\to\infty$. It thus follows from (9.35) that

$$\inf\{\exp[\int_{t-r}^{t} k(t-s,\mathfrak{z}_{x,y}(s))\,ds]:\, \mathfrak{z}_{x,y}\in A_{r,t}(x,y)\} \to e^r \tag{9.36}$$

uniformly over $y\geq N_r(t)$ as $t\to\infty$. In order to obtain a lower bound over the interval $[0,2r]$, we note that from (3.66) of Proposition 3.4 (with $x = 0$), for $y'\geq m_{1/2}(t)$ and $t-s\geq 1$,

$$u(t-s,y') \leq C_5\,e^{-s} \tag{9.37}$$

for some $C_5 > 0$. On account of (1.2'), this implies that

$$k(t-s,y') \geq 1 - C_6\,e^{-\rho s} \quad ,$$

where $\rho, C_6 > 0$. Consequently, for $x \geqq m_{1/2}(t)$,

$$(9.38) \qquad \inf\{\exp[\int_0^{2r} k(t-s,\mathfrak{z}_{x,y}(s))\,ds] : \mathfrak{z}_{x,y} \in A_{r,t}(x,y)\} \geqq C_3 e^{2r}$$

for some $C_3 > 0$. Application of (9.34), (9.36), and (9.38) to (9.33) shows that for $x \geqq m_{1/2}(t)$,

$$u(t,x) \geqq C_3(r,t)\, e^t \int_{N_r(t)}^{\infty} u(0,y) \frac{e^{-(x-y)^2/2t}}{\sqrt{2\pi t}} P[A_{r,t}(x,y)]\,dy ,$$

where

$$\lim_{r\to\infty} \lim_{t\to\infty} C_3(r,t) = C_3 > 0 . \qquad //$$

We are now finally in a position to demonstrate Proposition 9.3. The first result (9.39) reduces computation of $u(t,x)$ to computation of the hitting probability of $m_{1/2}(s)$ by Brownian motion. To demonstrate (9.39), we first strengthen Proposition 9.2 so that the bound C_3 given there is replaced by 1 . The desired result then follows by applying Proposition 7.5 in conjunction with Propositions 9.1 and 9.2, and a little work. The other estimate (9.40) is an upper bound on the tail of $u(t,x)$, and will be applied together with Propositions 3.2 and 3.3 to demonstrate convergence of $u(t,x+m_{1/2}(t))$ to $w^{2^{1/2}}(x)$. The bound is obtained by applying the corollary of Proposition 9.1 to the estimate given in (9.43). As previously, we will use the notation $z_{\cdot} = x_{\cdot} - m_{1/2}(t)$.

<u>Proposition 9.3</u>. Let $u(t,x)$ be a solution to (1.1), where f satisfies (1.2) and (1.2'), and is concave. Also, assume that the initial data $u(0,x)$ satisfy (1.19) and (9.1). Then for fixed $s_0 > 0$,

$$(9.39a) \quad u(t,x) = C_4(t,z)\, e^t E[u(0,\mathfrak{X}_x(t)) ; \mathfrak{X}_x(s) > m_{1/2}(t-s) ,\ 0 \leqq s \leqq t-s_0] ,$$

where

$$(9.39b) \qquad \lim_{t,z\to\infty} C_4(t,z) = 1 .$$

For each fixed $M_1 > 0$, this is at most

$$C_5(t,z_1)\frac{z}{z_1}e^{-2^{1/2}(z-z_1)}u(t,x_1) \tag{9.40a}$$

for $m_{1/2}(t) < x_1 \leq x \leq x_1 + M_1$, where

$$\lim_{z_1\to\infty}\lim_{t\to\infty} C_5(t,z_1) = 1 . \tag{9.40b}$$

<u>Proof</u>. We first recall from Propositions 9.1 and 9.2 that for $x \geq m_{1/2}(t)$ and large r,

$$\begin{aligned} &C_3(r,t)\,e^t\int_{N_r(t)}^{\infty} u(0,y)\frac{e^{-(x-y)^2/2t}}{\sqrt{2\pi t}} P[\mathfrak{z}_{x,y}(s) > \bar{\mathcal{M}}^{x,y}_{r,t}(t-s)\,,\ 0 \leq s \leq t]\,dy \\ &\leq u(t,x) \\ &\leq C_1(r)\,e^t\int_{N_r(t)}^{\infty} u(0,y)\frac{e^{-(x-y)^2/2t}}{\sqrt{2\pi t}} P[\mathfrak{z}_{x,y}(s) > \underline{\mathcal{M}}_{r,t}(t-s)\,,\ 2r \leq s \leq t-r]\,dy\ , \end{aligned} \tag{9.41}$$

where $\beta_r(t) \leq N_r(t) \to \infty$ as $t \to \infty$, $C_1(r) \to 1$ as $r \to \infty$, and

$$\lim_{r\to\infty}\lim_{t\to\infty} C_3(r,t) = C_3\ , \tag{9.32}$$

with $C_3 > 0$ satisfying (9.38). We set out to improve on (9.41) by replacing C_3 by 1 in (9.32). By Proposition 7.5,

$$\frac{P[\mathfrak{z}_{x,y}(s) > \bar{\mathcal{M}}^{x,y}_{r,t}(t-s) \quad \text{for} \quad 0 \leq s \leq t]}{P[\mathfrak{z}_{x,y}(s) > \underline{\mathcal{M}}_{r,t}(t-s) \quad \text{for} \quad 2r \leq s \leq t-r]} \to 1 \tag{9.42}$$

as $t, y_1, z \to \infty$ and then $r \to \infty$, where we set $z = x - m_{1/2}(t)$, and $y_1 = y$ if $y \geq \beta_r(t)$ and $y_1 = 0$ if $y < \beta_r(t)$. (Recall that $\underline{\mathcal{M}}'_{r,t}(s) = -\infty$ for $s \in [0,r) \cup (t-2r,t]$.) Together, (9.41) and (9.42) imply that

$$\begin{aligned} u(t,x) &= C_6(r,t,z)\,e^t \\ &\cdot \int_{N_r(t)}^{\infty} u(0,y)\frac{e^{-(x-y)^2/2t}}{\sqrt{2\pi t}} P[\mathfrak{z}_{x,y}(s) > \underline{\mathcal{M}}_{r,t}(t-s)\,,\ 2r \leq s \leq t-r]\,dy\ , \end{aligned} \tag{9.43}$$

where

$$(9.44)\qquad \liminf_{r\to\infty}\ \liminf_{t,z\to\infty} C_6(r,t,z) \geqq C_3 \ .$$

Application of the corollary of Proposition 9.1 to (9.43) then shows that for large enough s and z,

$$(9.45)\qquad u(s,m_{1/2}(s)+z) \leqq C_7(r)z\,e^{-z} + e^{-\delta_2 s} \ ,$$

where $C_7(r) > 0$ and $\delta_2 > 0$. But, by (3.66) of Proposition 3.4, for large enough r, $m_{1/2}(t-r) \leqq m_{1/2}(t)$. Application of (9.45) at time $s = t-r$ and a little work therefore shows that for each r and $\epsilon > 0$,

$$(9.46)\qquad u(s,m_{1/2}(t)+z) \leqq \epsilon \quad \text{for} \quad t-r \leqq s \leqq t$$

for large enough z and t. Consequently, for $A_{r,t}(x,y)$ as defined in (9.30),

$$(9.47)\qquad \inf\{\exp[\int_0^{2r} k(t-s,\mathfrak{z}_{x,y}(s))\,ds] : \mathfrak{z}_{x,y} \in A_{r,t}(x,y)\} \to e^{2r}$$

as $t \to \infty$ and $z \to \infty$, with convergence being uniform in y. We may now replace (9.38) of Proposition 9.2 with the improved estimate given in (9.47). Together with (9.34) and (9.36) of Proposition 9.2, the limit in (9.47) shows that

$$(9.48)\qquad u(t,x) = C_8(r,t,z)\,e^{t}\int_{N_r(t)}^{\infty} u(0,y)\,\frac{e^{-(x-y)^2/2t}}{\sqrt{2\pi t}}\,P[A_{r,t}(x,y)]\,dy \ ,$$

where

$$(9.49)\qquad \liminf_{r\to\infty}\ \liminf_{t,z\to\infty} C_8(r,t,z) \geqq 1 \ .$$

Formulas (9.48) and (9.49) present us with an improved version of the left inequality of (9.41) and of (9.32). Since we know that $C_1(r) \to 1$ as $r \to \infty$, a second application of (9.42) shows that in fact

$$(9.50)\qquad \lim_{r\to\infty}\ \overline{\lim_{t,z\to\infty}}\ C_8(r,t,z) = 1 \ .$$

The same reasoning also shows that

$$(9.51)\qquad \lim_{r\to\infty}\ \overline{\lim_{t,z\to\infty}}\ C_6(r,t,z) = 1 \ .$$

Furthermore, on account of (9.46), it is not difficult to show that for given s_o and large enough y ,

$$\underline{\mathcal{m}}'_{r,t}(s) \leq m_{1/2}(s) \leq \bar{\mathcal{m}}^{x,y}_{r,t}(s) \quad \text{on} \quad [s_o,r]\cup[t-2r,t] \ .$$

One may therefore repeat the reasoning employed in Proposition 7.5 to obtain the analogue of (7.46), but with either $\underline{\mathcal{m}}'_{r,t}(s)$ or $\bar{\mathcal{m}}^{x_o,y_o}_{r,t}(s)$ replaced by $m_{1/2}(s)$. It therefore follows from (9.48) that

$$(9.52)\qquad \begin{aligned} u(t,x) &= C_9(r,t,z)\, e^t \\ &\cdot \int_{N_r(t)}^{\infty} u(0,y)\, \frac{e^{-(x-y)^2/2t}}{\sqrt{2\pi t}}\, P[\mathfrak{z}_{x,y}(s) > m_{1/2}(t-s)\ ,\ 0 \leq s \leq t-s_o]\, dy \ , \end{aligned}$$

where $C_9(r,t,z)$ satisfies the equivalent of (9.50).

In order to reduce (9.52) to (9.39), we still need to justify replacement of $N_r(t)$ by $-\infty$ as the lower limit of integration. Recall that $N_r(t) = M_r(t) \vee \beta_r(t)$, where $M_r(t)$ was chosen so that $M_r(t) \to \infty$ as $t \to \infty$, with its growth being slow enough so as to allow application of Lemma 9.2 to (9.22) to replace the lower limit $\beta_r(t)$ by $N_r(t)$. Since Lemma 9.2 is also applicable to $m_{1/2}(t)$, one may also assume that $M_r(t)$ is chosen to grow slowly enough so as to allow substitution of $\beta_r(t)$ for $N_r(t)$ in (9.52). By doing a little estimation on the integrand at time $b_r(t)$ (defined in (7.59)), one can moreover justify extension of $\beta_r(t)$ to $-\infty$ in (9.52). One thus obtains

$$u(t,x) = C_4(r,t,z)\, e^t$$

$$\cdot \int_{-\infty}^{\infty} u(0,y) \frac{e^{-(x-y)^2/2t}}{\sqrt{2\pi t}} P[\mathfrak{z}_{x,y}(s) > m_{1/2}(t-s)\,,\, 0 \leqq s \leqq t-s_o]\, dy \quad ,$$

where $C_4(r,t,z)$ satisfies the analogue of (9.50). Finally, since the integral no longer depends on r, neither does C_4, and therefore (9.39) is valid.

In order to exhibit (9.40), one may apply the corollary of Proposition 9.1 to (9.43) to obtain

$$(9.53) \qquad u(t,x) \leqq C_{10}(t,z_1) \frac{z}{z_1} \exp[-(2^{1/2} - \delta_1)(z-z_1)]\, u(t,x_1) + e^{-\delta_2 t}$$

for $m_{1/2}(t) < x_1 \leqq x$, where $\delta_1, \delta_2 > 0$. On account of (9.51),

$$\lim_{t,z_1 \to \infty} C_{10}(t,z_1) = 1 \quad .$$

By Corollary 2 of Proposition 3.2, for fixed z_1, $u(t,x_1)$ remains bounded away from 0 as $t \to \infty$. By imposing the restriction $x \leqq x_1 + M_1$ and letting $\delta_1 \to 0$ as $t \to \infty$, it is therefore not difficult to see that (9.40) is also valid. //

Having established Proposition 9.3, we can now demonstrate Theorem 4 without difficulty. We show that for initial data satisfying (1.17), (1.19), and (9.1), $u(t,x)$ converges under appropriate translation to a travelling wave. The basic approach taken here has already been used repeatedly - first show that the right tail of $u(t,x)$ approaches that of $w^{2^{1/2}}(x)$, and then apply Proposition 3.3 to show that the entire solution approaches $w^{2^{1/2}}(x)$. In the case of concave f, (9.40) gives a sufficiently precise upper bound on the right tail of $u(t,x)$ for our needs. On the other hand, comparison of $u(t,x)$ to the solution with Heaviside initial data shows that the corresponding lower bound on the right tail of $u(t,x)$ also holds. Together, these bounds yield $u(t,x+m_{1/2}(t)) \sim Cxe^{-2^{1/2}x}$, which is asymptotically equivalent to the right tail of $w^{2^{1/2}}(x)$. To extend the theorem to nonconcave f, one applies the result to the solutions with forcing terms $\underline{f}$ and $\bar{f}$; these terms are concave, and satisfy $\underline{f} \leq f \leq \bar{f}$. One then uses (9.39) to squeeze the right tail of $u(t,x)$ between those corresponding to the concave forcing terms, from which the result in the general case follows.

Theorem 4. Let $u(t,x)$ be a solution of

$$(1.1) \qquad u_t = \frac{1}{2}u_{xx} + f(u) ,$$

where

$$(1.2) \qquad \begin{gathered} f \in C^1[0,1] , \ f(0) = f(1) = 0 , \ f(u) > 0 \ \text{for} \ 0 < u < 1 , \\ f'(0) = 1 , \ f'(u) \leq 1 \ \text{for} \ 0 < u \leq 1 , \end{gathered}$$

and

$$(1.2') \qquad 1 - f'(u) = O(u^\rho) \ \text{for some} \ \rho > 0 .$$

Assume that for some $h > 0$,

$$(1.19) \qquad \limsup_{t\to\infty} \frac{1}{t} \log [\int_t^{t(1+h)} u(0,y)\, dy] \leq -2^{1/2} ,$$

and for some $\eta, M, N > 0$,

$$(1.17) \qquad \int_x^{x+N} u(0,y)\, dy > \eta \ \text{for} \ x \leq -M .$$

Also, assume that

$$(9.1) \qquad \int_0^\infty y e^{2^{1/2}y} u(0,y)\, dy = \infty .$$

Then,

$$(9.54) \qquad u(t,x+m_{1/2}(t)) \to w^{2^{1/2}}(x) \ \text{uniformly in} \ x \ \text{as} \ t \to \infty ,$$

where $w^{2^{1/2}}(x)$ is the solution of (1.5) and (1.6) with $w^{2^{1/2}}(0) = 1/2$.

<u>Proof</u>. We first handle the case where f is concave, that is, where $f(u)/u$ is decreasing in u . In this case, we know from Proposition 9.3 that for each $M_1 > 0$ and for $m_{1/2}(t) < x_1 \leqq x \leqq x_1 + M_1$,

$$(9.40a) \qquad u(t,x) \leqq C_5(t,z_1)\,\frac{z}{z_1}\,e^{-2^{1/2}(z-z_1)}\,u(t,x_1) \quad ,$$

where $z_\cdot = x_\cdot - m_{1/2}(t)$ and

$$(9.40b) \qquad \lim_{z_1\to\infty}\ \lim_{t\to\infty} C_5(t,z_1) = 1 \quad .$$

On the other hand, in the notation of Section 3,

$$u(0,\cdot) \overset{s}{\geqq} u^H(0,\cdot) \quad ,$$

and therefore by Corollary 2 of Proposition 3.2,

$$(9.55) \qquad u(t,\cdot) \overset{s}{\geqq} u^H(t,\cdot) \quad \text{for each } t \quad ,$$

where u^H denotes the solution with Heaviside initial data. Moreover, by the Kolmogorov theorem, $u^H(t,z+m^H_{1/2}(t)) \to w^{2^{1/2}}(z)$, where we know that

$$(9.56) \qquad w^{2^{1/2}}(z) \sim C\,z\,e^{-2^{1/2}z} \quad \text{as } z\to\infty$$

with $C > 0$. One can also conclude from (9.40) and Proposition 7.1 that $u(t,z+m_{1/2}(t)) \to 0$ as $t,z \to \infty$. It therefore follows from (9.55) and (9.56) that for each $M_1 > 0$ and for $m_{1/2}(t) < x_1 \leqq x \leqq x_1 + M_1$,

$$(9.57a) \qquad u(t,x) \geqq C_6(t,z_1)\,\frac{z}{z_1}\,e^{-2^{1/2}(z-z_1)}\,u(t,x_1)$$

for some $C_6(t,z_1)$, where

$$(9.57b) \qquad \lim_{z_1\to\infty}\ \lim_{t\to\infty} C_6(t,z_1) = 1 \quad .$$

We will employ the upper and lower bounds (9.40) and (9.57) together with Proposition 3.3 to demonstrate convergence of $u(t,x)$ to a travelling wave. First of all, set $C_7(z_1) = C_5(z_1) \vee C_6^{-1}(z_1)$, where $C_5(z_1)$ and $C_6(z_1)$ are the limits of $C_i(t,z_1)$ as $t \to \infty$. One may of course assume here that $C_7(z_1) \downarrow 1$ as $z_1 \uparrow \infty$. It follows without difficulty from (9.40) and (9.57) that if $M_1(t) \to \infty$ sufficiently slowly as $t \to \infty$, then

$$C_7^{-3}(z_1)u(t,M_1(t)+m_{1/2}(t)) \leqq \frac{M_1(t)}{z_1} e^{2^{1/2}(z_1-M_1(t))} u(t,z_1+m_{1/2}(t)) \tag{9.58}$$
$$\leqq C_7^3(z_1)u(t,M_1(t)+m_{1/2}(t))$$

for $1 \leqq z_1 \leqq M_1(t)$. But it also follows from (9.40) and (9.57) that

$$u(t,M_1(t)+m_{1/2}(t)) = C M_1(t) e^{-2^{1/2}(M_1(t)-\zeta(t))} , \tag{9.59}$$

where $|\zeta(t)| \leqq N_1$ for large enough t and appropriate N_1. Together, (9.58) and (9.59) imply that if $1 \leqq z_1 \leqq M_1(t)$, then

$$C_7^{-3}(z_1) \leqq \frac{1}{Cz_1} e^{2^{1/2}(z_1-\zeta(t))} u(t,z_1+m_{1/2}(t)) \leqq C_7^3(z_1) . \tag{9.60}$$

Therefore,

$$C C_8^{-1}(z)\, z e^{-2^{1/2}z} \leqq u(t,z+m(t)) \leqq C C_8(z)\, z e^{-2^{1/2}z} \tag{9.61}$$

for $N_1+1 \leqq z \leqq M_1(t) - N_1$, where we set

$$m(t) = m_{1/2}(t) + \zeta(t) ,$$

and

$$C_8(z) = \left(1+\frac{N_1}{z-N_1}\right) C_7(z-N_1) .$$

Note that $C_8(z) \to 1$ as $z \to \infty$.

Now, it is an immediate consequence of (9.61) and (9.56) that

$$\gamma_1^{-1}(z)\, w^{2^{1/2}}(z) \leqq u(t,z+m(t)) \leqq \gamma_1(z)\, w^{2^{1/2}}(z) \tag{9.62}$$

on $[N_1+1, M_1(t)-N_1]$ for appropriate γ_1 satisfying

$$\gamma_1(z) \to 1 \quad \text{as} \quad z \to \infty \ .$$

By increasing N_1 if necessary, we may assume here that $\gamma_1(z) \leq 2$ for all z. If we set

$$\gamma_2(t) = \sup\{2u(t,z+m(t)) : z \geq M_1(t)-N_1\} \ ,$$

then

$$\gamma_1^{-1}(z)\, w^{2^{1/2}}(z) - \gamma_2(t) \leq u(t,z+m(t)) \leq \gamma_1(z)\, w^{2^{1/2}}(z) + \gamma_2(t)$$

for all $z \geq N_1+1$. It follows from Proposition 7.1 that

$$\gamma_2(t) \to 0 \quad \text{as} \quad t \to \infty \ .$$

Applying Proposition 3.3, we therefore deduce that

$$\sup_{z \geq c(t)} |u(t,z+m(t)) - w^{2^{1/2}}(z)| \to 0 \quad \text{as} \quad t \to \infty \ ,$$

where $c(t) \to -\infty$ as $t \to \infty$. Since (1.17) is assumed to hold, the second part of Proposition 7.1 shows that in fact

(9.63) $$u(t,z+m(t)) \to w^{2^{1/2}}(z) \quad \text{uniformly in } z \text{ as } t \to \infty \ .$$

In order to derive (9.54) from (9.63), we need only to replace $m(t)$ by $m_{1/2}(t)$ in the argument. This modification follows as a simple consequence of the uniform continuity of $w^{2^{1/2}}$. We have thus demonstrated (9.54) for concave f.

We now demonstrate (9.54) for general f. To do so, it suffices to show that (9.61) continues to be valid over $[N_2, M_2(t)]$ for some choice of $M_2(t)$ and N_2 with $M_2(t) \to \infty$ as $t \to \infty$; a quick check then shows that the remaining arguments make no use of the concavity of f, and may therefore be repeated as before. We first replace f in (1.1) by the concave functions $\underline{f}$ and $\bar{f}$, which are as defined in (3.50). Since $\underline{f} \leq f \leq \bar{f}$, by the maximum principle the corresponding solutions $\underline{u}(t,x)$ and $\bar{u}(t,x)$ and their medians $\underline{m}_{1/2}(t)$ and $\bar{m}_{1/2}(t)$ satisfy

$$\underline{u}(t,x) \leqq u(t,x) \leqq \bar{u}(t,x) \tag{9.64}$$

and

$$\underline{m}_{1/2}(t) \leqq m_{1/2}(t) \leqq \bar{m}_{1/2}(t) \tag{9.65}$$

for all t and x. On the other hand, since $\underline{f}$ and $\bar{f}$ are both concave, it follows from Proposition 9.3 that for fixed $s_o > 0$,

$$\underline{u}(t,x) = \underline{C}_4(t,\underline{z})\, e^t E[u(0,\mathfrak{X}_x(t))\, ;\, \mathfrak{X}_x(s) > \underline{m}_{1/2}(t-s)\, ,\, 0 \leqq s \leqq t-s_o] \tag{9.66a}$$

and

$$\bar{u}(t,x) = \bar{C}_4(t,\bar{z})\, e^t E[u(0,\mathfrak{X}_x(t))\, ;\, \mathfrak{X}_x(s) > \bar{m}_{1/2}(t-s)\, ,\, 0 \leqq s \leqq t-s_o]\ , \tag{9.66b}$$

where

$$\lim_{t,z\to\infty} \underline{C}_4(t,z) = \lim_{t,z\to\infty} \bar{C}_4(t,z) = 1\ ,$$

$\underline{z} = x - \underline{m}_{1/2}(t)$, and $\bar{z} = x - \bar{m}_{1/2}(t)$. Note that $\underline{z} \geqq \bar{z}$ for all t. By (9.64), the left side of (9.66a) is at most as great as the left side of (9.66b), whereas by (9.65),

$$E[u(0,\mathfrak{X}_x(t))\, ;\, \mathfrak{X}_x(s) > \underline{m}_{1/2}(t-s)\, ,\, 0 \leqq s \leqq t-s_o]$$

$$\geqq E[u(0,\mathfrak{X}_x(t))\, ;\, \mathfrak{X}_x(s) > \bar{m}_{1/2}(t-s)\, ,\, 0 \leqq s \leqq t-s_o]\ .$$

Since $u(t,x)$ is sandwiched in between the solutions $\underline{u}(t,x)$ and $\bar{u}(t,x)$, we conclude from (9.66) that

$$u(t,x) = \bar{C}_6(t,\bar{z})\, \bar{u}(t,x)\ , \tag{9.67a}$$

where

$$\lim_{t,z\to\infty} \bar{C}_6(t,z) = 1\ . \tag{9.67b}$$

One may now apply (9.61) to $\bar{u}(t,x)$ to obtain the bounds

$$\bar{C}\,\bar{C}_6(t,z)\,\bar{C}_8^{-1}(z)\,z\,e^{-2^{1/2}z} \leq u(t,z+\bar{m}(t)) \leq \bar{C}\,\bar{C}_6(t,z)\,\bar{C}_8(z)\,z\,e^{-2^{1/2}z}$$

over $[N_1+1,\ M_1(t)-N_1]$, where notation is as in (9.61). With a little work, one can show this implies that

$$\bar{C}\,C_9^{-1}(z)\,z\,e^{-2^{1/2}z} \leq u(t,z+m(t)) \leq \bar{C}\,\bar{C}_9(z)\,z\,e^{-2^{1/2}z}$$

over $[N_2,M_2(t)]$, where $C_9(z) \to 1$ as $z \to \infty$, $N_2 = N_1+1$, and $M_2(t)$ may be chosen so that $M_2(t) \to \infty$ as $t \to \infty$. We have thus extended (9.61) to nonconcave f. Continuing as before, we may reason as through (9.63) to derive (9.54) for general f. //

Before proceeding, we note that owing to (9.54) and (9.67),

(9.68) $\quad \bar{m}_{1/2}(t) - m_{1/2}(t)$ is bounded,

where $\bar{m}_{1/2}(t)$ denotes the median of the solution $\bar{u}(t,x)$ with forcing term $\bar{f}$. Therefore, (9.67) may be rephrased as

(9.69a) $\quad u(t,x) = \hat{C}_6(t,z)\,\bar{u}(t,x)$

with

(9.69b) $\quad \lim_{t,z\to\infty} \hat{C}_6(t,z) = 1$,

where $z = x - m_{1/2}(t)$. (It also follows that (9.39) is valid for nonconcave f.) Referring back to the proof of Theorem 4, a little thought shows that (9.69) is dependent only on the inequality (9.64), rather than our specific choice of $\bar{u}(t,x)$. (9.69) therefore also holds for any pair of solutions $u(t,x)$ and $\bar{u}(t,x)$ with the same initial data and with forcing terms which satisfy (1.2) and (1.2'). The asymptotic position of the wave $u(t,x)$ is thus essentially unaffected by the choice of its forcing term f as long as (1.2) and (1.2') hold. In the case of finite initial mass, it follows from Theorem 3 of Section 8 that for large t, $m_{1/2}(t)$ is only shifted over by a bounded quantity if f is altered. Here, in the case of infinite initial mass, more is in fact true. As z and t increase, $u(t,z+m_{1/2}(t))$ loses its dependence on f ; f thus affects the asymptotic shape of the wave (owing to (1.12)), but not its asymptotic position.

Having demonstrated convergence to the travelling wave $w^{2^{1/2}}(z)$ under the assumptions (1.17) and (1.19) on the initial data, we now check to see what can be said about the translation term $m(t)$ (or more specifically, $m_{1/2}(t)$). We of course already know that since $u(t,z+m_{1/2}(t)) \to w^{2^{1/2}}(z)$, $m(t)/t \to 2^{1/2}$ (Corollary 1 of Lemma 3.2). We also know from Section 8 that if the initial data has finite mass, then we may choose $m(t) = 2^{1/2}t - 3\cdot 2^{-3/2}\log t$. In the case where the initial data has infinite mass, we have so far only derived the implicit relation (9.39) (extended to nonconcave f). We know now, however, that as $t \to \infty$, $u(t,z+m(t)) \to w^{2^{1/2}}(z)$, which will enable us to simplify this expression. Employing estimates from Lemmas 9.3 and 9.4, we will show in Theorem 5 that the translation terms $m(t)$ for which $u(t,z+m(t)) \to w^{2^{1/2}}(z)$ are in fact the only solutions of the relation. Using this uniqueness, we will then derive $m(t)$ for a large class of initial data, which will shed light as to how a change in the initial data affects the relative position of the wave. The question of an explicit formula for $m^u(t)$ for $u(t,x)$ with general initial data is still unresolved.

The basic approach to be taken in deriving Theorem 5 will be to observe that under certain mild conditions on $m(s)$, the quantity

$$E[u(0,\mathfrak{X}_x(t)) ; \mathfrak{X}_x(s) > m(t-s) , 0 \leq s \leq t-s_o]$$

is quite stable under moderate perturbations of $m(s)$. On account of the relation $z = x - m(t)$, it will therefore only be possible for $E[\cdot] \sim w^{2^{1/2}}(z)$ for an asymptotically unique choice of $m(t)$. Spelling out the details now, we will assume here that $m(t)$, $t > 0$, is a function such that $m(t) < \infty$ (with possibly $m(t) = -\infty$) which satisfies

$$(9.70) \qquad m(t)/t \to 2^{1/2} \quad \text{as} \quad t \to \infty \ .$$

s_o may then be chosen so that

$$(9.71) \qquad \{m(s) : s_o \leqq s \leqq t\} \text{ is bounded for fixed } t \ .$$

We furthermore assume that $m(t)$ satisfies the lower bounds

$$(9.72a) \qquad m(t) \geqq 2^{1/2}t - t^{\delta}$$

and

$$(9.72b) \qquad m(t) - m(s) \geqq 2^{1/2}(t-s) - (t-s)^{\delta} - C_1$$

for $s_o \leqq s \leqq t$ and t chosen large enough, where $\delta < 1/2$ and C_1 are constants. From our work in previous sections, it is already clear that under (1.17) and (1.19), $m_{1/2}(t)$ satisfies the above conditions; (9.72) in particular follows from Theorem 3 and Proposition 3.4. Our first lemma states that under these conditions, if $m^2(s) - m^1(s)$ is uniformly bounded in $s_o \leqq s \leqq t$, then the ratio of the corresponding values of $E[\cdot]$ cannot be too far from one. The proof employs Proposition 6.3 to rewrite the ratio in a manner so as to allow explicit computation.

<u>Lemma 9.3</u>. Assume that $u(0,x)$ satisfies (9.1), and that $m^1(t)$ and $m^2(t)$ satisfy conditions (9.70) - (9.72). In addition assume that for some $M_t \geqq 0$,

$$(9.73) \qquad m^2(s) - m^1(s) \leqq M_t \quad \text{for} \quad s_o \leqq s \leqq t \ ,$$

where s_o is chosen as in (9.71) and (9.72). For each $c > 0$ and appropriate $N_c \geqq 1$, if t is chosen large enough, then

$$(9.74) \qquad \frac{E[u(0,\mathfrak{X}_x(t)) ; \mathfrak{X}_x(s) > m^2(t-s) \quad \text{for} \quad 0 \leqq s \leqq t-s_o]}{E[u(0,\mathfrak{X}_x(t)) ; \mathfrak{X}_x(s) > m^1(t-s) \quad \text{for} \quad 0 \leqq s \leqq t-s_o]} \geqq (1+M_t/y(t))^{-1}(1+M_t/z')^{-1}e^{-cM_t}$$

for $x > m^2(t) + N_c$, where $y(t) \to \infty$ as $t \to \infty$ and $z' = x - m^2(t) - N_c$.

Proof. We start off by observing that by (9.1), (9.72a), and Lemma 9.2, the left hand side of (9.74) is at least

$$(9.75)\qquad \frac{C(t) \int_{y_1(t)}^{\infty} u(0,y)\, e^{-(x-y)^2/2t}\, P[\mathfrak{z}_{x,y}(s) > m^2(t-s)\,,\ 0 \leq s \leq t-s_o]\,dy}{\int_{y_1(t)}^{\infty} u(0,y)\, e^{-(x-y)^2/2t}\, P[\mathfrak{z}_{x,y}(s) > m^1(t-s)\,,\ 0 \leq s \leq t-s_o]\,dy}\,,$$

where $C(t) \to 1$ if $y_1(t)$ is chosen to grow slowly enough to ∞ as $t \to \infty$. If we let $\underline{m}^1(s) \equiv m^2(s) - M_t \leq m^1(s)$ for $s_o \leq s \leq t$, then we may of course replace $m^1(t-s)$ by $\underline{m}^1(t-s)$ in (9.75). But, since $\underline{m}^1(s) \leq m^2(s)$ for $s_o \leq s \leq t$, by Proposition 6.3,

$$P[\mathfrak{z}_{x,y}(s) > m^2(t-s)\,,\ 0 \leq s \leq t-s_o] / P[\mathfrak{z}_{x,y}(s) > \underline{m}^1(t-s)\,,\ 0 \leq s \leq t-s_o]$$

is increasing in y. Therefore, (9.75) is at least

$$(9.76)\qquad C(t)\, \frac{P[\mathfrak{z}_{x,y_1}(s) > m^2(t-s) \ \text{ for } \ 0 \leq s \leq t-s_o]}{P[\mathfrak{z}_{x,y_1}(s) > \underline{m}^1(t-s) \ \text{ for } \ 0 \leq s \leq t-s_o]}\,,$$

where we suppress the argument t in $y_1(t)$. Now, one may show without too much difficulty from conditions (9.70) - (9.72) that for t and N_c chosen large enough,

$$(9.77)\qquad m^2(s) \leq \frac{s}{t}\, m^2(t) + \frac{c}{4}(s \wedge (t-s)) + N_c \quad \text{for} \quad s_o \leq s \leq t\,.$$

A second application of Proposition 6.3 therefore shows that (9.76) is at least

$$C(t)\, \frac{P[\mathfrak{z}_{z',y}(s) > \frac{c}{4}(s \wedge (t-s)) \ \text{ for } \ 0 \leq s \leq t]}{P[\mathfrak{z}_{z',y}(s) > \frac{c}{4}(s \wedge (t-s)) - M_t \ \text{ for } \ 0 \leq s \leq t]}\,,$$

where $z' = x - m^2(t) - N_c$ and $y(t) = y_1(t) - N_c$. The above probabilities are evaluated in Lemma 2.2(d). With a little work one can show that their ratio is at least

$$(9.78) \qquad (1+M_t/y(t))^{-1}(1+M_t/z')^{-1}\exp[-M_t(c+\epsilon(t))/2] \quad ,$$

where $\epsilon(t) \to 0$ and depends only on c. Incorporating $C(t)$ and $\epsilon(t)$ into c, one obtains (9.74) from (9.75)-(9.78). //

Lemma 9.4 furnishes an upper bound on the right tail of

$$E[u(0,\mathfrak{X}_x(t)) ; \mathfrak{X}_x(s) > m(t-s) \ , \ 0 \leq s \leq t-s_o]$$

considered as a function of x. The proof is analogous to that of Lemma 9.3.

Lemma 9.4. Assume that $u(0,x)$ satisfies (1.19) and (9.1), and that $m(t)$ satisfies (9.70)-(9.72) for fixed s_o. Then for each $b < 2^{1/2}$ and large enough t,

$$(9.79) \qquad \begin{aligned} &E[u(0,\mathfrak{X}_{x_2}(t)) ; \mathfrak{X}_{x_2}(s) > m(t-s) \ , \ 0 \leq s \leq t-s_o] \\ &\leq C_2 e^{-b(x_2-x_1)} E[u(0,\mathfrak{X}_{x_1}(t)) ; \mathfrak{X}_{x_1}(s) > m(t-s) \ , 0 \leq s \leq t-s_o] + e^{-t-bz_2} \end{aligned}$$

for $x_2 \geq x_1 \geq m(t) + N^b$, where $z_2 = x_2 - m(t)$, and C_2 and N^b are constants.

Proof. Application of Lemma 9.2 shows that for $x \geq m(t)+1$ and large enough t,

$$E[u(0,\mathfrak{X}_x(t)) ; \mathfrak{X}_x(s) > m(t-s) \ , \ 0 \leq s \leq t-s_o]$$

$$\leq 2E[u(0,\mathfrak{X}_x(t)) ; \mathfrak{X}_x(t) > y_1(t) \ , \mathfrak{X}_x(s) > m(t-s) \ , \ 0 \leq s \leq t-s_o]$$

for appropriate $y_1(t)$ with $y_1(t) \to \infty$ as $t \to \infty$. Also, owing to (1.19) and (9.70), one can show from (4.7) that for each $\delta > 0$,

$$(9.81) \qquad E[u(0,\mathfrak{X}_{x_2}(t)) ; \mathfrak{X}_{x_2}(t) > \delta t] \leq \tfrac{1}{2} e^{-t-bz_2}$$

for large enough t and $z_2 = x_2 - m(t) \geq 0$. In order to demonstrate (9.79), it therefore suffices to show that for some $\delta > 0$,

(9.80)
$$\frac{\int_{y_1(t)}^{\delta t} u(0,y)\, e^{-(x_2-y)^2/2t} P[\mathfrak{z}_{x_2,y}(s) > m(t-s) \text{ for } 0 \leq s \leq t-s_o]\, dy}{\int_{y_1(t)}^{\delta t} u(0,y)\, e^{-(x_1-y)^2/2t} P[\mathfrak{z}_{x_1,y}(s) > m(t-s) \text{ for } 0 \leq s \leq t-s_o]\, dy} \leq C_3\, e^{-b(x_2-x_1)}$$

for $x_2 \geq x_1 \geq m(t) + N^b$, large t, and appropriate C_3 and N^b.

The remainder of the proof is analogous to the proof of Lemma 9.3. Arguing as in Lemma 9.3, we see that by Proposition 6.3,

$$P[\mathfrak{z}_{x_2,y}(s) > m(t-s)\, ,\, 0 \leq s \leq t-s_o] / P[\mathfrak{z}_{x_1,y}(s) > m(t-s)\, ,\, 0 \leq s \leq t-s_o]$$

is decreasing in y since $x_2 \geq x_1$. Also, because of (9.70),

$$\max\{e^{-(x_2-y)^2/2t} / e^{-(x_1-y)^2/2t} : y \in [y_1(t), \delta t]\}$$
$$\leq \exp[(2^{1/2} - 2\delta)(x_1-x_2)]$$

for large enough t. To demonstrate (9.80), it therefore suffices to show that for large enough t,

(9.81)
$$\frac{P[\mathfrak{z}_{x_2,y_1}(s) > m(t-s)\, ,\, 0 \leq s \leq t-s_o]}{P[\mathfrak{z}_{x_1,y_1}(s) > m(t-s)\, ,\, 0 \leq s \leq t-s_o]} \leq C_3\, e^{c(x_2-x_1)},$$

where $c = 2^{1/2} - 2\delta - b$, and δ is chosen small enough so that $c > 0$. As in Lemma 9.3, for large enough t and N_c, (9.77) will hold. Applying Proposition 6.3 again, one therefore sees that the left side of (9.81) is at most

$$(9.82)\qquad \frac{P[\mathfrak{z}_{z_2',y}(s) > \frac{c}{4}(s\wedge(t-s)) \text{ for } 0 \leq s \leq t]}{P[\mathfrak{z}_{z_1',y}(s) > \frac{c}{4}(s\wedge(t-s)) \text{ for } 0 \leq s \leq t]},$$

with $z_i' = x_i - m(t) - N_c$ and $y(t) = y_1(t) - N_c$. By applying Lemma 2.2(d) and doing a little computation, one can show that (9.82) is at most

$$(9.83)\qquad \frac{z_2'}{z_1'} \exp[(z_2' - z_1')(c + \epsilon(t))/2] ,$$

where $\epsilon(t) \to 0$ as $t \to \infty$ and depends only on c. Since $z_2' - z_1' = x_2 - x_1$, (9.81) follows from (9.82) and (9.83) if one sets $N^b = N_c + 1$ and chooses C_2 appropriately. //

We now demonstrate Theorem 5. We show that under the conditions (9.70)-(9.72), if $m(t)$ satisfies (9.84), then $m(t) - m_{1/2}(t) \to C$ for some constant C. Thus, $u(t,x+m(t))$ will converge to a translate of the travelling wave; here we will let $w^{2^{1/2}}(x)$ denote the translate with $w^{2^{1/2}}(0) = 1/2$. The approach taken in the proof will be to first use Lemmas 9.3 and 9.4 to show that $m(t) - m_{1/2}(t)$ is bounded; in particular, inequality (9.74) does not allow this difference to grow beyond a certain limit. Having shown that $m(t) - m_{1/2}(t)$ is bounded, one can reapply the same line of reasoning to show that the ratio of the quantities

$$E[u(0,\mathfrak{X}_x(t)) ; \mathfrak{X}_x(s) > m_{\cdot}(t-s) , 0 \leq s \leq t-s_0]$$

corresponding to $m(t)$ and $m_{1/2}(t)$ in fact tends to 1. With a little further work, one then obtains the desired result that $m(t) - m_{1/2}(t) \to C$ for some C. Note that on account of Theorem 4, we may assume the conditions (1.17) and (1.19) rather than (9.54) in the statement of the theorem.

<u>Theorem 5</u>. Let $u(t,x)$ be a solution of

$$(1.1)\qquad u_t = \frac{1}{2}u_{xx} + f(u) ,$$

with $f(u)$ satisfying (1.2) and (1.2'). Assume that the initial data $u(0,x)$

satisfy (9.1), and that

(9.54) $u(t,x+m_{1/2}(t)) \to w^{2^{1/2}}(x)$ uniformly in x as $t \to \infty$.

Let $m(t)$ be chosen so that (9.70)-(9.72) hold, with $s_o > 0$ selected so that (9.71) is also valid for $m_{1/2}(t)$. Also, assume that

(9.84a) $e^t E[u(0,\mathfrak{X}_x(t)) : \mathfrak{X}_x(s) > m(t-s) \quad \text{for} \quad 0 \leqq s \leqq t-s_o] = D(t,z)\, v(z)$

for some function $v(z)$ with $z = x - m(t)$, and that

(9.84b) $\lim_{z \to \infty} \overline{\lim_{t \to \infty}} D(t,z) = 1$.

Then,

(9.85) $m(t) - m_{1/2}(t) \to C$ as $t \to \infty$,

and consequently

(9.86) $u(t,x+m(t)) \to w^{2^{1/2}}(x+C)$ uniformly in x as $t \to \infty$

for some constant C . One may in fact choose $v(z) = w^{2^{1/2}}(z+C)$.

<u>Proof</u>. We first of all note that both $m(t)$ and $m_{1/2}(t)$ satisfy (9.70)-(9.72). If one sets

$$M_t = \sup\{|m(s) - m_{1/2}(s)| : s_o \leqq s \leqq t\} ,$$

one may therefore apply Lemma 9.3 to conclude that for $c > 0$ and appropriate $N_c \geqq 1$,

(9.87)
$$\frac{E[u(0,\mathfrak{X}_x(t))\,;\,\mathfrak{X}_x(s) > m_{1/2}(t-s) \quad \text{for} \quad 0 \leqq s \leqq t-s_o]}{E[u(0,\mathfrak{X}_x(t))\,;\,\mathfrak{X}_x(s) > m(t-s) \quad \text{for} \quad 0 \leqq s \leqq t-s_o]} \geqq (1+M_t/y(t))^{-1}(1+M_t/z')^{-1} e^{-cM_t}$$

if $x > m_{1/2}(t) + N_c$, where $y(t) \to \infty$ and $z' = x - m_{1/2}(t) - N_c$. On the other hand, $v(z) > 0$ for $z > C_1$. If z is chosen large enough and then fixed, it follows from (9.84) that

$$(9.88)\quad e^{t} E[u(0,\mathfrak{X}_x(t))\,;\mathfrak{X}_x(s) > m(t-s)\ ,\ 0 \leqq s \leqq t-s_o] \geqq C(z) > 0$$

for large t. Together, (9.87) and (9.88) show that

$$(9.89)\quad \begin{aligned} &e^{t} E[u(0,\mathfrak{X}_x(t))\,;\mathfrak{X}_x(s) > m_{1/2}(t-s)\ ,\ 0 \leqq s \leqq t-s_o] \\ &\geqq C(z)(1+M_t/y(t))^{-1}(1+M_t/z')^{-1}\, e^{-cM_t} \end{aligned}$$

for z fixed, $z' > 0$, and large enough t.

We proceed now to show that $m(t) - m_{1/2}(t)$ is bounded in t. Assume first that

$$(9.90)\quad m_{1/2}(t) \leqq m(t) \quad \text{for a given large fixed value of } t\ ,$$

and choose $z \geqq N_c + 1$. Then $z' \geqq 1$, and the right side of (9.89) is at least

$$(9.91)\quad C(z)(1+M_t)^{-2}\, e^{-cM_t} \geqq C(z)\, e^{-2cM_t}$$

for large enough M_t. The left side of (9.89) with $x = z + m(t)$ is therefore decreasing at most at exponential rate c in M_t. On the other hand, Lemma 9.4 states that for $b < 2^{1/2}$ and large enough choice of t in (9.90), the left side of (9.89) is at most

$$(9.92)\quad C_3 e^{-bz'}(1+e^{t} E[u(0,\mathfrak{X}_x(t))\,;\mathfrak{X}_x(s) > m_{1/2}(t-s)\ ,0 \leqq s \leqq t-s_o])$$

for $x \geqq x_1 \equiv m_{1/2}(t) + N^b$. Owing to (9.39), for N^b chosen large enough, the quantity in parentheses above is bounded for large t. Therefore, if $z \geqq N^b + N_c$, (9.92) is at most

$$(9.93)\quad C_4\, e^{-bz'} = C_5\, e^{-bz} \exp\,[m(t) - m_{1/2}(t)]$$

for constants C_4 and C_5. Consequently, the left side of (9.89) is decreasing at least at exponential rate b in z'. Together, (9.89)-(9.93) imply that $m(t) - m_{1/2}(t)$ cannot be too large, and in fact, setting $b = 1$ and $c = 1/4$, one obtains that

$$(9.94)\qquad m(t) - m_{1/2}(t) \leqq \tfrac{1}{2}M_t + C_6$$

for appropriate C_6, if t is large enough. Reversing the roles of $m(t)$ and $m_{1/2}(t)$, and employing the same reasoning as in (9.87)-(9.94) (but with the roles of (9.84) and (9.39) interchanged), one may in fact conclude that

$$(9.95)\qquad |m(t) - m_{1/2}(t)| \leqq \tfrac{1}{2}M_t + C_7 \ .$$

Since M_t is increasing in t, it follows from (9.95) that $M_t \leqq 2C_7$, and is hence bounded for all t.

We need still to show that under (9.84), $m(t) - m_{1/2}(t)$ is not only bounded (for $t \geqq s_o$), but also converges. We proceed by returning to (9.87). If $M_t \leqq M$ for all t, then for fixed $c > 0$ and large enough t, the left side of (9.87) is at least

$$(9.96)\qquad (1 + M/y(t))^{-1}(1 + M/z')^{-1}\, e^{-cM}$$

on $z' > 0$, where $y(t) \to \infty$ as $t \to \infty$. As $t, z \to \infty$ (and hence $z' \to \infty$), and then $c \downarrow 0$, (9.96) $\to 1$. It follows that for $x = m(t) + z$,

$$(9.97)\qquad \frac{E[u(0,\mathfrak{X}_x(t))\, ; \mathfrak{X}_x(s) > m_{1/2}(t-s)\ ,\ 0 \leqq s \leqq t-s_o]}{E[u(0,\mathfrak{X}_x(t))\, ; \mathfrak{X}_x(s) > m(t-s)\ ,\ 0 \leqq s \leqq t-s_o]} \to 1$$

as $t, z \to \infty$. On the other hand, by (9.84),

$$(9.98)\qquad e^t E[u(0,\mathfrak{X}_x(t))\, ; \mathfrak{X}_x(s) > m(t-s)\ ,\ 0 \leqq s \leqq t-s_o]/v(z) \to 1$$

as $t \to \infty$ and then $z \to \infty$. Similarly, because of (9.39) and (9.54),

$$(9.99)\qquad \frac{e^t E[u(0,\mathfrak{X}_x(t))\, ; \mathfrak{X}_x(s) > m_{1/2}(t-s)\ ,\ 0 \leqq s \leqq t-s_o]}{w^{2^{1/2}}(z + m(t) - m_{1/2}(t))} \to 1$$

as $t \to \infty$ and then $z \to \infty$. Combining (9.97)-(9.99), we see that

$$(9.100)\qquad v(z)/w^{2^{1/2}}(z+m(t)-m_{1/2}(t)) \to 1$$

as $t \to \infty$ and then $z \to \infty$. Since $w^{2^{1/2}}(z)$ has an exponential right tail, (9.100) implies that

$$(9.85)\qquad m(t)-m_{1/2}(t) \to C \quad \text{as} \quad t \to \infty$$

for some constant C. Therefore, since $w^{2^{1/2}}(z)$ is uniformly continuous,

$$(9.86)\qquad u(t,z+m(t)) \to w^{2^{1/2}}(z+C) \quad \text{uniformly in } z \text{ as } t \to \infty .$$

Also,

$$v(z)/w^{2^{1/2}}(z+C) \to 1 \quad \text{as} \quad z \to \infty ,$$

and so (9.84) will continue to hold if $w^{2^{1/2}}(z+C)$ is substituted for $v(z)$. //

We conclude this section by showing that if $u(0,x)$ satisfies the growth condition (9.101), then $u(t,x+m(t)) \to w^{2^{1/2}}(x)$ and one may explicitly evaluate the translation term $m(t)$. This result is an application of Theorems 4 and 5; the major part of the argument here consists of verifying condition (9.84) of Theorem 5. The main point is that one is able to treat $m(s)$ in (9.84a) as if it were strictly linear. This approximation is justified by (9.103) in conjunction with Proposition 6.1. The remainder of the proof then consists of rather tedious computation which reduces the left side of (9.84a) to an explicit form.

Application of Theorems 4 and 5. Let $u(t,x)$ be a solution to the Kolmogorov equation (1.1), with $f(u)$ satisfying conditions (1.2) and (1.2'). Assume that $u(0,x)$ satisfies the left tail condition (1.17), that (9.1) holds, and moreover, that

$$(9.101)\qquad \int_0^\infty y\, e^{2^{1/2}y}\, u(0,y)\, e^{-y^2/2t}\, dy \le e^{t^\delta}$$

for some $\delta < 1/3$ and large enough t. Then

(1.18) $$u(t,x+m(t)) \to w^{2^{1/2}}(x) \text{ uniformly in } x \text{ as } t \to \infty ,$$

where

(9.102a) $$m(t) = 2^{1/2}t - 3\cdot 2^{-3/2}\log t + b(t)$$

with

(9.102b) $$b(t) = 2^{-1/2}\log [\int_0^\infty y\, e^{2^{1/2}y}\, u(0,y)\, e^{-y^2/2t}\, dy] ,$$

and $w^{2^{1/2}}(x)$ is the translate of (1.5) and (1.6) with

$$w^{2^{1/2}}(x) \sim (2/\pi)^{1/2}\, x\, e^{-2^{1/2}x} \quad \text{as } x \to \infty .$$

Proof. It is an easy consequence of (9.101) that

(4.39) $$\lim_{t\to\infty} \frac{1}{t} \log\varphi(t,2^{1/2}t) = 0 ,$$

where $\varphi(t,x)$ is defined in (4.2). Proposition 4.2 states that (4.39) is equivalent to (1.19). (1.18) therefore follows from Theorem 4 for some choice of $m(t)$. To conclude that $m(t)$ may be chosen as in (9.102), we show that $m(t)$ satisfies the hypotheses of Theorem 5. First observe that (9.70) follows from (9.101); (9.71) is therefore also satisfied with $s_0 = 1$. Since $b(t)$ is increasing in t, (9.72) is also clearly valid. Once the condition (9.84) is verified, we will be able to apply Theorem 5 to deduce (9.102).

Before exhibiting (9.84), we first show that for each δ_1 with $\delta < \delta_1 < 1/3$,

(9.103)
$$|m(s) - \tfrac{s}{t}m(t)| < s^{\delta_1} + C_{\delta_1} \quad \text{for } 1 \leq s \leq t/2$$
$$< (t-s)^{\delta_1} + C_{\delta_1} \quad \text{for } t/2 \leq s \leq t$$

for appropriate choice of C_{δ_1}. (9.103) may be broken up into four cases depending on whether $s \in [1,t/2]$ or $s \in [t/2,t]$, and whether $m(s) - \frac{s}{t}m(t)$

is being bounded above or below. First observe that on account of (9.101),

(9.104) $$C^1 \leq b(s) \leq 2^{-1/2}s^{\delta} + C^2$$

for $s \geq 1$ and suitable constants C^1 and C^2. (9.103) is thus obvious for $s \in [1,t/2]$. Since $b(s)$ is increasing, it is also easy to show from (9.101) that the upper bound in (9.103) is also valid for $s \in [t/2,t]$. In order to bound $m(s) - \frac{s}{t}m(t)$ from below on $[t/2,t]$, one can check that

(9.105) $$\int_0^{t^{\delta_2}} y\, e^{2^{1/2}y}\, u(0,y)\, e^{-y^2/2t}\, dy \Big/ \int_0^{t^{\delta_2}} y\, e^{2^{1/2}y}\, u(0,y)\, e^{-y^2/2s}\, dy \leq e^{(t-s)^{\delta_1}} ,$$

where $\delta_2 = (1+\delta_1)/2$. Also, by (9.101),

$$\int_{t^{\delta_2}}^{\infty} y\, e^{2^{1/2}y}\, u(0,y)\, e^{-y^2/4t}\, dy \leq e^{2t^{\delta}} ,$$

and so

(9.106) $$\int_{t^{\delta_2}}^{\infty} y\, e^{2^{1/2}y}\, u(0,y)\, e^{-y^2/2t}\, dy \leq \exp\,[2t^{\delta} - t^{\delta_1}/4] \to 0$$

as $t \to \infty$. Since $\int_0^{\infty} y\, e^{2^{1/2}y}\, u(0,y)\, dy = \infty$, it follows from (9.105) and (9.106) that for large enough t,

(9.107) $$\int_0^{\infty} y\, e^{2^{1/2}y}\, u(0,y)\, e^{-y^2/2t}\, dy \Big/ \int_0^{\infty} y\, e^{2^{1/2}y}\, u(0,y)\, e^{-y^2/2s}\, dy \leq 2\, e^{(t-s)^{\delta_1}} .$$

Plugging this bound into (9.102), we see that

$$m(s) - \frac{s}{t}m(t) > -(t-s)^{\delta_1} + C_{\delta_1} \quad \text{for } t/2 \leq s \leq t$$

for all t and an appropriate constant C_{δ_1}. This last estimate completes (9.103).

We will now demonstrate (9.84). The left side of (9.84a) may be written as

$$e^t \int_{-\infty}^{\infty} u(0,y) \frac{e^{-(x-y)^2/2t}}{\sqrt{2\pi t}} P[\mathfrak{z}_{x,y}(s) > m(t-s)\ ,\ 0 \leq s \leq t-1]\, dy\ ,$$

where we choose $s_0 = 1$. On account of Lemma 9.2, this equals

$$(9.108)\quad D_1(t,x)\, e^t \int_{M(t)}^{\infty} u(0,y) \frac{e^{-(x-y)^2/2t}}{\sqrt{2\pi t}} P[\mathfrak{z}_{x,y}(s) > m(t-s)\ ,\ 0 \leq s \leq t-1]\, dy\ ,$$

where $M(t) \to \infty$ and $D_1(t,x) \to 1$ uniformly in $x \geq m(t)+1$ as $t \to \infty$. Inequality (9.103) says that $m(t)$ is close to being linear. From Proposition 6.1 and a little estimation, we may therefore rewrite (9.108) as

$$D_2(t,z)\, e^t \int_{M(t)}^{\infty} u(0,y) \frac{e^{-(x-y)^2/2t}}{\sqrt{2\pi t}} P[\mathfrak{z}_{z,y}(s) > 0\ ,\ 0 \leq s \leq t]\, dy\ ,$$

where as usual $z = x - m(t)$ and where $D_2(t,z) \to 1$ as $t,z \to \infty$. By Lemma 2.2(a), this reduces to

$$(9.109)\quad D_2(t,z)\, e^t \int_{M(t)}^{\infty} u(0,y) \frac{e^{-(x-y)^2/2t}}{\sqrt{2\pi t}} (1 - \exp[-2yz/t])\, dy\ .$$

Now, since the initial data satisfy (1.19), one may show from (4.7) that the upper limit of the integral in (9.109) may be truncated at $2^{1/2}t$. Also,

$$(9.110a)\quad 1 - \exp[-2yz/t] \leq 2yz/t$$

and

$$(9.110b)\quad \begin{aligned} e^{-(x-y)^2/2t} &= \exp[2^{1/2}y - y^2/2t - t] \cdot \exp[-(z+m(t) - 2^{1/2}t)^2/2t] \\ &\quad \cdot \exp[\tfrac{1}{t}(y - 2^{1/2}t)(z+m(t) - 2^{1/2}t)]\ , \end{aligned}$$

with the last quantity being at most

$$t^{3/2} \exp[2^{1/2}y - y^2/2t - t]$$

for $0 \leq y \leq 2^{1/2}t$. Plugging these estimates into (9.109) while keeping (9.106) in mind, one may in fact truncate at t^{δ_2} , and rewrite (9.109) as

$$(9.111) \qquad D_2(t,z)\, e^t \int_{M(t)}^{t^{\delta_2}} u(0,y) \frac{e^{-(x-y)^2/2t}}{\sqrt{2\pi t}} (1 - \exp[-2yz/t])\, dy + D_3(t,z) ,$$

where $D_3(t,z) \to 0$ uniformly in $1 \leq z \leq N$ with N fixed.

We now further simplify the integrand of (9.111). We assume here that $M(t) \leq y \leq t^{\delta_2}$, $1 \leq z \leq N$, and recall the bounds on $b(t)$ (and hence on $m(t)$) given in (9.104). Clearly,

$$(9.112a) \qquad 1 - \exp[-2yz/t] \sim 2yz/t ,$$

and since $\delta_1 < 1/3$ and $\delta_2 < 2/3$, it follows from (9.110b) that

$$(9.112b) \qquad e^{-(x-y)^2/2t} \sim \exp[2^{1/2}y - y^2/2t + t - 2^{1/2}(z + m(t))] ,$$

where in both cases the estimates are uniform over the specified values of y and z as $t \to \infty$. Plugging (9.112) into (9.111), we obtain

$$(9.113) \qquad \begin{aligned} &D_4(t,z)\, z\, e^{-2^{1/2}z} (2/\pi t^3)^{1/2} \exp[-2^{1/2}(m(t) - 2^{1/2}t)] \\ &\cdot \int_{M(t)}^{t^{\delta_2}} y\, e^{2^{1/2}y} u(0,y)\, e^{-y^2/2t}\, dy + D_3(t,z) , \end{aligned}$$

where $D_4(t,z) \to 1$ as $t \to \infty$ and then $z \to \infty$. Since $M(t)$ may be chosen so as to increase to infinity as slowly as desired, another application of (9.106) allows us to rewrite (9.113) as

$$(9.114) \qquad \begin{aligned} &\hat{D}_4(t,z)\, z\, e^{-2^{1/2}z} (2/\pi t^3)^{1/2} \exp[-2^{1/2}(m(t) - 2^{1/2}t)] \\ &\cdot \int_0^{\infty} y\, e^{2^{1/2}y} u(0,y)\, e^{-y^2/2t}\, dy + \hat{D}_3(t,z) , \end{aligned}$$

where $\hat{D}_3(t,z)$ and $\hat{D}_4(t,z)$ have the same limiting behavior as $D_3(t,z)$ and $D_4(t,z)$.

We now conclude the proof. Substitution for $m(t)$ as specified by (9.102) reduces (9.114) to

$$(2/\pi)^{1/2}\hat{D}_4(t,z)\,z\,e^{-2^{1/2}z} + \hat{D}_3(t,z) \;,$$

or equivalently,

$$(9.115)\qquad (2/\pi)^{1/2} D(t,z)\,z\,e^{-2^{1/2}z} \;,$$

where $D(t,z) \to 1$ as $t \to \infty$ and then $z \to \infty$. Together, the estimates (9.108)-(9.115) imply that

$$(9.116)\qquad \begin{aligned} &e^{t} E[u(0,\mathfrak{X}_x(t)) ; \mathfrak{X}_x(s) > m(t-s) \quad \text{for} \quad 0 \leq s \leq t-1] \\ &= (2/\pi)^{1/2} D(t,z)\,z\,e^{-2^{1/2}z} \;. \end{aligned}$$

Condition (9.84) of Theorem 5 is therefore satisfied with

$$(9.117)\qquad v(z) = (2/\pi)^{1/2}\,z\,e^{-2^{1/2}z} \;.$$

Application of Theorem 5 enables us to conclude that $m(t) - m_{1/2}(t) \to C$, and hence that

$$(1.18)\qquad u(t,x+m(t)) \to w^{2^{1/2}}(x) \quad \text{uniformly in } x \text{ as } t \to \infty$$

for some translate $w^{2^{1/2}}(x)$ of (1.5) and (1.6). It moreover follows from (9.117) that

$$w^{2^{1/2}}(x) \sim (2/\pi)^{1/2}\,x\,e^{-2^{1/2}x} \quad \text{as} \quad x \to \infty \;,$$

which completes the proof. //

It is illustrative to compute the translation term $m(t)$ for certain explicit choices of initial data. In Examples 1 and 2, we observe how $m(t)$ changes as the right tail of $u(0,x)$ is increased. In both examples, we assume that (1.17) holds.

Example 1. Suppose that the initial data $u(0,x)$ satisfies

(1.25) $$\lim_{x\to\infty} e^{bx}u(0,x) = 0$$

for some $b > 2^{1/2}$. Then $\int_0^\infty x\,e^{2^{1/2}x}u(0,x)\,dx < \infty$, and we may apply Theorem 3 of Section 8 to conclude that

(1.18) $$u(t,x+m(t)) \to w^{2^{1/2}}(x) \text{ uniformly in } x \text{ as } t\to\infty$$

for some translate $w^{2^{1/2}}(x)$ of (1.5) and (1.6), where

(1.26) $$m(t) = 2^{1/2}t - 3\cdot 2^{-3/2}\log t \quad .$$

Note that if $u(0,x) \sim e^{-bx}$ as $x\to\infty$ with $b < 2^{1/2}$, then by Theorem 1, $u(t,x+m_{1/2}(t))$ will converge to the travelling wave $w^{\lambda}(x)$ moving at rate $\lambda = 1/b + b/2$.

Example 2. Let $h(x) = e^{2^{1/2}x}u(0,x)$ with $h(x) = O(x^{\gamma})$, $\gamma > 0$. Then

(9.118) $$\int_0^\infty x\,h(x)\,e^{-x^2/2t}\,dx = O(t^{1+\gamma/2}) \quad ,$$

and so (9.101) is satisfied. It may still be true that $\int_0^\infty x\,h(x)\,dx < \infty$, in which case (1.18) and (1.26) will continue to hold as they did in Example 1. If the integral is infinite, then (1.18) and (9.102) will instead hold.

In particular, consider the case where

(1.28) $$h(x) = x^{\alpha} \text{ for } x \geq 1 ,$$

where $\alpha > -2$. Then $\int_0^\infty x\,h(x)\,dx = \infty$, and application of (9.102) shows that one may choose

(1.29) $$m(t) = 2^{1/2}t + (\alpha-1)2^{-3/2}\log t \quad .$$

If $h(x) = x^{-2}$ for $x \geq 1$, then it is not difficult to compute from (9.102) that

(1.30) $$m(t) = 2^{1/2}t - 3\cdot 2^{-3/2}\log t + 2^{-1/2}\log\log t .$$

If $\alpha < -2$ in (1.28), then $\int_0^\infty x\,h(x)\,dx < \infty$, and so (1.26) holds. For $\int_0^\infty x\,h(x)\,dx = \infty$, it is possible to specify the translate $w^{2^{1/2}}(x)$, whereas if the integral is finite, it is not.

We also make the observation that (1.26) is not possible for $\int_0^\infty x\,h(x)\,dx = \infty$, where $h(x) = e^{2^{1/2}x}u(0,x)$. That is, if $u(t,x+m(t))$ converges to a travelling wave, then necessarily

(9.119) $$m_{1/2}(t) - 2^{1/2}t + 3\cdot 2^{-3/2}\log t \to \infty \quad \text{as} \quad t \to \infty .$$

To see this, choose $\hat{u}(0,x)$ and $\hat{h}(x) = e^{2^{1/2}x}\hat{u}(0,x)$ so that (1) $\hat{h}(x) \leq h(x)$ for all x, (2) $\int_0^y x\,\hat{h}(x)\,dx \leq y$ for all y, and (3) $\int_0^\infty x\,\hat{h}(x)\,dx = \infty$. Integration by parts shows that $\hat{h}(x)$ satisfies (9.118) with $\gamma = 1/2$, and so by (9.102),

$$m^{\hat{u}}(t) - 2^{1/2}t + 3\cdot 2^{-3/2}\log t \to \infty \quad \text{as} \quad t \to \infty .$$

On the other hand by the maximum principle, $\hat{u}(t,x) \leq u(t,x)$ for all t and x, and so $m^{\hat{u}}(t) \leq m(t)$ for all t. The limit in (9.119) follows.

REFERENCES

[1] Aronson, D.G., Remarks on McKean's paper, unpublished.

[2] Aronson, D.G., and Weinberger, H.F., Nonlinear diffusion in population genetics, combustion, and nerve propagation, in Partial Differential Equations and Related Topics, ed. J.A. Goldstein; Lecture Notes in Mathematics No. 446, pp. 5-49, Springer, New York, 1975.

[3] Aronson, D.G., and Weinberger, H.F., Multidimensional nonlinear diffusions arising in population genetics, Adv. in Math. 30, 1978, pp. 33-76.

[4] Bramson, M.D., Maximal displacement of branching Brownian motion, Comm. Pure Appl. Math. 31, 1978, pp. 531-581.

[5] Fife, P.C., and McLeod, J.B., The approach of solutions of nonlinear diffusion equations to travelling wave front solutions, Arch. Rational Mech. Anal. 65, 1977, pp. 335-362.

[6] Fisher, R.A., The advance of advantageous genes, Ann. of Eugenics 7, 1937, pp. 355-369.

[7] Friedman, A., Stochastic Differential Equations and Applications, Vol. 1, Academic Press, New York, 1975.

[8] Gidas, B., Ni, W.-M., and Nirenberg, L., Symmetry and related properties via the maximum principle, Comm. Math. Phys. 68, 1979, pp. 209-243.

[9] Gihman, I.T., and Skorohod, A.V., Stochastic Differential Equations, Springer-Verlag, New York, N.Y., 1972.

[10] Griffeath, D., A maximal coupling for Markov chains, Z. Wahrscheinlichkeitstheorie und verw. Gebiete 31, 1975, pp. 95-106.

[11] Jones, C., Spherically symmetric waves of a reaction-diffusion equation, submitted to J. Diff. Equations.

[12] Kac, M., Wiener and integration in function spaces, Bull. Amer. Math. Soc. 72, 1966, pp. 52-68.

[13] Kanel, Ya.I., The behavior of solutions of the Cauchy problem when the time tends to infinity in the case of quasilinear equations arising in the theory of combustion, Soviet Math. Dokl. 1, 1960, pp. 533-536.

[14] Kolmogorov, A., Petrovsky, I., and Piscounov, N., Étude de l'équation de la diffusion avec croissance de la quantité de matière et son application à un problème biologique, Moscou Universitet. Bull. Math. 1, 1937, pp. 1-25.

[15] McKean, H.P., Application of Brownian motion to the equation of Kolmogorov-Petrovskii-Piskunov, Comm. Pure Appl. Math. 28, 1975, pp. 323-331.

[16] Protter, M.H., and Weinberger, H.F., Maximum Principles in Differential Equations, Prentice-Hall, Englewood Cliffs, N.J., 1967.

[17] Uchiyama, K., The behavior of solutions of some non-linear diffusion equations for large time, J. Math. Kyoto Univ. 18, 1978, pp. 453-508.

School of Mathematics
University of Minnesota
Minneapolis, Minnesota

General instructions to authors for PREPARING REPRODUCTION COPY FOR MEMOIRS

For more detailed instructions send for AMS booklet, "A Guide for Authors of Memoirs." Write to Editorial Offices, American Mathematical Society, P. O. Box 6248, Providence, R. I. 02940.

MEMOIRS are printed by photo-offset from camera copy fully prepared by the author. This means that, except for a reduction in size of 20 to 30%, the finished book will look exactly like the copy submitted. Thus the author will want to use a good quality typewriter with a new, medium-inked black ribbon, and submit clean copy on the appropriate model paper.

Model Paper, provided at no cost by the AMS, is paper marked with blue lines that confine the copy to the appropriate size. Author should specify, when ordering, whether typewriter to be used has PICA-size (10 characters to the inch) or ELITE-size type (12 characters to the inch).

Line Spacing – For best appearance, and economy, a typewriter equipped with a half-space ratchet – 12 notches to the inch – should be used. (This may be purchased and attached at small cost.) Three notches make the desired spacing, which is equivalent to 1-1/2 ordinary single spaces. Where copy has a great many subscripts and superscripts, however, double spacing should be used.

Special Characters may be filled in carefully freehand, using dense black ink, or INSTANT ("rub-on") LETTERING may be used. AMS has a sheet of several hundred most-used symbols and letters which may be purchased for $5.

Diagrams may be drawn in black ink either directly on the model sheet, or on a separate sheet and pasted with rubber cement into spaces left for them in the text. Ballpoint pen is *not* acceptable.

Page Headings (Running Heads) should be centered, in CAPITAL LETTERS (preferably), at the top of the page – just above the blue line and touching it.

LEFT-hand, EVEN-numbered pages should be headed with the AUTHOR'S NAME;

RIGHT-hand, ODD-numbered pages should be headed with the TITLE of the paper (in shortened form if necessary).

Exceptions: PAGE 1 and any other page that carries a display title require NO RUNNING HEADS.

Page Numbers should be at the top of the page, on the same line with the running heads.

LEFT-hand, EVEN numbers – flush with left margin;

RIGHT-hand, ODD numbers – flush with right margin.

Exceptions: PAGE 1 and any other page that carries a display title should have page number, centered below the text, on blue line provided.

FRONT MATTER PAGES should be numbered with Roman numerals (lower case), positioned below text in same manner as described above.

MEMOIRS FORMAT

It is suggested that the material be arranged in pages as indicated below.
Note: Starred items (*) are requirements of publication.

Front Matter (first pages in book, preceding main body of text).

Page i – *Title, *Author's name.

Page iii – Table of contents.

Page iv – *Abstract (at least 1 sentence and at most 300 words).

*1980 Mathematics Subject Classifications represent the primary and secondary subjects of the paper. For the classification scheme, see Annual Subject Indexes of MATHEMATICAL REVIEWS beginning in December 1978.

Key words and phrases, if desired. (A list which covers the content of the paper adequately enough to be useful for an information retrieval system.)

Page v, etc. – Preface, introduction, or any other matter not belonging in body of text.

Page 1 – Chapter Title (dropped 1 inch from top line, and centered).

Beginning of Text.

Footnotes: *Received by the editor date.
Support information – grants, credits, etc.

Last Page (at bottom) – Author's affiliation.

ABCDEFGHIJ–AMS–89876543

Memoirs of the American Mathematical Society

Number 286

H. Peter Gumm

Geometrical methods in congruence modular algebras

Published by the
AMERICAN MATHEMATICAL SOCIETY
Providence, Rhode Island, USA

September 1983 · Volume 45 · Number 286

MEMOIRS of the American Mathematical Society

This journal is designed particularly for long research papers (and groups of cognate papers) in pure and applied mathematics. It includes, in general, longer papers than those in the TRANSACTIONS.

Mathematical papers intended for publication in the Memoirs should be addressed to one of the editors. Subjects, and the editors associated with them, follow:

Real analysis (excluding harmonic analysis) and **applied mathematics** to JOEL A. SMOLLER, Department of Mathematics, University of Michigan, Ann Arbor, MI 48109.

Harmonic and complex analysis to LINDA PREISS ROTHSCHILD, Department of Mathematics, University of California at San Diego, LaJolla, CA 92093

Abstract analysis to WILLIAM B. JOHNSON, Department of Mathematics, Ohio State Univeristy, Columbus, OH 43210

Algebra and number theory (excluding universal algebras) to LANCE W. SMALL, Department of Mathematics, University of California at San Diego, LaJolla, CA 92093

Logic, foundations, universal algebras and combinatorics to JAN MYCIELSKI, Department of Mathematics, University of Colorado, Boulder, CO 80309

Topology to WALTER D. NEUMANN, Department of Mathematics, University of Maryland, College Park, College Park, MD 20742

Global analysis and differential geometry to TILLA KLOTZ MILNOR, Department of Mathematics, Hill Center, Rutgers University, New Brunswick, NJ 08903

Probability and statistics to DONALD L. BURKHOLDER, Department of Mathematics, University of Illinois, Urbana, IL 61801

Combinatorics and number theory to RONALD GRAHAM, Mathematical Studies Department, Bell Laboratories, Murray Hill, NJ 07974

All other communications to the editors should be addressed to the Managing Editor, R. O. WELLS, JR., Department of Mathematics, University of Colorado, Boulder, CO 80309

MEMOIRS are printed by photo-offset from camera-ready copy fully prepared by the authors. Prospective authors are encouraged to request booklet giving detailed instructions regarding reproduction copy. Write to Editorial Office, American Mathematical Society, P.O. Box 6248, Providence, Rhode Island 02940. For general instructions, see last page of Memoir.

SUBSCRIPTION INFORMATION. The 1983 subscription begins with Number 272 and consists of six mailings, each containing one or more numbers. Subscription prices for 1983 are $104.00 list; $52.00 member. Each number may be ordered separately; *please specify number* when ordering an individual paper. For prices and titles of recently released numbers, refer to the New Publications sections of the **NOTICES** of the American Mathematical Society.

BACK NUMBER INFORMATION. For back issues see the AMS Catalogue of Publications.

TRANSACTIONS of the American Mathematical Society

This journal consists of shorter tracts which are of the same general character as the papers published in the MEMOIRS. The editorial committee is identical with that for the MEMOIRS so that papers intended for publication in this series should be addressed to one of the editors listed above.

Subscriptions and orders for publications of the American Mathematical Society should be addressed to American Mathematical Society, P. O. Box 1571, Annex Station, Providence, R. I. 02901. *All orders must be accompanied by payment.* Other correspondence should be addressed to P. O. Box 6248, Providence, R. I. 02940.

MEMOIRS of the American Mathematical Society (ISSN 0065-9266) is published bimonthly (each volume consisting usually of more than one number) by the American Mathematical Society at 201 Charles Street, Providence, Rhode Island 02904. Second Class postage paid at Providence, Rhode Island 02940. Postmaster: Send address changes to Memoirs of the American Mathematical Society, American Mathematical Society, P. O. Box 6248, Providence, RI 02940.

TABLE OF CONTENTS

ABSTRACT

We develop a geometric approach to algebras in congruence modular varieties. The idea of coordinatization of lines in affine geometry finds an almost perfect analog in the coordinatization of algebras.
The geometry is the congruence class geometry, i.e. the subspaces are the blocks of congruence relations.

We show that congruence modularity guarantees that the congruence class geometry behaves nicely, because the Desarguesian and the Pappian theorems are true, if interpreted correctly. The innocuously looking "Shifting Lemma" is the basic and powerful tool we need.
The obstacle to a perfect coordinatization is a congruence relation called the "commutator". The commutator is zero iff nonparallel lines have precisely one point of intersection.
This approach leads to a simple geometric development of commutator theory for arbitrary congruences. Results about affine algebras on the one hand and about distributive varieties on the other hand are tied together where only the commutator appears as a parameter. For the extreme values of this parameter we find theorems about affine, nilpotent and solvable congruences and varieties at one end and theorems generalizing Jónsson's lemma at the other end. A radical, $\sqrt{\underline{A}}$, is defined and we show that Jónsson's lemma is true for every algebra $\underline{A}/\sqrt{\underline{A}}$.

AMS (MOS) subject classification (1980): 08B10, 08B05, 08A30, 08A05.

Keywords and phrases: commutator, coordinatization, Desarguesian theorem, Shifting Lemma, Jónsson Lemma, radical, congruence class geometry, affine, solvable, Kronecker products, Jónsson Tarski algebras.

Library of Congress Cataloging in Publication Data

Gumm, H. Peter (Heinz Peter), 1951-
Geometrical methods in congruence modular algebras.

(Memoirs of the American Mathematical Society, ISSN 0065-9266 ; v. 45, no. 286 (Sept. 1983))
Revision of the author's Habilitationsschrift--Technische Hochschule Darmstadt, 1980.
Bibliography: p.
1. Algebra, Universal. 2. Algebraic varieties. I. Title. II. Title: Congruence modular algebras. III. Series: Memoirs of the American Mathematical Society ; no. 286.
QA3.A57 no. 286 [QA251] 510s [512] 83-11810
ISBN 0-8218-2286-1

PREFACE

Affine planes can be coordinatized by certain algebraic structures called planar ternary rings. If the Desarguesian theorem holds, then a (generally noncommutative) field is obtained. The algebraic structure can then be used in turn to prove results which may be translated back into geometrical theorems. The proof that every finite Desarguesian affine plane is Pappian, using Wedderburn's theorem is a prominent example.

By the same token geometrical reasoning may assist the mathematician working on algebraic structures. In theories related to linear algebra, as for example in ring - and module theory geometric intuition may suggest algebraic results and may be a guide to algebraic proofs. This method is not limited to fields that essentially originated from geometry. Here we deal with classes of algebraic structures, general enough to include groups, rings and modules on the one hand and lattices on the other hand, but they do not include semigroups, for example. We shall see that thereby we seem to have found the right level of abstraction where a geometrical language may reasonably be used and a geometrical intuition may be developped.

The key is that every (universal) algebra coordinatizes a pseudo-geometry. This geometry was investigated and characterized by Wille in [41], he called it the "Kongruenzklassengeometrie". It may be a geometry of a rather strange nature, but the fundamental notions like "points", "lines" and "incidence" make sense, so as to allow us to draw pictures of geometrical configurations, corresponding to algebraic contexts. These configurations may then lead to a deeper understanding of the theory, but also suggest proofs, which then have to be reformulated algebraically.

Here we make extensive use of this geometric visualization. We draw points and lines to express and explain algebraic situations. Previously known theories become clearer and new results are obtained. Indeed, by a consequent use of the geometric method, some deep algebraic results may seem more obvious because they are suggested by the geometry.

As we have mentioned before, the geometry may be very nasty in general, and it was long believed that only the class of "permutable" varieties was satisfactorily tractable by geometric methods. Those algebras share the fundamental property that "parallelograms" in their Kongruenzklassengeometrie can be completed, i.e. given three points, there exists a least one fourth point, so that the four points form a parallelogram. Those algebras comprise all classical structures mentioned above, so certainly they might seem to form a reasonable level of abstraction. On the other hand, one would like to include some more classes of structures into a systematic treatment. For example, the class of all lattices which does not have this property.

The framework we have chosen is modularity. We assume that the algebras in question have modular congruence lattices. This class is known since Birkhoff [2] to include permutable varieties, and clearly lattices are captured too, since in fact their congruence lattices are distributive.

Thus the difficulty is to find the right geometric properties common to both kinds of theories. A very simple property, the "Shifting Lemma" was discovered in [17] and its importance has become more apparent henceforth. In fact, more-dimensional analogues like the "Little Desarguesian theorem" and the "Escher-Cube" could be proved [18], showing the richness of structure in modular congruence class spaces.

There is a broad spectrum of modular varieties, reaching from abelian groups and modules at the one end to lattices or more generally to distributive varieties at the other end. Moreover, given a modular variety then after imposing appropriate conditions one often finds that either the variety is affine (i.e. polynomially equivalent to a variety of modules) or it has typical features of a distributive variety. This dichotomy was first noticed for permutable varieties.

Now, as we have said before, lattices and abelian groups seem to be modular out of different reasons and in fact they turn out as different ends in a spectrum of modular varieties which lie in between. In fact those ends could be isolated inside the subclass of permutable varieties. As an example we mention R. McKenzie's results [33] and also [16]. It was only when J.D.H. Smith [36] introduced "commutators" into general algebra, that an overview over the subspectrum of permutable varieties could be handled. Commutators in a way acted like a prism, making the spectrum visible.

J. Hagemann and C. Herrmann [24] managed to carry the commutator concept over to modular varieties, proving many of its properties that now seem to be fundamental for the theory. The sacrifice, however, was the loss of geometric intuition, the complicacy of the concept, which made it extremely hard to handle.

Using the Shifting Lemma, a very simplified definition could be given in [19], making it possible to give an elementary and geometrical development of commutator theory. The theory has since been proven extremely useful, pushing the theory of modular varieties forcefully ahead. Most prominently R. McKenzie's work with R. Freese [13] and with S. Burris [6] has to be mentioned here. Progress in other directions was also made in [21].

Let us now give a brief overview to the present treatise. After establishing the fundamental concepts and notions in chapter 0, we introduce the reader to the concept of modularity in chapter 1. The congruence class geometry is developed in chapter 2, we prove the fundamental configuration theorems, like the Little Desarguesian theorem, the Cube Lemma and the

Escher-Cube Lemma which will reappear throughout these notes. The syntactical description of modularity due to Day [9] follows in chapter 3, we try to explain the geometry behind it. Chapter 4 provides a term which plays a fundamental rôle in modular varieties. The door is opened to carry results from permutable varieties over to modular varieties. The construction of the aforementioned term is an outstanding example of how geometry provides the ideas for an algebraic proof. In chapter 5 we show that the classical idea of coordinatizing the affine plane with an abelian group has a direct counterpart in modular varieties. An analysis of the ingredients leads to the investigation of commutators in chapter 6, the theory is developed and the properties that nowadays seem to be fundamental are proved. The famous Jónsson Lemma turns out to be a special instance of a more general theorem for modular varieties. A radical $\sqrt{\underline{A}}$ is defined and it is shown that Jónssons Lemma is true for every algebra of the form $\underline{A}/\sqrt{\underline{A}}$ in a modular varietiey. A formula which shows that congruences permute modulo some commutators allows us to give a Mal'cev type description of modular varieties using ternary terms only in chapter 7. More evidence is gathered that modularity lives between the poles of permutability and distributivity. Theorems that refer to permutability are proved in chapter 8. Abelian congruences and corresponding affine algebras are then examined in chapter 9 and nilpotent and solvable varieties characterized in chapter 10. In chapter 11 we look at the possibility of yet extending the framework of modular varieties. FP-varieties seem to be appropriate for many results. Their congruence class geometry has special properties only in direct products of algebras. Terms being "n-ary homomorphisms" with respect to other terms are studied in chapter 12, a theme which is intimately connected with commutators and coordinatization. Unitary groupoid objects in modular varieties (and in FP-varieties) are shown to be abelian group objects.

ACKNOWLEDGEMENTS

This paper formed the "Habilitationsschrift" of the author and was presented to the Fachbereich Mathematik of the Technische Hochschule Darmstadt in August 1980. Theorems 6.18 - 6.21 and 9.2 - 9.6 were added in the fall of 1981. The results included were obtained at the Arbeitsgruppe Allgemeine Algebra of the Technische Hochschule Darmstadt, at the Math. Dept's of Lakehead University and of McMaster University. They were presented in a series of lectures at the above institutions as well as in a lecture at the AMS meeting in Boulder, CO in 1980 and at the University of Hawaii in 1981. The author expresses his gratitude to R. Wille, A. Day, G. Bruhns, W. Taylor and R. Freese. Many thanks also to Traudel Ridder for the excellent preparation of the typed version of this manuscript.

0. FUNDAMENTAL CONCEPTS

This chapter summarizes fundamental notions and elementary results of general algebra. We omit proofs since they can be found in most elementary textbooks, such as GRÄTZER [14], or PIERCE [43].

Regarding our terminology we mainly follow [14]. Our word "polynomial", however, stands for "algebraic function" as in [14]. We will in fact use both words simultaneously.

Let $\Delta := (n_i)_{i \in I}$ be a family of natural numbers. An <u>algebra of type Δ</u> is a pair $\underline{A} := (A, (f_i)_{i \in I})$ where A is a nonempty set and every f_i is an n_i-ary operation on A, i.e. a map $f_i \colon A^{n_i} \to A$.

From now on we tacitly assume we have specified a type Δ so that all algebras we deal with are of type Δ.

Let $\underline{B}$ be another algebra, i.e. $\underline{B} = (B, (g_i)_{i \in I})$.
A map $\phi \colon A \to B$ is a <u>homomorphism</u> from $\underline{A}$ to $\underline{B}$ if for every $i \in I$ and elements $a_1, a_2, \ldots, a_{n_i} \in A$ we have

$$\phi(f_i(a_1, \ldots, a_{n_i})) = g_i(\phi(a_1), \ldots, \phi(a_{n_i})).$$

If ϕ is injective it is called an embedding. If ϕ is bijective then ϕ^{-1} is a homomorphism from $\underline{B}$ to $\underline{A}$. ϕ is then called an isomorphism. We say $\underline{A}$ and $\underline{B}$ are isomorphic and write $\underline{A} \cong \underline{B}$.

A subalgebra $\underline{C}$ of $\underline{A}$ is an algebra $\underline{C} = (C, (h_i)_{i \in I})$ where $C \subseteq A$ and each operation h_i is the restriction of the corresponding f_i to C^{n_i}.

If D is a nonempty subset of A and for all $d_1, \ldots, d_{n_i} \in D$ we have that $f_i(d_1, \ldots, d_{n_i}) \in D$ then $\underline{D} = (D, (h_i)_{i \in I})$ with $h_i := f_i|_{D^{n_i}}$ is a subalgebra of $\underline{A}$. By abuse of language we sometimes phrase shortly: D is a subalgebra of $\underline{A}$.

The <u>product</u> $\prod_{j \in J} \underline{C}_j$ of the algebras $\underline{C}_j$ has as underlying set the cartesian product of the sets C_j and the operations are defined componentwise. A subalgebra $\underline{S}$ of $\prod_{j \in J} \underline{C}_j$ is called a <u>subdirect product</u> of the $\underline{C}_j$, if the restrictions of the canonical projections π_j to $\underline{S}$ are still onto.

A <u>congruence relation</u> Θ on the algebra $\underline{A}$ is an equivalence relation on A which is at the same time a subalgebra of $\underline{A} \times \underline{A}$. For $(x,y) \in \Theta$ we frequently write $x \Theta y$ or $x \equiv y \pmod{\Theta}$. The fact that Θ is a subal-

Received by the editors December 15, 1981.

gebra of $\underline{A} \times \underline{A}$ can be expressed by the implication

$$x_1 \theta y_1, \ldots, x_{n_i} \theta y_{n_i} \Rightarrow f_i(x_1,\ldots,x_{n_i}) \,\theta\, f_i(y_1,\ldots,y_{n_i}).$$

This property is often referred to as the compatibility of θ with the operation f_i.

If θ is a congruence relation on $\underline{A}$ and $a \in \underline{A}$ we define $[a]\theta := \{x \in A \mid x\,\theta\,a\}$ and call it the θ-block of a.

The set $A/\theta := \{[a]\theta \mid a \in A\}$ can be given the structure of an algebra again by defining

$$f_i([a_1]\theta, \ldots, [a_{n_i}]\theta) := [f(a_1,\ldots,a_{n_i})]\theta.$$

The resulting algebra is called the factor of $\underline{A}$ by θ and

π_θ: $A \to A/\theta$

$a \to [a]\theta$ is a surjective homomorphism.

Note that for an arbitrary homomorphism ϕ: $\underline{A} \to \underline{B}$ the relation $\ker \phi := \{(x,y) \mid \phi(x) = \phi(y)\}$ is a congruence relation and every congruence relation arises this way, namely $\theta = \ker \pi_\theta$. Moreover, if ϕ: $\underline{A} \to \underline{B}$ is a surjective homomorphism then $\underline{B}$ and $\underline{A}/\ker \phi$ are isomorphic, in symbols $\underline{B} \cong \underline{A}/\ker \phi$.

Let now $\underline{V}$ be a class of algebras. $\underline{V}$ is called a variety if $\underline{V}$ is closed under the formation of subalgebras, homomorphic images and direct products of any of its members. G. BIRKHOFF's theorem says that a class of algebras is definable by equations if and only if it is a variety.

In a variety $\underline{V}$ there exist free algebras $\underline{F}_{\underline{V}}(X)$ for every set X. $\underline{F}_{\underline{V}}$ is a functor, left adjoint to the forgetful functor into sets, in particular: For every algebra $\underline{A}$ in $\underline{V}$ and every map α: $X \to A$ there exists precisely one homomorphism $\bar{\alpha}$: $\underline{F}_{\underline{V}}(X) \to \underline{A}$ extending α.

If X is finite, say $X = \{x_1,\ldots,x_k\}$ then $\underline{F}_{\underline{V}}(X)$ can be considered as the algebra of all k-ary V-terms with variables from X in the following manner:

For $p \in \underline{F}_{\underline{V}}(X)$ and $\underline{A} \in \underline{V}$ define a k-ary operation $p^{\underline{A}}$ on A by $p^{\underline{A}}(a_1,\ldots,\bar{a}_k) := \bar{\alpha}(p)$, where $\bar{\alpha}$ is the unique homomorphism from $\underline{F}_{\underline{V}}(X)$ to $\underline{A}$ with $\bar{\alpha}(x_i) = a_i$.

The k-ary operations thus arising on A are called term-functions. They can also be characterized as those k-ary operations on A which can be built up by superposition from the fundamental operations f_i and the projection operations π_i^n: $A^n \to A$, $(a_1,\ldots,a_n) \to a_i$.

If n-k places in an n-ary term function are frozen with fixed elements of A, we obtain a k-ary polynomial of $\underline{A}$, frequently also called an algebraic function of $\underline{A}$.

We will have to take a closer look at congruence relations. If θ is a congruence relation on $\underline{A}$ and τ a polynomial of $\underline{A}$ then Θ is compatible with τ. (Similarly the equality characterizing homomorphisms remains true if f_i is replaced by any term-function.)

There are two trivial congruences 0 and 1 (sometimes subscripted as $0_{\underline{A}}$ and $1_{\underline{A}}$) on every algebra $\underline{A}$, given by

$$0 = \{(x,x) \mid x \in \underline{A}\} \text{ and}$$

$$1 = \{(x,y) \mid x,y \in \underline{A}\}.$$

$\underline{A}$ is called simple if there are no other congruences on $\underline{A}$.
$\underline{A}$ is called subdirectly irreducible, if $\underline{A}$ possesses a smallest nontrivial congruence relation μ, called the monolith of $\underline{A}$. G. BIRKHOFF's theorem asserts that every algebra is a subdirect product of subdirectly irreducible factors.

The intersection of an arbitrary family of congruences is a congruence again, thus for a subset $T \subseteq A \times A$ there exists a smallest congruence relation containing T which we will denote by $\langle T\rangle_{\underline{A}}$ or $\Theta_{(a,b)}$ for $T = \{(a,b)\}$.

A.I. MAL'CEV in [32] gave an explicit description of $\langle T\rangle_{\underline{A}}$. This description specializes as follows:

0.1 Theorem: (i) If T is a reflexive symmetric relation, then $(a,b) \in \langle T\rangle_{\underline{A}}$ iff there exist unary algebraic functions $\tau_0,\ldots,\tau_n$ and $(s_0,t_0),\ldots,(s_n,t_n) \in T$ such that

$$a = \tau_0(s_0)$$

$$\tau_i(t_i) = \tau_{i+1}(s_{i+1}) \quad \text{for } 0 \le i < n,$$

$$\tau_n(t_n) = b$$

(ii) For $T = \{(x,y)\}$: $(a,b) \in \Theta_{(x,y)}$ iff there exist unary algebraic functions $\tau_0,\ldots,\tau_{n-1}$ and elements $a = c_0,\ldots,c_{n-1}, c_n = b$ such that $\{c_i,c_{i+1}\} = \{\tau_i(x),\tau_i(y)\}$ for $0 \le i < n$.

(iii) A subset $S \subseteq A$ is a class of some congruence relation on $\underline{A}$ iff for all algebraic functions τ on $\underline{A}$ we have either $\tau(S) \subseteq S$ or $\tau(S) \cap S = \emptyset$.

The congruences on an algebra form a complete algebraic lattice, in fact a sublattice of the lattice of all equivalence relations on the set A. We shall denote this lattice by $\mathrm{Con}(\underline{A})$. The join of two congruences Θ and Ψ always contains the relational product $\Theta\circ\Psi$ and can be described as

$$\Theta\vee\Psi = \bigcup\{\overbrace{\Theta\circ\Psi\circ\Theta\circ\Psi\circ \ldots \circ\Theta\circ\Psi}^{n \text{ times}} \mid n \in \mathbb{N}\}.$$

Since $\Theta\circ\Theta = \Theta$ by transitivity, it follows that $\Theta\vee\Psi = \Theta\circ\Psi$ just in case that $\Theta\circ\Psi = \Psi\circ\Theta$. We shall then say that Θ and ψ are permutable.

Let now $\phi: \underline{A} \to \underline{B}$ be a homomorphism. If β is a congruence relation on $\underline{B}$ then

$$\overset{\leftarrow}{\phi}\beta := \{(x,y) \in A\times A \mid (\phi(x),\phi(y)) \in \beta\}$$

is a congruent relation on $\underline{A}$. Incidentally $\ker\phi = \overset{\leftarrow}{\phi}0$ and $\overset{\leftarrow}{\phi}$ is a lattice isomorphism from $\mathrm{Con}(\underline{B})$ to the sublattice of $\mathrm{Con}(\underline{A})$ consisting of all congruence relations on $\underline{A}$ which contain $\ker\phi$.

For $\alpha \in \mathrm{Con}(\underline{A})$ we get a congruence

$$\overset{\rightarrow}{\phi}\alpha := \langle\{(\phi(x),\phi(y)) \mid x\,\alpha\,y\}\rangle_{\underline{B}} \quad \text{on } \underline{B}.$$

One checks the relations

$$\overset{\rightarrow}{\phi}\overset{\leftarrow}{\phi}\beta \le \beta \quad \text{and}$$

$$\overset{\leftarrow}{\phi}\overset{\rightarrow}{\phi}\alpha \ge \alpha \vee \ker\phi$$

with equality holding in both formulas if ϕ is onto.

If $\prod_{j\in J} \underline{C}_j$ is a direct product, then the kernels of the canonical projections will also be denoted as π_j. Those congruences are often called factor congruences. They are mutually permutable. Given a filter $\mathcal{D}$ on the indexing set J, a congruence relation $\Theta_{\mathcal{D}}$ arises on defining: $a\,\Theta_{\mathcal{D}}\,b$ iff $\{i\in J \mid \pi_i(a) = \pi_i(b)\} \in \mathcal{D}$. If $\mathcal{D}$ is an ultrafilter, the important construction of an ultraproduct $\prod_{i\in J} \underline{C}_j/\Theta_{\mathcal{D}}$ is obtained. It has important modeltheoretic properties, see e.g. [1].

1. MODULARITY

1.1 Definition: A lattice $\underline{L}$ is modular if for every $x,y,z \in L$ the implication

$$x \geq z \Rightarrow x \wedge (y \vee z) \leq (x \wedge y) \vee z$$

holds.

An algebra $\underline{A}$ is called congruence modular if $\mathrm{Con}(\underline{A})$ is a modular lattice. Similarly a variety $\underline{V}$ is called modular if every $\underline{A} \in \underline{V}$ is congruence modular.

The following theorems will be used to provide us with a sufficiently large class of examples.

The first theorem is again due to MAL'CEV. It describes congruence permutable varieties, i.e. varieties all of whose algebras are congruence permutable.

1.2 Theorem [32]: A variety $\underline{V}$ of algebras is congruence permutable if and only if there exists a ternary $\underline{V}$-term $p(x,y,z)$, such that the equations $p(x,y,y) = x$ and $p(x,x,y) = y$ are true in $\underline{V}$.

To generalize the concept of permutability of congruences we define for a natural number k and congruences Θ_o and Θ_1:

Definition: Congruences Θ_o and Θ_1 are k-permutable if $\Theta_o \circ \Theta_1 \circ \Theta_o \circ \ldots \circ \Theta_\varepsilon \subseteq \Theta_1 \circ \Theta_o \circ \Theta_1 \circ \ldots \circ \Theta_{1-\varepsilon}$, where ε is 1 or o, depending on whether k is even or odd, and both sides are k-fold relational products. An algebra is called k-permutable, if any two congruences on $\underline{A}$ are k-permutable. Similarly a variety consisting of k-permutable algebras only will be called a k-permutable variety.

Thus permutability is just 2-permutability and k-permutability implies (k+1)-permutability.

J. HAGEMANN and A. MITSCHKE obtained a characterization of k-permutable varieties reminiscent of MAL'CEV's theorem, and including this for the case $k = 2$ [25]:

1.3 Theorem: A variety $\underline{V}$ of algebras is k-permutable if and only if there exist ternary $\underline{V}$-terms $p_o,\ldots,p_k$ such that the equations

$$p_o(x,y,z) = x$$
$$p_i(x,x,y) = p_{i+1}(x,y,y) \quad \text{for} \quad 0 \leq i < k$$
$$p_k(x,y,z) = z$$

are true in $\underline{V}$.

To see that 3-permutable algebras are congruence modular, first shown by B. JONSSON [29], we note that joins of two congruences Θ and Ψ in this case are computed as $\Theta\circ\Psi\circ\Theta$. Here the modular law reduces to $\alpha \geq \gamma \Rightarrow \alpha \wedge (\gamma\circ\beta\circ\gamma) \subseteq (\alpha \wedge \beta) \vee \gamma$. For (x,y) from the left hand side there exist u,v with $x\ \gamma\ u\ \beta\ v\ \gamma\ y$ and consequently $x\ \alpha\ u$ and $v\ \alpha\ y$. α is transitive, yielding $u\ \alpha\ v$. Finally $x\ \gamma\ u\ (\alpha\wedge\beta)\ v\ \gamma\ y$ hence $(x,y) \in (\alpha \wedge \beta) \vee \gamma$.

Many examples of congruence modular algebras are in fact <u>congruence distributive</u>. B. JONSSON characterized varieties containing congruence distributive algebras only, as follows:

1.4 <u>Theorem ([28]):</u> A variety $\underline{V}$ of algebras is congruence distributive if and only if for some natural number n there exist ternary $\underline{V}$-terms $q_o,\ldots,q_n$ such that the equations

$$
\begin{aligned}
q_o(x,y,z) &= x \\
q_i(x,y,x) &= x && \text{for all } 0 \leq i \leq n, \\
q_i(x,x,y) &= q_{i+1}(x,x,y) && \text{for } 0 \leq i < n, \ i \text{ even.} \\
q_i(x,y,y) &= q_{i+1}(x,y,y) && \text{for } 0 < i < n, \ i \text{ odd} \\
q_n(x,y,z) &= z
\end{aligned}
$$

hold in $\underline{V}$.

Now we are able to list many varieties which are congruence modular:

<u>Groups:</u> Groups are permutable according to MAL'CEV's theorem. $P(x,y,z) := x\cdot y^{-1}\cdot z$ is the term witnessing permutability.

<u>Rings:</u> Same as above with $p(x,y,z) = x-y+z$.

<u>Quasigroups:</u> $p(x,y,z) := (x/(y\backslash y))(y\backslash z)$ is a term for MAL'CEV's theorem.

<u>Median algebras:</u> Those are algebras with a ternary "majority term" i.e. a term $m(x,y,z)$ satisfying the equations $m(x,x,y) = m(x,y,x) = m(y,x,x) = x$. Median algebras are congruence distributive, JONSSON's theorem applies with $n = 2$.

<u>Lattices:</u> Lattices are median algebras. Take $m(x,y,z) := (x \vee y) \wedge (y \vee z) \wedge (z \vee x)$.

<u>Implication algebras:</u> They are groupoids $(G,\to)$ satisfying $(x\to y)\to x = x$; $(x\to y)\to y = (y\to x)\to x$; $x \to (y\to z) = y \to (x\to z)$. Implication algebras are 3-permutable and congruence distributive. See MITSCHKE [34] and HAGEMANN, MITSCHKE [25].

<u>Boolean algebras:</u> Boolean algebras are lattices, hence congruence distributive. But they are also rings, hence permutable.

So far all of our examples were already either permutable or distributive. For an example of a congruence modular variety which is neither permutable, nor congruence distributive we introduce:

Generalized right complemented semigroups: These algebras have two binary operations $\cdot$ and $*$, satisfying:

$$x\cdot(x*y) = y\cdot(y*x)$$
$$x\cdot(y*y) = x.$$

Generalized right complemented semigroups have 3-permutable congruences. HAGEMANN and MITSCHKE showed that their theorem applies with $p_1(x,y,z) = x\cdot(y*z)$ and $p_2(x,y,z) = z\cdot(y*x)$.

To see that generalized right complemented semigroups are in general non-distributive, take a ring $\underline{R}$ with unit, in which 2 has an inverse. Then define $x\cdot y := x+y$ and $x*y := \frac{1}{2}(y-x)$ as operations on any module over $\underline{R}$.

Implication algebras are also models of the above equations. If we define $x\cdot y := y\to x$ and $x*y := y\to x$ then the above equations hold. See MITSCHKE [34] for an example of a (three-element) implication algebra with non permutable congruences.

If a further equation

$$(x\cdot y)*z = y*(x*z)$$

is added, we obtain the class of right complemented semigroups. See BOSBACH [3] for the fact that right complemented semigroups are congruence distributive.

Also see the remarks at the end of chapter 7.

2. CONGRUENCE CLASS GEOMETRY

If V is a vector space then the blocks of congruence relations on V are precisely the affine subspaces of V. With this example in mind, geometrical terms suggest themselves for the study of congruence relations. Indeed the system of congruence classes of an algebra $\underline{A}$ can be considered as a geometry, the so called "Kongruenzklassengeometrie". This geometry was introduced and investigated by R. WILLE in [41].

We do not use any results from this approach, yet we adopt and heavily exploit the geometrical viewpoint, by using a pseudogeometrical language and by drawing geometrical figures.

Thus we draw points for elements of a given algebra $\underline{A}$ and we connect two points, say x and y with a line, if x and y are congruent modulo some congruence relation (or some compatible relation), say Θ. In this case we label the line connecting x and y with the symbol Θ and think of it as representing all points from $[x]\Theta$ when Θ is a congruence relation. Two lines will be drawn parallel, just in case they are classes of the same congruence relation.

As an example, the following picture expresses the relation $x \,\Theta\, y \,\Psi\, z$:

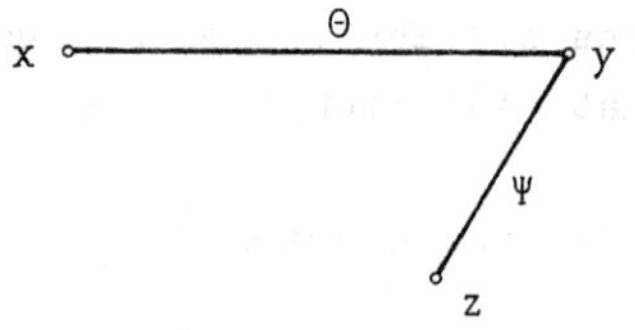

Moreover we will have $\Theta \circ \Psi = \Psi \circ \Theta$ if and only if for every $x,y,z \in \underline{A}$, the above picture can be completed to

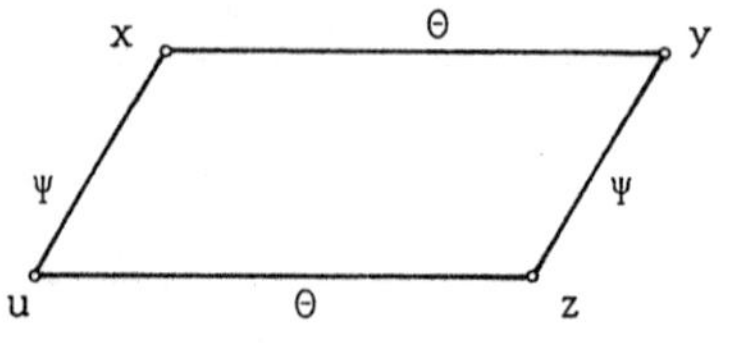

for some $u \in \underline{A}$.

Thus permutability of congruences can be expressed geometrically by the existence of the "4-th parallelogram point". (x,y,u,z) in this case would be called a <u>Θ-Ψ-parallelogram</u>. MAL'CEV's term $p(x,y,z)$ always provides us with one 4-th parallelogram point. From the equations $p(x,y,y) = x$ and $p(y,y,z) = z$ and the relations $x\ \Theta\ y\ \Psi\ z$ we infer $p(x,y,z)\ \Theta\ p(y,y,z) = z$ and $p(x,y,z)\ \Psi\ p(x,y,y) = x$.

In congruence modular algebras such a strong geometrical tool cannot be expected, however the following "Shifting Lemma" is still powerful enough to replace the modular equation in everything that follows.

2.1 <u>Shifting Lemma:</u> Let α,β, and γ be congruences on a congruence modular algebra $\underline{A}$ and let x,y,z,u be elements of $\underline{A}$. If $\alpha \wedge \beta \leq \gamma$ then

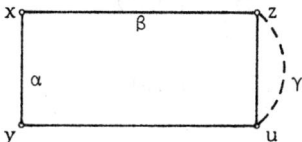

<u>implies</u>

<u>Proof:</u> We have $(x,y) \in \alpha \wedge (\beta \circ (\alpha \wedge \gamma) \circ \beta) \subseteq \alpha \wedge (\beta \vee (\alpha \wedge \gamma)) \subseteq (\alpha \wedge \beta) \vee (\alpha \wedge \gamma)$ by modularity. From the assumption that $\alpha \wedge \beta \leq \gamma$ it follows that $(x,y) \in \gamma$. □

If the condition $\alpha \wedge \beta \leq \gamma$ is dropped in the Shifting Lemma the hypothesis still guarantees that $(x,y) \in (\alpha \wedge \beta) \vee \gamma$; simply replace γ by $\gamma' := (\alpha \wedge \beta) \vee \gamma$ and apply 2.1. Thus both versions are equivalent. We will find that within a variety the validity of the Shifting Lemma is equivalent to modularity. In fact, the Shifting Lemma will replace the modular law in everything that follows. In particular for the rest of this chapter we will assume that the Shifting Lemma holds in $\underline{A} \times \underline{A}$. Thus we are able to obtain "higher-dimensional" configuration theorems which are important for later chapters. Those theorems were found in GUMM [18]. Our original proof used the DAY-terms which we shall introduce in the following chapter. The proof was shortened by TAYLOR and by WOLF, to the form we present it in here.

2.2 <u>Theorem:</u> Let Θ,α_1,α_2 and Ψ be congruences with $\Theta\wedge\alpha_1 \leq \Psi \geq \Theta\wedge\alpha_2$. If x,y,z,u,x',y',z',u' are elements of $\underline{A}$ then

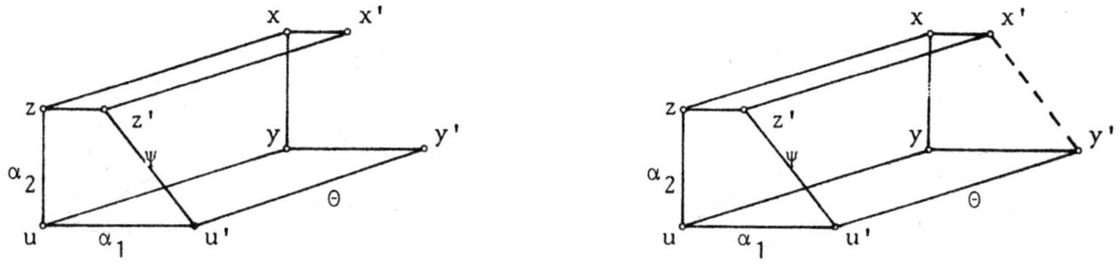

<u>implies</u>

Proof: The pairs (x',x), (y',y), (z',z), (u',u) are elements of α_1 which is a subalgebra of $\underline{A} \times \underline{A}$. Defining congruences $1 \times \alpha_2$ and $\Psi \times \alpha_2$ and $\Theta \times \Theta$ on α_1 by

$$(u,v)\ 1 \times \alpha_2\ (r,s) \quad \text{iff} \quad v\ \alpha_2\ s$$

$$(u,v)\ \Psi \times \alpha_2\ (r,s) \quad \text{iff} \quad u\ \Psi\ r \ \text{and}\ v\ \alpha_2\ s$$

$$(u,v)\ \Theta \times \Theta\ (r,s) \quad \text{iff} \quad u\ \Theta\ r \ \text{and}\ v\ \Theta\ s$$

we obtain

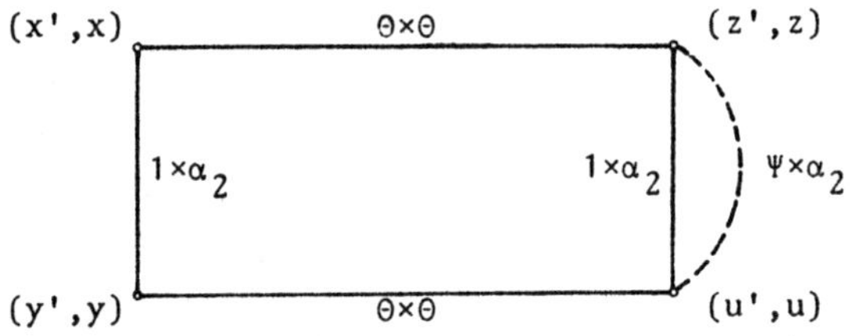

If we are allowed to apply the Shifting Lemma, then from $(x',x)\ \Psi \times \alpha_2\ (y',y)$ we get immediately $x'\ \Psi\ y'$ and we are done. However, we have to check that $\Theta \times \Theta \wedge 1 \times \alpha_2 \leq \Psi \times \alpha_2$. Thus let $(a,b)\ \Theta\times\Theta \wedge 1\times\alpha_2\ (c,d)$. First of all $(a,b) \in \alpha_1$ and $(c,d) \in \alpha_1$. Moreover $(b,d) \in \Theta\wedge\alpha_2$ hence $(b,d) \in \Psi$ by assumption. Hence

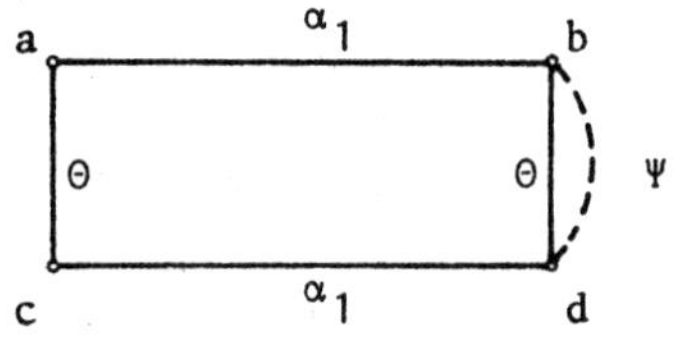

The other assumption, $\Theta\wedge\alpha_1 \leq \Psi$ allows us to apply the Shifting Lemma in this situation, yielding $(a,c) \in \Psi$.

Mainly we are interested in the following two special cases. Firstly, letting $x = x'$ and $z = z'$ we obtain

2.3 The Little Desarguesian Theorem: Let Θ, α_1, α_2 and Ψ be congruences with $\Theta\wedge\alpha_1 \leq \Psi \geq \Theta\wedge\alpha_2$. Then

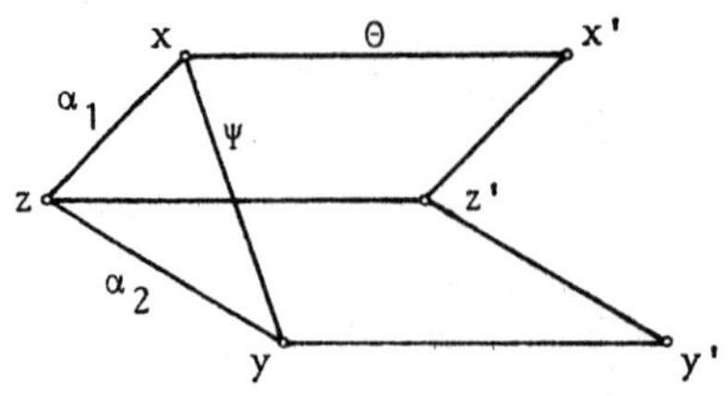

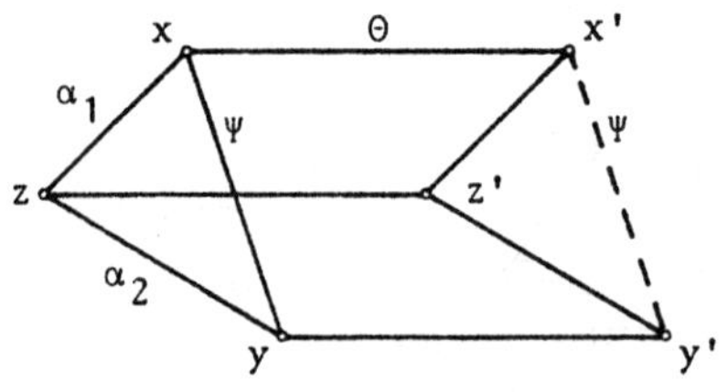

implies

This theorem is particularly interesting, since R. FREESE and B. JONSSON [11] have shown that the congruence lattices of algebras in modular varieties are arguesian. Thus our theorem may be considered as an affine counterpart to their result. The existence of such an affine counterpart is surprising since no transition between the projective geometry of algebras (as manifested in their congruence lattices) and the affine geometry (Kongruenzklassengeometrie) is known.

Our next specialization of 2.2 has many applications in what follows. In particular it guarantees the closure of the "REIDEMEISTER configuration" which was first shown and applied in [16] and [17].

2.4 Cube Lemma: Let Θ_o, Θ_1 and Ψ be congruences with $\Theta_o \wedge \Theta_1 \leq \Psi$ and let x,y,z,u,x',y',z',u' be elements of $\underline{A}$, then

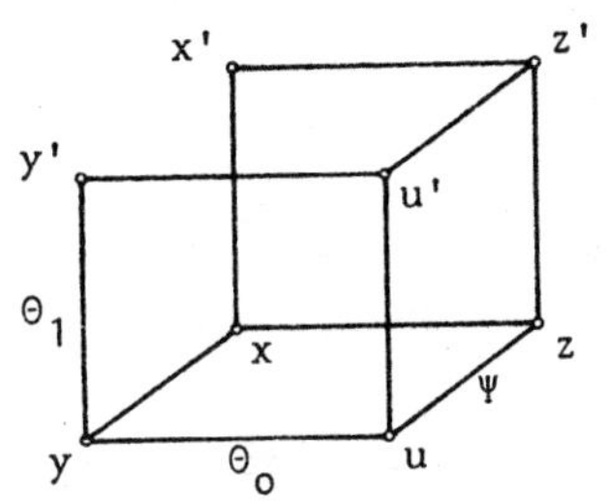

implies

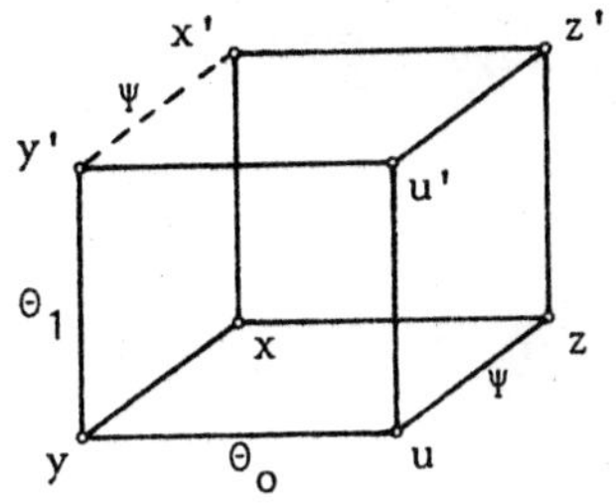

Proof: Set $\Psi = \alpha_2$, $\Theta_o := \Theta$, $\Theta_1 := \alpha_1$ in 2.2. □

A "twisted" form of 2.2 will be needed to give us the closure of the "Desarguesian configuration", as termed in [17]:
Because of apparent common features of the configuration with pictures of M.C. ESCHER we call it

2.5 The ESCHER Cube: Let Θ, α_1, α_2 and Ψ be congruences with $\Theta\wedge\alpha_1 \leq \Psi \geq \Theta\wedge\alpha_2$ and let x,y,z,u,x',y',z',u' be elements of $\underline{A}$. Then

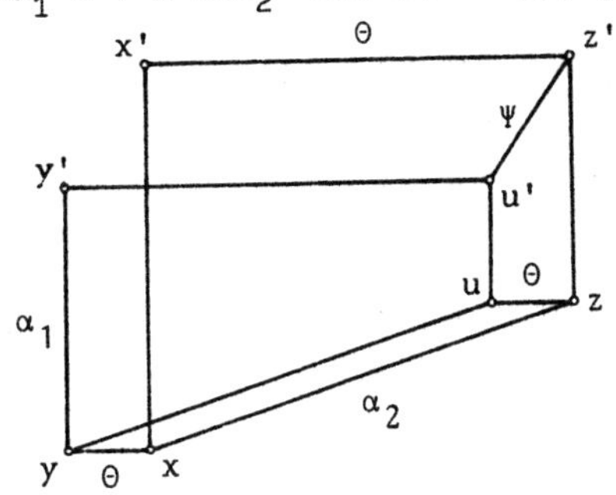

implies

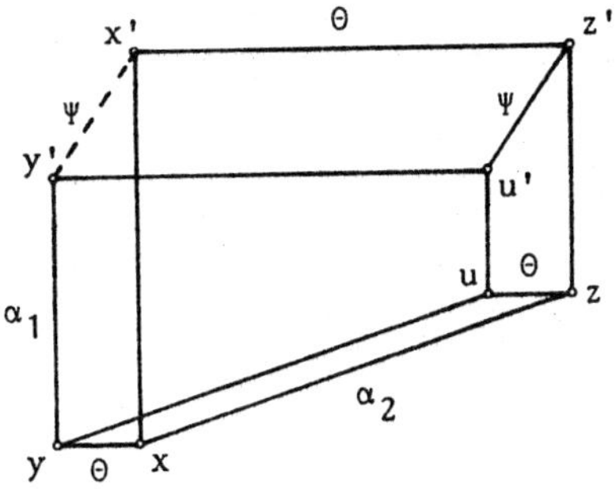

Proof: Look at α_1 as a subalgebra of $\underline{A}\times\underline{A}$ and define congruences $\Theta\times\alpha_2$, $1\times\Theta$ and $\Psi\times\Theta$ on the algebra α_1 in the obvious way to obtain

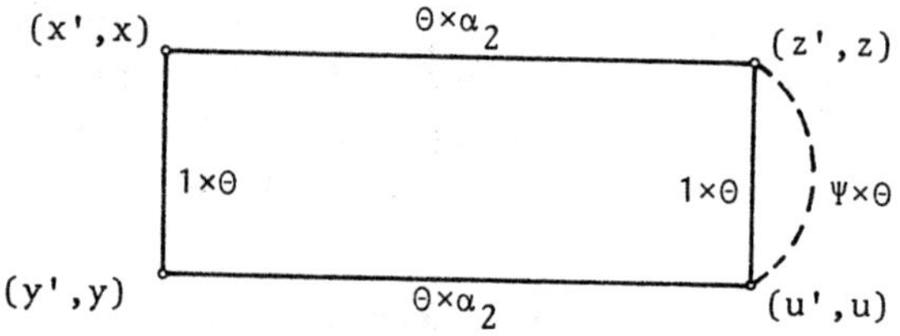

If the conditions for the Shifting Lemma are satisfied we get $(x',x)\ \Psi\times\Theta\ (y',y)$ and in particular $x'\ \Psi\ y'$. Thus to show that $\Theta\times\alpha_2 \wedge 1\times\Theta \leq \Psi\times\Theta$ we take $((a,b),(c,d))$ from the left hand side and obtain:

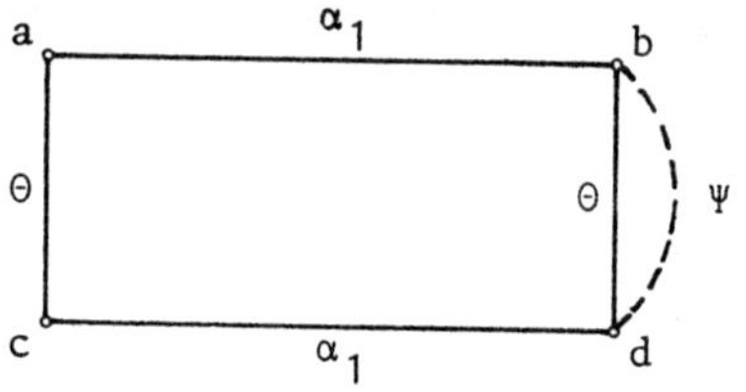

To see that $(b,d) \in \Psi$ we use $(b,d) \in \Theta$ and $(b,d) \in \alpha_2$ and our hypothesis. Thus the latter Shifting Lemma provides for $(a,c) \in \Psi$, i.e. $((a,b),(c,d))$ from the right hand side.

Again we are interested in two special cases. Firstly, on identifying z with z' and y with y' we obtain

2.6 <u>The Little Pappian Theorem:</u> If Θ, α_1, α_2 and Ψ are congruences with $\Theta\wedge\alpha_1 \leq \Psi \geq \Theta\wedge\alpha_2$ and $x,y,z,u,x',u' \in \underline{A}$ then

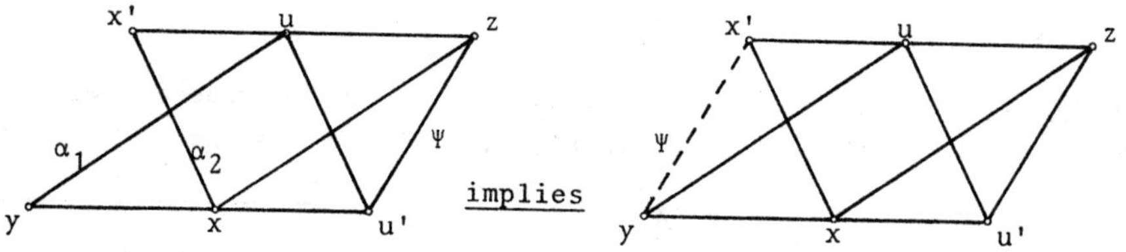

Similarly, on identifying α_2 with Ψ and u with u' we obtain

2.7 <u>Lemma:</u> Let Θ_o, Θ_1 and Ψ be congruences with $\Theta_o\wedge\Theta_1 \leq \Psi$ and x,y,z,u,x',y',z' elements of $\underline{A}$ then

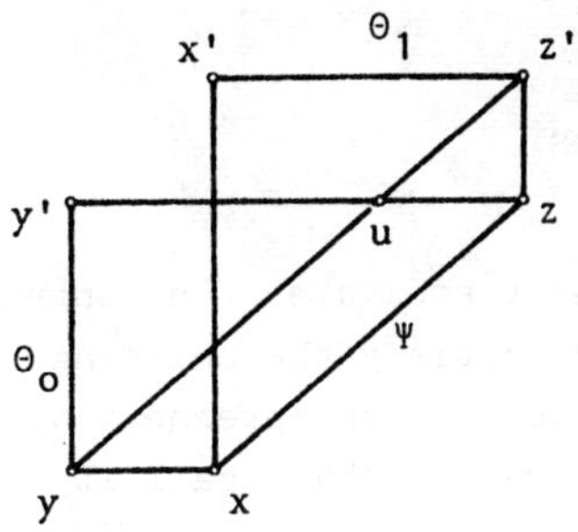

<u>implies</u>

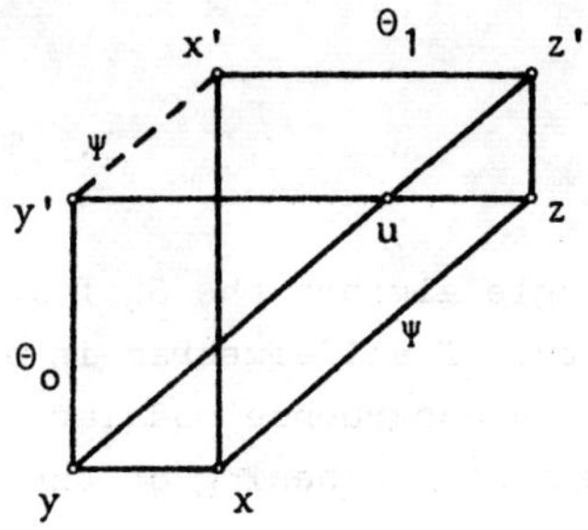

This lemma, as will turn out later is precisely what is needed to have groups in modular varieties being abelian.

3. THE DAY-TERMS

For a single algebra the Shifting Lemma is not equivalent to congruence modularity. If all algebras in a variety V satisfy the Shifting Lemma then V is congruence modular as we shall see. As an intermediate step we will use a strengthening of the Shifting Lemma. We shall call it

3.1 The Shifting Principle: Let α and γ be congruences and Λ a reflexive, symmetric and compatible relation on $\underline{A}$ with $(\alpha \cap \Lambda) \leq \gamma \leq \alpha$. For any elements $x,y,z,u \in \underline{A}$

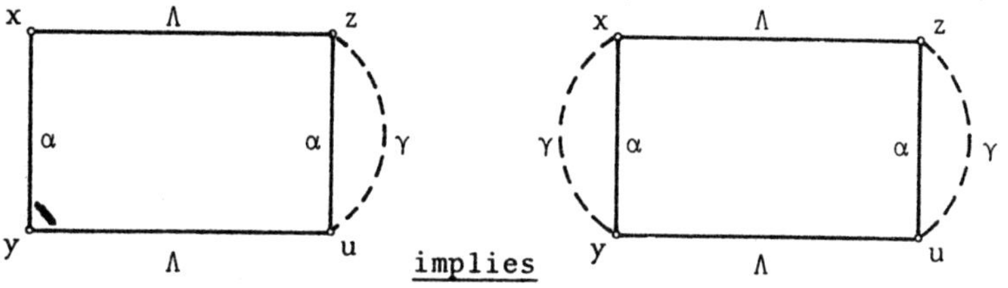

The proof that the Shifting Principle holds in a modular variety will have to be postponed for a few pages. We shall first show that the Shifting Principle implies congruence modularity.

3.2 Lemma: If the Shifting Principle holds for any algebra $\underline{A}$ then $\underline{A}$ is congruence modular.

Proof: The proof is based on an idea of A. DAY [9]. Suppose $\alpha \geq \gamma$ and β are congruences. Then

$$\alpha \wedge (\beta \vee \gamma) = \bigcup_{n \in \mathbb{N}} \alpha \cap \Lambda_n$$

where

$$\Lambda_0 := \beta \quad \text{and}$$

$$\Lambda_{n+1} := \Lambda_n \circ \gamma \circ \beta$$

Thus it suffices to show that

$$\alpha \cap \Lambda_n \subseteq (\alpha \wedge \beta) \vee \gamma \quad \text{for every} \quad n \in \mathbb{N}.$$

For $n = 0$ this is trivial. Assuming then that $\alpha \cap \Lambda_k \subseteq (\alpha \wedge \beta) \vee \gamma$, $(x,y) \in \alpha \cap \Lambda_{k+1}$ implies that $(x,y) \in \alpha \cap (\Lambda_k \circ \gamma \circ \beta) \subseteq \alpha \cap (\Lambda_k \circ \gamma \circ \Lambda_k)$. Thus there exist $u,v \in \underline{A}$ with

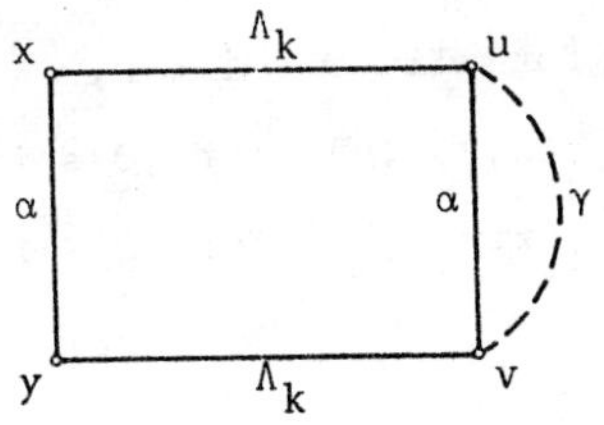

Replacing γ by $(\alpha\wedge\beta)\vee\gamma$ then the inductive hypothesis makes the Shifting Principle applicable, yielding $(x,y)\in(\alpha\wedge\beta)\vee\gamma$.

To prepare an immediate application, let us define:

3.3 Definition: An algebra $\underline{A}$ has regular congruences, if every congruence is uniquely determined by any of its classes.

J. HAGEMANN used R. WILLE's Mal'cev-type characterization of regularity [41] to show that varieties of regular algebras are congruence modular (and even n-permutable for some n) [22]. Refining this theorem, S. BULMAN-FLEMING, A. DAY and W. TAYLOR proved [4]:

3.4 Theorem: If all subalgebras of $\underline{A}\times\underline{A}$ have regular congruences then $\underline{A}$ is congruence modular.

Proof: We prove the Shifting Principle, namely, with notation as in 3.1 we take Λ any reflexive subalgebra of $\underline{A}\times\underline{A}$ and define congruences $\alpha\times\gamma$ and $\gamma\times\gamma$ on Λ by

$$(a,b)\ \alpha\times\gamma\ (c,d) \quad \text{iff} \quad a\ \alpha\ c \quad \text{and} \quad b\ \gamma\ d$$

and $$(a,b)\ \gamma\times\gamma\ (c,d) \quad \text{iff} \quad a\ \gamma\ c \quad \text{and} \quad b\ \gamma\ d.$$

Now for an arbitrary $a\in\underline{A}$ look at the (a,a)-class of $\alpha\times\gamma$. Namely, if $(a,a)\ \alpha\times\gamma\ (r,s)$ we get $(r,s)\in\alpha$ since $\gamma\leq\alpha$ and α is transitive. Hence $(r,s)\in\Lambda\cap\alpha$, therefore $(r,s)\in\gamma$, yielding $(a,a)\ \gamma\times\gamma\ (r,s)$ by transitivity of γ. We have just shown that $[(a,a)]\ \alpha\times\gamma = [(a,a)]\ \gamma\times\gamma$ and may now infer $\alpha\times\gamma=\gamma\times\gamma$ by regularity. Thus $(x,z)\ \gamma\times\gamma\ (y,u)$ in 3.1 hence $x\ \gamma\ y$.

Note that in fact we have shown slightly more, namely that c-regularity implies modularity. Here $\underline{c}$-regularity is a weakening of the notion of regularity. Algebras are supposed to have a constant $\underline{c}$, and every congruence is supposed to be determined by the class containing the constant $\underline{c}$.

We can now state and prove A. DAY's Mal'cev type characterization of congruence modular varieties [9]. The following chapter will very much depend on a deeper understanding of the geometrical meaning of those terms.

3.5 Theorem: A variety $\underline{V}$ is congruence modular if and only if for some natural number n there exist quaternary terms $m_o,\dots,m_n$ such that the following equations hold in $\underline{V}$:

(MO) $m_o(x,y,z,u) = x$

(M1) $m_i(x,x,y,y) = x$ for all $0 \leq i \leq n$,

(M2) $m_i(x,y,x,y) = m_{i+1}(x,y,x,y)$ for $0 \leq i < n$, i even

(M3) $m_i(x,y,z,z) = m_{i+1}(x,y,z,z)$ for $0 \leq i < n$, i odd

(M4) $m_n(x,y,z,u) = y$.

Proof: (existence):
Let $\underline{F}_V(\{x,y,z,u\})$ be the free algebra in $\underline{V}$ freely generated by the set $X = \{x,y,z,u\}$. For $a,b \in X$ let $\Theta_{(a,b)}$ be the smallest congruence relation on $\underline{F}_V(X)$ containing the pair (a,b). Thus for $\alpha := \Theta_{(x,y)} \vee \Theta_{(z,u)}$, $\beta := \Theta_{(x,z)} \vee \Theta_{(y,u)}$ and $\gamma := \Theta_{(z,u)}$ we set $\delta := (\alpha \wedge \beta) \vee \gamma$ to have the situation

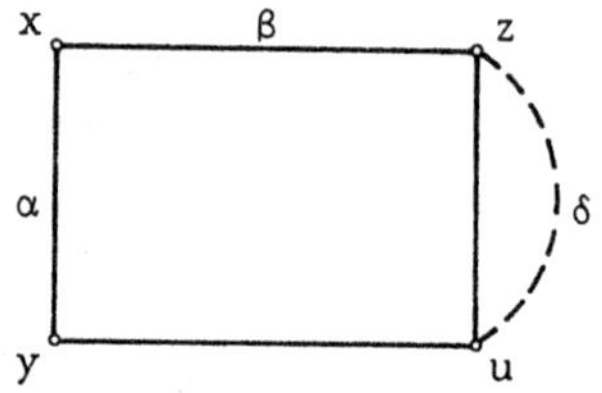

The Shifting Lemma hence yields $(x,y) \in (\alpha\wedge\beta) \vee \gamma$ i.e. there exists a number n such that (x,y) is in the n-fold relational product of $\alpha\wedge\beta$ and γ. More precisely, for n there exist elements $m_o,\ldots,m_n$ of $\underline{F}_V(X)$ such that the following relations hold:

(mO) $m_o = x$

(m2') $m_i (\Theta_{(x,y)} \vee \Theta_{(z,u)}) \wedge (\Theta_{(x,z)} \vee \Theta_{(y,u)}) m_{i+1}$ for i even, $0 \leq i < n$

(m3') $m_i \; \Theta_{(z,u)} m_{i+1}$ for i odd, $0 \leq i < n$

(m4) $m_n = y$.

(m2') together with (m3') may be replaced by

(m1) $m_i \; \Theta_{(x,y)} \vee \Theta_{(z,u)} x$ for all $0 \leq i \leq n$

(m2) $m_i \; \Theta_{(x,z)} \vee \Theta_{(y,u)} m_{i+1}$ for i even, $0 \leq i < n$

(m3) $m_i \; \Theta_{(z,u)} m_{i+1}$ for i odd, $0 < i < n$.

Since every element of $\underline{F}_V(X)$ may be written as a quaternary $\underline{V}$-term, and using MAL'CEV's argument [32] we obtain terms $m_o,\ldots,m_n$ satisfying the equations (MO),...,(M4) as they correspond to the relations (mO),...,(M4).

Sufficiency: We prove the Shifting Principle 3.1.
With the notation of 3.1 we define elements $\underline{m}_i := m_i(x,y,z,u)$ and

$\underline{m}_i' := m_i(x,y,x,y)$. Then the equations for the m_i ensure us of the relations

S1: $\underline{m}_i = m_i(x,y,z,u) \ \alpha \ m_i(x,x,z,z) = x$ for all i

S2: $\underline{m}_i' = m_i(x,y,x,y) \ \alpha \ m_i(x,x,x,x) = x$ as well as

S3: $\underline{m}_i' = m_i(x,y,x,y) \ \Lambda \ m_i(x,y,z,u) = \underline{m}_i$

S1, S2 and S3 jointly imply $\underline{m}_i' \ \gamma \ \underline{m}_i$ for all i.
For <u>i even</u> we have $\underline{m}_i' = \underline{m}_{i+1}'$, therefore with the above
S4: $\underline{m}_i \ \gamma \ \underline{m}_{i+1}$ for i even. For i odd the corresponding relation follows from
$\underline{m}_i = m_i(x,y,z,u) \ \gamma \ m_i(x,y,z,z) = m_{i+1}(x,y,z,z) \ \gamma \ m_{i+1}(x,y,z,u) = \underline{m}_{i+1}$.
Therefore with transitivity of γ we obtain
$x = \underline{m}_o \ \gamma \ \underline{m}_1 \ \gamma \ \underline{m}_2 \ \ldots \ \underline{m}_{n-1} \ \gamma \ \underline{m}_n = y$, i.e. $(x,y) \in \gamma$.
Now 3.2 completes the proof.

As a corollary we obtain

3.6 <u>Corollary:</u> In a variety of algebras the Shifting Lemma is equivalent to modularity, and both are equivalent to the Shifting Principle.

It is worthwile, to look more closely at the points $\underline{m}_i$ and $\underline{m}_i'$ constructed in the above proof. Let therefore β and γ be congruences and x,y,z,u be points with

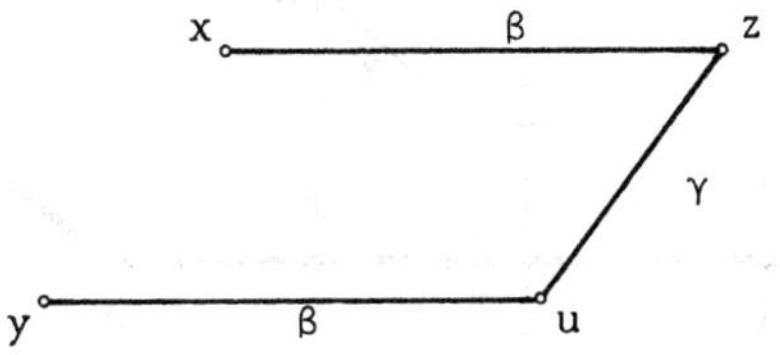

Under which conditions do these points form a β-γ-parallelogram, i.e. when do we have $x \equiv y (\mathrm{mod}\ \gamma)$?
As above we will define again $\underline{m}_i := m_i(x,y,z,u)$. Then we get:

(a) $x = m_o$ by (M0)

(b) $\underline{m}_i \ \beta \ \underline{m}_{i+1}$ for i even by (M2)

(c) $\underline{m}_i \ \gamma \ \underline{m}_{i+1}$ for i odd by (M3)

(d) $\underline{m}_n = y$ by (M4)

This situation is shown in the following figure (using $n = 7$).

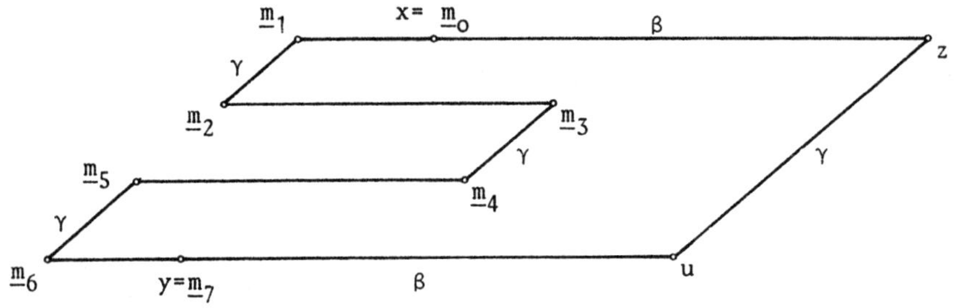

One direction of the following lemma is obvious now:

3.7 <u>Lemma:</u> $x \gamma y$ if and only if $\underline{m}_i \gamma \underline{m}_{i+1}$ for i even.

Notice that we had no chance so far, to use the powerful equations (M1). Only if we start with a completed parallelogram we can use (M1) and obtain $\underline{m}_i(x,y,z,u) \gamma \underline{m}_i(x,x,z,z) = x$. In particular $\underline{m}_i \gamma \underline{m}_{i+1}$ follows, finishing the proof of lemma 3.7.

Putting together a completed parallelogram and an uncompleted parallelogram, so that corresponding points are congruent modulo some congruence relation α, we obtain the following familiar figure:

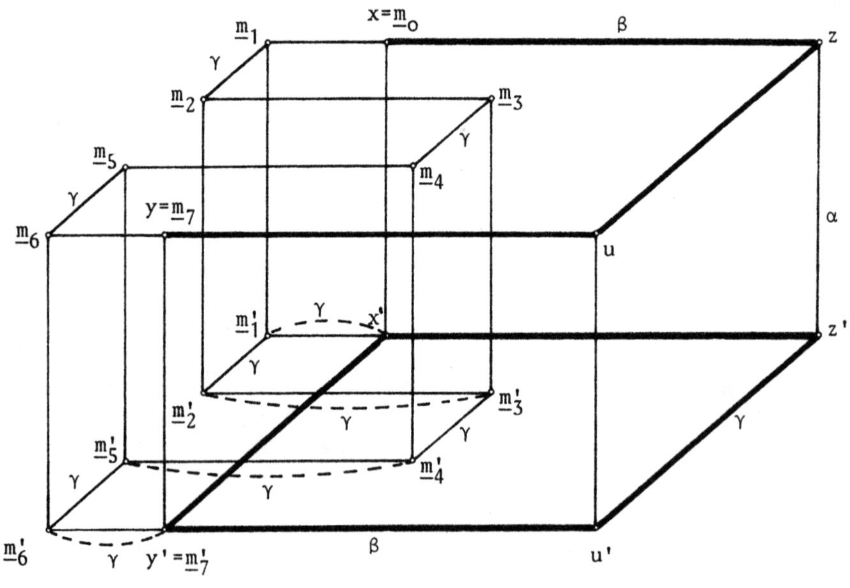

Thus all we need for showing $x \gamma y$ is, to apply the Shifting Lemma to the situation

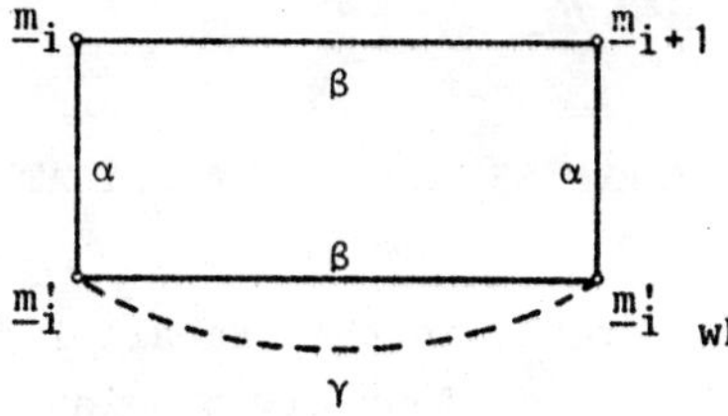

whenever i is even.

Thus all we need is the condition $\alpha\wedge\beta \leq \gamma$. This is precisely the original proof for the cube lemma.

Clearly we can easily formulate a theorem having as special cases all the configuration theorems of the preceding chapter. We call it the Parallelogram Principle:

3.8 Parallelogram Principle: To prove that (x,y,z,u) with $x\ \beta\ z\ \gamma\ u\ \beta\ y$ forms a γ-β-parallelogram, find a β'-γ'-parallelogram (x',y',z',u') such that (x,x'), (y,y'), (z,z') and (u,u') are from some congruence α. If $(\alpha \vee (\beta'\wedge\gamma')) \wedge \beta \leq \gamma$ holds then (x,y,z,u) is a γ-β-parallelogram.

Loosely spoken: "Look along some congruence α onto a completed parallelogram".

To prove it, just follow the reasoning after 3.7 with β and γ replaced by β' and γ' in the completed parallelogram. The only crucial step is the application of the Shifting lemma. In the picture of the preceding page replace β by $\phi := \beta \vee (\beta'\wedge\gamma')$ and γ by $\Psi := (\alpha\wedge\phi) \vee (\beta'\wedge\gamma')$. Thus the Shifting Lemma yields $\underline{m}_i'\ \phi\wedge\Psi\ \underline{m}'_{i+1}$ hence

$$\begin{aligned}(\underline{m}_i,\underline{m}_{i+1}) &\in [([\beta \vee (\beta'\wedge\gamma')] \wedge \alpha) \vee (\beta'\wedge\gamma')] \wedge \beta = \\ &= [\beta \vee (\beta'\wedge\gamma')] \wedge [\alpha \vee (\beta'\wedge\gamma')] \wedge \beta = \\ &= [\alpha \vee (\beta'\wedge\gamma')] \wedge \beta \leq \gamma.\end{aligned}$$

4. A SIXARY TERM AND ITS APPLICATIONS

In the last two chapters, notably in the "Parallelogram Principle" we developed a method for showing that four given points form a parallelogram. There is, however, no method visible to construct a fourth parallelogram point from three given ones. Clearly this is not possible in general, since it would imply congruence permutability. It turns out though, that, given some auxiliary points and congruences, parallelograms can be completed. In particular, parallelograms with one pair of sides being lines of a factor congruence can always be completed (Corollary 4.5). This chapter will in fact provide a term p, doing this uniformly throughout the variety.

The construction of p relies almost exclusively on the geometrical visualization developed in former chapters and provides an excellent example of how geometry can inspire and lead algebraic computations. Thus instead of just writing down p and proving the characteristic relations, we will include the geometric reasoning which necessarily leads to the discovery of p.

The usefulness of p will become apparent in later chapters. In many instances this term simulates MAL'CEV's term from 1.2, thus allowing us to carry many results about permutable varieties from [16] over to modular varieties. In particular the methods of coordinatizing algebras in permutable varieties as developed in [16] need precisely Corollary 4.5 as an additional ingredient for being valid in modular varieties. This has been worked out in [17].

4.1 __Theorem:__ In a modular variety $\underline{V}$ there exists a 6-ary term $p(x_1,\ldots,x_6)$ with the following property: Let Θ_0, Θ_1 and Ψ be congruences on $\underline{A} \in \underline{V}$ with $\Theta_0 \wedge \Theta_1 \leq \Psi$ and let a,b,c,d,e,f be elements of $\underline{A}$. Then the relations

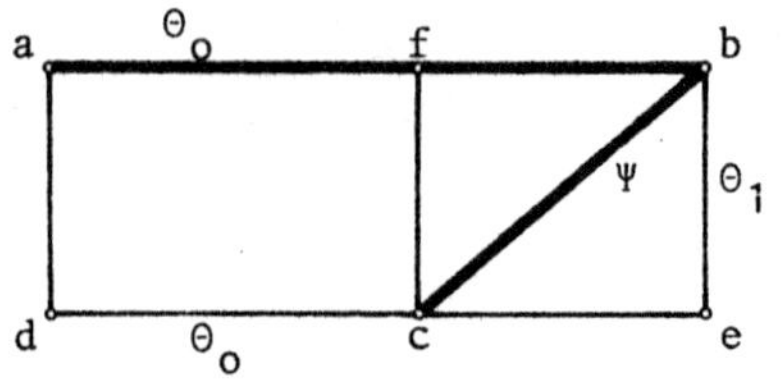

imply that $\underline{p} := p(a,b,c,d,e,f)$ satisfies

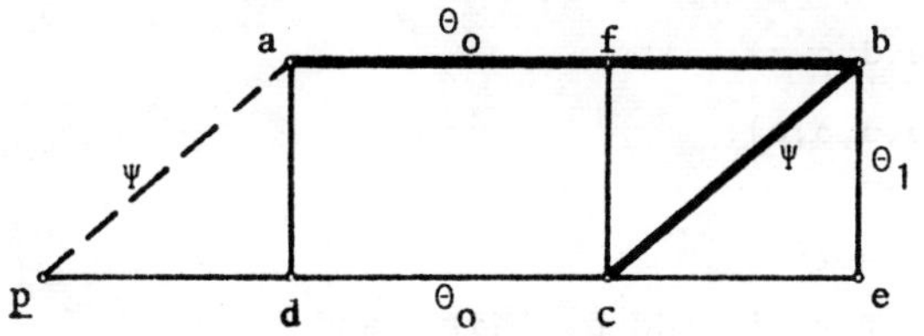

This term p thus gives a fourth parallelogram point $\underline{p}$ for a,b and c, with d,e,f being auxiliary points.

Let us then start with the hypothesis of the theorem. For the first part of the proof we can do without the point f. Since we also assumed that $\Theta_o \wedge \Theta_1 \leq \Psi$ we do not loose generality if we set $\Theta_o \wedge \Theta_1 = 0$. Firstly let us apply the DAY-terms to find the points $\underline{m}_i := m_i(a,d,b,c)$. Then the equations (M0), (M2), (M3) and (M4) yield the relations familiar from 3.7:

(1) $\underline{m}_o = a$

(2) $\underline{m}_i \; \Theta_o \; \underline{m}_{i+1}$ for i even

(3) $\underline{m}_i \; \Psi \; \underline{m}_{i+1}$ for i odd and

(4) $\underline{m}_n = d$.

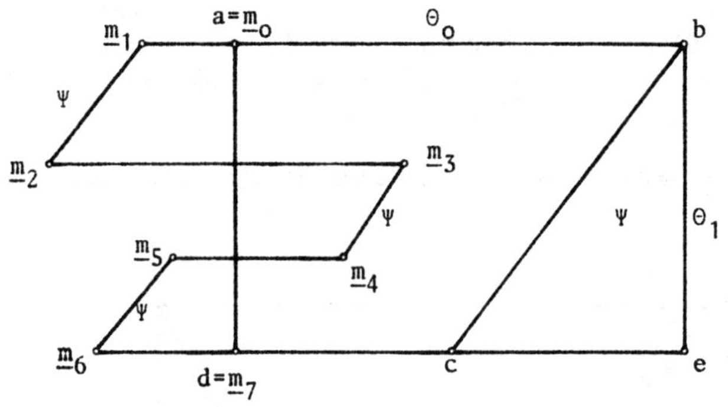

Now geometrically we have to shift the Ψ-line $\overline{b,c}$ to join up with the point a so, that it intersects the Θ_o-line $\overline{d,e}$ in a point, which will be $\underline{p}$. This is not possible in just one step, hence we will proceed by shifting the line-segments $\overline{\underline{m}_i,\underline{m}_{i+1}}$ for i odd and then "add them together" to make them form the desired Ψ-line starting at a. The problem is that we have no control about as to where the $\overline{\underline{m}_i,\underline{m}_{i+1}}$ line segments are positioned "horizontally" (modulo Θ_o).

A first step in overcoming this difficulty is the observation that the segments $\overline{\underline{m}_i,\underline{m}_{i+1}}$ do have points of intersection with the Θ_1-line through a.

Namely define:

$$\underline{\hat{m}}_i := m_i(a,d,c,c) \quad \text{and}$$
$$\underline{\tilde{m}}_i := m_i(a,d,b,e).$$

Then

(5) $\underline{\hat{m}}_i = \underline{\hat{m}}_{i+1}$ for i odd by (M3) and

(6) $\underline{\hat{m}}_i \;\Theta_1\; a$ by (M1).

From the definition and using (5) we find

(7) $\underline{m}_i \;\Psi\; \underline{\hat{m}}_i \;\Psi\; \underline{m}_{i+1}$ for i odd.

Similarly for the $\underline{\tilde{m}}_i$ we find the relations

(8) $\underline{\tilde{m}}_i \;\Theta_o\; \underline{m}_i$ for every i by definition and

(9) $\underline{\tilde{m}}_i \;\Theta_1\; a$ by (M1).

Using (9), (8) and (2) we conclude

(10) $\underline{\tilde{m}}_i \;\Theta_o \wedge \Theta_1\; \underline{\tilde{m}}_{i+1}$ if i is even, hence with our assumption:

(11) $\underline{\tilde{m}}_i = \underline{\tilde{m}}_{i+1}$ if i is even.

Now the relations (6) and (7) show that for i odd the line segment $\overline{\underline{m}_i,\underline{m}_{i+1}}$ intersects $\overline{a,d}$ in $\underline{\hat{m}}_i$, for i even, the corresponding fact follows from the relations (8), (9) and (11).

Our last picture therefore needs to be corrected as follows:

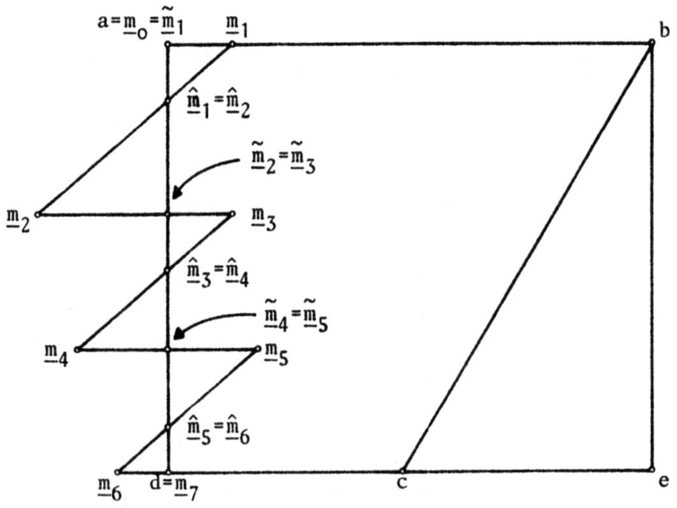

At this point we observe that the Θ_o-Ψ-parallelograms given by three points $\underline{\tilde{m}}_i$, $\underline{m}_i$ and $\underline{\hat{m}}_i$ for i odd can be completed. For this we define

$$\underline{\ddot{m}}_i := m_i(a,d,c,e).$$

The relations

(12) $\underline{\tilde{m}}_i \; \Psi \; \underline{\ddot{m}}_i \; \Theta_o \; \underline{\hat{m}}_i$ are obvious from the definition.

The points and relations collected so far provide a proof of the following lemma which is an important intermediate step in the proof of 4.1.

4.2 Lemma: There exist terms $s_o,\dots,s_{n-1}$ and $t_1,\dots,t_{n-1}$ in every modular variety such that with the hypothesis of 4.1 we have for $\underline{t}_i := t_i(a,b,c,d,e)$ and $\underline{s}_i := s_i(a,b,c,d,e)$ that $a = \underline{s}_o$, $d = \underline{s}_{n-1}$, and $\underline{s}_i \; \Psi \; \underline{t}_{i+1} \; \Theta_o \; \underline{s}_{i+1} \; \Theta_1 \; \underline{s}_i$ for every $i \le n-2$.

Proof: Define

$$t_i(x_1,x_2,x_3,x_4,x_5,x_6) := \begin{cases} m_i(x_1,x_4,x_3,x_5), & i \text{ odd} \\ m_i(x_1,x_4,x_2,x_3), & i \text{ even.} \end{cases}$$

and

$$s_i(x_1,x_2,x_3,x_4,x_5,x_6) := \begin{cases} m_i(x_1,x_4,x_3,x_3), & i \text{ odd} \\ m_i(x_1,x_4,x_2,x_5), & i \text{ even.} \end{cases}$$

then in our previous notation

$$\underline{t}_i = \begin{cases} \underline{\ddot{m}}_i & \text{for } i \text{ odd} \\ \underline{m}_i & \text{for } i \text{ even} \end{cases}$$

and

$$\underline{s}_i = \begin{cases} \underline{\hat{m}}_i & \text{for } i \text{ odd} \\ \underline{\tilde{m}}_i & \text{for } i \text{ even.} \end{cases}$$

Now the obvious idea, motivated by the geometry is to apply Lemma 4.2 onto itself. This will become clear by looking at the picture on the next page.

Applying Lemma 4.2 first yields us points $\underline{t}_1,\dots,\underline{t}_{n-1}$ and $\underline{s}_o,\dots,\underline{s}_{n-1}$. Then we shall construct new points a^1 and d^1 in the shown position. After having done that we shall apply Lemma 4.2 again but now replacing a by a^1 and d by d^1. This way another collection of points $\underline{t}^1_1$, $\underline{t}^1_2$, ..., $\underline{t}^1_{n-1}$ and $\underline{s}^1_o$, $\underline{s}^1_1$, ..., $\underline{s}^1_{n-1}$ is obtained. Now the choice of the points a^1 and d^1 will guarantee that $\underline{s}^1_1 = \underline{t}_1$. This way we have prolonged the short segment $\overline{a,\underline{t}_1}$ by the segment $\overline{\underline{t}_1,\underline{t}^1_2}$, which is just a parallel shift of $\overline{\underline{s}_1,\underline{t}_1}$. Thus continuing with the new points a^2 and d^2, and so on, we manage to connect all segments $\overline{\underline{s}_i,\underline{t}_{i+1}}$ to finally end up with the desired line $\overline{a,\underline{p}}$.

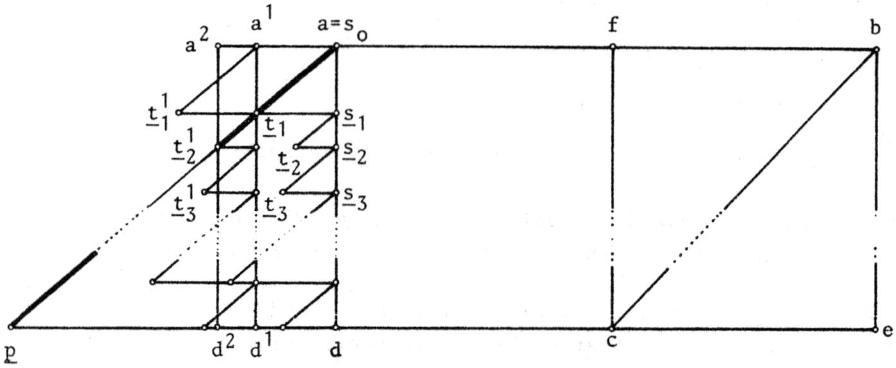

We have promised to construct $\underline{p}$ as the value of a term. But our plan, presented above can easily be modified this way. Thus for the proof of 4.1 we define terms

$$a^o(x_1, \dots, x_6) := x_1$$
$$b^o(x_1, \dots, x_6) := x_2$$
$$\vdots \qquad \vdots$$
$$f^o(x_1, \dots, x_6) := x_6$$

and $t_i^o(x_1, \dots, x_6) := t_i(x_1, \dots, x_6)$ as well as

$s_i^o(x_1, \dots, x_6) := s_i(x_1, \dots, x_6)$ where the t_i and s_i are the terms from Lemma 4.2. Further define recursively for $1 \le k \le n-1$:

$$a^k(x_1,\dots,x_6) := t_k^{k-1}(x_1,x_2,x_6,x_1,x_2,x_6)$$
$$d^k(x_1,\dots,x_6) := t_k^{k-1}(x_4,x_5,x_3,x_4,x_5,x_3)$$
$$t_i^k(x_1,\dots,x_6) := t_i(a^k(x_1,\dots,x_6),x_2,x_3,d^k(x_1,\dots,x_6),x_5,x_6)$$
$$s_i^k(x_1,\dots,x_6) := s_i(a^k(x_1,\dots,x_6),x_2,x_3,d^k(x_1,\dots,x_6),x_5,x_6)$$

Let us agree that with $\underline{a}^i$, $\underline{d}^i$, $\underline{s}_i^k$, $\underline{t}_i^k$ we mean the result from applying the term with the corresponding name onto the points a,b,c,d,e,f (in that order).

We need to show five relations:

(13) $\underline{a}^k \ \Theta_1 \ \underline{d}^k$

(14) $\underline{a}^k \ \Theta_o \ a$ and $\underline{d}^k \ \Theta_o \ d$

(15) $\underline{t}_i^k \ \Theta_o \ \underline{t}_i$ for $1 \le i \le n-1$

(16) $\underline{s}_i^k \ \Theta_o \ \underline{s}_i$ for $i \le n-1$

(17) $\underline{t}_{k+1}^k \ \Psi \ a.$

The first four relations are easy namely

$$\underline{a}^{k+1} = t^k_{k+1}(a,b,f,a,b,f)\ \Theta_1\ t^k_{k+1}(d,e,c,d,e,c) = \underline{d}^{k+1}$$
$$\underline{a}^{k+1} = t^k_{k+1}(a,b,f,a,b,f)\ \Theta_o\ t^k_{k+1}(a,a,a,a,a,a) = a$$
$$\underline{d}^{k+1} = t^k_{k+1}(d,e,c,d,e,c)\ \Theta_o\ t^k_{k+1}(d,d,d,d,d,d) = d$$

since all the terms m_i and hence all their composites are idempotent. To prove (15) and (16) we need (14):

$\underline{t}^k_i = t_i(a^k,b,c,d^k,e,f)\ \Theta_o\ t_i(a,b,c,d,e,f) = \underline{t}_i$ and

$\underline{s}^k_i = s_i(a^k,b,c,d^k,e,f)\ \Theta_o\ s_i(a,b,c,d,e,f) = \underline{s}_i$.

Clearly (17) is the crucial relation. We use induction on k. The case $k=0$ comes from 4.2. In the induction step we first use (15), (16) and Lemma 4.2 to show that

(18) $\underline{t}^k_{k+1}\ \Theta_o\ \underline{t}_{k+1}\ \Theta_o\ \underline{s}_{k+1}\ \Theta_o\ \underline{s}^{k+1}_{k+1}$.

Now we apply 4.2 again with a replaced by $\underline{a}^{k+1}$ and d replaced by $\underline{d}^{k+1}$. Thus we obtain $\underline{s}^{k+1}_{k+1}\ \Theta_1\ \underline{a}^{k+1}$. On the other hand $\underline{a}^{k+1} = t^k_{k+1}(a,b,f,a,b,f)\ \Theta_1\ t^k_{k+1}(a,b,c,d,e,f) = \underline{t}^k_{k+1}$. Thus $\underline{s}^{k+1}_{k+1}\ \Theta_1\ \underline{t}^k_{k+1}$ and $\underline{s}^{k+1}_{k+1}\ \Theta_o\ \underline{t}^k_{k+1}$ from above, yielding

(19) $\underline{s}^{k+1}_{k+1} = \underline{t}^k_{k+1}$

according to our assumption that $\Theta_o \wedge \Theta_1 = 0$. Using the inductive hypothesis and Lemma 4.2 we finally arrive at

$\underline{t}^{k+1}_{k+2} = t_{k+2}(a^{k+1},b,c,d^{k+1},e,f)\ \Psi\ s_{k+1}(a^{k+1},b,c,d^{k+1},e,f) = \underline{s}^{k+1}_{k+1} = \underline{t}^k_{k+1}\ \Psi\ a$,

proving (17).

Hence defining $p(x_1,\ldots,x_6) := t^{n-2}_{n-1}(x_1,\ldots,x_6)$ we combine (15) and (17) to have for $k = n-2$:

$$p(a,b,c,d,e,f) = \underline{t}^{n-2}_{n-1}\ \Theta_o\ \underline{t}_{n-1}\ \Theta_o\ d \quad \text{and}$$
$$p(a,b,c,d,e,f) = \underline{t}^{n-2}_{n-1}\ \Psi\ a.$$

This finishes the proof of 4.1.

We can use the techniques from the above proof to make the following improvement:

4.3 Theorem: There exists a ternary term $t(x,y,z)$ in every modular variety such that given congruences Θ_o, Θ_1 and Ψ with $\Theta_o \wedge \Theta_1 \le \Psi$ and elements a,b,c,d,e,f with

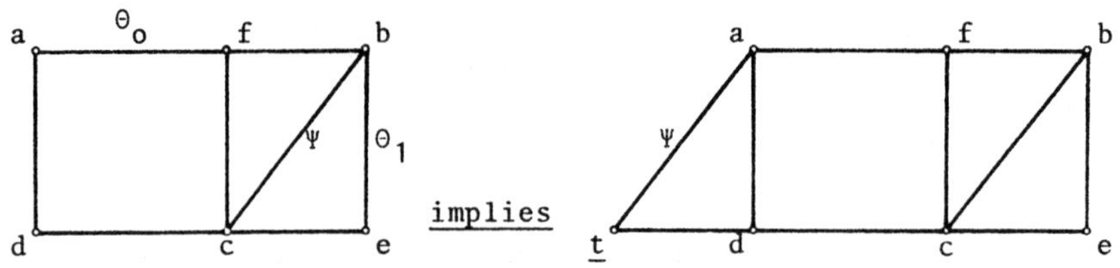

with $\underline{t} := t(d,e,c)$. Moreover, the equation $t(x,y,y) = x$ holds in $\underline{V}$.

Proof: Define $t(x,y,z) := p(x,y,z,x,y,z)$, then
$t(d,e,c) = p(d,e,c,d,e,c)\ \Theta_1\ p(a,b,c,d,e,f)$ and
$t(d,e,c)\ \Theta_0\ t(d,d,d) = d\ \Theta_0\ p(a,b,c,d,e,f)$.
Thus $t(d,e,c)\ \Theta_0 \wedge \Theta_1\ p(a,b,c,d,e,f)$.
To see that the equation $t(x,y,y) = x$ holds in $\underline{V}$, take an algebra $\underline{A} \in \underline{V}$ and $x,y \in \underline{A}$. Consider the direct product $\underline{A} \times \underline{A}$ and set $\Theta_0 := \pi_1$, $\Theta_1 := \pi_2$ and $\Psi := \pi_1$. With $a = (x,y)$, $b = f = (y,y)$, $e = c = (y,x)$ and $d = (x,x)$ the geometric conditions demand that $t((x,x),(y,x),(y,x))$ has to be (x,x). Evaluating the first component gives us $t(x,y,y) = x$.

4.4 Corollary: Let Θ_0, Θ_1 and Ψ be congruences with $\Theta_0 \wedge \Theta_1 \leq \Psi \leq \Theta_0 \vee \Theta_1$. If Θ_0 permutes with Θ_1 then Ψ permutes with Θ_0 and with Θ_1.

Proof: Suppose $a\ \Theta_0\ b\ \Psi\ c$ then $(b,c) \in \Theta_0 \circ \Theta_1$, $(b,c) \in \Theta_1 \circ \Theta_0$ and $(a,c) \in \Theta_0 \circ \Psi \subseteq \Theta_0 \circ \Theta_0 \circ \Theta_1 \subseteq \Theta_0 \circ \Theta_1 \subseteq \Theta_1 \circ \Theta_0$ imply the existence of further points d,e,f with the configuration of 4.3. Applying 4.3 therefore we get $a\ \Psi\ p(d,e,c)\ \Theta_0\ c$.

The form in which this corollary usually will be applied is as follows:

4.5 Corollary: Let $\underline{B} = \prod_{i \in I} \underline{A}_i$ be the direct product of the algebras $\underline{A}_i$, i I. Then every congruence relation on $\underline{B}$ permutes with every factor congruence.

This corollary has many useful applications in GUMM, HERRMANN [21] where it guarantees that certain lattice theoretical decompositions (in the congruence lattices) actually yield algebraic decompositions.

More applications are provided in the following chapter. The following corollary appears in WERNER [39] under the assumption that $\underline{B}$ has 3-permutable congruences:

4.6 Corollary: Let $\underline{B} = \prod_{i \in I} \underline{A}_i$ be the direct product of the algebras $\underline{A}_i$, $i \in I$. For a subset $S \subseteq I$ let π_S be the canonical projection onto $\underline{C} := \prod_{i \in S} \underline{A}_i$. For any congruence Θ on $\underline{B}$, the image of Θ under π_S $(= \{(\pi_S(a),\pi_S(b)) \mid (a,b) \in \Theta\})$ is a congruence relation on $\underline{C}$.

5. COORDINATIZATION

To coordinatize a line ℓ in Desarguesian affine geometry we would embed ℓ in a plane and choose two more lines, ℓ' and ℓ'' to obtain a set of three mutually nonparallel lines. After choosing an element o from ℓ arbitrarily, two points x and y on ℓ are added according to the following figure

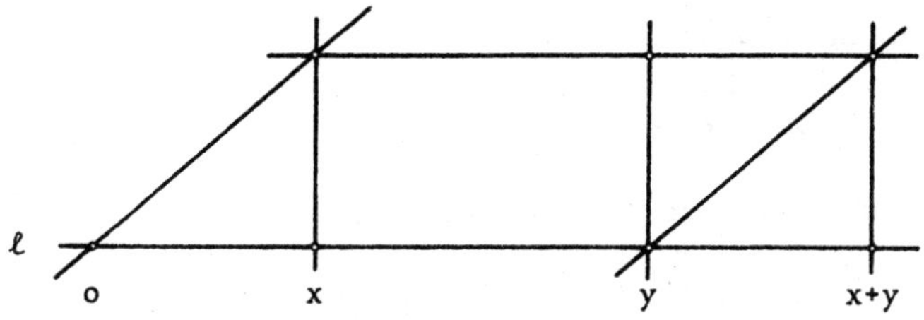

where horizontal, vertical and skew lines are lines parallel to ℓ, ℓ' and ℓ''.

This construction is known to yield an abelian group $\underline{G} = (\ell,+,o)$. Moreover the particular choice of o is irrelevant because, treating o as a variable, we actually have defined the ternary operation $x-o+y$.

In this chapter we will do precisely the same thing as above with an algebra $\underline{A}$ in a modular variety $\underline{V}$ playing the rôle of the line ℓ. The details we know about the congruence class geometry will provide us with the necessary tools making this process work. In particular, we will assume throughout this chapter that the algebra $\underline{A}$ is contained in a modular variety $\underline{V}$.

Thus we let the algebra $\underline{A}$ play the rôle of the line ℓ. Naturally our "plane" will be $\underline{A}\times\underline{A}$. There are two canonical congruences π_1 and π_2 on $\underline{A}\times\underline{A}$. Let $\ell := [(a,b)]\pi_2 = \{(x,b) \mid x \in \underline{A}\}$ be a class of π_2 for some $(a,b) \in \underline{A}\times\underline{A}$ chosen at will. Clearly A may be identified with the points of ℓ. For our line ℓ' an obvious candidate is found in $[(a,b)]\pi_1 = \{(a,x) \mid x \in A\}$. Now for ℓ'' we would like to choose the "diagonal", i.e. $\mathrm{diag}(\underline{A}) := \{(x,x) \mid x \in A\}$. Unfortunately $\mathrm{diag}(\underline{A})$ need not be a "line" in our sense, because lines are congruence classes. Thus we are forced to consider the smallest line containing $\mathrm{diag}(\underline{A})$, i.e. we set

$$\Delta := \langle\{((x,x),(y,y)) \mid (x,y) \in \underline{A}\times\underline{A}\}\rangle_{\underline{A}\times\underline{A}}.$$

Then ℓ'' has to be $[(x,x)]\Delta$ for some (any) x from $\underline{A}$.

We will go through this chapter however, assuming that ℓ'' is a line. This assumption is equivalent to: ℓ'' intersects ℓ and ℓ' in at most one point (see 5.1 below).

At this stage it is not clear whether we loose any generality by taking this particular choice for ℓ''. Of course for a "third line" ℓ'', in order to qualify for the geometric construction explained above two requirements are essential:

(L1) ℓ'' intersects ℓ and ℓ' in at most one point, i.e. if (x,y) and (x,z) are on ℓ'' then $y = z$ and symmetricall if (y,x) and (z,x) are on ℓ'' then $y = z$.

(L2) ℓ'' is long enough, i.e. for any $x \in \underline{A}$ there exist $y,z \in \underline{A}$ with $(x,y) \in \ell''$ and $(z,x) \in \ell''$.

The status of these requirements is made clear in the following observation:

5.1 Lemma: Let Θ be a congruence relation on $\underline{A} \times \underline{A}$. Then the following are equivalent:

(i) Some class of Θ is a line satisfying (L1) and (L2).

(ii) Every class of Θ is a line satisfying (L1) and (L2).

(iii) Θ is a common complement to the factor congruences on $\underline{A} \times \underline{A}$, i.e. $\Theta \vee \pi_1 = \Theta \vee \pi_2 = 1_{\underline{A}\times\underline{A}}$ and

$$\Theta \wedge \pi_1 = \Theta \wedge \pi_2 = 0_{\underline{A}\times\underline{A}}.$$

Proof: (i) → (iii): $\Theta \vee \pi_1 = \Theta \vee \pi_2 = 1_{\underline{A}\times\underline{A}}$ comes from (L2). Suppose $\Theta \wedge \pi_1 \neq 0_{\underline{A}\times\underline{A}}$, then for some $x,y,z \in \underline{A}$ with $y \neq z$ we have $(x,y)\ \Theta\ (x,z)$. Choose $u \in A$ with $(u,z) \in \ell''$ then (u,z) and (u,y) are both on ℓ'' by an application of the Shifting Lemma.

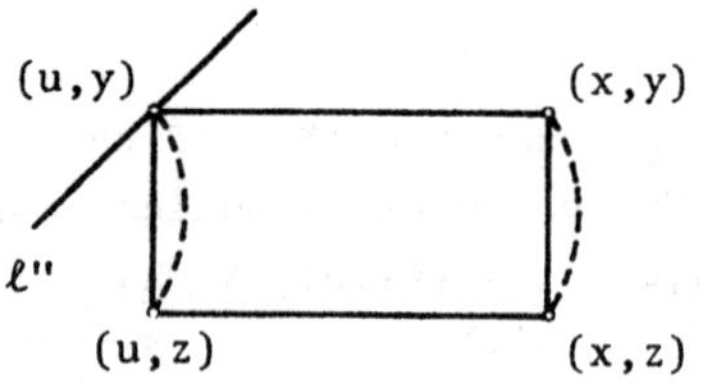

This contradiction proves (iii).

(iii) → (ii): (L1) is trivial for every class of Θ. Let ℓ be a class of Θ, $x \in A$ and $(a,b) \in \ell$. Then $(x,x)\ \pi_1 \vee \Theta\ (a,b)$ and, since Θ permutes with π_1 there exists a y from A with $(x,x)\ \pi_1\ (x,y)\ \Theta\ (a,b)$, hence $(x,y) \in \ell$.

In other words, Θ together with π_1 and π_2 form a 0-1-sublattice (named $\underline{M}_3$) of $\mathrm{Con}(\underline{A}\times\underline{A})$ as in the following lattice diagram:

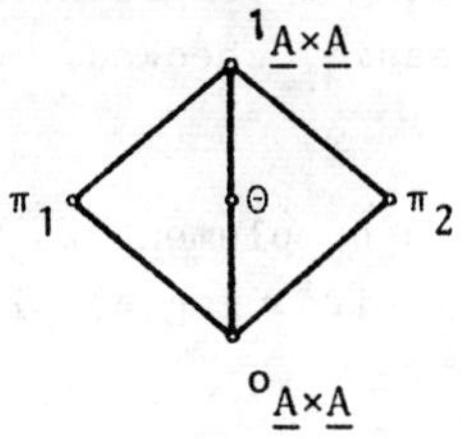

Next we show that we might as well assume that diag($\underline{A}$) is a class of some congruence Δ. How to define Δ?

Suppose $(x,y)\ \Delta\ (u,v)$. Then certainly we want $(y,y)\ \Delta\ (v,v)$. By the properties of Θ there exists a $z \in \underline{A}$ with $(x,y)\ \Theta\ (u,z)$. Now the little Desarguesian Theorem forces $(y,y)\ \Theta\ (v,z)$.

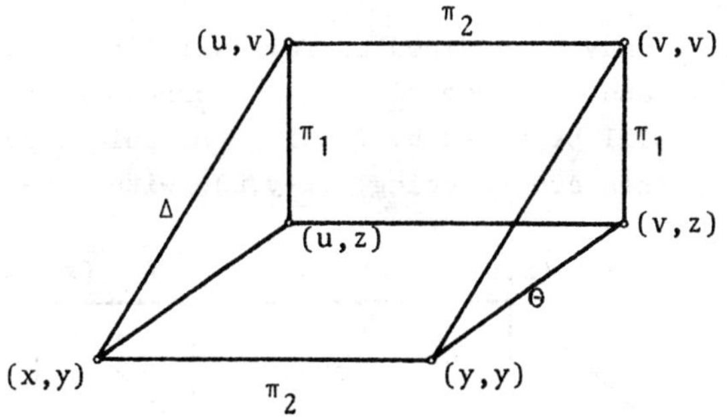

This reasoning goes back and forth between Θ and Δ thus we are forced to define:

$$(x,y)\ \Delta\ (u,v) \quad :\underline{\text{iff}}\ \exists z \in A \quad (x,y)\ \Theta\ (u,z) \text{ and } (y,y)\ \Theta\ (v,z).$$

We claim that this definition is equivalent to

$$(x,y)\ \Delta\ (u,v) \quad :\underline{\text{iff}}\ \exists a,b \in A \quad (x,a)\ \Theta\ (u,b) \text{ and } (y,a)\ \Theta\ (v,b).$$

Namely, if the right side is true then, since $\pi_1 \circ \Theta = 1_{\underline{A}\times\underline{A}}$ there exists a $z \in A$ with $(y,y)\ \Theta\ (v,z)$ hence the Cube Lemma yields: $(x,y)\ \Theta\ (u,z)$.

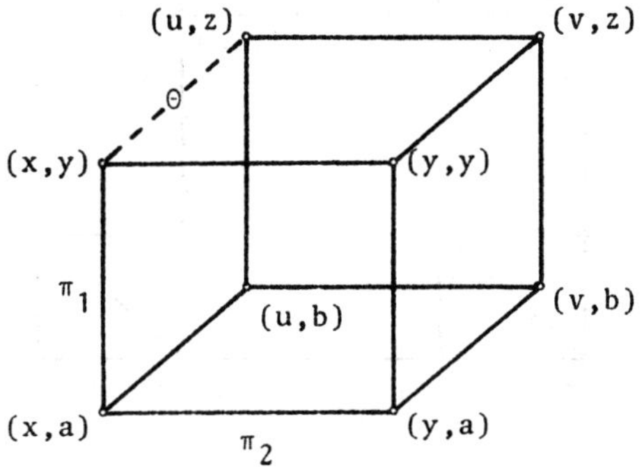

Now the properties of Δ, being a congruence relation on $\underline{A} \times \underline{A}$ with diag($\underline{A}$) a class of Δ are easily checked.
Thus we obtain:

5.2 Lemma: There is a common complement to the factor congruences π_1 and π_2 on $\underline{A} \times \underline{A}$ if and only if diag($\underline{A}$) is a class of some congruence on $\underline{A} \times \underline{A}$.

From now on we may assume diag($\underline{A}$) being a line. Geometrically we know then:

(11) any two lines (parallel to ℓ, ℓ' or ℓ'') intersect in at most one point and

(12) any two nonparallel lines (of the above) intersect in at least one point.

Thus addition on $\underline{A}$ may be defined as follows: First choose some element $o \in \underline{A}$. Elements x and y from $\underline{A}$, which correspond to points (x,b) and (y,b) on ℓ will be added by finding the unique point (x,u) with $(o,b)\ \Delta\ (x,u)$ and then constructing $(x+y,u)$ with $(x+y,u)\ \Delta\ (y,b)$.

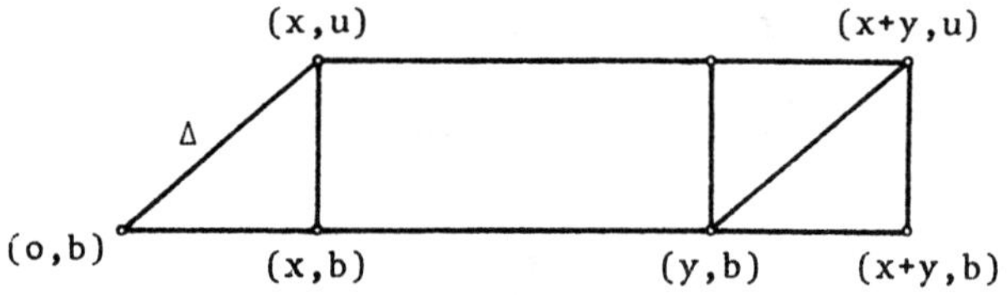

(11) and (12) from above guarantee that this process works and defines a binary operation + on $\underline{A}$. First we show independence of the choice of b. Thus suppose we had used $c \in \underline{A}$ instead of b. Clearly we are done if we can show $(y,c)\ \Delta\ (x+y,v)$ in the picture below.

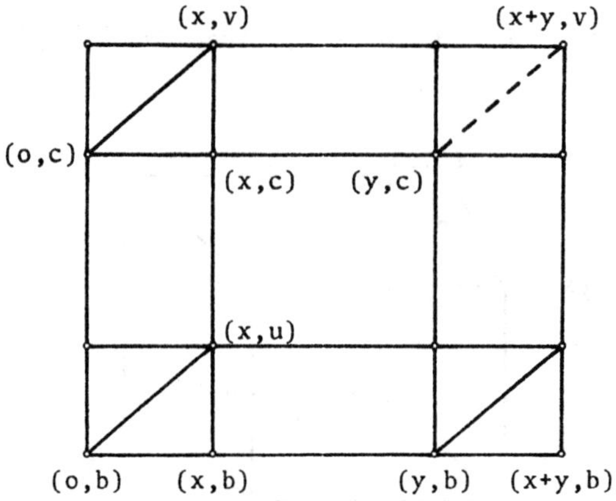

So the Cube Lemma gives the needed result. It is involved again in showing associativity of $+$. This is a simple exercise. For commutativity we need $(x+y,v)\ \Delta\ (x,b)$ in the situation below. 2.7 is all we need here:

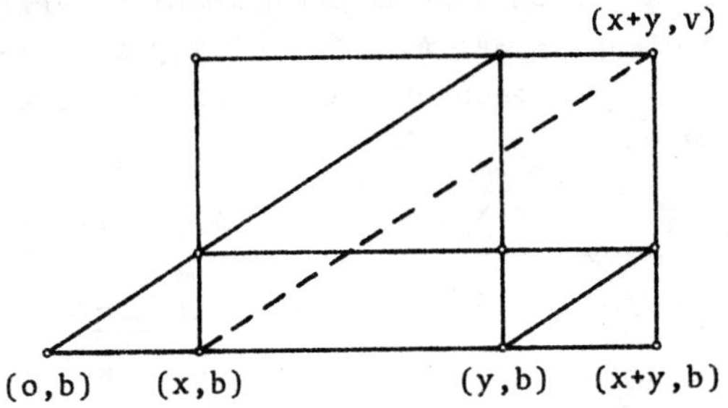

Since $o+x = x+o = x$ is obvious from the definition we only have to find $-x$. But, given x, $-x$ is found as indicated below:

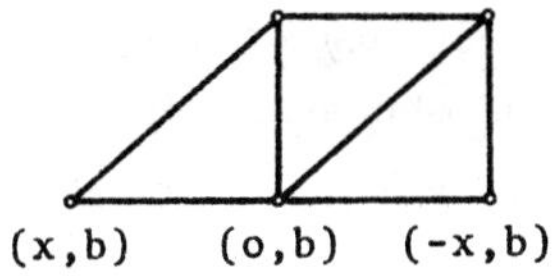

Clearly we have that $(A,+,-,o)$ is an abelian group by now. Had we chosen a different element $e \in A$, we would have obtained an isomorphic group $(A,+_e,-_e,e)$ with the isomorphism defined by $x \to x+e$.

The arbitrariness of the choice of the neutral element is removed if we consider the ternary operation $x-y+z$. Comparing the construction of $x-y+z$ with Theorem 4.3, we find that it agrees with our ternary term $t(x,y,z)$. In other words, $x-y+z$ is given as a term function on $\underline{A}$.

Let us now see how the other operations of $\underline{A}$ behave with respect to $x-y+z$. To this end consider an algebraic function τ on $\underline{A} \times \underline{A}$. Since $x-y+z = w$ if and only if for some elements u and v the configuration

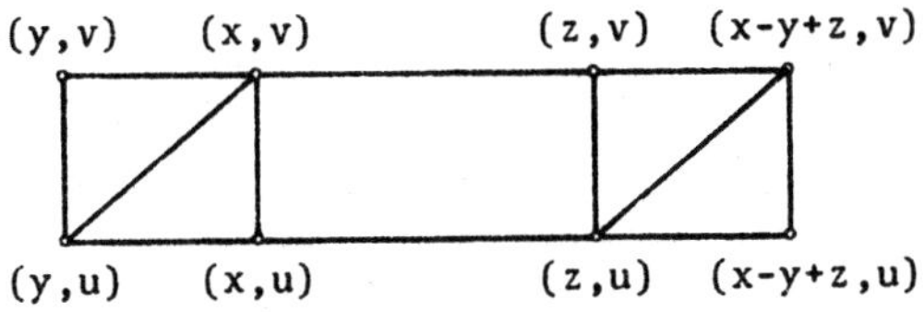

is given and since τ preserves congruences on $\underline{A} \times \underline{A}$, the image of this configuration will be a configuration of the same kind. In particular, looking at the first component τ_1 of τ only, we find the equation

$$\tau_1(x-y+z) = \tau_1(x) - \tau_1(y) + \tau_1(z).$$

Clearly this extends to τ being n-ary.

Let us collect the results achieved so far in a theorem:

5.3 Theorem: Let $\underline{A}$ be an algebra in a modular variety with $\mathrm{diag}(\underline{A})$ being a congruence class on $\underline{A} \times \underline{A}$. Then $t(x,y,z) = x-y+z$ for some abelian group defined on $\underline{A}$. Moreover every n-ary algebraic function τ on $\underline{A}$ satisfies:

$$\tau(x_1-y_1+z_1,\ldots,x_n-y_n+z_n) = \tau(x_1,\ldots,x_n) - \tau(y_1,\ldots,y_n) + \tau(z_1,\ldots,z_n).$$

Algebras as in the above theorem are called affine.

Affine algebras are almost modules. Let us work this out now. Again we look at a special case first. Suppose that $\underline{A}$ has a one-element subalgebra $\{o\}$. We then let o be the neutral element of our abelian group defined on $\underline{A}$ by $x+y := x-o+y$. Let R be the set of all unary algebraic functions τ on $\underline{A}$ which have no other constant than o in their representation. (If o is given by some constant, R is just the set of all unary term functions). The property of affineness clearly states that $\underline{R}$ can be viewed as a subring of the endomorphism ring of $(A,+,-,o)$. Moreover, if $f(x_1,\ldots,x_n)$ is a term-function on $\underline{A}$ we may write

$$\begin{aligned} f(x_1,\ldots,x_n) &= f(x_1-o+o,o-o+x_2,\ldots,o-o+x_n) = \\ &= f(x_1,o,\ldots,o) + f(o,x_2,\ldots,x_n) = \\ &= \ldots \\ &= f(x_1,o,\ldots,o) + f(o,x_2,o,\ldots,o) + \ldots + f(o,\ldots,o,x_n) \\ &= \tau_1(x_1) + \tau_2(x_2) + \ldots + \tau_n(x_n) \\ &= \sum_{i=1}^{n} \tau_i x_i \end{aligned}$$

where product is taken in the ring $\underline{R}$.

Thus there is a module structure defined on $\underline{A}$ such that every operation on A is linear. Moreover, if o is an algebraic constant then the linear operations are precisely the term-functions of $\underline{A}$.

Now if $\underline{A}$ does not have a one-element subalgebra, then, after choosing an element o, we let $R(\underline{A})$ be the set of all unary algebraic functions τ on $\underline{A}$ having o as only constant and satisfying $\tau(o) = o$. The above arguments again tell us that $R(\underline{A})$ is a unitary ring and every term function $f(x_1,\ldots,x_n)$ can be written as

$$f(x_1,\ldots,x_n) = \sum_{i=1}^{n} \tau_i x_i + a,$$

where a is a fixed element of $\underline{A}$, depending only on o and f, in fact $a = f(o,\ldots,o)$ and $\tau_i(x) = f(o,\ldots,o,x,o,\ldots,o) - f(o,\ldots,o)$.

Thus, if $\underline{A}$ is affine, then $\underline{A}$ is polynomially equivalent, i.e. has the same polynomial functions as the module ${}_{R(\underline{A})}A$.

There is another way to obtain the abelian group structure on $\underline{A}$. Suppose again that $\mathrm{diag}(\underline{A})$ is a congruence class for a congruence relation Δ on $\underline{A}\times\underline{A}$. From 5.1 Δ is a complement of π_1 and of π_2. Moreover, since for any $x,y,z \in \underline{A}$ we have $(y,y)\ \Delta\ (z,z)$ it follows that $t((x,z),(y,z),(z,z))\ \Delta\ (x,y)$, i.e. $(t(x,y,z),z)\ \Delta\ (x,y)$. From the properties of Δ the equations $t(x,x,z) = z$ and $t(y,y,z) = z$ follow

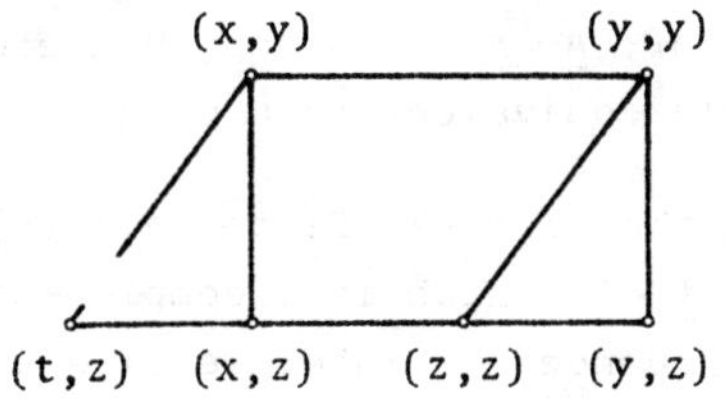

Hence t is a Mal'cev-term (as in 1.2), so $\underline{A}$ generates a permutable variety. Let f be any n-ary operation of $\underline{A}$, then in particular $(t(f(\vec{x}),f(\vec{y}),f(\vec{z})),\ f(\vec{z}))\ \Delta\ (f(\vec{x}),f(\vec{y}))$. On the other hand, since $(t(x_i,y_i,z_i),z_i)\ \Delta\ (x_i,y_i)$ and from the compatibility of Δ we conclude $(f(t(x_1,y_1,z_1),\dots,t(x_n,y_n,z_n)),\ f(\vec{z}))\ \Delta\ (f(\vec{x}),f(\vec{y}))$. Since the right sides are equal, the left sides are congruent. Since their second components coincide, so do their first components and we find:
$t(f(\vec{x}),f(\vec{y}),f(\vec{z})) = f(t(x_1,y_1,z_1),\dots,t(x_n,y_n,z_n))$.

5.4 __Proposition:__ An algebra $\underline{A}$ is affine if and only if there is a terrary term $t(x,y,z)$ on $\underline{A}$ satisfying $t(x,y,y) = x$, $t(x,x,z) = z$ and if every operation f of $\underline{A}$ sommutes with t.

__Proof:__ t commutes with every operation means that $t\colon \underline{A}^3 \to \underline{A}$ is a homomorphism, so t commutes with every term function of $\underline{A}$ as well. In particular it commutes with itself, i.e.

a) $t(x,y,z) = t(t(x,y,y),t(x,x,y),t(z,x,x))$
$\qquad = t(t(x,x,z),t(y,x,x),t(y,y,x)) = t(z,y,x)$

b) $t(t(x,y,z),y,v) = t(t(x,y,z),t(y,y,y),t(y,y,v))$
$\qquad = t(t(x,y,y),t(y,y,y),t(z,y,v)) = t(x,y,t(z,y,v))$

and similarly

c) $t(x,y,t(y,x,y)) = y$.

Choosing an arbitrary element $o \in \underline{A}$ and setting $y = o$ in the above equations we find that $x+y := t(x,o,y)$ defines an abelian group structure on $\underline{A}$ with $t(x,y,z) = x-y+z$ and neutral element o. The rest follows with 5.3.

Let us keep a record now of the several characterizations of affine algebras. (More of them will result from the next chapters.)

5.5 Theorem: Let $\underline{A}$ be an algebra in a modular variety, then the following are equivalent:

(i) $\underline{A}$ is affine.

(ii) $\underline{A}$ has a Mal'cev-term t and every operation of $\underline{A}$ commutes with t.

(iii) There is a congruence relation Θ on $\underline{A} \times \underline{A}$ which is a common complement of π_1 and of π_2.

(iv) diag($\underline{A}$) is a congruence class on $\underline{A} \times \underline{A}$.

(v) Every subalgebra of $\underline{A} \times \underline{A}$ is a congruence class.

(vi) $\underline{A}$ is polynomially equivalent to a module over a ring with unit.

Theorem 5.3 deserves another remark: Since diag($\underline{A}$) is a congruence class of a congruence Δ on $\underline{A} \times \underline{A}$ which is a complement of π_1 and of π_2, and since all those congruences permute, we find that $\underline{A} \times \underline{A} \cong \underline{A} \times (\underline{A} \times \underline{A})/_{\Delta}$. Since a subalgebra of $\underline{A} \times \underline{A}$, namely diag($\underline{A}$) is identified by Δ, $(\underline{A} \times \underline{A})/_{\Delta}$ has a one-element subalgebra whilst this need not be so for $\underline{A}$. In particular, $\underline{A}$ and $\underline{A}/_{\Delta}$ need not be isomorphic. $\underline{A}/_{\Delta}$ can also be defined by changing each operation f into a new operation f^{∇} by $f^{\nabla}(x_1,\ldots,x_n) := f(x_1,\ldots,x_n) - f(o,\ldots,o)$ yielding the "linearization" $\underline{A}^{\nabla}$ of $\underline{A}$. This situation is examined more closely and in greater generality in section 9.

6. COMMUTATORS

So far we have assumed that $\mathrm{diag}(\underline{A})$ be a congruence class on $\underline{A} \times \underline{A}$. Suppose now this was not the case. We would then consider the smallest congruence relation Δ such that $\mathrm{diag}(\underline{A})$ is contained in a Δ-class. Thus we must set

$$\Delta := \langle\{((x,x),(y,y)) \mid (x,y) \in \underline{A} \times \underline{A}\}\rangle_{\underline{A}\times\underline{A}} .$$

If we do now try our coordinatization we have the difficulty that intersections of Δ-lines with π_1-lines or with π_2-lines need not be unique, hence the construction of $x-y+z$ is not unique.

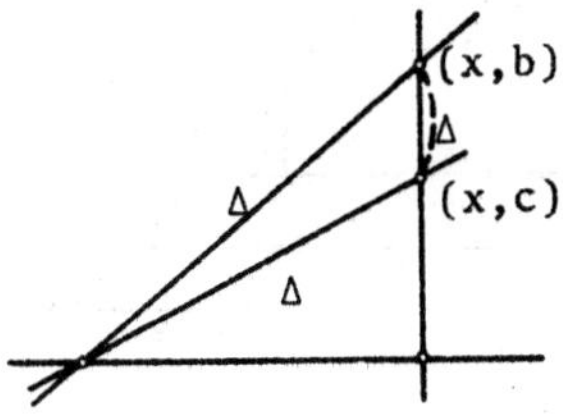

Thus we have $(x,b)\ \Delta\ (x,c)$ <u>for some</u> $x \in \underline{A}$. Fortunately the Shifting Lemma tells us that in this case

$$(y,b)\ \Delta\ (y,c) \quad \underline{\text{for every}} \quad y \in \underline{A}.$$

Thus the following is a congruence relation on $\underline{A}$:

$$[1,1] := \{(b,c) \mid (x,b)\ \Delta\ (x,c) \quad \text{for some} \quad x \in A\}$$

and it is equal to

$$\{(b,c) \mid (b,b)\ \Delta\ (b,a)\}.$$

Factoring by this congruence relation we will indeed get an algebra $B = \underline{A}/[1,1]$ where now $\mathrm{diag}(\underline{B})$ is a congruence class in $\underline{B} \times \underline{B}$ and hence $\underline{B}$ is polynomially equivalent to a module. This will become clear later in this chapter in a more general setting.

Let us now generalize the above notions to obtain the important definition of a <u>commutator</u> of congruences. Let α and β be congruences on the algebra $\underline{A}$. We think of α now as a subalgebra of $\underline{A} \times \underline{A}$ and define a congruence relation on α by

$$\Delta_\alpha^\beta := \langle\{((x,x),(y,y)) \mid (x,y) \in \beta\}\rangle_\alpha .$$

Then $[\alpha,\beta] := \{(b,c) \mid (b,b)\ \Delta_{\alpha}^{\beta}\ (b,c)\}$ is called the <u>commutator of α and β</u>.

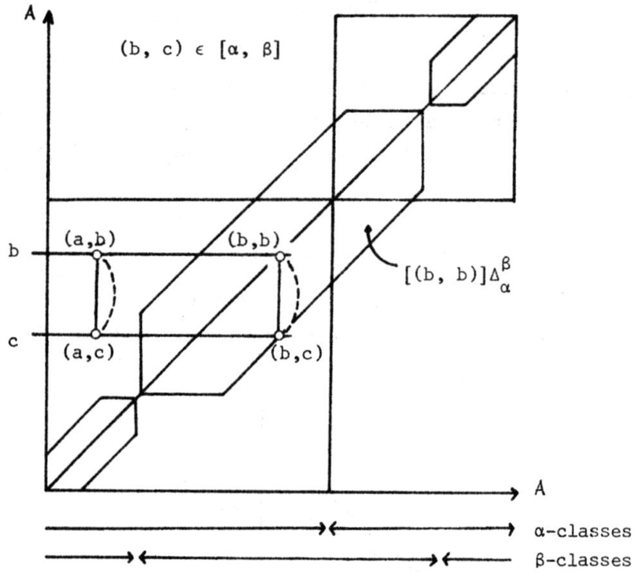

Just as [1,1] measures the "thickness" of the smallest line containing diag($\underline{A}$), the commutator $[\alpha,\beta]$ measures the "thickness" of the diagonal pieces given by β in the algebra α. This is shown in the preceding figure where the two squares represent two blocks of α (as subsets of $\underline{A} \times \underline{A}$!).

Commutators were introduced into General Algebra by J.D.H. SMITH in [36]. He used them as a major tool for studying permutable varieties (which he calls Mal'cev varieties). His fundamental concept is "centrality". Here two congruences centralize each other if their commutator is 0. Starting with this concept he develops commutators. We believe that his approach is less direct and harder to work with than the approach we suggest.

J. HAGEMANN and C. HERRMANN [24] managed to carry the whole concept over to modular varieties. They analyized the lattice theoretical properties of Con(α) that made SMITH's concept work and gave a new definition of the commutator in modular varieties which coincided with SMITH's concept in the permutable case. Although many results true in permutable varieties could be proved again in modular varieties and an impressing list of new results could be added, the simplicity and clarity of the concept is lost and consequently their methods are unusually difficult to comprehed and to work with.

In our approach, the vehicle for transferring the concept from permutable to modular varieties is, of course, geometry.

Let us first see what this notion amounts to in some familiar varieties as in groups, rings (not necessarily associative), and lattices.

To facilitate the computations, we will freely use a result which we are going to prove later, namely that $[\alpha,\beta] \leq \gamma$ if and only if $[\bar{\phi}\alpha,\bar{\phi}\beta] = 0$, where ϕ is the canonical homomorphism from $\underline{A}$ onto $\underline{A}/\gamma$, see 6.17 below.

6.1 Groups: Via the obvious translation between congruences and normal subgroups given by

$$x \ \Theta \ y \iff x\cdot y^{-1} \in N(\Theta) \quad \text{and}$$

$$x \in N \iff x \ \Theta(N) \ 1$$

we claim that the above definition captures precisely the notion of commutator (normal) subgroup of two normal subgroups.

Thus let $\underline{N}$ and $\underline{M}$ be normal in $\underline{G}$. Take $x \in N$, $y \in M$. Then $(y,1) \in \Theta(M) =: \beta$ hence $(y,y) \ \Delta_\alpha^\beta \ (1,1)$ with $\alpha := \Theta(N)$. Multiplication by $(x,1)$ and $(1,x)$ (from α) gives $(x,1)\cdot(y,y)\cdot(1,x) \ \Delta_\alpha^\beta \ (x,1)\cdot(1,1)\cdot(1,x)$ i.e. $(x\cdot y,y\cdot x) \ \Delta_\alpha^\beta \ (x,x)$. The fact that $(x,x\cdot y) \in \beta$ and consequently $(x,x) \ \Delta_\alpha^\beta \ (x\cdot y,x\cdot y)$ together with transitivity give us $(x\cdot y,x\cdot y) \ \Delta_\alpha^\beta \ (x\cdot y,y\cdot x)$, showing that $x\cdot y \ [\alpha,\beta] \ y\cdot x$. Therefore the above commutator contains the usual group theoretic commutator.

Equality is seen using 6.17. We factor our given group $\underline{G}$ by the group theoretic commutator $[N,M]$. ($\gamma = \Theta[N,M]$ in 6.17). Hence we may assume that N centralizes M, i.e. every $x \in N$ commutes with every $y \in M$.

It follows that $N_\alpha^\beta := \{(x,x) \mid x \in M\}$ is a normal subgroup of α. Hence defining a congruence relation Θ_α^β on α by

$$(a,b) \ \Theta_\alpha^\beta \ (c,d) \ :\iff (a\cdot c^{-1},b\cdot d^{-1}) \in N_\alpha^\beta, \ \text{i.e.}$$
$$\iff a\cdot c^{-1} = b\cdot d^{-1} \ \text{and} \ a\cdot c^{-1} \in M$$

yields $\Theta_\alpha^\beta \geq \Delta_\alpha^\beta$.

Hence $(a,a) \ \Delta_\alpha^\beta \ (a,b) \Rightarrow (a,a) \ \Theta_\alpha^\beta \ (a,b) \Rightarrow (1,a\cdot b^{-1}) \in N_\alpha^\beta \Rightarrow a = b$. Thus $[\alpha,\beta] = 0$ finishing the claim.

6.2 Rings: For ideals I,J we get $[I,J] = (I\cdot J + J\cdot I)$, i.e. the ideal generated by all sums $i\cdot j + j\cdot i$ with $i \in I$ and $j \in J$.
The proof is analogous to the preceding one.
Firstly for $i \in I$ and $j \in J$ we get with $\alpha = \Theta(I)$ and $\beta = \Theta(J)$ that $(j,j) \ \Delta_\alpha^\beta \ (o,o)$, hence $(j,j)\cdot(i,o) \ \Delta_\alpha^\beta \ (o,o)\cdot(i,o)$ whence $(j\cdot i,o) \ \Delta_\alpha^\beta \ (o,o)$. This proves $J\cdot I \subseteq [I,J]$. Similarly $I\cdot J \subseteq [I,J]$, therefore $(I\cdot J + J\cdot I) \subseteq [I,J]$.

For the reserve inclusion 6.17 again permits us to assume that $(I\cdot J + J\cdot I) = 0$. Therefore $N_\alpha^\beta := \{(x,x) \mid x \in J\}$ is an ideal of (the ring) α. As above we define

$$(a,b)\ \Theta_\alpha^\beta\ (c,d) \quad :\Longleftrightarrow \quad (a-c,b-d) \in N_\alpha^\beta$$

and find $\Theta_\alpha^\beta \geq \Delta_\alpha^\beta$ implying $[I,J] = 0$, proving the claim.

6.3 Lattices: If α and β are congruences on a lattice $\underline{L}$ then $[\alpha,\beta] = \alpha\wedge\beta$.

(This is more generally true for every congruence distributive variety).

Namely consider $(x,y) \in \alpha\wedge\beta$ and look at Δ_α^β, π_1 and π_2, all thought of as congruences on α. Then

$$(x,x)\ \Delta_\alpha^\beta\ (y,y)\ \pi_2\ (x,y) \quad \text{and}$$

$$(x,x)\ \Delta_\alpha^\beta \vee \pi_1\ (x,y) \quad \text{hence}$$

$$(x,x)\ \Delta_\alpha^\beta \vee (\pi_1 \wedge \pi_2)\ (x,y)$$

by distributively and consequently $x\ [\alpha,\beta]\ y$ (since $\pi_1\wedge\pi_2 = 0$). On the other hand $[\alpha,\beta] \leq \alpha\wedge\beta$ follows from the definition of the commutator.

Thus we have seen that adding operations to our algebras increases the commutator until it is maximal when the algebras become congruence distributive. Congruence distributive varieties are in fact characterized amongst other modular varieties by this property as a corollary to 6.12 below.

Let us now return to develop the general theory of commutators.

6.4 Properties of Δ_α^β:

(i) $(a,b)\ \Delta_\alpha^\beta\ (c,d)$ implies

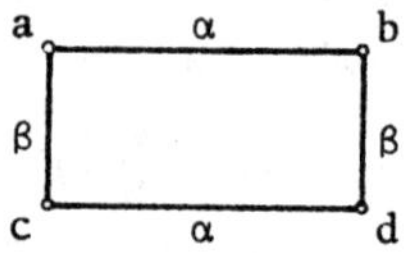

(ii) $(a,b)\ \Delta_\alpha^\beta\ (c,d)$ implies $(b,a)\ \Delta_\alpha^\beta\ (d,c)$

(iii) $a\ \beta\ b$ implies $(a,a)\ \Delta_\alpha^\beta\ (b,b)$.

Proof: (iii) being part of the definition, (ii) follows immediately from the symmetry of α, or fancier, note that $(x,y) \to (y,x)$ yields an automorphism of α, leaving invariant the generating set of Δ_α^β. For (i) note that $\Delta_\alpha^\beta \leq \beta\times\beta|_\alpha$ where $\beta\times\beta|_\alpha$ is the congruence on α given by $(x,y)\ \beta\times\beta|_\alpha\ (z,u)$ if $x\ \beta\ z$ and $y\ \beta\ u$.

6.5 Properties of $[\alpha,\beta]$:

(i) $[\alpha,\beta] = \{(x,y) \mid (x,x)\ \Delta_\alpha^\beta\ (y,x)\}$

(ii) $[\alpha,\beta] = \{(x,y) \mid \exists z\ ((z,x)\ \Delta_\alpha^\beta\ (z,y))\}$

$= \{(x,y) \mid \exists z\ ((x,z)\ \Delta_\alpha^\beta\ (y,z))\}$

$= \{(x,y) \mid \exists z\ ((z,z)\ \Delta_\alpha^\beta\ (x,y))\}$.

(iii) $[\alpha,\beta]$ is a congruence relation on $\underline{A}$.

(iv) $[\alpha,\beta] \le \alpha \wedge \beta$.

Proof: (i) follows from 6.4 (ii).
(ii) follows with the Shifting Lemma applied to

The last equality follows from 6.4 (i) and (iii). (iv) is immediate from the definition and from 6.4 (i). For (iii): All properties of a congruence relation are immediate with 6.4. For transitivity we use 6.5, namely $x\ [\alpha,\beta]\ y\ [\alpha,\beta]\ z$ implies $(x,y)\ \Delta_\alpha^\beta\ (y,y)\ \Delta_\alpha^\beta\ (z,y)$ hence $x\ [\alpha,\beta]\ z$ with 6.5.

From MAL'CEV's description of congruences generated by a binary (symmetric) relation (0.1) we readily obtain, using 6.5:

6.6 An alternative description of the commutator:

$(x,y) \in [\alpha,\beta] \iff$ there exist unary algebraic functions $\tau_0,\dots,\tau_n$ on α and $(s_0,u_0),\dots,(s_n,u_n) \in \beta$ with

$\exists z$ with $\tau_0(s_0,s_0) = (z,z)$ [or: $\tau_0(s_0,s_0) = (x,x)$]

$\tau_i(u_i,u_i) = \tau_{i+1}(s_{i+1},s_{i+1}),\quad 0 \le i < n$

$\tau_n(u_n,u_n) = (x,y)$.

This description may of course be formulated coordinatewise. A unary algebraic function τ_i on α is nothing else than a pair $(t_i(x,\tilde{a}^i), t_i(x,\tilde{b}^i))$, where the t_i are term functions on $\underline{A}$ and $\tilde{a}^i = (a_1^i,\dots,a_n^i)$, $\tilde{b}^i = (b_1^i,\dots,b_n^i)$ are vectors componentwise congruent modulo α.

Note that the first line says no nore than

$$t_0(s_0,\tilde{a}^1) = t_0(s_0,\tilde{b}^1).$$

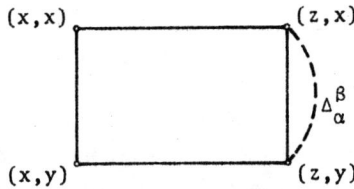

In particular it follows:

$$t_o(u_o,\tilde{a}^1)\ [\alpha,\beta]\ t_o(u_o,\tilde{b}^1).$$

This comes from 6.6 for $n = 0$. Since n may be arbitrary, this process can be iterated, defining $[\alpha,\beta]$ again as in 6.8 below. The case where we don't get started is

6.7 <u>A syntactical description of $[\alpha,\beta] = 0$:</u>

$[\alpha,\beta] = 0 \iff$ for all term functions $p(x_1,\ldots,x_n)$ on $\underline{A}$ and $(a_2,b_2),\ldots,(a_n,b_n) \in \alpha$ and $(x,y) \in \beta$ we have

$p(x,a_2,\ldots,a_n) = p(x,b_2,\ldots,b_n)$

<u>implies</u>

$p(y,a_2,\ldots,a_n) = p(y,b_2,\ldots,b_n)$.

The condition on the right hand side is called the <u>"term condition"</u> by several authors. The above result was independently found by R. McKENZIE who proposes to use it to define the commutator in nonmodular varieties (see 6.8 below). The first place where a similar condition is studied in the connection with congruences on direct products seems to be in WERNER [40], theorem 9. His assumptions nevertheless are far too strong and treat only the special case of simple algebras.

The condition also comes up in a totally different (so it seems) setting. We refer the reader to the interesting paper by FREESE, LAMPE and TAYLOR [12]. W. TAYLOR in [38] also showed that semigroups satisfying the above condition for $\alpha = \beta = 1$ are medial (which is only a minor part of [38] and not hard to show).

As we have indicated before, 6.7 actually can be turned into a definition of the commutator as

6.8 <u>Theorem:</u> $[\alpha,\beta]$ is the smallest congruence relation on $\underline{A}$ such that for all term functions t, $(x,y) \in \beta$ $(a_2,b_2),\ldots,(a_n,b_n) \in \alpha$ we have

$t(x,a_2,\ldots,a_n)\ [\alpha,\beta]\ t(x,b_2,\ldots,b_n)$

<u>implies</u>

$t(y,a_2,\ldots,a_n)\ [\alpha,\beta]\ t(y,b_2,\ldots,b_n)$.

There is also a more geometric way to see 6.7. For this recall that by 0.1 a set $S \subseteq A$ is a congruence class for some congruence Θ on $\underline{A}$, iff for all unary algebraic functions τ of $\underline{A}$, we have: If one element of S is mapped back into S by τ then this is true for every element of S. Thus the right hand side of 6.7 states precisely that the sets $\delta_x^\beta = \{(y,y) \mid x\ \beta\ y\}$ are classes of some congruence on the algebra α. If so, they are certainly classes of Δ_α^β, hence $[\alpha,\beta] = 0$. On the other hand, if it was not a class of some congruence, we could assume $(x,x)\ \Delta_\alpha^\beta\ (u,v)$ for some $u \neq v$, yielding $(u,u)\ \Delta_\alpha^\beta\ (u,v)$ with 6.5 and hence $(u,v) \in [\alpha,\beta]$. This was our original proof of 6.7 in [19].

Notice that replacing equality signs in 6.7 by $\equiv \pmod{[\alpha,\beta]}$ is another way to 6.8. The left hand side then becomes a tautology, making the right hand side universally true. The justification for this will be given in 6.17 below.

Another conclusion is immediate from 6.6:

6.9 Proposition: Let $\phi\colon \underline{A} \to \underline{B}$ be a homomorphism and α,β congruences on $\underline{A}$. Then $\vec{\phi}[\alpha,\beta] \leq [\vec{\phi}\alpha,\vec{\phi}\beta]$.

Proof: Consider the homomorphism $\phi\times\phi\colon \alpha \to \vec{\phi}\alpha$ and apply it to the equations of 6.6. Each τ_i, which is an algebraic function of (the algebra) α will be transformed by $\phi\times\phi$ into an algebra function $\bar{\tau}_i$ of (the algebra) $\vec{\phi}\alpha$. Thus we get

$$\bar{\tau}_0(\phi(s_0),\phi(s_0)) = (\phi(x),\phi(x))$$
$$\bar{\tau}_i(\phi(u_i),\phi(u_i)) = \bar{\tau}_{i+1}(\phi(s_{i+1}),\phi(s_{i+1})), \quad \text{for } 0 \leq i < n$$
$$\bar{\tau}_n(\phi(u_n),\phi(u_n)) = (\phi(x),\phi(y)).$$

Hence with 6.6 we have $(\phi(x),\phi(y)) \in [\vec{\phi}\alpha,\vec{\phi}\beta]$.

Well known for groups, we get the following corollary:

6.10 Corollary: The commutator of fully invariant congruences is again fully invariant.

As we go on we need the following technical result:

6.11 Theorem: Let $\underline{D}$ be a subalgebra $\underline{A}\times\underline{A}$ (with $\mathrm{Con}(\underline{D})$ modular). Let κ_i, $i \in I$ be a family of congruence relations on $\underline{D}$ with the property $(x,y)\ \kappa_i\ (z,u) \Rightarrow (x,x)\ \kappa_i\ (z,z)$. Then for all $x,y,z \in \underline{D}$ we have:

$$(x,x) \bigvee_{i\in I} \kappa_i\ (y,z) \Rightarrow (y,y) \bigvee_{i\in I} (\kappa_i \wedge \pi_1)\ (y,z).$$

Proof: If $(x,x) \bigvee \kappa_j\ (y,z)$ then there exist w.l.o.g. $\kappa_0,\ldots,\kappa_{n-1}$ and $(u_0,v_0),\ldots,(u_n,v_n) \in \underline{D}$ with

$$(u_0,v_0) = (x,x), \quad (u_n,v_n) = (y,z) \quad \text{and}$$
$$(u_i,v_i)\ \kappa_i\ (u_{i+1},v_{i+1}) \quad \text{for } 0 \leq i < n.$$

By induction we show that

$$(u_i,u_i) \bigvee (\kappa_j \wedge \pi_1)\ (u_i,v_i).$$

Indeed this is trivial for $i=0$. In passing from i to $i+1$ we note that $(u_i,u_i)\ \kappa_i\ (u_{i+1},u_{i+1})$ and, using the induction hypothesis we have the situation:

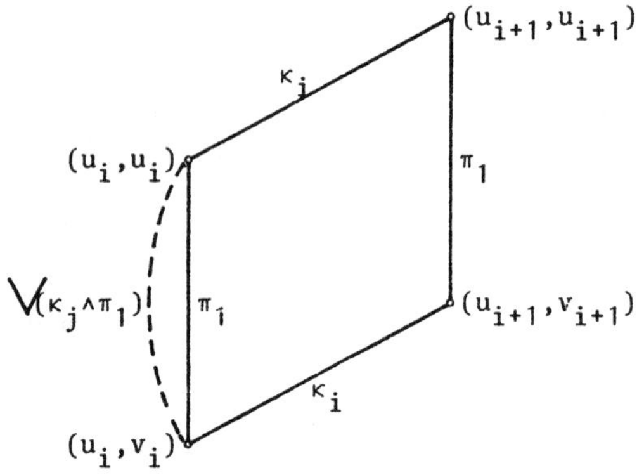

Thus the induction step is achieved with the Shifting Lemma. Setting now $i=n$ the theorem is proved.

As a corollary we have one of the most important properties of the commutator, namely join-distributivity. This was discovered by HAGEMANN and HERRMANN [24].

6.12 <u>Corollary:</u> $[\alpha,\bigvee\beta_i]=\bigvee[\alpha,\beta_i]$.

<u>Proof:</u> $\geq$ is clear since $\beta_i\leq\bigvee\beta_i$.

Trivially $\Delta_\alpha^{\bigvee\beta_i}=\bigvee\Delta_\alpha^{\beta_i}$. Hence supposing $(x,y)\in[\alpha,\bigvee\beta_i]$, i.e.

$(x,x)\ \bigvee\Delta_\alpha^{\beta_i}\ (x,y)$ we conclude with 6.11 the relation

$(x,x)\ \bigvee(\Delta_\alpha^{\beta_i}\wedge\pi_1)\ (x,y)$ which clearly means $(x,y)\in\bigvee[\alpha,\beta_i]$.

A second application of 6.11 yields a result of R. FREESE and R. McKENZIE [13]:

6.13 <u>Theorem:</u> Let $\phi:\underline{A}\twoheadrightarrow\underline{B}$ be an onto homomorphism and α,β congruences on $\underline{B}$. Then $\overleftarrow{\phi}[\alpha,\beta]=[\overleftarrow{\phi}\alpha,\overleftarrow{\phi}\beta]\vee\ker\phi$.

<u>Proof:</u> Using 6.9 we get that $(x,y)\in[\overleftarrow{\phi}\alpha,\overleftarrow{\phi}\beta]$ implies $(\phi(x),\phi(y))\in[\vec{\phi}\overleftarrow{\phi}\alpha,\vec{\phi}\overleftarrow{\phi}\beta]=[\alpha,\beta]$ because ϕ is onto. For the reverse inclusion suppose $(a,b)\in\overleftarrow{\phi}[\alpha,\beta]$ i.e. $(x,y)\in[\alpha,\beta]$ with $x=\phi(a)$ and $y=\phi(b)$. The last relation can be written down as in 6.6. Since ϕ is onto there exist $(\bar{s}_i,\bar{t}_i)\in\overleftarrow{\phi}\beta$ with $(\bar{s}_i)=s_i$ and $(\bar{t}_i)=t_i$ and there are similarly algebraic functions $\bar{\tau}_i$ on $\overleftarrow{\phi}\alpha$ which arise from the given τ_i by replacing any constant (i.e. an element of α) by an arbitrary preimage under $\phi\times\phi$ (i.e. an element of $\overleftarrow{\phi}\alpha$). Since $\phi\times\phi$ is a homomorphism we obtain:

$$\bar\tau_0(\bar s_0,\bar s_0)\ \ker\phi\times\phi\ (a,a)$$
$$\bar\tau_i(\bar t_i,\bar t_i)\ \ker\phi\times\phi\ \bar\tau_{i+1}(\bar s_{i+1},\bar s_{i+1})\quad \text{for } 0\le i<n$$
$$\bar\tau_n(\bar t_n,\bar t_n)\ \ker\phi\times\phi\ (a,b)$$

Hence $(a,a)\ \Delta^{\overleftarrow{\phi}\beta}_{\overleftarrow{\phi}\alpha} \vee \ker\phi\times\phi\ (a,b)$. Application of 2.6 with $\kappa_1 = \Delta^{\overleftarrow{\phi}\beta}_{\overleftarrow{\phi}\alpha}$ and $\kappa_2 = \ker\phi\times\phi$ yields

$$(a,a)\ (\Delta^{\overleftarrow{\phi}\beta}_{\overleftarrow{\phi}\alpha}\wedge\pi_1)\vee(\ker\phi\times\phi\wedge\pi_1)\ (a,b)$$

which immediately gives the missing inclusion.

Another important property of the commutator is commutativity, see [24]. To prove it we use the Cube Lemma to imitate SMITH's proof of the permutable case.

6.14 Theorem: $[\alpha,\beta] = [\beta,\alpha]$.

Proof: Let us define $\bar\Delta^\beta_\alpha := \{((x,y),(u,v)) \mid (x,u)\ \Delta^\beta_\alpha\ (y,v)\}$. Clearly $(x,y)\in[\alpha,\beta]$ implies $(x,x)\ \bar\Delta^\beta_\alpha\ (x,y)$, hence we are done if we can show that $\bar\Delta^\beta_\alpha = \Delta^\alpha_\beta$.
Obviously $\bar\Delta^\beta_\alpha$ is a binary relation on β containing $((x,x),(y,y))$ whenever $(x,y)\in\alpha$. Reflexivity and symmetry are precisely properties 6.4 (iii) and 6.4 (ii).
For transitivity suppose $(x,u)\ \bar\Delta^\beta_\alpha\ (y,v)\ \bar\Delta^\beta_\alpha\ (z,w)$ which with the aid of 6.4 provides the following relations for Δ^β_α:

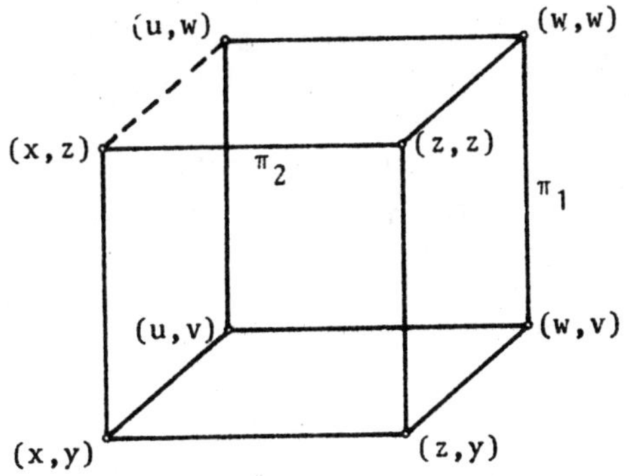

The Cube Lemma thus yields $(x,z)\ \Delta^\beta_\alpha\ (u,w)$ i.e. $(x,u)\ \bar\Delta^\beta_\alpha\ (z,w)$. Compatibility of $\bar\Delta^\beta_\alpha$ is trivially seen, hence $\bar\Delta^\beta_\alpha$ is a congruence relation on β, containing $((x,x),(y,y))$ whenever $x\ \alpha\ a$. Therefore $\bar\Delta^\beta_\alpha \ge \Delta^\alpha_\beta$. We conclude $\Delta^\alpha_\beta = \bar{\bar\Delta}^\alpha_\beta \ge \bar\Delta^\beta_\alpha \ge \Delta^\alpha_\beta$ which results in $\Delta^\alpha_\beta = \bar\Delta^\beta_\alpha$.

6.15 Theorem ([24]): Let α, β and γ be congruences on $\underline{B}$. Then $[\alpha,\beta]\le\gamma\le\alpha\wedge\beta$ if and only if there exists an algebra $\underline{A}$, a homomorphism $\phi\colon \underline{A}\twoheadrightarrow\underline{B}$ and congruences σ and τ on $\underline{A}$ such that

(1) $\sigma \wedge \tau \leq \overleftarrow{\phi}\gamma$

(2) $\sigma \vee \overleftarrow{\phi}\gamma \geq \overleftarrow{\phi}\alpha$ and

(3) $\tau \vee \overleftarrow{\phi}\gamma \geq \overleftarrow{\phi}\beta$.

Proof: "→": $[\sigma \vee \overleftarrow{\phi}\gamma, \tau \vee \overleftarrow{\phi}\gamma] \leq [\sigma,\tau] \vee \overleftarrow{\phi}\gamma \leq \overleftarrow{\phi}\gamma$ applying 6.12, 6.14 and 6.5. Hence $[\overleftarrow{\phi}\alpha\ \overleftarrow{\phi}\beta] \leq \overleftarrow{\phi}\gamma$. From 6.13 we get $\overleftarrow{\phi}[\alpha,\beta] = [\overleftarrow{\phi}\alpha,\overleftarrow{\phi}\beta] \vee \ker \phi \leq \overleftarrow{\phi}\gamma$ and after applying $\overrightarrow{\phi}$ the result follows.

For the other direction take $\underline{A} := \alpha$ and $\phi := \pi_1 : \underline{A} \to \underline{B}$ as given by $\pi_1(x,y) := x$. Define $\sigma := \ker \pi_2$ and $\tau := \Delta_\alpha^\beta$. Then $(x,y)\ \sigma \wedge \tau\ (u,v)$ implies $y = v$ and hence $x\ [\alpha,\beta]\ u$, hence $((x,y),(u,v)) \in \overleftarrow{\phi}[\alpha,\beta] \leq \overleftarrow{\phi}\gamma$.

$(x,y)\ \overleftarrow{\phi}\alpha\ (u,v)$ implies $(x,u) \in \alpha$ hence $(x,v) \in \alpha$. Thus $(x,y)\ \overleftarrow{\phi}0\ (x,v)\ \sigma\ (u,v)$ hence $\overleftarrow{\phi}\alpha \leq \overleftarrow{\phi}0 \vee \sigma \leq \overleftarrow{\phi}\gamma \vee \sigma$. $(x,y)\ \overleftarrow{\phi}\beta\ (u,v)$ implies $x\ \beta\ u$ hence $(x,y)\ \overleftarrow{\phi}0\ (x,x)\ \Delta_\alpha^\beta\ (u,u)\ \overleftarrow{\phi}0\ (u,v)$ so $\overleftarrow{\phi}\beta \leq \overleftarrow{\phi}0 \vee \Delta_\alpha^\beta \leq \overleftarrow{\phi}\gamma \vee \tau$.

Inspection of the above proof leads to the following corollary which could also be used to define the commutator in nonmodular varieties. This corollary is due to HERMANN (unpublished).

6.16 Corollary: The commutator operation is the biggest binary operation $\langle , \rangle$ on the congruence lattices of algebras in a modular variety satisfying:

1. $\langle\alpha,\beta\rangle \leq \alpha \wedge \beta$
2. $\langle\alpha,\beta\vee\gamma\rangle = \langle\alpha,\beta\rangle \vee \langle\alpha,\gamma\rangle$
3. $\langle\alpha\vee\beta,\gamma\rangle = \langle\alpha,\gamma\rangle \vee \langle\beta,\gamma\rangle$
4. $\overleftarrow{\phi}\langle\alpha,\beta\rangle \leq \langle\overleftarrow{\phi}\alpha,\overleftarrow{\phi}\beta\rangle \vee \overleftarrow{\phi}0$.

Proof: Use 6.15 with $\gamma := [\alpha,\beta]$ and construct $\underline{A}$, ϕ, σ and τ. Then repeat the proof of "←" with $\gamma = [\alpha,\beta]$ and elsewhere $[,]$ replaced by $\langle , \rangle$.

6.17 Corollary: For congruences α, β, γ of the algebra $\underline{A}$ we have that $[\alpha,\beta] \leq \gamma$ if and only if $[\overrightarrow{\phi}\alpha,\overrightarrow{\phi}\beta] = 0$ where ϕ is the canonical homomorphism from $\underline{A}$ onto $\underline{A}/\gamma$.

Proof: $[\overrightarrow{\phi}\alpha,\overrightarrow{\phi}\beta] = 0$ implies with 6.9 that $\overrightarrow{\phi}[\alpha,\beta] = 0$ which is equivalent to $[\alpha,\beta] \leq \ker \phi \leq \gamma$. On the other hand $\overleftarrow{\phi}[\overrightarrow{\phi}\alpha,\overrightarrow{\phi}\beta] = [\overleftarrow{\phi}\overrightarrow{\phi}\alpha,\overleftarrow{\phi}\overrightarrow{\phi}\beta] \vee \ker \phi = [\alpha \vee \ker \phi, \beta \vee \ker \phi] \vee \ker \phi \leq$ $\leq [\alpha,\beta] \vee \ker \phi$. Assuming $[\alpha,\beta] \leq \gamma = \ker \phi$ we get that $[\overrightarrow{\phi}\alpha,\overrightarrow{\phi}\beta] = 0$.

SMITH [36] says that "α centralizes β" if $[\alpha,\beta] = 0$. 6.17 shows that it is enough to know the relation of centralizing to describe the commutator operation. Moreover the results obtained so far allow the notion of the "centralizer" of a congruence α, which is defined to be the largest congruence $\zeta(\alpha)$ which centralizes α. The existence is guaranteed by

6.12 and

$$\zeta(\alpha) = \{\beta \mid [\alpha,\beta] = 0\}.$$

$\zeta := \zeta(1)$ is called the center of $\underline{A}$.

Recently the importance of the centralizer concept became clear through a beautiful theorem by HRUSHOVSKII, which generalized the important Jonsson Lemma [28] and its subsequent generalizations due to HAGEMANN, HERRMANN [24] and FREESE, McKENZIE [13].

6.18 Theorem (HRUSHOVSKII): Let $\underline{S} \in HSP(\mathcal{K})$ be subdirectly irreducible with monolith μ. Then $\underline{S}/_{\zeta(\mu)} \in HSP_u(\mathcal{K})$.

Here $P_u(\mathcal{K})$ is the class of all ultraproducts of members of $\mathcal{K}$. Note that in congruence-distributive varieties $\zeta(\mu) = 0$ so $\underline{S} \in HSP_u(\mathcal{K})$ as stated in Jonsson's Lemma. See [1] for the definition and the important properties of ultraproducts. We prove the theorem here in a slightly more general form. Recall that an algebra $\underline{S}$ is called finitely subdirectly irreducible if the smallest congruence $0_{\underline{S}}$ is not an intersection of finitely many congruences above $0_{\underline{S}}$.

6.19 Theorem: Let $\underline{S} \in HSP(\mathcal{K})$ be finitely subdirectly irreducible and define $\xi := \bigvee\{\zeta(\alpha) \mid \alpha \neq 0\}$. Then $\underline{S}/_{\xi} \in HSP_u(\mathcal{K})$.

Note that in 6.18 ξ coincides with $\zeta(\mu)$, so 6.19 implies 6.18. The proof is very similar to Jonsson's proof in [28] and it is based on W. LAMPE's proof of HRUSHOVSKII's theorem.
Since $\underline{S} \in HSP(\mathcal{K})$ there is a subalgebra $\underline{U}$ of a product $\prod_{i\in I} \underline{A}_i$ with $\underline{A}_i \in \mathcal{K}$ and a surjective homomorphism $\phi: \underline{U} \twoheadrightarrow \underline{S}$. Let Θ be the kernel of ϕ in $Con(\underline{U})$. Θ is finitely meet-irreducible. First we show:

(§) Suppose $\alpha,\beta \in Con(U)$ with $\alpha \wedge \beta \leq \Theta$ and $\alpha \not\leq \Theta$ and $\beta \not\leq \Theta$
Then there exists a $\gamma \not\leq \Theta$ with $[\alpha\vee\beta,\gamma] \leq \Theta$.

Take $\gamma := (\alpha\vee\Theta) \wedge (\beta\vee\Theta)$, then γ is properly above Θ since Θ is finitely meet-irreducible. Now
$[\alpha,\gamma] \leq [\alpha,\beta\vee\Theta] = [\alpha,\beta] \vee [\alpha,\Theta] \leq (\alpha\wedge\beta) \vee \Theta \leq \Theta$ and similarly $[\beta,\gamma] \leq \Theta$ so $[\alpha\vee\beta,\gamma] \leq \Theta$.

For subsets D of I define a congruence relation η_D on $\underline{U}$ by $x \,\eta_D\, y$ iff $\{i \mid x(i) = y(i)\} \supseteq D$.
Let $\mathcal{F}$ be a filter on I, maximal with respect to the condition that $\eta_D \leq \Theta$ for all $D \in \mathcal{F}$. Let $\mathcal{U}$ be any ultrafilter extending $\mathcal{F}$. We claim:

$$\forall_{E\in\mathcal{U}} \quad \eta_E \leq \overset{\leftarrow}{\phi}(\xi).$$

If $E \in \mathcal{F}$ then this is clear since $\eta_E \leq \Theta \leq \overset{\leftarrow}{\phi}(\xi)$. If $E \notin \mathcal{F}$ then by the maximality of $\mathcal{F}$ there exists a $G \in \mathcal{F}$ with $\eta_{E\cap G} \not\leq \Theta$ and

$\eta_{E'\cap G} \not\leq \theta$. But obviously $\eta_{E\cap G} \wedge \eta_{E'\cap G} = \eta_G \leq \theta$. Now with (§) we find a γ properly above θ with $\eta_E \leq \eta_{E\cap G} \vee \eta_{E'\cap G} \leq \zeta(\gamma:\theta)$ (where $\zeta(\gamma:\theta)$ denotes the greatest congruence whose commutator with γ is below θ). Since finally $\zeta(\gamma:\theta) \leq \overleftarrow{\phi}(\xi)$ we have that $\eta_E \leq \overleftarrow{\phi}(\xi)$. So the ultrafilter congruence η given by $\bigvee\{\eta_E \mid E \in \mathfrak{U}\}$ is below $\overleftarrow{\phi}(\xi)$. With standard arguments (see [28]) now $\underline{S}/_\xi \in HSP_u(\mathcal{K})$.

Let us introduce <u>prime</u> and <u>semiprime</u> congruences.

6.20 <u>Definition:</u> A congruence relation θ on the algebra $\underline{A}$ is called <u>prime</u> if for any two congruences α,β on $\underline{A}$ we have: $[\alpha,\beta] \leq \theta$ implies $\alpha \leq \theta$ or $\beta \leq \theta$. θ is called <u>semiprime</u>, if the above implication holds in the special cases where $\alpha = \beta$.

Clearly prime congruences are finitely meet irreducible moreover they are precisely the fixed points of the following mapping ξ: $\mathrm{Con}(\underline{A}) \to \mathrm{Con}(\underline{A})$:

$$\theta \to \xi(\theta) := \bigvee\{\alpha \mid \exists_{\beta \not\leq \theta}\ [\alpha,\beta] \leq \theta\}.$$

Note that for (nonassociative) rings R and ideals I,J,K the above notion of primeness is equivalent to the usual notion using products of ideals, i.e. K is prime iff for all I,J ideals of R we have: $I\cdot J \leq K$ implies $I \leq K$ or $J \leq K$. This is not hard to show.

The intersection of prime congruences is semiprime. The converse is due to KEIMEL [44]: Every semiprime congruence is the intersection of prime congruences. It is useful to introduce the notation $\sqrt{\theta}$ for the intersection of all prime congruences above θ. Then KEIMEL's theorem says that θ is semiprime iff $\sqrt{\theta} = \theta$. We write $\sqrt{\underline{A}}$ for $\sqrt{0_{\underline{A}}}$ and call $\sqrt{\underline{A}}$ the <u>prime radical</u> of $\underline{A}$. From 6.19 we obtain

6.21 <u>Theorem:</u> Let $\underline{A}$ be an algebra in a modular variety $HSP(\mathcal{K})$. Then $\underline{A}/\sqrt{\underline{A}} \in P_sHSP_u(K)$.

7. TERNARY TERMS FOR MODULARITY

In the last chapter we have not fully made use of the fact that we are in a modular variety. We could have done with the hypothesis that congruences on $\underline{A}$, when considered as subalgebras of $\underline{A} \times \underline{A}$ have modular congruence lattices.

In this chapter we will use the ternary term $t(x,y,z)$ as constructed in theorem 4.3 to obtain an important permutability formula for congruences. This in turn will be the key for the construction of ternary terms which may replace the quaternary DAY-terms.

The discovery of those terms was surprising for various reasons. Firstly, since ternary terms are describing properties of three generated algebras, the existence of ternary terms seems to contradict the fact that there exist varieties, all of whose three-generated algebras are modular, but the variety as a whole is not congruence modular. The solution however is, that our terms are in fact describing a property which is strictly stronger than modularity and which is shared automatically by every algebra which is contained in a modular variety.

Secondly, the terms we produce are just B. JONSSON's terms for distributivity (thm. 1.4) and A.I. MAL'CEV's term for permutability (thm. 1.2) "glued" together. Thus giving account of the fact that modular varieties are somewhere in between (and including) permutable varieties on the one hand and distributive varieties on the other hand.

This general principle will come up time and again in the later chapters. Investigation of the commutators will usually indicate in which direction to go.

Let us now start with investigating the rôle of the ternary term $t(x,y,z)$ from 4.3 within commutator theory.

To this end let α and β be congruences on $\underline{A}$ with $\alpha \geq \beta$. Suppose x,y,z are elements from $\underline{A}$ with $x \;\alpha\; y \;\beta\; z$. We set $\Psi := \Delta_{\alpha}^{\beta}$ and apply the term t to the situation (inside the algebra α):

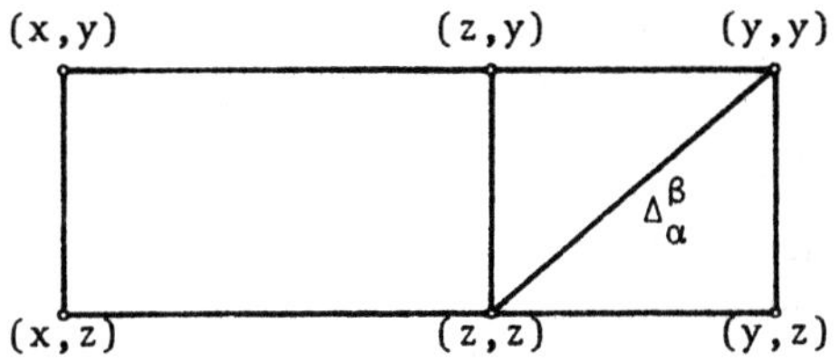

Using 4.3 we find

(§) $(x,y)\ \Delta_{\alpha}^{\beta}\ (t(x,y,z),z)$.

In particular, setting $x = y$, we obtain:

$$t(x,x,z)\ [\alpha,\beta]\ z.$$

The interesting case is where $\alpha = \beta$:

7.1 <u>Lemma:</u> There is a term t in every modular variety $\underline{V}$ such that

$t(x,y,y) = x$ is an equation in $\underline{V}$

and $t(a,a,b)\ [\alpha,\alpha]\ b$ holds for $a\ \alpha\ b$.

An important application, which has also been found independently by W. TAYLOR is:

7.2 <u>Lemma:</u> For any congruences α and β the formulas

$$\Theta \circ \Psi \subseteq [\Theta,\Theta] \circ \Psi \circ \Theta$$

and $\Theta \circ \Psi \subseteq \Psi \circ \Theta \circ [\Psi,\Psi]$

are true.

<u>Proof:</u> For $(x,z) \in \Theta \circ \Psi$ there exists a y with $x\ \Theta\ y\ \Psi\ z$. Then $x\ [\Theta,\Theta]\ t(y,y,x)\ \Psi\ t(z,y,x)\ \Theta\ t(z,y,y) = z$ and $x = t(x,y,y)\ \Psi\ t(x,y,z)\ \Theta\ t(y,y,z)\ [\Psi,\Psi]\ z$ by 7.1.

Our plan is now, to use 7.2 for obtaining a new Mal'cev-type condition for congruence modularity. To this end, the commutator has to be removed from the formulas in 7.2, since we have not defined it for nonmodular varieties. We find the desired version of 7.2 in the following lemma:

7.3 <u>Lemma:</u> For congruences α, β, γ and δ with $\alpha \leq \gamma \vee \delta$ we have $\alpha \circ \beta \subseteq ((\alpha \wedge \gamma) \vee (\alpha \wedge \delta)) \circ \beta \circ \alpha$.

<u>Proof:</u> $\alpha \circ \beta \subseteq [\alpha,\alpha] \circ \beta \circ \alpha \subseteq$
$\subseteq [\alpha,\gamma \vee \delta] \circ \beta \circ \alpha \subseteq$
$\subseteq ([\alpha,\gamma] \vee [\alpha,\delta]) \circ \beta \circ \alpha \subseteq$
$\subseteq ((\alpha \wedge \gamma) \vee (\alpha \wedge \delta)) \circ \beta \circ \alpha$.

Finally then the characterization theorem [20]:

7.4 <u>Theorem:</u> For a variety $\underline{V}$ the following are equivalent:

(i) $\underline{V}$ is congruence modular.

(ii) For all congruences α, β, γ, δ on $\underline{A} \in \underline{V}$ with $\gamma \vee \delta \geq \alpha$ the formula $\alpha \circ \beta \subseteq ((\alpha \wedge \gamma) \vee (\alpha \wedge \delta)) \circ \beta \circ \alpha$ holds.

(iii) For all congruences α, β, γ on $\underline{A} \in \underline{V}$ with $\gamma \vee \beta \geq \alpha$ the formula $\alpha \circ \beta \subseteq ((\alpha \wedge \beta) \vee (\alpha \wedge \gamma)) \circ \beta \circ \alpha$ holds.

(iv) For some $n \in \mathbb{N}$ there exist ternary terms $q_o, \ldots, q_n$ and p such that the following equations are true in $\underline{V}$:

(1) $q_o(x,y,z) = x$

(2) $q_i(x,y,x) = x$ for all $0 \leq i < n$

(3) $q_i(x,x,y) = q_{i+1}(x,x,y)$ for i even

(4) $q_i(x,y,y) = q_{i+1}(x,y,y)$ for i odd

(5) $q_n(x,y,y) =$
$= p(x,y,y)$

(6) $p(x,x,y) = y$

(1)–(4): JONSSON-terms; (5)–(6): MAL'CEV-term

Comment: It may be interesting to notice that p could be trivial (i.e. a projection). In this case p would be the third projection and $q_n(x,y,y)$ would be equal to y. We may suppose that n is odd, for otherwise we have $q_{n-1}(x,y,y) = y$ as well. Now define $q_{n+1}(x,y,z) = z$. Now the equations we are left with are precisely B. JONSSON's equations showing that $\underline{V}$ is congruence distributive (1.4).

On the other hand, if all the q_i's are projections, this would imply $q_i(x,y,z) = x$ for every i. Hence what we are left with are the equations which are precisely those of MAL'CEV witnessing permutability (1.2).

Proof of 7.4: (i) $\rightarrow$ (ii) $\rightarrow$ (iii) is lemma 7.3.
For (iii) $\rightarrow$ (iv) let $\underline{F}_{\underline{V}}(3)$ be the free algebra in $\underline{V}$ generated by $X = \{x,y,z\}$. Consider the congruences $\Theta_{(x,y)}, \Theta_{(y,z)}$ and $\Theta_{(x,z)}$ which are generated by the nontrivial partitions of X.

Clearly $\Theta_{(y,z)} \vee \Theta_{(x,z)} \geq \Theta_{(x,y)}$. Hence (iii) tells us:

$$\Theta_{(x,y)} \circ \Theta_{(y,z)} \leq ((\Theta_{(x,y)} \wedge \Theta_{(x,z)}) \vee (\Theta_{(x,y)} \wedge \Theta_{(y,z)})) \circ \Theta_{(y,z)} \circ \Theta_{(x,y)}.$$

(x,z) is therefore in the right hand side, which implies that there exist elements $t_o, \ldots, t_n$, and r in $\underline{F}_{\underline{V}}(3)$ such that

(0) $x = t_o$

(2') $t_i \; \Theta_{(x,y)} \wedge \Theta_{(x,z)} \; t_{i+1}$ for i even

(3') $t_i \; \Theta_{(x,y)} \wedge \Theta_{(y,z)} \; t_{i+1}$ for i odd

(4) $t_n \; \Theta_{(y,z)} \; r$ and

(5) $r \; \Theta_{(x,y)} \; z$.

We rewrite (2') and (3') by

(1) $t_i \; \Theta_{(x,y)} \; x$ for all i

(2) $t_i \; \Theta_{(x,z)} \; t_{i+1}$ for i even

(3) $t_i \; \Theta_{(y,z)} \; t_{i+1}$ for i odd.

By the usual arguments then the t_i and r do correspond to ternary terms $\overline{q}_i$ and p such that (in accordance with (0), ..., (5)) the following equations are satisfied in $\underline{V}$:

(0) $x = \overline{q}_o(x,y,z)$

(1) $\overline{q}_i(x,x,y) = x$ for all $0 \le i < n$

(2) $\overline{q}_i(x,y,x) = \overline{q}_{i+1}(x,y,x)$ for i even

(3) $\overline{q}_i(x,y,y) = \overline{q}_{i+1}(x,y,y)$ for i odd

(4) $\overline{q}_n(x,y,y) = p(x,y,y)$

(5) $p(x,x,y) = y$.

By simply redefining $q_i(x,y,z) := \overline{q}_i(x,z,y)$ we get the desired terms.

For (iv) → (i) it is enough to prove the Shifting Lemma according to 3.6. So we start with congruences α, β, γ with $\alpha \wedge \beta \le \gamma$ and elements x, y, z, u such that $x \; \alpha \; z \; (\beta \wedge \gamma) \; u \; \alpha \; y \; \beta \; x$. Again we might as well assume that $\alpha \wedge \beta = 0$, otherwise we would have to replace equality signs by $\equiv (\text{mod } \alpha \wedge \beta)$.

Consider the following points of $\underline{A}$:

$$\overline{p} := p(z,u,y)$$

$$\hat{q}_i := q_i(x,u,y)$$

$$\overline{q}_i := q_i(x,z,y) \quad \text{and}$$

$$\tilde{q}_n := q_n(z,y,u).$$

We obtain the following relations:

I $x \; \beta \; \overline{q}_i$ for all i

II $x \; \beta \; \hat{q}_i$ for all i, by using equation 2.

Equation 6 yields:

III $y \; (\beta \wedge \gamma) \; \overline{p}$.

Thus the $\overline{q}_i$, $\hat{q}_i$ and $\overline{p}$ lie on the β-line connecting x and y, in particular they are mutually β-congruent. Equations 3 and 4 now provide for

IV $\overline{q}_i \; \alpha \; \overline{q}_{i+1}$ for i even and

V $\hat{q}_i \; \alpha \; \hat{q}_{i+1}$ for i odd.

Hence with I and II we have

VI $\overline{q}_i = \overline{q}_{i+1}$ for i even and

VII $\hat{q}_i = \hat{q}_{i+1}$ for i odd.

But notice that by definition we have also

VIII $\overline{q}_i \; \gamma \; \hat{q}_i$ for every i.

This, together with VI and VII gives us:

$$x = \bar{q}_o = \bar{q}_1 \ \gamma \ \hat{q}_1 = \hat{q}_2 \ \gamma \ \bar{q}_2 = \bar{q}_3 \ \gamma \ \hat{q}_3 \ \ldots \ \hat{q}_n, \quad \text{i.e.}$$

IX $x \ \gamma \ \hat{q}_n$. (No matter whether n is odd or even!)

Hence in view of III, all we have to do is to show that $\hat{q}_n = \bar{p}$. For this reason notice that

$$\hat{q}_n = q_n(x,u,y) \ \alpha \ q_n(z,y,u) = \tilde{q}_n \quad \text{and}$$

$$\bar{p} = p(z,u,y) \ \alpha \ p(z,y,y) = q_n(z,y,y) \ \alpha \ q_n(z,y,u) = \tilde{q}_n.$$

Hence $\hat{q}_n \ \alpha \ \bar{p}$ and $\hat{q}_n \ \beta \ \bar{p}$ by I and III. Thus $\hat{q}_n = \bar{p}$ and we are finished.

To see what is really going on, a look at the following picture is worthwile. (Notice, that $\tilde{q}_n \ \gamma \wedge \beta \ u$ by equation 2.)

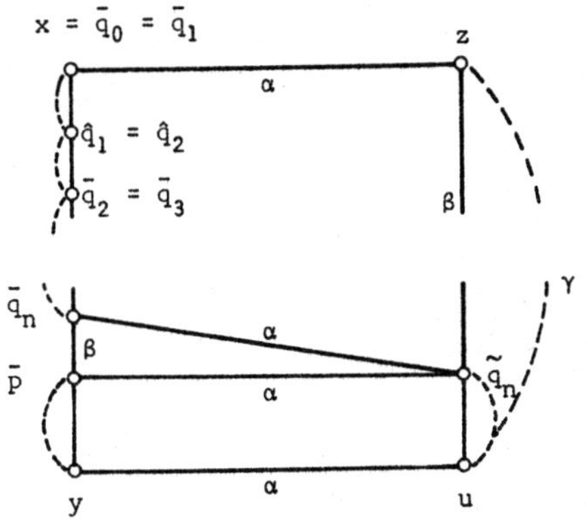

Of course, as a consequence of 7.4 one should be able to construct our terms $q_o,\ldots,q_n$ and p from the DAY-terms $m_o,\ldots,m_k$. In fact a constructive proof could be worked out following the ribbon backwards through the theory of commutators and through the construction of the sixary term from chapter 4 to the DAY-terms. This would be an enormous and tedious task, so messy that it is inconceivable that a proof of 7.4 would ever have been found without the theory of commutators at disposal.

For the simplest nontrivial case, i.e. if $\underline{V}$ is 3-permutable we have worked out the terms:

7.5 <u>Proposition:</u> Suppose $\underline{V}$ is a 3-permutable variety, where $r(x,y,z)$ and $s(x,y,z)$ are the terms for 3-permutability (1.3) i.e. satisfying

$$\begin{aligned} x &= r(x,y,y) \\ r(x,x,y) &= s(x,y,y) \\ s(x,x,y) &= y. \end{aligned}$$

Then the following are terms satisfying the equations of 7.4:

$$q_1(x,y,z) = s(r(s(x,z,z),y,x),r(x,y,z),x)$$

$$q_2(x,y,z) = r(r(s(x,z,z),z,x),r(s(x,z,z),y,x),r(x,y,z))$$

$$q_3(x,y,z) = r(s(x,z,z),z,x)$$

$$p(x,y,z) = r(s(x,y,z),y,x).$$

We can do better, namely have 7.4 hold with $n = 2$ because of 3-permutability. But the terms will get deeper nested as

$$\bar{q}_1(x,y,z) = r(x,q_1(x,y,z),q_2(x,y,z))$$

$$\bar{q}_2(x,y,z) = s(q_1(x,y,z),q_2(x,y,z),q_3(x,y,z))$$

$$p(x,y,z) = r(s(x,y,z),y,x).$$

Is there any shorter way?

Very beautifully the terms of 7.4 come up when we form the join of two independent varieties $\underline{V}_1$ and $\underline{V}_2$ where $\underline{V}_1$ is distributive and $\underline{V}_2$ is permutable. Independence of $\underline{V}_1$ and $\underline{V}_2$ means that there is a binary term $f(x,y)$ such that

$f(x,y) = x$ holds in $\underline{V}_1$ and

$f(x,y) = y$ holds in $\underline{V}_2$.

Clearly modularity is the common denominator of $\underline{V}_1$ and $\underline{V}_2$ and should hence be shared by their join. (Here independence is indispensable as a consequence of [15].)

Let $t_o,\ldots,t_n$ be the JONSSON-terms of $\underline{V}_1$ and m the MAL'CEV-term of $\underline{V}_2$ then

$q_i(x,y,z) := f(t_i(x,y,z),x)$ and

$p(x,y,z) := f(z,m(x,y,z))$

are the terms for theorem 7.4.

There is, also an interesting converse to this construction. Namely, given the terms q_i and p of Theorem 7.4, define subvarieties of the given variety $\underline{V}$ by

$$\underline{V}_d := \text{Mod}\ \{p(x,y,y) = y\} \cap \underline{V}$$

$$\underline{V}_p := \text{Mod}\ \{p(x,y,y) = x\} \cap \underline{V}.$$

Then we obtain:

7.6 <u>Proposition:</u> $\underline{V}_d$ is congruence distributive, $\underline{V}_p$ is permutable and $\underline{V}_d$ and $\underline{V}_p$ are independent subvarieties of $\underline{V}$.

If $\underline{V}$ as an example is the variety of generalized right complemented semigroups then $r(x,y,z) := x\cdot(y*z)$ and $s(x,y,z) := z\cdot(y*x)$ show that $\underline{V}$ is three-permutable. Thus the term $p(x,y,z)$ from theorem 7.4 is

$$p(x,y,z) = r(s(x,y,z),y,x) = $$
$$= (z\cdot(y*x))\cdot(y*x).$$

Thus $\underline{V}_d = \text{Mod}\ \{(y\cdot(y*x))\cdot(y*x) = y\} \cap \underline{V}$ and

$\underline{V}_p = \text{Mod}\ \{(y\cdot(y*x))\cdot(y*x) = x\} \cap \underline{V}$.

Let us add a remark about the connections between the different terms we have been using so far. It will become obvious in the following chapter (see thm. 8.5 below) that, on defining

$$t(x,y,z) := p(z,y,x)$$

we obtain a term satisfying all properties from our originally constructed ternary term from chapter 4. On the other hand, for every term $t(x,y,z)$ with the properties described in theorem 4.3,

$$p(x,y,z) := t(z,y,x)$$

yields a term for which there exist $q_o,\dots,q_n$ with the properties of theorem 7.4. This is a consequence of the proof of 7.2.

8. PERMUTABILITY RESULTS

In this chapter we give criteria for congruences Θ and Ψ to permute. Some of the results improve corresponding findings from chapter 4. The formula on which most of this is based is the one from 7.2.

We start with

8.1 <u>Theorem:</u> Let $\underline{A}$ be an algebra in a modular variety and Θ and Ψ congruences on $\underline{A}$. Then the following statements are equivalent:

(i) Θ permutes with Ψ

(ii) $\Theta^{(n)}$ permutes with $\Psi^{(m)}$ for all $n,m \in \mathbb{N}$

(iii) $\Theta^{(n)}$ permutes with $\Psi^{(m)}$ for some $n,m \in \mathbb{N}$.

Here we use the following definition:

$$\Theta^{(o)} := \Theta, \quad \Theta^{(n+1)} := [\Theta^{(n)},\Theta^{(n)}].$$

Θ is called <u>solvable</u>, if $\Theta^{(n)} = 0$ for some natural number n.

Note that as a corollary to 8.1 we have a result from [21]:

8.2 <u>Corollary:</u> A solvable congruence relation permutes with every congruence relation.

<u>Proof of 8.1:</u> Iterating the formula $\Theta \circ \Psi \subseteq [\Theta,\Theta] \circ \Psi \circ \Theta$ and its symmetric form we obtain:

$$\Theta \circ \Psi \subseteq \Psi \circ \Theta^{(n)} \circ \Psi^{(m)} \circ \Theta \quad \text{for any } n,m \in \mathbb{N}.$$

Namely, by symmetry we are done if we show the induction step from n to $n+1$. Assuming the above formula we obtain

$$\begin{aligned} \Theta \circ \Psi &\subseteq \Psi \circ \Theta^{(n)} \circ \Psi^{(m)} \circ \Theta \subseteq \\ &\subseteq \Psi \circ [\Theta^{(n)},\Theta^{(n)}] \circ \Psi^{(m)} \circ \Theta^{(n)} \circ \Theta \subseteq \\ &\subseteq \Psi \circ \Theta^{(n+1)} \circ \Psi^{(m)} \circ \Theta. \end{aligned}$$

Now the above formula gives us (iii) → (i) since $\Theta^{(n)} \leq \Theta$ and $\Psi^{(m)} \leq \Psi$. It remains to prove (i) → (ii).

Again by induction we may assume we have already proven that $\Theta^{(n-1)}$ permutes with $\Psi^{(m)}$ hence have to show $\Theta^{(n)}$ permutes with $\Psi^{(m)}$. Changing notation, we have to show that $[\Theta,\Theta]$ permutes with Ψ in case Θ permutes with Ψ. Suppose $(x,z) \in [\Theta,\Theta] \circ \Psi$, i.e. for some y we have

$$x \; [\Theta,\Theta] \; y \; \Psi \; z.$$

In particular $(x,y) \in \Theta$ hence $x \,\Theta\, y \,\Psi\, z$. Since Θ and Ψ permute we find a u with

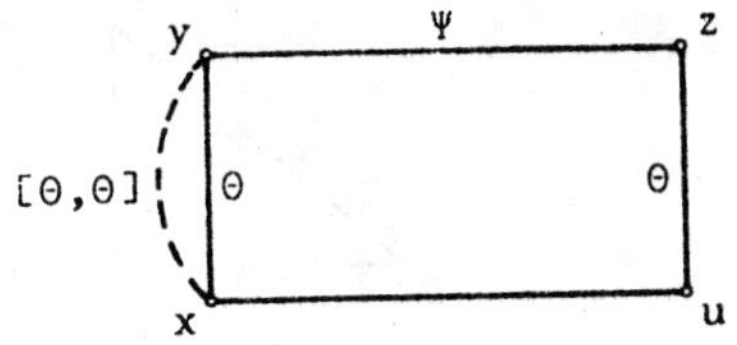

The Shifting Lemma gives us $(u,z) \in (\Theta \wedge \Psi) \vee [\Theta,\Theta]$. But since $[\Theta \wedge \Psi, \Theta \wedge \Psi] \leq [\Theta,\Theta]$, $(\Theta \wedge \Psi)^{(1)}$ permutes with $[\Theta,\Theta]$ which implies that $\Theta \wedge \Psi$ permutes with $[\Theta,\Theta]$ by the direction (iii) → (i).

Hence there exists an element w with $u \,\Theta \wedge \Psi\, w \,[\Theta,\Theta]\, z$, thus $x \,\Psi\, w \,[\Theta,\Theta]\, z$ which was to be shown.

As a corollary to the proof we get a stronger kind of Shifting Lemma, namely:

8.3 <u>Corollary:</u> Let α, β, γ be congruences on an algebra $\underline{A}$ in a modular variety such that $(\alpha \wedge \beta)^{(n)}$ permutes with $\gamma^{(m)}$ for some $n,m \in \mathbb{N}$. Then

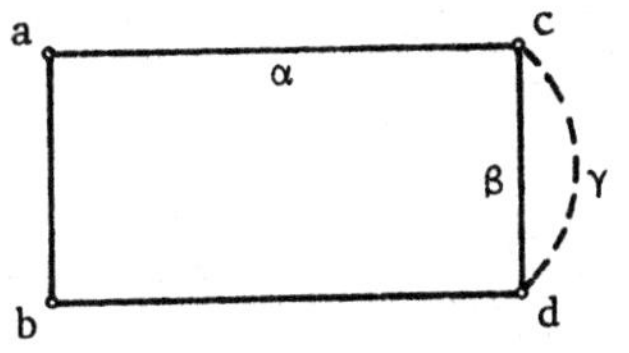

implies $(a,b) \in (\alpha \wedge \beta) \circ \gamma$.

Iterating the formula of 7.2 in a way similarly as in the proof of 8.1 we find that joins of congruences may be easily computed, once the join of some of there iterated commutators are known:

8.4 <u>Lemma:</u> If Θ and Ψ are congruences and $n,m \in \mathbb{N}$, then

$$\begin{aligned}\Theta \vee \Psi &= \Theta \circ (\Theta^{(n)} \vee \Psi^{(m)}) \circ \Psi = \\ &= (\Theta^{(n)} \vee \Psi^{(m)}) \circ \Theta \circ \Psi = \\ &= \Theta \circ \Psi \circ (\Theta^{(n)} \vee \Psi^{(m)}).\end{aligned}$$

An application will be given in 8.8 below.

Theorem 4.3 may also be improved as follows:

8.5 <u>Theorem:</u> In every modular variety $\underline{V}$ there exists a ternary term $p(x,y,z)$ such that $p(x,x,y) = y$ is an equation of $\underline{V}$ and for congruences α, β and γ with $\alpha \wedge \beta$ permuting with γ we obtain

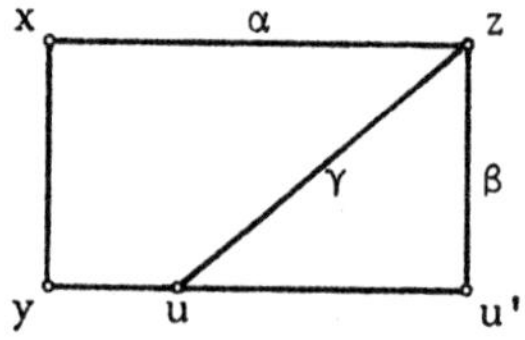

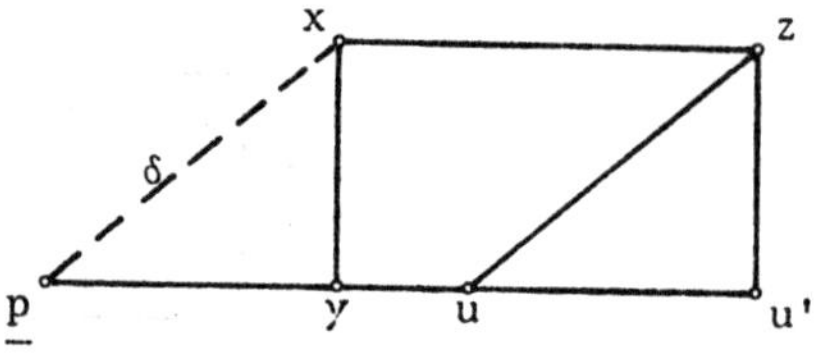

with $\delta = \gamma \circ (\alpha \wedge \beta)$ and $\underline{p} = p(u,u',y)$.

<u>Proof:</u> Define $\hat{q}_i$ and $\underline{p}$ precisely as in the proof of 7.4. Hence $x \ \gamma \vee (\alpha \wedge \beta) \ \hat{q}_n$ and finally $p(u,u',y) \ \gamma \ p(z,u',y)$ $p(u,u',y) \ \gamma \ p(z,u',y) \ (\alpha \wedge \beta) \ \hat{q}_n$ with $p(u,u',y) \ \alpha \ y$ yielding $x \ \gamma \circ (\alpha \wedge \beta) \ \underline{p}$.

With the help of the above finally 4.4 can be strengthened:

8.6 <u>Corollary:</u> Let α, β and γ be congruences such that $\gamma \leq \alpha \circ \beta = \beta \circ \alpha$ and $\gamma^{(n)}$ permutes with $(\alpha \wedge \beta)^{(m)}$ for some $n,m \in \mathbb{N}$. Then γ permutes with α (and with β).

<u>Remark:</u> The special case where $\alpha \wedge \beta \leq \gamma$ was the first result in this direction. It was proven in [17]. Combining this with the result that solvable congruences permute with every other congruence (8.2), A. WOLF gave a short argument to replace the condition $\alpha \wedge \beta \leq \gamma$ by $(\alpha \wedge \beta)^{(n)} \leq \gamma$.

Corollary 8.5 is its present form subsumes all those versions as well as 8.2, just set $\beta = 1$ and $m = 0$.

If α and β are congruences then modularity of the lattice of congruence relations implies that the maps $x \longrightarrow \alpha \vee x$ and $x \longrightarrow \beta \wedge x$ are isomorphisms between the intervals $\alpha \wedge \beta \leq x \leq \beta$ and $\alpha \leq x \leq \alpha \vee \beta$.

In [21] it was proved that

8.7 <u>Lemma:</u> If α and β permute, then the above isomorphisms preserve permutability of congruences.

<u>Proof:</u> For $\alpha \wedge \beta \leq \Theta, \Psi \leq \beta$ we find by 4.4 that Θ and Ψ permute with α. Hence $\alpha \vee \Theta$ $(= \alpha \circ \Theta)$ permutes with $\alpha \vee \Psi$ $(= \alpha \circ \Psi)$.

If $\alpha \leq \Theta, \Psi \leq \alpha \vee \beta$ and Θ permutes with Ψ then notice that by repeated use of 4.4 $\Theta \wedge \Psi$ permutes with $\beta \wedge (\Theta \vee \Psi)$. Thus we might as well assume that $\alpha \wedge \beta = 0$ and $\Theta \wedge \Psi = \alpha$. Moreover, instead of looking at each class of $\Theta \vee \Psi$ separately, we may assume $\Theta \vee \Psi = 1$.

Thus Θ and Ψ give a factor decomposition of $\underline{A}/\alpha$. Since α permutes with β, we have a direct decomposition of $\underline{A}$.

Therefore, all the congruences considered so far were factor-congruences, in particular, $\Theta \wedge \beta$ and $\Psi \wedge \beta$ permute.

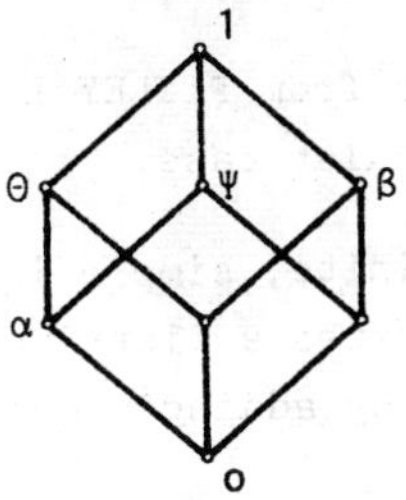

Combining the preceding lemma with 8.4 we obtain a short proof of a theorem which is due to A. WOLF [42]:

8.8 Theorem: If $\underline{A}$ and $\underline{B}$ are algebras in a modular variety such that $\underline{A}$ and $\underline{B}$ have permutable congruences, then so has $\underline{A} \times \underline{B}$.

Proof: For Θ, Ψ congruences on $\underline{A} \times \underline{B}$ we have by 8.4:

$$\begin{aligned}
\underline{\Theta \vee \Psi} &= \Theta \circ ([\Theta,\Theta] \vee [\Psi,\Psi]) \circ \Psi \subseteq \\
&\subseteq \Theta \circ ([\Theta,1] \vee [\Psi,1]) \circ \Psi \subseteq \\
&\subseteq \Theta \circ ([\Theta,\pi_1 \vee \pi_2] \vee [\Psi,\pi_1 \vee \pi_2]) \circ \Psi \subseteq \\
&\subseteq \Theta \circ ([\Theta,\pi_1] \vee [\Theta,\pi_2] \vee [\Psi,\pi_1] \vee [\Psi,\pi_2]) \circ \Psi \subseteq \\
&\subseteq \Theta \circ ((\Theta \wedge \pi_1) \vee (\Psi \wedge \pi_1) \vee (\Theta \wedge \pi_2) \vee (\Psi \wedge \pi_2)) \circ \Psi \subseteq \\
&\subseteq \Theta \circ (((\Theta \wedge \pi_1) \circ (\Psi \wedge \pi_1)) \vee ((\Theta \wedge \pi_2) \circ (\Psi \wedge \pi_2))) \circ \Psi \subseteq \\
&\subseteq \Theta \circ ((\Theta \wedge \pi_1) \circ (\Psi \wedge \pi_1) \circ (\Theta \wedge \pi_2) \circ (\Psi \wedge \pi_2)) \circ \Psi \subseteq \\
&\subseteq \Theta \circ (\Psi \wedge \pi_1) \circ (\Theta \wedge \pi_2) \circ \Psi \subseteq \\
&\subseteq \Theta \circ (\Theta \circ \pi_2) \circ (\Psi \circ \pi_1) \circ \Psi \subseteq \underline{\Theta \circ \Psi}
\end{aligned}$$

We have twice used the fact that every congruence below π_1 permutes with every congruence below π_2. This follows from 4.4 because π_1 and π_2 permute.

Quite useful is the following instance of Corollary 8.2:

8.9 Corollary: If $\mathrm{Con}(\underline{A})$ has M_3, (the five-element modular nondistributive lattice) as a sublattice, then any two elements of this sublattice permute.

Proof: Clearly if

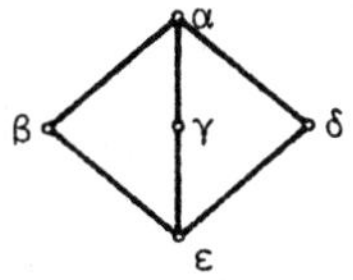

is a sublattice, then $[\alpha,\alpha] \leq \varepsilon$ by successively applying the rules of 6.16. Hence $\alpha, \beta, \gamma, \delta$ are solvable in $\underline{A}/\varepsilon$. Thus in fact $\alpha, \beta, \gamma, \delta$ permute with everything above ε.

As an example, using arguments from PIXLEY [35], a result of R. McKENZIE [33] can be adapted to the modular case:

8.10 Corollary: If $\underline{A}$ is finite, simple in a modular variety, then $\underline{A}$ is either affine or $\underline{A}^+$ generates a distributive variety. (Here $\underline{A}^+$ is the algebra obtained from $\underline{A}$ by adding a constant $\underline{a}$ with value a for every element $a \in A$.)

Proof: Let $\underline{F} := \underline{F}_{V(\underline{A}^+)}(3)$ be freely three-generated in the variety $V(\underline{A}^+)$ generated by $\underline{A}^+$. Clearly $\underline{F}$ is a subdirect power of $\underline{A}^+$. If $Con(\underline{F})$ is distributive, then $V(\underline{A}^+)$ is congruence distributive as a consequence of JONSSON's theorem (thm. 1.4). Otherwise there are coatoms α,β in $Con(\underline{F})$ with $\{x \mid \alpha \wedge \beta \leq x \leq 1\}$ being a nondistributive interval; α and β may be chosen to be part of the subdirect decomposition for $\underline{F}$ (see BURRIS [5]). Since $\underline{F}/\alpha \cong \underline{F}/\beta \cong \underline{A}^+$, and, by corollary 8.9, $\underline{F}/\alpha\wedge\beta \cong \underline{A}^+ \times \underline{A}^+$ for $\alpha \neq \beta$. Hence $\underline{A}^+ \times \underline{A}^+$ has a congruence which is a complement of the canonical factor congruences, since $\underline{A}^+$ is simple.

9. ABELIAN CONGRUENCES AND AFFINE ALGEBRAS

In the preceding chapters we have mainly made use of the commutators of the form $[\alpha,\alpha]$. Here we look at a case, slightly more general, namely we assume $\alpha \geq \beta$ and consider $[\alpha,\beta]$.

If x, y, z are elements with $x \; \alpha \; y \; \beta \; z$ then we have already found the important relation

$$(x,y) \; \Delta_\alpha^\beta \; (t(x,y,z),z). \qquad (\S)$$

Therefore, since Δ_α^β is a congruence we find for $\vec{x} := (x_1,\dots,x_n)$, $\vec{y} := (y_1,\dots,y_n)$, $\vec{z} := (z_1,\dots,z_n)$ with $x_i \; \alpha \; y_i \; \beta \; z_i$ and any n-ary operation f:

$$(f(\vec{x}),f(\vec{y})) \; \Delta_\alpha^\beta \; (f(t(x_1,y_1,z_1),\dots,t(x_n,y_n,z_n)),f(\vec{z})).$$

Using (§) again we get

$$(f(\vec{x}),f(\vec{y})) \; \Delta_\alpha^\beta \; (t(f(\vec{x}),f(\vec{y}),f(\vec{z})),f(\vec{z}))$$

and hence

$$f(t(x_1,y_1,z_1),\dots,t(x_n,y_n,z_n)) \; [\alpha,\beta] \; t(f(\vec{x}),f(\vec{y}),f(\vec{z})). \qquad (\S\S)$$

This yields one direction of an equational description of $[\alpha,\beta]$:

9.1 <u>Theorem:</u> Suppose $\alpha \geq \beta$. Then $[\alpha,\beta] = 0$ if and only if for all $x_i \; \alpha \; y_i \; \beta \; z_i$ with $x_i, y_i, z_i \in A$ the equations

$$t(y_i,y_i,z_i) = z_i \quad \text{and}$$

$$f(t(x_1,y_1,z_1),\dots,t(x_n,y_n,z_n)) = t(f(\vec{x}),f(\vec{y}),f(\vec{z}))$$

are satisfied.

<u>Proof:</u> For the proof of the missing direction we define a congruence relation $\equiv$ on α by

$$(x,y) \equiv (u,z) \;:\Leftrightarrow\; x \; \alpha \; y \; \beta \; z \quad \text{and} \quad t(x,y,z) = u.$$

To show symmetry we suppose $t(x,y,z) = u$ and $x \; \alpha \; y \; \beta \; z$ and compute:

$$\begin{aligned} t(u,z,y) &= t(t(x,y,z),t(y,y,z),t(y,y,y)) = \\ &= t(t(x,y,y),t(y,y,y),t(z,z,y)) = \\ &= t(x,y,y) = x. \end{aligned}$$

Clearly $x\ \beta\ u$, hence $u\ \alpha\ z\ \beta\ y$.

For transitivity suppose $x\ \alpha\ y\ \beta\ z$, $u\ \alpha\ z\ \beta\ s$, $t(x,y,z) = u$, $t(u,z,s) = r$ and compute:

$$\begin{aligned} t(x,y,s) &= t(t(x,y,y),t(y,y,y),t(z,z,s)) = \\ &= t(t(x,y,z),t(y,y,z),t(y,y,s)) = \\ &= t(u,z,s) = r. \end{aligned}$$

Again $x\ \alpha\ y\ \beta\ s$ trivially.

Using that $t(x,x,y) = y$ for $x\ \beta\ y$ we find $(x,x) \equiv (y,y)$ whenever $x\ \beta\ y$ and consequently $\equiv\ \geq \Delta_\alpha^\beta$.

Hence suppose $x\ [\alpha,\beta]\ y$, then $(y,x)\ \Delta_\alpha^\beta\ (x,x)$, therefore $(y,x) \equiv (x,x)$ hence $t(y,x,x) = x$ which implies $x = y$. Thus $[\alpha,\beta] = 0$.

We have created a situation similar to the hypothesis of Proposition 5.4. Indeed, theorem 9.1 permits us to associate affine algebras with congruences α which are abelian, i.e. for which $[\alpha,\alpha] = 0$. If additionally α is contained in the center, then all affine algebras associated with α are isomorphic. Moreover, the congruences below the center ξ correspond uniquely to the subalgebras of the affine algebra associated with ξ. This reflects the group theoretic situation where subalgebras of the center of $\underline{G}$ are normal in $\underline{G}$. One of the difficulties we have here is, that our algebras need not have any one-element subalgebras.

We have to use 9.1 over and over again. The approach is rather naturally and intuitively clear, but we have to be careful in setting up the right equations so that the conditions in 9.1 remain satisfied.

Choose an arbitrary element a from $\underline{A}$ and a congruence relation $\beta \in \mathrm{Con}(\underline{A})$. Let f be a fundamental operation of $\underline{A}$, or the ternary operation $t(x,y,z)$. For $x_1,\ldots,x_n \in [a]\beta$ we define

$$f^\nabla(x_1,\ldots,x_n) := t(a,f(a,\ldots,a),f(x_1,\ldots,x_n)).$$

Let $\underline{A}^\nabla[\beta]_a$ be the algebra with base set $[a]\beta$ and the operations of the form f^∇. Clearly $[a]\beta$ is closed under the new operations so that the definition makes sense. Note that in case $[\beta,\beta] = 0$ idempotent operations remain unchanged, in particular, if $\{a\}$ happened to be a one-element subalgebra of $\underline{A}$ then $\underline{A}^\nabla[\beta]_a$ is the subalgebra $[a]\beta$ of $\underline{A}$. This is immediate from 7.1.

First we need:

9.2 <u>Theorem:</u> Let $\alpha \geq \beta$ with $[\alpha,\beta] = 0$. Then $\underline{A}^\nabla[\beta]_a$ is an affine algebra and for $(a,b) \in \alpha$ we get $\underline{A}^\nabla[\beta]_a \cong \underline{A}^\nabla[\beta]_b$.

9.3 <u>Corollary:</u> If $\alpha \geq \beta$ and $[\alpha,\beta] = 0$ then β is uniform with respect to α, i.e. all β-classes within a fixed α-class have the same size.

Defining $\alpha^1 := \alpha$ and $\alpha^{n+1} := [\alpha^n, 1]$ then by induction:

9.4 Corollary: If $\alpha^n = 0$ for some $n \in \mathbb{N}$ then all classes of α have the same size (α is a uniform congruence).

Proof of 9.2: a) $\underline{A}^\nabla[\beta]_a$ is affine with respect to t $(= t^\nabla)$. We use the notation $\bar{a}$ for the constant sequence $(a,\dots,a)$. $\vec{x}$ denotes $(x_1,\dots,x_n)$, similarly we use $\vec{y}$ and $\vec{z}$.
So, given $x_i, y_i, z_i \in [a]\beta$ we compute

$$\begin{aligned} t(f^\nabla(\vec{x}),f^\nabla(\vec{y}),f^\nabla(\vec{z})) &= t(t(a,f(\bar{a}),f(\vec{x})),t(a,f(\bar{a}),f(\vec{y})),t(a,f(\bar{a}),f(\vec{z}))) \\ &= t(t(a,a,a),t(f(\bar{a}),f(\bar{a}),f(\bar{a})),t(f(\vec{x}),f(\vec{y}),f(\vec{z}))) \\ &= t(a,f(\bar{a}),f(t(x_1,y_1,z_1),\dots,t(x_n,y_n,z_n))) \\ &= f^\nabla(t(x_1,y_1,z_1),\dots,t(x_n,y_n,z_n)). \end{aligned}$$

b) For $a\,\alpha\,b$ we show $\underline{A}^\nabla[\beta]_a \cong \underline{A}^\nabla[\beta]_b$:
Define $\xi_{a,b}\colon [a]\beta \twoheadrightarrow [b]\beta$ by $\xi_{a,b}(x) := t(b,a,x)$, then with 9.1 again

$$\begin{aligned} \xi_{b,a}\circ\xi_{a,b}(x) &= t(a,b,t(b,a,x)) = t(t(a,a,a),t(b,a,a),t(b,a,x)) = \\ &= t(t(a,b,b),t(a,a,a),t(a,a,x)) = t(a,a,x) = x. \end{aligned}$$

Hence the $\xi_{a,b}$ are bijective for $a\,\alpha\,b$.
Furthermore, (denoting the operations of $\underline{A}^\nabla[\beta]_a$ and of $\underline{A}^\nabla[\beta]_b$ with the same symbol) we get for $x_i \in [a]\beta$:

$$\begin{aligned} &\xi_{a,b}(f^\nabla(x_1,\dots,x_n)) = t(b,a,t(a,f(\bar{a}),f(\vec{x}))) = \\ &= t(t(b,f(\bar{b}),f(\bar{b})),t(a,f(\bar{a}),f(\bar{a})),t(a,f(\bar{a}),f(\vec{x}))) = \\ &= t(t(b,a,a),t(f(\bar{b}),f(\bar{a}),f(\bar{a})),t(f(\bar{b}),f(\bar{a}),f(\vec{x}))) = \\ &= t(b,f(\bar{b}),f(t(b,a,x_1),\dots,t(b,a,x_n))) \\ &= f^\nabla(\xi_{a,b}(x_1),\dots,\xi_{a,b}(x_n)). \end{aligned}$$

Now from the above it is clear that in the special case where $[1,\alpha] = 0$, i.e. $\alpha \le \zeta$ (the center of $\underline{A}$) there is <u>one</u> affine algebra $\underline{A}[\alpha]$ associated with α.
$\underline{A}[\alpha]$ even is contained in the variety generated by $\underline{A}$ and the relation between $\underline{A}$, $\underline{A}[\alpha]$ and (the algebra) $\underline{\alpha}$ is given by

9.5 Theorem: If $[1,\alpha] = 0$ then

(i) $A^\nabla[\alpha] \cong \underline{\alpha}/\Delta^1_\alpha$ and

(ii) $\underline{\alpha} \cong \underline{A} \times \underline{A}^\nabla[\alpha]$.

9.5(ii) is a generalization of the situation which has first appeared in Sec. 5, namely for A an affine algebra, i.e. setting $\alpha = 1$ we get $\underline{A} \times \underline{A} = \underline{A} \times \underline{A}^\nabla$ where now $\underline{A}^\nabla$ (by 9.5(i)) is the factor of $\underline{A} \times \underline{A}$ by the congruence $\Delta = \Delta_1^1$ on $\underline{A} \times \underline{A}$ (see Sec. 5). Note that $\underline{A}^\nabla[\alpha]$ always has a one-element subalgebra, as (possibly) opposed to $\underline{A}$.

Proof of 9.5: (i) Define a map $\delta_a: \underline{\alpha} \twoheadrightarrow [a]\alpha$ by $\delta_a(x,y) := t(a,x,y)$, then δ_a is a map from α to $[a]\alpha$ and $\ker \delta_a = \Delta_\alpha^1$ because suppose $t(a,x,y) = t(a,x',y') =: u$ then $(a,x)\ \Delta_1^\alpha\ (u,y)$ and $(a,x')\ \Delta_1^\alpha\ (u,y')$ so $(a,u)\ \Delta_\alpha^1\ (x,y)$ and $(a,u)\ \Delta_\alpha^1\ (x',y')$ according to the beginning of this section, hence $(x,y)\ \Delta_\alpha^1\ (x',y')$, i.e. $\ker \delta_a \le \Delta_\alpha^1$.
Conversely $(x,y)\ \Delta_\alpha^1\ (x',y')$ leads to $(x,x')\ \Delta_1^\alpha\ (y,y')$ and for $v := t(a,x,y)$ we find $(a,x)\ \Delta_1^\alpha\ (v,y)$ hence $(a,x')\ \Delta_1^\alpha\ (v,y')$ and $t(a,x,y) = v = t(a,x',y')$.

For the homomorphism condition we calculate

$$\begin{aligned}
&\delta_a(f((x_1,y_1),\dots,(x_n,y_n))) = t(a,f(\vec{x}),f(\vec{y})) = \\
&= t(a,a,t(a,f(\vec{x}),f(\vec{y}))) \\
&= t(t(a,a,a),t(a,f(\bar{a}),f(\bar{a})),t(a,f(\vec{x}),f(\vec{y}))) \\
&= t(a,f^\nabla(\vec{x}),f^\nabla(\vec{y})) \\
&= f^\nabla(t(a,x_1,y_1),\dots,t(a,x_n,y_n)) \\
&= f^\nabla(\delta_a(x_1,y_1),\dots,\delta_a(x_n,y_n)).
\end{aligned}$$

For the proof of (ii) we find in the congruence lattice of the algebra $\underline{\alpha}$ the congruences Δ_α^1, π_1 and π_2 (the kernels of the projections onto A). $[1,\alpha] = 0$ means $\Delta_\alpha^1 \wedge \pi_1 = 0$ and, equivalently $\Delta_\alpha^1 \wedge \pi_2 = 0$. Moreover $\Delta_\alpha^1 \vee \pi_1 = \Delta_\alpha^1 \vee \pi_2 = 1$. Thus Δ_α^1 and π_1 are complements in $\mathrm{Con}(\underline{\alpha})$. If they permute they will give the desired decomposition of $\underline{\alpha}$. But since $[\pi_1,\pi_1] = 0$ follows from the above relations, π_1 permutes with every congruence (recall 8.2).

Next we are going to look at congruences below the center ζ and we shall show that they correspond uniquely to subalgebras of $\underline{A}[\zeta]$.

Fix an element a from $\underline{A}$. For a subalgebra $\underline{S}$ of $\underline{A}^\nabla[\zeta]$ containing the element a, define $\Psi(\underline{S}) := \{(x,y) \in \zeta \mid t(a,x,y) \in \underline{S}\}$ and for a congruence relation Θ of $\underline{A}$ with $\Theta \le \zeta$ define $U(\Theta) := [a]\Theta$. We show

9.6 Proposition: $U(-)$ and $\Psi(-)$ are mutually inverse lattice isomorphisms between the interval $[0,\zeta]$ of $\mathrm{Con}(\underline{A})$ and $\mathrm{Sub}(\underline{A}^\nabla[\zeta])$, the lattice of subalgebras of $\underline{A}^\nabla[\zeta]$.

Remark: The point is here that subalgebras of $\underline{A}^\nabla[\zeta]$ correspond to congruences on $\underline{A}$. For congruences on $\underline{A}^\nabla[\zeta]$ the corresponding property is trivial from affineness. Another way to phrase 9.6 would be: Congruences on $\underline{A}^\nabla[\zeta]$ can be extended to congruences on $\underline{A}$.

Proof: We show that $\Psi(\underline{S})$ is a congruence on $\underline{A}$.
Symmetry: If $x\ \Psi(\underline{S})\ y$ then $a\ \zeta\ t(a,y,x)$ hence

$$\begin{aligned} t(a,y,x) &= t(a,a,t(a,y,x)) \\ &= t(t(a,a,a),t(a,x,x),t(a,y,x)) \\ &= t(a,t(a,x,y),a) \in \underline{S}. \end{aligned}$$

Transitivity: $t(a,x,y) \in S$ and $t(a,y,z) \in \underline{S}$, $x\zeta y\zeta z$ imply

$$\begin{aligned} t(a,x,z) &= t(t(a,a,a),t(x,y,y),t(y,y,z)) \\ &= t(t(a,x,y),a,t(a,y,z)) \in \underline{S}. \end{aligned}$$

Compatibility: Given $x_i \zeta y_i$ and $t(a,x_i,y_i) \in \underline{S}$ then $f(\vec{x})\zeta f(\vec{y})$ and

$$\begin{aligned} & t(a,f(\vec{x}),f(\vec{y})) = \\ & = t(t(a,f(\bar{a}),f(\bar{a})),t(f(\vec{x}),f(\vec{x}),f(\vec{x})),t(f(\vec{x}),f(\vec{x}),f(\vec{y}))) \\ & = t(a,f(\bar{a}),f(t(a,x_1,y_1),\ldots,t(a,x_n,y_n))) \\ & = f^{\nabla}(t(a,x_1,y_1),\ldots,t(a,x_n,y_n)) \in \underline{S}. \end{aligned}$$

Finally for $\Theta \le \zeta$, $\Theta \in \mathrm{Con}(\underline{A})$:
$x\ \Psi\ U(\Theta)\ y$ <u>iff</u> $t(a,x,y) \in U(\Theta)$ & $x\zeta y$ <u>iff</u>
$t(a,x,y)\ \Theta\ t(a,x,x) = a$ & $x\zeta y$ <u>iff</u> $x\Theta y$.

The last equivalence here is due to

$$\begin{aligned} x = t(x,a,a)\ \Theta\ t(x,a,t(a,x,y)) &= t(t(x,x,x),t(a,x,x),t(a,x,y)) \\ &= t(x,x,y) = y. \end{aligned}$$

With the preceding results the congruences below the center and the corresponding affine algebras seem to be well understood. Next abelian congruences β i.e. those for which $[\beta,\beta] = 0$ should be studied for $\beta \not\le \xi$. In general the corresponding affine algebras $\underline{A}^{\nabla}[\beta]_a$ do not lie in the variety generated by $\underline{A}$. In the case of groups though they do generate equivalent varieties. A description of those affine algebras might lead to improving the bounds for the cardinality of subdirectly irreducible algebras in residually finite varieties, given by FREESE, McKENZIE [13].

10. VARIETIES OF AFFINE ALGEBRAS

Analogously to the theory of groups we define for a congruence α:

$$\alpha_{(o)} := \alpha; \qquad \alpha^{(o)} := \alpha$$

$$\alpha_{(n+1)} := [\alpha, \alpha_{(n)}]; \qquad \alpha^{(n+1)} := [\alpha^{(n)}, \alpha^{(n)}].$$

Then α is called nilpotent (solvable) of degree $\leq k$ if $\alpha_{(k)} = 0$ ($\alpha^{(k)} = 0$). α is called nilpotent (solvable) if for some $k \in \mathbb{N}$ α is nilpotent (solvable) of degree $\leq k$. Clearly $\underline{A}$ is called nilpotent (solvable) if $1_{\underline{A}}$ is.

If $\underline{V}$ is a variety then with $\underline{V}_{(k)}$ (resp. $\underline{V}^{(k)}$) we denote the class of all algebras which are nilpotent (resp. solvable) of degree $\leq k$.

Starting with $t(x,y,z)$ from chapter 4 (which may of course be $p(z,y,x)$ from chapter 7) we define recursively:

$$t_1(x,y,z) := t(x,y,z)$$

$$t_{n+1}(x,y,z) := t(t_n(x,y,z), t_n(y,y,z), z).$$

Then we get:

10.1 Observation: $\underline{V}_{(k)}$ and $\underline{V}^{(k)}$ are permutable varieties with Mal'cev term t_k.

Proof: The fact that $\underline{V}_{(k)}$ and $\underline{V}^{(k)}$ are varieties can be proven just as in the case of groups. In other words, the class of algebras with a solvable (nilpotent) series of length $\leq k$ for some fixed k, is closed under taking homomorphic images, subalgebras and direct products. It has been clear from 8.2, that $\underline{V}_{(k)}$ and $\underline{V}^{(k)}$ are permutable varieties. The fact that t_k is a Mal'cev term on every solvable algebra (degree $\leq k$) is easy by induction using 7.1.

Now we are going to use 9.1 to give an equational description for $\underline{V}_{(k)}$ and for $\underline{V}^{(k)}$.

To this end let us define by recursion:

$$N_o := \{x \equiv y\}$$

$$N_{k+1} := \{t(\sigma,\sigma,\tau) \equiv \tau \mid N_k \vdash \sigma \equiv \tau\} \cup$$

$$\cup \{f(t(x_1,\sigma_1,\tau_1),\ldots,t(x_n,\sigma_n,\tau_n)) \equiv$$

$$\equiv t(f(x_1,\ldots,x_n), f(\sigma_1,\ldots,\sigma_n), f(\tau_1,\ldots,\tau_n)) \mid$$

$$\mid f \text{ is n-ary operation and } N_k \vdash \sigma_i \equiv \tau_i \text{ for } 0 \leq i \leq n\}$$

and

$$\begin{aligned} S_0 &:= \{x \equiv y\} \\ S_{k+1} &:= \{t(\sigma,\sigma,\tau) \equiv \tau \mid S_k \vdash \sigma \equiv \tau\} \cup \\ &\quad \cup \{f(t(\gamma_1,\sigma_1,\tau_1),\dots,t(\gamma_n,\sigma_n,\tau_n)) \equiv \\ &\quad \equiv t(f(\gamma_1,\dots,\gamma_n),f(\sigma_1,\dots,\sigma_n),f(\tau_1,\dots,\tau_n)) \mid \\ &\quad \mid f \text{ n-ary operation and } S_k \vdash \gamma_i \equiv \sigma_i \equiv \tau_i,\ 0 \le i \le n\}. \end{aligned}$$

9.2 yields by induction:

10.2 Corollary: Relatively to $\underline{V}$ the varieties $\underline{V}_{(k)}$ (resp. $\underline{V}^{(k)}$) are defined by the equation N_k (resp. S_k).

For the case $k = 1$ nilpotency and solvability coincide, leaving us with the variety of affine algebras.

We have seen before that affine algebras are just modules in disguise. Indeed $\underline{V}^{(1)}$ is polynomially equivalent to a variety of modules over a ring $R(\underline{V})$. Let us have a closer look at $R(\underline{V})$.

Clearly we should expect $R(\underline{V})$ to be the free module on one generator. However, instead of looking at $\underline{F}_{\underline{V}}(1)$, where we have to make up our mind which element to pick for a o-element, we just adjoin a new free generator, which we call o. Thus technically we look at the free $\underline{V}$-algebra with free generators x and o. (This idea goes back at least to CSAKANY [7].) Then we do the construction of § 5, i.e. $R(\underline{V})$ has an underlying set the set of all binary idempotent terms $r(o,x)$. Addition and multiplication are given by

$$\begin{aligned} r(o,x) + s(o,x) &:= t(r(o,x),o,s(o,x)) \\ r(o,x) \cdot s(o,x) &:= r(o,s(o,x)). \end{aligned}$$

Clearly, since $[1,1] = 0$, by chapter 5 we have defined an abelian group with an associative multiplication. The one distributive law which is nontrivial, is a consequence of 9.1.

Obviously, for every $\underline{A} \in \underline{V}$, the ring $R(\underline{A})$ from § 5 is a homomorphic image of $R(\underline{V})$.

11. GENERALIZATIONS: FP-VARIETIES

The crucial idea that eventually led us to the consideration of commutators was the idea of coordinatizing the congruence class geometry. The important results about modular varieties which were needed were:

1. that congruences on direct products permute with the factor congruences and

2. the Shifting Lemma.

Clearly, as we have seen in Corollary 3.6, the Shifting Lemma in its general form is equivalent to modularity. The first property however, is strictly weaker. Thus let us define:

11.1 Definition: A variety $\underline{V}$ is called factor permutable (or in short: FP-variety) if every congruence relation on a direct product $\underline{A} \times \underline{B}$ of algebras $\underline{A}, \underline{B} \in \underline{V}$ permutes with the canonical factor congruences π_1 and π_2.

Recall that in general a factor congruence α is a congruence for which there exists another congruence β with $\alpha \wedge \beta = 0$ and $\alpha \circ \beta = 1$.

As an example let us look at the variety $\underline{W}$ given by one ternary term $p(x,y,z)$ and a constant 0, satisfying the equations

$$p(x,x,y) = y$$
$$p(x,0,0) = x.$$

Then $\underline{W}$ is not a modular variety, indeed no equation is satisfied in all congruences lattices of algebras in $\underline{W}$. To witness, let S be any nonempty set and $0 \notin S$. On $A := S \cup \{0\}$ define the ternary operation $p(x,y,z)$ by

$$p(x,y,z) := \begin{cases} x & \text{if } y = z = 0 \\ z & \text{else.} \end{cases}$$

Then clearly we obtain an algebra in $\underline{W}$. Moreover, every partition of S together with the singleton class $\{0\}$, is a congruence relation of $\underline{A}$.

On the other hand, to see that $\underline{W}$ is an FP-variety, let a congruence relation Θ be given on $\underline{A} \times \underline{B} \in \underline{W}$. Suppose $x \; \pi_1 \; y \; \Theta \; z$, then for some appropriate elements $x = (a_1,b_1)$, $y = (a_1,b_2)$ and $z = (a_2,b_3)$. It follows

$$p((a_1,b_2),(0,b_2),(0,b_1)) \; \Theta \; p((a_2,b_3),(0,b_2),(0,b_1))$$

hence

$$(a_1,b_1)\ \Theta\ (a_2,p(b_3,b_2,b_1)),$$

thus with $u := (a_2,p(b_3,b_2,b_1))$ we get $x\ \Theta\ u\ \pi_1\ z$.

Other examples of FP-varieties which are not already modular include those varieties studied by FRASER and HORN [10] and by HU [27]. Those are varieties where congruences on direct products are products of congruences on the factors.

A simple example from [10] is defined by the equations $x + 0 = 0 + x = x \cdot 1 = x$, $x \cdot 0 = 0$. FRASER and HORN have given a Mal'cev-type description of their varieties. A similar description of FP-varieties is as easy:

11.2 Theorem: A variety $\underline{W}$ is an FP-variety, if and only if there exist natural numbers $m,n \geq 1$, a map $k: \{1,\dots,n\} \to \{0,1\}$, $(m+1)$-ary terms $p_1,\dots,p_n$, binary terms r_{ij} and ternary terms s_{ij} with $1 \leq i \leq n$, $1 \leq j \leq m$ such that the following equations hold in $\underline{W}$:

(1) $x_0 = p_1(x_{k(1)},r_{11},\dots,r_{1m})$

(2) $x_1 = p_n(x_{1-k(n)},r_{n1},\dots,r_{nm})$

(3) $x_2 = p_n(x_{1-k(n)},s_{n1},\dots,s_{nm})$

(4) $p_i(x_{1-k(i)},r_{i1},\dots,r_{im}) = p_{i+1}(x_{k(i+1)},r_{i+1,1},\dots,r_{i+1,m})$ for all $i < n$

(5) $p_i(x_{1-k(i)},s_{i1},\dots,s_{im}) = p_{i+1}(x_{k(i+1)},s_{i+1,1},\dots,s_{i+1,m})$ for all $i < n$.

Proof: The proof is by looking at the congruence relation Θ on the direct product $\underline{F}_W(\{x_0,x_1\}) \times \underline{F}_W(\{x_0,x_1,x_2\})$ generated by the pair $((x_0,x_0),(x_1,x_1))$. The arguments are routine, see [23].

Note that adding the equation

$$x_2 = p_1(x_{k(1)},s_{n1},\dots,s_{nm})$$

to the above equations yields the condition of FRASER and HORN.

Remark: The case $n = 1$ is particularly interesting. If $k(1) = 1$ then $\underline{W}$ is a permutable variety with $p(x,r_1(y,z),\dots,r_m(y,z))$ being the MAL'CEV term.

If $k(1) = 0$ and $m = 1$ we are left with the equations

(Σ1) $x_0 = p(x_0,r(x_0,x_1))$

(Σ2) $x_1 = p(x_1,r(x_0,x_1))$

(Σ3) $x_2 = p(x_1,s(x_0,x_1,x_2))$

11.3 <u>Proposition:</u> The equations $\Sigma 1$, $\Sigma 2$, $\Sigma 3$ jointly imply that every finite algebra in $\underline{W}$ generates a permutable variety.

<u>Proof:</u> $\Sigma 3$ implies that the map $p(a,-)$ is onto for every $a \in \underline{A}$. If A is finite then $p(a,-)$ must be 1-1. Equations $\Sigma 1$ and $\Sigma 2$ yield $p(x,r(x,y)) = p(x,r(y,x))$, hence $r(x,y) = r(y,x)$ be the above.

$\Sigma 1$ alone yields $p(x,r(x,y)) = p(x,r(x,z))$ hence $r(x,y) = r(x,z)$. Combining this, $r(x,y) = r(x,z) = r(z,x) = r(z,u)$ thus $r(x,y)$ is a constant, $0 := r(x,y)$.

Similarly, $\Sigma 3$ gives $s(x_o,x_1,x_2) = s(y,x_1,x_2)$ hence s does not depend on the first place, i.e. $s(x_o,x_1,x_2) =: u(x_1,x_2)$. Furthermore $p(x_o,0) = x_o = p(x_o,u(x_o,x_o))$ so $u(x_o,x_o) = 0$. Combining this we define

$$d(x_o,x_1,x_2) := p(x_o,u(x_1,x_2)),$$

then the above arguments show that d is a Mal'cev term on every finite algebra.

The term $d(x,y,z)$ as constructed above plays a central rôle in FP-varieties:

11.4 <u>Proposition:</u> In every FP-variety $\underline{W}$ there exists a term $d(x,y,z)$ such that $d(x,x,y) = y$ is an equation in $\underline{W}$ and for every congruence relation Θ on $\underline{A} \times \underline{B}$ with $(a_o,b_o)\ \Theta\ (a_1,b_1)$ and for every $a_2 \in \underline{A}$, $(d(a_o,a_1,a_2),b_o)$ completes the following parallelogram:

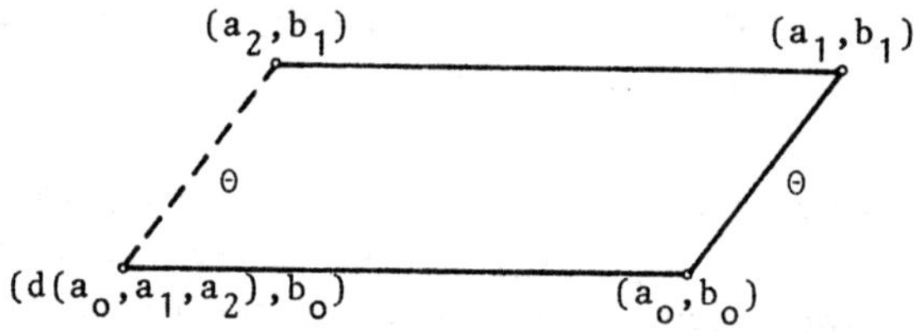

<u>Proof:</u> Depending on whether $k(1)$ is 0 or 1 in the Mal'cev condition we define

$$d(x,y,z) := p_1(x,s_{11}(x,y,z),\dots,s_{1m}(x,y,z))$$

or

$$d(x,y,z) := p_1(y,s_{11}(x,y,z),\dots,s_{1m}(x,y,z)).$$

Now for $1 \le i \le n$, $j \in \{0,1\}$ we set

$$s_i^j := p_i(a_j,s_{i1}(a_o,a_1,a_2),\dots,s_{im}(a_o,a_1,a_2))$$

and

$$t_i^j := p_i(b_j,r_{i1}(b_o,b_1),\dots,r_{im}(b_o,b_1)).$$

Then $(s_i^j,t_i^j)\ \Theta\ (s_i^k,t_i^k)$ for $j,k \in \{0,1\}$ and
$d(a_o,a_1,a_2),b_o) = (s_1^{k(1)},t_1^{k(1)})$ by equation (1),
$(s_i^{1-k(i)},t_i^{1-k(i)}) = (s_{i+1}^{k(i+1)},t_{i+1}^{k(i+1)})$ by equations (5) and (4) and
$(s_n^{1-k(n)},t_n^{1-k(n)}) = (a_2,b_1)$ by equations (3) and (2). Thus
$(d(a_o,a_1,a_2),b_o)\ \Theta\ (a_2,b_1)$ by transitivity.
To see that $d(x,x,y) = y$, set $\Theta := \pi_1$ on $\underline{A} \times \underline{A}$ and $b_1 = a_1 = a_o = x$ and $a_2 = b_o = y$. Then the above implies that
$d(x,x,y) = d(a_o,a_1,a_2) = a_2 = y$.

We do also have a weak replacement for the Shifting Lemma in an FB-variety, i.e. it is clear that the Shifting Lemma 2.1 holds, provided β is a factor congruence.

As we have seen in the examples, we cannot expect the congruence class geometry of algebras in FP-varieties to behave nicely on every algebra. However, direct products are still comparatively well behaved, and counterparts for closure theorems in the modular case can be found. Thus e.g. the Cube lemma will have to be replaced by

11.5 The REIDEMEISTER-theorem: Let Θ be a congruence relation on $\underline{A} \times \underline{B}$ with $\Theta \wedge \pi_2 = 0$. Then

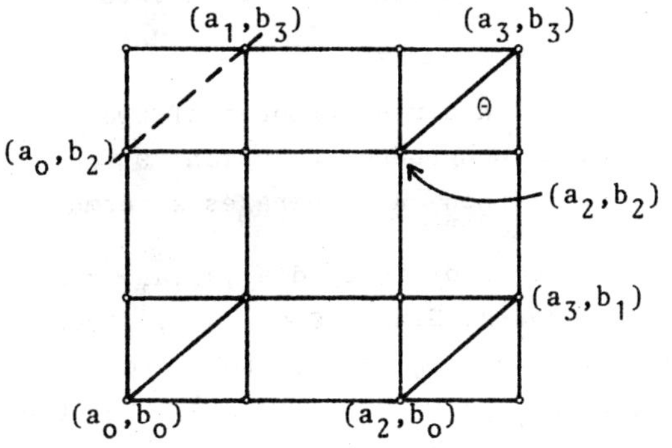

implies $(a_o,b_2)\ \Theta\ (a_1,b_3)$.

Proof: According to 11.4 we have $(d(a_2,a_3,a_1),b_o)\ \Theta\ (a_1,b_1)$ hence $(d(a_2,a_3,a_1),b_o)\ \Theta \wedge \pi_2\ (a_o,b_o)$ whence $d(a_2,a_3,a_1) = a_o$.
It follows then from $(d(a_2,a_3,a_1),b_2)\ \Theta\ (a_1,b_3)$ that $(a_o,b_2)\ \Theta\ (a_1,b_3)$ which we claimed.

There is also a pendant to the Escher Cube which we will need later:

11.6 Lemma: Let Θ be a congruence relation on the direct product $\underline{A} \times \underline{B}$. If $(a_2,b_o)\ \Theta\ (a_o,b_2)$, $(a_1,b_2)\ \Theta\ (a_2,b_1)$ then $(a_1,b_o)\ \Theta\ (a_o,b_1)$.

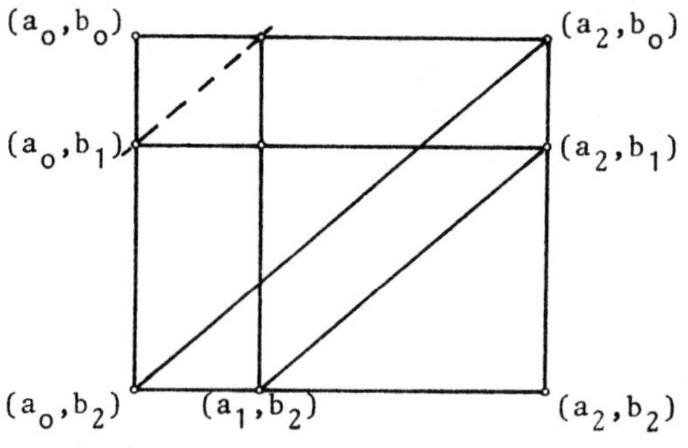

Proof: Factor permutability accounts for the existence of a_3, b_3 with $(a_3,b_1)\ \theta\ (a_1,b_3)\ \theta\ (a_o,b_2)$. We apply d from 11.4 to obtain

$d((a_3,b_1),(a_2,b_1),(a_1,b_1)) = (d(a_3,a_2,a_1),b_1)$ and

$d((a_1,b_3),(a_1,b_2),(a_1,b_1)) = (a_1,d(b_3,b_2,b_1))$.

Since corresponding entries in d are filled with congruent elements, we get, using the geometrical properties of d from 11.4:

$$(a_1,b_o)\ \theta\ (d(a_3,a_2,a_1),b_1)\ \theta\ (a_1,d(b_3,b_2,b_1))\ \theta\ (a_o,b_1).$$

The following lemma is needed to throw us back into permutable varieties in certain circumstances:

11.7 Lemma: Let $\underline{A}\times\underline{B}$ be a direct product of two algebras in an FP-variety. If there exists a congruence relation θ on $\underline{A}\times\underline{B}$ with $\theta\vee\pi_1 = 1$ and $\theta\wedge\pi_2 = 0$ then $\underline{A}$ generates a permutable variety.

Proof: It is sufficient to show that $d(a_o,a_1,a_1) = a_o$ holds for $a_o,a_1 \in \underline{A}$ arbitrarily chosen. Since $\theta\vee\pi_1 = 1$, there exist $b_o,b_1 \in \underline{B}$ with $(a_o,b_o)\ \theta\ (a_1,b_1)$. Set $a_2 := a_1$, then $(d(a_o,a_1,a_1),b_o)\ \theta\ (a_1,b_1)$ by 11.4, hence $(d(a_o,a_1,a_1),b_o)\ \theta\wedge\pi_2\ (a_o,b_o)$. Now the condition $\theta\wedge\pi_2 = 0$ forces $d(a_o,a_1,a_1) = a_o$.

Thus the characterization theorem for affine algebras in permutable varieties from [16] carries over unchanged to FP-varieties. The commutator machinery was used in [26] to carry the original result ([16], Theorem 4.7) over to modular varieties. That it could be done without was shown in [17]. The above lemma shows that the original proof still works in the FP case.

11.8 Theorem: Let $\underline{A}$ be an algebra in an FP-variety. Then the following conditions are equivalent:

(i) $\underline{A}$ is affine

(ii) There exists a congruence relation θ on $\underline{A}\times\underline{A}$ which is a common complement of π_1 and of π_2.

(iii) $\mathrm{diag}(\underline{A}) = \{(x,x) \mid x \in \underline{A}\}$ is a congruence class on $\underline{A} \times \underline{A}$.

Proof: Only (iii) → (ii) needs a proof. The rest follows with 11.7 and chapter 5.

Let Ψ be the congruence relation of which $\mathrm{diag}(\underline{A})$ is a class. $\Psi \vee \pi_i = 1$ is clear for $i \in \{0,1\}$. If $\Psi \wedge \pi_1$, say, is different from 0, i.e. $(a,b)\ \Psi\ (a,c)$ then the Shifting Lemma for π_1, π_2, and Ψ provides $(b,b)\ \Psi\ (b,c)$, hence $(b,c) \in \mathrm{diag}(\underline{A})$ i.e. $b = c$.

As a particular example how to use the above theorem let us ask, in which algebras in an FP-variety, the set of solutions of a family of equations can be described by congruence classes (as in the familiar example of vector spaces). Simply look at the trivial equation $x = y$ whose set of solutions just consists of $\mathrm{diag}(\underline{A})$. Thus only modules over a commutative ring R remain, as in the modular case [17].

Similarly many results from CSAKANY [7] or [8], can directly be restated for FP-varieties.

Let us define an algebra to be hamiltonian, if every subalgebra is a class of some congruence relation.

Since $\mathrm{diag}(\underline{A})$ is always a subalgebra of $\underline{A} \times \underline{A}$ we get

11.9 Corollary: An algebra A in an FP-variety is affine, if and only if $\underline{A} \times \underline{A}$ is hamiltonian.

If we look at varieties of affine algebras, then we may even step outside the framework of FP-varieties. For this sake let us define:

An algebra $\underline{A}$ is called Jonsson-Tarski algebra, if it has a binary term $+$ and a constant term 0 such that the equations $x + 0 = 0 + x = x$ hold.

Thus Jonsson-Tarski algebras are just groupoids with unit with possibly some more operations added. A deep theory of decompositions was developed by JONSSON and TARSKI for those algebras in [30] (they required $\{0\}$ to be a subalgebra).

Clearly the examples of FP-varieties, cited at the beginning of this chapter are Jonsson-Tarski-varieties on defining

$$x + y := p(x,o,y).$$

We need a theorem due to KLUKOVITS [31]:

11.10 Theorem: A variety $\underline{V}$ is hamiltonian, if and only if for every term $f(x_1,\dots,x_n)$ there exists a ternary term h_f, such that the equation

$$f(x_1,\dots,x_n) = h_f(x_o, f(x_o,x_2,\dots,x_n), x_1)$$

holds.

Proof: Given f, look at the free $\underline{V}$-algebras over the generating set $\{x_o,x_1,\ldots,x_n\}$. Since the subalgebra $\underline{U}$, generated by $\{x_o,x_1,f(x_o,x_2,\ldots,x_n)\}$ must be a congruence class, and $f(x_o,x_2,\ldots,x_n)$, considered as unary algebraic function $\tau(x_o)$, throws an element of $\underline{U}$, namely x_o back into $\underline{U}$, we need $\tau(x_1)$ to be inside $\underline{U}$. Thus $\tau(x_1)$ has to be the result of a term h_f applied to the generators of $\underline{U}$.

For the other direction suppose $\underline{S}$ to be a subalgebra of $\underline{A}$ and $\tau(x)$ a unary algebraic function on $\underline{A}$, i.e. $\tau(x) = f(x,a_2,\ldots,a_n)$. If $\tau(u) \in \underline{S}$ for some $u \in \underline{S}$, then for any other $u' \in \underline{S}$ we get $\tau(u') = h_f(u,\tau(u),u')$, an element of $\underline{S}$, thus $\underline{S}$ is a congruence class.

11.11 Theorem: Let $\underline{V}$ be a hamiltonian variety of Jonsson-Tarski algebras. Then $\underline{V}$ is polynomially equivalent to a variety of modules.

Proof: Set $p(x,y,z) := h_+(y,x,z)$.
Then

$$\underline{p(x,x,z)} = h_+(x,x,z) = h_+(x,x+0,z) = z+0 = \underline{z}$$

and

$$\underline{p(x,0,0)} = h_+(0,x,0) = h_+(0,0+x,0) = 0+x = \underline{x},$$

bringing us back inside an FP-variety.

12. KRONECKER PRODUCTS

We have frequently been encountered with situations where some (or all) operations of an algebra $\underline{A}$ have properties which are commonly characteristic for homomorphisms. A typical example is theorem 9.1. The equation just says that every n-ary operation $f(x_1,\ldots,x_n)$ is an "n-ary homomorphism" with respect to $t(x,y,z)$. This statement is symmetric in the sense, that it can be read as $t(x,y,z)$ being a ternary homomorphism with regard to $f(x_1,\ldots,x_n)$. The importance of this property is evident in the proof of 9.2.

Secondly, the kernel of a homomorphism is a congruence relation. In particular, suppose $\phi: \underline{A}^n \longrightarrow \underline{A}$ is an n-ary homomorphism and consider the kernel of ϕ. Suppose $\phi(x,z_2,\ldots,z_n) = \phi(x,y_2,\ldots,y_n)$ for some $x,z_2,\ldots,z_n,y_2,\ldots,y_n$. Then the Shifting Lemma along factor congruence implies: $\phi(a,z_2,\ldots,z_n) = \phi(a,y_2,\ldots,y_n)$ for <u>every</u> $a \in \underline{A}$.
Thus in an FP-variety this property is formally the same as the term condition from 6.7.

To put the discussion into a more general surrounding, let us look at general categories $\mathbf{C}$ having finite products and let us define an <u>algebra</u> (of type Δ) <u>in the category</u> $\mathbf{C}$. This is a $\mathbf{C}$-object $\underline{A}$ together with $\mathbf{C}$-morphisms $f_i \in \mathrm{Hom}_{\mathbf{C}}(\underline{A}^{n_i},\underline{A})$.

A $\underline{V}$-algebra <u>in</u> $\mathbf{C}$ must also satisfy the equations given gy the variety $\underline{V}$. Equations have to be expressed by commutative diagram unless the category is concrete.

For example, if $\mathbf{T}$ is the category of topological spaces, a $\underline{V}$-algebra <u>in</u> $\underline{\mathbf{T}}$ is a topological $\underline{V}$-algebra where every operation is continuous. If $\underline{V}$ is idempotent then homotopy groups are <u>Groups in V</u>, see TAYLOR [37].

If $\mathbf{P}$ is the category of posets then algebras <u>in P</u> are supposed to have all their operations order preserving. We will however only be concerned with the case where $\mathbf{C}$ is a variety $\underline{W}$ of algebras.

Thus let $\underline{A}$ be a $\underline{V}$-algebra <u>in W</u>, f an n-ary $\underline{V}$-operation and g an m-ary $\underline{W}$-operation. Both f and g are defined on $\underline{A}$ but moreover f is a homomorphism (with respect to the $\underline{W}$-structure) from $\underline{A}^n$ to $\underline{A}$. Thus for elements $x_{ij} \in A$ with $1 \le i \le m$, $1 \le j \le n$ we have

$$f(g(x_{11},\ldots,x_{m1}),\ldots,g(x_{1n},\ldots,x_{mn})) = $$
$$= g(f(x_{11},\ldots,x_{1n}),\ldots,f(x_{m1},\ldots,x_{mn})).$$

Obviously the $\underline{V}$-algebras <u>in W</u> form a variety of type $\Delta_V \cup \Delta_W$ which we denote by $\underline{V} \otimes \underline{W}$ and which we call the Kronecker product of $\underline{V}$ and $\underline{W}$.

Clearly $\underline{V} \otimes \underline{W} = \underline{W} \otimes \underline{V}$ and $V \otimes (\underline{W} \otimes \underline{U}) = (\underline{V} \otimes \underline{W}) \otimes \underline{U}$. As an example how to deal with Kronecker products let us show

12.1 Proposition: (a) The Kronecker product of two permutable varieties is affine and (b) the Kronecker product of a permutable variety with a distributive variety is trivial.

Proof: (a) Let p, resp. q be the Mal'cev terms. Then

$$\begin{aligned} p(x,y,z) &= p(q(x,y,y),q(z,z,y),q(z,z,z)) = \\ &= q(p(x,z,z),p(y,z,z),q(y,y,z)) = \\ &= q(x,y,z). \end{aligned}$$

Hence $p = q$ is a Mal'cev term commuting with itself. Thus theorem 4.7 in [16] may be applied. Another way is by 9.1. For (b) let p be the Mal'cev term for $\underline{V}$ and q_i be the Jonsson terms for $\underline{W}$, $1 \le i \le n$. Let n be as small as possible such that Jonsson terms $q_o,\ldots,q_n$ exist. Depending on whether n is odd or even we calculate:

$$\begin{aligned} q_{n-1}(x,y,z) &= q_{n-1}(p(x,x,x),p(y,x,x),p(x,x,z)) = \\ &= p(q_{n-1}(x,y,x),q_{n-1}(x,x,x),q_{n-1}(x,x,z)) = \\ &= q_{n-1}(x,x,z) = z \end{aligned}$$

or

$$\begin{aligned} q_{n-1}(x,y,z) &= q_{n-1}(p(x,x,x),p(y,z,z),p(x,x,z)) = \\ &= p(q_{n-1}(x,y,x),q_{n-1}(x,z,x),q_{n-1}(x,z,z)) = \\ &= q_{n-1}(x,z,z) = z, \end{aligned}$$

in both cases contradicting the minimality of n.

If the terms defining the varieties $\underline{V}$ and $\underline{W}$ become more complicated, this method of determining $\underline{V} \otimes \underline{W}$ is too circumstantial. For example, how to set up the "right" equations for the Kronecker product of two modular varieties? How about two FP-varieties?

Here the geometric methods help us to circumvent the problem of setting up the appropriate equations.

The first example considers Jonsson-Tarski-algebras in FP-varieties and shows that they are abelian groups.
For modular varieties instead of FP-varieties the proof could have been given right after chapter 4. It illuminates the usefulness of "thinking in pictures". We make this clear by pointing out in every step, which geometric picture is responsible for which algebraic property.

12.2 Proposition: Let $+: \underline{A}^2 \longrightarrow \underline{A}$ be a homomorphism such that for some $o \in \underline{A}$, $o + x = x + o = x$. If $\underline{A}$ generates an FP-variety then $+$ is an abelian group operation.

Proof: (i) + is cancellative (Shifting Lemma). If $a + x = a + y$, then (a,x) ker+ (a,y), hence (o,x) ker+ (o,y) by the Shifting Lemma, i.e. $x = o + x = o + y = y$. Similarly $x + a = y + a$ implies $x = y$.
Note that (i) is equivalent to ker+ $\wedge \pi_1$ = ker+ $\wedge \pi_2 = 0$. We need this for

(ii) + is associative (Reidemeister Configuration, Cube Lemma).
Since (y,z) ker+ $(o,y+z)$, (y,o) ker+ (o,y) and (x,y) ker+ $(o,x+y)$ we get from 11.5 that $(x+y,z)$ ker+ $(x,y+z)$ i.e. $(x+y) + z = x + (y+z)$.

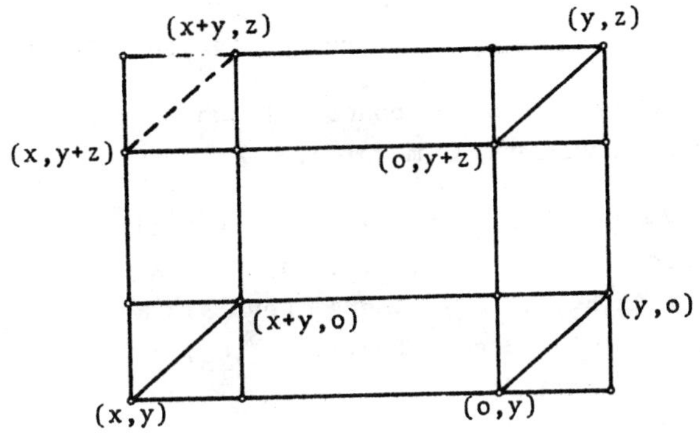

(iii) + is commutative (Escher Cube, resp. 11.6).
Since (o,y) ker+ (y,o), (o,x) ker+ (x,o) we get from 11.6 that (x,y) ker+ (y,x) i.e. $x + y = y + x$.

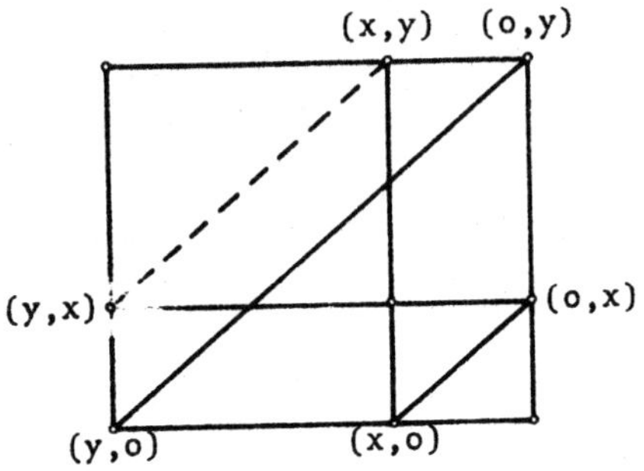

(iv) Existence of an inverse (Factor permutability).
Given (x,o) ker+ (o,x), factor permutability provides an element a with (o,o) ker+ (a,x) i.e. $a + x = o$.

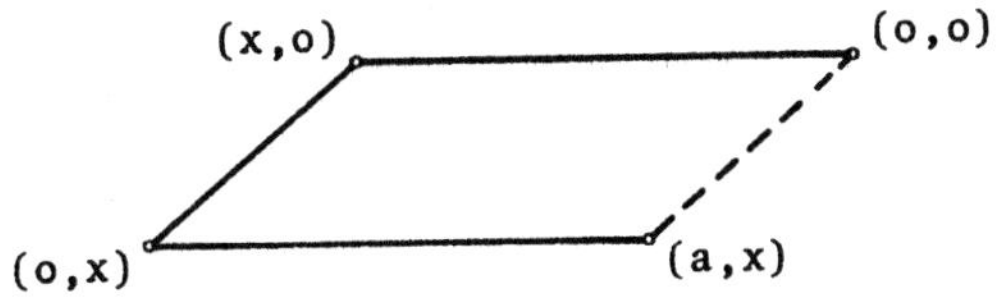

As a corollary:

12.3 Corollary: The Kronecker product of a Jonsson-Tarski-variety with an FP-variety is polynomially equivalent to a variety of modules.

Proof: Suppose $\underline{V}$ is the Jonsson-Tarski-variety and $\underline{W}$ is the FP-variety. We know that $+$ is an abelian group operation commuting with every $\underline{W}$-operation. From the geometrical property of the W-terms $d(x,y,z)$ we see that in fact $d(x,y,z) = x - y + z$. Hence every other $\underline{V}$-operation commutes with $x - y + z$ too.

With the help of 11.7 we get the same result for two FP-varieties:

12.4 Proposition: The Kronecker-product of two FP-varieties is polynomially equivalent to a variety of modules.

Proof: Take $\underline{A} \in \underline{V} \circledast \underline{W}$ and let f be an n-ary fundamental operation. Say, f is a $\underline{V}$-operation. If $f(x,a_2,\ldots,a_n) = f(x,b_2,\ldots,b_n)$ then $(x,a_2,\ldots,a_n)$ kerf $(x,b_2,\ldots,b_n)$ where kerf is a $\underline{W}$-congruence relation. Hence for all $y \in \underline{A}$ we get $f(y,a_2,\ldots,a_n) = f(y,b_2,\ldots,b_n)$ because of the Shifting Lemma for $\underline{W}$.

Therefore $\text{diag}(\underline{A})$ is a congruence class for the $\underline{V}$-operations and similarly for the $\underline{W}$-operations. With 11.7 and 11.8 the result follows.

BIBLIOGRAPHY

1 J.L. BELL, A.B. SLOMSON: Models and ultraproducts. North Holland, Amsterdam, 1969.

2 G. BIRKHOFF: Lattice theory, A.M.S. Colloquium Publ., Providence, R.I. 1967.

3 B. BOSBACH: Rechtskomplementäre Halbgruppen. Math. Z. 124 (1972), 273-288.

4 S. BULMAN-FLEMING, A. DAY, W. TAYLOR: Regularity and modularity of congruences. Algebra Universalis 4 (1974), 58-60.

5 S. BURRIS: Separating sets in modular lattices with applications to congruence lattices. Algebra Universalis 5 (1975), 213-223.

6 S. BURRIS, R. McKENZIE: Decidable varieties with modular congruence lattices. Memoirs of the A.M.S., 32 (1981), Number 246.

7 B. CSÁKÁNY: Varieties of affine modules. Acta Sci. Math., Szeged, 37 (1975), 3-10.

8 B. CSÁKÁNY: Varieties of modules and affine modules. Acta Math. Acad. Sci. Hung. 26 (1975), 263-266.

9 A. DAY: A characterization of modularity for congruence lattices of algebras. Canad. Math. Bull. 12 (1969), 167-173.

10 G.A. FRASER, A. HORN: Congruence relations in direct products. Proc. Amer. Math. Soc. 26 (1970), 390-394.

11 B. FREESE, B. JONSSON: Congruence modularity implies the Arguesian identity. Algebra Universalis 7 (1977), 191-194.

12 R. FREESE, W.A. LAMPE, W. TAYLOR: Congruence lattices of algebras of fixed similarity type, I. Pacific J. Math. 82 (1979), 59-68.

13 R. FREESE, R. McKENZIE: Residually small varieties with modular congruence lattices. Transaction of the A.M.S., 246 (1981), 419-430.

14 G. GRÄTZER: Universal algebra. Second edition. Springer Verlag, New York - Heidelberg - Berlin, 1979.

15 H.P. GUMM: Mal'cev conditions in sums of varieties and a new Mal'cev condition. Algebra Universalis 5 (1975), 56-64.

16 H.P. GUMM: Algebras in permutable varieties: Geometrical properties of affine algebras. Algebra Universalis 9 (1979), 8-34.

17 H.P. GUMM: Über die Lösungsmengen von Gleichungssystemen über allgemeinen Algebren. Math. Z. 162 (1978), 51-62.

18 H.P. GUMM: The Little Desarguesian Theorem for modular varieties. Proc. of the A.M.S. 80 (1980), 393-397.

19 H.P. GUMM: An easy way to the commutator in modular varieties. Arch. d. Math. 34 (1980), 220-228.

20 H.P. GUMM: Congruence modularity is permutability composed with distributivity. Arch. d. Math. 36 (1981), 569-576.

21 H.P. GUMM, C. HERRMANN: Algebras in modular varieties: Baer refinements, cancellation and isotopy. Houston J. of Math. 5 (1979), 503-523.

22 J. HAGEMANN: On regular and weakly regular congruences. Preprint 1973.

23 J. HAGEMANN: Congruences on products and subdirect products of algebras. Preprint 1975.

24 J. HAGEMANN, C. HERRMANN: A concrete ideal multiplication for algebraic systems and its relation to congruence distributivity. Arch. d. Math. 32 (1979), 234-245.

25 J. HAGEMANN, A. MITSCHKE: On n-permutable congruences. Algebra Universalis 3 (1973), 8-12.

26 C. HERRMANN: Affine algebras in congruence modular varieties. Acta Sci. Math. Szeged 41 (1979), 119-125.

27 T.K. HU: On equational classes of algebras in which congruences on finite products are induced by congruences on their factors. Manuscript 1970.

28 B. JÓNSSON: Algebras whose congruence lattices are distributive. Math. Scand. 21 (1967), 110-121.

29 B. JÓNSSON: On the representation of lattices. Math. Scand. 1 (1953), 193-206.

30 B. JÓNSSON, A. TARSKI: Direct decompositions of finite algebraic systems. Notre Dame Math. Lectures No. 5, University of Notre Dame, Ind. 1947.

31 L. KLUKOVITS: Hamiltonian varieties of universal algebras. Acta Sci. Math., Szeged, 37 (1975), 11-15.

32 A.I. MAL'CEV: On the general theory of algebraic systems (Russian). Mat. Sb. (New Series), 35, (77),195

33 R. McKENZIE: On minimal, locally finite varieties with permuting congruence relations. Preprint 1976.

34 A. MITSCHKE: Implication algebras are 3-permutable and 3-distributive. Algebra Universalis 1 (1971), 182-186.

35 A.F. PIXLEY: Completeness in arithmetical algebras.
Algebra Universalis 2 (1972), 179-196.

36 J.D.H. SMITH: Mal'cev varieties.
Lecture Notes in Math., Springer-Verlag, Berlin - New York, 1976.

37 W. TAYLOR: Varieties obeying homotopy laws.
Canad. J. Math. 29 (1977), 498-527.

38 W. TAYLOR: Some applications of the term condition.
Algebra Universalis 14 (1982), 11-24.

39 H. WERNER: Produkte von Kongruenzklassengeometrien universeller Algebren. Math. Z. 121 (1971), 111-140.

40 H. WERNER: Congruences on products of algebras and functionally complete algebras. Algebra Universalis 4 (1974), 99-105.

41 R. WILLE: Kongruenzklassengeometrien. Lecture Notes in Math., Vol. 113, Springer-Verlag, Berlin - New York, 1970.

42 A. WOLF: Finite direct products in congruence modular varieties preserve congruence permutability. Manuscript 1980.

43 R.S. PIERCE: Introduction to the theory of abstract algebras.
Holt, Rinehart and Winston, New York, N.Y., 1968.

44 K. KEIMEL: A unified theory of minimal prime ideals.
Acta Math. Acad. Sci. Hung. 23 (1972), 51-69.

Technische Hochschule Darmstadt
Fachbereich Mathematik
Arbeitsgruppe 1
D - 6100 Darmstadt
West-Germany

starting 1983:

Ges. f. Strahlen u. Umweltforschung
Inst. f. Medizinische Informatik
Ingolstädter Landstr. 1
D-8042 Neuherberg
West-Germany

General instructions to authors for PREPARING REPRODUCTION COPY FOR MEMOIRS

For more detailed instructions send for AMS booklet, "A Guide for Authors of Memoirs." Write to Editorial Offices, American Mathematical Society, P. O. Box 6248, Providence, R. I. 02940.

MEMOIRS are printed by photo-offset from camera copy fully prepared by the author. This means that, except for a reduction in size of 20 to 30%, the finished book will look exactly like the copy submitted. Thus the author will want to use a good quality typewriter with a new, medium-inked black ribbon, and submit clean copy on the appropriate model paper.

Model Paper, provided at no cost by the AMS, is paper marked with blue lines that confine the copy to the appropriate size. Author should specify, when ordering, whether typewriter to be used has PICA-size (10 characters to the inch) or ELITE-size type (12 characters to the inch).

Line Spacing – For best appearance, and economy, a typewriter equipped with a half-space ratchet – 12 notches to the inch – should be used. (This may be purchased and attached at small cost.) Three notches make the desired spacing, which is equivalent to 1-1/2 ordinary single spaces. Where copy has a great many subscripts and superscripts, however, double spacing should be used.

Special Characters may be filled in carefully freehand, using dense black ink, or INSTANT ("rub-on") LETTERING may be used. AMS has a sheet of several hundred most-used symbols and letters which may be purchased for $5.

Diagrams may be drawn in black ink either directly on the model sheet, or on a separate sheet and pasted with rubber cement into spaces left for them in the text. Ballpoint pen is *not* acceptable.

Page Headings (Running Heads) should be centered, in CAPITAL LETTERS (preferably), at the top of the page – just above the blue line and touching it.

LEFT-hand, EVEN-numbered pages should be headed with the AUTHOR'S NAME;
RIGHT-hand, ODD-numbered pages should be headed with the TITLE of the paper (in shortened form if necessary).
Exceptions: PAGE 1 and any other page that carries a display title require NO RUNNING HEADS.

Page Numbers should be at the top of the page, on the same line with the running heads.

LEFT-hand, EVEN numbers – flush with left margin;
RIGHT-hand, ODD numbers – flush with right margin.
Exceptions: PAGE 1 and any other page that carries a display title should have page number, centered below the text, on blue line provided.

FRONT MATTER PAGES should be numbered with Roman numerals (lower case), positioned below text in same manner as described above.

MEMOIRS FORMAT

It is suggested that the material be arranged in pages as indicated below.
Note: Starred items (*) are requirements of publication.

Front Matter (first pages in book, preceding main body of text).

Page i – *Title, *Author's name.

Page iii – Table of contents.

Page iv – *Abstract (at least 1 sentence and at most 300 words).

*1980 Mathematics Subject Classifications represent the primary and secondary subjects of the paper. For the classification scheme, see Annual Subject Indexes of MATHEMATICAL REVIEWS beginning in December 1978.

Key words and phrases, if desired. (A list which covers the content of the paper adequately enough to be useful for an information retrieval system.)

Page v, etc. – Preface, introduction, or any other matter not belonging in body of text.

Page 1 – Chapter Title (dropped 1 inch from top line, and centered).
Beginning of Text.
Footnotes: *Received by the editor date.
Support information – grants, credits, etc.

Last Page (at bottom) – Author's affiliation.

ABCDEFGHIJ–AMS–89876543

Memoirs of the American Mathematical Society
Number 287

Bernard R. McDonald

R-linear endomorphisms of $(R)_n$ preserving invariants

Published by the
AMERICAN MATHEMATICAL SOCIETY
Providence, Rhode Island, USA

November 1983 · Volume 46 · Number 287 (first of 2 numbers)

MEMOIRS of the American Mathematical Society

This journal is designed particularly for long research papers (and groups of cognate papers) in pure and applied mathematics. It includes, in general, longer papers than those in the TRANSACTIONS.

Mathematical papers intended for publication in the Memoirs should be addressed to one of the editors. Subjects, and the editors associated with them, follow:

MEMOIRS are printed by photo-offset from camera-ready copy fully prepared by the authors. Prospective authors are encouraged to request booklet giving detailed instructions regarding reproduction copy. Write to Editorial Office, American Mathematical Society, P.O. Box 6248, Providence, Rhode Island 02940. For general instructions, see last page of Memoir.

SUBSCRIPTION INFORMATION. The 1983 subscription begins with Number 272 and consists of six mailings, each containing one or more numbers. Subscription prices for 1983 are $104.00 list; $52.00 member. Each number may be ordered separately; *please specify number* when ordering an individual paper. For prices and titles of recently released numbers, refer to the New Publications sections of the **NOTICES** of the American Mathematical Society.

BACK NUMBER INFORMATION. For back issues see the AMS Catalogue of Publications.

TRANSACTIONS of the American Mathematical Society

This journal consists of shorter tracts which are of the same general character as the papers published in the MEMOIRS. The editorial committee is identical with that for the MEMOIRS so that papers intended for publication in this series should be addressed to one of the editors listed above.

Subscriptions and orders for publications of the American Mathematical Society should be addressed to American Mathematical Society, P. O. Box 1571, Annex Station, Providence, R. I. 02901. *All orders must be accompanied by payment.* Other correspondence should be addressed to P. O. Box 6248, Providence, R. I. 02940.

MEMOIRS of the American Mathematical Society (ISSN 0065-9266) is published bimonthly (each volume consisting usually of more than one number) by the American Mathematical Society at 201 Charles Street, Providence, Rhode Island 02904. Second Class postage paid at Providence, Rhode Island 02940. Postmaster: Send address changes to Memoirs of the American Mathematical Society, American Mathematical Society, P. O. Box 6248, Providence, RI 02940.

CONTENTS

ABSTRACT

Let R be a commutative ring and $(R)_n$ denote the $n \times n$ matrix ring over R. In this paper we classify the R-linear mappings $T : (R)_n \to (R)_n$ which preserve rank one matrices. This classification gives as a corollary those R-linear mappings which preserve the determinant. Other invariant preserving maps are also determined. These maps are invertible and we describe the groups that they generate.

1980 Mathematics Subject Classification

15A33, 15A72, 13C10

Key Words

Matrices over commutative rings,

Invariant preserving maps.

Library of Congress Cataloging in Publication Data

McDonald, Bernard R.
R-linear endomorphisms of (R) [subscript n] preserving invariants.

(Memoirs of the American Mathematical Society, ISSN 0065-9266 ; no. 287)
"November 1983."
Bibliography: p.
1. Matrix rings. 2. Commutative rings.
3. Invariants. I. Title. II. Series.
QA3.A57 no. 287 [QA188] 510s [512'.5] 83-15648
ISBN 0-8218-2287-X

(A) INTRODUCTION

Let R denote a commutative ring with identity and $(R)_n$ denote the ring of $n \times n$ matrices over R. For 90 years much effort has been devoted to the following question. Suppose $\rho(A)$ is an invariant defined on matrices A in $(R)_n$. Then, determine the set of R-linear mappings $T : (R)_n \to (R)_n$ which preserve the invariant, i.e., find the linear mappings T satisfying $\rho(T(A)) = \rho(A)$ for all A in $(R)_n$. For example, the invariant might be the determinant, i.e., $\rho(A) = \det(A)$, and here we would seek all linear mappings T with $\det(T(A)) = \det(A)$.

The case of the determinant was first studied where $R = \mathbb{C}$, the complex numbers, by Frobenius in 1897 [1]. Frobenius showed that if $T : (\mathbb{C})_n \to (\mathbb{C})_n$ preserved determinants, then either $T(A) = PAQ$ for all A in $(\mathbb{C})_n$ or that $T(A) = PA^tQ$ for all A in $(\mathbb{C})_n$ where P and Q are invertible matrices with $\det(PQ) = 1$.

Many of the questions regarding linear maps $T : (R)_n \to (R)_n$ which preserve invariants can be reduced to the problem of determining the set of linear maps which carry the rank one matrices into themselves. This fact was noted by M. Marcus in a survey article [10] in 1971 on this subject where R was assumed to be a field. In 1959, assuming that k is an algebraically closed field of characteristic zero, then Marcus and Moyls [12] proved that, if $T : (k)_n \to (k)_n$ is a k-linear mapping with the property that $\operatorname{rank}(T(A)) = 1$ whenever $\operatorname{rank}(A) = 1$, then T has the form $T(A) = PAQ$ for all A in $(k)_n$ or $T(A) = PA^tQ$ for all A in $(k)_n$ where P and Q are invertible matrices. If one examines carefully the proofs of Marcus and Moyls [12], one finds that their proof carries over to *any field of any characteristic.* The Marcus and Moyls result was proven by the use of multilinear algebra. An elementary matrix theoretic proof of the same result was given by Minc [17]

Received by Editor Dec. 7, 1981

in 1977.

There has been considerable interest in this problem. In addition to the above-noted survey by Marcus [10] of this and related problems, there is an earlier survey by Marcus [9] and, more recently, a survey in the Ph.D. thesis of Robert Grone [3] which lists 103 related papers on linear mapping problems over fields. Even here, Grone fails to list the extensive literature concerning the automorphisms of the classical linear groups which is also relevant to these problems.

In 1980 D. G. James [8] classified the linear mappings of $(R)_n$ which preserved determinant where R was in integral domain. It was this paper that initiated our interest in this problem for the case where R is a commutative ring; and, in Section (H) we give the solutions of determinant preservers for an arbitrary commutative ring. W. Waterhouse [21] also began to work on this problem at that time and has communicated to me a separate solution of the determinant preservers over a commutative ring by the use of group scheme techniques. Many problems of the above type are related to the study of an affine group scheme and Waterhouse in [21] provides an excellent exposition of this point of view.

On the other hand, the classification of rank one preservers has also been extended to division rings by W. J. Wong [22]. Thus, there exist analogs for noncommutative rings.

In this paper, we extend the theory to an arbitrary commutative ring. The approach is to adopt the thesis of Marcus and attack initially the problem of rank one preservers. When this is complete, we deduce from it the form of the determinant preservers and several other invariant preserving linear mappings.

The outline of the paper is as follows:

In Section (B) we discuss the decomposition of $(R)_n$ when idempotents are present in R . This is due to the transpose mapping which is significant throughout this discussion; however, for a splitting of R into two subrings induced by an idempotent, the "transpose" may affect only one summand of the induced splitting of $(R)_n$ and leave the other summand fixed. Thus, we have many "transposes", and they form a group which may be identified with the set of idempotents of R.

In Section (C) we summarize the standard theory of equivalence transformations of $(R)_n$.

The heart of this paper is Section (D). Here we develop the theory of invertible R-submodules of $(R)_n \oplus (R)_n$. The work is patterned after the development of the theory for $(R)_n$ by M. Issacs [6] in his classification and discussion of the R-algebra automorphisms of $(R)_n$ in 1980. His work was a reformulation of earlier and more general work by Rosenberg and Zelinsky [19] in 1961. Here we identify "generalized" equivalence transformations of $(R)_n$ certain "twisted" lines, i.e., rank one projectives, in "twisted" planes, i.e., direct sums of two rank one projectives, which sit in $(R)_n \oplus (R)_n$.

In Section (E) we generalize the concept of a rank one matrix over a field. Here we say that a matrix A in $(R)_n$ has rank one if its range, as a linear mapping, is a rank one projective R-module; equivalently, if the columns of A generate a rank one projective. In this section, we develop the theory of these rank one mappings.

In Section (F), we prove the Marcus-Moyls Theorem for a <u>local</u> commutative ring R. Then, in Section (G), we develop the general case for an arbitrary commutative ring by the use of localization techniques from commutative algebra. In this section we complete the classification of rank one preservers and our results are summarized in the paper's main theorem - Theorem (G.4).

Section (H) concerns applications of Theorem (G.4) to other invariant preserving questions. In particular, the determinant preserving linear maps are classified by Theorem (H.4), thus giving a modern formulation of Frobenius' original theorem.

It remains to express appreciation to those who contributed in some fashion to the writing of this paper. I was fortunate to have initial written communications from D. James and W. Waterhouse which excited my interest in the problem. During the writing of this paper, I was a guest of the University of California, Santa Barbara as a member of their "Year in Algebra" at the invitation of J. Zelmanowitz. While at Santa Barbara I enjoyed conversations with H. Minc, M. Marcus, E. Formanek and C. Squier concerning this problem. Finally, I was encouraged by correspondence with R. Loeffler while most of the research was being done. This research was supported in part by National Science Foundation grant MCS-8023735.

(B) IDEMPOTENTS AND LINEAR MAPS OF $(R)_n$

In this section we summarize certain facts from commutative algebra concerning idempotents of a commutative ring R and utilize them to describe "idempotent" mappings of $(R)_n$.

Let R denote a commutative ring and $B(R)$ denote the set $\{e \text{ in } R \mid e^2 = e\}$ of idempotents of R. Recall that $B(R)$ is a Boolean algebra where meet, join, and complement are given by $e \wedge \bar{e} = e\bar{e}$, $e \vee \bar{e} = e + \bar{e} - e\bar{e}$, and $e' = 1 - e$, respectively. A <u>partition of one</u> is a finite set of idempotents $\{e_1, \ldots, e_t\}$ in R satisfying $e_i e_j = 0$ for $i \neq j$ (orthogonality) and $1 = e_1 + e_2 + \cdots + e_t$.

The <u>prime spectrum</u> of R, denoted $\mathrm{Spec}(R)$, is the collection of prime ideals q of R. It is well known that $\mathrm{Spec}(R)$ is a compact topological space under the Zariski topology where, if E is a subset of R, then

$$\Gamma(E) = \{q \text{ in } \mathrm{Spec}(R) \mid E \not\subset q\}$$

is an open set. If e is an idempotent, then the open set $\Gamma(e)$ is also closed and partitions $\mathrm{Spec}(R)$ as a disjoint union $\mathrm{Spec}(R) = \Gamma(e) \cup \Gamma(1 - e)$. Further, the idempotent e also decomposes R as a direct product of rings $R = Re \times R(1 - e)$. It is well known ([20], page 140) that this determines natural bijections between each of the following sets.

(a) Direct factors of R,

(b) Idempotents of R,

(c) Open and closed subsets of $\mathrm{Spec}(R)$.

Suppose q is a prime ideal and e is an idempotent of R. Then localizing R at q, we have $(e)_q = 0$ if e is in q and $(e)_q = 1$ if e is not in q. Thus, define a mapping

$$\rho_e : \mathrm{Spec}(R) \to \{\pm 1\}$$

by

$$\rho_e(q) = \begin{cases} 1 & \text{if } (e)_q = 1 \\ -1 & \text{if } (e)_q = 0 \end{cases} .$$

Then $\rho_e = 1$ on $\Gamma(1 - e)$ and $\rho_e = -1$ on $\Gamma(e)$. For e an idempotent of R, we call ρ_e an <u>involution</u> on Spec(R) and denote the set $\{\rho_e \mid e \text{ an idempotent}\}$ by Inv(R). Observe that $\rho_0 = -1$, $\rho_1 = 1$ and Inv(R) is a multiplicative Abelian group of exponent 2 where $\rho_e\rho_f = \rho_{e\wedge f + e'\wedge f'}$.

Let e be an idempotent in R, $R_1 = Re$ and $R_2 = Re'$. Then the direct product decomposition of the ring $R = R_1 \times R_2$ induces a direct product decomposition of the matrix ring $(R)_n = (R_1)_n \oplus (R_2)_n$. If A is a matrix in $(R)_n$, then we will denote the decomposition of A by $A = \langle A_1, A_2 \rangle$ where A_i is in $(R_i)_n$. Applying the transpose to the second coordinate A_2 of A determines an invertible R-linear map

$$\Omega_e : (R)_n \to (R)_n$$

defined by

$$\Omega_e(A) = \Omega_e(\langle A_1, A_2 \rangle) = \langle A_1, A_2^t \rangle .$$

If e and f are idempotents of R, then it is easy to see that $\Omega_e\Omega_f = \Omega_{e\wedge f + e'\wedge f'}$, Ω_1 = identity map, Ω_0 = transpose map, and $\Omega_e \to \rho_e$ is a group isomorphism. Thus, the group $\{\Omega_e \mid e \text{ an idempotent}\}$ is isomorphic to the group Inv(R) of involutions on Spec(R). Due to this isomorphism, we will also denote $\{\Omega_e \mid e \text{ an idempotent}\}$ by Inv(R), and call this group the <u>group of involutions on the spectrum of</u> R.

(C) EQUIVALENCE TRANSFORMATIONS AND INVOLUTIONS

Suppose that P and Q are invertible $n \times n$ matrices over R. The mapping

$$E_{(P,Q)} : (R)_n \to (R)_n$$

given by

$$E_{(P,Q)}(A) = PAQ^{-1}$$

is called a <u>standard equivalence transformation</u> of $(R)_n$. The standard equivalence transformations of $(R)_n$ are invertible R-linear mappings of $(R)_n$, form a multiplicative group, and satisfy

(a) $E_{(P,Q)}E_{(\bar{P},\bar{Q})} = E_{(P\bar{P},Q\bar{Q})}$,

(b) $E_{(P,Q)} = E_{(\bar{P},\bar{Q})}$ if and only if $\bar{P} = \alpha P$ and $\bar{Q} = \alpha Q$ where α is a unit in R.

Let R^0 denote the group of units of R and let $GL_n(R)$ denote the general linear group of invertible $n \times n$ matrices. Thus, R^0 is $GL_1(R)$. If $\mathrm{Equiv}_n(R)$ denotes the group of standard equivalence transformations of $(R)_n$, then $\mathrm{Equiv}_n(R) \approx [GL_n(R) \times GL_n(R)]/R^0$. This follows from the exact sequence of groups

$$1 \longrightarrow R^0 \xrightarrow{\tau} GL_n(R) \times GL_n(R) \xrightarrow{E} \mathrm{Equiv}_n(R) \longrightarrow 1 ,$$

where $\tau(\alpha) = \langle \alpha I, \alpha I \rangle$ and $E(\langle P,Q \rangle) = E_{(P,Q)}$.

Suppose e is an idempotent of R, $R_1 = Re$ and $R_2 = Re'$. Then $R = R_1 \times R_2$ induces a decomposition of the general linear group $GL_n(R) = GL_n(R_1) \times GL_n(R_2)$ and, analogously, decompositions of

$GL_n(R) \times GL_n(R)$ and $Equiv_n(R)$. If $<P,Q>$ is in $GL_n(R) \times GL_n(R)$, then let $P = <P_1,P_2>$ and $Q = <Q_1,Q_2>$ denote the decompositions of P and Q relative to $R = R_1 \times R_2$. If A is any invertible matrix, let $A^* = (A^{-1})^t$ (inverse-transpose). Note that $A \to A^*$ is a group automorphism of the general linear group. The idempotent e together with $*$ determine an automorphism Λ_e of $GL_n(R) \times GL_n(R)$ given by

$$\Lambda_e : <P,Q> \to \ll P_1,Q_2^*>,<Q_1,P_2^*\gg = \Lambda_e(P,Q) .$$

Observe that the automorphism Λ_e is the composition of two automorphisms: (1) the $*$ automorphism on the coordinate ring R_2, (2) an interchange of the R_2 coordinates. Further, $\Lambda_1(<P,Q>) = <P,Q>$, $\Lambda_0(<P,Q>) = <Q^*,P^*>$ and $\Lambda_e(\tau(\alpha)) = \tau(\bar{\alpha})$ where if $\alpha = <\alpha_1,\alpha_2>$ is in $R^0 = R_1{}^0 \times R_2{}^0$, then $\bar{\alpha}$ is given by $\bar{\alpha} = <\alpha_1,\alpha_2{}^{-1}>$. Thus, Λ_e carries $\tau(R^0)$ onto $\tau(R^0)$ and induces an automorphism of $Equiv_n(R)$ which we also denote by Λ_e. The image of $E_{(P,Q)}$ under Λ_e will be written as $E_{\Lambda_e(P,Q)}$ where $\Lambda_e(P,Q)$ is given above.

We are concerned with the invertible R-endomorphisms of the matrix ring $(R)_n$. The group of all invertible R-linear mappings of $(R)_n$ may be identified with the general linear group $GL_{n^2}(R)$ of dimension n^2. Both $Inv(R)$ (the group of involutions on $Spec(R)$ introduced in (B)) and $Equiv_n(R)$ are subgroups of $GL_{n^2}(R)$. The subgroup generated by $Inv(R)$ and $Equiv_n(R)$ will be denoted by $Rank_n(R)$.

It is now straight-forward to show $Inv(R) \cap Equiv_n(R) = 1$. Further, if A is in $(R)_n$, then

$$\begin{aligned}
\Omega_e E_{(P,Q)}(A) &= \Omega_e(PAQ^{-1}) \\
&= <P_1A_1Q_1^{-1},(P_2A_2Q_2^{-1})^t> \\
&= <P_1A_1Q_1^{-1},Q_2^*A_2^tP_2^t> \\
&= <P_1,Q_2^*> <A_1,A_2^t> <Q_1^{-1},(P_2^*)^{-1}> \\
&= E_{\Lambda_e(P,Q)}\Omega_e(A) .
\end{aligned}$$

Hence, $\Omega_e E_{(P,Q)} = E_{\Lambda_e(P,Q)}\Omega_e$. This motivates the definition of Λ_e . From this, it is easy to see that $\mathrm{Rank}_n(R)$ is a semidirect product of $\mathrm{Inv}(R)$ and $\mathrm{Equiv}_n(R)$. This discussion is summarized in the following Proposition.

(C.1) PROPOSITION.

(a) The sequence

$$1 \to R^0 \to GL_n(R) \times GL_n(R) \xrightarrow{E} \mathrm{Equiv}_n(R) \to 1$$

is exact.

(b) The sequence

$$1 \to \mathrm{Equiv}_n(R) \to \mathrm{Rank}_n(R) \to \mathrm{Inv}(R) \to 1$$

is split exact.

(D) INVERTIBLE SUBMODULES OF $(R)_n$ AND EQUIVALENCE TRANSFORMATIONS

This section concerns a certain specified set of R-submodules of the direct sum $(R)_n \oplus (R)_n$ - the "invertible" submodules - and the mappings of $(R)_n$ which these submodules induce. These modules are defined below and some of their theory is summarized. M. Isaacs in [6] provides a good introduction to the basic theory of the invertible submodules of the matrix ring $(R)_n$. Our initial remarks summarize Isaacs' results.

Let $\mathcal{M}$ denote the collection of all R-submodules of $(R)_n$. If U and V are R-submodules in $\mathcal{M}$, define a multiplication by

$$UV = \left\{ \sum_{\text{finite}} uv \mid u \text{ in } U,\ v \text{ in } V \right\}$$

where the summation extends over all finite sets of products uv where u is in U and v is in V. Thus, UV is the additive subgroup of $(R)_n$ generated by the set $\{uv \mid u \text{ in } U, v \text{ in } V\}$. Let $E = \{\alpha I \mid \alpha \text{ in } R\}$ where I denotes the $n \times n$ identity matrix. An R-submodule U in $\mathcal{M}$ is called <u>invertible</u> if there is an R-submodule V in $\mathcal{M}$ satisfying $UV = VU = E$. Let $\mathcal{G}$ denote the subset of invertible R-submodules in $\mathcal{M}$.

The following are useful in this paper and are given either explicitly or implicitly in Isaac's paper [6]:

(a) If U is in $\mathcal{G}$ and V is such that $UV = VU = E$, then V is unique, called the <u>inverse</u> of U, and denoted by U^{-1}. If $UV = E$ then $VU = E$. The set $\mathcal{G}$ is a group. If U and V are in and $U \subseteq V$, then $U = V$.

(b) If U is in $\mathcal{G}$, u, $\bar{u}$ are in U, and v, $\bar{v}$ are in U^{-1}, then $uv = vu$, $u\bar{u} = \bar{u}u$, and $v\bar{v} = \bar{v}v$.

(c) If P is an invertible matrix in $(R)_n$, then $RP = \{\alpha P \mid \alpha \text{ in } R\}$

is an invertible R-submodule of $(R)_n$ which is free of R-dimension one. An invertible R-module U contains an invertible matrix P if and only if $U = RP$. If U and V are in G, then $U \simeq V$ as R-modules if and only if $U = PV$ for a suitable invertible matrix P.

(d) Recall that an R-module U is said to be a rank one projective R-module if

(1) U is a finitely generated projective R-module and

(2) For each prime ideal q of R, the localization U_q of U at q is a free R_q-module of dimension one.

If U is in G, then U is a rank one projective R-module.

(e) An R-module U is an element of G if and only if $\oplus \sum_{i=1}^{n} U \simeq R^{(n)}$.

(f) If U and V are in G then $UV \simeq U \otimes_R V$ as R-modules.

We need a remark about localization at a prime ideal q of R. Suppose U is in G. Localizing the R-module U at q, we obtain U_q which is a free R_q-submodule of $(R_q)_n$. Over the local ring R_q, since U_q is free of R_q-dimension one, we have $U_q = \{\alpha P \mid \alpha \text{ in } R_q\}$ where P is an $n \times n$ matrix over R_q. Since U_q is invertible, the matrix P is invertible. Thus, in $(R_q)_n$ an R_q-module $\bar{U}$ is invertible if and only if $\bar{U} = \{\alpha P \mid \alpha \text{ in } R_q\}$ where P is an invertible matrix over R_q.

We consider next the generalization of the previous remarks to R-submodules of the R-algebra $(R)_n \oplus (R)_n$. Let $S = S_1 \oplus S_2$ where $S_i = (R)_n$ for $i = 1,2$. Let $\pi_i : S \to S_i$ denote the natural projection and $\lambda_i : S_i \to S$ denote the natural injection for $i = 1,2$. If I denotes the $n \times n$ identity matrix of $(R)_n$, then let $\bar{E} = \{\alpha \langle I,I \rangle \mid \alpha \text{ in } R\}$. Note that $\langle I,I \rangle$ is the identity of S.

An R-submodule $\bar{U}$ of S is called invertible if there is an R-submodule $\bar{V}$ of S with $\bar{U}\bar{V} = \bar{E}$. (The product $\bar{U}\bar{V}$ of two R-submodules in $S = (R)_n \oplus (R)_n$ is the coordinate-wise extension of the product in $(R)_n$.)

If $\bar{U}\bar{V} = \bar{E}$, then

$$E = \pi_i\bar{E} = \pi_i(\bar{U}\bar{V}) = \pi_i\bar{U}\pi_i\bar{V}$$

for $i = 1,2$. Hence, the projection $U_i = \pi_i\bar{U}$, $i = 1,2$, of an invertible R-submodule $\bar{U}$ of S is an invertible R-submodule U_i of S_i . Set

$$\bar{U}_1 = \lambda_1\pi_1\bar{U} , \quad \bar{U}_2 = \lambda_2\pi_2\bar{U} .$$

Then $\bar{U}$ is a submodule of the direct sum $\bar{U}_1 \oplus \bar{U}_2$.

(D.1) LEMMA. Let R be a local commutative ring with maximal ideal m . Let $P = L_1 \oplus L_2$ be a free R-module of R-dimension 2 , i.e., a "plane," with $L_1 = Rb_1$, $L_2 = Rb_2$ free direct summands of P of R-dimension 1 , i.e., "lines." Suppose that $L \subset P$ is a finitely generated submodule with $\pi_iL = L_i$ where the $\pi_i : P \to L_i$ for $i = 1,2$, are the natural projections. Then, either $L = P$ or L is a free summand of P of R-dimension 1 , that is, L is a line.

PROOF. Let the natural maps $R \to R/m$, $P \to P/mP$, etc., be denoted by "bar." Then $\overline{\pi_iL} = \bar{L}_i = \pi_i(\bar{L}) \neq 0$, so L is not in mP . Suppose $L \neq P$. Since $\bar{L}$ is a nonzero subspace of $\bar{P}$ and $\bar{L} \neq \bar{P}$ (note: if $\bar{L} = \bar{P}$ then $L = P$), then $\bar{L}$ is a line in $\bar{P}$. Let $\bar{b}$ be a basis vector for $\bar{L}$ with b in L . Then $L = Rb$. Let $b = \alpha b_1 + \beta b_2$. Since $\overline{\pi_iL} = \bar{L}_i$, both α and β are units. Hence b is a unimodular vector and $L = Rb$ is a free summand of P , i.e., L is a line.

We apply the above lemma to the localization of $\bar{U}\bar{V} = \bar{E}$ where $\bar{U}$ is an invertible R-submodule of S . Shortly, in Lemma (D.2)(c), we show that $\bar{U}$ is finitely generated. Let q be a prime ideal in R . Then, localizing at a prime q of R , $\bar{U}_q\bar{V}_q = \bar{E}_q$ where $\bar{E}_q = \{\alpha\langle I,I\rangle \mid \alpha \text{ is in } R_q\}$. Since for $i = 1,2$, $\bar{U}_i = \lambda_i\pi_i\bar{U}$ is invertible in $S_i = (R)_n$, Isaac's results imply that $(\bar{U}_i)_q$ is a free R_q-module of dimension 1 . Let $(\bar{U}_i)_q = \{\alpha P_i \mid \alpha \text{ in } R_q, P_i \text{ invertible in } (R_q)_n\}$. Then $(\bar{U}_1)_q \oplus (\bar{U}_2)_q$ is a plane with basis $\{P_1,P_2\}$. The invertible R_q-submodule $\bar{U}_q$ satisfies the hypothesis of (D.1) where $(\bar{U}_1)_q$ and $(\bar{U}_2)_q$ are the lines L_1 and L_2 . Therefore, $\bar{U}_q$ is a line since it cannot locally be the plane and remain invertible. Thus, there exist units α and β in R_q such that

$$\bar{U}_q = \{\delta(\alpha P_1 + \beta P_2) \mid \delta \text{ in } R_q\}$$

or, in terms of "coordinates,"

$$\bar{U}_q = \{\delta < \alpha P_1, \beta P_2 > \mid \delta \text{ in } R_q\} .$$

Since α and β are units, we may set $P = \alpha P_1$, $Q = \beta P_2$ and write

$$\bar{U}_q = \{\delta < P,Q > \mid \delta \text{ in } R_q\}$$

where P and Q are invertible matrices in $(R_q)_n$.

Motivation for the above arises from Hua's geometry of matrices [4]. We regard $S = (R)_n \oplus (R)_n$ as a free R-space of dimension $2n^2$. Suppose that $P = Ra \oplus Rb$ is a plane, i.e., a free R-summand of S of R-dimension 2 and a basis $\{a,b\}$ where a and b are invertible matrices. We seek lines L in P of the form $L = R(\alpha a + \beta b)$ where α and β are units in R in order that the line L will project both onto Ra and onto Rb .

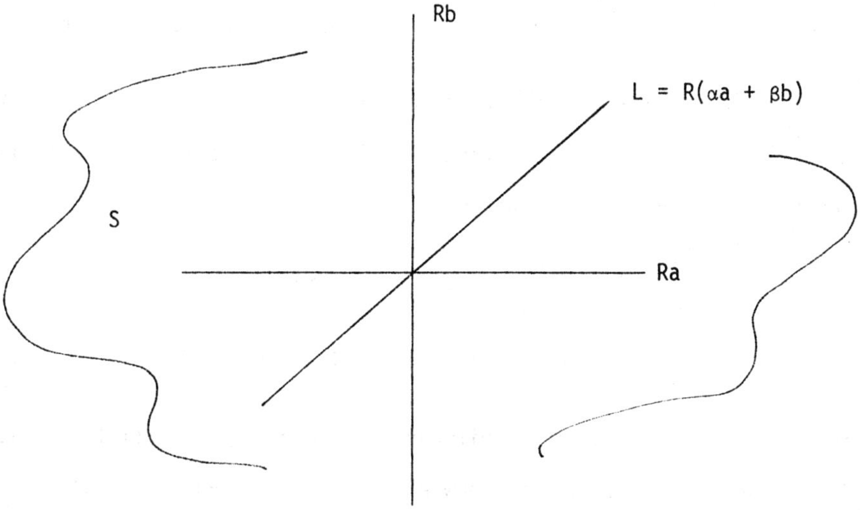

The invertible R-submodules of S (which will prove to be rank one projectives and, consequently, locally free of dimension one, i.e., locally "lines") will play the role of "lines" in the "planes" which occur as direct sums of two rank one projectives. More appropriately, these should be called "twisted" lines and "twisted" planes.

The idea of "twisting" is known to experts in commutative algebra and arises in our setting as follows: if $M = U_1 \oplus \cdots \oplus U_n$ is a direct sum of rank one projective R-modules with $U_i \simeq U_j$, then, since $\mathrm{End}_R(U_i) \simeq R$ for $1 \leqslant i \leqslant n$, we have $\mathrm{End}_R(M) \simeq (R)_n \simeq \mathrm{End}_R(F)$ where $F \simeq R^n$ is a free R-module of R-dimension n. When this is the case (see Lemma 6.4, p. 69 of [5]) there is a rank one projective R-module U such that $U \otimes F \simeq M$, that is, the rank one projective module U "distorts" or "twists" F through tensoring to produce M.

Let $\mathfrak{A}$ denote the set of invertible R-submodules of $S = (R)_n \oplus (R)_n$. If $\bar{U}$ and $\bar{V}$ in $\mathfrak{A}$ satisfy $\bar{U}\bar{V} = \bar{E}$, then $\bar{V}$ is denoted by $\bar{U}^{-1}$.

(D.2) LEMMA. Let $\bar{U}$ be in $\mathfrak{A}$. Then

(a) $\bar{U}\bar{U}^{-1} = \bar{U}^{-1}\bar{U} = \bar{E}$.

(b) If $\bar{u}_1, \bar{u}_2$ are in $\bar{U}$ and $\bar{v}_1, \bar{v}_2$ are in $\bar{U}^{-1}$ then

$\bar{u}_1\bar{v}_1 = \bar{v}_1\bar{u}_1$, $\bar{u}_1\bar{u}_2 = \bar{u}_2\bar{u}_1$, and $\bar{v}_1\bar{v}_2 = \bar{v}_2\bar{v}_1$.

(c) $\bar{U}$ is a finitely generated, rank one, projective R-module.

PROOF. If the formulas in (b) are true for each localization at each prime ideal q of R, then they are true globally. Localizing at a prime ideal q and applying the above discussion, one has

$$\bar{U}_q = \{\delta < \alpha P_1, \beta P_2 > \mid \delta \text{ in } R_q\},$$

$$\bar{U}_q^{-1} = \{\delta < (\alpha P_1)^{-1}, (\beta P_2)^{-1} > \mid \delta \text{ in } R_q\}$$

where α and β are suitable units and P_1 and P_2 invertible matrices in $(R_q)_n$. The formulas in (b) are now easy to verify. Part (a) follows from (b).

To show (c) it is necessary to show that $\bar{U}$ is finitely generated and projective. Recall that a finitely generated rank one projective module is a finitely generated projective module which is locally free of dimension one. We first show that $\bar{U}$ is finitely generated and projective. We do this by exhibiting a "dual basis" for $\bar{U}$. Since $\bar{U}\bar{V} = \bar{E}$ there are $\bar{u}_1,\ldots,\bar{u}_m$ in $\bar{U}$ and $\bar{v}_1,\ldots,\bar{v}_m$ in $\bar{V}$ with $\sum_i \bar{u}_i\bar{v}_i = 1$, $1 = \langle I,I\rangle$, where we identify $\bar{E} = \{\alpha\langle I,I\rangle \mid \alpha \in R\}$ with R by the ring isomorphism $\alpha\langle I,I\rangle \to \alpha$. If $\bar{u}$ is in $\bar{U}$, define a linear form $f_i : \bar{U} \to R$ by $f_i(\bar{u}) = \bar{u}\bar{v}_i$ for $1 \leqslant i \leqslant m$. Consider

$$\begin{aligned} \bar{u} = \bar{u}1 &= \bar{u} \textstyle\sum_i \bar{u}_i\bar{v}_i \\ &= \textstyle\sum_i \bar{u}_i\bar{u}\bar{v}_i \qquad \text{(by (b))} \\ &= \textstyle\sum_i \bar{u}_i f_i(\bar{u}) . \end{aligned}$$

Thus, the sets $\{\bar{u}_1,\ldots,\bar{u}_m\}$, $\{f_1,\ldots,f_m\}$ form a dual basis for $\bar{U}$ and it is well known that the existence of a dual basis is equivalent to $\bar{U}$ being finitely generated and projective. Since $\bar{U}$ is finitely generated and projective the localization discussion prior to the lemma shows that $\bar{U}_q$ is free of dimension one for each prime q of R. Thus, $\bar{U}$ has rank one. This completes the proof.

We have that $\bar{U}$ is a submodule of $\bar{U}_1 \oplus \bar{U}_2$ and finitely generated projective of rank one. We show next that $\bar{U}$ is actually a summand of $\bar{U}_1 \oplus \bar{U}_2$, thus justifying our thinking of $\bar{U}$ as a "twisted line" in the "twisted plane" $\bar{U}_1 \oplus \bar{U}_2$. Note that the projection maps are locally isomorphisms, thus $\bar{U}_1 \simeq \bar{U}_2$. There is an exact sequence

$$0 \longrightarrow \bar{U} \xrightarrow{\lambda} \bar{U}_1 \oplus \bar{U}_2 \xrightarrow{\pi} \bar{U}_1 \oplus \bar{U}_2/\lambda\bar{U} \longrightarrow 0$$

where λ is the natural injection. Also there exists a projective R-module P such that $\bar{U}_1 \oplus \bar{U}_2 \oplus P = F_0$ is a finitely generated free R-module. Thus, we have an exact sequence

$$0 \longrightarrow \bar{U} \oplus P \xrightarrow{\lambda \oplus 1} F_0 \xrightarrow{\pi \oplus 0} \bar{U}_1 \oplus \bar{U}_2/\lambda\bar{U} \longrightarrow 0 .$$

Since $\bar{U} \oplus P$ is a finitely generated projective, there is a finitely generated free module F_1 and a surjection $F_1 \to \bar{U} \oplus P$. Thus, we have a finite free presentation

$$F_1 \longrightarrow F_0 \longrightarrow \bar{U}_1 \oplus \bar{U}_2/\lambda\bar{U} \longrightarrow 0 .$$

An R-module which has a finite free presentation is finitely generated and projective provided the module is locally free. Thus, examine the localization of the initial exact sequence.

For a prime ideal q of R, the localization of

$$0 \longrightarrow \bar{U} \xrightarrow{\lambda} \bar{U}_1 \oplus \bar{U}_2 \xrightarrow{\pi} \bar{U}_1 \oplus \bar{U}_2/\lambda\bar{U} \longrightarrow 0$$

at q gives the exact sequence

$$0 \longrightarrow \bar{U}_q \xrightarrow{\lambda_q} (\bar{U}_1 \oplus \bar{U}_2)_q \xrightarrow{\pi_q} (\bar{U}_1 \oplus \bar{U}_2)_q/\lambda\bar{U}_q \longrightarrow 0$$

which, by the previous discussion, has the general form of

$$0 \longrightarrow R_q(b_1 + b_2) \longrightarrow R_q b_1 \oplus R_q b_2 \xrightarrow{\pi} M \longrightarrow 0$$

where b_1, b_2 and $b_1 + b_2$ are unimodular[1] vectors in the plane $P = R_q b_1 \oplus R_q b_2$. It is easy to see that $M = R_q \pi(b_1)$ is a cyclic module over R_q. Suppose $\alpha\pi(b_1) = 0$ for some α in R_q. Then $\pi(\alpha b_1) = 0$ and thus $\alpha b_1 = \beta(b_1 + b_2)$ for some β in R_q. This implies $(\alpha - \beta)b_1 = \beta b_2$ and, since $\{b_1, b_2\}$ are R_q-independent, $\beta = 0$ and thus $\alpha = 0$. Hence, $M = R_q\pi(b_1) \simeq R_q$ is free. Thus, by the above remarks, $\bar{U}_1 \oplus \bar{U}_2/\lambda\bar{U}$ is a projective R-module; hence, the sequence splits and $\bar{U}$ is a summand of $\bar{U}_1 \oplus \bar{U}_2$.

(D.3) LEMMA. (For the notation of this section.) The invertible R-module $\bar{U}$ of $\mathfrak{A}$ is a direct summand of rank one of $\bar{U}_1 \oplus \bar{U}_2$ where $\bar{U}_1$ and $\bar{U}_2$ are the projections of $\bar{U}$ onto $S_1 = (R)_n$ and $S_2 = (R)_n$, respectively.

---

[1] "Unimodular" is defined in Section (E) before Proposition (E.1).

(D.4) THEOREM. A projective module X is isomorphic to an invertible R-submodule U of $S = (R)_n \oplus (R)_n$ if and only if

$$R^n \simeq X \oplus \cdots \oplus X \qquad (n \text{ summands}) .$$

PROOF. The proof is patterned on the proof of Theorem 16 of Isaac's [6] for $(R)_n$.

If U is an R-module, then let U^n denote the direct sum $U \oplus \cdots \oplus U$ of n copies of U.

For $1 \leqslant i \leqslant n$, define R-morphisms $\rho_i : S \to R^{2n}$ by $\rho_i(\langle A_1, A_2 \rangle) =$ i-th row of $\langle A_1, A_2 \rangle$.

Let $\bar{U}$ be in $\mathfrak{U}$. Define

$$\tau : \bar{U}^n \to R^{2n}$$

by

$$\tau(u_1, \ldots, u_n) = \textstyle\sum_{i=1}^{n} \rho_i(u_i) .$$

Clearly, τ is an R-morphism. Let b_i denote the i-th row of $\langle I, I \rangle$.

We claim first that τ is injective. If $\langle A_1, A_2 \rangle$ and $\langle B_1, B_2 \rangle$ are in S, then $\rho_i(\langle A_1B_1, A_2B_2 \rangle) = \rho_i(\langle A_1, A_2 \rangle)\langle B_1, B_2 \rangle$, i.e., ρ_i is a right S-morphism. For u in $\bar{U}$ and v in $\bar{U}^{-1}$, we have $uv = \alpha\langle I, I \rangle$ for some α in R. Then,

$$\rho_i(u)v = \rho_i(uv) = \rho_i(\alpha\langle I, I \rangle) = \alpha b_i .$$

Now suppose $x = \langle u_1, u_2, \ldots, u_n \rangle$ is in $\bar{U}^n$ and $\tau(x) = 0$. Select v in $\bar{U}^{-1}$ and let $u_i v = \alpha_i \langle I, I \rangle$. Then,

$$\begin{aligned} 0 = \tau(x)v &= \tau(\langle u_1, \ldots, u_n \rangle)v \\ &= \textstyle\sum_i \rho_i(u_i)v = \sum_i \rho_i(u_i v) \\ &= \textstyle\sum_i \alpha_i b_i . \end{aligned}$$

Since the set $\{b_1, \ldots, b_n\}$ is R-independent, each $\alpha_i = 0$, $1 \leqslant i \leqslant n$. Since v was arbitrary, we have $u_i \bar{U}^{-1} = 0$. Select $\bar{u}_j$ in $\bar{U}$ and $\bar{v}_j$ in

$\bar{U}^{-1}$ with $\sum_j \bar{u}_j\bar{v}_j = 1$. Then $u_i = u_i 1 = \sum_j u_i\bar{u}_j\bar{v}_j = \sum_j \bar{u}_j u_i \bar{v}_j = 0$. If u is in $\bar{U}$, then $\rho_i(u) = \tau(<0,\ldots,u,\ldots,0>)$ (where u occupies the i-th position) is in the image $Im(\tau)$ of τ , i.e., each row of U is in $Im(\tau)$. Since $Im(\tau)$ is an R-module, the image $Im(\tau)$ is the set of all R-linear combinations of all rows of all elements u in $\bar{U}$. In particular, if A is a matrix in $(R)_n$, then all rows of Au are in $Im(\tau)$ for u in $\bar{U}$. Select u_i in $\bar{U}$ and v_i in $\bar{U}^{-1}$ with $\sum v_i u_i = <I,I>$. Set $v_i = <A_i^1, A_i^2>$ and $u_i = <B_i^1, B_i^2>$ in $S = (R)_n \oplus (R)_n$. Then,

$$<I,I> = \sum v_i u_i = \sum_i <A_i^1 B_i^1, A_i^2 B_i^2> .$$

By our observations, the rows of

$$A_i^1 u_i = [I,*]$$

are in $Im(\tau)$. These rows generate a free R-module of R-dimension n . Let F denote this free R-module. Clearly $F \subset Im(\tau)$. Further each row, which is a basis vector of F , is unimodular as an element of $Im(\tau)$ and hence splits from $Im(\tau)$ making F a direct summand of $Im(\tau)$. Thus, we have

$$U^n \xrightarrow[\tau]{\simeq} F \oplus (*) = Im(\tau) .$$

Since τ is an isomorphism of U^n onto $Im(\tau)$, $Im(\tau)$ is a projective module which is locally free of dimension $n = \dim(F)$. Thus, for each prime ideal q , $(*)_q = 0$ and, consequently $(*) = 0$ and $F = Im(\tau)$.

Thus U determines an element of $Pic_n(R)$.

If X is an R-module with $X^n \simeq R^n$, then Isaacs' [6] shows on pp. 228-229 that there is an invertible R-submodule U in $(R)_n$ with $X \simeq U$.

Then one would set $\bar{U} = \{<u,u> \mid u \text{ in } U\}$. It is easy to see that $\bar{U}$ is invertible in $S = (R)_n \oplus (R)_n$ and that $X \simeq \bar{U}$. This would complete the proof; however, it is of interest to see the construction of U , so for completeness we provide the proof. The proof basically follows Isaacs' [6].

Suppose X is an R-module and $\sigma : X^n \to R^n$ is an R-module isomorphism. We first construct the "matrix of σ". Let $\lambda_j : X \to X^n$ be the natural j-th injection morphism and $\pi_i : R^n \to R$ denote the natural i-th projection. Define $\sigma_{ij} : X \to R$ by $\sigma_{ij} = \pi_i \sigma \lambda_j$ for $1 \leq i,j \leq n$. Then $[\sigma_{ij}]$ is the "standard matrix" for σ.

For each x in X, we have a matrix $m_x = [\sigma_{ij} x]$. Set $U = \{m_x \mid x \text{ in } X\}$. We claim that $m : X \to U$ by $m : x \to m_x$ is an R-module isomorphism and that U is an invertible R-submodule of $(R)_n$.

Certainly, m is surjective and it is also easily seen to be injective. Hence X is isomorphic to U and U is a finitely generated rank one projective R-module. It remains to show that U is invertible. Since $\sigma : X^n \to R^n$ is an isomorphism, select x_i in X^n with $\sigma(x_i) = [0,0,\ldots,1,0,\ldots,0]^t$ (1 in i-th position). Write x_i as $x_i = [x_{1i}, x_{2i}, \ldots, x_{ni}]^t$. For each f in $\mathrm{Hom}_R(X,R)$, let n_f denote the matrix $[f(x_{ij})]$. Then $n : \mathrm{Hom}_R(X,R) \to (R)_n$. Let $V = \{n_f \mid f \text{ is in } \mathrm{Hom}_R(X,R)\}$. Then, V is an R-submodule of $(R)_n$. It remains to show that $UV = E$. This may be verified directly as was done in Isaacs' ([6], p. 229) or by a straight-forward localization argument, since locally at each prime of R, it is easily seen to be true. This finishes the proof of (D.4).

The above theorem gives as a corollary the following useful result.

(D.5) <u>COROLLARY</u>. Let U be a R-submodule of $(R)_n$ (or $(R)_n \oplus (R)_n$). The following are equivalent:

(a) U is invertible.

(b) U_q is invertible for each prime ideal q of R.

PROOF. We prove the result for $(R)_n$. Suppose U is invertible in $(R)_n$. Then there is a V with $UV = E$. Since $(UV)_q = U_q V_q = E_q$ we have that (a) implies (b).

Suppose that U is a R-submodule of $(R)_n$ and U_q is invertible for each prime ideal q of R. Let $\{E_{ij} \mid 1 \leq i,j \leq n\}$ be the standard matrix units. Then, analogous to the proof of (D.4), define $\rho_i : (R)_n \to R^n$ by $\rho_i(A) = i$-th row of A. Define

$$\tau : U^n \to R^n$$

by

$$\tau(u_1,\ldots,u_n) = \textstyle\sum_{i=1}^n \rho_i(u_i) .$$

But, both ρ_i, $1 \leq i \leq n$, and τ localize and locally τ is an isomorphism since U_q is invertible. Hence, τ is an isomorphism and thus U is invertible.

The proof of the next lemma is easy and completely analogous to the proof of Lemma 1 of [6].

(D.6) LEMMA. Let $\bar{U}_1$ and $\bar{U}_2$ be in $\mathfrak{U}$. If $\bar{U}_1 \subseteq \bar{U}_2$ then $\bar{U}_1 = \bar{U}_2$.

It is clear that the set $\mathfrak{U}$ of invertible submodules of $S = (R)_n \oplus (R)_n$ forms a group. Let $\mathfrak{U}_0$ denote the subset of $\mathfrak{U}$ consisting of those invertible $\bar{U}$ which contain an element u of the form $u = \langle P,Q \rangle$ where P and Q are invertible matrices in $(R)_n$. If $\bar{U}$ contains a u of the form $u = \langle P,Q \rangle$, set $\hat{U} = \{\alpha\langle P,Q\rangle \mid \alpha \text{ in } R\}$. Then $\hat{U}$ is an invertible R-submodule of S and $\hat{U} \subseteq U$. By the above lemma, $\bar{U} = \hat{U}$ and $\bar{U} = \{\alpha\langle P,Q\rangle \mid \alpha \text{ in } R\}$ is a free R-module. It is now easy to see that $\mathfrak{U}_0$ is a subgroup of $\mathfrak{U}$.

We denote the quotient group $\mathfrak{U}/\mathfrak{U}_0$ by $\mathrm{Pic}_n(R)$.

We have developed the theory of invertible R-submodules of $S = (R)_n \oplus (R)_n$ sufficiently so that easy adaptations of Issacs' proofs of analogous results in [6] for invertible submodules of $(R)_n$ carry over. A modification of the proof of Lemma 4 of [6] gives part (a) of the next theorem, the proof of Theorem 5 of [6] gives part (b), and the proof of Lemma 17 of [6] gives part (c). All of these arguments are straight-forward.

(D.7) THEOREM. Let $\bar{U}$ and $\bar{V}$ be in $\mathfrak{H}$.

(a) The following are equivalent:

(1) $\bar{U} \simeq \bar{V}$ as R-modules.

(2) $\bar{U} = \langle P,Q\rangle \bar{V}$ for a pair $\langle P,Q\rangle$ of invertible matrices in $(R)_n$.

(3) The cosets $\mathfrak{H}_0\bar{U}$ and $\mathfrak{H}_0\bar{V}$ are equal.

(b) $\mathrm{Pic}_n(R) = \mathfrak{H}/\mathfrak{H}_0$ is an Abelian group.

(c) $\bar{U}\bar{V} \simeq \bar{U} \otimes \bar{V}$.

For completeness, we sketch part of the proof of (a) above. (See Isaacs' [6], p. 220).

Suppose $\bar{U} \simeq \bar{V}$ and $\sigma : \bar{U} \to \bar{V}$ is the isomorphism. Let u_i be in $\bar{U}$ and $\bar{u}_i$ be in $\bar{U}^{-1}$ with $\langle I,I\rangle = \sum u_i\bar{u}_i$. Let $x = \sum \sigma(u_i)\bar{u}_i$ be in $\bar{V}\bar{U}^{-1}$. Then for u in U , $xu = \sum \sigma(u_i)\bar{u}_i u = \sum \sigma(u_i\bar{u}_i u)$ since $\bar{u}_i u$ may be regarded as a scalar. Thus $xu = \sigma(u)$ and thus $x\bar{U} = \bar{V}$. Finally, since $x\bar{V}\bar{U}^{-1} = \bar{E}$, we have $xy = \langle I,I\rangle$ for some y in $\bar{V}\bar{U}^{-1}$. Thus, x is invertible and has the form $x = \langle P,Q\rangle$ for invertible matrices P and Q in $(R)_n$.

The Picard group Pic(R) of R , as noted in the next section, is the Abelian group of R-isomorphism classes [U] of rank one projective R-modules U where multiplication is given by tensor product. Suppose U is a rank one projective R-module and $U \oplus U \oplus \cdots \oplus U$ (n copies of U) is isomorphic to R^n . Then, computing exterior powers,

$$\begin{aligned} U \otimes \cdots \otimes U &\simeq \Lambda^n[U \oplus \cdots \oplus U] \\ &\simeq \Lambda^n[R^n] \\ &\simeq R . \end{aligned}$$

Thus, in Pic(R) , $[U]^n = [U \otimes \cdots \otimes U] = [R] = 1$ and [U] is an n-torsion element. With (D.6) complete, we may identify $\mathrm{Pic}_n(R)$ with a subgroup of the n-torsion subgroup of Pic(R) . For certain commutative rings, e.g., Dedekind domains ([6], Corollary 18), $\mathrm{Pic}_n(R)$ is the full n-torsion subgroup of Pic(R) .

If Pic(R) is trivial, for example, if R is a principal ideal domain, a local ring, a semilocal ring, a von Neumann regular ring, a commutative ring "with many units" [15], or a polynomial ring $k[X_1,\ldots,X_n]$ (Serre's Theorem) where k is a field, then $\mathrm{Pic}_n(R)$ is also trivial and, of course, equal to Pic(R) . In this case all invertible R-submodules of $S = (R)_n \oplus (R)_n$ have the form $U = \{\alpha\langle P,Q\rangle \mid \alpha \text{ in } R\}$ where P and Q are invertible matrices.

With this lengthy preliminary on invertible R-submodules $\bar{U}$ of $S = (R)_n \oplus (R)_n$ complete, we return to our central idea of "equivalence transformations." The key idea is that the invertible R-submodules of S provide the generalization of the standard equivalence transformations.

Recall that, if $\bar{U}$ is in $\mathfrak{U}$ and q is a prime ideal of R , then $\bar{U}_q$ has the form

$$\bar{U}_q = \{\alpha\langle P,Q\rangle \mid \alpha \text{ in } R_q\}$$

where P and Q are fixed invertible matrices of $(R_q)_n$. Similarly,

$$\bar{U}_q^{-1} = (\bar{U}^{-1})_q = (\bar{U}_q)^{-1} = \{\beta\langle P^{-1},Q^{-1}\rangle \mid \beta \text{ in } R_q\} .$$

Select u_i in $\bar{U}$ and v_i in $\bar{V} = \bar{U}^{-1}$ with $\sum_i u_i v_i = \langle I,I\rangle$. Set $u_i = \langle u_1^{(i)},u_2^{(i)}\rangle$ and $v_i = \langle v_1^{(i)},v_2^{(i)}\rangle$ in $S = (R)_n \oplus (R)_n$. Then, localizing $\langle I,I\rangle$ at a prime ideal q of R ,

$$\begin{aligned}\langle I,I\rangle &= \langle I,I\rangle_q \\ &= [\textstyle\sum_i u_i v_i]_q \\ &= [\textstyle\sum_i \langle u_1^{(i)}v_1^{(i)},u_2^{(i)}v_2^{(i)}\rangle]_q \\ &= [\textstyle\sum \langle u_1^{(i)},u_2^{(i)}\rangle_q \langle v_1^{(i)},v_2^{(i)}\rangle_q] \\ &= \textstyle\sum \alpha_i\langle P,Q\rangle\beta_i\langle P^{-1},Q^{-1}\rangle \\ &= (\textstyle\sum_i \alpha_i\beta_i)\langle I,I\rangle .\end{aligned}$$

Thus, $\sum_i \alpha_i\beta_i = 1$. This is used below.

Continuing the above notation, we define for $\bar{U}$ in $\mathfrak{U}$ a mapping

$$E_{\bar{U}} : (R)_n \to (R)_n$$

by

$$E_{\bar{U}}(A) = \sum_i u_1^{(i)} A v_2^{(i)} .$$

Clearly, $E_{\bar{U}}$ is a linear mapping. Consider the localization of $E_{\bar{U}}$ at the above prime ideal q :

$$\begin{aligned} [E_{\bar{U}}(A)]_q &= \sum_i (u_1^{(i)})_q A_q (v_2^{(i)})_q \\ &= \sum_i \alpha_i P A_q \beta_i Q^{-1} \\ &= \left(\sum_i \alpha_i\beta_i\right) P A_q Q^{-1} \\ &= P A_q Q^{-1} \end{aligned}$$

since, as noted above, $\sum \alpha_i\beta_i = 1$.

Thus, the transformations $E_{\bar{U}} : (R)_n \to (R)_n$ localize to equivalence transformations. Further, the above localization argument shows

(1) $E_{\bar{U}}$ is an isomorphism since it is locally an isomorphism.

(2) $E_{\bar{U}}$ depends only on $\bar{U}$ and not on the choice of elements satisfying

$$\sum_i \langle u_1^{(i)} v_1^{(i)}, u_2^{(i)} v_2^{(i)} \rangle = \langle I, I \rangle .$$

This follows since two distinct choices will everywhere locally produce the same (local) equivalence transformation and linear mappings which are locally identical are identical.

Let

$$EQUIV_n(R) = \{E_{\bar{U}} \mid \bar{U} \text{ in } \mathfrak{U}\} .$$

The multiplication in $\mathfrak{U}$ induces a multiplication in $EQUIV_n(R)$ by

$$E_{\bar{U}} E_{\bar{V}} = E_{\bar{U}\bar{V}}$$

and $\mathrm{EQUIV}_n(R)$ forms a group. Since $E_{\bar{U}} = E_{\bar{V}}$ if and only if $\bar{U} = \bar{V}$, the group $\mathrm{EQUIV}_n(R) \simeq \mathfrak{A}$.

There is a natural group morphism

$$\mathrm{Equiv}_n(R) \to \mathrm{EQUIV}_n(R)$$

given by

$$E_{(P,Q)} \to E_{\bar{U}}$$

where $\bar{U} = \{\alpha\langle P,Q\rangle \mid \alpha \text{ in } R\}$. By the initial remarks in Section (C), this map does not depend on the choice of the pair $\langle P,Q\rangle$ representing $E_{(P,Q)}$. Further, it is easy to see that $\mathfrak{A}_0 \simeq \mathrm{Equiv}_n(R)$. This gives an exact sequence

$$1 \longrightarrow \mathrm{Equiv}_n(R) \longrightarrow \mathrm{EQUIV}_n(R) \longrightarrow \mathrm{Pic}_n(R) \longrightarrow 1 \ .$$

As a concluding remark, the reader has perhaps noted that $S = (R)_n \oplus (R)_n$ may be replaced by $S = (R)_n \oplus \cdots \oplus (R)_n$ (m summands) and the results of this section will carry over with modest modification.

These comments are summarized in the following theorem.

(D.8) THEOREM. (For the notation in this section.) Let $\bar{U}$ be an invertible R-submodule of $S = (R)_n \oplus (R)_n$.

(a) $E_{\bar{U}}$ (defined above) is an R-module isomorphism of $(R)_n$. We call $E_{\bar{U}}$ an *equivalence transformation of* $(R)_n$.

(b) For each prime ideal q of R, the localization of $E_{\bar{U}}$, $(E_{\bar{U}})_q = E_{\bar{U}_q}$, is a standard equivalence transformation of $(R_q)_n$.

(c) Suppose $T : (R)_n \to (R)_n$ is an R-linear transformation such that at each localization at a prime q the map T_q is a standard equivalence transformation. Then there is a unique invertible R-submodule $\bar{U}$ of $S = (R)_n \oplus (R)_n$ with $T = E_{\bar{U}}$.

(d) There is a natural exact sequence

$$1 \longrightarrow \mathrm{Equiv}_n(R) \longrightarrow \mathrm{EQUIV}_n(R) \longrightarrow \mathrm{Pic}_n(R) \longrightarrow 1$$

where $\mathrm{EQUIV}_n(R) \approx \mathfrak{U}$ and $\mathrm{Equiv}_n(R) \approx \mathfrak{U}_0$.

PROOF. It is only necessary to show part (c). Suppose $T : (R)_n \to (R)_n$ is an R-linear map with $T_q = E_{(P,Q)}$ at the prime ideal q of R where P and Q are suitable invertible matrices in $(R_q)_n$. Set

$$\bar{U} = \{<A_1,A_2> \text{ in } S \mid T(X)A_2 = A_1X \text{ for all } X \text{ in } (R)_n\} .$$

It is easy to see that $\bar{U}$ is a non-empty R-submodule of S . Consider the localization of $\bar{U}$ at the prime q .

$$\begin{aligned} \bar{U}_q &= \{<A_1,A_2>_q \mid [T(X)A_2]_q = [A_1X]_q\} \\ &= \{<\bar{A}_1,\bar{A}_2> \mid PX_qQ^{-1}\bar{A}_2 = \bar{A}_1X_q\} \end{aligned}$$

where $\bar{A}_1 = (A_1)_q$ and $\bar{A}_2 = (A_2)_q$. Thus,

$$X_qQ^{-1}\bar{A}_2 = P^{-1}\bar{A}_1X_q$$

for all X_q in $(R_q)_n$. Then, there is an α in R_q with

$$Q^{-1}\bar{A}_2 = \alpha I = P^{-1}\bar{A}_2 .$$

Hence $<A_1,A_2>_q = <\bar{A}_1,\bar{A}_2> = \alpha<P,Q>$ and

$$(\bar{U})_q = \{\alpha<P,Q> \mid \alpha \text{ in } R_q\}$$

is invertible. Hence, by (D.5) $\bar{U}$ is invertible. Clearly, $E_{\bar{U}} = T$ since $(E_{\bar{U}})_q = T_q$ for each prime ideal q of R .

Finally, suppose $T = E_{\bar{U}} = E_{\bar{V}}$, then it is easy to check by localization that $\bar{U} = \bar{V}$. This completes (D.8).

It is worth noting the special case of a standard inner automorphism or similarity transformation. Suppose P is an invertible matrix in $(R)_n$, then the map $(R)_n \to (R)_n$ by $A \to PAP^{-1}$ is called a <u>standard similarity</u>

transformation and is precisely the standard equivalence transformation $E_{(P,P)}$.

Often, the double subscript is deleted and $E_{(P,P)}$ is written as E_P . Let

$$\mathrm{Inn}_n(R) = \{E_{(P,P)} \mid P \text{ invertible matrix in } (R)_n\}$$

denote the set of standard similarity transformations. Then $\mathrm{Inn}_R(R)$ is a subgroup of $\mathrm{Equiv}_n(R)$.

Let U be an invertible R-submodule of $(R)_n$. Then

$$\bar{U} = \{\langle u,u\rangle \mid u \text{ is in } U\}$$

is an invertible R-submodule of $S = (R)_n \oplus (R)_n$. We call $\bar{U}$ the diagonal of U and write $\bar{U} = \mathrm{diag}(U)$. Then, the inner automorphisms or similarity transformations of $(R)_n$ is the group

$$\mathrm{INN}_n(R) = \{E_{\bar{U}} \mid \bar{U} = \mathrm{diag}(U),\ U \text{ invertible R-submodule of } (R)_n\} .$$

This is discussed again in Section (H).

(E) RANK ONE ENDOMORPHISMS

Let R be a commutative ring and P be a finitely generated projective R-module. Recall that, a module P is said to be a <u>rank one</u> projective R-module if the localized module P_q is free of R_q-dimension one for each prime ideal q in R. For background see [5], [19].

It is known [5] that the following are equivalent for a finitely generated projective R-module P:

(a) P has rank one,

(b) $P^* = \mathrm{Hom}_R(P,R)$ has rank one,

(c) $R \approx \mathrm{End}_R(P)$ under $\phi : R \to \mathrm{End}_R(P)$ by $\phi(r)(p) = rp$.

Let $[P]$ denote the R-isomorphism class of P and let $\mathrm{Pic}(R)$ denote the set of R-isomorphism classes of rank one projective R-modules. The set $\mathrm{Pic}(R)$ is a multiplicative Abelian group under $[P] \circ [\bar{P}] = [P \otimes_R \bar{P}]$ where $1 = [R]$ and $[P]^{-1} = [P^*]$. This group, as noted in the last section, is called the <u>Picard group</u> of R and is a trivial group if R has the property that all its rank one projective modules are free.

Suppose k is a field and A is an $n \times n$ matrix over k. One of the standard definitions of A to be a <u>rank one</u> matrix is that the column space of A has k-dimension one.

Suppose A is an $n \times n$ matrix over a commutative ring R. We generalize the above idea by defining A to be a <u>rank one</u> matrix if the R-module generated by the columns of A is a rank one projective.

R. Gilmer and R. Heitmann [2] recently characterized rank one matrices. Let $A = [a_{ij}]$ be an $n \times n$ matrix over R. Then, they show the following are equivalent:

(a) A has rank one,

(b) (1) The ideal generated by the elements a_{ij} of A is R, and

(2) All 2×2 submatrices of A have determinant 0 .

A direct proof of the Gilmer-Heitmann result can be given by a localization argument. The above can be stated invariantly in terms of the determinantal ideals of A . That is, let $F_k(A)$ denote the ideal generated by the determinants of all $k \times k$ submatrices of A . Then, A has rank one if and only if $F_1(A) = R$ and $F_2(A) = 0$.

By letting A act on the left of $n \times 1$ columns, we may view A as an R-endomorphism of R^n where R^n is the free R-module of $n \times 1$ columns over R . If A has rank one, then since the image of A as an endomorphism is the module generated by the columns of A , we have an exact sequence $R^n \xrightarrow{A} P \longrightarrow 0$ where P is a rank one projective. Since P is projective, the sequence splits and $R^n = \bar{P} \oplus Q$ where $\bar{P} \simeq P$ and $Q = \ker(A)$. Relative to the decomposition $R^n = \bar{P} \oplus Q$, we can find a 2×2 matrix for A : (Recall that:

$$
\begin{aligned}
(R)_n &\simeq \mathrm{End}_R(R^n) \\
&= \mathrm{End}_R(\bar{P} \oplus Q) \\
&\simeq \begin{bmatrix} \mathrm{End}_R(\bar{P}) & \mathrm{Hom}_R(Q,\bar{P}) \\ \mathrm{Hom}_R(\bar{P},Q) & \mathrm{End}_R(Q) \end{bmatrix} . \;)
\end{aligned}
$$

Let $x = \bar{p} + q$ be an element in $R^n = \bar{P} \oplus Q$. Then $A(x) = A\bar{p} + Aq = A\bar{p} = p$ in P where $p = p_1 + q_1$ in $\bar{P} \oplus Q$. Then, the matrix of A is

$$
\mathrm{Mat}(A) = \begin{bmatrix} \alpha & 0 \\ \delta & 0 \end{bmatrix}
$$

where

α in $\mathrm{End}_R(\bar{P})$ is given by $\alpha : \bar{p} \to p_1$,

δ in $\mathrm{Hom}_R(\bar{P},Q)$ is given by $\delta : \bar{p} \to q_1$,

0 in $\mathrm{Hom}_R(Q,\bar{P})$ is the zero morphism, and

0 in $\mathrm{End}_R(Q)$ is the zero morphism.

Suppose R is a commutative ring for which

(1) $\mathrm{Pic}(R) = 1$.

(2) If $Q \oplus R^n \simeq R^m$, then $Q \simeq R^{m-n}$.†

The second condition is referred to as "stably-free projectives are free." Local ring, semilocal rings, von Neuman regular rings, rings with "many units" [15], principal ideal domains, and $k[X_1,\ldots,X_n]$ (k a field) are examples of commutative rings satisfying (1) and (2). Then, by (1), we have $\bar{P}$ free of dimension one and, by (2), Q is free of dimension $n-1$.

Then, for the above endomorphism A of rank one, we have a matrix

$$\mathrm{Mat}(A) = \begin{bmatrix} \alpha_1 & \\ \alpha_2 & \\ \vdots & 0 \\ \alpha_n & \end{bmatrix} \qquad (\alpha_i \text{ in } R)$$

where $(\alpha_1,\alpha_2,\ldots,\alpha_n) = R$. This matrix corresponds to the standard column reduced matrix of A over a field.

We will call properties (1) and (2) above, the <u>stability conditions for rank</u>. Recall that a row (or column) is called <u>unimodular</u> if its coordinates generate R as an ideal. Then, in the presence of the above stability conditions,

$$\mathrm{Mat}(A) = \begin{bmatrix} \alpha_1 \\ \alpha_2 \\ \vdots \\ \alpha_n \end{bmatrix} [1 \quad 0 \cdots\cdots 0]$$

is the product of a unimodular column and a unimodular row.

If F is a free R-module of finite dimension, then an element x in F is <u>unimodular</u> if its coordinates relative to any basis of F generate R as an ideal. It is easy to see that x being unimodular is independent of the

---

†We assume (2) for $m = 1,2,3,\ldots$ and $1 \leqslant n \leqslant m$.

choice of basis. Further, if x in F is unimodular, then Rx splits F as $F = Rx \oplus G$, i.e., Rx is a free direct summand of dimension 1 .

(E.1) PROPOSITION. Let R satisfy the stability conditions for rank. Let A be an $n \times n$ matrix over R . Then, the following are equivalent:

(a) A has rank one.

(b) $A = xy$ where x is a unimodular $n \times 1$ column and y is a unimodular $1 \times n$ row.

Further, if in (b), $A = xy$ and $A = \bar{x}\bar{y}$, then there is a unit α in R with $x = \alpha\bar{x}$ and $y = \alpha^{-1}\bar{y}$.

PROOF. The above discussion shows that (a) implies (b). Suppose $A = xy$ as in (b) where $x = [p_1,\ldots,p_n]^t$ and $y = [q_1,\ldots,q_n]$. Then $A = [p_i q_i]$ and

$$\det \begin{bmatrix} p_i q_j & p_i q_k \\ p\ q_j & p\ q_k \end{bmatrix} = 0 .$$

Thus, determinants of all 2×2 submatrices of A are 0 . Let m be a maximal ideal of R and consider the reduction of R modulo m , i.e., the ring morphism $R \to R/m$. Both x and y are nonzero modulo m and, thus A is not zero modulo m . Thus, the elements of A are not contained in m . Since this is true for all maximal ideals m , the elements of A generate R as an ideal.[1]

To show the final remark, recall that (2) of the stability conditions for rank is equivalent to: for each positive integer m , $GL_m(R)$ acts transitively on unimodular vectors in R^m . (For example, see Theorem (IV.41) of [14].) Hence, for x and y above, there exist invertible matrices P and Q satisfying $Px = [1\ 0 \cdots 0]^t$ and $yQ = [1\ 0 \cdots 0]$. Thus, after modifying $xy = \bar{x}\bar{y}$ by P and Q , we may assume $xy = \bar{x}\bar{y}$ has the form

---

[1]This shows that (b) implies (a) for any commutative ring R .

$$\begin{bmatrix}1\\0\\\vdots\\0\end{bmatrix}[1\ 0\ \cdots\ 0] = \begin{bmatrix}\bar{p}_1\\\bar{p}_2\\\vdots\\\bar{p}_n\end{bmatrix}[\bar{q}_1,\bar{q}_2,\ldots,\bar{q}_n] .$$

Then $\bar{p}_1\bar{q}_1 = 1$, $\bar{p}_1\bar{q}_j = 0$ for $2 \leqslant j \leqslant n$, and $\bar{p}_k\bar{q}_1 = 0$ for $2 \leqslant k \leqslant n$. Then, $\bar{p}_1 = \alpha$ is a unit, $\bar{q}_1 = \alpha^{-1}$, and $\bar{q}_j = \bar{p}_j = 0$ for $2 \leqslant j \leqslant n$. Thus, $x = \alpha\bar{x}$ and $y = \alpha^{-1}\bar{y}$.

Let F be a free R-module of dimension n and basis $\{b_1,b_2,\ldots,b_n\}$. Let $F^* = \mathrm{Hom}_R(F,R)$ be the dual of F with dual basis $\{b_1^*,\ldots,b_n^*\}$ where $b_j^*(b_i) = \delta_{ij}$ (Kronecker delta) for $1 \leqslant i,j \leqslant n$. Then, the mapping $[\ ,\ ] : \langle x,f\rangle \to [x,f]$ and $[x,f](y) = f(y)x$ is R-bilinear and induces an $(\mathcal{S},\mathcal{S})$-module isomorphism

$$[\ ,\ ] : F \otimes_R F^* \to \mathcal{S} \qquad (\mathcal{S} = \mathrm{End}_R(F))$$

$$\text{by } [\ ,\ ] : x \otimes f \to [x,f] .$$

Under this map $b_j \otimes b_i^* \to [b_j,b_i^*] = E_{ji}$ where $E_{ji} : F \to F$ given by $E_{ji}(b_k) = \delta_{ik}b_j$ is the standard elementary matrix unit.

If $x = p_1b_1 + p_2b_2 + \cdots + p_nb_n$ is written, relative to this basis, as a column $x = [p_1\ p_2\ \cdots\ p_n]^t$ and $y = q_1b_1^* + \cdots + q_nb_n^*$ as a row $y = [q_1\ q_2\ \cdots\ q_n]$, then

$$x \otimes y \to [x,y] = xy = \begin{bmatrix}p_1\\p_2\\\vdots\\p_n\end{bmatrix}[q_1\ q_2\ \cdots\ q_n] .$$

(E.2) <u>COROLLARY</u>. Let R satisfy the stability conditions for rank and let A be an $n \times n$ matrix over R . Then, the following are equivalent.

(a) A has rank one.

(b) A is the image under $[\ ,\ ] : F \otimes F^* \to \mathrm{End}_R(F)$ of $x \otimes y$ where x is a unimodular vector in F and y is a unimodular vector in F^*.

As noted in (E.1), condition (b) will imply condition (a) for any commutative ring R.

Let $\mathrm{RANK}_n(R)$ denote the set of R-linear morphisms $T : (R)_n \to (R)_n$ with the property that, if A has rank one, then $T(A)$ has rank one. We will show that a rank one preserver T is always invertible. It will then be easy to see that $\mathrm{RANK}_n(R)$ is a subgroup of the invertible R-linear endomorphisms of $(R)_n$. Further, it is easy to verify that the elements of the semidirect product $\mathrm{Rank}_n(R) = \mathrm{Equiv}_n(R) \times \mathrm{Inv}(R)$ form a subgroup of $\mathrm{RANK}_n(R)$. If k is a field, then $\mathrm{RANK}_n(k) = \mathrm{Rank}_n(k)$; but, in general, $\mathrm{Rank}_n(R)$ is only a subgroup of $\mathrm{RANK}_n(R)$. In Section (G) we will examine exactly how $\mathrm{Rank}_n(R)$ sits in $\mathrm{RANK}_n(R)$. We first treat the case where R is a local ring in Section F.

Our approach will be to use (E.2) and to translate the question to R-linear mappings $T : F \otimes F^* \to F \otimes F^*$ with the property that, if x in F and y in F^* are unimodular, then $T(x \otimes y) = u \otimes v$ where u in F and v in F^* are unimodular.

Let M be an R-module. A submodule F of M will be called an R-<u>subspace</u> of M if F is an R-direct summand of M <u>and</u> F is a free R-module.

(E.3) <u>LEMMA</u>. Let M and N be finitely generated, projective R-modules. Suppose $x_1,\ldots,x_r$, $w_1,\ldots,w_s$ are in M and $y_1,\ldots,y_r$, $z_1,\ldots,z_s$ are in N. Suppose also that $\sum_{i=1}^r x_i \otimes y_i = \sum_{j=1}^s w_j \otimes z_j$ are in $M \otimes_R N$. If $x_1,\ldots,x_r$ are a basis for a subspace of M, then each y_i is in $\sum_{j=1}^s Rz_j$. Similarly, if $y_1,\ldots,y_r$ are a basis for a subspace of N, then each x_i is in $\sum_{j=1}^s Rw_j$.

<u>PROOF</u>. The proof follows the proof over a field given in Lemma 1 of [12]. Let $F = Rx_1 \oplus \cdots \oplus Rx_r$ and $M = F \oplus \bar{F}$. Define $\phi : F \to R$ by $\phi(x_1) = 1$, $\phi(x_i) = 0$ for $2 \leq i \leq r$, and $\phi(\bar{F}) = 0$. Let $\alpha : N \to R$ be arbitrary. Define an R-bilinear form $\beta : M \otimes N \to R$ by $\beta(m,n) = \phi(m)\,\alpha(n)$. Then β

induces a linear map $\beta : M \otimes N \to R$ and

$$\alpha(y_1) = \beta(\sum x_i \otimes y_i) = \beta(\sum w_j \otimes z_j) = \sum \phi(w_j)\,\alpha(z_j)\ .$$

Thus, $\alpha(y_1) = \alpha(\sum \phi(w_j) z_j)$.

If M is a finitely generated projective module and $f(m) = 0$ for all f in $M^* = \mathrm{Hom}_R(M,R)$, then $m = 0$. Thus, since α was arbitrary, we have $y_1 = \sum \phi(w_j) a_j$. A similar argument will show that y_i is in $\sum Rz_j$ for $2 \leqslant i \leqslant r$.

Let F and $\bar{F}$ be free R-modules of finite dimension. An element m in $F \otimes_R \bar{F}$ will be said to have <u>rank</u> <u>one</u> if $m = x \otimes y$ where x and y are unimodular vectors in F and $\bar{F}$, respectively.

(E.4) <u>COROLLARY</u>. Let F and $\bar{F}$ be free R-modules of finite dimension. Let x , w be unimodular in F and y , z be unimodular in $\bar{F}$. Then, $x \otimes y = w \otimes z$ if and only if there is a unit α in R with $x = \alpha w$ and $y = \alpha^{-1} z$.

(F) THE LOCAL CASE

In this section, we assume that R is a local commutative ring with maximal ideal m and residue class field $k = R/m$. For a local ring, we will completely classify the linear maps T which carry rank one vectors to rank one vectors. The method of proof is a direct lifting of the theory from the known results over the field k. The classification of such T over a field which is algebraically closed of characteristic 0 was given by Marcus and Moyls in [12]. An examination of their proofs in [12] shows that characteristic 0 is nowhere used. Thus, their proofs work in any characteristic. Further, we will see that the condition of algebraic closure is unnecessary.

Let F and $\bar{F}$ be free R-modules of dimension n. A line in F (or $\bar{F}$) is a free summand of dimension one. Unimodular vectors are the basis vectors of lines.

Let $T : F \otimes_R \bar{F} \to F \otimes_R \bar{F}$ be an R-linear map having the property that T takes rank one vectors to rank one vectors. Since R is local, R satisfies the stability conditions for rank. (See (E.1) and (E.2)).

Let $\pi : R \to R/m = k$ denote the natural ring morphism of R onto k. We also denote by π the natural induced morphisms $F \to F/mF$ and $\bar{F} \to \bar{F}/m\bar{F}$. It is easy to see that unimodular vectors in F correspond under π to nonzero vectors in F/mF. Further, a set of vectors $\{x_1,\ldots,x_n\}$ is a basis for F if and only if $\{\pi x_1,\ldots,\pi x_n\}$ is a basis for F/mF. The next result is straight-forward.

(F.1) LEMMA. The linear map $T : F \otimes \bar{F} \to F \otimes \bar{F}$ given above induces a k-linear mapping $\pi T : F/mF \otimes_k \bar{F}/m\bar{F} \to F/mF \otimes_k \bar{F}/m\bar{F}$ which also preserves rank one vectors.

Let $\{b_1,\ldots,b_n\}$ and $\{\bar{b}_1,\ldots,\bar{b}_n\}$ be bases for F and $\bar{F}$, respectively. Then, for $1 \leq i,j \leq n$,

$$T(b_i \otimes \bar{b}_j) = u_{ij} \otimes v_{ij}$$

where u_{ij} and v_{ij} are unimodular in F and $\bar{F}$, respectively. We use this notation throughout the remainder of this section.

All the arguments in the proof of the classification of T over a local ring will be based on the proofs in [12]. These depend only on the action of T on $\{b_i \otimes \bar{b}_j \mid 1 \leq i,j \leq n\}$ and on linear combinations of the $b_i \otimes \bar{b}_j$. To obtain the theory for an arbitrary commutative ring, we will employ localization arguments. A linear map T in the general setting clearly extends to a linear mapping on the localization at each prime ideal. But, it is <u>not</u> obvious the localized linear map still preserves rank one vectors. However, bases and the action of T on these bases will localize and, the point we wish to make in this paragraph, is that the proofs indicate the calculations of T on these bases is sufficient to determine T.

(F.2) <u>THEOREM</u>. (For the above setting.)

(a) For each i, $1 \leq i \leq n$, either

(1) $Ru_{i1} + \cdots + Ru_{in} = F$ and $Rv_{i1} + \cdots + Rv_{in} = \bar{L}$ where $\bar{L}$ is a line in $\bar{F}$, or

(2) $Ru_{i1} + \cdots + Ru_{in} = L$ and $Rv_{i1} + \cdots + Rv_{in} = \bar{F}$ where L is a line in F.

(b) For each j, $1 \leq j \leq n$, either

(3) $Ru_{1j} + \cdots + Ru_{nj} = F$ and $Rv_{1j} + \cdots + Rv_{nj} = \bar{L}$ where $\bar{L}$ is a line in $\bar{F}$, or

(4) $Ru_{1j} + \cdots + Ru_{nj} = L$ and $Rv_{1j} + \cdots + Rv_{nj} = \bar{F}$ where L is a line in F.

<u>PROOF</u>. This is precisely Lemma 2 of [12] restated for local rings. The proof of this lemma does not utilize characteristic 0 or algebraic closure. It is

perhaps worthwhile to show initially how the fact that one of the sets of unimodular vectors forms a basis (this fact lifts directly from the case over a field) implies that the other corresponding set of unimodular vectors will generate a line.

Consider the vectors u_{i1} and u_{i2} and v_{i1} and v_{i2}. By (F.1) the mapping πT preserves rank one vectors. Then either πu_{i1} and πu_{i2} are k-independent or πv_{i1} and πv_{i2} are k-independent. Suppose πu_{i1} and πu_{i2} are k-independent. Then u_{i1} and u_{i2} are R-independent. But

$$\begin{aligned} T\left(b_i \otimes (\bar{b}_1 + \bar{b}_2)\right) &= u_{i1} \otimes v_{i1} + u_{i2} \otimes v_{i2} \\ &= u \otimes v \quad \text{(by (E.2))} \end{aligned}$$

for some unimodular vectors u and v. Then, by (E.3), $Rv_{i1} = Rv_{i2} = Rv$.

We sketch the remainder of the proof. It is based on the proof of Marcus and Moyls ([12], Lemma 2).

Suppose $\alpha \neq 1,2$. Then $u_{i\alpha}$ must be R-independent of at least one of u_{i1} and u_{i2} since this is true of $\pi u_{i\alpha}$, πu_{i1}, and πu_{i2}. Thus, by the above argument, $Rv = Rv_{i\alpha}$ for $1 \leqslant \alpha \leqslant n$. Thus, if $\dim(k\pi u_{i1} + \cdots + k\pi u_{in}) \geqslant 2$, then there is a line $\bar{L} = Rv$ with $\bar{L} = Rv_{i\alpha}$, $1 \leqslant \alpha \leqslant n$.

Suppose that we have a line $\bar{L}$ with $Rv_{i\alpha} = \bar{L}$ for $1 \leqslant \alpha \leqslant n$. We show next that $F = Ru_{i1} + \cdots + Ru_{i1}$. Here it suffices to show that $F/mF = k\pi u_{i1} + \cdots + k\pi u_{in}$. So suppose not. Then, the vectors $\{\pi u_{i1}, \ldots, \pi u_{in}\}$ are k-dependent. Assume $\sum_j a_j \pi u_{ij} = 0$. Further, for $j = 1, \ldots, n$, let $\pi v_{ij} = c_j \pi v_{i1}$. Then

$$\begin{aligned} &\pi T\left(\left(\sum_j (a_j/c_j)\pi b_j\right) \otimes \pi \bar{b}_i\right) \\ &= \left(\sum_j (a_j/c_j)\pi u_{ij}\right) \otimes c_j \pi v_{i1} \\ &= \left(\sum_j a_j \pi u_{ij}\right) \otimes \pi v_{i1} \\ &= 0 \end{aligned}$$

which violates the fact that πT preserves rank one vectors. Hence,

$\{\pi u_{i1},\ldots,\pi u_{in}\}$ generate F/mF and thus $\{u_{i1},\ldots,u_{in}\}$ generate F .

(F.3) THEOREM. (For the setting of (F.1).) For all i and j , either (1) and (4) hold uniformly; or (2) and (3) hold uniformly.

PROOF. This is Lemma 3 of [12] when R is a field. Using the first part of the proof of Lemma 3 [12] for πT and then lifting to the local ring, it is easy to see that (1) or (2) will hold uniformly for $1 \leqslant i \leqslant n$; and, in a similar fashion, (3) or (4) will hold uniformly for $1 \leqslant j \leqslant n$. We do not sketch the proof since the previous proof illustrates how to adapt the proof of Lemma 3 of [12]. Of crucial importance is the fact that a set of vectors $\{x_1,\ldots,x_n\}$ is a basis for F if and only if $\{\pi x_1,\ldots,\pi x_n\}$ is a basis for F/mF .

It remains to show that (1) and (3) cannot hold simultaneously. An analogous proof will show (2) and (4) cannot hold simultaneously.

Assume that both (1) and (3) are true. Then, over the field k , both (1) and (3) hold for the rank one preserving linear map

$$\pi T : F/mF \otimes_k \bar{F}/m\bar{F} \to F/mF \otimes_k \bar{F}/m\bar{F} .$$

We extend to the algebraic closure $\hat{k}$ of k . By extension of scalars, πT extends to $\widehat{\pi T}$ over the algebraic closure. Further, $\widehat{\pi T}$ is also a rank one preserver and, since bases extend to bases, $\widehat{\pi T}$ will also satisfy both (1) and (3). However, the concluding remarks of the proof of Lemma 3 of [12] show that this is not possible over an algebraically closed field. This completes the proof.

The next result for a field is Theorem 1 of [12]. The proof over a local ring can be obtained immediately by directly mimicking the proof of Theorem 1 [12]. However, for completeness, we will sketch this proof.

(F.4) THEOREM. Let F and $\bar{F}$ be n-dimensional free modules over a local ring R . Let

$$T : F \otimes \bar{F} \to F \otimes \bar{F}$$

be a linear transformation which maps elements of rank one to elements of rank one. Let $\tau : \bar{F} \otimes F \to F \otimes \bar{F}$ be the twist mapping where $\tau : y \otimes x \to x \otimes y$. Then, there exist invertible R-linear mappings

$$\sigma : F \to F$$

and

$$\beta : \bar{F} \to \bar{F}$$

such that either

$$T = \sigma \otimes \beta$$

or

$$T = \tau(\psi\sigma \otimes \psi^{-1}\beta)$$

where $\psi : F \to \bar{F}$ is an invertible R-linear mapping.

<u>PROOF</u>. Let $T(b_i \otimes \bar{b}_j) = u_{ij} \otimes v_{ij}$ for $1 \leqslant i,j \leqslant n$. By Theorems (F.2) and (F.3) we assume that both (1) and (4) of (F.2) hold uniformly. The other case is similar and is done in [12].

Thus, there is a fixed line L in F and $\bar{L}$ in $\bar{F}$ such that for each i, $1 \leqslant i \leqslant n$,

$$F = Fu_{i1} + \cdots + Ru_{in},$$
$$L = Rv_{i\alpha} \quad \text{for } 1 \leqslant \alpha \leqslant n,$$

and for each j, $1 \leqslant j \leqslant n$,

$$\bar{F} = Rv_{1j} + \cdots + Rv_{nj},$$
$$L = Ru_{\alpha j} \quad \text{for } 1 \leqslant \alpha \leqslant n.$$

Then, there exist scalars s_{ij} and t_{ij} with

$$v_{ij} = t_{ij}v_{i1},$$
$$u_{ij} = s_{ij}u_{1j}.$$

Set $u_j = u_{1j}$ and $v_i = v_{i1}$. Then,

$$\begin{aligned} T(b_i \otimes \bar{b}_j) &= u_{ij} \otimes v_{ij} \\ &= (s_{ij}u_{1j}) \otimes (t_{ij}v_{i1}) \\ &= c_{ij}(u_j \otimes v_i) \end{aligned}$$

where $c_{ij} = s_{ij}t_{ij}$.

Then, for $i \geq 2$,

$$\begin{aligned} &T\left[\left(\textstyle\sum_{j=1}^n b_j\right) \otimes (\bar{b}_1 + \bar{b}_i)\right] \\ &= \sum_j c_{j1}u_1 \otimes v_j + \sum_j c_{ji}u_i \otimes v_j \\ &= u_1 \otimes \left(\sum_j c_{j1}v_j\right) + u_i \otimes \left(\sum_j c_{ji}v_j\right) . \end{aligned}$$

On the other hand,

$$T\left[\left(\textstyle\sum_{j=1}^n b_j\right) \otimes (\bar{b}_1 + \bar{b}_i)\right] = w \otimes z$$

where w and z are unimodular vectors in F and $\bar{F}$, respectively. Applying Lemma (E.3), we have that

$$x = \sum_j c_{j1}v_j \quad \text{and} \quad y = \sum_j c_{ji}v_j$$

are in Rz . Reduction modulo the maximal ideal will show that both x and y are unimodular vectors. Hence $Rx = Ry = Rz$ and there is a unit scalar α_i with $\alpha_i x = y$, i.e.,

$$\alpha_i\left(\sum_j c_{j1}v_j\right) = \sum_j c_{ji}v_j .$$

Therefore, $\alpha_i c_{j1} = c_{ji}$. Now set $x_j = c_{j1}u_j$ and $y_i = \alpha_i v_i$. Then, for the above equation,

$$\begin{aligned} T(b_i \otimes \bar{b}_j) &= u_{ij} \otimes v_{ij} \\ &= c_{ij}(u_j \otimes v_i) \\ &= (\alpha_i c_{j1})(u_j \otimes v_i) \\ &= (c_{j1}u_j) \otimes (\alpha_i v_i) \\ &= x_j \otimes y_i . \end{aligned}$$

Thus, there exist R-linearly independent bases $\{x_1,\ldots,x_n\}$ and $\{y_1,\ldots,y_n\}$ spanning F and $\bar{F}$, respectively, such that

$$T(b_i \otimes \bar{b}_j) = x_j \otimes y_i .$$

Let $\psi : F \to \bar{F}$ be any invertible R-linear mapping. Then there exist invertible R-linear mappings σ and β of F and $\bar{F}$ such that

$$\sigma b_i = \psi^{-1} y_i , \quad 1 \leq i \leq n , \text{ and}$$

$$\beta \bar{b}_j = \psi x_j , \quad 1 \leq j \leq n .$$

Then,

$$\begin{aligned} &\tau(\psi\sigma \otimes \psi^{-1}\beta)(b_i \otimes \bar{b}_j) \\ &= \tau(y_i \otimes x_j) \\ &= x_j \otimes y_i \\ &= T(b_i \otimes \bar{b}_j) \end{aligned}$$

and this shows that

$$T = \tau(\psi\sigma \otimes \psi^{-1}\beta) .$$

The other case is similar.

In the language of matrices, we have the following corollary.

(F.5) <u>COROLLARY</u>. Let R be a local ring and $T : (R)_n \to (R)_n$ a linear mapping which carries rank one matrices to rank one matrices. Then, there exist invertible matrices P and Q such that either

$$T(A) = PAQ^{-1} \text{ for all } A \text{ in } (R)_n$$

$$\text{or } T(A) = PA^tQ^{-1} \text{ for all } A \text{ in } (R)_n .$$

Thus, for a local ring R, $\mathrm{Inv}(R) = Z_2$ (the cyclic group of order 2) and $\mathrm{RANK}_n(R) = \mathrm{Rank}_n(R)$ is a semidirect product of $\mathrm{Equiv}_n(R)$ and Z_2. In the local ring case, $\mathrm{Equiv}_n(R) = \mathrm{EQUIV}_n(R)$ and $\mathrm{Pic}_n(R) = 1$.

The results of Marcus and Moyls [12] were formulated for mappings $T : F \otimes \bar{F} \rightarrow F \otimes \bar{F}$ where $\dim(F) = m$ and $\dim(\bar{F}) = n$, and it was not necessary for $m = n$. The results in this paper can be formulated for unequal dimension in a straightforward fashion. It was our thought that $(R)_n$ or $F \otimes F^*$ provided the most important example of the theory and thus was our focus. These arguments, with modest changes, will also apply to the case of unequal dimension, i.e., rectangular matrices.

(G) THE GLOBAL CASE

In this section R denotes an arbitrary commutative ring with identity. Let $(R)_n$ denote the $n \times n$ ring over R and let $T : (R)_n \to (R)_n$ be an R-linear mapping.

Suppose q is a prime ideal of R. Then R_q denotes the localization of R at q. Let $k_q = R_q/qR_q$ be the residue class field and let $\pi : R_q \to k_q$ denote the natural ring morphism. Let $\rho : R \to R/q$ denote the natural morphism onto the domain R/q. If q is a maximal ideal, then R/q is a field naturally isomorphic to k_q. Consequently, when useful, and when q is a maximal ideal, we will identify R/q and k_q, and consider the diagram

$$\begin{array}{ccc} R & \xrightarrow{(\ )_q} & R_q \\ & {\scriptstyle\rho}\searrow \quad \swarrow{\scriptstyle\pi} & \\ & R/q = k_q & \end{array}$$

as commutative. If q is a prime ideal, then k_q can be identified with the field of quotients of R/q. We will also use ρ and π for the natural extension of these maps to modules and matrix rings.

Let $T_q : (R_q)_n \to (R_q)_n$ be the localization of T at q. Note that, if A is in R, then $[T(A)]_q = T_q(A_q)$.

(G.1) THEOREM. The following are equivalent:

(a) T preserves rank one matrices.

(b) T_q preserves rank one matrices for each prime ideal q of R.

(c) T_q preserves rank one matrices for each maximal ideal q of R.

PROOF. It is immediate that (b) implies (c). To show (c) implies (a), assume that A is a rank one matrix in $(R)_n$. Then, computing determinantal ideals, we have $F_1(A) = R$ and $F_2(A) = 0$. Then, $[F_1(A)]_q = F_1(A_q) = R_q$ and $[F_2(A)]_q = F_2(A_q) = 0$ for q a maximal ideal. Thus, A_q has rank one. By (c), $T_q(A_q)$ has rank one. Then $F_1(T_q(A_q)) = [F_1(T(A))]_q = R_q$ and $F_2(T_q(A_q)) = [F_2(T(A))]_q = 0$ for each maximal ideal q. Therefore, $F_1(T(A)) = R$ and $F_2(T(A)) = 0$ and, consequently, $T(A)$ has rank one.

To show that (a) implies (b) is more difficult. Let $(R)_n$ be identified with $\mathrm{End}_R(F)$ where F is a free R-module with basis $\{b_1,b_2,\ldots,b_n\}$. Let F^* denote the dual space of F having the natural dual basis $\{b_1^*,b_2^*,\ldots,b_n^*\}$. As noted after (E.1), we may identify the modules $F \otimes F^*$ and $\mathrm{End}_R(F)$ and, thus, also $(R)_n$. Via this identification, we will consider T as an R-linear mapping $T : F \otimes F^* \to F \otimes F^*$. Since T maps rank one matrices to rank one matrices,

$$T : b_i \otimes b_j^* \to y_{ij} = T(b_i \otimes b_j^*)$$

where y_{ij} is an element of $F \otimes F^*$ of rank one. Elements of rank one localize at a prime q to elements of rank one. For simplicity we will "bar" elements which are localized at the ideal q. Thus, at q,

$$T_q : \bar{b}_i \otimes \bar{b}_j^* \to \bar{y}_{ij} .$$

(Recall that a basis and its dual are preserved under localization.) Since a local ring satisfies the stability conditions for rank, by (E.2),

$$\bar{y} = u_{ij} \otimes v_{ij}$$

for unimodular vectors u_{ij} in F_q and v_{ij} in F_q^*. That is, in the localization at q,

$$T_q : \bar{b}_i \otimes \bar{b}_j^* \to u_{ij} \otimes v_{ij} .$$

We pause to comment on the philosophy of the proof. If one examines the arguments of Marcus and Moyls' in [12] and the discussion in Section (F), then one realizes there is no need to evaluate T at <u>all</u> rank one matrices in order to characterize its form. One needs only to compute the images of rank one matrices or elements of the form $b_i \otimes \bar{b}_j$ where $\{b_1, b_2, \ldots, b_n\}$ is a basis for F and $\{\bar{b}_1, \bar{b}_2, \ldots, \bar{b}_n\}$ is a basis for $\bar{F}$ and $b_i \otimes \bar{b}_j$ is in $F \otimes \bar{F}$. This idea was exploited previously in [13] in the characterizations of the automorphisms of $GL_n(R)$ where there it sufficed to compute the effect of an automorphism on only elementary transvections having a unit in their off-diagonal position, rather than to compute the images of all elementary transvections, in order to determine the form of the group automorphism. Of course, since we are dealing with R-linear mappings, the maps are uniquely determined by their actions on the R-bases of the endomorphism ring, i.e. sets of rank one matrices. Thus, by examinating the arguments in [12] and (F), we will determine the form of T_q by studying $T_q(\bar{b}_i \otimes \bar{b}_j^*) = u_{ij} \otimes v_{ij}$ where $\{\bar{b}_1, \ldots, \bar{b}_n\}$ and $\{\bar{b}_1^*, \ldots, \bar{b}_n^*\}$ are bases of F_q and F_q^*, respectively, which lift from bases of F and F^*. Indeed, as noted above, this is reasonable since T is characterized as a linear mapping by its action on the basis $\{b_i \otimes b_j^* \mid 1 \leqslant i,j \leqslant n\}$ of $F \otimes F^*$.

To show that T_q preserves rank one matrices in $(R_q)_n$ when T preserves rank one matrices in $(R)_n$, we actually show that sufficient information is available to show T has the form given by (F.5). To do this we illustrate how the arguments of Marcus and Moyls [12] may be extended.

We begin by proving the first part of Theorem (F.2); equivalently, Lemma 2 of [12], in the above context. Namely, we show for each i, $1 \leqslant i \leqslant n$, either

(1) $R_q u_{i1} + \cdots + R_q u_{in} = F_q$ and $R_q v_{i1} + \cdots + R_q v_{in} = \bar{L}$ where $\bar{L}$ is a line in F_q^*, or

(2) $R_q u_{i1} + \cdots + R_q u_{in} = L$ and $R_q v_{i1} + \cdots + R_q v_{in} = F_q^*$ where L is a line in F_q.

We have that $T_q(\bar{b}_i \otimes \bar{b}_j^*) = u_{ij} \otimes v_{ij}$, $1 \leqslant i,j \leqslant n$. In the k_q-vector space $\pi F_q = F_q/qF_q$ either

$$\dim_{k_q}(k_q\pi u_{i1} + \cdots + k_q\pi u_{in}) \geq 2 \quad \text{or}$$

$$\dim_{k_q}(k_q\pi u_{i1} + \cdots + k_q\pi u_{in}) = 1 .$$

(a) Suppose $\dim_{k_q}(k_q\pi u_{i1} + \cdots + k_q\pi u_{in}) \geq 2$. Then, in πF_q at least two of the vectors in the set $\{\pi u_{i1},\ldots,\pi u_{in}\}$ are k_q-independent. Assume $\pi u_{i\alpha}$ and $\pi u_{i\beta}$ are k_q-independent. Since $u_{i\alpha}$ and $u_{i\beta}$ are unimodular in F_q, we have that $u_{i\alpha}$ and $u_{i\beta}$ are R_q-independent and form a basis for a two-dimensional plane in F_q. Then,

$$\begin{aligned}[T(b_i \otimes (b^*_\alpha + b^*_\beta))]_q &= u_{i\alpha} \otimes v_{i\alpha} + u_{i\beta} \otimes v_{i\beta} \\ &= u \otimes v .\end{aligned}$$

By (E.3), $v_{i\alpha}$ and $v_{i\beta}$ are in $R_q v$. Hence, since $v_{i\alpha}$ and $v_{i\beta}$ are unimodular in F^*_q, $R_q v = R_q v_{i\alpha} = R_q v_{i\beta}$ and $v_{i\alpha}$ and $v_{i\beta}$ generate the same line. Next, select $\gamma \neq \alpha,\beta$. Then $\pi u_{i\gamma}$, which is nonzero, must be k_q-independent of either $u_{i\alpha}$ or $u_{i\beta}$. Thus, $u_{i\gamma}$ will be R_q-independent of either $u_{i\alpha}$ or $u_{i\beta}$. By repeating the argument, we can then conclude that $v_{i1},\ldots,v_{in}$ all generate the same line.

(b) Now assume that $\dim_{k_q}(k_q\pi u_{i1} + \cdots + k_q\pi u_{in}) = 1$. The proof of (a) mimicked the proof of Lemma 2 of [12]. The second part requires more modification since the unit c_α appearing in the proof of Lemma 2 [12] does not necessarily exist.

Let m be a maximal ideal of R containing q. Recall that the localization R_q can be obtained in a transitive manner by first localizing at m and then localizing at the extension $\bar{q} = qR_m$ of q in R_m, i.e., $R_q = (R_m)_{\bar{q}}$ (for example, see Proposition 19, p. 165 of [18]). Since m is also prime, R_m is a local ring. Without loss of generality, we may assume the u_{ij} and v_{ij} "appear" at the localization at m and then extend to R_q.

Thus, suppose $\dim_{k_m}(k_m\pi u_{i1} + \cdots + k_m\pi u_{in}) = 1$ where $\pi : R_m \to k_m$. Also assume that

$$\dim_{k_m}(k_m\pi v_{i1} + \cdots + k_m\pi v_{in}) = t < n .$$

We may suppose that the v_{ij} are so ordered that $\{\pi v_{i1},\dots,\pi v_{it}\}$ are k_m-independent. Then, there exist $\bar{\alpha}_i$ in k_m with

$$\pi v_{i,t+1} = -\sum_{j=1}^{t} \bar{\alpha}_j \pi v_{ij}$$

and β_j in k_m with

$$\beta_j \pi u_{i,t+1} = \pi u_{ij}, \quad 1 \leqslant j \leqslant t .$$

Then, since each β_j is a unit, there is an α_j in k_m with $\alpha_j\beta_j = \bar{\alpha}_j$ for $1 \leqslant j \leqslant t$. In the diagram (*) below, select δ_j in R with $\rho(\delta_j) = \alpha_j$, $1 \leqslant j \leqslant t$, under $\rho : R \to R/m$. In $F \otimes F^*$ the element

$$x = b_i \otimes [b^*_{t+1} + \sum_j \delta_j b^*_j]$$

has rank one. Then,

$$[T(b_i \otimes [b^*_{t+1} + \sum_j \delta_j b^*_j])]_m$$

has rank one and so does its image under $\pi : F \otimes F^* \to \bar{F} \otimes \bar{F}^*$. We compute this image:

$$\begin{aligned}
&\pi\big([T(b_i \otimes [b^*_{t+1} + \textstyle\sum_{j=1}^{t} \delta_j b^*_j])]_m\big) \\
&= \pi\big(u_{i,t+1} \otimes v_{i,t+1} + \textstyle\sum_j (\delta_j)_m u_{ij} \otimes v_{ij}\big) \\
&= \pi u_{i,t+1} \otimes \pi v_{i,t+1} + \textstyle\sum_j \pi(\delta_j)_m \pi u_{ij} \otimes \pi v_{ij} \\
&= \pi u_{i,t+1} \otimes \pi v_{i,t+1} + \textstyle\sum_j \alpha_j \beta_j \pi u_{i,t+1} \otimes \pi v_{ij}
\end{aligned}$$

by the above and the commutativity of

(*)

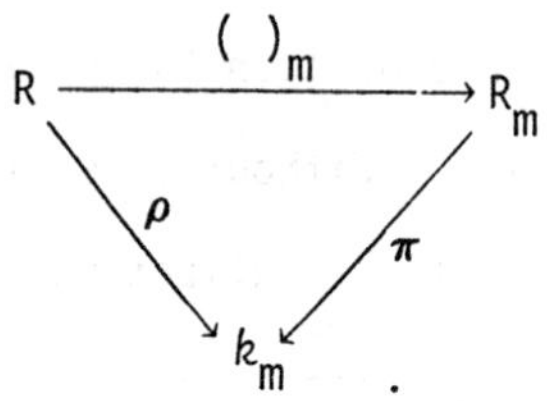

.

Thus,

$$\pi([T(x)]_m) = \pi u_{i,t+1} \otimes [- \textstyle\sum_j \bar{\alpha}_j \pi v_{ij} + \sum_j \alpha_j \beta_j \pi v_{ij}]$$
$$= u_{i,t+1} \otimes 0$$
$$= 0$$

which is not of rank one - a contradiction. Hence, $\dim_{k_m}(k_m \pi v_{i1} + \cdots + k_m \pi v_{in}) = n$. Thus, $\{v_{i1},\ldots,v_{in}\}$ is an R_m-basis for F_m^* and extends to an R_q-basis for F_q^* by transitivity of localizations.

The analogues of Lemma 3, Theorem 1, and Corollary of [12] are now straightforward. (See Section (F).) The above proof was more involved since the Marcus-Moyls proof needed to invert a nonzero element and this had to be avoided in the general case. Thus, $T_q(A) = PAQ^{-1}$ for all A in $(R_q)_n$ or $T_q(A) = PA^tQ^{-1}$ for all A in $(R_q)_n$ and, in particular, T_q preserves rank one matrices. This completes the proof of Theorem (G.1).

We assume for the remainder of this section that $T : (R)_n \to (R)_n$ preserves rank one matrices. Thus, for each prime ideal q of R, T_q also preserves rank one matrices and, consequently, has the form given by (F.5).

This also shows that T is an invertible linear mapping since at each localization T_q is invertible for each q. Further, $(T^{-1})_q = (T_q)^{-1}$ so the inverse T^{-1} of T will also preserve rank one matrices. Let E_{ij} denote the elementary matrix unit which has 1 in the (i,j)-positions and zeros elsewhere.

(G.2) THEOREM. (For the above setting.) There exist disjoint open-closed sets X_1 and X_2 of prime ideals such that $\mathrm{Spec}(R) = X_1 \cup X_2$ and

$$[T(E_{ij})]_q = \begin{cases} PE_{ij}Q^{-1} & \text{if } q \text{ is in } X_1 \\ PE_{ji}Q^{-1} & \text{if } q \text{ is in } X_2 \end{cases}$$

for $1 \leqslant i,j \leqslant n$.

PROOF. First observe that, since T is linear, if $A = [a_{ij}] = \sum_{i,j} a_{ij}E_{ij}$, then $[T(A)]_q = [T(\sum_{i,j} a_{ij}E_{ij})]_q = \sum_{i,j} \bar{a}_{ij}T_q(E_{ij})$. Thus, we know T by knowing its action on $\{E_{ij} \mid 1 \leq i,j \leq n\}$.

Since it is not possible (as is easily checked) for $P_1XQ_1 = P_2X^tQ_2$ for all X over a commutative ring, it is clear that $X_1 \cap X_2 = \emptyset$. Further, by (G.1) and (F.5) each q in $\mathrm{Spec}(R)$ belongs to either X_1 or X_2 . Hence, $\mathrm{Spec}(R) = X_1 \cup X_2$. It remains to show that each X_i is both open and closed. Here it suffices to show each X_i is open.

Suppose q is in X_1 . (A similar argument will show X_2 is open.) Then, if $A = T(E_{ij})$, we have $A_q = PE_{ij}Q^{-1}$ for P and Q invertible matrices in $(R_q)_n$. By "clearing denominators" we may assume $P = (\bar{P})_q$ and $Q^{-1} = (\bar{Q})_q$ for $\bar{P}$ and $\bar{Q}$ in $(R)_n$. Then, in $(R)_n$,

$$A = \bar{P}E_{ij}\bar{Q} + E$$

where E is in $(R)_n$ and $\lambda E = 0$ for some λ in $R - q$. Set $\hat{Q} = \lambda\bar{Q}$. Then

$$\begin{aligned}\bar{P}(E_{ij} + E)\hat{Q} &= \bar{P}E_{ij}\hat{Q} \\ &= \lambda\bar{P}E_{ij}\bar{Q} \\ &= \lambda A .\end{aligned}$$

Set

$$\Omega = \{\bar{q} \text{ in } \mathrm{Spec}(R) \mid \lambda\det(\bar{P}\bar{Q}) \text{ is not in } \bar{q}\} .$$

Thus, by Section (B), Ω is a basic open set in $\mathrm{Spec}(R)$:

$$\Omega = \Gamma(\lambda\det(\bar{P})\det(\bar{Q}))$$

where $\Gamma(a) = \{\bar{q} \text{ in } \mathrm{Spec}(R) \mid a \text{ not in } \bar{q}\}$. Further, q is in Ω . If $\bar{q}$ is in Ω , then

$$A_q = (\bar{P}E_{ij}\bar{Q})_q = \bar{P}_qE_{ij}\bar{Q}_q$$

where $\bar{P}_q$ and $\bar{Q}_q$ are invertible. Thus $\Omega \subset X_1$. Hence X_1 is open and

$X_2 = \mathrm{Spec}(R) - X_1$ is closed. This completes the proof.

It is well known, for example see Swan ([20], Theorem 7.12 and proof) and Section (B), that if $\mathrm{Spec}(R) = X_1 \cup X_2$ (disjoint) where X_i is both open and closed, then there are idempotents e and f in R with $1 = e+f$, $ef = 0$, and

$$X_1 = \Gamma(e) \ , \qquad X_2 = \Gamma(f) \ .$$

For this setting and the hypothesis of (G.2), let Ω_e be the idempotent map defined in Section (B).

Then, for $\Omega_e T$ we have

$$[(\Omega_e T)(E_{ij})]_q = PE_{ij}Q^{-1}$$

for all q in $\mathrm{Spec}(R)$ where P and Q are invertible matrices in $(R_q)_n$.

That is, by modifying T by a suitable Ω_e we may assume uniform behavior locally across the total spectrum of R.

Thus, without loss of generality, we may now <u>assume that</u> $T : (R)_n \to (R)_n$ is a linear map preserving rank one matrices and

$$[T(A)]_q = T_q(A_q) = PA_qQ^{-1}$$

for P, Q invertible in $(R_q)_n$ and for all prime ideals q of R.

For the above linear mapping T, set

$$U_T = \left\{ \langle A_1, A_2 \rangle \text{ in } (R)_n \oplus (R)_n \mid T(X)A_2 = A_1 X \text{ for all } X \text{ in } (R)_n \right\} .$$

It is easy to see that U_T is nonempty since it contains $\langle 0,0 \rangle$ and that U_T is an R-submodule of $(R)_n \oplus (R)_n$. We will show that U_T is an invertible R-submodule of $(R)_n \oplus (R)_n$.

(G.3) <u>PROPOSITION</u>. Let T and U_T be described as above. Then,

(a) U_T is an invertible R-submodule of $(R)_n \oplus (R)_n$.

(b) $(U_T)^{-1} = U_{T^{-1}}$.

(c) If $\bar{T} : (R)_n \to (R)_n$ is another rank one preserving linear map which is locally "uniform" across Spec(R) , then $U_T U_{\bar{T}} = U_{T\bar{T}}$.

PROOF. Consider the localization of U_T at a prime ideal q of R :

$$(U_T)_q = \{ < A_1, A_2 >_q \mid [T(X)A_2]_q = [A_1 X]_q \}$$
$$= \{ < \bar{A}_1, \bar{A}_2 > \mid P X_q Q^{-1} \bar{A}_2 = \bar{A}_1 X \}$$

where $\bar{A}_1 = (A_1)_q$ and $\bar{A}_2 = (A_2)_q$. Thus,

$$X_q Q^{-1} \bar{A}_2 = P^{-1} \bar{A}_1 X_q$$

for all X in $(R)_n$. Over a commutative ring, if $\bar{Y}A = B\bar{Y}$ for all $\bar{Y} = E_{ij}$, $1 \leqslant i,j \leqslant n$, then $A = \alpha I = B$. Thus,

$$Q^{-1} \bar{A}_2 = \alpha I = P^{-1} \bar{A}_1$$

for some α in R_q . Hence,

$$< A_1, A_2 >_q = < \bar{A}_1, \bar{A}_2 > = \alpha < P, Q >$$

and

$$(U_T)_q = \{ \alpha < P, Q > \mid \alpha \text{ in } R_q \text{ , } P \text{ and } Q \text{ fixed} \} .$$

It is easy to see that

$$(U_{T^{-1}})_q = \{ \alpha < P^{-1}, Q^{-1} > \mid \alpha \text{ in } R_q \} .$$

Suppose that $\bar{T}$ is another rank one preserving linear mapping of $(R)_n$ which is locally "uniform" across Spec(R) . Consider the product of the modules $U_{\bar{T}} U_T$. If $< A_1, A_2 >$ is in U_T and $< B_1, B_2 >$ is in $U_{\bar{T}}$, then for X in $(R)_n$

$$(\bar{T}T)(X) B_2 A_2 = T[T(X)] B_2 A_2$$
$$= B_1 T(X) A_2$$
$$= B_1 A_1 T(X) .$$

Thus $U_{\bar{T}}U_T \subseteq U_{\bar{T}T}$. In particular, $U_T U_{T^{-1}} \subseteq U_{TT^{-1}} = \bar{E}$. But, locally, it is an easy calculation to show

$$(U_T U_{T^{-1}})_q = (U_T)_q (U_{T^{-1}})_q = \bar{E}_q = (\bar{E})_q$$

for each q in $\mathrm{Spec}(R)$. Hence $U_T U_{T^{-1}} = E$ and U_T is invertible. Then, by (D.5), $U_{\bar{T}}U_T = U_{T\bar{T}}$.

(G.4) PROPOSITION. Let $T : (R)_n \to (R)_n$ and $\bar{T} : (R)_n \to (R)_n$ be as above. Then, the following are equivalent:

(a) $T = \bar{T}$.

(b) $U_T = U_{\bar{T}}$.

PROOF. Certainly (a) implies (b). Assume that $U_T = U_{\bar{T}}$. Then, for all X in $(R)_n$,

$$\{\langle A_1, A_2\rangle \mid T(X)A_2 = A_1 X\} = \{\langle A_1, A_2\rangle \mid \bar{T}(X)A_2 = A_1 X\} .$$

Then, $T(X)A_2 - \bar{T}(X)A_2 = 0$ for all X in $(R)_n$. Thus, for each element A_2 which is a projection of $\langle A_1, A_2\rangle$ of U_T on the second coordinate,

$$[T(X) - \bar{T}(X)]A_2 = 0 .$$

But, by Section (D), the projection U_2 of $U_T = U_{\bar{T}}$ onto its second coordinate produces an invertible R-submodule of $(R)_n$. Hence

$$[T(X) - \bar{T}(X)]U_2 = 0 .$$

Select U_1 with $U_2 U_1 = E = RI = \{\alpha I \mid \alpha \text{ in } R\}$. Then $[T(X) - \bar{T}(X)]R = 0$, that is, $T(X) - \bar{T}(X) = 0$ for all X in $(R)_n$. Hence $T = \bar{T}$.

Thus, each rank one preserving linear map $T : (R)_n \to (R)_n$ which is "uniform" across $\mathrm{Spec}(R)$ determines a unique invertible R-submodule U_T of $(R)_n \oplus (R)_n$.

On the other hand, each invertible R-submodule U of $(R)_n \oplus (R)_n$ determines an element E_U of $\mathrm{EQUIV}_n(R)$ (see Section (D)) where $(E_U)_q$ is an equivalence transformation for each q in $\mathrm{Spec}(R)$. Thus, $(E_U)_q$ preserves rank one matrices and, by (G.1), E_U preserves rank one matrices. By checking locally at each prime ideal, it is now easy to show that

$$E_{U_T} = T .$$

Thus, the rank one preserving linear maps of $(R)_n$ which are locally "uniform" across $\mathrm{Spec}(R)$ are exactly $\mathrm{EQUIV}_n(R)$. If T is any element of $\mathrm{RANK}_n(R)$, then there is an idempotent e (by (G.2)) such that $\Omega_e T$ is in $\mathrm{EQUIV}_n(R)$. Thus, $\mathrm{RANK}_n(R) = \mathrm{Inv}(R)\,\mathrm{EQUIV}_n(R)$.

The results of this and the previous sections are summarized in the next theorem which is the central result of this section and the paper. The proof of this theorem is now complete.

(G.5) THEOREM. (Rank One Preservers) The following diagram having exact rows and split exact columns is commutative:

$$\begin{array}{ccccccccc}
 & & 1 & & 1 & & 1 & & \\
 & & \downarrow & & \downarrow & & \downarrow & & \\
1 & \longrightarrow & \mathrm{Equiv}_n(R) & \xrightarrow{\text{inj}} & \mathrm{EQUIV}_n(R) & \xrightarrow{\rho} & \mathrm{Pic}_n(R) & \longrightarrow & 1 \quad \text{(exact)} \\
 & & \text{inj}\downarrow & & \text{inj}\downarrow & & \text{id}\downarrow & & \\
1 & \longrightarrow & \mathrm{Rank}_n(R) & \xrightarrow{\text{inj}} & \mathrm{RANK}_n(R) & \xrightarrow{\bar{\rho}} & \mathrm{Pic}_n(R) & \longrightarrow & 1 \quad \text{(exact)} \\
 & & \downarrow & & \downarrow & & \downarrow & & \\
1 & \longrightarrow & \mathrm{Inv}(R) & \xrightarrow{\text{id}} & \mathrm{Inv}(R) & \longrightarrow & 1 & & \text{(exact)} \\
 & & \downarrow & & \downarrow & & & & \\
 & & 1 & & 1 & & & & \\
 & & \text{(split exact)} & & \text{(split exact)} & & & &
\end{array}$$

The groups appearing are as follows:

(a) $\mathrm{Equiv}_n(R) = \{E_{(P,Q)} \mid P, Q \text{ invertible in } (R)_n\}$, i.e., the group of "standard" equivalence transformations. (Section (C).)

(b) $\mathrm{EQUIV}_n(R) = \{E_U \mid U \text{ an invertible } R\text{-submodule of } (R)_n \oplus (R)_n\}$, i.e., the group of "generalized" equivalence transformations. (Section (D).)

(c) $\mathrm{Pic}_n(R) = \mathfrak{H}/\mathfrak{H}_0$ = the subgroup of the n-torsion subgroup of $\mathrm{Pic}(R)$ consisting of all $[P]$ with $P^n \simeq R^n$. (Section (D).)

(d) $\mathrm{Inv}(R) = \{\Omega_e \mid e$ an idempotent of $R\}$. (Section (B).)

(e) $\mathrm{Rank}_n(R) = \{E_{(P,Q)}\Omega_e \mid E_{(P,Q)}$ in $\mathrm{Equiv}_n(R)$, Ω_e in $\mathrm{Inv}(R)\}$, i.e., the group of "standard" rank one preserving linear mappings of $(R)_n$. (Section (E).)

(f) $\mathrm{RANK}_n(R) = \{E_U\Omega_e \mid E_U$ in $\mathrm{EQUIV}_n(R)$, Ω_e in $\mathrm{Inv}(R)\}$, i.e., the group of all rank one preserving linear mappings of $(R)_n$. (Section (G).)

The mappings ρ and $\bar{\rho}$ are:

$$\rho : E_U \to [U] \quad \text{(isomorphism class)}$$

and

$$\bar{\rho} : E_U\Omega_e \to [U] \quad \text{(isomorphism class)} .$$

(H) APPLICATIONS

In this section, we apply Theorem (G.5) to obtain descriptions of various groups which preserve specified invariants of $(R)_n$. The central idea is that preservation of rank one matrices is a feature possessed by other invariant preserving linear transformations. Once this is shown, then the groups of invariant preservers may be deduced from Theorem (G.4). If R is a field this approach is discussed by Marcus in [10] and Minc in [17].

(H.1) THEOREM. Suppose that $T : (R)_n \to (R)_n$ is an R-linear mapping that preserves any of the following:

(a) Determinants, i.e., $\det(T(A)) = \det(A)$

(b) Determinantal ideals, i.e., $F_K(T(A)) = F_K(A)$ for $K = 1,2,\ldots$

(c) Characteristic polynomials, i.e., $\chi(T(A),X) = \chi(A,X)$ where $\chi(A,X)$ is the characteristic polynomial of A, or

(d) T is an R-algebra automorphism.

Then T preserves rank one matrices.

PROOF. Part (a) is the most difficult part of the above result. We begin a proof of (a). Assume $T : (R)_n \to (R)_n$ is an R-linear mapping with the property that $\det(T(A)) = \det(A)$ for all A in $(R)_n$.

(H.2) LEMMA. Let $T : (R)_n \to (R)_n$ be an R-linear mapping. The following are equivalent:

(a) T preserves determinants.

(b) T_q preserves determinants for each prime ideal q of R.

(c) T_m preserves determinants for each maximal ideal m of R.

PROOF. Clearly (b) implies (c). To show (c) implies (a), suppose T_q preserves determinants for each maximal ideal q. Then

$$[\det(T(A)) - \det(A)]_q = \det(T_q(A_q)) - \det(A_q) = 0$$

for each q. Hence $\det(T(A)) = \det(A)$ and T preserves determinants.

To show (a) implies (b), assume T preserves determinants. Let $\bar{A}$ be in $(R_q)_n$ for q a prime ideal. Then "clearing denominators" there is a scalar λ in $R-q$ with $\lambda\bar{A} = A$ for A in $(R)_n$.

Then,

$$\begin{aligned} \lambda^n \det(\bar{A}) &= \det(\lambda\bar{A}) \\ &= \det(A) \\ &= \det(T(A)) \\ &= \det(T(\lambda\bar{A})) \\ &= \lambda^n \det(T(\bar{A})). \end{aligned}$$

Then, locally, $\lambda^n \det(\bar{A}) = \lambda^n \det(T_q(\bar{A}))$ and λ^n is a unit in R_q. Hence, $\det(\bar{A}) = \det(T_q(\bar{A}))$.

We now return to the proof of Theorem (H.1).

By (H.2) and (G.1), it suffices to show that if $T_q : (R_q)_n \to (R_q)_n$ preserves determinants, then it preserves rank one matrices. Thus, assume R is a local commutative ring with maximal ideal m and residue field $k = R/m$. Assume $T : (R)_n \to (R)_n$ is an R-linear map preserving determinants. Let $\pi : R \to R/m = k$ be the natural mapping. Define $\pi T : (k)_n \to (k)_n$ by $\pi T(\pi A) = \pi(T(A))$. Then

$$\begin{aligned} \det[\pi T(\pi A)] &= \det \pi(T(A)) \\ &= \pi \det(T(A)) \\ &= \pi \det(A) \\ &= \det(\pi A). \end{aligned}$$

Thus, πT preserves determinants.

We claim that πT is invertible. This is the first part of the proof of Theorem 2 of [17]. For completeness, we present the proof here. Suppose

$\bar{A}$ is in $(k)_n$ and $\pi T(\bar{A}) = 0$. Then, there is a matrix $\bar{B}$ such that $\bar{A} + \bar{B}$ is invertible in $(k)_n$ and $\text{rank}(\bar{B}) = n - \text{rank}(\bar{A})$. Then,

$$\begin{aligned} \det(\pi T(\bar{B})) &= \det(\pi T(\bar{A}) + \pi T(\bar{B})) \\ &= \det(\pi T(\bar{A} + \bar{B})) \\ &= \det(\bar{A} + \bar{B}) \\ &\neq 0 . \end{aligned}$$

Thus, $\det(\bar{B}) \neq 0$ and $\bar{B}$ is invertible. Then, $\text{rank}(\bar{A}) = 0$ and thus $\bar{A} = 0$. Therefore, the k-linear map πT is injective and thus surjective, i.e., πT is invertible.

Suppose A in $(R)_n$ has rank one. Then πA has rank one in $(k)_n$. Therefore $\pi T(\pi A) \neq 0$ since πT is invertible and $\pi T(\pi A) = \pi(T(A))$ has nonzero rank. Thus, over the local ring R there exist invertible matrices U and V such that

$$T(A) = U \begin{bmatrix} 1 & 0 & \cdots & 0 \\ 0 & & & \\ \vdots & & \bar{A} & \\ 0 & & & \end{bmatrix} V .$$

Set

$$B = U \begin{bmatrix} 0 & 0 & \cdots & 0 \\ 0 & & & \\ \vdots & & \bar{B} & \\ 0 & & & \end{bmatrix} V .$$

Then, for an indeterminate X,

$$\begin{aligned} \det(XT(A) + B) &= \det(UV) \det \begin{bmatrix} X & 0 & \cdots & 0 \\ 0 & & & \\ \vdots & & X\bar{A} + \bar{B} & \\ 0 & & & \end{bmatrix} \\ &= \det(UV) \det(X\bar{A} + \bar{B}) X . \end{aligned}$$

On the other hand, if

$$PAQ = \begin{bmatrix} 1 & 0 & \cdots & 0 \\ 0 & & & \\ \vdots & & 0 & \\ 0 & & & \end{bmatrix}$$

then,

$$\begin{aligned} \det(XT(A) + B) &= \det(T(XA + T^{-1}(B))) \\ &= \det(XA + T^{-1}(B)) \\ &= \det(PQ)\det(XE_{11} + P^{-1}T^{-1}(B)Q^{-1}) \\ &= \alpha X + \beta \end{aligned}$$

which is a linear polynomial in X. Thus, $\det(X\bar{A} + \bar{B})$ is in R for all choices of the $(n-1) \times (n-1)$ block $\bar{B}$. Take $\bar{B}$ to be invertible. Then $\det(X\bar{A}\bar{B}^{-1} + I)$ is in R for all $\bar{B}$ in $GL_{n-1}(R)$. Equivalently, $\det(X\bar{A}\bar{P} + I)$ is in R for all $\bar{P}$ in $GL_{n-1}(R)$. Recall that the "trace polynomial" of a matrix D in $(R)_{n-1}$ is

$$\det(XD + I) = \sum_{i=0}^{n-1} Tr(\Lambda^i(D))X^i$$

where Tr denotes the trace and Λ^i the i-th exterior power. Thus, in our setting,

$$\sum_{i=0}^{n-1} Tr(\Lambda^i(\bar{A}\bar{P}))X^i$$

is a scalar in R for all invertible $\bar{P}$. Therefore,

$$0 = Tr(\Lambda^1(\bar{A}\bar{P})) = Tr(\bar{A}\bar{P})$$

for all $\bar{P}$ in $GL_{n-1}(R)$. But the trace mapping is a nondegenerate R-bilinear form on $(R)_{n-1} \oplus (R)_{n-1}$. Hence $\bar{A} = 0$ and $T(A)$ has rank one. Thus, if T preserves determinants then T preserves rank one matrices.

Part (b) is immediate since rank one matrices are characterized by their determinantal ideals.

Part (c) is also immediate since, if characteristic polynomials are preserved, then their constant coefficients are preserved and these constant coefficients are precisely the determinants of the given matrix.

Finally, suppose T is an R-algebra automorphism of $(R)_n$. Then T_q, for q a prime ideal of R, is an R_q-algebra automorphism of $(R_q)_n$. If A has rank one in $(R_q)_n$, then $PE_{11}Q = A$ for invertible P and Q. Then $T_q(A) = T_q(PE_{11}Q) = T_q(P)T_q(E_{11})T_q(Q)$ and it is an easy exercise to show that over a local ring R_q an R_q-algebra automorphism carries a set of matrix units to a set of matrix units. Hence, $T_q(E_{11})$ is a matrix unit and, consequently, has rank one.

In order to illustrate our remark before (H.1), we examine determinant preservers in greater detail. This is an old problem which dates at least to Frobenius [1] in 1897. Indeed, our interest in linear mappings of $(R)_n$ arose because of D. G. James' classification [8] in 1980 of determinant preserving mappings of $(R)_n$ where R is an integral domain. W. Waterhouse [21] by different techniques also recently classified determinant preserving linear mappings of $(R)_n$ where R is a commutative ring.

Let

$$\mathrm{DET}_n(R) = \left\{T : (R)_n \to (R)_n \mid T \text{ is R-linear and preserves determinants}\right\}.$$

If T is in $\mathrm{DET}_n(R)$, then the proof of (H.1) shows that T is invertible since T_q is invertible for each prime ideal q of R. Further, (H.1) indicates that $\mathrm{DET}_n(R)$ is a subset of $\mathrm{RANK}_n(R)$ - this also implies that T is invertible.

Thus, we examine the elements of $\mathrm{RANK}_n(R)$ to determine which preserve determinants.

First, note that $\det[\Omega_e(A)] = \det(A)$ for any A in $(R)_n$ so that $\mathrm{Inv}(R)$ is contained in $\mathrm{DET}_n(R)$.

If $E_{(P,Q)}$ from $\mathrm{Equiv}_n(R)$ is in $\mathrm{DET}_n(R)$, then

$$\begin{aligned}\det[E_{(P,Q)}(A)] &= \det(PAQ^{-1})\\ &= \det(PQ^{-1})\det(A)\\ &= \det(A)\end{aligned}$$

for all A implies that $\det(PQ^{-1}) = 1$ or that $\det(P) = \det(Q)$. Let

$$\mathrm{Det}_n(R) = \left\{\Omega_e E_{(P,Q)} \mid e \text{ idempotent}, \det(P) = \det(Q)\right\} .$$

Then $\mathrm{Det}_n(R)$ is a subgroup of $\mathrm{DET}_n(R)$.

Suppose U is an invertible R-submodule of $(R)_n \oplus (R)_n$. Then, there exist $u_i, \bar{u}_i$ in U and $v_i, \bar{v}_i$ in U^{-1} with $\sum_i \langle u_i,\bar{u}_i\rangle\langle v_i,\bar{v}_i\rangle = \langle I,I\rangle$. The mapping E_U in $\mathrm{EQUIV}_n(R)$ was given by

$$E_U(A) = \sum_i u_i A\bar{v}_i$$

for A in $(R)_n$. Let q be a prime ideal in R . Then, by Section (D), $(u_i)_q = \alpha_i P$, $(\bar{v}_i)_q = \beta_i Q^{-1}$ where $\sum_i \alpha_i\beta_i = 1$ and P and Q are invertible matrices in $(R_q)_n$. For E_U to be in $\mathrm{DET}_n(R)$, we must have $(E_U)_q$ in $\mathrm{DET}_n(R_q)$. Thus, $\det(P) = \det(Q)$. Hence, for each q ,

$$[\det(\textstyle\sum_i u_i\bar{v}_i)]_q = \det(PQ^{-1}) = 1 .$$

Hence, $\det(\sum_i u_i\bar{v}_i) = 1$. Also, since $\det(\sum_i u_i\bar{v}_i) = 1$ is locally sufficient to place $(E_U)_q$ in $\mathrm{DET}_n(R_q)$, it is sufficient.

(H.3) <u>LEMMA</u>. (For the above notation.) The mapping E_U of $\mathrm{EQUIV}_n(R)$ is in $\mathrm{DET}_n(R)$ if and only if $\det(\sum u_i\bar{v}_i) = 1$.

It is also easy to check that this condition is independent of the sets $\{u_i\}$ $\{\bar{v}_i\}$ representing E_U .

Thus,

$$\mathrm{DET}_n(R) = \{\Omega_e E_U \mid e \text{ idempotent and } \det(\textstyle\sum u_i\bar{v}_i) = 1\}$$

where $E_U(A) = \sum u_i A\bar{v}_i$. In particular, $\mathrm{DET}_n(R)$ is a subgroup of $\mathrm{RANK}_n(R)$ and, it is easily checked, a normal subgroup.

Also, $\mathrm{Det}_n(R)$ is a normal subgroup of $\mathrm{DET}_n(R)$. We next compute $\mathrm{DET}_n(R)/\mathrm{Det}_n(R)$.

Suppose E_U is in $\mathrm{EQUIV}_n(R)$. Then, we have a natural map $E_U \to [U]$

(isomorphism class of U) which takes $\mathrm{EQUIV}_n(R)$ to $\mathrm{Pic}_n(R)$. Invertible R-submodules U and V of $(R)_n \oplus (R)_n$ determine the same isomorphism class of $\mathrm{Pic}_n(R)$ if and only if $U = \langle P_1, P_2 \rangle V$ where P_1 and P_2 are invertible (Theorem (D.6)). Suppose E_U is given by

$$E_U(A) = \textstyle\sum_i u_i A \bar{v}_i$$

where $\det(\sum_i u_i \bar{v}_i) = \alpha$. Note that α is a unit in R . Since $\det : GL_n(R) \to R^0$ is surjective, select an invertible matrix P with $\det(P) = \alpha^{-1}$. Consider the invertible R-submodule $V = \langle P, I \rangle U$. Then

$$E_V(A) = \textstyle\sum_i P u_i A \bar{v}_i$$

and $\det(\sum_i P u_i \bar{v}_i) = \det P(\sum u_i \bar{v}_i) = \det(P)\det(\sum_i u_i \bar{v}_i) = \alpha^{-1}\alpha = 1$. Thus, E_V is in $\mathrm{DET}_n(R)$ and $[V] = [U]$. Hence, for every $[U]$ in $\mathrm{Pic}_n(R)$, there is an E_V in $\mathrm{DET}_n(R)$ with $E_V \to [V] = [U]$.

This discussion gives the following theorem.

(H.4) THEOREM. (Determinant Preservers.) The following diagram, having exact rows and split exact columns, is commutative:

$$\begin{array}{ccccccccc}
 & & 1 & & 1 & & 1 & & \\
 & & \downarrow & & \downarrow & & \downarrow & & \\
1 & \longrightarrow & \mathrm{Equiv}_n^1(R) & \xrightarrow{\text{inj}} & \mathrm{EQUIV}_n^1(R) & \xrightarrow{\rho} & \mathrm{Pic}_n(R) & \longrightarrow & 1 \quad \text{(exact)} \\
 & & \text{inj}\downarrow & & \text{inj}\downarrow & & \text{id}\downarrow & & \\
1 & \longrightarrow & \mathrm{Det}_n(R) & \xrightarrow{\text{inj}} & \mathrm{DET}_n(R) & \xrightarrow{\bar{\rho}} & \mathrm{Pic}_n(R) & \longrightarrow & 1 \quad \text{(exact)} \\
 & & \downarrow & & \downarrow & & \downarrow & & \\
1 & \longrightarrow & \mathrm{Inv}(R) & \xrightarrow{\text{id}} & \mathrm{Inv}(R) & \longrightarrow & 1 & & \\
 & & \downarrow & & \downarrow & & & & \\
 & & 1 & & 1 & & & & \\
 & & \text{(split exact)} & & \text{(split exact)} & & & &
\end{array}$$

where

(a) $\mathrm{Equiv}_n^1(R) = \{E_{(P,Q)} \mid \det(P) = \det(Q)\}$

(b) $\mathrm{EQUIV}_n^1(R) = \{E_U \mid \det(\sum u_i\bar{v}_i) = 1 \text{ where } E_U(A) = \sum_i u_iA\bar{v}_i\}$.

It is easy to see that $\mathrm{Equiv}_n^1(R)$ is a normal subgroup of $\mathrm{Equiv}_n(R)$. Further, there is a natural group morphism $\tau : \mathrm{Equiv}_n(R) \to R^0$ (units of R) given by $\tau(E_{(P,Q)}) = \det(PQ^{-1})$. The kernel of τ is $\mathrm{Equiv}_n^1(R)$ which gives an exact sequence

$$1 \longrightarrow \mathrm{Equiv}_n^1(R) \longrightarrow \mathrm{Equiv}_n(R) \xrightarrow{\tau} R^0 \longrightarrow 1$$

which splits under $\alpha \to E_{(\alpha I, I)}$.

If E_U is in $\mathrm{EQUIV}_n(R)$ with $E_U(A) = \sum_i u_iA\bar{v}_i$ for all A in $(R)_n$, then we have

$$\bar{\tau} : \mathrm{EQUIV}_n(R) \to R^0$$

by

$$\bar{\tau}(E_U) = \det(\textstyle\sum_i u_i\bar{v}_i) .$$

This mapping has kernel $\mathrm{EQUIV}_n^1(R)$ and is split. Thus, it is easy to see how $\mathrm{DET}_n(R)$ sits in $\mathrm{RANK}_n(R)$. We have the following corollary describing the above remarks. This corollary connects the diagram of Theorem (H.5) to that of Theorem (G.5).

(H.5) COROLLARY. (For the above notation.) The following sequences of groups are split exact:

(a) $1 \longrightarrow \mathrm{Equiv}_n^1(R) \xrightarrow{\text{inj}} \mathrm{Equiv}_n(R) \xrightarrow{\tau} R^0 \longrightarrow 1$,

(b) $1 \longrightarrow \mathrm{EQUIV}_n^1(R) \xrightarrow{\text{inj}} \mathrm{EQUIV}_n(R) \xrightarrow{\bar{\tau}} R^0 \longrightarrow 1$,

(c) $1 \longrightarrow \mathrm{DET}_n(R) \xrightarrow{\text{inj}} \mathrm{RANK}_n(R) \xrightarrow{\bar{\tau}} R^0 \longrightarrow 1$,

(d) $1 \longrightarrow \mathrm{Det}_n(R) \xrightarrow{\text{inj}} \mathrm{Rank}_n(R) \xrightarrow{\bar{\tau}} R^0 \longrightarrow 1$,

where R^0 denotes the units of R and

$$\tau(E_{(P,Q)}) = \det(PQ^{-1}) ,$$

$$\bar{\tau}(E_U) = \det(\textstyle\sum_i u_i\bar{v}_i) \quad \text{if} \quad E_U(A) = \sum u_iA\bar{v}_i ,$$

$$\bar{\tau}(\Omega_e E_U) = \bar{\tau}(E_U) , \text{ and}$$

$$\bar{\tau}(\Omega_e E_{(P,Q)}) = \tau(E_{(P,Q)}) = \det(PQ^{-1}) .$$

We next examine the preservation of determinantal ideals.

Let A be in $(R)_n$. The j-th determinantal ideal $F_j(A)$, $1 \leq j \leq n$, of A is the ideal generated by the determinants of all $j \times j$ submatrices of A. Suppose $T : (R)_n \to (R)_n$ preserves determinantal ideals. Since a rank one matrix A is characterized by the fact that $F_1(A) = R$ and $F_2(A) = 0$ (see Section (E)), it is clear that T also preserves rank one matrices. Thus, it is only necessary to see if preservation of determinantal ideals imposes additional constraints on T as was the case with preservation of determinants.

However, the elements which generate $F_j(A)$ are precisely the elements of $C_j(A)$ = the j-th compound matrix of A. (See, for example, [11], pp. 16-17.) The Binet-Cauchy theorem, stating that $C_j(AB) = C_j(A)C_j(B)$ together with the fact that $C_j(P)^{-1} = C_j(P^{-1})$ indicate that equivalence transformations preserve determinantal ideals. Further, for the transpose we have $C_j(A)^t = C_j(A^t)$. Using these facts and straightforward localization arguments, one sees that all elements in $\mathrm{RANK}_n(R)$ preserve determinantal ideals.

Before we examine the linear mappings $T : (R)_n \to (R)_n$ which preserve the characteristic polynomial of a matrix, we discuss the R-algebra automorphisms of $(R)_n$. Here we generalize the question slightly and place the formulation of the solution in our context. If R is a commutative ring, then the R-algebra automorphisms of $(R)_n$ are well known. (In the case that R is a field, a consequence of the Skolem-Noether theorem is that every algebra automorphism is inner.) Indeed, if P is a finitely generated faithful projective R-module, then Rosenberg and Zelinsky [19] classified in 1961 the R-algebra automorphisms of $\mathrm{End}_R(P)$. Issacs in [6] provides an excellent discussion and solution of this problem for $(R)_n$ in terms of invertible R-submodules of $(R)_n$. Isaacs' formulation heavily influenced this work.

Using Isaacs' description, we have an exact sequence

$$1 \to \mathrm{Inn}_n(R) \to \mathrm{Aut}_n(R) \to \mathrm{Pic}_n(R) \to 1$$

where $\mathrm{Inn}_n(R)$ is the group of inner R-automorphisms of $(R)_n$ of the form I_P where $I_P(A) = PAP^{-1}$ for P invertible; and, $\mathrm{Aut}_n(R)$ is the group I_U for U an invertible R-submodule <u>of</u> $(R)_n$ where $I_U(A) = \sum_i u_i A \bar{u}_i$ for u_i in U and $\bar{u}_i$ in U^{-1} with $\sum u_i \bar{u}_i = \bar{1}$.

In the context of this paper, a standard inner-automorphism I_P would be identified with $E_{(P,P)}$ where $E_{(P,P)}(A) = PAP^{-1}$. Thus,

$$\mathrm{Inn}_n(R) = \{E_{(P,P)} \mid P \text{ is invertible in } (R)_n\} .$$

To retain a uniform notation, we will denote $\mathrm{Aut}_n(R)$ by $\mathrm{INN}_n(R)$. In our context, if U is an invertible R-submodule of $(R)_n$, then U determines a twisted-plane $U \oplus U$. The element of $\mathrm{INN}_n(R)$ corresponding to I_U of $\mathrm{Aut}_n(R)$ is determined by the R-submodule $\bar{U} = \{\langle u,u\rangle \mid u \text{ in } U\}$ of $(R)_n \oplus (R)_n$ which is the <u>diagonal</u> of the twisted plane $U \oplus U$. Thus, translating [6], we find an exact sequence

$$1 \to \mathrm{Inn}_n(R) \to \mathrm{INN}_n(R) \to \mathrm{Pic}_n(R) \to 1$$

where

$$\mathrm{Inn}_n(R) = \{E_{(P,P)} \mid P \text{ invertible in } (R)_n\}$$

is a normal subgroup of $\mathrm{Equiv}_n(R)$ and

$$\mathrm{INN}_n(R) = \left\{E_{\bar{U}} \mid \bar{U} = \{\langle u,u\rangle \mid u \text{ in } U,\ U \text{ invertible R-submodule of } (R)_n\}\right\}$$

is a subgroup of $\mathrm{EQUIV}_n(R)$.

To generalize the context slightly, and in doing so place it in a setting which utilizes $\mathrm{Inv}(R)$, recall that an algebra anti-isomorphism of $(k)_n$ (k a field) has the form $A \to PA^tP^{-1}$ (for example, see Jacobson [7], pp. 23-24). Thus, over a commutative ring, E_U (for U an invertible R-submodule of $(R)_n \oplus (R)_n$) determines an R-algebra automorphism of $(R)_n$ and $\Omega_e E_U$, for $e = 0$, produces an R-algebra anti-automorphism of $(R)_n$. Recall that Ω_0 is the transpose map.

Thus, we define the <u>R-skew automorphisms</u>, denoted $SAUT_n(R)$, of $(R)_n$ as the subgroup $Inv(R)INN_n(R)$ of $RANK_n(R)$. Similarly, we denote by $Saut_n(R)$ the subgroup $Inv(R)Inn_n(R)$. Then, those elements of the form $\Omega_1 E_U$ of $SAUT_n(R)$ determine the subgroup $INN_n(R)$ and those elements of the form $\Omega_0 E_U$ determine the R-algebra anti-automorphisms of $(R)_n$. We have

$$1 \to Saut_n(R) \to SAUT_n(R) \to Pic_n(R) \to 1$$

exact. This is summarized in the following result.

(H.6) PROPOSITION. (Automorphisms) The following diagram, having exact rows and split exact columns, is commutative:

$$\begin{array}{ccccccccc}
 & & 1 & & 1 & & 1 & & \\
 & & \downarrow & & \downarrow & & \downarrow & & \\
1 & \longrightarrow & Inn_n(R) & \longrightarrow & INN_n(R) & \longrightarrow & Pic_n(R) & \longrightarrow & 1 \\
 & & \downarrow & & \downarrow & & \downarrow & & \\
1 & \longrightarrow & Saut_n(R) & \longrightarrow & SAUT_n(R) & \longrightarrow & Pic_n(R) & \longrightarrow & 1 \\
 & & \downarrow & & \downarrow & & \downarrow & & \\
1 & \longrightarrow & Inv(R) & \longrightarrow & Inv(R) & \longrightarrow & 1 & & \\
 & & \downarrow & & \downarrow & & & & \\
 & & 1 & & 1 & & & &
\end{array}$$

The purpose for the above discussion is to determine the subgroup of $RANK_n(R)$ whose elements preserve the characteristic polynomial $\chi(A,X)$ of a matrix A. Recall that

$$\chi(A,X) = X^n - Tr(A)X^{n-1} + \cdots + (-1)^n \det(A) ,$$

so that, if $T : (R)_n \to (R)_n$ and $\chi(T(A),x) = \chi(A,X)$, then T preserves the determinant and the trace. Hence, T belongs to $RANK_n(R)$. Over a field k, a linear mapping $T : (k)_n \to (k)_n$ preserves both trace and determinant if and only if either $T(A) = PAP^{-1}$ for all A or $T(A) = PA^tP^{-1}$ for all A [16].

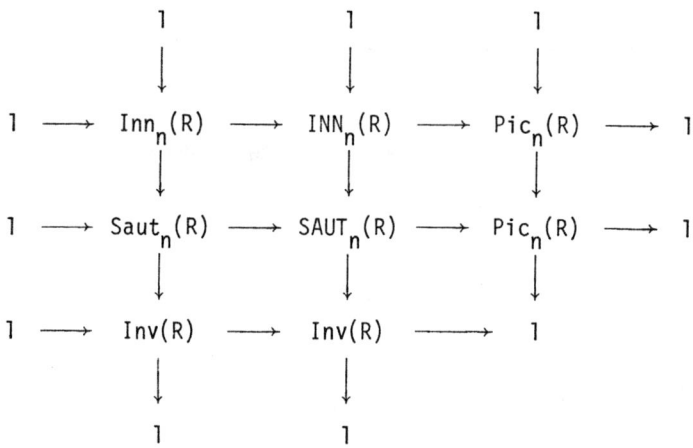

Since the trace map is R-linear, it localizes nicely and, using (H.2), it is not difficult to see that T preserves trace and determinant if and only if T_q preserves trace and determinant for each prime ideal q of R. Then it is straightforward to verify that those linear mappings in $RANK_n(R)$ which preserve characteristic polynomials are exactly $SAUT_n(R)$.

REFERENCES

1. G. Frobenius, Über die Darstellung der endlichen Gruppen durch lineare Substitutionen, S.-B. Preuss. Akad. Will. Berlin (1897), 994-1015.

2. R. Gilmer and R. Heitmann, On Pic(R[X]) for R Seminormal, J. Pure Appl. Alg. 16 (1980), 251-257.

3. R. Grone, Isometries of Matrix Algebras, Ph.D. Thesis, Univ. of California, Santa Barbara.

4. L. K. Hua, A theorem on matrices over a field and its applications, J. Chinese Math. Soc. (N. S.) 1 (1951), 110-163.

5. E. Ingraham and F. DeMeyer, Separable Algebras Over Commutative Rings. Vol. 181, Lecture Notes in Math., Springer-Verlag, 1971.

6. I. M. Issacs, Automorphisms of matrix algebras over commutative rings, Linear Alg. and Appl. 31 (1980), 215-231.

7. N. Jacobson, The Theory of Rings. Math. Surveys, Number II, Amer. Math. Soc. (1943).

8. D. James, On the automorphisms of $\det(X_{ij})$, Math. Chronicle 9 (1980), 35-40.

9. M. Marcus, Linear operations on matrices, Amer. Math. Monthly 69 (1962), 837-847.

10. ______, Linear transformations on matrices, J. Nat. Bureau Standards 75B (1971), 107-113.

11. M. Marcus and H. Minc, A Survey of Matrix Theory and Matrix Inequalities. Allyn and Bacon, Boston (1964).

12. M. Marcus and B. Moyls, Transformations on tensor product spaces, Pacific J. of Math. 9 (1959), 1215-1221.

13. B. McDonald, Automorphisms of $GL_n(R)$, Trans. Amer. Math. Soc. 246 (1978), 155-171.

14. B. McDonald, Linear Algebra Over Commutative Rings. To appear: Monographs in Pure and Applied Math., Marcel Dekker, Inc.

15. B. McDonald and W. Waterhouse, Projective modules over rings with many units, Proc. Amer. Math. Soc. 83 (1981), 455-458.

16. H. Minc, The invariance of elementary symmetric functions, Linear and Multilinear Alg. 4 (1976), 209-215.

17. _______, Linear transformations on matrices: Rank 1 preservers and determinant preservers, Linear and Multilinear Alg. 4 (1977), 265-272.

18. D. Northcott, Lessons on Rings, Modules and Multiplicities. Cambridge Univ. Press, Cambridge (1968).

19. A. Rosenberg and D. Zelinsky, Automorphisms of separable algebras, Pacific J. Math. 11 (1961), 1109-1117.

20. R. G. Swan, Algebraic K-theory. Lecture Notes in Math., #76, Springer-Verlag (1968).

21. W. C. Waterhouse, Automorphisms of $\det(X_{ij})$: The Group Scheme Approach, to appear: Adv. in Math.

22. W. Wong, Maps on simple algebras preserving zero products I: The associative case, Pacific J. Math. 89 (1980), 229-247.

Author's permanent address:

Department of Mathematics
The University of Oklahoma
Norman, Oklahoma 73019

General instructions to authors for
PREPARING REPRODUCTION COPY FOR MEMOIRS

For more detailed instructions send for AMS booklet, "A Guide for Authors of Memoirs." Write to Editorial Offices, American Mathematical Society, P. O. Box 6248, Providence, R. I. 02940.

MEMOIRS are printed by photo-offset from camera copy fully prepared by the author. This means that, except for a reduction in size of 20 to 30%, the finished book will look exactly like the copy submitted. Thus the author will want to use a good quality typewriter with a new, medium-inked black ribbon, and submit clean copy on the appropriate model paper.

Model Paper, provided at no cost by the AMS, is paper marked with blue lines that confine the copy to the appropriate size. Author should specify, when ordering, whether typewriter to be used has PICA-size (10 characters to the inch) or ELITE-size type (12 characters to the inch).

Line Spacing – For best appearance, and economy, a typewriter equipped with a half-space ratchet – 12 notches to the inch – should be used. (This may be purchased and attached at small cost.) Three notches make the desired spacing, which is equivalent to 1-1/2 ordinary single spaces. Where copy has a great many subscripts and superscripts, however, double spacing should be used.

Special Characters may be filled in carefully freehand, using dense black ink, or INSTANT ("rub-on") LETTERING may be used. AMS has a sheet of several hundred most-used symbols and letters which may be purchased for $5.

Diagrams may be drawn in black ink either directly on the model sheet, or on a separate sheet and pasted with rubber cement into spaces left for them in the text. Ballpoint pen is *not* acceptable.

Page Headings (Running Heads) should be centered, in CAPITAL LETTERS (preferably), at the top of the page – just above the blue line and touching it.

LEFT-hand, EVEN-numbered pages should be headed with the AUTHOR'S NAME;
RIGHT-hand, ODD-numbered pages should be headed with the TITLE of the paper (in shortened form if necessary).
Exceptions: PAGE 1 and any other page that carries a display title require NO RUNNING HEADS.

Page Numbers should be at the top of the page, on the same line with the running heads.

LEFT-hand, EVEN numbers – flush with left margin;
RIGHT-hand, ODD numbers – flush with right margin.
Exceptions: PAGE 1 and any other page that carries a display title should have page number, centered below the text, on blue line provided.

FRONT MATTER PAGES should be numbered with Roman numerals (lower case), positioned below text in same manner as described above.

MEMOIRS FORMAT

It is suggested that the material be arranged in pages as indicated below.
Note: Starred items (*) are requirements of publication.

Front Matter (first pages in book, preceding main body of text).

Page i – *Title, *Author's name.

Page iii – Table of contents.

Page iv – *Abstract (at least 1 sentence and at most 300 words).

*1980 Mathematics Subject Classifications represent the primary and secondary subjects of the paper. For the classification scheme, see Annual Subject Indexes of MATHEMATICAL REVIEWS beginning in December 1978.

Key words and phrases, if desired. (A list which covers the content of the paper adequately enough to be useful for an information retrieval system.)

Page v, etc. – Preface, introduction, or any other matter not belonging in body of text.

Page 1 – Chapter Title (dropped 1 inch from top line, and centered).
Beginning of Text.
Footnotes: *Received by the editor date.
Support information – grants, credits, etc.

Last Page (at bottom) – Author's affiliation.

ABCDEFGHIJ–AMS–89876543

Memoirs of the American Mathematical Society
Number 288

R. H. Cameron and D. A. Storvick

A simple definition of the Feynman integral, with applications

Published by the
AMERICAN MATHEMATICAL SOCIETY
Providence, Rhode Island, USA

November 1983 · Volume 46 · Number 288 (end of volume)

MEMOIRS of the American Mathematical Society

This journal is designed particularly for long research papers (and groups of cognate papers) in pure and applied mathematics. It includes, in general, longer papers than those in the TRANSACTIONS.

Mathematical papers intended for publication in the Memoirs should be addressed to one of the editors. Subjects, and the editors associated with them, follow:

Real analysis (excluding harmonic analysis) and **applied mathematics** to JOEL A. SMOLLER, Department of Mathematics, University of Michigan, Ann Arbor, MI 48109.

Harmonic and complex analysis to LINDA PREISS ROTHSCHILD, Department of Mathematics, University of California at San Diego, LaJolla, CA 92093

Abstract analysis to WILLIAM B. JOHNSON, Department of Mathematics, Ohio State Univeristy, Columbus, OH 43210

Algebra and number theory (excluding universal algebras) to LANCE W. SMALL, Department of Mathematics, University of California at San Diego, LaJolla, CA 92093

Logic, foundations, universal algebras and combinatorics to JAN MYCIELSKI, Department of Mathematics, University of Colorado, Boulder, CO 80309

Topology to WALTER D. NEUMANN, Department of Mathematics, University of Maryland, College Park, College Park, MD 20742

Global analysis and differential geometry to TILLA KLOTZ MILNOR, Department of Mathematics, University of Maryland, College Park, MD 20742

Probability and statistics to DONALD L. BURKHOLDER, Department of Mathematics, University of Illinois, Urbana, IL 61801

Combinatorics and number theory to RONALD GRAHAM, Mathematical Studies Department, Bell Laboratories, Murray Hill, NJ 07974

All other communications to the editors should be addressed to the Managing Editor, R. O. WELLS, JR., Department of Mathematics, University of Colorado, Boulder, CO 80309

MEMOIRS are printed by photo-offset from camera-ready copy fully prepared by the authors. Prospective authors are encouraged to request booklet giving detailed instructions regarding reproduction copy. Write to Editorial Office, American Mathematical Society, P.O. Box 6248, Providence, Rhode Island 02940. For general instructions, see last page of Memoir.

SUBSCRIPTION INFORMATION. The 1983 subscription begins with Number 272 and consists of six mailings, each containing one or more numbers. Subscription prices for 1983 are $104.00 list; $52.00 member. Each number may be ordered separately; *please specify number* when ordering an individual paper. For prices and titles of recently released numbers, refer to the New Publications sections of the **NOTICES** of the American Mathematical Society.

BACK NUMBER INFORMATION. For back issues see the AMS Catalogue of Publications.

TRANSACTIONS of the American Mathematical Society

This journal consists of shorter tracts which are of the same general character as the papers published in the MEMOIRS. The editorial committee is identical with that for the MEMOIRS so that papers intended for publication in this series should be addressed to one of the editors listed above.

Subscriptions and orders for publications of the American Mathematical Society should be addressed to American Mathematical Society, P. O. Box 1571, Annex Station, Providence, R. I. 02901. *All orders must be accompanied by payment.* Other correspondence should be addressed to P. O. Box 6248, Providence, R. I. 02940.

MEMOIRS of the American Mathematical Society (ISSN 0065-9266) is published bimonthly (each volume consisting usually of more than one number) by the American Mathematical Society at 201 Charles Street, Providence, Rhode Island 02904. Second Class postage paid at Providence, Rhode Island 02940. Postmaster: Send address changes to Memoirs of the American Mathematical Society, American Mathematical Society, P. O. Box 6248, Providence, RI 02940.

TABLE OF CONTENTS

ABSTRACT

This memoir presents a simple sequential definition of the Feynman integral which is applicable to a rather large class of functionals. The existence theorem shows that this sequential Feynman integral exists and equals the analytic Feynman for all elements of a Banach algebra of functionals expressable as Fourier transforms of measures of finite variation on L_2^ν. This integral has good translation and rotation properties and permutes with other integrals and sums in a reasonable way. Applications to the Schroedinger equation are given and the relationship to other sequential definitions are discussed.

1980 Mathematics Subject Classification 28C20

KEY WORDS AND PHRASES

Sequential Feynman Integral,Analytic Feynman Integral,Feynman Path Integral, Integration in Function Space,Fourier Transforms in Function Space,Schroedinger Equation.

Library of Congress Cataloging in Publication Data

Cameron, Robert Horton, 1908-
A simple definition of the Feynman integral, with applications.

(Memoirs of the American Mathematical Society, ISSN 0065-9266 ; no. 288)
"November 1983."
Bibliography: p.
1. Feynman integral. 2. Function spaces.
I. Storvick, David Arne, 1929- . II. Series.
QA3.A57 no. 288 510s [515.4'2] 83-15605
[QA312]
ISBN 0-8218-2288-8

1. INTRODUCTION: DEFINITION OF THE SEQUENTIAL FEYNMAN INTEGRAL

It is the purpose of this paper to give a simple sequential definition of the Feynman integral which is applicable to a rather large class of functionals and which can be conveniently manipulated. We shall show that our integral has good translation and rotation properties and permutes with other integrals and sums in a reasonable way. It is defined as the limit of a sequence of finite dimensional Lebesgue integrals, (or rather as the common limit of a set of such sequences.) The definition involves no statistics, functional analysis, analytic continuation, repeated limits, or integrals in function space.

In order to introduce our definition, we first present some necessary notation.

NOTATION: Let $C \equiv C[a,b]$ be the space of continuous functions $x(t)$ on $[a,b]$ such that $x(a) = 0$ and let $C^{\nu}[a,b] = \overset{\nu}{\underset{j=1}{\times}} C[a,b]$.

Let a subdivision σ of $[a,b]$ be given:

$$\sigma: [a = \tau_0 < \tau_1 < \tau_2 < \dots < \tau_k < \dots < \tau_m = b] .$$

Let $\vec{X} \equiv \vec{X}(t)$ be a polygonal curve in C^{ν} based on a subdivision σ and the matrix of real numbers $\vec{\vec{\xi}} \equiv \{\xi_{j,k}\}$, and defined by

$$(1.1) \qquad \vec{X}(t) \equiv \vec{X}(t,\sigma,\vec{\vec{\xi}}) = [X_1(t,\sigma,\vec{\vec{\xi}}),\dots,X_{\nu}(t,\sigma,\vec{\vec{\xi}})]$$

where

$$X_j(t,\sigma,\vec{\vec{\xi}}) = \frac{\xi_{j,k-1}(\tau_k - t) + \xi_{j,k}(t - \tau_{k-1})}{\tau_k - \tau_{k-1}}$$

when $\tau_{k-1} \leqq t \leqq \tau_k$, $k = 1,2,\dots,m$, $\xi_{j,0} \equiv 0$.

(We note that as $\vec{\vec{\xi}}$ ranges over all of νm dimensional real space, the polygonal functions $\vec{X}((\cdot),\sigma,\vec{\vec{\xi}})$ range over all polygonal approximations to the functions in $C^{\nu}[a,b]$ based on the subdivision σ . Specifically if $\vec{x}$ is a particular element of $C^{\nu}[a,b]$ and we set $\xi_{j,k} = x_j(\tau_k)$, the function $\vec{X}((\cdot),\sigma,\vec{\vec{\xi}})$ is the polygonal approximation of $\vec{x}$ based on the subdivision σ .)

Received by the editors June 24, 1981.
Presented to the Society April 16, 1982.

DEFINITION: Let $q \neq 0$ be a given real number and let $F(x)$ be a functional defined on a subset of $C^{\nu}[a,b]$ containing all the polygonal elements of $C^{\nu}[a,b]$. Let $\sigma_1, \sigma_2, \ldots$ be a sequence of subdivisions such that norm $\sigma_n \to 0$ and let $\{\lambda_n\}$ be a sequence of complex numbers with $\Re e\, \lambda_n > 0$ such that $\lambda_n \to -iq$. Then if the integral in the right hand side of (1.2) exists for all n and if the following limit exists and is independent of the choice of the sequences $\{\sigma_n\}$ and $\{\lambda_n\}$, we say that the sequential Feynman integral with parameter q exists and is given by

$$(1.2) \qquad \int^{sf_q} F(\vec{x})\,d\vec{x} \equiv \lim_{n\to\infty} \gamma_{\sigma_n,\lambda_n} \int_{\mathbb{R}^{\nu m_n}} \exp\{-\frac{\lambda_n}{2}\int_a^b \|\frac{d\vec{X}}{dt}(t,\sigma_n,\vec{\xi})\|^2 dt\} F(\vec{X}((\cdot),\sigma_n,\vec{\xi}))\,d\vec{\xi},$$

where

$$(1.3) \qquad \gamma_{\sigma,\lambda} = (\frac{\lambda}{2\pi})^{\nu m/2} \prod_{k=1}^{m} (\tau_k - \tau_{k-1})^{-\nu/2}.$$

We note that m depends on σ and m is the number of subintervals in σ. We emphasize that the Lebesgue integral on the right side of (1.2) exists for all n.

In section 3 we shall show that the sequential Feynman integral exists for a broad class of functionals $\hat{S}$ which we now define.

DEFINITION: Let $D[a,b]$ be the class of elements $x \in C[a,b]$ such that x is absolutely continuous on $[a,b]$ and $x' \in L_2[a,b]$. Let $D^{\nu} = \overset{\nu}{\underset{1}{X}} D$.

DEFINITION: Let $\mathcal{M} \equiv \mathcal{M}(L_2^{\nu}[a,b])$ be the class of complex measures of finite variation defined on $B(L_2^{\nu})$, the Borel measurable subsets of $L_2^{\nu}[a,b]$. We set $\|\varkappa\| = \text{var}\,\varkappa$. (In this paper, L_2 always means <u>real</u> L_2).

DEFINITION: The functional F defined on a subset of C^{ν} that contains D^{ν} is said to be an element of $\hat{S} \equiv \hat{S}(L_2^{\nu})$ if there exists a measure $\varkappa \in \mathcal{M}$ such that for $\vec{x} \in D^{\nu}$,

$$(1.4) \qquad F(\vec{x}) \equiv \int_{L_2^{\nu}} \exp\{i \sum_{j=1}^{\nu} \int_a^b v_j(t) (\frac{dx_j(t)}{dt}) dt\} d\varkappa(\vec{v}).$$

We shall show in Corollaries I and II of our fundamental existence theorem, Theorem 3.1, that the sequential Feynman integral of each functional $F \in \hat{S}$ exists and satisfies

$$\int^{sf_q} F(\vec{x})\,d\vec{x} = \int_{L_2^{\nu}} \exp\{\frac{1}{2qi} \sum_{j=1}^{\nu} \int_a^b [v_j(t)]^2 dt\} d\varkappa(\vec{v}).$$

In section 2 we shall show that with the proper norm, $\hat{S}$ is a Banach algebra, and shall show its relationship to other Banach algebras. In particular we shall show that $\hat{S}$ contains the Banach algebras S' and S" defined in [4]. (The definition of S" involves no integration in function space).

In section 4 we shall present a translation theorem and show that the sequential Feynman integral remains invariant under orthogonal transformations of $\mathbb{R}^{\nu}$.

In section 5 we present a Fubini theorem giving conditions for permuting sequential Feynman integrals with Lebesgue type integrals or with infinite series.

In section 6 we give applications of the sequential Feynman integral to the Schroedinger equation and to the quadratic potentials of Johnson and Skoug [9].

In section 7 we shall show the relationship to other sequential definitions of the Feynman integral. In particular the sequential Wiener integral [3] and the Truman integral [13], [14]. Our work and that of Truman are closely related to that of Albeverio and Høegh - Krohn [1] and [3]. Johnson [7] shows the relationship between our space S (given in [4]) to the space $\mathfrak{F}(H)$ of Fresnel integrable functionals of Albeverio and Høegh-Krohn. Indeed, in a recent communication, Johnson has pointed out that our space $\hat{S}$ defined above is identical (and not merely isometrically isomorphic) to the space $\mathfrak{F}(H)$.

2. THE SPACES OF FUNCTIONALS $\hat{S}$ AND S^*

In this section we shall show that $\hat{S}$ (with the proper norm) is a Banach algebra and show how it is related to the Banach algebras S and S' introduced in [4]. We shall also define a new Banach algebra S^* which is intermediate between S' and S, and which is closely related to $\hat{S}$.

TERMINOLOGY: We shall say that two functionals $F(\vec{x})$ and $G(\vec{x})$ are equal s-almost everywhere if for each $\rho > 0$ the equation $F(\rho\vec{x}) = G(\rho\vec{x})$ holds for almost all $\vec{x} \in C^\nu[a,b]$, in other words, if $F(\vec{x}) = G(\vec{x})$ except for a scale-invariant null set. We denote this equivalence relation between functionals by $F \overset{s}{\approx} G$. (Our measure in C^ν is Wiener measure.)

The definition of S also involves the P.W.Z. (Paley-Wiener-Zygmund) integral [12] which is defined as follows.

DEFINITION: Let $\varphi_1, \varphi_2, \ldots$ be a C.O.N. (complete orthonormal) set of real functions of bounded variation on $[a,b]$. Let $v \in L_2[a,b]$ and $v_n(t) = \sum_{j=1}^{n} \varphi_j(t) \int_a^b v(s)\varphi_j(s)ds$. Then the P.W.Z. integral is defined by

$$\int_a^b v(s)\tilde{d}x(s) \equiv \overset{\{\varphi\}}{\int_a^b} v(s)\tilde{d}x(s) \equiv \lim_{n\to\infty} \int_a^b v_n(s)dx(s)$$

for all $x \in C[a,b]$ for which the above limit exists.

NOTE: It was shown in [12] that this integral exists for almost all $x \in C[a,b]$ and is essentially independent of the choice of $\varphi_1, \varphi_2, \ldots$ Moreover if v is of bounded variation, it is essentially equivalent to the Riemann - Stieltjes integral. Clearly "almost all" may be replaced by "s-almost all" in this statement.

DEFINITION: Let $S \equiv S(L_2^\nu)$ be the space of functionals expressable in the form

$$(2.1) \qquad F(\vec{x}) \equiv \int_{L_2^\nu} \exp\{i \sum_{j=1}^{\nu} \int_a^b v_j(x)\tilde{d}x_j(t)\} d\mu(\vec{v})$$

for s-almost all $\vec{x} \in C^\nu[a,b]$, where $\mu \in \mathcal{M} \equiv \mathcal{M}[L_2^\nu]$. (Note, it is assumed that the P.W.Z. integral $\int_a^b v_j(t)\tilde{d}x_j(t)$ is based on the same C.O.N. sequence $\{\varphi_n\}$ for all choices of $\vec{v}$ and $\vec{x}$).

NOTATION: If $F(\vec{x}) = G(\vec{x})$ s-almost everywhere on $C^\nu[a,b]$ and also for every $\vec{x} \in D^\nu[a,b]$, we shall write $F(\vec{x}) \stackrel{\approx}{=} G(\vec{x})$.

From Remark 2 of [6] we have that if $v \in L_2[a,b]$ and $x \in D[a,b]$ then

$$(2.2) \qquad \int_a^b v(t)\tilde{d}x(t) = \int_a^b v(t)x'(t)dt .$$

Thus if $v \in L_2[a,b]$ and $\{\varphi_n\}$ and $\{\Psi_n\}$ are two C.O.N. sequences of B.V., then

$$\overset{\{\varphi\}}{\int_a^b} v(t)\tilde{d}x(t) = \overset{\{\Psi\}}{\int_a^b} v(t)\tilde{d}x(t)$$

for $x \in D[a,b]$, and hence

$$\int_{L_2} \exp\{i \overset{\{\varphi\}}{\int_a^b} v(t)\tilde{d}x(t)\}d\varkappa(v) \stackrel{\approx}{=} \int_{L_2} \exp\{i \overset{\{\Psi\}}{\int_a^b} v(t)\tilde{d}x(t)\}d\varkappa(v) .$$

We now introduce the space S^*. We shall prove later that

$$S \cap \hat{S} \underset{\neq}{\supset} S^* \underset{\neq}{\supset} S' \supset S'' .$$

DEFINITION: We say $F \in S^* \equiv S^*[L_2^\nu]$ iff there exists a $\varkappa \in \mathfrak{M}$ such that

$$(2.3) \qquad F(\vec{x}) \stackrel{\approx}{=} \int_{L_2^\nu} \exp\{i \sum_{j=1}^{\nu} \int_a^b v_j(t)\tilde{d}x_j(t)\}d\varkappa(\vec{v}) .$$

NOTE: It follows from the definitions of $\hat{S}$ and S^* and the properties of the P.W.Z. integral on D that $\hat{S} \supset S^*$.

LEMMA 2.1. If $F \in S^*$ and F is given by (2.3) with $\varkappa \in \mathfrak{M}$ it follows that $\varkappa$ is uniquely determined by F.

PROOF: Since $F \in S^*$, it follows that $F \in S$. Hence from Theorem 2.1 of [4], $\varkappa \in \mathfrak{M}$ is uniquely determined by the equation

$$F(\vec{x}) \approx \int_{L_2^\nu} \exp\{i \sum_{j=1}^{\nu} \int_a^b v_j(t)\tilde{d}x_j(t)\}d\varkappa(\vec{v}) .$$

But since $F \in S^*$, (2.3) holds for some $\varkappa \in \mathfrak{M}$ with the stronger equivalence " $\stackrel{\approx}{=}$ " and thus this $\varkappa$ is uniquely defined. In the Banach algebra S we defined $\|F\| \equiv \|\varkappa\|$ and so for $F \in S^*$, we define $\|F\| \equiv \|\varkappa\|$.

LEMMA 2.2. The space $S^* \equiv S^*(L_2^\nu)$ is a Banach algebra.

The proof is identical with the proof of Theorems 2.2 and 2.3 of [4] except that " $\approx$ " must be replaced by " $\stackrel{\approx}{=}$ ".

NOTATION:

A binary polygonal function in $C[a,b]$ is one which is linear on each interval

$$[a + (\frac{j-1}{2^m})(b-a)\ ,\ a + \frac{j}{2^m}(b-a)] \quad \text{for} \quad j = 1,\dots,2^m .$$

The mth binary polygonal approximation $[x]_m$ to the function $x \in C[a,b]$ is defined by

$$\begin{cases} [x]_m(t) = x(t) \quad \text{when} \quad t = a + \dfrac{k(b-a)}{2^m}\ , \quad k = 0,1,\dots,2^m \\ [x]_m(t) \quad \text{is linear on} \quad [a + \dfrac{(k-1)(b-a)}{2^m}\ , \quad a + \dfrac{k(b-a)}{2^m}]\ , \ k = 1,2,\dots,2^m . \end{cases}$$

DEFINITION: A functional F will be called continuous at x with respect to binary polygonal approximation if

$$\lim_{m \to \infty} F([x]_m) = F(x)$$

where $[x]_m$ is the mth binary polygonal approximation of x .

The Haar functions are a C.O.N. set on $[0,1]$, and are defined as follows [11; p 44]:

$$\chi_0^{(0)}(t) \equiv 1 \quad \text{on} \quad [0,1]$$

$$\chi_0^{(1)}(t) = \begin{cases} 1 & \text{on} \quad [0,1/2) \\ -1 & \text{on} \quad (1/2,1] \\ 0 & \text{at} \quad 1/2\ , \end{cases}$$

$$\chi_1^{(1)}(t) = \begin{cases} \sqrt{2} & \text{on} \quad [0,1/4) \\ -\sqrt{2} & \text{on} \quad (1/4,1/2] \\ 0 & \text{elsewhere on} \quad [0,1] \end{cases}$$

$$\chi_1^{(2)}(t) = \begin{cases} \sqrt{2} & [1/2,3/4) \\ -\sqrt{2} & (3/4,1] \\ 0 & \text{elsewhere on} \quad [0,1] \end{cases}$$

and in general for $k = 1,2,\dots,2^n$; $n = 1,2,\dots$

$$\chi_n^{(k)}(t) = \begin{cases} +\sqrt{2^n} & \text{on} \quad [\dfrac{2k-2}{2^{n+1}}\ ,\ \dfrac{2k-1}{2^{n+1}}) \\ -\sqrt{2^n} & \text{on} \quad (\dfrac{2k-1}{2^{n+1}}\ ,\ \dfrac{2k}{2^{n+1}}] \\ 0 & \text{elsewhere on} \quad [0,1]\ . \end{cases}$$

REMARK 1: Let $h(t)$ be a step-function on $[a,b]$ and let $f(t)$ be continuous on $[a,b]$, and let $f(t) = 0$ when $t = a$ and when $t = b$ and whenever $h(t)$ has a discontinuity at t . Then

$$\int_a^b h(t)df(t) = 0 .$$

For by integration by parts $\int_a^b h(t)df(t) = - \int_a^b f(t)dh(t) = 0$, since f is zero whenever h has a jump.

REMARK 2: Let $h(t)$ be a step-function on $[a,b]$ and let $f(t)$ and $g(t)$ be continuous on $[a,b]$. Let $f(t) = g(t)$ when $t = a$ and when $t = b$ and whenever $h(t)$ has a discontinuity at t . Then

$$\int_a^b h(t)df(t) = \int_a^b h(t)dg(t) .$$

REMARK 3: Let $h(t)$ be a step-function on $[0,1]$ whose discontinuities occur only at the points $0, \frac{1}{2^m}, \frac{2}{2^m}, \frac{3}{2^m}, \ldots, \frac{2^m}{2^m}$. Then

$$(2.4)\qquad h(t) = \int_0^1 h(s)ds + \sum_{n=0}^{m-1} \sum_{k=1}^{2^n} \chi_n^{(k)}(t) \int_0^1 \chi_n^{(k)}(s)h(s)ds$$

for all t on $[0,1]$ except at the points $t = 0 , \frac{1}{2^m}, \frac{2}{2^m}, \ldots, \frac{2^m}{2^m}$.

This can be easily seen because the Haar functions are a C.O.N. set. Consequently if all the terms of the orthogonal development of h are included, it converges in the L_2 mean to h . But all the non-vanishing terms in the development are included in the right member of (2.4), since h is orthogonal to $\chi_n^{(k)}$ when $n \geq m$. Then (2.4) holds for almost all t in $[a,b]$, and since both members of (2.4) are continuous except at $\frac{k}{2^m}$, $k = 0,1,\ldots,2^m$, it follows that (2.4) holds on $[0,1]$ except at these points.

LEMMA 2.3. Let $v \in L_2[a,b]$. If

$$(2.5)\qquad F(x) \equiv \int_a^b v(t)\tilde{d}\, x(t) ,$$

then for s-almost every $x \in C[a,b]$ and every $x \in D[a,b]$, $F(x)$ is continuous with respect to binary polygonal approximation (continuous B.P.A.).

PROOF Case I. Let $a = 0$ and $b = 1$ and assume that the P.W.Z. integral is given in terms of the Haar Functions

Let

$$v_m(t) \equiv \int_0^1 v(s)ds + \sum_{n=0}^{m-1} \sum_{k=1}^{2^n} \chi_n^{(k)}(t) \int_0^1 \chi_n^{(k)}(s)v(s)ds ,$$

$$\hat{x}_m(t) \equiv x(1) + \sum_{n=0}^{m-1} \sum_{k=1}^{2^n} \chi_n^{(k)}(t) \int_0^1 \chi_n^{(k)}(s)dx(s) ,$$

then

$$\int_0^1 v_m(t)dx(t) = x(1)\int_0^1 v(s)ds + \sum_{n=0}^{m-1}\sum_{k=1}^{2^n}\int_0^1 \chi_n^{(k)}(t)dx(t)\int_0^1 \chi_n^{(k)}(s)v(s)ds$$

$$= \int_0^1 v(s)[x(1) + \sum_{n=0}^{m-1}\sum_{k=1}^{2^n}\chi_n^{(k)}(s)\int_0^1 \chi_n^{(k)}(t)dx(t)]ds$$

$$= \int_0^1 v(s)\hat{x}_m(s)ds .$$

Thus, we have

$$\int_0^1 v_m(t)dx(t) = \int_0^1 v(s)\hat{x}_m(s)ds . \tag{2.6}$$

We now show for $x \in C[0,1]$ that

$$\frac{d[x]_m(s)}{ds} = \hat{x}_m(s) ,$$

for $s \in [0,1]$, $s \neq a + \frac{k}{2^m}$, $k = 0,1,2,\ldots,2^m$.

Now by Remark 2,

$$\hat{x}_m(s) = x(1) + \sum_{n=0}^{m-1}\sum_{k=1}^{2^n}\chi_n^{(k)}(t)\int_0^1 \chi_n^{(k)}(s)dx(s)$$

$$= x(1) + \sum_{n=0}^{m-1}\sum_{k=1}^{2^n}\chi_n^{(k)}(t)\int_0^1 \chi_n^{(k)}(s)d[x]_m(s)$$

$$= x(1) + \sum_{n=0}^{m-1}\sum_{k=1}^{2^n}\chi_n^{(k)}(t)\int_0^1 \chi_n^{(k)}(s)\frac{d[x]_m(s)}{ds}ds ,$$

and by Remark 3 we have

$$\hat{x}_m(s) = \frac{d[x]_m(s)}{ds} \tag{2.7}$$

for $s \in [0,1]$ except for $s = \frac{k}{2^m}$, $k = 0,1,\ldots,2^m$.

Since for s-almost all x in $C[0,1]$ and all $x \in D[0,1]$, $\int_0^1 v(t)\tilde{d}x(t)$ exists with respect to the orthogonal development in Haar functions, then by the definition of the P.W.Z. integral,

$$F(x) \equiv \int_0^1 v(t)\tilde{d}x(t)$$

$$\overset{\approx}{=} \lim_{m\to\infty}\int_0^1 v_m(t)dx(t) .$$

Thus by equation (2.6),(2.7), and (2.2) we have

$$F(x) = \lim_{n\to\infty} \int_0^1 v(t)\hat{x}_n(t)dt$$

$$= \lim_{n\to\infty} \int_0^1 v(t) \frac{d[x]_n(t)}{dt} dt$$

$$= \lim_{n\to\infty} \int_0^1 v(t)\tilde{d}[x]_n(t)$$

$$= \lim_{n\to\infty} F([x]_n) \quad \text{for s-almost all } x \in C[a,b] \text{ and all } x \in D[a,b] .$$

Case II: Let $a = 0$ and $b = 1$, and let the P.W.Z. integral be defined by any C.O.N. sequence $\{\varphi_n\}$ of functions of bounded variation. Then by the theorem of Paley, Wiener and Zygmund, the integral

$$\overset{\{\varphi_n\}}{\int_0^1} v(t)\tilde{d}x(t) = \overset{\{\chi_n^{(k)}\}}{\int_0^1} v(t)\tilde{d}x(t)$$

for almost all $x \in C[0,1]$, and since the integral is linear, this holds for s-almost all $x \in C[0,1]$. Moreover by (2.2) the above equation holds for all $x \in D[0,1]$.

Case III: Let $[a,b]$ be any interval, and the C.O.N. sequence be unrestricted. By a translation and a change of scale from $[0,1]$ to $[a,b]$ the result follows, since scale invariant Wiener null sets go into scale invariant Wiener null sets by such transformations, and the Lemma is proved.

THEOREM 2.1 If $F(x) \in S^*$, then $F(\vec{x})$ is continuous with respect to binary polygonal approximation s-almost everywhere in $C^\nu[a,b]$ and everywhere in $D^\nu[a,b]$.

PROOF: Since $F \in S^*$, there exists $\varkappa \in \mathfrak{M}$ such that

$$(2.8) \qquad F(\vec{x}) \approx \int_{L_2^\nu} \exp\{i \sum_{j=1}^{\nu} \int_a^b v_j(t)\tilde{d}\, x_j(t)\}d\mu(\vec{v}) .$$

(It is assumed that the P.W.Z. integrals in the exponential are based on the same C.O.N. sequence $\{\varphi_k\}$ of functions of bounded variation for all $v_j \in L_2$ and all $x_j \in C[a,b]$ $(j = 1,2,\ldots,\nu))$.

Then substituting $[\vec{x}_m]$ for $\vec{x}$, we have

$$F([\vec{x}]_m) = \int_{L_2^\nu} \exp\{i \sum_{j=1}^{\nu} \int_a^b v_j(t)\tilde{d}[x_j]_m(t)\}d\varkappa(\vec{v}) .$$

By Lemma 2.3, the above exponential approaches the exponential in (2.8) as $m\to\infty$, so by bounded convergence and because the exponential is measurable in $\vec{v}\times\vec{x}$ on $L_2^\nu\times C^\nu$, we have $F([\vec{x}]_m)\to F(\vec{x})$ for s-almost all $\vec{x}\in C^\nu[a,b]$ and every $\vec{x}\in D^\nu[a,b]$, and the theorem is proved.

COROLLARY 1: If $F, G \in S^*$ and $F(\vec{x}) = G(\vec{x})$ for all binary polygonal functions in $C^\nu[a,b]$, then

$$F(\vec{x}) \stackrel{\approx}{=} G(\vec{x}) .$$

COROLLARY 2: If $F \in \hat{S}$ and F is defined only on D^ν, then there exists an extension $F^* \in S^*$ such that $F^*(\vec{x}) = F(\vec{x})$ on D^ν. Moreover F^* is essentially unique in the sense that if $F^*, F^{**} \in S^*$ and $F^*(\vec{x}) = F^{**}(\vec{x}) = F(\vec{x})$ on D^ν, then

$$F^*(\vec{x}) \stackrel{\approx}{=} F^{**}(\vec{x}) .$$

Finally, if $\varkappa$ is associated with F by (1.4), it follows that $\varkappa$ is associated with F^* by (2.3).

PROOF: Let

$$F^*(\vec{x}) \equiv \int_{L_2^\nu} \exp\{i \sum_{j=1}^{\nu} \int_a^b v_j(t)\tilde{d}x_j(t)\}d\varkappa(\vec{v})$$

for every $\vec{x} \in C^\nu[a,b]$ for which the right member exists. By the definition of S^*, we have $F^* \in S^*$. By (2.2) we have $F^*(\vec{x}) = F(\vec{x})$ for $\vec{x} \in D^\nu[a,b]$. Thus the existence of F^* satisfying the required conditions has been established. The uniqueness follows from Corollary 1.

LEMMA 2.4. If $F \in \hat{S}$ then the measure $\varkappa$ is uniquely determined by equation (1.4) on D^ν.

PROOF: Let F_1 be the restriction of F to D^ν. Then $F_1 \in \hat{S}$ and by Corollary 2 of Theorem 2.1, there exists an essentially unique $F^* \in S^*$ such that $F_1(\vec{x}) = F^*(\vec{x})$ on D^ν. Since the measure defining F^* is unique, the measure $\varkappa \in \mathfrak{M}$ satisfying (1.4) is unique.

NOTATION: If $F \in \hat{S}$, we define $\|F\| \equiv \|\varkappa\|$, where $\varkappa$ is associated with $F \equiv F(\vec{x})$ by (1.4) for $\vec{x} \in D^\nu$.

It follows from Lemma (2.4) that for $F \in \hat{S}$, the measure $\varkappa$ is uniquely determined by F and it is clear that $\|F\|$ is a norm for $\hat{S}$ if we identify elements of $\hat{S}$ which are equal on D^ν.

LEMMA 2.5: The space $\hat{S} \equiv \hat{S}(L_2^\nu)$ is a Banach algebra where elements of $\hat{S}$ that are equal on D^ν are considered equivalent.

The proof is identical with the proofs of Theorems 2.2 and 2.3 of [4].

OBSERVATION: Since each functional F in S or S^* or $\hat{S}$ determines a unique measure $\varkappa \in \mathfrak{M}$ and each measure $\varkappa \in \mathfrak{M}$ determines a unique equivalence class of functionals in each of the three spaces S, S^*, and $\hat{S}$, the three Banach algebras are pairwise isomorphic.

REMARK 4: Let $v \in BV[a,b]$ and let x be absolutely continuous on $[a,b]$. Then the following Riemann-Stieltjes and Lebesgue integrals are equal:

$$\int_a^b v(t)dx(t) = \int_a^b v(t) \frac{dx(t)}{dt} dt .$$

This follows from the definition of the Riemann-Stieltjes integral and the fact that $\operatorname{var}_{[a,t]} x$ is a uniformly continuous function of t on $[a,b]$.

LEMMA 2.6. $S' \subsetneq S^* \subsetneq S \cap \hat{S}$.

PROOF: For $F \in S'$, there exists $\mu' \in \mathfrak{M}'$ (definitions given in [4]) so that

$$F(\vec{x}) = \int_{(BV)^\nu} \exp\{i \sum_{j=1}^{\nu} \int_a^b v_j(t)dx_j(t)\}d\varkappa'(\vec{v})$$

for $x \in C^\nu[a,b]$. Just as in the proof of Theorem 3.0 in [4], we define a measure $\varkappa$ on $L_2[a,b]$ as follows. Let $E \in B(L_2^\nu)$, then set $\varkappa(E) \equiv \varkappa'(E \cap (BV)^\nu)$. Let $x \in D[a,b]$, then by Remark 2 of [6] and Remark 4 we have for $v \in BV$,

$$\int_a^b v(t)dx(t) = \int_a^b v(t) \frac{dx}{dt} dt = \int_a^b v(t)\tilde{d}x(t) .$$

Thus for $\vec{x} \in D^\nu$,

$$F(\vec{x}) = \int_{(BV)^\nu} \exp\{i \sum_{j=1}^{\nu} \int_a^b v_j(t)dx_j(t)\}d\varkappa'(\vec{v})$$

$$= \int_{(BV)^\nu} \exp\{i \sum_{j=1}^{\nu} \int_a^b v_j(t) \frac{dx_j}{dt} dt\}d\varkappa'(\vec{v})$$

$$= \int_{L_2^\nu} \exp\{i \sum_{j=1}^{\nu} \int_a^b v_j(t) \frac{dx_j}{dt} dt\}d\varkappa(\vec{v})$$

$$= \int_{L_2^\nu} \exp\{i \sum_{j=1}^{\nu} \int_a^b v_j(t)\tilde{d}x_j(t)\}d\varkappa(\vec{v}) .$$

By Theorem 3.0 of [4], the first and last members above are equal for s-almost all $x \in C^\nu[a,b]$, so that $S' \subset S^*$; $S^* \subset S$ by definition.

We now present an example which shows that $S^* \neq S \cap \hat{S}$.

EXAMPLE: Let

$$(2.9)\qquad F(\vec{x}) = \begin{cases} 0 & \text{if } \vec{x} \in D^\nu \\ 1 & \text{if } \vec{x} \in C^\nu - D^\nu . \end{cases}$$

Then since D^ν is a scale invariant null set, we have $F(\vec{x}) \approx 1$, and hence $F \in S$. Clearly F is also an element of $\hat{S}$. By Theorem 2.1,if F were an element of S^*, it would be continuous (B.P.A.) s-almost everywhere on $C^\nu[a,b]$, and it would therefore be zero s-almost everywhere on $C^\nu[a,b]$, since it is zero for all elements of D^ν.

This contradicts the fact that it is unity s-almost everywhere and so $F \notin S^*$ and hence $S^* \neq S \cap \hat{S}$.

Now finally, we shall show that $S' \neq S^*$. From Proposition 2 of [8], when $\nu = 1$ we have that $\mathfrak{M}' \neq \mathfrak{M}$, and if σ is the measure generated by the unit mass concentrated in an element v_0 of $L_2 - BV$, then $\sigma \in \mathfrak{M} - \mathfrak{M}'$.

Let

$$F(x) \equiv \int_{L_2} \exp\{i \int_a^b v(t)\tilde{d}x(t)\} d\sigma(v) = \exp\{i \int_a^b v_0(t)\tilde{d}x(t)\} \in S^* .$$

By Propositions I & II of [8], we have $F \notin S'$ and the lemma is proved.

LEMMA 2.7. If $F, G \in S^*$ and $F(\vec{x}) = G(\vec{x})$ almost everywhere on $C^\nu[a,b]$, then $F(\vec{x}) \stackrel{\approx}{=} G(\vec{x})$.

PROOF. By definition of S^*, there exist $\varkappa_1, \varkappa_2 \in \mathfrak{M}$ such that

$$(2.10)\qquad F(\vec{x}) \stackrel{\approx}{=} \int_{L_2^\nu[a,b]} \exp\{i \sum_{j=1}^{\nu} \int_a^b v_j(t)\tilde{d}x_j(t)\} d\mu_1(\vec{v}) .$$

and

$$(2.11)\qquad G(\vec{x}) \stackrel{\approx}{=} \int_{L_2^\nu[a,b]} \exp\{i \sum_{j=1}^{\nu} \int_a^b v_j(t)\tilde{d}x_j(t)\} d\mu_2(\vec{v}) .$$

Thus $F(\vec{x})$ is almost everywhere equal to both the right member of (2.10) and the right member of (2.11). Since $F \in S^* \subset S$, it follows from Theorem 2.1 in [4], that $\varkappa_1$ and $\varkappa_2$ are identical. Thus $F(\vec{x}) \stackrel{\approx}{=} G(\vec{x})$ follows from (2.10) and (2.11), and the Lemma is proved.

REMARK: If $F \in S^*$, then the values of F on $D^\nu[a,b]$ determine the values of F s-almost everywhere on $C^\nu[a,b]$; and conversely, the values of F s-almost everywhere on $C^\nu[a,b]$ determine the values of F everywhere on $D^\nu[a,b]$.

LEMMA 2.8. If $F \in S$ there exists $F^* \in S^*$ such that $F^* \approx F$ on $C^\nu[a,b]$.

PROOF. By the definition of S, there exists a unique μ such that

$$F(\vec{x}) \approx \int_{L_2^\nu} \exp\{i \sum_{j=1}^{\nu} \int_a^b v_j(t)\tilde{d}x_j(t)\} d\mu(\vec{v}) \equiv F^*(\vec{x}) .$$

Thus F^* is defined whenever the integral exists, and so F^* exists s-almost everywhere on $C^\nu[a,b]$ and everywhere on $D^\nu[a,b]$. Thus $F^* \in S^*$ and $F^* \approx F$ and the Lemma is proved.

3. EXISTENCE OF THE SEQUENTIAL FEYNMAN INTEGRAL ON $\hat{S}$ AND S^*

In preparation for our fundamental existence theorem, we now introduce some necessary notion and establish two lemmas. Let

$$W_\lambda(\sigma,\vec{\xi}) \equiv \gamma_{\sigma,\lambda}\exp\{-\frac{\lambda}{2}\int_a^b \|\frac{d}{dt}\vec{X}(t,\sigma,\vec{\xi})\|^2 dt\}$$

(3.1)

$$= (\frac{\lambda}{2\pi})^{\frac{\nu m}{2}} \prod_{k=1}^{m} (\tau_k - \tau_{k-1})^{-\nu/2} \exp\{-\frac{\lambda}{2}\sum_{k=1}^{m}\frac{\|\vec{\xi}_k - \vec{\xi}_{k-1}\|^2}{(\tau_k - \tau_{k-1})}\} .$$

(By the notation $\lambda^{\frac{\nu m}{2}}$ we mean $(\sqrt{\lambda})^{\nu m}$ where $\mathrm{Re}\sqrt{\lambda} > 0$, and $\vec{\xi}_k$ is the vector $[\xi_{1,k},\ldots,\xi_{m,k}]$ and $\|\vec{\xi}_k\|^2 = \sum_{j=1}^{\nu} \xi_{j,k}^2$.) Thus in terms of W, the sequential Feynman integral (defined in (1.2)) can be written

$$\int^{sf_q} F(\vec{x})d\vec{x} = \lim_{n\to\infty} \int_{\mathbb{R}^{\nu m_n}} W_{\lambda_n}(\sigma_n,\vec{\xi})F(\vec{X}((\cdot),\sigma_n,\vec{\xi}))d\vec{\xi} . \tag{3.2}$$

REMARK: Since the $\{\sigma_n\}$ and $\{\lambda_n\}$ were chosen arbitrarily and independently in the definition, the single limit may also be expressed as a double limit, thus,

$$\int^{sf_q} F(x)dx = \lim_{\substack{n\to\infty \\ k\to\infty}} I_{n,k} \tag{3.3}$$

where

$$I_{n,k} \equiv \int_{\mathbb{R}^{\nu m_k}} W_{\lambda_n}(\sigma_k,\vec{\xi})F(\vec{X}((\cdot),\sigma_k,\vec{\xi}))d\vec{\xi} . \tag{3.4}$$

NOTATION: Let $v(t) \in L_2[a,b]$ and let σ be any subdivision $a = \tau_0 < \tau_1 < \ldots < \tau_m = b$. We define the averaged function $v_\sigma(t)$ for v on σ by

$$\begin{cases} v_\sigma(t) \equiv \dfrac{1}{\tau_j - \tau_{j-1}}\displaystyle\int_{\tau_{j-1}}^{\tau_j} v(s)ds \quad \text{when } \tau_{j-1} \le t < \tau_j ,\ j = 1,\ldots,m \\ v_\sigma(t) = 0 \quad \text{when } t = b . \end{cases} \tag{3.5}$$

Where there is a sequence of subdivisions, $\sigma_1,\sigma_2,\ldots$ then σ, m, and τ_j will be replaced by $\sigma_n, m_n, \tau_{n,j}$.

LEMMA 3.1: Let $\sigma_1,\sigma_2,\ldots$ be a sequence of subdivisions of $[a,b]$ such that $\|\sigma_n\| \to 0$, and let $v(t)$ be integrable on $[a,b]$. Then for almost every $t \in [a,b]$, the sequence of averaged functions converges to the function:

(3.6) $$\lim_{n\to\infty} v_{\sigma_n}(t) = v(t) .$$

PROOF: By the fundamental theorem of integral calculus for Lebesgue integrals,

(3.7) $$\frac{d}{dt}\int_a^t v(s)ds = v(t)$$

for almost all $t \in [a,b]$. Let $t^* \in [a,b)$ be a value of t where (3.7) holds, and hence

$$\lim_{h\to 0} \frac{1}{h}\int_{t^*}^{t^*+h} v(s)ds = v(t^*) .$$

Let $\epsilon > 0$ be given. Choose $\delta > 0$ such that, when $0<|h|<\delta$,

$$\left| \frac{1}{h}\int_{t^*}^{t^*+h} v(s)ds - v(t^*)\right| < \epsilon ;$$

then

(3.8) $$\left|\int_{t^*}^{t^*+h} v(s)ds - hv(t^*)\right| \leq \epsilon|h|$$

when $|h| < \delta$. Choose M such that when $n>M$, $\|\sigma_n\| < \delta$. For j_n such that $\tau_{n,j_n-1} \leq t^* < \tau_{n,j_n}$. Let us first put $h = \tau_{n,j_n} - t^*$ in (3.8), thus

$$\left|\int_{t^*}^{\tau_{n,j_n}} v(s)ds - (\tau_{n,j_n} - t^*)v(t^*)\right| \leq \epsilon(\tau_{n,j_n} - t^*) .$$

Next put $h = \tau_{n,j_n} - t^*$ in (3.8), thus

$$\left|\int_{t^*}^{\tau_{n,j_n-1}} v(s)ds - (\tau_{n,j_n-1} - t^*)v(t^*)\right| \leq \epsilon|t^* - \tau_{n,j_n-1}|$$

or

$$\left|\int_{\tau_{n,j_n-1}}^{t^*} v(s)ds - (t^* - \tau_{n,j_n-1})v(t^*)\right| \leq \epsilon(t^* - \tau_{n,j_n-1})$$

and we have

$$\left|\int_{\tau_{n,j_n-1}}^{\tau_{n,j_n}} v(s)ds - (\tau_{n,j_n} - \tau_{n,j_n-1})v(t^*)\right| \leq \epsilon(\tau_{n,j_n} - \tau_{n,j_n-1})$$

so

$$\left|\frac{1}{\tau_{n,j_n} - \tau_{n,j_n-1}} \int_{\tau_{n,j_n-1}}^{\tau_{n,j_n}} v(s)ds - v(t^*)\right| \leq \epsilon .$$

Therefore we've established

$$\lim_{n\to\infty} v_{\sigma_n}(t) = v(t^*) \ .$$

LEMMA 3.2. Let $\sigma_1, \sigma_2, \ldots$ be a sequence of subdivisions such that $\|\sigma_n\| \to 0$, and let $v \in L_2[a,b]$; then

$$\lim_{n\to\infty} \int_a^b [v_{\sigma_n}(t)]^2 dt = \int_a^b [v(t)]^2 dt \ .$$

PROOF: By Lemma 3.1, $\lim_{n\to\infty} v_{\sigma_n}(t) = v(t)$ for almost every $t \in [a,b]$, and by Fatou's lemma,

$$\liminf_{n\to\infty} \int_a^b [v_{\sigma_n}(t)]^2 dt \geqq \int_a^b [v(t)]^2 dt \ .$$

Now

$$v_{\sigma_n}(t) = \frac{1}{\tau_{n,j} - \tau_{n,j-1}} \int_{\tau_{n,j-1}}^{\tau_{n,j}} v(s)ds \ , \text{ when } \tau_{n,j-1} \leq t < \tau_{n,j} \ ;$$

and by the Schwarz inequality,

$$\text{(3.9)} \qquad [\int_{\tau_{n,j-1}}^{\tau_{n,j}} v(s)ds]^2 \leqq \int_{\tau_{n,j-1}}^{\tau_{n,j}} [v(s)]^2 ds \cdot (\tau_{n,j} - \tau_{n,j-1}) \ .$$

Therefore we have

$$\text{(3.10)} \qquad \int_a^b [v_{\sigma_n}(t)]^2 dt = \sum_{j=1}^{m_n} \left[\frac{\int_{\tau_{n,j-1}}^{\tau_{n,j}} v(s)ds}{\tau_{n,j} - \tau_{n,j-1}} \right]^2 (\tau_{n,j} - \tau_{n,j-1})$$

$$\leqq \sum_{j=1}^{m_n} \int_{\tau_{n,j-1}}^{\tau_{n,j}} [v(s)]^2 ds$$

$$= \int_a^b [v(t)]^2 dt \ .$$

Therefore we have

$$\limsup_{n\to\infty} \int_a^b [v_{\sigma_n}(t)]^2 dt \leqq \int_a^b [v(t)]^2 dt \ .$$

We have proved that the lim sup is less than or equal to the lim inf and so we have

$$\lim_{n\to\infty} \int_a^b [v_{\sigma_n}(t)]^2 dt = \int_a^b [v(t)]^2 dt \ .$$

At this time, we introduce the definition of the analytic Feynman integral given in [4] page 20, see also [3].

DEFINITION: Let F be a functional such that the ν-dimensional Wiener integral

$$(3.11)\qquad J(\lambda) \equiv \int_{C^\nu[a,b]} F(\lambda^{-1/2}\vec{x})\,d\vec{x}$$

exists for all real $\lambda > 0$. If there exists a function $J^*(\lambda)$ analytic in the half-plane $\mathrm{Re}\,\lambda > 0$ such that $J^*(\lambda) = J(\lambda)$ for all real $\lambda > 0$, then we define J^* to be the analytic Wiener integral of F over $C^\nu[a,b]$ with parameter λ, and for $\mathrm{Re}\,\lambda > 0$ we write

$$(3.12)\qquad \int_{C^\nu[a,b]}^{\mathrm{anw}_\lambda} F(\vec{x})\,d\vec{x} \equiv J^*(\lambda) .$$

DEFINITION: Let q be a real parameter $(q \neq 0)$ and let F be a functional whose analytic Wiener integral exists for $\mathrm{Re}\,\lambda > 0$. Then if the following limit exists, we call it the analytic Feynman integral of F over $C^\nu[a,b]$ with parameter q, and we write

$$(3.13)\qquad \int^{\mathrm{anf}_q} F(\vec{x})\,d\vec{x} \equiv \lim_{\substack{\lambda \to -iq \\ \mathrm{Re}\,\lambda > 0}} \int_{C^\nu[a,b]}^{\mathrm{anw}_\lambda} F(\vec{x})\,d\vec{x} .$$

THEOREM 3.1. If $F \in S^*$ and q is real $(q \neq 0)$, then F is sequentially Feynman integrable and its sequential Feynman integral is equal to its analytic Feynman integral.

Since $F \in S^*(L_2^\nu)$ there exists $\mu \in \mathcal{M}(L_2^\nu)$ such that

$$(3.14)\qquad F(\vec{x}) \approx \int_{L_2^\nu} \exp\{i \sum_{j=1}^{\nu} \int_a^b v_j(t)\tilde{d}x_j(t)\}\,d\mu(\vec{v}) .$$

In particular, this equality holds for all polygonal functions $\vec{x}$. Let $\{\sigma_n\}$ and $\{\lambda_n\}$ be such that $\|\sigma_n\| \to 0$, and $\mathrm{Re}\,\lambda_n > 0$, and $\lambda_n \to -iq$. Then

$$\begin{aligned} \mathcal{J}_n &\equiv \int_{\mathbb{R}^{\nu m_n}} W_{\lambda_n}(\sigma_n,\vec{\xi})F(\vec{X}(\cdot,\sigma_n,\vec{\xi}))\,d\vec{\xi} \\ &= \int_{\mathbb{R}^{\nu m_n}} W_{\lambda_n}(\sigma_n,\vec{\xi}) \int_{L_2^\nu} \exp\{i \sum_{j=1}^{\nu} \int_a^b v_j(t)\tilde{d}_t X_j(t,\sigma_n,\vec{\xi})\}\,d\mu(\vec{v})\,d\vec{\xi} . \end{aligned}$$

By the Fubini Theorem and the properties of the P.W.Z. integral, (see equation (2.2) above), we have

$$\begin{aligned} \mathcal{J}_n &= \int_{L_2^\nu}\int_{\mathbb{R}^{\nu m_n}} W_{\lambda_n}(\sigma_n,\vec{\xi})\exp\{i \sum_{j=1}^{\nu} \int_a^b v_j(t)\frac{d}{dt} X_j(t,\sigma_n,\vec{\xi})\,dt\}\,d\vec{\xi}\,d\mu(\vec{v}) \\ &= \gamma_{\sigma_n,\lambda_n}\int_{L_2^\nu}\int_{\mathbb{R}^{\nu m_n}} \exp\{-\frac{\lambda_n}{2}\sum_{k=1}^{m_n}\frac{\sum_{j=1}^{\nu}(\xi_{j,k}-\xi_{j,k-1})^2}{(\tau_{n,k}-\tau_{n,k-1})}\} \\ &\times \exp\{i \sum_{j=1}^{\nu}\sum_{k=1}^{m_n}\int_{\tau_{n,k-1}}^{\tau_{n,k}} v_j(t)\,\frac{(\xi_{j,k}-\xi_{j,k-1})}{(\tau_{n,k}-\tau_{n,k-1})}\,dt\}\,d\vec{\xi}\,d\mu(\vec{v}) \end{aligned}$$

$$= \gamma_{\sigma_n,\lambda_n} \int_{L_2^\nu} \int_{\mathbb{R}^{\nu m_n}} \exp\{-\frac{\lambda_n}{2} \sum_{k=1}^{m_n} \frac{\sum_{j=1}^{\nu} \eta_{j,k}^2}{(\tau_{n,k}-\tau_{n,k-1})}\}$$

$$\exp\{i \sum_{j=1}^{\nu} \sum_{k=1}^{m_n} \int_{\tau_{n,k-1}}^{\tau_{n,k}} v_j(t)dt \frac{\eta_{j,k}}{(\tau_{n,k}-\tau_{n,k-1})}\} d\vec{\eta}\, d\mu(\vec{v})$$

where $\eta_{j,k} = \xi_{j,k} - \xi_{j,k-1}$. Thus

$$\mathcal{J}_n = \gamma_{\sigma_n,\lambda_n} \int_{L_2^\nu} \int_{\mathbb{R}^{\nu m_n}} \prod_{k=1}^{m_n} \prod_{j=1}^{\nu} \exp\{\frac{-\lambda_n \eta_{j,k}^2}{2(\tau_{n,k}-\tau_{n,k-1})} +$$

$$i(\int_{\tau_{n,k-1}}^{\tau_{n,k}} v_j(t)dt)\frac{\eta_{j,k}}{(\tau_{n,k}-\tau_{n,k-1})}\} d\vec{\eta} d\mu(\vec{v})$$

$$= \gamma_{\sigma_n,\lambda_n} \int_{L_2^\nu} \prod_{k=1}^{m_n} \prod_{j=1}^{\nu} \Bigg[\int_{\mathbb{R}} \exp\{\frac{-\lambda_n \eta_{j,k}^2}{2(\tau_{n,k}-\tau_{n,k-1})} +$$

$$i(\int_{\tau_{n,k-1}}^{\tau_{n,k}} v_j(t)dt)\frac{\eta_{j,k}}{(\tau_{n,k}-\tau_{n,k-1})}\} d\eta_{j,k} \Bigg] d\mu(\vec{v}) .$$

Now if $\operatorname{Re} \alpha > 0$,

$$\int_{-\infty}^{\infty} e^{-\alpha\eta^2 + i\beta\eta}\, d\eta = e^{-\beta^2/4\alpha} \sqrt{\frac{\pi}{\alpha}} ,$$

so

$$\int_{-\infty}^{\infty} \exp\{\frac{-\lambda_n \eta_{j,k}^2}{2(\tau_{n,k}-\tau_{n,k-1})} + i(\int_{\tau_{n,k-1}}^{\tau_{n,k}} v_j(t)dt)\frac{\eta_{j,k}}{\tau_{n,k}-\tau_{n,k-1}}\} d\eta_{j,k}$$

$$= \exp\{\frac{-(\int_{\tau_{n,k-1}}^{\tau_{n,k}} v_j(t)dt)^2}{2(\tau_{n,k}-\tau_{n,k-1})\lambda_n}\} \left[\frac{2\pi(\tau_{n,k}-\tau_{n,k-1})}{\lambda_n}\right]^{1/2} .$$

Thus we have

$$\mathcal{J}_n = \gamma_{\sigma_n,\lambda_n} \int_{L_2^\nu} \prod_{k=1}^{m_n} \prod_{j=1}^{\nu} (\exp\{\frac{-(\int_{\tau_{n,k-1}}^{\tau_{n,k}} v_j(t)dt)^2}{2(\tau_{n,k}-\tau_{n,k-1})\lambda_n}\} \left[\frac{2\pi(\tau_{n,k}-\tau_{n,k-1})}{\lambda_n}\right]^{1/2}) d\mu(\vec{v})$$

$$= \int_{L_2^\nu} \exp\{-\sum_{j=1}^{\nu} \sum_{k=1}^{m_n} \frac{(\int_{\tau_{n,k-1}}^{\tau_{n,k}} v_j(t)dt)^2}{2(\tau_{n,k}-\tau_{n,k-1})\lambda_n}\} d\mu(\vec{v}) .$$

We now set

$$V_{j,n}(t) = \frac{1}{(\tau_{n,k}-\tau_{n,k-1})} \int_{\tau_{n,k-1}}^{\tau_{n,k}} v_j(s)ds$$

where $\tau_{n,k-1} \leqq t < \tau_{n,k}$ for $k = 1,\dots,m_n$ and

$$V_{j,n}(b) = 0 ,$$

so

$$\int_a^b [V_{j,n}(t)]^2 dt = \sum_{k=1}^{m_n} \left[\frac{\int_{\tau_{n,k-1}}^{\tau_{n,k}} v_j(s)ds}{(\tau_{n,k} - \tau_{n,k-1})} \right]^2 (\tau_{n,k} - \tau_{n,k-1})$$

$$= \sum_{k=1}^{m_n} \frac{[\int_{\tau_{n,k-1}}^{\tau_{n,k}} v_j(s)ds]^2}{(\tau_{n,k} - \tau_{n,k-1})} .$$

Thus we have

$$\mathcal{J}_n = \int_{L_2^\nu} \exp\{-\frac{1}{2\lambda_n} \sum_{j=1}^{\nu} \int_a^b [V_{j,n}(t)]^2 dt\} d\varkappa(\vec{v})$$

$$\to \int_{L_2^\nu} \exp\{\frac{1}{2qi} \sum_{j=1}^{\nu} \int_a^b [v_j(t)]^2 dt\} d\varkappa(\vec{v}) \quad \text{as} \quad n \to \infty$$

by Lemma 3.2 and the bounded convergence theorem. Therefore we have proved that

$$\lim_{n\to\infty} \mathcal{J}_n \equiv \int^{sf_q} F(\vec{x}) d\vec{x} = \int_{L_2^\nu} \exp\{\frac{1}{2qi} \sum_{j=1}^{\nu} \int_a^b [v_j(t)]^2 dt\} d\varkappa(\vec{v}) .$$

Since $F \in S^* \subset S$ we have by Theorem 5.1 of [4] that the last member above is equal to the analytic Feynman integral of F and the theorem is proved.

COROLLARY 1: If $F \in S^*$ and q is real $(q \neq 0)$, and F is given by (2.3) where $\varkappa \in \mathcal{M}(L_2^\nu)$, then

$$\int^{sf_q} F(\vec{x}) d\vec{x} = \int_{L_2^\nu} \exp\{\frac{1}{2qi} \sum_{j=1}^{\nu} \int_a^b [v_j(t)]^2 dt\} d\varkappa(\vec{v}) = \int_{C^\nu[a,b]}^{anf_q} F(\vec{x}) d\vec{x} . \tag{3.15}$$

COROLLARY 2: Let $F \in \hat{S}$. Then F is sequentially Feynman integrable and the first two members of (3.15) are equal.

These corollaries follow immediately from the proof of Theorem 3.1.

We now present an example to show that there are sequentially Feynman integrable functionals which do not lie in any of the Banach algebras that we have presented above.

EXAMPLE We present a sequentially Feynman integrable functional which is not in any S class. Let $\nu = 1$. Let $F(x) \equiv f(x(b))$, where $f \in L_1(-\infty,\infty)$, and f is not essentially bounded: e.g. let

$$f(u) \equiv (1+u^2)^{-1} u^{-1/3} .$$

Then f is unbounded in an essential way. Since F is unbounded on C in an essential way, $F \notin S$, and since F is unbounded on $D, F \notin \hat{S}$. However we shall now establish that F is sequentially Feynman integrable.

Let $\sigma_1, \sigma_2, \ldots$ be a sequence of subdivisions of $[a,b]$ such that $\|\sigma_n\| \to 0$ as $n \to \infty$ and let $\lambda_1, \lambda_2, \ldots$ be a sequence of complex numbers with $\operatorname{Re} \lambda_n > 0$ and such that $\lambda_n \to -iq$ as $n \to \infty$. Now

$$\int_{\mathbb{R}^{m_n}} W_{\lambda_n}(\sigma_n, \vec{\xi}) F(X(\cdot, \sigma_n, \vec{\xi}))\, d\vec{\xi}$$

$$= \left(\frac{\lambda_n}{2\pi}\right)^{m_n/2} \left[\prod_{k=1}^{m_n} (\tau_k - \tau_{k-1})^{-1/2}\right] \int_{-\infty}^{\infty} \cdots \int_{-\infty}^{\infty} \exp\left\{-\frac{\lambda_n}{2} \sum_{k=1}^{m_n} \frac{(\xi_k - \xi_{k-1})^2}{(\tau_k - \tau_{k-1})}\right\} f(\xi_{m_n})\, d\xi_1 \cdots d\xi_{m_n}$$

$$= \left(\frac{\lambda_n}{2\pi(b-a)}\right)^{1/2} \int_{-\infty}^{\infty} \exp\left\{-\frac{\lambda_n}{2} \frac{(\xi_{m_n})^2}{(b-a)}\right\} f(\xi_{m_n})\, d\xi_{m_n}$$

$$= \left(\frac{\lambda_n}{2\pi(b-a)}\right)^{1/2} \int_{-\infty}^{\infty} \exp\left\{-\frac{\lambda_n u^2}{2(b-a)}\right\} f(u)\, du ,$$

and

$$\int^{sf_q} F(x)\, dx = \left(\frac{q}{2\pi i(b-a)}\right)^{1/2} \int_{-\infty}^{\infty} \exp\left\{\frac{iqu^2}{2(b-a)}\right\} f(u)\, du .$$

Thus there are functionals which are not in any S class, which are sequentially Feynman integrable. This particular functional is also analytic Feynman integrable and the sequential Feynman integral equals the analytic Feynman integral.

NOTE. If $F \approx G$ and the sequential Feynman integrals of both functionals exist, it does not follow that they are equal. However if $F(\vec{x}) = G(\vec{x})$ for all $\vec{x} \in D^{\nu}[a,b]$ and the sequential Feynman integral of one functional exist, the sequential Feynman integrals of both functionals exist and they are equal.

With the analytic Feynman integral the situation is reversed. Thus if $F \approx G$ and the analytic Feynman integral of one functional exists, the analytic Feynman integral of the other functional exists and they are equal. However if $F(\vec{x}) = G(\vec{x})$ for all $\vec{x} \in D^{\nu}[a,b]$ and the analytic Feynman integrals of both functionals exist, it does not follow that they are equal.

NOTE. Each equivalence class (based on $\approx$) of elements of S contains a unique equivalence class (based on $\stackrel{\approx}{=}$) of elements of S^*, and each equivalence $(\stackrel{\approx}{=})$ class of elements of S^* is contained in a unique equivalence $(\approx)$ class of elements of S.

COUNTEREXAMPLE: Finally we note that the sequential and analytic Feynman integrals are not always equal, even when both exist. The example given in equation (2.9) has this property. Thus it is easy to see that the sequential Feynman integral of this functional is zero while its analytic Feynman integral is unity.

§4. TRANSFORMATION THEOREMS

In this section we prove a translation theorem for the sequential Feynman integral, and show how the integral can be modified so as to be translation invariant. We also prove that the sequential Feynman integral is invariant under orthogonal transformations.

THEOREM 4.1. Let $F \in S^*$, and let $\vec{y} \in D^\nu$. Then $F((\cdot)+\vec{y}) \in S^*$, and for each real $q, q \neq 0$, the following sequential Feynman integrals exist and the equality below holds:

$$\int^{sf_q} F(\vec{x}+\vec{y})\,d\vec{x} = \tag{4.0}$$
$$\exp\{\tfrac{qi}{2}\|\vec{y}'\|^2\} \int^{sf_q} F(\vec{x}) \exp\{-iq \sum_{j=1}^{\nu} \int_a^b y_j'(t)\tilde{d}x_j(t)\}\,d\vec{x} .$$

PROOF: To establish that $F((\cdot)+y) \in S^*$, we use the same proof as was used to prove equation (11) of Theorem 3 of [6], except that "$\approx$" is replaced by "$\stackrel{\approx}{=}$". Since $F \in S$, we apply Thm 4 of [6] to obtain equation (12) of [6]. Since the integrands of both sides of (12) are in S^*, the analytic Feynman integrals equal the sequential Feynman integrals and the theorem is proved.

COROLLARY 1. Under the hypotheses of Theorem 4.1, the following sequential Feynman integrals exist and the equality below holds:

$$\int^{sf_q} F(\vec{x})\,d\vec{x} = \exp\{\tfrac{qi}{2}\|\vec{y}'\|^2\}$$
$$\int^{sf_q} F(\vec{x}+\vec{y})\exp\{iq \sum_{j=1}^{\nu} \int_a^b y'_j(t)\tilde{d}x_j(t)\}\,d\vec{x} .$$

COROLLARY 2. Theorem 4.1 and Corollary 1 remain true if S^* is replaced by $\hat{S}$.

This follows from Corollary 2 of Theorem 2.1.

While the above translation theorem is true for $F \in S^*$ and $y \in D$, it does not hold for all sf_q-integrable functionals.

EXAMPLE: Let $\nu = 1$ and let

$$F(x) \equiv \begin{cases} \int_a^b \left[\frac{d^2x}{dt^2}\right]^2 dt & \text{when this exists} \\ 0 & \text{otherwise} \end{cases}$$

and let $y(t) \equiv (t-a)^2$.

We shall show that both members of (4.0) exist, but they are not equal, for

$$F(X((\cdot),\sigma_n,\vec{\xi})) \equiv 0 \text{ , so}$$

$$\int^{sf_q} F(x)dx = 0 \text{ .}$$

Again, $F(X((\cdot),\sigma_n,\vec{\xi}) + y) \equiv 4(b-a)$, so

$$\int^{sf_q} F(x+y)dx = \lim_{n\to\infty} 4(b-a) \int_{\mathbb{R}^{m_n}} W_{\lambda_n}(\sigma_n,\vec{\xi})d\vec{\xi} = 4(b-a) \text{ .}$$

DEFINITION: Let $F(\vec{x})$ be sf_q-integrable for some real $q \neq 0$. Let

$$(4.1) \qquad G(\vec{x}) \equiv \exp\{\frac{iq}{2}\int_a^b \|\frac{d\vec{x}(t)}{dt}\|^2 dt\}F(\vec{x}) \quad \text{on } D^\nu \text{ .}$$

Then we define the translation invariant sequential Feynman integral of G with parameter q by

$$(4.2) \qquad \int^{tisf_q} G(\vec{x})d\vec{x} \equiv \int^{sf_q} F(\vec{x})d\vec{x} \text{ .}$$

THEOREM 4.2: Let q be real, $q \neq 0$, let $F \in S^*$, and let G be given by (4.1). Let $y \in D^\nu[a,b]$; then $G((\cdot)+\vec{y})$ is $tisf_q$-integrable and

$$(4.3) \qquad \int^{tisf_q} G(\vec{x}+\vec{y})d\vec{x} = \int^{tisf_q} G(\vec{x})d\vec{x} \text{ .}$$

PROOF: By (4.1) we have for $\vec{x} \in D^\nu$,

$$(4.4) \qquad \begin{aligned} G(\vec{x}+\vec{y}) &= \exp\{\frac{iq}{2}\int_a^b \|\frac{d\vec{x}(t)}{dt} + \frac{d\vec{y}(t)}{dt}\|^2 dt\}F(\vec{x}+\vec{y}) \\ &= \exp\{\frac{iq}{2}\int_a^b \|\frac{d\vec{x}(t)}{dt}\|^2 dt\} \exp\{iq\int_a^b (\frac{d\vec{x}}{dt}, \frac{d\vec{y}}{dt})dt\} \exp\{\frac{iq}{2}\int_a^b \|\frac{d\vec{y}}{dt}\|^2 dt\} \cdot F(\vec{x}+\vec{y}) \text{ .} \end{aligned}$$

Now by Theorem 4.1, $F((\cdot)+\vec{y}) \in S^* \subset \hat{S}$. If we define a measure $\mu \in \mathcal{M}$ for sets $E \in B(L_2^\nu)$ by

$$\mu(E) \equiv \begin{cases} 1 & \text{if } \frac{d\vec{y}(t)}{dt} \in E \\ 0 & \text{otherwise ,} \end{cases}$$

then

$$\exp\{iq\sum_{j=1}^{\nu} \frac{dy_j(t)}{dt}\frac{dx_j(t)}{dt} dt\} = \int_{L_2^\nu} \exp\{iq\sum_{j=1}^{\nu}\int_a^b v_j(t)\frac{dx_j(t)}{dt} dt\}d\mu(\vec{v}) \in \hat{S} \text{ .}$$

Thus each factor on the right hand side of (4.4) except the first is an element of $\hat{S}$, and so $G(\vec{x}+\vec{y})$ is translation invariant sequential Feynman integrable.

Applying Corollary 2 of Theorem 4.1 and equations (4.1) and (4.2) we have

$$\int^{\text{tisf}_q} G(\vec{x}+\vec{y})\,d\vec{x}$$

$$= \int^{\text{tisf}_q} \exp\{\frac{iq}{2}\int_a^b \|\frac{d\vec{x}(t)}{dt}\|^2 dt\}\exp\{iq\int_a^b (\frac{d\vec{x}}{dt}, \frac{d\vec{y}}{dt})dt\}\exp\{\frac{iq}{2}\int_a^b \|\frac{d\vec{y}}{dt}\|^2 dt\}F(\vec{x}+\vec{y})\,d\vec{x}$$

$$= \int^{\text{sf}_q} \exp\{iq\int_a^b (\frac{d\vec{x}}{dt}, \frac{d\vec{y}}{dt})dt\}\exp\{\frac{iq}{2}\int_a^b \|\frac{d\vec{y}}{dt}\|^2 dt\}F(\vec{x}+\vec{y})\,d\vec{x}$$

$$= \int^{\text{sf}_q} F(\vec{x})\,d\vec{x}$$

$$= \int^{\text{tisf}_q} G(\vec{x})\,d\vec{x} ,$$

and the theorem is proved.

THEOREM 4.3. (Orthogonal Transformation Theorem:) If $F(\vec{x})$ is sequentially Feynman integrable, then $F(T_R\vec{x})$ is sequentially Feynman integrable and

$$\int^{\text{sf}_q} F(\vec{x})\,d\vec{x} = \int^{\text{sf}_q} F(T_R\vec{x})\,d\vec{x} ,$$

where T_R is an orthogonal transformation of $\mathbb{R}^\nu$ given by an orthogonal matrix A .

PROOF:

Now

$$A\vec{X}((\cdot),\sigma_n,\vec{\vec{\xi}}) = \vec{X}((\cdot),\sigma_n,A\vec{\vec{\xi}}) ,$$

since $A\vec{X}(\tau_k,\sigma_n,\vec{\vec{\xi}})$ is a vector whose j^{th} component is

$$\sum_{\ell=1}^{\nu} a_{j,\ell}X_\ell(\tau_k,\sigma_n,\vec{\vec{\xi}}) = \sum_{\ell=1}^{\nu} a_{j,\ell}\xi_{\ell,k} ;$$

and $\vec{X}(\tau_k,\sigma_n,A\vec{\vec{\xi}})$ is a vector whose j^{th} component is

$$X_j(\tau_k,\sigma_n,A\vec{\vec{\xi}}) = \sum_{\ell=1}^{\nu} a_{j,\ell}\xi_{\ell,k} .$$

Thus

$$\mathcal{J} \equiv \int_{\mathbb{R}^{\nu m_n}} W_{\lambda_n}(\sigma_n,\vec{\vec{\xi}})F(A\vec{X}((\cdot),\sigma_n,\vec{\vec{\xi}}))\,d\vec{\vec{\xi}}$$

$$= \int_{\mathbb{R}^{\nu m_n}} W_{\lambda_n}(\sigma_n,\vec{\vec{\xi}})F(\vec{X}((\cdot),\sigma_n,A\vec{\vec{\xi}}))\,d\vec{\vec{\xi}} .$$

Let us make the transformation $\vec{\vec{\eta}} = A\vec{\vec{\xi}}$; so $\vec{\vec{\xi}} = A^{-1}\vec{\vec{\eta}}$ and the determinant of A is unity. But A^{-1} is an orthogonal transformation and hence leaves $W_{\lambda_n}(\sigma_n,(\cdot))$ invariant and so

$$\mathcal{J} = \int_{\mathbb{R}^n} \nu m_n \, W_{\lambda_n}(\sigma_n, A^{-1}\vec{\eta})\vec{F}(\vec{X}((\cdot),\sigma_n,\vec{\eta}))d\vec{\eta}$$

$$= \int_{\mathbb{R}^n} \nu m_n \, W_{\lambda_n}(\sigma_n,\vec{\eta})F(\vec{X}((\cdot),\sigma_n,\vec{\eta}))d\vec{\eta} \, .$$

By taking limits as $n \to \infty$ we have

$$\int^{sf_q} F(\vec{x})d\vec{x} = \int^{sf_q} F(T_R\vec{x})d\vec{x} \, .$$

COROLLARY TO THEOREM 4.3:

Let q be real, $q \neq 0$. Let $G(\vec{x})$ be $tisf_q$-integrable. Then $G(T_R\vec{x})$ is $tisf_q$-integrable and

$$\int^{tisf_q} G(\vec{x})d\vec{x} = \int^{tisf_q} G(T_R\vec{x})d\vec{x}$$

where T_R is an orthogonal transformation of $\mathbb{R}^n$.

§5. ORDER REVERSAL THEOREMS

In order to facilitate the use of the sequential Feynman integral, we present two theorems giving conditions under which sequential Feynman integration may be interchanged with generalized Lebesgue integration or with infinite summation.

NOTE: Whenever the term "measurable on C^ν" occurs, it refers to the uncompleted (Borel) Wiener measure.

THEOREM 5.1. Let σ be a complex measure of B.V. on a measure space $\mathcal{Y}$. For every $\lambda > 0$, let $F(\lambda x, y)$ be measurable on $C^\nu \times \mathcal{Y}$. For each $y \in \mathcal{Y}$ let $F(\cdot, y) \in S^*$, and let $\|F(\cdot, y)\|$ be dominated by a σ-integrable function of y on $\mathcal{Y}$. Then for real $q \neq 0$, the following iterated integrals exist and are equal:

$$(5.1)\qquad \int_{\mathcal{Y}} \Big[\int^{sf_q} F(\vec{x}, y)\, d\vec{x}\Big] d\sigma(y) = \int^{sf_q} \Big[\int_{\mathcal{Y}} F(\vec{x}, y)\, d\sigma(y)\Big] d\vec{x} .$$

PROOF: Since $F(\cdot, y) \in S^* \subset S$, we have by Theorem 6.2 of [4] that the following integrals exist and are equal:

$$\int_{C^\nu}^{anf_q} \Big[\int_{\mathcal{Y}} F(\vec{x}, y)\, d\sigma(y)\Big] d\vec{x}$$

$$(5.2)\qquad = \int_{\mathcal{Y}} \Big[\int_{C^\nu}^{anf_q} F(\vec{x}, y)\, d\vec{x}\Big] d\sigma(y)$$

$$= \int_{\mathcal{Y}} \Big[\int^{sf_q} F(\vec{x}, y)\, d\vec{x}\Big] d\sigma(y) .$$

Here the second equality follows from Theorem 3.1.

We need to show that $\int_{\mathcal{Y}} F(x, y)\, d\sigma(y) \in S^*$. Now, by hypothesis, for each $y \in \mathcal{Y}$ there exists $\mu_y \in \mathcal{M}$ such that

$$(5.3)\qquad F(\vec{x}, y) \approx \int_{L_2^\nu} \exp\Big\{ i \sum_{j=1}^{\nu} \int_a^b v_j(t)\, \tilde{d}x_j(t) \Big\} d\mu_y(\vec{v}) .$$

By hypothesis the left member is integrable in y for s-almost every $\vec{x}$, and we have by Theorem 6.1 of [4] with the Corollary to Thm 2.1 of [4] that

$$\int_{\mathcal{Y}} F(\vec{x},y)\,d\sigma(y) \approx \int_{\mathcal{Y}}\int_{L_2^\nu} \exp\{i \sum_{j=1}^{\nu} \int_a^b v_j(t)\,\tilde{d}x_j(t)\}\,d\mu_y(\vec{v})\,d\sigma(y)$$

$$(5.4) \qquad = \int_{L_2^\nu} \exp\{i \sum_{j=1}^{\nu} \int_a^b v_j(t)\,\tilde{d}x_j(t)\}\,d[\int_{\mathcal{Y}} \mu_y(\vec{v})\,d\sigma(y)]$$

$$= \int_{L_2^\nu} \exp\{i \sum_{j=1}^{\nu} \int_a^b v_j(t)\,\tilde{d}x_j(t)\}\,dm(\vec{v}) \quad \text{where} \quad m \in \mathcal{M} .$$

Here the last two equalities hold when the second member of (5.4) exists. Hence

$\int_{\mathcal{Y}} F(\vec{x},y)\,d\sigma(y) \in S^*$.

Consequently, by Theorem 3.1,

$$\int^{sf_q} \int_{\mathcal{Y}} F(\vec{x},y)\,d\sigma(y)\,d\vec{x}$$

$$= \int_{C^\nu}^{anf_q} \int_{\mathcal{Y}} F(\vec{x},y)\,d\sigma(y)\,d\vec{x} ,$$

and hence by (5.2) the theorem is proved.

In Theorem 5.4 of [4] we proved that under appropriate conditions

$$\sum_{n=0}^{\infty} \int_{C^\nu[a,b]}^{anf_q} F_n(\vec{x})\,d\vec{x} = \int_{C^\nu[a,b]}^{anf_q} \sum_{n=0}^{\infty} F_n(\vec{x})\,d\vec{x} .$$

We now present a similar theorem for the sequential Feynman integral.

THEOREM 5.2: Let $F_n \in S^*$ for $n = 1,2,\ldots$, and let

$$(5.5) \qquad \sum_{n=1}^{\infty} \|F_n\| < \infty .$$

Then $F \in S^*$ where

$$(5.6) \qquad F(\vec{x}) \approx \sum_{n=1}^{\infty} F_n(\vec{x}) ,$$

and

$$(5.7) \qquad \int^{sf_q} F(\vec{x})\,d\vec{x} = \sum_{n=1}^{\infty} \int^{sf_q} F_n(\vec{x})\,d\vec{x} .$$

The proof follows the same pattern as the proof of Theorem 5.4 of [4].

§6. APPLICATIONS:

We first consider the relationship of our sequential Feynman integral to an integral equation which is formally equivalent to the Schroedinger equation. For this application we make the temporary restriction that $\nu = 1$.

Since $S'' \subset S^*$, and the sequential Feynman integral exists and equals the analytic Feynman integral for $F \in S^*$, each of the seven theorems of Analytic Feynman integral solutions of an integral equation related to the Schroedinger equation, [5], remains true when we replace "analytic Feynman integral" by "sequential Feynman integral" and replace "$\int^{\text{anf}_q}$" by "$\int^{\text{sf}_q}$" . Here we state a new combined form of Theorems 4 and 6 of [5].

DEFINITION: Let $\mathfrak{F}_0$ be the class of Fourier-Stieltjes transforms of right continuous functions of bounded variation on $\mathbb{R}$.

DEFINITION: Let $\mathcal{V}$ be the class of functions $h(\cdot,\cdot)$ Borel measurable on $[0,t_0] \times \mathbb{R}$ such that for each $t \in [0,t_0]$, $h(t,\cdot)$ is a right-continuous function of bounded variation on $\mathbb{R}$ such that the variation of $h(t,\cdot)$ over $\mathbb{R}$ is bounded in t .

DEFINITION: Let $\mathfrak{F}$ be the class of Fourier-Stieltjes transforms H of elements $h \in \mathcal{V}$, thus

$$H(t,u) \equiv \int_{-\infty}^{\infty} e^{iuv} d_v h(t,v) .$$

THEOREM 6.1. Let $\Theta \in \mathfrak{F}$ and $\Psi \in \mathfrak{F}_0$, let q be a real number, $q \neq 0$, and let

$$\Gamma(t,\xi) \equiv \int^{\text{sf}_q} F(x;t,\xi)dx$$

where for $x \in C[0,t_0]$, $t \in (0,t_0]$ and $\xi \in \mathbb{R}$,

$$F(x;t,\xi) \equiv \exp\{\int_0^t \Theta(t-s,x(s)+\xi)ds\}\Psi(x(t)+\xi) .$$

Then $\Gamma(t,\xi)$ is the unique solution in $\mathfrak{F}$ of the integral equation

$$\Gamma(t,\xi) = \left(\frac{q}{2\pi i t}\right)^{1/2} \int_{\to -\infty}^{\to\infty} \Psi(u)\exp\left\{\frac{iq(\xi-u)^2}{2t}\right\}du$$

$$+ \left(\frac{q}{2\pi i}\right)^{1/2} \int_0^t (t-s)^{-1/2} \int_{\to -\infty}^{\to\infty} \Theta(s,u)\Gamma(s,u)\exp\left\{\frac{iq(\xi-u)^2}{2(t-s)}\right\}duds$$

for $(t,\xi) \in [0,t_0] \times \mathbb{R}$.

Recently the authors have learned that G. Johnson and D. Skoug (see [10]) have obtained a similar result for the analytic Feynman integral without the restriction that $\nu = 1$.

In [8] and [9], Johnson and Skoug have given a number of results which show that certain types of functionals are in S and consequently their analytic Feynman integrals exist for all real values of the parameter q . Straight forward modifications of their proofs show that these functionals are also in S^* and consequently their sequential Feynman integrals exist and are equal to the analytic Feynman integrals. We now state some of these modified results.

THEOREM 6.2. (cf. the Thm in [9]) Let

$$F(\vec{x}) \approx \exp\{-\int_a^b (\vec{A}(s)\vec{x}(s),\vec{x}(s))ds\} \equiv \exp\{-\int_a^b \sum_{i=1}^{\nu} \sum_{j=1}^{\nu} a_{i,j}(s)x_i(s)x_j(s)ds\} ,$$

where $\{\vec{A}(s) = (a_{i,j}(s)) : a \leq s \leq b\}$ is a commutative family of $\nu \times \nu$ real symmetric positive definite matrices such that the (necessarily positive) eigenvalues $\{p_1(s), p_2(s), \ldots, p_\nu(s)\}$ have square roots which are of bounded variation on $[a,b]$. Then F is in S^* and so possesses a sequential Feynman integral for all real values of the parameter q , $q \neq 0$.

THEOREM 6.3. (cf. Thm 1 in [8]) Let $\Theta(t,\vec{u})$ be a complex valued function defined on $(a,b] \times \mathbb{R}^\nu$ which for each $t \in (a,b]$ is the Fourier-Stieltjes transform of a complex-valued countably-addive Borel measure σ_t on $\mathbb{R}^\nu$, that is

$$\Theta(t,\vec{u}) = \int_{\mathbb{R}^\nu} \exp\{i(\vec{u},\vec{v})\}d\sigma_t(\vec{v}) ,$$

and assume that $\|\sigma_t\|$ is dominated by a function $h(t)$ in $L_1[a,b]$ and that for each Borel set E in $(a,b] \times \mathbb{R}^\nu$, $\sigma_t(E^{(t)})$ is a Borel measurable function of t . (Here $E^{(t)}$ denotes the t-section of E .) Then

$$G(\vec{x}) \equiv \exp\{\int_a^b \Theta(t,\vec{x}(t)dt\} , \quad \vec{x} \in C^\nu[a,b]$$

is an element of S^* and so possesses a sequential Feynman integral for all real values of the parameter q , $q \neq 0$.

Theorem 1 of [8] was proved for the case $\nu = 1$, but the authors state in their introduction that their proof works just as well for general ν .

THEOREM 6.4. (cf. Corollary 4 in [9]) Let $\{\vec{\vec{A}}(s)\}$ satisfy the hypotheses of Theorem 6.2 and let $\Theta(t,\vec{u})$ satisfy the hypotheses of Theorem 6.3. Then

$$H(\vec{x}) \approx \exp\{-\int_a^b (\vec{\vec{A}}(s)\vec{x}(s),\vec{x}(s))ds + \int_a^b \Theta(t,\vec{x}(t))dt\}$$

belongs to S^* and so possesses a sequential Feynman integral for all real values of the parameter q , $q \neq 0$.

§7. RELATIONSHIPS TO OTHER SEQUENTIAL DEFINITIONS OF THE FEYNMAN INTEGRAL

In this section we shall show some of the relationships between the sf_q-integral defined above and other forms of the Feynman integral.

DEFINITION: Let $\text{Re}\,\lambda > 0$ and $F(\vec{x})$ be defined on a set of functions that contains all polygonal functions in $C^\nu[a,b]$. Let $\{\sigma_k\}$ be a sequence of subdivisions of $[a,b]$ such that norm $\sigma_k \to 0$. Then if the following limit exists and is independent of the choice of the sequence $\{\sigma_k\}$, we say that the sequential Wiener integral with parameter λ exists and is given by

$$(7.0) \qquad \int^{sw[\lambda]} F(\vec{x})\,d\vec{x} \equiv \lim_{k\to\infty} \int_{R^{\nu m_k}} W_\lambda(\sigma_k,\vec{\xi})F(\vec{X}(\cdot,\sigma_k,\vec{\xi}))\,d\vec{\xi}\ .$$

(Here it is of course implied that the functional F is such that the integral on the right of (7.0) exists as a Lebesgue integral for each k.)

We first compare with the integrals defined in [3]. The definition we present is a generalization from C to C^ν and is written in a sequential form rather than as a limit as norm $\sigma \to 0$. Moreover the λ in our definition corresponds to σ^{-2} in [3]. Thus when $\nu = 1$ the

$$\int^{sw[\lambda]} F(\vec{x})\,d\vec{x} = \int_{C[a,b]}^{sw_\sigma} F(\vec{x})\,d\vec{x}$$

where $\lambda = \sigma^{-2}$.

THEOREM 7.1. Let $\text{Re}\,\lambda > 0$ and $F \in \hat{S}$, then the following integrals exist and are equal

$$(7.1) \qquad \int^{sw[\lambda]} F(\vec{x})\,d\vec{x} = \int_{L_2^\nu} \exp\{\frac{-1}{2\lambda}\sum_{j=1}^{\nu}\int_a^b [v_j(t)]^2\,dt\}\,d\varkappa(\vec{v})\ ,\ \text{where}$$

$\varkappa$ is the element of $\mathcal{M}(L_2^\nu)$ associated with F by (2.3).

The proof is identical with that of Theorem 3.1 except that λ remains constant throughout.

THEOREM 7.2. Let q be real, $q \neq 0$, and let F be a functional which is sf_q-integrable and is also $sw_{[\lambda]}$ integrable for each λ in $\text{Re}\,\lambda > 0$, $|\lambda + iq| < \delta$ for some $\delta > 0$. Then

(7.2) $$\int^{sf_q} F(\vec{x})\,d\vec{x} = \lim_{\substack{\lambda\to -iq \\ \mathrm{Re}\,\lambda>0}} \int^{sw[\lambda]} F(\vec{x})\,d\vec{x} .$$

PROOF:

Let $\{\sigma_k\}$ be a sequence of subdivisions of $[a,b]$ such that norm $\sigma_k \to 0$ as $k\to\infty$ and let $\{\lambda_n\}$ be a sequence of complex numbers such that $\mathrm{Re}\,\lambda_n > 0$ and $\lambda_n \to -iq$ as $n\to\infty$. Let

(7.3) $$I_{n,k} \equiv \int_{\mathbb{R}^{\nu m_k}} W_{\lambda_n}(\sigma_k,\vec{\xi})F(\vec{X}((\cdot),\sigma_k,\vec{\xi}))\,d\vec{\xi} .$$

Then by hypothesis and (7.0) and (7.3),

(7.4) $$\lim_{k\to\infty} I_{n,k} \equiv \int^{sw[\lambda_n]} F(\vec{x})\,d\vec{x}$$

exists for each n such that $|\lambda_n + iq| < \delta$. By equation (3.3), the double limit of $I_{n,k}$ as $n,k\to\infty$ exists and equals the sequential Feynman integral. Hence by the Moore-Osgood theorem we have

(7.5) $$\lim_{\substack{n\to\infty \\ k\to\infty}} I_{n,k} = \lim_{n\to\infty} \lim_{k\to\infty} I_{n,k}$$

and

(7.6) $$\int^{sf_q} F(\vec{x})\,d\vec{x} = \lim_{n\to\infty} \int^{sw[\lambda_n]} F(\vec{x})\,d\vec{x} .$$

and the theorem is proved.

COROLLARY 1: If $F \in \hat{S}$, then both members of (7.2) exist and equality holds.

COROLLARY 2: Let $F \in \hat{S}$, let $\sigma_1,\sigma_2,\ldots$ be a sequence of subdivisions of $[a,b]$ such that $\|\sigma_k\| \to 0$ as $k\to\infty$, and let $\lambda_1,\lambda_2,\ldots$ be a sequence of complex numbers such that $\mathrm{Re}\,\lambda_n > 0$ and $\lambda_n \to -iq$ where q is real, $q \neq 0$. Then

(7.7) $$\int^{sf_q} F(\vec{x})\,d\vec{x} = \lim_{\substack{n\to\infty \\ k\to\infty}} I_{n,k} = \lim_{n\to\infty}\lim_{k\to\infty} I_{n,k}$$

where $I_{n,k}$ is given by (7.3).

We next show the relationship of the complex Wiener integral to the integral of Aubrey Truman [13; p. 77].

DEFINITION: Let s be a complex number such that $\mathrm{Im}\, s \leqq 0$, $s \neq 0$. Let σ_k^* be the subdivision of $[a,b]$ which has k equal subdivisions. Let F be a functional defined on D^ν for which the following integral exists as a Lebesgue integral

(7.8) $$\mathfrak{F}_k^s[F] \equiv \int_{\mathbb{R}^{\nu k}} W_{-i/s}(\sigma_k^*,\vec{\xi})F(\vec{X}((\cdot),\sigma_k^*,\vec{\xi}))d\vec{\xi} \; .$$

Then if the following limit exists, we call it <u>the Truman integral</u> of F and write

(7.8a) $$\mathfrak{F}^s[F] = \lim_{k\to\infty} \mathfrak{F}_k^s[F] \; .$$

NOTE: Truman actually interprets the above integral, $\mathfrak{F}_k^s[F]$, in a generalized sense, at least in the case when s is real. We shall make his generalization specific when we take up the case: s a real number. If the functional F is such that $\mathfrak{F}^s[F]$ exists in the above sense, i.e. with the right member of (7.8) existing as a Lebesgue (and hence absolutely convergent) integral for each integer k, we shall say that F <u>is strongly Truman integrable</u>.

We next give conditions under which (7.7) is true with the order of the iterated limits reversed.

It turns out that the Truman integral is related to the (complex) sequential Wiener integral when $\operatorname{Im} s < 0$ and to the sequential Feynman integral when s is real. In the former case we have the following relationship which is an immediate consequence of the two definitions.

OBSERVATION: Let $\operatorname{Im} s < 0$, let $\lambda = \frac{-i}{s}$ and let F be sequentially Wiener integrable with parameter λ. Then the strong Truman integral of F exists and

$$\mathfrak{F}^s[F] = \int^{sw[\lambda]} F(\vec{x})d\vec{x} \; .$$

The following theorem follows immediately from (3.3) and the Moore-Osgood theorem:

THEOREM 7.3. Let q be real, $q \neq 0$, and let F be a functional which is sf_q-integrable on C^ν. Let $\sigma_1,\sigma_2,\ldots$ be a sequence of subdivisions of $[a,b]$ such that norm $\sigma_k \to 0$ as $k\to\infty$ and let $\lambda_1,\lambda_2,\ldots$ be a sequence of complex numbers such that $\operatorname{Re}\lambda_n > 0$ and $\lim \lambda_n = -iq$ as $n\to\infty$. Then if $\lim_{n\to\infty} I_{n,k}$ exists for each k, the following repeated limit exists and

(7.9) $$\int^{sf_q} F(\vec{x})d\vec{x} = \lim_{k\to\infty}\lim_{n\to\infty} I_{n,k} \; .$$

NOTATION: Let σ be the subdivision $a = t_0 < t_1 < \ldots < t_{n-1} < t_n = b$. Then P_σ denotes the transformation of C^ν into C^ν that takes $\vec{x}$ into $\vec{x}_\sigma$,

$$x_\sigma(t) = \frac{(\tau_k - t)\vec{x}(\tau_{k-1}) + (t-\tau_{k-1})\vec{x}(\tau_k)}{\tau_k - \tau_{k-1}} \quad \text{on } \tau_{k-1} < t \leq \tau_k \, ;$$

thus $P_\sigma(\vec{x}) = \vec{x}_\sigma$, and $P_\sigma(\vec{x})(t) = \vec{x}_\sigma(t)$.

NOTATION: If the set of subdivisions points of a subdivision σ_1 is contained in the set of subdivision points of a subdivision σ_2 , we shall write $\sigma_1 \subset \sigma_2$.

REMARK: Let $\sigma_k \subset \sigma_n$. Then

$$P_{\sigma_k} P_{\sigma_n} = P_{\sigma_n} P_{\sigma_k} = P_{\sigma_k} \tag{7.10}$$

and

$$P_{\sigma_n}(\vec{X}((\cdot),\sigma_k,\vec{\xi})) = \vec{X}((\cdot),\sigma_k,\vec{\xi}) .$$

LEMMA 7.1. If σ is a subdivision of $[a,b]$, and if $\lambda_1,\lambda_2,\ldots$ is a sequence of complex numbers such that $\operatorname{Re}\lambda_n > 0$, and $\lambda_n \to -iq$ where q is real, $q \neq 0$ and if $F(P_\sigma(\vec{x}))$ is sf_q integrable and σ has m subintervals then

$$\int^{sf_q} F(P_\sigma(\vec{x}))d\vec{x} = \lim_{n\to\infty} \int_{\mathbb{R}^{\nu m}} W_{\lambda_n}(\sigma,\vec{\xi})F(\vec{X}((\cdot),\sigma,\vec{\xi}))d\vec{\xi} . \tag{7.11}$$

PROOF: Let $\sigma_1,\sigma_2,\ldots$ be a sequence of subdivisions such that $\sigma_k \supset \sigma$ for each k and $\lim \operatorname{norm} \sigma_k \to 0$ as $k \to \infty$. Then

$$\int^{sf_q} F(P_\sigma(\vec{x}))d\vec{x}$$

$$= \lim_{\substack{n\to\infty\\k\to\infty}} \int_{\mathbb{R}^{\nu m_k}} W_{\lambda_n}(\sigma_n,\vec{\xi})F(P_\sigma\vec{X}((\cdot),\sigma_k,\vec{\xi}))d\vec{\xi} \tag{7.12}$$

$$= \lim_{\substack{n\to\infty\\k\to\infty}} \int_{\mathbb{R}^{\nu m_k}} W_{\lambda_n}(\sigma,\vec{\xi})F(\vec{X}((\cdot),\sigma,\vec{\xi}_{\ell_1},\ldots,\vec{\xi}_{\ell_m}))d\vec{\xi} ,$$

where ℓ_j is the subscript of the subdivision point of σ_k corresponding to the jth sub script of σ . If we perform the integration on each of the $\xi_{j,k}$ which does not appear in $X((\cdot),\sigma,\xi_{\ell_1},\ldots,\xi_{\ell_m})$, we obtain

$$\int^{sf_q} F(P_\sigma(\vec{x})d\vec{x} = \lim_{n\to\infty} \int_{\mathbb{R}^{\nu m}} W_{\lambda_n}(\sigma,\vec{\xi})F(\vec{X}((\cdot),\sigma,\vec{\xi})d\vec{\xi} ,$$

and the lemma is proved.

THEOREM 7.4. Let q be a real number, $q \neq 0$, and let $\sigma_1,\sigma_2,\ldots$ be a sequence of subdivisions of $[a,b]$ such that $\lim_{k\to\infty} \operatorname{norm}(\sigma_k) = 0$. Then if F is a functional which is sequential Feynman integrable, and for each $k = 1,2,\ldots,$ $F(P_{\sigma_k})$ is also sequential Feynman integrable, then it follows that

$$(7.13)\qquad \int^{sf_q} F(\vec{x})\,d\vec{x} = \lim_{k\to\infty} \int^{sf_q} F(P_{\sigma_k}(\vec{x}))\,d\vec{x}\ .$$

PROOF: Let $\lambda_1,\lambda_2,\ldots$ be a sequence of complex numbers with $\operatorname{Re}\lambda_n > 0$ for each n and $\lim_{n\to\infty}\lambda_n = -iq$. Then by Lemma 7.1 and equation (3.4), we have for each k

$$\int^{sf_q} F(P_{\sigma_k}(\vec{x}))\,d\vec{x} = \lim_{n\to\infty} I_{n,k}$$

and from Theorem 7.3, our theorem follows.

REMARK: If σ is any subdivision of $[a,b]$ and $\lambda_1,\lambda_2,\ldots$ is a sequence of complex numbers such that $\operatorname{Re}\lambda_n > 0$, $\lim \lambda_n = -iq$ as $n\to\infty$ and if $F(\vec{X}((\cdot),\sigma,\vec{\xi}))$ is integrable on $\mathbb{R}^{\nu m}$ we have by dominated convergence that

$$(7.14)\qquad \lim_{n\to\infty} \int_{\mathbb{R}^{\nu m}} W_{\lambda_n}(\sigma,\vec{\xi})F(\vec{X}((\cdot),\sigma,\vec{\xi}))\,d\vec{\xi}$$
$$= \int_{\mathbb{R}^{\nu m}} W_{-iq}(\sigma,\vec{\xi})F(\vec{X}((\cdot),\sigma,\vec{\xi}))\,d\vec{\xi}\ .$$

(We note that the functional $F\equiv 1$ fails to satisfy the hypotheses of this remark because $F\equiv 1$ is not integrable over $\mathbb{R}^{\nu m}$. Since the left member may exist when the right member does not, the left member may be used to define the right member in a generalized sense.)

We next consider Aubrey Truman's integral [13] (see page 77, equation (3.2) and (3.3)) for the case s real, $s\neq 0$, and we set $q = \frac{1}{s}$. Using the definitions of the two integrals, we obtain the following relationship.

OBSERVATION: If F is sf_q-integrable and σ_k^* is the subdivision of $[a,b]$ with k equal subdivisions and if $F(\vec{X}((\cdot),\sigma_k,\vec{\xi}))$ is integrable over $\mathbb{R}^{\nu k}$ for each k and $qs=1$; q,s real , it follows that F is strongly Truman integrable and

$$(7.15)\qquad \mathfrak{Z}^s[F] = \int^{sf_q} F(\vec{x})\,d\vec{x}\ .$$

To establish equation (7.15), we note that (7.14) holds by hypothesis. Then from the definition of the sequential Feynman integral, equation (7.14), the Moore-Osgood Theorem and the definition of the Truman integral, we have

$$\int^{sf_q} F(\vec{x})d\vec{x} = \lim_{\substack{n\to\infty \\ k\to\infty}} \int_{\mathbb{R}^{\nu k}} W_{\lambda_n}(\sigma_k^*,\vec{\xi})F(\vec{X}((\cdot),\sigma_k^*,\vec{\xi}))d\vec{\xi}$$

$$= \lim_{k\to\infty} \lim_{n\to\infty} \int_{\mathbb{R}^{\nu k}} W_{\lambda_n}(\sigma_k^*,\vec{\xi})F(\vec{X}((\cdot),\sigma_k^*,\vec{\xi}))d\vec{\xi}$$

$$= \lim_{k\to\infty} \int_{\mathbb{R}^{\nu k}} W_{-i/s}(\sigma_k^*,\vec{\xi})F(\vec{X}((\cdot),\sigma_k^*,\vec{\xi}))d\vec{\xi}$$

$$= \lim_{k\to\infty} \mathfrak{F}_k^s[F] = \mathfrak{F}^s[F] .$$

NOTATION: If $\vec{\eta} \equiv \{\eta_{j,\ell}\}_{\substack{j=1,\dots,\nu \\ \ell=1,\dots,k}}$ is a matrix of real numbers, we shall use the notation

$$\Sigma\vec{\eta} \equiv \{\sum_{\ell'=1}^{\ell} \eta_{j,\ell'}\}_{\substack{j=1,\dots,\nu \\ \ell=1,\dots,k}} .$$

NOTATION: (Cauchy Principal Value). We shall write

$$\int_{\mathbb{R}^n}^{(p)} f(\eta_1,\dots,\eta_n)d\vec{\eta} \equiv \lim_{R\to\infty} \int_{-R}^{R}\dots\int_{-R}^{R} f(\eta_1,\dots,\eta_n)d\vec{\eta}$$

if it exists.

Note that the definition of the Truman integral given in (7.8) can be transformed by the transformation

$$\eta_{j,\ell} = \xi_{j,\ell} - \xi_{j,\ell-1} \quad (\text{where } \xi_{j,0} \equiv 0)$$

into the form

$$(7.16) \qquad \mathfrak{F}_k^s[F] = \left[\frac{k}{2s\pi(b-a)i}\right]^{\nu k/2} \int_{\mathbb{R}^{\nu k}} \exp\{\frac{ik}{2s(b-a)} \sum_{j=1}^{\nu}\sum_{\ell=1}^{k} \eta_{j,\ell}^2\}F(\vec{X}((\cdot),\sigma_k^*,\Sigma\vec{\eta}))d\vec{\eta} .$$

Here we are assuming that $F(\vec{X}((\cdot),\sigma_k^*,\Sigma\vec{\eta}))$ is Lebesgue integrable (and hence absolutely integrable) over $\mathbb{R}^{\nu k}$. However Truman also applies his definition to various functionals (for example $F \equiv 1$), where $F(\vec{X}((\cdot),\sigma_k^*,\Sigma\vec{\eta}))$ is not absolutely integrable. In such cases he extends the definition of his integral by evaluating the integral in the right hand member of (7.16) as a Cauchy principal value (see Truman [13] J. Math Phy. page 1744 column 2, line 12) thus:

$$(7.17)\qquad \mathfrak{J}_k^s[F] \equiv \left[\frac{k}{2s\pi(b-a)i}\right]^{\nu k/2} (p)\int_{\mathbb{R}^{\nu k}} \exp\{\frac{ik}{2s(b-a)} \sum_{j=1}^{\nu}\sum_{\ell=1}^{k} \eta_{j,\ell}^2\} F(\vec{X}((\cdot),\sigma_k^*,\Sigma\vec{\eta}))d\vec{\eta}\ .$$

Using this definition of $\mathfrak{J}_k^s[F]$, the extended definition of the Truman integral is

$$(7.17a)\qquad \mathfrak{J}^s[F] \equiv \lim_{k\to\infty} \mathfrak{J}_k^s[F]\ .$$

REMARK: When the Truman integral exists in the strong sense given by (7.8) and (7.8a) it also exists in the extended sense given by (7.17) and (7.17a) and has the same value, so that functionals which are strongly Truman integrable are also Truman integrable in the extended sense. However the converse is not true, since the functional $F(\vec{x}) \equiv 1$ is Truman integrable but not strongly Truman integrable when s is real, $s \neq 0$.

In order to show the relationship of the sequential Feynman integral to the extended Truman integral, we shall now prove three lemmas.

LEMMA 7.2. Let $\alpha > 0$, let $f(x)e^{-\alpha x}$ be integrable on $(0,\infty)$, and let

$$(7.18)\qquad g(x) \equiv \int_0^x f(s)ds\ .$$

Then

$$(7.19)\qquad \int_0^\infty |f(x)|e^{-\alpha x}dx \geqq \alpha \int_0^\infty |g(x)|e^{-\alpha x}dx$$

and

$$(7.20)\qquad \int_0^\infty f(x)e^{-\alpha x}dx = \alpha \int_0^\infty g(x)e^{-\alpha x}dx\ .$$

PROOF: If $0 < R < \infty$, we have by integration by parts

$$(7.21)\qquad \int_0^R f(x)e^{-\alpha x}dx = g(R)e^{-\alpha R} + \alpha \int_0^R g(x)e^{-\alpha x}dx\ ,$$

and similarly

$$\int_0^R |f(x)|e^{-\alpha x}dx = \int_0^R |f(s)|ds\,e^{-\alpha R} + \alpha \int_0^R \int_0^x |f(s)|ds\,e^{-\alpha x}dx$$
$$\geqq \alpha \int_0^R \left|\int_0^x f(s)ds\right| e^{-\alpha x}dx = \alpha \int_0^R |g(x)|e^{-\alpha x}dx\ .$$

Taking the limit as $R \to \infty$, we have

$$\int_0^\infty |f(x)|e^{-\alpha x}dx \geqq \alpha \int_0^\infty |g(x)|e^{-\alpha x}dx\ .$$

Thus $g(x)\exp(-\alpha x)$ is integrable on $(0,\infty)$, and hence we can choose a sequence $R_1, R_2, \ldots$ such that $R_n \to \infty$ and $g(R_n)\exp(-\alpha R_n) \to 0$ as $n \to \infty$. Replacing R by R_n in (7.21) and taking the limit as $n \to \infty$, we obtain (7.20), and the lemma is proved.

LEMMA 7.3: For each $\alpha > 0$, let $f(x_1,\ldots,x_p)\exp\{-\alpha \sum_{j=1}^{p} x_j\}$ be integrable on $(0,\infty)^p$. Let $\int_0^{R_p}\cdots\int_0^{R_1} f(x_1,\ldots,x_p)dx_1\ldots dx_p$ exist and be bounded on $\vec{R} \in (0,\infty)^p$. Let the following limit exist,

$$(7.22)\qquad L \equiv \lim_{\substack{R_1\to\infty\\ \vdots \\ R_p\to\infty}} \int_0^{R_p}\cdots\int_0^{R_1} f(x_1,\ldots,x_p)dx_1\ldots dx_p\ .$$

Then

$$(7.23)\qquad \lim_{\alpha\to 0+} \int_0^{\infty}\cdots\int_0^{\infty} f(x_1,\ldots,x_p)\exp\{-\alpha \sum_{j=1}^{p} x_j\}dx_1\ldots dx_p = L\ .$$

PROOF. Let

$$g_1(x_1,\ldots,x_p) \equiv \int_0^{x_1} f(s,x_2,\ldots,x_p)ds$$

and

$$g_i(x_1,\ldots,x_p) \equiv \int_0^{x_i} g_{i-1}(x_1,\ldots,x_{i-1},s,x_{i+1},\ldots,x_p)ds \quad \text{for } i = 2,3,\ldots,p\ ,$$

so that

$$g_p(x_1,\ldots,x_p) = \int_0^{x_p}\cdots\int_0^{x_1} f(s_1,\ldots,s_p)ds_1\ldots ds_p\ .$$

Thus g_p is bounded on $(0,\infty)^p$ and

$$L = \lim_{\substack{R_1\to\infty\\ \vdots \\ R_p\to\infty}} g_p(R_1,\ldots,R_p)\ .$$

Let

$$|g_p(x_1,\ldots,x_p)| \leqq M\ .$$

Now for almost all $(x_2,\ldots,x_p) \in (0,\infty)^{p-1}$, $f(x_1,x_2,\ldots,x_p)\exp(-\alpha x_1)$ is integrable on $(0,\infty)$, and hence by Lemma 7.2 we have

$$(7.24)\qquad \int_0^{\infty} f(x_1,\ldots,x_p)e^{-\alpha x_1}dx_1 = \alpha\int_0^{\infty} g_1(x_1,\ldots,x_p)e^{-\alpha x_1}dx_1$$

and

$$\int_0^\infty |f(x_1,\ldots,x_p)| e^{-\alpha x_1} dx_1 \geqq \alpha \int_0^\infty |g_1(x_1,\ldots,x_p)| e^{-\alpha x_1} dx_1 .$$

Thus

$$\infty > \int_0^\infty \overset{(p-1)}{\cdots} \int_0^\infty [\int_0^\infty |f(x_1,\ldots,x_p)| e^{-\alpha x_1} dx_1] e^{-\alpha(x_2+\ldots+x_p)} dx_2\ldots dx_p$$

$$\text{(7.25)} \qquad \geqq \int_0^\infty \overset{(p-1)}{\cdots} \int_0^\infty [\alpha \int_0^\infty |g_1(x_1,\ldots,x_p)| e^{-\alpha x_1} dx_1] e^{-\alpha(x_2+\ldots+x_p)} dx_2\ldots dx_p$$

$$= \alpha \int_0^\infty \overset{(p-1)}{\cdots} \int_0^\infty [\int_0^\infty |g_1(x_1,x_2,\ldots,x_p) e^{-\alpha x_2} dx_2] e^{-\alpha(x_1+x_3+\ldots+x_p)} dx_1 dx_3\ldots dx_p$$

and then g_1 considered as a function of x_2 satisfies the hypotheses of Lemma 7.2 for almost every fixed value of $x_1,x_3,\ldots,x_p$. Thus we have for almost all $x_1,x_3,\ldots,x_p$,

$$\text{(7.26)} \qquad \int_0^\infty g_1(x_1,x_2,\ldots,x_p) e^{-\alpha x_2} dx_2 = \alpha \int_0^\infty g_2(x_1,x_2,\ldots,x_p) e^{-\alpha x_2} dx_2 ,$$

and

$$\int_0^\infty |g_1(x_1,\ldots,x_p)| e^{-\alpha x_2} dx_2 \geqq \alpha \int_0^\infty |g_2(x_1,\ldots,x_p)| e^{-\alpha x_2} dx_2 .$$

Substituting in (7.25) we have

$$\infty > \int_0^\infty \overset{(p)}{\cdots} \int_0^\infty |g_1(x_1,\ldots,x_p)| e^{-\alpha(x_1+\ldots+x_p)} dx_1\ldots dx_p$$

$$\geqq \alpha \int_0^\infty \overset{(p)}{\cdots} \int_0^\infty |g_2(x_1,\ldots,x_p)| e^{-\alpha(x_1+\ldots+x_p)} dx_1\ldots dx_p .$$

Multiplying (7.24) by $e^{-\alpha x_2}$ and (7.26) by $e^{-\alpha x_1}$, we have for almost all $x_3,\ldots,x_p$,

$$\int_0^\infty \int_0^\infty f(x_1,\ldots,x_p) e^{-\alpha(x_1+x_2)} dx_1 dx_2 = \alpha \int_0^\infty\int_0^\infty g_1(x_1,\ldots,x_p) e^{-\alpha(x_1+x_2)} dx_1 dx_2$$

$$= \alpha^2 \int_0^\infty\int_0^\infty g_2(x_1,\ldots,x_p) e^{-\alpha(x_1+x_2)} dx_1 dx_2 .$$

Continuing the process, we obtain

$$\int_0^\infty \overset{(p)}{\cdots} \int_0^\infty f(x_1,\ldots,x_p) e^{-\alpha(x_1+\ldots+x_p)} dx_1\ldots dx_p$$

$$= \alpha^p \int_0^\infty \overset{(p)}{\cdots} \int_0^\infty g_p(x_1,\ldots,x_p) e^{-\alpha(x_1+\ldots+x_p)} dx_1\ldots dx_p ,$$

and the fact that the integrand on the right is integrable. Now let $\epsilon > 0$ be given. Choose R such that when $x_1 \geqq R$, $x_2 \geqq R$, ..., $x_p \geqq R$,

$$|g_p(x_1,\ldots,x_p) - L| < \epsilon .$$

Then

$$\int_0^\infty\cdots\int_0^\infty f(x_1,\ldots,x_p)e^{-\alpha(x_1+\ldots+x_p)}dx_1\ldots dx_p$$

$$= \alpha^p\int\cdots\int_{(0,\infty)^p\setminus(R,\infty)^p} g_p(x_1,\ldots,x_p)e^{-\alpha(x_1+\ldots+x_p)}dx_1\ldots dx_p$$

$$+ \alpha^p\int_R^\infty\cdots\int_R^\infty [g_p(x_1,\ldots,x_p) - L]e^{-\alpha(x_1+\ldots+x_p)}dx_1\ldots dx_p$$

$$+ L\,\alpha^p\int_R^\infty\cdots\int_R^\infty e^{-\alpha(x_1+\ldots+x_p)}dx_1\ldots dx_p .$$

Evaluating the integral in the last term above and transposing it to the left hand side, we have

$$\int_0^\infty\cdots\int_0^\infty f(x_1,\ldots,x_p)e^{-\alpha(x_1+\ldots+x_p)}dx_1\ldots dx_p - Le^{-p\alpha R}$$

$$= \alpha^p\int\cdots\int_{(0,\infty)^p\setminus(R,\infty)^p} g_p(x_1,\ldots,x_p)e^{-\alpha(x_1+\ldots+x_p)}dx_1\ldots dx_p$$

$$+ \alpha^p\int_R^\infty\cdots\int_R^\infty [g_p(x_1,\ldots,x_p) - L]e^{-\alpha(x_1+\ldots+x_p)}dx_1\ldots dx_p .$$

Thus we have

$$\left|\int_0^\infty\cdots\int_0^\infty f(x_1,\ldots,x_p)e^{-\alpha(x_1+\ldots+x_p)}dx_1\ldots dx_p - Le^{-p\alpha R}\right|$$

$$\leqq \alpha^p\int\cdots\int_{(0,\infty)^p\setminus(R,\infty)^p} Me^{-\alpha(x_1+\ldots+x_p)}dx_1\ldots dx_p$$

$$+ \alpha^p\int_R^\infty\cdots\int_R^\infty \epsilon\, e^{-\alpha(x_1+\ldots+x_p)}dx_1\ldots dx_p$$

$$= M[1 - e^{-p\alpha R}] + \epsilon\, e^{-p\alpha R} .$$

Taking the upper limit as $\alpha\to 0^+$, we have

$$\limsup_{\alpha\to 0^+}\left|\int_0^\infty\cdots\int_0^\infty f(x_1,\ldots,x_p)e^{-\alpha(x_1+\ldots+x_p)}dx_1\ldots dx_p - Le^{-p\alpha R}\right| = 0 ,$$

so (7.23) holds and the lemma is established.

LEMMA 7.4. Let $f(u_1,\ldots,u_p)\exp\{-\alpha\sum_{j=1}^p u_j^2\}$ be integrable on $\mathbb{R}^p$ for each $\alpha > 0$. Let $\int_{-R_p}^{R_p}\cdots\int_{-R_1}^{R_1} f(u_1,\ldots,u_p)du_1\ldots du_p$ exist and be bounded for $\vec{R}\in(0,\infty)^p$. Let

$$L \equiv \lim_{\substack{R_1\to\infty\\ \vdots\\ R_p\to\infty}} \int_{-R_p}^{R_p}\cdots\int_{-R_1}^{R_1} f(u_1,\ldots,u_p)du_1\ldots du_p \quad \text{exist.}$$

Then

$$\lim_{\alpha\to 0^+} \int\cdots\int_{\mathbb{R}^p} f(u_1,\ldots,u_p)\exp\{-\alpha \sum_{j=1}^{p} u_j^2\}du_1\ldots du_p = L .$$

PROOF. Let

$$f^*(u_1,\ldots,u_p) = \Sigma\, f(\pm u_1,\pm u_2,\ldots,\pm u_p)$$

where the summation is taken over the 2^p terms obtained by taking all possible combinations of plus and minus signs. Then our hypotheses imply that $f^*(u_1,\ldots,u_p)$ $\exp\{-\alpha \sum_{j=1}^{p} u_j^2\}$ is integrable over $(0,\infty)^p$, and also that $\int_0^{R_p}\cdots\int_0^{R_1} f^*(u_1,\ldots,u_p)$ $du_1\ldots du_p$ exists and is bounded on $(0,\infty)^p$, and $L = \lim_{\substack{R_1\to\infty\\ \vdots \\ R_p\to\infty}} \int_0^{R_p}\cdots\int_0^{R_1} f^*(u_1,\ldots,u_p)du_1\ldots du_p$ exists. If we make the change of variables $u_j^2 = x_j$, $j=1,\ldots,p$ and set

$f^{**}(x_1,\ldots,x_p) = \dfrac{f^*(\sqrt{x_1},\ldots,\sqrt{x_p})}{2^p\sqrt{x_1\ldots x_p}}$, we have for $\alpha \geq 0$,

$$\int_0^{R_p}\cdots\int_0^{R_1} f^*(u_1,\ldots,u_p)\exp\{-\alpha\sum_{j=1}^{p} u_j^2\}du_1\ldots du_p$$

$$= \int_0^{R_p^2}\cdots\int_0^{R_1^2} f^{**}(x_1,\ldots,x_p)\exp\{-\alpha\sum_{j=1}^{p} x_j\}dx_1\ldots dx_p .$$

Therefore for $\alpha > 0$,

$$f^{**}(x_1,\ldots,x_p)\exp\{-\alpha\sum_{j=1}^{p} x_j\}$$

is integrable on $(0,\infty)^p$ and

$\int_0^{\rho_p}\cdots\int_0^{\rho_1} f^{**}(x_1,\ldots,x_p)dx_1\ldots dx_p$ exists and is bounded for $\vec{\rho}\in(0,\infty)^p$,

and $L = \lim_{\substack{\rho_1\to\infty\\ \vdots \\ \rho_p\to\infty}} \int_0^{\rho_p}\cdots\int_0^{\rho_1} f^{**}(x_1,\ldots,x_p)dx_1\ldots dx_p$ exists. Thus by Lemma 7.3

$$\lim_{\alpha\to 0^+}\int_0^{\infty}\cdots\int_0^{\infty} f^{**}(x_1,\ldots,x_p)\exp\{-\alpha\sum_{j=1}^{p} x_j\}dx_1\ldots dx_p = L ,$$

and by reversing the change of variable we have

$$\lim_{\alpha\to 0^+}\int_0^{\infty}\cdots\int_0^{\infty} f^*(u_1,\ldots,u_p)\exp\{-\alpha\sum_{j=1}^{p} u_j^2\}du_1\ldots du_p = L$$

and Lemma 7.4 is proved.

THEOREM 7.5. Let F be sf_q-integrable and σ_k^* be the subdivision of $[a,b]$ with k equal subdivisions. Let $s = \frac{1}{q}$ (s and q real). Let

$$\Gamma(\vec{R}) \equiv \int_{-R_{\nu,k}}^{R_{\nu,k}} \cdots \int_{-R_{1,1}}^{R_{1,1}} \exp\{\frac{ikq}{2(b-a)} \sum_{j=1}^{\nu} \sum_{\ell=1}^{k} \eta_{j,\ell}^2\} F(\vec{X}((\cdot),\sigma_k^*,\Sigma\vec{\eta}))\, d\eta_{1,1} \cdots d\eta_{\nu,k}$$

(which exists as a Lebesgue integral by the first hypothesis) be bounded for $\vec{R} \in (0,\infty)^{\nu k}$ and let

$$\lim_{\vec{R}\to\vec{\infty}} \Gamma(\vec{R}) \text{ exist.}$$

(It is understood that the components of $\vec{R}$ approach ∞ independently.) Then the Truman integral of F exists and

$$\mathfrak{J}^s[F] = \int^{sf_q} F(\vec{x})\, d\vec{x} \,.$$

To prove this theorem we shall use Lemma 7.4 with $p = k\nu$ and

$$f(\vec{\eta}) \equiv F(\vec{X}(\cdot,\sigma_k^*,\Sigma\vec{\eta})) \exp\{\frac{ikq}{2(b-a)} \sum_{j=1}^{\nu} \sum_{\ell=1}^{k} \eta_{j,\ell}^2\} \,.$$

Since F is sf_q-integrable, integrals of the form

$$\int_{\mathbb{R}^{\nu k}} W_{\lambda_k}(\sigma_k^*,\vec{\xi}) F(\vec{X}(\cdot,\sigma_k^*,\vec{\xi}))\, d\vec{\xi}$$

exist for each k and each choice of λ_k with $\operatorname{Re}\lambda_k > 0$. Thus setting $\eta_{j,\ell} = \xi_{j,\ell} - \xi_{j,\ell-1}$ we have

$$\begin{aligned}
\infty &> \int_{\mathbb{R}^{\nu k}} |W_{\lambda_k}(\sigma_k^*,\vec{\xi}) F(\vec{X}((\cdot),\sigma_k^*,\vec{\xi}))|\, d\vec{\xi} \\
&= \left(\frac{|\lambda_k|k}{2\pi(b-a)}\right)^{\nu k/2} \int_{\mathbb{R}^{\nu k}} \exp\{-\frac{(\operatorname{Re}\lambda_k)k}{2(b-a)} \sum_{j=1}^{\nu} \sum_{\ell=1}^{k} \eta_{j,\ell}^2\} |F(\vec{X}((\cdot),\sigma_k^*,\Sigma\vec{\eta}))|\, d\vec{\eta} \\
&= \left(\frac{|\lambda_k|k}{2\pi(b-a)}\right)^{\nu k/2} \int_{\mathbb{R}^{\nu k}} \exp\{-\frac{(\operatorname{Re}\lambda_k)k}{2(b-a)} \sum_{j=1}^{\nu} \sum_{\ell=1}^{k} \eta_{j,\ell}^2\} |f(\vec{\eta})|\, d\vec{\eta} \,,
\end{aligned}$$

and so

$$\exp\{-\alpha \sum_{j=1}^{\nu} \sum_{\ell=1}^{k} \eta_{j,\ell}^2\} |f(\vec{\eta})|$$

is integrable over $\mathbb{R}^{\nu k}$ for all $\alpha > 0$ and hence f satisfies all the hypotheses of Lemma 7.4. Hence the following limit exists and we have

$$L \equiv \lim_{\alpha \to 0^+} \int_{\mathbb{R}^{\nu k}} F(\vec{X}((\cdot),\sigma_k^*,\Sigma\vec{\eta}))\exp\{\frac{ikq}{2(b-a)} \sum_{j=1}^{\nu} \sum_{\ell=1}^{k} \eta_{j,\ell}^2\}$$

$$\exp\{-\alpha \sum_{j=1}^{\nu} \sum_{\ell=1}^{k} \eta_{j,\ell}^2\} d\vec{\eta}$$

$$= \lim_{\vec{R} \to \vec{\infty}} \int_{-R_{\nu,k}}^{R_{\nu,k}} \overset{(\nu k)}{\cdots\cdot} \int_{-R_{1,1}}^{R_{1,1}} F(\vec{X}((\cdot),\sigma_k^*,\Sigma\vec{\eta}))$$

$$\exp\{\frac{ikq}{2(b-a)} \sum_{j=1}^{\nu} \sum_{\ell=1}^{k} \eta_{j,\ell}^2\} d\vec{\eta} .$$

Thus $\mathfrak{F}_k^s(F)$ exists and $\mathfrak{F}_k^s(F) = [\frac{k}{2s\pi(b-a)i}]^{\nu k/2} L$. Now if we set $\alpha = \frac{1}{n}$, $\lambda_n = -iq + 2(b-a)/nk$, we obtain

$$L = \lim_{\alpha \to 0^+} \int_{\mathbb{R}^{\nu k}} F(X((\cdot),\sigma_k^*,\vec{\xi}))\exp\{(\frac{iq}{2} - \frac{\alpha(b-a)}{k}) \sum_{j=1}^{\nu} \sum_{\ell=1}^{k} \frac{(\xi_{j,\ell} - \xi_{j,\ell-1})^2}{(b-a)k^{-1}}\} d\vec{\xi}$$

$$= \lim_{\lambda_n \to iq} (\frac{2\pi(b-a)}{\lambda_n k})^{\nu k/2} \int_{\mathbb{R}^{\nu k}} F(X((\cdot),\sigma_k^*,\vec{\xi}))W_{\lambda_n}(\sigma_k^*,\vec{\xi})\, d\vec{\xi}$$

$$= (\frac{2\pi(b-a)}{-ikq})^{\nu k/2} \lim_{n \to \infty} I_{n,k} ,$$

where we have set $\sigma_k \equiv \sigma_k^*$. Thus $\mathfrak{F}_k^s(F) = \lim_{n\to\infty} I_{n,k}$, and from Theorem 7.3, equation 7.9, we have

$$\lim_{k\to\infty} \lim_{n\to\infty} I_{n,k} = \lim_{k\to\infty} \mathfrak{F}_k^s(F) = \mathfrak{F}^s(F) = \int^{sf_q} F(\vec{x})\, d\vec{x}$$

and the theorem is proved.

In [5] we stated some elementary facts about Fresnel integrals, namely: if ρ and σ are real and $\rho \neq 0$

$$(7.27) \qquad \int_{\to -\infty}^{\to \infty} \exp\{i\rho u^2 + i\sigma u\} du = \sqrt{\frac{\pi i}{\rho}} \exp\{\frac{-i\sigma^2}{4\rho}\}$$

and

$$(7.28) \qquad |\int_{-B}^{B'} \exp\{i\rho u^2 + i\sigma u\}| < \frac{3}{\sqrt{|\rho|}} .$$

THEOREM 7.6. If $F \in S^*$ and q is real, $q \neq 0$, and $s = \frac{1}{q}$, then the sequential Feynman integral and the Truman integral exist and we have

$$\int^{sf_q} F(\vec{x})\, d\vec{x} = \mathfrak{F}^s[F] .$$

PROOF. Since $F \in S^*$, F is sf_q-integrable. Moreover there exists a $\varkappa \in \mathcal{M}(L_2^\nu)$ such that

$$F(x) \approx \int_{L_2^\nu} \exp\{i \sum_{j=1}^{\nu} \int_a^b v_j(t)\tilde{d}x_j(t)\} d\mu(\vec{v}) .$$

Hence if $\Gamma(\vec{\vec{R}})$ is defined as in Theorem 7.5, we have for $\tau(k,\ell) = a + (b-a)\ell/k$,

$$\Gamma(\vec{\vec{R}}) \equiv \int_{-R_{\nu,k}}^{R_{\nu,k}} \cdots \int_{-R_{1,1}}^{R_{1,1}} \exp\{\frac{ikq}{2(b-a)} \sum_{j=1}^{\nu} \sum_{\ell=1}^{k} \eta_{j,\ell}^2\} .$$

$$\int_{L_2^\nu} \exp\{i \sum_{j=1}^{\nu} \sum_{\ell=1}^{k} \int_{\tau(k,\ell-1)}^{\tau(k,\ell)} v_j(t) \frac{\eta_{j,\ell}}{(b-a)k^{-1}} dt\} d\mu(\vec{v}) d\eta_{1,1} \cdots d\eta_{\nu,k}$$

$$= \int_{L_2^\nu} \int_{-R_{\nu,k}}^{R_{\nu,k}} \cdots \int_{-R_{1,1}}^{R_{1,1}} \exp\{\frac{ik}{2(b-a)} \sum_{j=1}^{\nu} \sum_{\ell=1}^{k} [q\eta_{j,\ell}^2 + 2\eta_{j,\ell} \int_{\tau(k,\ell-1)}^{\tau(k,\ell)} v_j(t)dt]\}$$

$$d\eta_{1,1} \cdots d\eta_{\nu,k} d\mu(\vec{v})$$

$$= \int_{L_2^\nu} \{ \prod_{j=1}^{\nu} \prod_{\ell=1}^{k} \int_{-R_{j,\ell}}^{R_{j,\ell}} \exp\{\frac{ik}{2(b-a)} [q\eta_{j,\ell}^2 + 2\eta_{j,\ell} \int_{\tau(k,\ell-1)}^{\tau(k,\ell)} v_j(t)dt]\} d\eta_{j,\ell}\} d\mu(\vec{v}) .$$

By (7.28) we have

$$\left| \int_{-R_{j,\ell}}^{R_{j,\ell}} \exp\{\frac{ik}{2(b-a)} [q\eta_{j,\ell}^2 + 2\eta_{j,\ell} \int_{\tau(k,\ell-1)}^{\tau(k,\ell)} v_j(t)dt]\} d\eta_{j,\ell} \right| < \frac{3}{\sqrt{\frac{k|q|}{2(b-a)}}} .$$

Thus we have

$$|\Gamma(\vec{\vec{R}})| < \left(\frac{18(b-a)}{k|q|}\right)^{\nu k/2} \text{var } \mu .$$

By (7.27) we see that

$$\int_{-R_{j,\ell}}^{R_{j,\ell}} \exp\{\frac{ik}{2(b-a)} [q\eta_{j,\ell}^2 + 2\eta_{j,\ell} \int_{\tau(k,\ell-1)}^{\tau(k,\ell)} v_j(t)dt\} d\eta_{j,\ell}$$

has a limit as $R_{j,\ell} \to \infty$. Therefore we may apply the dominated convergence theorem to conclude that

$$\lim_{\vec{\vec{R}} \to \vec{\vec{\infty}}} \Gamma(\vec{\vec{R}})$$

exists. Therefore by Theorem 7.5 the Truman integral exists and

$$\int^{sf_q} F(\vec{x}) d\vec{x} = \mathfrak{F}^s[F] .$$

It has been pointed out by G.W. Johnson that a shorter proof of Theorem 7.6 can be given by combining Corollary 1 to Theorem 3.1 above and Theorem 1 of Truman [13].

REMARK: We note that Theorem 7.5 applies to some functionals (such as $F(x) \equiv 1$) to which the Observation following Theorem 7.4 does not apply. Moreover Theorem 7.5 also applies to the functional

$$F(x) = \begin{cases} \exp\{-\operatorname{var}_{[a,b]}[x(\cdot)]\} \cdot [x(b)]^{-1/3} & \text{if it exists} \\ 0 & \text{otherwise,} \end{cases}$$

though this functional is unbounded on $D[a,b]$ and is therefore not in S^* and Theorem 7.6 does not apply to it. Thus Theorem 7.5 is stronger than theorem 7.6.

BIBLIOGRAPHY

[1] S. Albeverio and R. Høegh-Krohn, Mathematical theory of Feynman path integrals, Lecture notes in Mathematics 523, Springer-Verlag Berlin-Heidleberg-New York 1976.

[2] S. Albeverio and R. Høegh-Krohn, Feynman path integrals and the corresponding method of stationary phase, Feynman path integrals, Lecture notes in physics 106, Springer-Verlag, Berlin, Heidelberg, New York (1979), 3-57.

[3] R.H. Cameron, A family of integrals serving to connect the Wiener and Feynman integrals, J. Math. Phys. XXXIX July 1960, 126-140.

[4] R.H. Cameron and D.A. Storvick, Some Banach algebras of analytic Feynman integrable functionals, Analytic Functions Kozubnik 1979, Lecture Notes in Mathematics 798, Springer-Verlag Berlin, Heidelberg, New York (1980) 18-67.

[5] R.H. Cameron and D.A. Storvick, Analytic Feynman integral solutions of an integral equation related to the Schroedinger equation, J. Analyse Math. 38 (1980), 34-66.

[6] R.H. Cameron and D.A. Storvick, A new translation theorem for the analytic Feynman integral, Rev. Roumaine Math. Pures Appl. vol 27, No 9 (1982), 937-944.

[7] G.W. Johnson, the equivalence of two approaches to the Feynman integral, J. Math. Phys. 23 (11) Nov. (1982), 2090-2096.

[8] G.W. Johnson and D.L. Skoug, Notes on the Feynman integral, I, to appear in Pacific J. of Math.

[9] G.W. Johnson and D.L. Skoug, Notes on the Feynman integral, II, to appear in J. Funct. Anal.

[10] G.W. Johnson and D.L. Skoug, Notes on the Feynman integral, III: The Schroedinger equation, to appear in Pacific J. of Math.

[11] S. Kaczmarz and H. Steinhaus, Theorie der Orthogonalreihen, z subvencjifunduszu kultury narodoweu, Warszawa-Lwow, (1935).

[12] R.E.A.C. Paley, N. Wiener, and A. Zygmund, Notes on random functions, Math. Z. 37 (1933), 647-688.

[13] A. Truman, The Feynman maps and the Wiener integral, J. Math Phys. 19 (8), August 1978.

[14] A. Truman, The polygonal path formulation of the Feynman path integral, Feynman path integrals, Lecture notes in physics 106, Springer-Verlag, Berlin, Heidelberg, New York (1979), 73-102.

Department of Mathematics, University of Minnesota, Minneapolis, Minnesota 55455, U.S.A.

General instructions to authors for
PREPARING REPRODUCTION COPY FOR MEMOIRS

For more detailed instructions send for AMS booklet, "A Guide for Authors of Memoirs." Write to Editorial Offices, American Mathematical Society, P. O. Box 6248, Providence, R. I. 02940.

MEMOIRS are printed by photo-offset from camera copy fully prepared by the author. This means that, except for a reduction in size of 20 to 30%, the finished book will look exactly like the copy submitted. Thus the author will want to use a good quality typewriter with a new, medium-inked black ribbon, and submit clean copy on the appropriate model paper.

Model Paper, provided at no cost by the AMS, is paper marked with blue lines that confine the copy to the appropriate size. Author should specify, when ordering, whether typewriter to be used has PICA-size (10 characters to the inch) or ELITE-size type (12 characters to the inch).

Line Spacing – For best appearance, and economy, a typewriter equipped with a half-space ratchet – 12 notches to the inch – should be used. (This may be purchased and attached at small cost.) Three notches make the desired spacing, which is equivalent to 1-1/2 ordinary single spaces. Where copy has a great many subscripts and superscripts, however, double spacing should be used.

Special Characters may be filled in carefully freehand, using dense black ink, or INSTANT ("rub-on") LETTERING may be used. AMS has a sheet of several hundred most-used symbols and letters which may be purchased for $5.

Diagrams may be drawn in black ink either directly on the model sheet, or on a separate sheet and pasted with rubber cement into spaces left for them in the text. Ballpoint pen is *not* acceptable.

Page Headings (Running Heads) should be centered, in CAPITAL LETTERS (preferably), at the top of the page – just above the blue line and touching it.

LEFT-hand, EVEN-numbered pages should be headed with the AUTHOR'S NAME;

RIGHT-hand, ODD-numbered pages should be headed with the TITLE of the paper (in shortened form if necessary).

Exceptions: PAGE 1 and any other page that carries a display title require NO RUNNING HEADS.

Page Numbers should be at the top of the page, on the same line with the running heads.

LEFT-hand, EVEN numbers – flush with left margin;

RIGHT-hand, ODD numbers – flush with right margin.

Exceptions: PAGE 1 and any other page that carries a display title should have page number, centered below the text, on blue line provided.

FRONT MATTER PAGES should be numbered with Roman numerals (lower case), positioned below text in same manner as described above.

MEMOIRS FORMAT

It is suggested that the material be arranged in pages as indicated below.
Note: Starred items (*) are requirements of publication.

Front Matter (first pages in book, preceding main body of text).

Page i – *Title, *Author's name.

Page iii – Table of contents.

Page iv – *Abstract (at least 1 sentence and at most 300 words).

*1980 Mathematics Subject Classifications represent the primary and secondary subjects of the paper. For the classification scheme, see Annual Subject Indexes of MATHEMATICAL REVIEWS beginning in December 1978.

Key words and phrases, if desired. (A list which covers the content of the paper adequately enough to be useful for an information retrieval system.)

Page v, etc. – Preface, introduction, or any other matter not belonging in body of text.

Page 1 – Chapter Title (dropped 1 inch from top line, and centered).

Beginning of Text.

Footnotes: *Received by the editor date.
Support information – grants, credits, etc.

Last Page (at bottom) – Author's affiliation.

ABCDEFGHIJ–AMS–89876543